ALFONSINE ASTRONOMY

ALFONSINE ASTRONOMY

STUDIES AND SOURCES

ALFONSINE ASTRONOMY

This collection is created under the auspices of the European Research Council project ALFA: Shaping a European scientific scene, Alfonsine astronomy, CoG 723085.

Alfonsine Astronomy

Computational Practices

Edited by
MATTHIEU HUSSON, JOSÉ CHABÁS †, AND
RICHARD L. KREMER

BREPOLS

D/2026/0095/145
ISBN 978-2-503-61077-1
e-ISBN 978-2-503-61078-8
DOI 10.1484/M.ALFA-EB.5.136820

Printed in the E.U. on acid-free paper

Table of Contents

MATTHIEU HUSSON, JOSÉ CHABÁS, AND
RICHARD L. KREMER

Introduction*

Alfonsine astronomers at work

According to the ongoing research of the ALFA project, from the end of the thirteenth century and well into the sixteenth century, no fewer than 120 astronomers produced more than 330 works of mathematical astronomy, including texts, tables, and instruments in Latin and related vernacular languages.[1] Unlike the Arabic zijes built according to differing parameters, these Alfonsine works generally produce the same celestial positions and share a common tropical reference frame. These works are now known in more than 900 manuscript codices, most of which were produced in the fourteenth and fifteenth centuries. Together, they constitute the written record of Alfonsine astronomy.

In an initial collective publication the ALFA Project analysed this vast corpus at the 'manuscript level', considering individual codices, rather than texts, through case studies.[2] Instead of focusing primarily on questions of astronomical content, we posed questions about the types of manuscripts assembled, the milieus in which the manuscripts were produced, and the agency of scribes, artists, binders, compilers, commentators, patrons, astronomers, and collectors, all of whom assembled the archive on which we are working today. This research demonstrates that Alfonsine astronomy was discussed and fostered within universities, of course, but also within courts, religious orders, printing shops, and humanist and other intellectual circles. These multiple venues have consequences for the type of works produced and the way they were transmitted. Although some works, such as the *Parisian Alfonsine Tables* or John of Saxony's canons for them (beginning *Tempus est mensura*) circulate in many more manuscript witnesses than any other works in this tradition, they never became completely authoritative or systematically entered university *curricula*. They continued to be widely treated as tools, subject to rearrangement, abridgment, expansion, or reformatting. The core works are in some cases commented on, but more frequently they were reconfigured with other materials, often original, in a dynamic process in which authorship was distributed among many different historical actors. In this situation, the boundaries of text and table collections and the circulation and function of diagrams and images challenged genres within

* Research presented here was supported by the ERC project *ALFA, Shaping a European Scientific Scene: Alfonsine Astronomy* (CoG 723085, PI Matthieu Husson).

1 The ALFA Survey' of 900+ medieval Latin manuscripts containing Alfonsine astronomical material is available at https://dishas.obspm.fr/alfa/survey. A collective paper exposing the methodology, choices, and main results of this research is also in preparation. The quantitative information on the Alfonsine corpus presented in this introduction comes from this survey.

2 Richard L. Kremer, Matthieu Husson José Chabás, eds, *Alfonsine Astronomy: The Written Record* (Turnhout: Brepols, 2022).

the tradition or led to efforts to establish new genres. In other words, although often static from the perspective of astronomical theory, the written corpus of Alfonsine astronomy attests to a very diverse and dynamic set of manuscript and textual practices.

Significantly, the manuscript evidence shows that astronomers' practices did not include systematic observations of celestial motions, the determination of new values for key astronomical parameters, or the consideration of alternative geometrical or arithmetic models for celestial motion during the Alfonsine era. Of course, isolated observation reports can be found in the Alfonsine corpus. Parameter values are sometimes discussed (e.g., the lengths of the sidereal and tropical years); parameters are often cited explicitly and geometrical models elaborated on, especially in texts on planetary equatoria. However, none of these efforts led to even tentative, let alone critical, challenges to Ptolemaic astronomy, or attempts to adjust the parameters of the Parisian Alfonsine Tables. Rather, Alfonsine astronomers centred their activity around the elaboration of procedures, addressed to ever-growing audiences. These procedures were expressed by a complex web of procedural texts, tables, diagrams, and instruments and aimed at computing astronomical positions. Thus, the design and comparison of computational procedures for practitioners with different levels of mathematical skill was at the heart of many original works by Alfonsine authors. Even the less specialised practitioners had to select from different available procedures as they computed celestial positions. To make these choices, given the accessible manuscripts and instruments, each with its own potential imperfections and scribal idiosyncrasies, they needed to evaluate the computational recipes, numerical accuracy and precision, difficulty and time required for execution or even the soundness and consistency of procedures with respect to the underlying astronomical and astrological theories. Such choices about computational practices were at the centre of Alfonsine astronomers' activities.

Following the evidence of the sources, this volume thus focuses on computational practices. It suggests that they include not only the rules of sexagesimal arithmetic, but also approaches to interpolation, attention to technical vocabulary for procedures, the incorporation of worked quantitative examples in canons for students, the invention of instruments for specialised tasks, the design of tables and table sets, and much more besides. In other words, we suggest broadening the definition of the term 'computation' so that it describes a set of mathematical practices including different objects and forms of reasoning, all geared towards the production of numerical results through meaningful procedures. Hence, analysing the astronomical practices documented in the corpus of Alfonsine texts is essentially the study of computational practices.[3]

The focus of this book on computational practices is part of a broader interest in the history of science that has often been designed as a 'practical turn', also a growing trend in the history of astronomy over the last several decades. This practical turn entails a view of science as a set of epistemic activities rather than primarily a set of propositions expressing theories. These activities are conducted by individuals and collectives in various milieus forming 'cultures' characterised by, for example, the sharing of certain tools and instruments, ways of working, distinct forms of training, or institutions regulating the

3 Several contributions in this collection take inspiration from recent works in the history of mathematics with a particular focus on sources and practices, as exemplified in the book series *Why Sciences of the Ancient World Matter*, directed by Karine Chemla, Agathe Keller, and Christine Proust.

development of the practices in connection with other areas of the cultural and social worlds in which they unfold. This shift of attention from propositions to practices is often described as a way to disclose the meanings and values historical actors themselves attributed to their activities and the bodies of knowledge they produce and transmit. Together with the results of the first ALFA collective book, these approaches contribute to our analyses in this volume.

In the history of mediaeval European astronomy, this practical turn probably began *c.* 2000 with the study of the computational practices linked to the determination of the times of true syzygy.[4] This focus on computational practices was followed by studies of interpolation, uses of instruments, and processes of translating technical terms from Latin to the vernacular.[5] ALFA seeks to add to such work the evidence from individual manuscripts in the Alfonsine corpus. Instead of considering tables, instruments, and various types of texts separately, it is now possible, using tools from the digital humanities, to anchor our analyses in the manuscript evidence, where, more often than not, these different elements have been assembled by the mediaeval astronomers themselves. Thus, in addition to a theoretical and technical analysis of astronomical tables and instruments and their contextualisation within intellectual traditions, it is now possible to consider a range of new questions concerning the history of mediaeval astronomy:

- How are the various elements of a codex assembled?
- What means are used to articulate the elements and to organise complementarity and redundancy among them?
- When considered together, what type of practices do these elements enable that would not be accessible were the elements considered separately?
- Do different combinations of material in given manuscripts imply differing understandings of the astral sciences?
- Do we see the same sorts of tool assemblages around varied topics such as longitudes, syzygies, latitudes, parallaxes, or eclipses?
- How are these different combinations distributed chronologically, geographically, and even socially within the corpus?
- Do such combinations suggest differing views on the relation of mathematical astronomers to the natural world, to natural philosophy, or to mathematics?

Such general questions have motivated many of the case studies presented here.

To unfold the broader meaning of 'computation' adopted in this book, the contributions are presented in four sections. A first group of chapters considers different examples of procedural texts oriented either towards table sets or instruments and analyses how astronomers might have construed them with computational, geometrical, theoretical, or even historiographical meanings. The second series of chapters analyses tables or table

4 José Chabás and Bernard R. Goldstein, 'Computational Astronomy: Five Centuries of Finding True Syzygy', *Journal for the History of Astronomy*, 28 (1997), 93–105.

5 José Chabás and Bernard R. Goldstein, *The Alfonsine Tables of Toledo* (Dordrecht: Kluwer, 2003); Richard L. Kremer, 'Experimenting with Paper Instruments in Fifteenth-and Sixteenth-Century Astronomy: Computing Syzygies with Isotemporal Lines and Salt Dishes', *Journal for the History of Astronomy*, 42 (2011), 223–58; Matthieu Husson, 'Ways to Read a Table: Reading and Interpolation Algorithms in Some Early Fourteenth Century Double Entry Tables', *Journal for the History of Astronomy*, 43 (2012), 299–319. Similar questions were pursued in James Evans, *The History and Practice of Ancient Astronomy* (New York: Oxford University Press, 1998).

sets and explores their geometrical and theoretical grounding as well as their purposes and numerical derivation. Contributions in the third section more specifically address questions directly related to numbers, accuracy, rounding and precision, or interpolation. The last set considers two common genres at the border between mathematical astronomy and its wider public: ephemerides and horoscopes.

Part 1: Procedural texts

A major result of ALFA's survey of the Alfonsine corpus is the precise quantification of the importance of procedural texts for table sets (canons) or for instruments (construction and use) in the production of Alfonsine astronomers. According to the ongoing research, tables are the most represented category of work in the corpus (around forty percent). Procedural texts for tables and the use of instruments come in second, almost on a par, with about twenty-five percent of the works each. The importance of procedural texts, comprising one-quarter of the corpus, is also apparent qualitatively in the great level of variety they present. These procedural texts may be focused on a specific astronomical question, such as solar and lunar velocities, or they may be encyclopaedic and deal with all aspects of mathematical astronomy. They could be directed at an advanced audience and present a wealth of alternative procedures, discussing their respective merits; they can take the form of detailed computed examples; or, they might be simple and didactic, in some cases providing little more than a glossary of terms. They may push reflection on the procedures towards, among other non-mutually exclusive possibilities such as astronomical theory, concrete computations, or astrological issues. They can be presented as independent texts or as commentaries and expansions on previous works. They can be more or less firmly attached to a given table set or instrument. Relevant table sets or instruments are often the object of several procedural texts redacted at different periods and for different purposes. In sum, Alfonsine astronomers considered procedural texts as a major arena of creativity within mathematical astronomy. They thought about new procedures for a diversity of purposes and often with multiple layers of meaning. They also developed their reflection by stretching the genres of texts with which they worked.

Andriani, **Gessner**, and **Jacobson** consider the case of the *Arabes maxime*, a text uniquely preserved in a mid-fifteenth-century manuscript produced by Conrad Heingarter (fl. 1450s-70s). They show that several features of this rather comprehensive procedural text indicate an audience primarily trained in Alfonsine astronomy. However, the text is also constructed as an expansion and commentary, including diagrams and different geometrical reasonings, on the most diffused canons to the eleventh-century Toledan Tables. In several instances, the text explicitly compares the Toledan Tables and Alfonsine mathematical astronomy. Although it is difficult to propose with any level of certitude a context of composition for the text, these authors note that it was copied by a first-rank astrologer at a time nearly a century after the Toledan Tables had long ceased to be the main computational tool for mathematical astronomers. Perhaps, the authors suggest, Heingarter and before him the author of this treatise were in fact interested in the Toledan canons for the procedures they contain. A substantial excerpt of a critical edition and translation of this text is included as part of this contribution.

Andriani, Bui, and **Jacobson** take an interest in procedures for planetary latitudes and their grounding in theory. Their chapter deals with an original treatise on planetary latitudes composed by the same Conrad Heingarter. This text, redacted as a letter to a fellow master at the Paris university, is presented as a commentary to the *Theorica planetarum*. The text integrates characteristics of several genres: the letter, the treatise, and the commentary. Two institutional contexts are evoked in the text: the court and the university. It is, moreover, interesting that this canon treatise is not presented in direct connection with any specific table set, but rather as a commentary to the *Theorica planetarum*. This illustrates again that procedural texts could be, for these historical actors, much more than simple instructional texts that one could easily forget after learning how to manipulate tables. The authors also analyse the long-term trends that overlap mathematical astronomy and astrology by which Heingarter is driven towards the topic of planetary latitude. A critical edition and translation of this short treatise is provided by the authors.

Gessner and **Saby** consider a procedural text dedicated to an instrument rather than a table. The text was composed by the early-fourteenth-century Parisian astronomer John Vimond, whose table set is the first known attestation of Parisian Alfonsine parameters. The text describes a variant of the astrolabe designed for specialised astrological computations like houses and directions which require projections of various circles onto the equator and the ecliptic. Like the texts analysed in the two preceding contributions, this work is preserved in a single copy. The authors provide a critical edition and translation of the treatise as well as a tentative reconstruction of the instrument itself and an analysis of its unique technical terminology. The authors also consider the problem of the motion of the solar apogee, a topic of primary concern for early-fourteenth-century Parisian astronomers. Analysing the instructions, they argue that the instrument was not only intended as a computational aid but also as a tool for theoretical reflection, of interest in itself but of little numerical consequence given the accuracy of the instrument's scales.

Finally, **Jacobson** studies two fourteenth-century commentaries to the most frequently copied canons on Alfonsine planetary astronomy: John of Saxony's *Tempus est mensura* and John of Saxony's master John of Lignères's *Priores astrologi*. These commentaries propose another sort of expansion of procedural texts in which procedures are linked to both theoretical diagrams and computed examples. In this context, Jacobson analyses the function of diagrams that are used both as icons of the geometrical models grounding the tables and computations and as ways of visualizing the computation and different steps and branching points of the procedures. This mathematical practice with diagrams has a trajectory in the Alfonsine tradition and, as the author shows, may also be found in Arabic and Persian sources stretching into the fifteenth century.

Part 2: Reconfiguring tables

A second major area of engagement for Alfonsine astronomers was in the design, configuration, and visual layout of tables and table sets (without modification of the underlying planetary models and corresponding parameters). The basic computational task as defined by Ptolemaic astronomy, combining mean motions and corrections (or equations) to obtain true positions of celestial bodies, could be performed via multiple paths. The layout of the table usually shapes the precision of the computation, the need to rely on interpolation, and the means of doing so. Deeper layers of the procedures also emerge from the design of tables, for instance, the exploitation of symmetries. Tables also select those astronomical quantities that are manipulated in the procedures and those that are not; they select the operations that must be performed by users and those that have been precomputed in the tables themselves, thus determining the meaning of the procedure. Of course, tables also necessarily contain quantitative information, that is, implicitly at least, information about astronomical theory and the relevant parameters such as the obliquity of the ecliptic, radius of planetary epicycles, or the length of the synodic month. Since Alfonsine astronomy is structured around the circulation of the *Parisian Alfonsine Tables*, all of its distinctive theoretical features were first expressed in a tabular format. In some cases, for example the motion of the eighth sphere, astronomical theory that would ground the tables remained a subject of discussion throughout the entire period. By analysing how Alfonsine astronomers worked with their tables in different contexts, the contributions in this section shift attention from parameters alone to the actual scenarios by which the tables themselves were computed or with which they are to be used. This allows us to understand the mutual relations among tables and their multiple roles in a mathematical astronomy centred on procedures and computational practices in more detail.

Hadrava studies a small but fundamental subset of the *Parisian Alfonsine Tables* dedicated to the motion of the apogees and fixed stars. These three tables propose a double motion, combining new quantitative parameters with the trepidation commonly ascribed to Thābit ibn Qurra in the Middle Ages and Ptolemy's constant motion of precession. This peculiar combination is characteristic of Alfonsine astronomy. It was one of the features that fostered much discussion and work from astronomers of the period; they explored its numerical consequences, tried to reformulate it with different procedures and tables, or speculated about possible geometrical models that would account for tables and be consistent with Aristotelian physics. Building on previous publications, Hadrava explores possible scenarios for geometrical models and their associated analytical expression that could generate the attested tabular data. He concludes that models following the hypothesis of a fixed obliquity best fit the historical data. In order to feed the different layers of information embedded in the analytical models he uses, Hadrava draws alternatively on tabular, diagrammatic, and textual sources.

Van Brummelen considers the emergence of a different category of tables in Latin sources. Auxiliary tables provide computational help for astronomers. They display relations between quantities that are more complex than primitive trigonometric quantities but not specific to a particular astronomical problem and thus can be used in a variety of astronomical contexts. In Europe, this type of table first appears in the works of the fifteenth-century astronomer Giovanni Bianchini as he elaborates the second version of his *Tabule primi mobilis*. Bianchini offers unique access to the primitive building blocks from which he composed this influential table set for spherical astronomy. By publishing his auxiliary tables under the title *Tabule magistrales*, Bianchini initiated a tradition that would be picked up by his followers, starting with Regiomontanus, and would culminate in the introduction of the logarithm about 150 years later. Auxiliary tables, however, had long existed in the context of Persian and Arabic astronomy, and Van Brummelen explores the contrasts and similarities between the Latin sources and those of these two traditions.

Chabás and **Goldstein** focus on an early Alfonsine Parisian astronomer, John of Genoa (fl. 1332), who was the compiler of a remarkable table for the radii and velocities of the Sun and the Moon at syzygies. Although this table was widely diffused in the Middle Ages, it has barely been addressed by modern scholars. Chabás and Goldstein explain the table and recompute its entries. The algorithms underlying this table turn out to be similar to those used by previous astronomers who compiled tables for the same purpose, except for the lunar velocity. In this case, John of Genoa computed the lunar velocity for the first time using Ptolemy's second or complete lunar model; he was followed in this by his contemporary John of Montfort and, a century later, by John of Gmunden. Analysing the computational practices underlying these tables shows how mediaeval astronomers made consistent use of an increasingly refined terminology along with the resources of the grid format to convey complex relations between astronomical quantities. This study shows once again that close examination of mediaeval astronomical tables can be a useful and efficient approach to understanding computational practices and the transmission of astronomical knowledge.

Penin also explores the making of a particular set of tables: Jean Fusoris's trigonometric tables composed between 1432 and 1436. Fusoris had planned to compute a full set of astronomical tables, but his work was interrupted after he completed the chord and sine tables. He nonetheless had the time to write, along with the tables themselves, a very interesting text explaining some aspects of his endeavour. Relying on this text and the only extant manuscript in which the tables are now accessible, Penin reconstructs the reasoning and computational scenarios that were probably used and designed by Fusoris to construct his tables. Not only are Fusoris's tables of an unmatched level of precision for his time in Europe, but he also developed a sort of 'tabular mathematics'. Working with differences up to the third order and a grid presentation of the numbers, Fusoris introduced a procedure that we would describe today as third-order polynomial interpolation. This allowed him to circumvent the delicate problem of the computation of the sine of one degree by using original geometrical lemmas. Thanks to Penin's analysis, the work of Fusoris may well receive greater attention in the context of the general history of trigonometry. As far as the history of Alfonsine astronomy is concerned, Penin's study also illustrates the case of an astronomer with great concern for the accuracy of the fundamental computation tools of mathematical astronomy. This is another manifestation of Alfonsine astronomical practices centered on procedures.

Finally, **Andriani** and **Zieme** investigate an original text by Regiomontanus, which they edit and translate in order to understand how the Viennese astronomer of the 1450s conceived the computation of the equation of time (the difference between apparent time observed on a sundial and mean time used in astronomical computation). This difference is an essential component of mathematical astronomy and, at least in theory, required occasional updates because of the secular motion of the solar apogee. Andriani and Zieme show that Regiomontanus's approach to the computation of the equation of time depended heavily on the uses of astronomical tables rather than raw trigonometric relationships. Although he undoubtedly had a precise grasp of the spherical configurations that define the equation of time, Regiomontanus addressed it not as geometrical problem to be solved with tools such as Menalaeus's theorem, but rather through more proximate relations found in, for example, right ascension tables. In particular, Regiomontanus's method is inspired by the analysis in the *Almagesti minor* (*c.* 1200), but he also relied on Bianchini's astronomical tables of the 1440s. Andriani and Zieme also reflect on the differences between computation scenarios of the kind presented by Regiomontanus in this text and the modern analytical formulae that are commonly used to examine ancient and mediaeval astronomical tables of the equation of time.

Part 3: Operating with numbers

Although, as explained above, procedures and tables were injected with multiple layers of meaning by Alfonsine astronomers, their central goal was to describe and manage the long and complex chains of computation that are required in order to yield quantitative descriptions of past or future celestial phenomena such as the true positions of celestial bodies. In this work, mathematical relations between multiple quantities are displayed. Some of these quantities are more important than others due to their astronomical meaning or their role as nodes which will impact subsequent stages in computation. Such quantities are often well identified by the computing astronomers and receive specific attention in procedures and tables. Astronomical computations manipulate different kinds of numeration systems (sexagesimal and decimal), measuring units (minutes of time and angle), and quantities to which the usual arithmetical procedures designed primarily for simple integers must be adapted. Much of the computational labor of astronomers is dedicated to the fundamental operations of managing units in multiplication or division and in converting a number from one format to another in order to allow its manipulation within the context of a particular tabular format or with a given instrument, etc. Finally, the study of astronomical computation also offers historians the opportunity to analyse how a celestial phenomenon, which was primarily understood as a continuous motion in the heavens, is managed via numbers associated with the category of discrete quantities. This fundamental tension explains many features of the numbers, units, and quantities used by mathematical astronomers and their need to make decisions regarding accuracy and precision.

Hadravová and **Hadrava** concentrate on a set of canons to the *Tabule resolute* that were one of the main formats under which Alfonsine astronomy circulated in fifteenth-century Prague and Cracow. Unlike the *Parisian Alfonsine Tables*, the *Tabule resolute* employ a calendrical format for the mean motions. They also set the radices to places and dates that are closer geographically and chronologically to the actual users of the tables. Setting these fundamental quantities in this way not only reduced computational labor but also facilitated better control over the potential scribal corruption of these values. Shortening the time spans over which mean motions were computed also allowed the use of less precise numbers in the table and in computation without significantly reducing the accuracy of the computation. Moreover, the different versions of the canons to the *Tabule resolute* propose several procedures that enable users to compute their own radices. Analysing these procedures and comparing them to computations of the *Parisian Alfonsine Tables* for the same purposes, Hadravová and Hadrava conclude that the *Tabule resolute* are designed for a different kind of user than are the *Parisian Alfonsine Tables.*

Chabás presents an extensive analysis and survey of two sets of computations attributed to the fourteenth-century astronomer Hermann of Saxony, who has remained mostly unknown to date. This genre of procedural texts, often referred to as *practica* by mediaeval scribes, is known from previous uses (e.g., John of Saxony for spherical astronomy or John of Genoa for eclipse computations). Hermann of Saxony provides full computations for planetary positions, following the procedures outlined by John of Saxony in his *Tempus est mensura*. Present in several manuscripts, these practical canons slowly established a genre that persisted well after the medieval period. The analysis of Hermann of Saxony's texts consisting of worked examples provides us with hints as to how medieval astronomers managed long astronomical computations, handled precision and rounding, and interpolated from their tables. At the same time, Hermann of Saxony's long and complex chains of computations highlight the need felt by late medieval astronomers for precise specialised terminology.

Husson's contribution focuses on a critical element in most astronomical computations, especially those relying on numerical tables: interpolation. He argues that what we categorise today under this term was, in the context of Alfonsine astronomy, more a family of procedures linked by their common use of proportional part computation. Husson analyses how proportional part computation is extended in different ways to handle diverse situations for single- or double-argument tables and forms various interpolation procedures. In order to elaborate on proportional part procedures, astronomers worked on the terminology, the sequence of operations, and the design and layout of specific tables or diagrammatic elements. This family of proportional part procedures in turn presents fundamental intuitions or implicit postulates about the smoothness of change in given astronomical quantities.

Part 4: On the border of mathematical astronomy

Alfonsine astronomers collected and copied materials that they assembled into toolbox manuscripts, developed new procedures and tabular formats to facilitate their work with these toolboxes, and computed astronomical positions and celestial events such as eclipses. Their astronomical manuscripts and practices would spread throughout Latin Europe during most of the fourteenth and fifteenth centuries. However, the celestial positions they computed mostly circulated in non-specialist circles, communicated in two related but distinct formats that can be understood as genres sitting on the border of mathematical astronomy: horoscopes and ephemerides (e.g., a list of serial positions, often daily, of the luminaries and the planets, generally set for a year). Analysing these genres can help us understand how the needs of non-specialists might shape the interface between mathematical astronomy and other disciplines such as medicine and astrology. For instance, it is interesting to notice that both ephemerides and horoscopes present lists of numbers in specific page layouts that enable lay viewers to quickly grasp the distribution of the planets and luminaries on the ecliptic at a given time. Ephemerides also offer the possibility of diachronic views of these celestial phenomena, and with their house divisions horoscopes situate the celestial position at a given time on a local horizon. Conversely, horoscopes and ephemerides also display the results of astronomical computation and thus provide clues about how late mediaeval astronomers and astrologers of different milieus and skills practised their art. Like numerical tables or computed examples, horoscopes and ephemerides can be recomputed. If we deploy their tools and seek to follow their procedures, perhaps we can learn something about their computational habits.

Kremer's chapter considers ephemerides. He asks what these documents might reveal about the computational practices of the astronomers who constructed them. Kremer first establishes a distinction between two levels of agency that can be unravelled. An explicit agency is visible in features such as the layout of the documents, the precision of the position proposed, the refinement of the temporal grid on which the position is given for each planet, the time span for which the ephemerides are computed, and so on. An implicit agency is accessible only through a closer analysis of the quantitative data, aiming to reconstruct the computation scenarios by which the ephemerides were originally made. The chapter considers the meridians used for the computations, the underlying tables producing these positions, the possible presence of interpolation schemes, and the computation signature or personal 'noise' of the astronomer computing the positions. Focusing on two well-known and large-scale ephemerides by John of Saxony and Regiomontanus, Kremer discloses some aspects of the computational practices of these two astronomers that are not accessible through other means or by reading their procedural texts.

Tur analyses, in five examples from the fifteenth century, the structure of what we usually refer to as 'horoscope squares', often designated in historical sources as *figurae caeli*. In a manner somewhat similar to Kremer's analysis, Tur examines explicit features of these diagrams as well as more implicit practices accessible only through the reconstruction of the possible computation scenarios by which the quantitative data recorded in these diagrams were obtained. He identifies the multiple ways the horoscope makers articulated mathematical astronomy procedures and astrological interpretations. More generally, the chapter points to the cultural meaning of these diagrams, which turns out to be an 'aestheticized symbol of the scientific status of astrology'. Tur collects evidence of the accuracy of the computation underlying the quantitative data in these diagrams, which can then be fruitfully compared with those assembled by Kremer and other authors in this book.

Future studies will continue to expand the sets of manuscripts, authors, and texts discussed in the literature and to deepen our understanding of how Alfonsine astronomers, at work with tables, diagrams, instruments, and texts, developed their own trends in the broader context of the Ptolemaic tradition. It might be useful in this respect to point out three complementary aspects of Alfonsine astronomers' practices emerging from the studies assembled in this volume.

Conceptually, tables, diagrams, instruments and texts are all implicated in the computational procedures of medieval astronomers. On a linguistic level, different contributions point to the importance of technical terminology in connecting the various elements used to express, compare, and elaborate procedures. Materially, these different elements were assembled in codices, which can be considered the essential tools of the astronomers at work. Understanding the interplay between these material, linguistic, and conceptual aspects is one of the promising venues of future research on this corpus.

PART 1

Procedural Texts

ELEONORA ANDRIANI, NICHOLAS JACOBSON and
SAMUEL GESSNER

A Note on *Arabes Maxime*

A Testimony of Alfonsine Astronomers' Interest in the Toledan Tables[*]

The Bibliothèque nationale de France (BnF) preserves eleven manuscript volumes related to Conrad Heingarter, a Swiss physician and astrologer active in the second half of the fifteenth century.[1] One of these manuscripts, Paris, BnF lat. 7295A, contains what is perhaps the only witness of an anonymous astronomical work beginning with the expression 'Arabes maxime'. This work has so far slipped through the net of investigation.[2] The aim of

* This chapter has been written collectively. The three authors have equally contributed to the writing process and revision work. Eleonora Andriani wrote the preamble and section 1, Samuel Gessner wrote section 2.1 and 2.2, Nicholas Jacobson wrote 2.3 and the conclusion. The edition was established collectively.
We extend our thanks to the members of the entire ALFA team for their participation in discussions and seminars on our work. Our thanks go particularly to Charles Burnett, José Chabás, Alena Hadravová, Matthieu Husson, Richard L. Kremer, Sophie Serra, and Stefan Zieme.
This work was supported by the ERC project *ALFA: Shaping a European scientific scene, Alfonsine astronomy*, CoG 723085, PI Matthieu Husson.

1 'The BnF collection now holds eight manuscripts related to Conrad Heingarter (lat. 7197, lat. 7273 (?), lat. 7295A (?), lat. 7314 (?), lat. 7432, lat. 7446, lat. 7447, lat. 7450)'. Matthieu Husson, 'Mathematical Astronomy and the Production of Multiple-Texts Manuscripts in Late Medieval Europe: A Comparison of BnF lat. 7197 and BnF lat. 7432', in *The Emergence of Multiple-Text Manuscripts*, ed. by Alessandro Bausi, Michael Friedrich and Marilena Maniaci (Berlin: De Gruyter, 2019) <https://doi.org/10.1515/9783110645989> [accessed: 14.04.2023], p. 249. The following manuscripts should be added to the list above: lat. 7333, lat. 11232, lat. 7305.

2 An exception should be made for some marginal or rather short references in the secondary literature and catalogues. See Matthieu Husson, 'The Art of Astrological Computations: Conrad Heingarter and the Manuscript Paris, BnF latin 7295A', in *Exploring Written Artefacts: Objects, Methods, and Concepts*, ed. by Jörg B. Quenzer (Berlin: De Gruyter, 2021) <https://doi.org/10.1515/9783110753301> [accessed: 14.04.2023], p. 526; this article focuses on the manuscript 7295A and highlights its specific characteristics of a personal 'toolbox' for computation. See also, Fritz S. Pedersen, *The Toledan Tables: A Review of the Manuscripts and the Textual Versions with an Edition* (Copenhagen: C.A. Reitzels Forlag, 2002), p. 157; Lynn Thorndike and Pearl Kibre, *A Catalogue of Incipits of Mediaeval Scientific Writings in Latin*, rev. ed. (London: Mediaeval Academy of America, 1963), p. 124; Francis S. Benjamin Jr. and Gerald J. Toomer, *Campanus of Novara and Medieval Planetary Theory:* Theorica planetarum (Madison: University of Wisconsin Press, 1971), pp. 83-85. See also the description prepared by Laure Rioust in the BnF catalogue (June 2018); see 'Latin 7295A' <https://archivesetmanuscrits.bnf.fr/ark:/12148/cc664943> [accessed: 14.04.2023].

Eleonora Andriani • Centre national de la recherche scientifique, Paris
Nicholas Jacobson • University of Wisconsin-Madison
Samuel Gessner • Universidade de Lisboa

Alfonsine Astronomy: Computational Practices, ed. by Matthieu Husson, José Chabás, and Richard L. Kremer, Turnhout, 2026 (*Alfonsine Astronomy: Studies and Sources*, 3), p. 19-51
© BREPOLS PUBLISHERS 10.1484/M.ALFA-EB.5.137907

this contribution — besides providing introductory information about the work at issue covering folios 182^{r}-193^{v} — is to highlight some of its striking characteristics. Although the work is anonymous and its attribution to Heingarter doubtful, we argue that a full edition and study of this text is justified, not only for its intrinsic interest, but also in the hope that it will help refine our understanding of the history of Alfonsine astronomy. The first section of this contribution focuses on the manuscript context and the text itself, including a treatment of its sources, the time frame of its composition, and its genre. The second section comments upon some of the most intriguing aspects of this work, namely its explicitly comparative approach in which the author weighs the Toledan Tables and canons against the Alfonsine procedures and the work's use of diagrams when 'demonstrating' the Toledan procedures. This section also discusses some relevant terminology and concepts related to mathematical reasoning that we have found in the text. The Appendix provides a selection of chapters excerpted from the critical edition in preparation.[3]

1. Manuscript context

The collection of works in Paris, BnF lat. 7295A (former shelfmarks: Colbert 979; Regius 5073. A) was produced by Conrad Heingarter at least partly during his university years in Paris (third quarter of the fifteenth century).[4] The manuscript codex entered the Royal Library in 1732 at the same time as BnF lat. 7197, another manuscript produced by Heingarter.[5]

Heingarter seems to have written most of the folios 1^{r}-32^{v}, as well as folios 35^{r}-45^{r}, 49^{r}-93^{r}, 145^{r}-52^{r}, and 181^{r}-193^{v}, including some passages in German (ff. 64^{v}-92^{r}) and in an explicit Greek lettering (f. 45^{r}).[6] He also annotated certain texts written by other copyists (e.g. ff. 99^{r}, 100^{r}, 164^{r}, 179^{r}-80^{v}). The manuscript consists of seventeen quires and contains several works, including the following: *Theorica planetarum Campani*, the *De armillis* by Profatius Judaeus, Ps.-Messahala's *De astrolabio* (only the part on construction), the Parisian Alfonsine Tables (augmented by the Oxford double-entry tables for latitude) accompanied by the canons of John of Saxony, and the tables of 1322 and related canons by John of Lignères.[7]

3 For the Appendix, see pp. 37-48 in the present contribution.

4 The watermarks identified in this manuscript are attested in Paris in the third quarter of the fifteenth century, between 1450 and 1463, which corresponds to the years when Heingarter was studying there. See Husson, 'The Art of Astrological Computations', p. 515 and the section 'présentation du contenu' in Laure Rioust's notice.

5 See Husson, 'Mathematical Astronomy', p. 255.

6 If transliterated into Latin, it would read: 'Explicit seu finis per Konradum Heingarter'. Paris, BnF lat. 7295A, f. 45^{r}. The explicit is bordered by three strokes drawn in red ink and is written in Greek letters. For general information on the use of Greek script in Latin manuscripts, see Walter Berschin, 'Greek Script in Latin Manuscripts', in *The Oxford Handbook of Latin Palaeography*, ed. by Frank T. Coulson and Robert G. Babcock (New York: Oxford University Press, 2020), pp. 115-17. A notable example of the practice of writing information in Latin but using Greek script is found in the manuscript Toledo, Archivo y Biblioteca Capitulares, 104 6, a Boccaccio autograph manuscript consisting of an anthology of Dante's works. Here, Giovanni Boccaccio seems to have written the following phrase in Greek script as a caption to a portrait found in the manuscript: 'Ioannes de Certaldo p[inx]it', meaning 'John of Certaldo painted it'. See Stefano Martinelli Tempesta and Marco Petoletti, 'Il ritratto di Omero e la firma greca di Boccaccio', *Italia Medioevale e Umanistica* 54 (2013), 399-409 (p. 401).

7 Although the last quire bears the number 18 in the bottom left margin, the quires constituting BnF lat. 7295A are only 17. The quire structure and a table of contents is available in the notice produced by Laure Rioust. For a complete overview of the works included in BnF lat. 7295A, see Husson, 'The Art of Astrological

The seventeenth and final quire is the one that contains our text, *Arabes maxime* (preceded by some shorter works), all written in the hand identified as Heingarter's. Heingarter is likely copying a preexisting work. Among the hints for this there are several inconsistencies between the text and the lettering of the diagrams. Another one is the text's reference to 'red circles' in a diagram that are not rubricated in this witness. This quire starts on folio 181^{r}, with a text on the colours of lunar eclipses, which Fritz S. Pedersen identified as an addition to the Toledan canons (version Cb) and defined as 'Alkhwarizmian'.[8] This paragraph on eclipse colours is followed by another short text bearing the incipit 'Nota quod eclipsis lune', which mostly corresponds to a passage from the chapter on the direct, retrograde, and stationary positions of the planets from the *Theorica planetarum communis*.[9] This text ends with the explicit *Expliciunt canones secundi* (!) *mobilis magistri Iohannes de Lineriis finiti Wienensi studio* indicating that the work was copied in the city of Vienne in France.[10] Indeed, the previous quire (i.e. the sixteenth) includes the canons *Priores astrologi* (ff. 174^{r}–80^{v}) by John of Lignères. These canons were not copied by Heingarter, but include some notes in his hand, and bear traces of his careful reading. Lignères' canons end abruptly, and a remark in the lower margin, *Hic deficiunt canones quinque, scilicet 41us, 42us, 43us, 44us et quadragesimus quintus*, indicates that the canons missing from the work had not gone unnoticed to the attentive reader.

The rest of this first folio in the seventeenth quire (f. 181^{r}–181^{v}) includes a text with the incipit *Nota secundum Alfraganum centrum ecentrici solis*. Small paragraphs starting with 'nota' and providing different kinds of definitions or clarifications show that the text is not an organic work, but rather a set of collected notes, or possibly expansions on various topics.

These short accounts defining or elucidating different subjects include the following topics: the distance of the centre of the eccentric of the Sun, the Moon, and the planets from the centre of the Earth; the semi-diameter of the epicycles of the planets; the diameter of the Sun, the Earth, and the Moon and their proportion to one another; parallaxes; planetary motion in longitude; planetary motion in latitude; lunar mean motion in latitude; the equated motion of lunar latitude; the equated argument of the Moon; the true latitude of the Moon; the apparent latitude of the Moon; the declination of the Sun; the *sinus rectus*, *sinus versus*, and the *kardaga*; the place of the stars according to longitude and latitude; a method for identifying the variation in the lengths of the daylight referring to the arcs *portio diei* and *portio noctis*; the definition of the conditions necessary for an eclipse to take place; and various details on sexagesimal computations.[11] This rather heterogeneous text

Computations', pp. 523-27.

8 'The text was popular later: part (1) is in the Alfonsine Tables, and the incipit of part (2) is quoted by Steinschneider (Études, 1881+, p. 41) from the 1518 printing of John of Saxony's rules for the Alfonsine Tables, folio 23^{v}, where it is ascribed to John of Lignères'. Pedersen, *The Toledan Tables*, p. 535. See Moritz Steinschneider, 'Études sur Zarkali, astronome arabe du XIe siècle, et ses ouvrages', offprint from *Bullettino di bibliografia e di storia delle scienze matematiche e fisiche* (ed. Boncompagni) 17 (1884), p. 781.

9 See *Theorica planetarum Gerardi* VI. 84-86, ed. by Francis J. Carmody (Berkeley, CA: University of California Press, Berkeley, 1942), pp. 39-40.

10 BnF lat. 7295A, f. 181ra.

11 Interestingly, some passages of the text beginning with 'Nota secundum Alfraganum centrum ecentrici solis' exhibit verbatim correspondences with certain passages included in the *Almagesti Minor*, books V-VI; see Henry Zepeda, *The First Latin Treatise on Ptolemy's Astronomy: The* Almagesti minor *(c. 1200)* (Turnhout: Brepols, 2018), pp. 330 and 434.

continues on the unruled page in a single column (f. 181^{v}) and is followed by a diagram, the caption of which reads 'The figure of the solar eclipse'.[12]

This section of the manuscript with the incipit *Nota secundum Alfraganum*, is displayed as a single text and consists of a continuous chain of notes. Significantly, these paragraphs introduced by the word 'nota' suggest that the genre of this text is some sort of *catena* commentary. *Catena* commentaries are glosses that 'break away from the immediacy of the text [...] and claim the status of independent expositions, which in turn may inspire their own glossing activity'.[13]

Opening with the incipit *Arabes maxime secundum motum lune*, the content of the manuscript from folio 182ra until the end of the quire on folio 193vb appears as a single work divided into nearly fifty unnumbered chapters of various lengths and touching upon a variety of astronomical and trigonometrical topics.[14] An overview of the chapters and the topics, some of which are explored in this contribution, is displayed in the table below. The table provides the order of the chapters, the folio reference to where they begin, the incipit, and the subject area they treat. The chapter beginnings are usually marked by spaces reserved for initials.[15]

Chapter	Folio	Incipit	Topic
1	182ra	<A>rabes maxime secundum motum lune tempora distinguentes	Chronology
2	182ra	<C>um volueris scire annos, mense<s> et dies Arabum ex annis domini nostri	
3	182rb	<S>i exordium anni vel mensis Arabici scire cupis	
4	182rb	<C>um alicuius arcus sinum rectum sive equalem quesieris	Trigonometry and *primum mobile*
5	182vb	<Q>ualiter invenitur per numerum portio circuli equalis	
6	183ra	Et est sciendum quod licet inter sinum et portionem non sit proportio	
7	183ra	<Q>uamvis in hoc opere de practica principaliter cogitemus	
8	183vb	<E>t quoniam in premissa demonstratione	
9	183vb	<I>tem in eadem demonstratione	
10	184ra	<S>i aliquot gradus ecliptice cuiuscumque signi multiplicaveris	
11	184ra	<Q>ualiter per altitudinem solis inveniantur puncta umbre	
12	184rb	<H>actenus de motu orbis primi et pertinentibus ad ipsum tractavimus	Planetary longitudes
13	184vb	<N>otandum quod centrum epicicli tam Veneris quam Mercurii	
14	185rb	<S>ecundum Alfraganum luna cum fuerit in superiori parte epicycli movetur	
15	185rb	<M>ercurius, Venus et sol in suis motibus mediis eodem modo se habent	
16	185va	<M>ars in die movetur in epicyclo 28 minutis et percurrit eum in duobus annis	
17	185vb	<E>t quoniam planete moventur in ecentricis ut visum est	

12 BnF lat. 7295A, f. 181^{v}: 'Figura eclipsis solaris'.

13 Greti Dinkova-Bruun, 'Text and Gloss', in *The Oxford Handbook of Latin Palaeography*, ed. by Frank T. Coulson and Robert G. Babcock (New York: Oxford University Press, 2020), pp. 933-34.

14 From this point on, we will refer to this work by its incipit, that is, *Arabes maxime*.

15 Only six chapters out of 49 include initials, five of which are rubricated; additional four chapters include guide letters and only one chapter includes a heading. See the Appendix, p. 37 n. 64.

Chapter	Folio	Incipit	Topic
18	186ra	<N>ota in tabulis Tholetanis utimur annis Arabum et ponuntur ibi radices	Planetary longitudes
19	186ra	<A>d inveniendum medios motus planetarum in tabulis tholetanis	
20	186rb	<S>olem equabis per omnem motum sicut in Alphontio	
21	187rb	<A>d habendum verum locum capitis draconis	
22	187rb	<H>oc placuit extrahere de equatione Saturni	
23	187vb	<V>enus et Mercurius equantur sicut tres superiores omnino hoc accepto	
24	188ra	<C>um aliquis quinque planetarum fuerit in superiori parte epicycli	
25	188rb	<S>i de aliquo planetarum scire volueris utrum sit stationarius	Planetary passions
26	188vb	<C>um fuerit centrum planete equatum ab uno gradu in 6 signa	
27	188vb	<O>rtum et occasum elaycum [*sic*] planetarum sic invenies in 3 superioribus	
28	189ra	<U>trum planeta sit sub radiis vel videri possit per tabulas ad hoc constitutas	
29	189rb	<S>i noscere desideras qua hora solis vel luna vel quilibet alius planeta	
30	189rb	<P>ostquam de motu longitudinis planetarum dictum est	Planetary latitudes
31	189va	<L>atitudinem lune sic inveneris	
32	189va	<A>d cognoscendum diversitatem que accidit in latitudine 5 planetarum	
33	190rb	<E>t si per tabulas invenire volueris quanta sit latitudo	
34	190va	De Venere. Veneris autem et Mercurii latitudines per similes tabulas invenies	
35	190va	Nota quando est coniunctio vel oppositio solis et lune	Syzygies
36	190va	Cum itaque sciveris in quo mense potest fieri eclipsis solis vel lune	
37	190va	Certum tempus et verum locum coniunctionis et oppositionis invenies hoc modo	
38	190vb	Et quoniam in operatione premissa docuimus ut longitudinem	
39	191ra	<C>um eclipsim lune absque errore prescire et sine dubio predicere volueris	Lunar eclipses
40	191va	<Q>uantum autem obscurabitur de superficie lunaris corporis	
41	191va	<L>une dyametrum ad horam eclipsis sic invenias	
42	191va	<D>yametrum circuli umbre in loco transitus invenies hoc modo	
43	191vb	<S>i eclipsim lune in figura geometrica representare volueris	
44	192ra	<N>unc videndum est de hiis que circa eclipsim solis consideranda occurrunt	Solar eclipses
45	192rb	<S>i diversitatem lune invenire volueris ad horam eclipsis	
46	192vb	<C>um scire volueris qua die et qua hora diei eclipsis solaris ventura sit	
47	193rb	<S>i scire volueris quantum de superficie solis obfuscari debeat	
48	193rb	<C>um solis dyametrum ad horam eclipsis invenire volueris	
49	193va	<S>i solaris eclipsis figuram depingere volueris	

Arabes maxime (i.e. 'The Arabs chiefly'), the initial words of this work, immediately provide the reader with a frame of reference for the main tradition underpinning it. The expression derives from the Toledan canon (canon Cb7 in the version Cb as established by Pedersen) that begins *Arabes vero, a lunae maxime motu ipsius temporis metientes spatia*.[16] Though most of the text appears to be clearly indebted to the Cb canons, the work also relies upon other sources that could be said to belong to an 'Arabic tradition'.[17] Hence, the text is remarkable in that it establishes this interest in the Arabic authorities at the very outset, and this initial marker of interest is matched by the stress put in the text on the approach to astronomical tables that is presented as 'Arabic' by the author.[18] As appears from the table above, the work includes forty-nine chapters, which can be distributed into several topics. These are: chronology, trigonometry and the *primum mobile*, planetary longitudes, planetary passions, planetary latitudes, syzygies, lunar eclipses, and solar eclipses. The author of the text has consciously chosen a limited set of topics, excluding some treated in the Toledan canons or other works. At times this is made explicit, such as when the author remarks that the solar and lunar motion would not be discussed in a certain passage, since they had been sufficiently treated in the *Theorica planetarum*.[19]

A rough temporal context for the work can be inferred from the references it includes to Toledan and Alfonsine frameworks. A few hints to define the time period are offered by the first lines of Chapters 18, 19, and 20.[20] Indeed, the incipit of Chapter 18 reads, 'Note that in the Toledan Tables we use the years of the Arabs, and the radices of their epoch are put there'. The incipit of Chapter 19 reads, 'The operation to find the mean motion of the planets in the Toledan Tables is the same as that in the Alfonsine Tables'. The incipit of Chapter 20 reads, 'You will equate the Sun through all the motion as it is taught in Alfonso'.[21] Some further hints from the work will be offered to narrow down the span of time in which it may have been composed (see Section 2.1). While the time of the composition of this work was posterior to Alfonso (1270s), it appears that there was an ongoing interest in the 'Toledan tradition', since this work refers to the Toledan Tables quite frequently and depends largely on the Cb canons.[22]

The bond that connects *Arabes maxime* to the Toledan canons is close enough to raise the question of its status as an independent work. Given its great dependency, it is possible to consider the text as an 'Alfonsine exposition' on the Cb canons. In his edition of the Toledan Tables and canons, Pedersen lists eleven texts that he considers as cognate to or

16 See Pedersen, *The Toledan Tables*, p. 382.

17 The sources that have been identified so far, besides the canons Cb, are Euclid, al-Farghānī, and Ps.-Thābit ibn Qurra. However, the ongoing work of editing the text will enable us to determine more sources. This is certainly one of the important advantages that a complete edition of this text will offer.

18 '<N>ota: in tabulis tholetanis utimur annis Arabum et ponuntur ibi radices ere Arabum'. See Appendix, Ch. 18, 44.

19 'de motu solis et lune satis habetur in theorica planetarum'. BnF lat. 7295A, Ch. 12, f. 184vb.

20 Note that an edition of these chapters is included in the Appendix of this contribution. See Appendix, pp. 44-45.

21 '<N>ota: in tabulis tholetanis utimur annis Arabum et ponuntur ibi radices ere Arabum'. See Appendix, Ch. 18, pp. 44; 'Ad inveniendum medios motus planetarum in tabulis tholetanis eadem est operatio que in tabulis Auphontii'. See Appendix, Ch. 19, pp. 45; '<S>olem equabis per omnem motum sicut in Alphontio docetur'. See Appendix, Ch. 20, pp. 45.

22 For a definition of the 'Toledan tradition' including the works and authors that it encompasses, see Pedersen, *The Toledan Tables*, pp. 44-45.

derived from the texts of Cb.[23] He provides a short characterization of each of these texts, denoting them in a variety of ways, although his categories seem somewhat loose. The resulting classification comprises the following characterizations: '[a] series of excerpts similar to rules from Cb, with lengthy explanations added'; '[an] expanded paraphrase of Cb'; '[a] literal commentary on Cb'; '[an] extract of texts from a literal commentary on Cb [i.e. John of Sicily]'; 'a set of lemmatized glosses to Cb'; 'a series of canons on chronology and trigonometry, lightly paraphrased from Cb'; 'a comprehensive set of mixed glosses covering trigonometry (including sines of *kardagas*), planetary equations, eclipses, and risings'; and 'trigonometrical canons [which] might be paraphrases of Cb'.

Arabes maxime seems to be one of these kinds of texts. Precisely what category best defines it, however, remains to be determined. The following paragraphs will further explore the dependency of this work upon the Toledan tradition, while also noting the subtle and sometimes sophisticated ways in which the author supplemented the mathematical reasoning of the canons through his own commentary.

2

2.1 Comparative viewpoint on the Toledan and Alfonsine Tables

As noted above, the incipit of our text is a true catchphrase. *Arabes maxime* seems to appeal to readers curious about the way the 'Arabs' understood and practiced astronomy. The text is clearly addressed to an audience used to and conversant with the Alfonsine way, but who wanted to understand the Toledan Tables as well. Indeed, several of the *Arabes maxime* chapters (19, 20, 22) make an explicit comparison between the Alfonsine approach and the one associated with the Toledan Tables.

One can imagine that this interest in comparing the Toledan Tables to the Alfonsine ones was possibly common practice from the 1270s onwards, when some astronomers were using the Alfonsine Tables (either the Castilian or the Parisian Latin version), while others continued to favour the (Latin translated) Toledan Tables. Chapter 20 of *Arabes maxime*, opening with the sentence 'You will equate the Sun through all the motion as it is taught in Alfonso', contains a tantalizing hint that our text *Arabes maxime* might even refer to the Castilian version of the Alfonsine Tables before they were reformatted in Paris in the 1320s.[24] However, let it be said that the interpretation of the chapter is ambiguous, and no definite conclusion can be reached yet. If this view were confirmed, though, it would constitute a significant argument for an early dating of the text. In any case, to understand this hint, it is important to remember that mean motions are counted in a sidereal reference

23 See Pedersen, *The Toledan Tables*, pp. 340-41. The texts mentioned by Pedersen are 'Diversi astrologi', 'Canones Luce' — 'Ut annos Arabum', 'Kardaga est portio', 'Sicut dicit Hermes in libro de natura dierum', John of Sicily's 'Inter cetera veritatis philosophicae documenta', 'Scripta Marsiliensis super canones Azarchelis', 'Cum in singulis scientiis', 'Quia ad notitiam caelestis motus secundum doctrinam magistri Arzachelis perutile est scire annos Arabum', 'Ad maiorem manifestationem eorum quae dicuntur de umbra supra in canonibus Arzachelis, folio 4, est istud notandum…', 'Circulus quilibet potest dividí in 360 partes', and 'De canonibus Arzachel tractaturus'. Note that Pedersen states that he mostly refrains from considering texts of the fourteenth century or later.

24 'Solem equabis per omnem motum sicut in Alphontio docetur'. See Appendix, Ch. 20, p. 45.

frame in the Castilian version of the Alfonsine Tables, while in the Latin version mean motions are reckoned with respect to a tropical reference frame.

The crucial sentence in this chapter translates as follows: '... having obtained the true place of the Sun, count it (*computa eum*) from the beginning of the movable [zodiac sign] Aries, to which certain position add the equated motion of the 8th sphere. Thus, you will have its [the Sun's] position, by counting most accurately from the intersection of the circles'.[25] Opposition is made here between measuring a position in the eighth sphere (typically from the first point of the movable Aries) on the one hand, and measuring it from the vernal point of the prime mover, defined as the rising intersection of the (moveable) ecliptic and the celestial equator, on the other.

In combination with the opening of the chapter, the straightforward reading of this passage is that while equating the Sun is done for all motions as taught by the Alfonsine astronomers, this process implies two equations: that of the argument and that of the motion of access and recess of the eighth sphere. Adding the latter equation only at the end (and not before computing the equation) agrees with the Castilian Alfonsine canons (as with the Toledan Tables) and is clearly different from the procedure for the Parisian Alfonsine Tables.[26]

It is not entirely clear from the text, however, whether the passage provided represents an operation that is 'like it is taught in Alfonso', or whether it is given to precisely highlight one of the points of divergence of the Alfonsine from the tradition of the Toledan Tables. In general, one observes that *Arabes maxime* typically addresses the cases where the Toledan tradition is different from Alfonsine doctrine, while it does not repeat notions from the Alfonsine that are assumed (by the author of *Arabes*) to be known by the reader.

The cited passage echoes a chapter of the Toledan Cb canons on the equation of the Sun: '... and then you will get the true place of the Sun, and this is reckoned from the beginning of Aries in the eighth sphere. To this true place, add the place of the eighth sphere, and you will get the truer place'.[27] Yet, even in the Castilian Alfonsine canons one finds some similarity in the wording, namely in Chapter 16, which reads 'And what remains of the mean motion after its [the equation's] subtraction or addition, will be the certain position of the Sun'.[28] Note that the Castilian *el lugar çierto* dovetails with the Latin expression of *cui certo loco*, which one finds in both the Toledan canons Cb and in *Arabes maxime*. In all cases it refers to the position of the Sun in the eighth sphere and appears to have been a way to distinguish it from the *verus locus*, the true position with respect to the vernal point on the fixed zodiac. Both the Toledan and the Castilian canons instruct the reader to add or subtract the equation of access and recess to the already equated positions of the planets.

---

25 'Habito ergo vero loco solis, computa eum a principio arietis mobilis, cui certo loco superadde motum 8ve spere equatum, et habebis locum eius certissime computando a coniunctione circulorum'. See Appendix, p. 45.

26 With the PAT the motion of the eighth sphere must be introduced as a first step to define the position of the apogee. The apogee is needed to determine the mean anomaly and then compute the solar equation. See José Chabás and Bernard R. Goldstein, *The Alfonsine Tables of Toledo* (Dordrecht: Kluwer, 2003) pp. 217–20.

27 'et tunc habebis locum solis certissime, cuius initium erit ab initio arietis 8'a sphaera. Cui certo loco addes locum 8'ae sphaerae, et habebis certiorem', Pedersen, *The Toledan Tables*, Cb142, p. 436.

28 'E lo que fuere el medio curso despues del menguamiento del, o despues del añadimiento sobre el, eso sera el lugar çierto del sol'. In the chapter 'De la equaçion del Sol'. See Chabás and Goldstein, *The Alfonsine Tables of Toledo*, Ch. 16, sentence 6, p. 37.

It remains uncertain whether the author of *Arabes maxime* was referring to the Castilian version of the Alfonsine doctrine or whether he was drawing from the older Toledan Cb canons. The difficulty here is to interpret the comparative perspective of the text. It is not obvious in this case whether the text describes a common or a divergent feature. One thing is clear: the text assumes that the readers were aware of two different reference frames. In the case of the Toledan and the Castilian Alfonsine traditions, planetary positions are located with respect to the eighth sphere (sidereal). In the case of the Parisian Alfonsine tradition, they are located with respect to the vernal point (tropical). The reader is further instructed to find a planetary position, *locum certissime computando*, which is the position most accurately computed from the vernal point.

2.2. Invoking Euclid in clarifying the notion of kardage

Beyond comparing Toledan material with the Alfonsine Tables, the author of *Arabes maxime* also consistently complements the arithmetic and tabular reasoning of the canons with geometrical diagrams and commentary. This is a distinctive feature found among the various categories of commentary texts to canons.[29] In several places the text justifies the addition of a diagram by arguing that it clarifies the argument. The first topic treated in this way is the derivation of trigonometric values. Indeed, much space is given to the explanation and the demonstration of the sine table in the form of what is known as a table of *kardage*: a way of dividing the unit circle into 24 equal arcs that goes back to Sanskrit sources and is also present in the canons of the Toledan tradition. The beginning of the passage on the *kardage* in *Arabes maxime* expresses the content of the Toledan canons numbered in Pedersen's edition as canons Cb52, Cb58 (explanation) and the canons Cb238, Cb242, and Cb244 (demonstration) treated by Pedersen as 'parallel canons' mainly derived from al- Khwārizmī.[30] The proof for the values of the *kardage* can be found in Chapter 7 of *Arabes maxime*.[31] It is based on the principles also invoked in Cb244 and the following canons. These include the Pythagorean theorem with right triangles to derive the right and versed sines.[32] A feature that distinguishes *Arabes maxime* from the original Toledan source, however, is the invocation of propositions from Euclid's *Elements* for many of the steps.[33]

29 The practice of adding glosses *with figures* to explain canons occurs of course in earlier works: for example, in Ibn al-Muthannā's commentary of al-Khwārizmī's *Ta 'līl zīj al-Khwārizmī*. Bernard R. Goldstein, *Ibn al-Muthannâ's Commentary on the Astronomical Tables of al-Khwârizmî* (New Haven: Yale University Press, 1967).

30 Pedersen refers to the specific version of al-Khwārizmī's astronomical tables in its 'non-Maslama' version used by Ibn al-Muthannā whose commentary survives only in two Hebrew versions and one Latin version. Goldstein, *Ibn al-Muthannâ's Commentary*. Pedersen, *The Toledan Tables*, p. 51, (2.5). According to this author, Cb244 is clearly a text that was added to the end of the canons. See Pedersen, *The Toledan Tables*, p. 488.

31 Appendix, pp. 41-43.

32 To be exact, the *kardage* table provides the differences of the right and versed sines for six arcs 15°, 30°, 45°, 60°, 75°, 90°. The 'right sine' (*sinus rectus*) corresponds to the modern notion of sine. The 'versed sine' (*sinus versus*) corresponds to the difference of the radius and the cosine. Both are expressed with respect to the length of the radius sometimes taken, according to our text, as 150 or 60. By approximation integers are used for the sine values.

33 This might be indicative of an arts faculty context. In medieval Latin astronomical literature, Euclid is invoked in some high-level works such as the *Almagesti minor*, which presents the material of the Almagest *modo geometrico*. Also, in treating the *kardage* and trigonometry, Richard of Wallingford's *Quadripartitum* makes proficient use of Euclid's *Elements* to prove the computations of sines to a great level of precision. See

For the derivation of the sine and versed sine tables, the *Arabes maxime* goes further than the edited version of the Toledan canons does. The text has two more chapters (8 and 9) showing which geometric constructions correspond to the arithmetic operations used in Chapter 7 when the Pythagorean theorem is used.[34] Indeed, it repeatedly involves extracting the root of a sum or a difference of squares. The constructive steps, to join the area of two squares in one — or to construct a square equal to the difference of two squares — are justified step by step, invoking propositions from Euclid's *Elements*.[35] An examination of the text shows that it contains important passages parallel to John of Gmunden's treatise *De sinibus et arcubus* composed in 1437.[36] The following table contains three parallel passages excerpted from Chapters 7, 8, and 9, which reveal unmistakable similarities between the two works.[37]

Arabes maxime	*De sinibus, chordis et arcubus*
[Ch. 7, pp. 42-43, lines 41-46]	p. 92
Ex hiis que dicta sunt, manifesta est quantitas sinus tam recti quam versi cuiuslibet kardage et quarumlibet simul sumptarum, quia sinus rectus prime est sinus versus 6^{e} et sinus rectus 2^{e} est sinus versus 5^{e}. Item sinus rectus duarum scilicet prime et 2^{e} est sinus versus duarum scilicet 5^{e} et 6^{e}, et versus illarum est rectus istarum. Item sinus rectus 3^{e} est sinus versus 4^{e}, et versus 3^{e} est rectus quarte. Similiter sinus rectus trium primarum est sinus versus trium ultimarum et sinus versus illarum est rectus istarum.	Ex hiis igitur quae dicta sunt manifesta est quantitas tam sinus recti quam versi cuiuslibet cardagae et quarumlibet simul sumptarum quia sinus rectus primae cardagae est sinus versus sextae cardagae et sinus rectus secundae est sinus versus quintae. Item sinus rectus duarum cardagarum scilicet primae et secundae est sinus versus duarum ultimarum scilicet quintae et sextae et sinus versus duarum primarum est sinus rectus duarum ultimarum. Item sinus rectus tertiae cardagae est sinus versus quartae et sinus versus tertiae est sinus rectus quartae. Similiter sinus rectus trium primarum cardagarum est sinus versus trium ultimarum et sinus versus trium primarum est sinus rectus trium ultimarum.

John D. North, *Richard of Wallingford. An Edition of his Writings with Introduction, English Translation, and Commentary*, vol. 1 (Oxford: Oxford University Press, 1976), pp. 21-178. It must be kept in mind that our text is of a mathematically less advanced level than those cited examples. As we show below, the use of Euclid here is more akin to introductory mathematical texts. See note 38.

34 Appendix, pp. 41-43.

35 Conspicuously, though, the style of the justifications is not 'Euclidean'.

36 The sources of these passages were not indicated in detail in Busard's 1971 edition. Hubertus. L. L. Busard, *Der Traktat* De sinibus, chordis et arcubus *von Johannes von Gmunden* (Vienna: Springer Verlag, 1971), edited from the manuscript in Vienna, Österreichische Nationalbibliothek (ÖNB) Cod. 5268, ff. 44ra-90va. Busard notes, however, that John of Gmunden did 'more than just use' John of Lignères' *Cuiuslibet arcus* canons, that is, the *Canones tabularum primi mobilis*. See also Glenn van Brummelen, *The Mathematics of the Heavens and the Earth* (Princeton: Princeton University Press, 2009), p. 218.

37 The left column includes selected passages from the text of *Arabes maxime* excerpted from the Appendix. Variant readings and conjectures in these passages can be found in the edition included in the Appendix pp. 41-43. The right column provides the text of *De sinibus, chordis et arcubus* exactly as it is found in Busard's edition. Page changes in the manuscript BnF lat. 7295A and in Busard's edition are noted by vertical bars. As far as the diagrams are concerned, their proportions show a similarity between those supporting Gmunden's treatise and those of *Arabes maxime*. However, the similarity does not appear in the diagrams edited by Busard, but only when examining those of the ÖNB manuscript.

[Ch. 8, p. 43, lines 52-55]
<E>t quoniam in premissa demonstratione alicui docuimus ut quadrata iungerentur in unum, qualiter hoc secundum artem geometrie fieri potest, considera, si duo quadrata fuerint equalia, tunc quadretur dyameter et habetur propositum scilicet unum quadratum equipollens duobus.

p. 84
Quod unum quadratum iungitur altero tunc quando describitur unum quadratum aequale ambobus. Hoc autem secundum artem geometriae fit taliter: considera si unum quadratum alteri fuerit aequale, tunc enim quadra diametrum unius eorum et habebis propositum scilicet unum quadratum aequipollens duobus.

[Ch. 9, p. 44, lines 63-74]
<I>tem in eadem demonstratione, quoniam diximus quadratum a quadrato esse tollendum, quod qualiter per geometriam fieri debeat, patet sic. Sit quadratum ***hklm*** equale duobus quadratis suprascriptis quemadmodum ostensum est supra. ***Abcd*** quadratum subtrahitur ab eo hoc modo: per 3^{am} primi Euclidis ex linea ***hm*** abscindatur linea ***ab*** et linee equalis sit ***nm***. Item ex linea ***ml*** abscindatur alia eidem ***nm*** equalis et sit ***mo***.
Item per 31^{am} ex puncto ***n*** trahatur eque distans ipsi ***hk*** et sit linea ***np***. Item ex puncto ***o*** per eandem 31^{am} ducatur eque distans ipsi ***kl*** et sit linea ***oe***. Ad hoc per primam petitionem Euclidis trahatur ***mk*** | dyagonaliter que transit per punctum ***r***.
Tunc argumentatur sic: cum ***nrom*** quadratum ratione compositionis figure sit equale ***abcd*** quadrato et ipsum totale quadratum ***hklm*** ex ypotesi sit equale duobus quadratis supra descriptis, patet quod gnomo residuus erit equalis ***efgh*** et sic est propositum.

pp. 85-86
In sequenti etiam demonstratione dicetur unum quadratum ab alio esse minuendum. Hoc autem per geometriam fit taliter: sit quadratum ***hlkm*** aequale duobus quadratis suprascriptis quemadmodum ostensum est. Subtrahe ergo ab eo quadratum ***abcd*** hoc modo: abscinde per 3^{am} primi Euclidis ex linea ***hm*** lineam aequalem ***ac*** et sit linea ***hn***. Item ex | linea ***hk*** abscinde aliam aequalem eidem ***hn*** et sit linea ***hp***.
Deinde per 31^{am} primi Euclidis ex puncto ***n*** trahe lineam aequidistantem lineae ***hk*** et sit linea ***no***. Similiter ex puncto ***p*** per eandem 31^{am} duc lineam aequidistantem lineae ***kl*** et sit linea ***pq***. Adhuc per (f. 85vb) primam petitionem Euclidis trahe lineam ***hl*** diagonaliter quae transit per punctum ***r***.
Cum igitur quadratum ***hnrp*** ratione compositionis figurae sit aequale quadrato ***abcd*** et ipsum totale quadratum ***hklm*** ex hypothesi sit aequale duobus quadratis supradescriptis patet, quod gnomo residuus erit aequalis quadrato ***efgh*** et hoc est propositum.

One may ask oneself whether the *Arabes maxime* pre-dates or post-dates Gmunden's treatise, or whether they share a common source. The relationship between these very close passages in Gmunden's text and the *Arabes maxime* is not clear yet. We suspect that they both derive from a common source, but one could also be a reworking of the other. In any case, the quick to-and-fro we find in the *kardage* chapters between abstract arguments about given and known quantities, arithmetic determinations, and geometricizing elaborations are characteristic of introductory texts of mathematics.[38]

38 For instance, the approach of the influential *Liber embadorum* by Abraham Bar Hiyya, called Savasorda (early twelfth century), translated by Plato of Tivoli, is comparable. After giving numerical solutions to the problems, the *Liber* then provides a geometrical justification with recourse to lettered diagrams.

2.3 Imagination and demonstration in the planetary canons

The chapters on the *kardage* are not the only places in the *Arabes maxime* that contain geometric diagrams elaborating on arithmetic canon procedures. The reader will encounter several more, for instance, accompanying the chapters on planetary motion. In these canons, the diagrams are deployed by the author as a kind of gloss on parallel material found in the Cb canons.[39] While basing his chapters on the Toledan corpus, however, it is worth noting that the author does not include verbatim instructions for using tables drawn from these texts.[40] Instead the contents of the chapters move beyond a mere description of the tabular steps necessary for finding true positions to include comparisons between different canon traditions, as well as illustrations of planetary models. Each illustration is accompanied by directions for composing the diagram, which effectively translate the arithmetic computations of tabular values into geometrical constructions.

In Chapter 20, for example, the reader is given a diagram of solar motion and is told to let the point ***o*** be the centre of the Earth, point ***p*** the centre of an eccentric circle, and point ***a*** the solar apogee or *aux* (Figure 1.1). The body of the Sun is placed in two possible positions on its eccentric circle in the diagram: the first position at ***b*** and the second position at ***c***.[41] The body of the Sun in each instance is then projected onto the zodiac circle at ***d*** and ***m***. Given that the Sun's position is measured from the beginning of Aries at point ***r*** in the zodiac circle, the reader can then easily see why the arcs ***de*** and ***mr***, which each visually represent the solar equation, must be, respectively, subtracted or added to the Sun's mean motion to obtain the true position as viewed from the Earth at point ***o***.[42]

It is not uncommon to find worked examples in witnesses to Toledan and Alfonsine canons. In these the authors would apply the general instructions of a canon to a specific date and time in a specific place for which a planetary position is desired.[43] In the *Arabes maxime*, however, we find the author has included diagrams that belong to the *theorica planetarum* tradition. In doing so the author has adapted the convention of the worked example to the diagram provided in the chapter. He writes:

> For example, when the Sun appears as it is placed at ***b***, the mean motion is the arc ***rqe***, but the true motion is the arc ***rqd***, [and] lying between is the arc of the equation ***de***. Therefore, if the *aux*, that is arc ***rq***, should be subtracted from the mean motion, the argument will remain, namely the arc ***qe***, which is smaller than 6 signs. And for this

39 For an introductory discussion of his treatment of the Cb canons, see Pedersen, *The Toledan Tables*, 'Part 2 Cb: Canons "Quoniam cuiusque actionis" ', pp. 329-42.

40 For the chapter on the true Sun, see Appendix, pp. 45-47; compare Pedersen, *The Toledan Tables*, Part 2, pp. 435-37. For the chapter on the true Moon, see Appendix, pp. 47-48, compare Pedersen, *The Toledan Tables*, Part 2, pp. 437-39. For the chapter on the superior planets (BnF lat. 7295A, f. 187rb, line 16–f. 187vb, line 2), compare Pedersen, *The Toledan Tables*, Part 2, pp. 439-41. For the chapter on the inferior planets (BnF 7295A, f. 187vb, lines 3-43), compare Pedersen, *The Toledan Tables*, Part 2, pp. 440-41.

41 Appendix, p. 46, lines 107-9.

42 Appendix, p. 46, lines 119-25.

43 In the case of the Cb version of the Toledan canons, see Pedersen, *The Toledan Tables*, Part 2, pp. 434-35 where Cremona is taken as the meridian position for a worked example for mean planetary motions. In Oxford, Hertford College 4, on ff. 140^{r}-47^{v}, John of Murs gives worked examples for his canons on the Alfonsine Tables throughout. See also the chapters of Chabás and Jacobson in this volume.

> reason, the arc of the equation is subtracted from the mean motion, and the arc ***rqd*** remains, which is the equated motion of the Sun, and the true position of the Sun is the point ***d***.[44]

The order of this example follows the instructions found in the Toledan canon, but it has replaced the tabular values with their counterparts found in the diagram.[45] These steps, organized as they were diagrammatically, conform to a common technique taught in the universities by which a student would fix concepts and operations to memory by means of figures.[46] These images contained strategically labelled elements of interest, or *loci*, designed to help users recollect crucial points of comparison, connection, or distinction in their mind's eye.[47] This became a commonplace in instruction across various disciplines in the arts curriculum.[48] It would thus appear that the diagram was intended to help users to recall the procedures of the canon into the faculty of their imagination as an *aide-mémoire* for future computations.

Furthermore, in several instances the instructions for the canons of the *Arabes maxime* appeal to the reader's power of imagination — a cognitive faculty, which, according to medieval accounts of human psychology, was uniquely suited to tackling the subtleties of mathematical relations.[49] Moreover, this appears to have been, not merely for the sake of memorization, but also to clarify the canons in a way that the author believed would be demonstrative.[50] With reference to the image of the solar motion, the author explains why the Sun must be equated in the first place. Even though 'the Sun runs through the zodiac in the same [time] in which it goes through the [whole] eccentric', he explains that he does not mean it does so proportionally throughout the two circles. Instead, 'on the upper half of the eccentric, which is ***cab***, it moves more on the eccentric than in the zodiac', and 'on the lower half of the eccentric, which is ***bnc***, it moves more in the zodiac than on the eccentric.'[51] The diagram stands as a demonstration of this.

---

44 Appendix, p. 46, lines 120-24.

45 Compare the language of the Toledan canon (Cb141a): '... obtain [the Sun's] mean motion ... from [it] subtract the apogee of the Sun ... the remainder will be the 'argument' of the Sun ... write out the solar equation ... subtract it if the argument is less than 6 signs ... then you will get the true place of the Sun'. Pedersen, *Toledan Tables*, Part 2, pp. 435-37.

46 See Francis Yates, *The Art of Memory* (Chicago: University of Chicago Press, 1966), pp. 70-81.

47 Thomas Aquinas provides a detailed explication of the human capacity for recollection through geometric images in the seventh lecture of his commentary to Aristotle's *On Memory and Recollection*. See Thomas Aquinas, *Commentary on* Memory and Recollection, trans. by John Burchill, 'Thomas Aquinas, *Commentary* on Aristotle, *On Memory and Recollection*', in *The Medieval Craft of Memory: An Anthology of Texts and Pictures*, ed. by Mary Carruthers and Jan M. Ziolkowski (Philadelphia: University of Pennsylvania Press, 2002), pp. 181-85.

48 As Francis Yates wrote in her seminal work on the subject of memory, Thomas Aquinas likened the strategies of the *loci* found in Cicero's *Rhetorica ad Herennium* to the stepwise memorization of mathematical instructions as described in Aristotle's base text. See Yates, *The Art of Memory*, pp. 71-72.

49 The term *imaginari* is first introduced in the chapter on the eighth sphere (BnF lat. 7295A, f. 184rb-84va).

50 The literature on this subject is vast, but for two introductory articles on the subject, see Andrew G. Molland, 'Mathematics in the Thought of Albertus Magnus', in *Albertus Magnus and the Sciences: Commemorative Essays*, ed. by James A. Weisheipl (Toronto: Pontifical Institute of Mediaeval Studies, 1980), pp. 463-78; Edith Sylla, 'Mathematical Physics and Imagination in the Work of the Oxford Calculators: Roger Swineshead's *On Natural Motions*', in *Mathematics and Its Applications to Science and Natural Philosophy in the Middle Ages: Essays in Honour of Marshall Clagett*, ed. by Edward Grant and John E. Murdoch (Cambridge: Cambridge University Press, 1987), pp. 69-103.

51 Appendix, p. 46, lines 113-17.

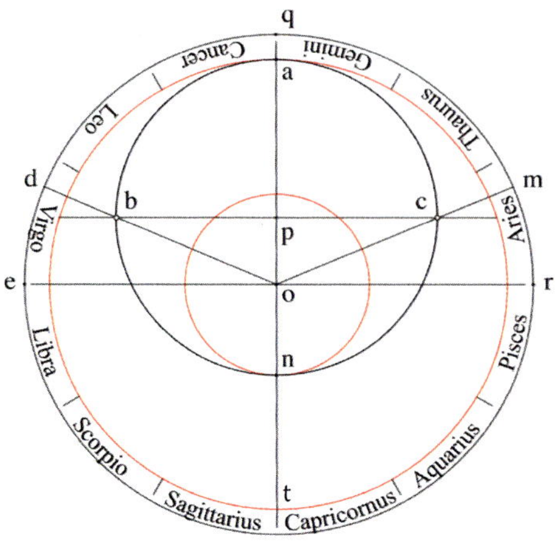

Figure 1.1. Theory of the Sun. Illustration of the theory of the Sun found in Chapter 20 (Appendix, pp. 45-46, line 99-130) where ***qetr*** represents the ecliptic circle with its centre at ***o***, which is the Earth. The circle ***abnc*** is the deferent carrying the body of the Sun with uniform angular velocity about ***p***, which is drawn eccentrically to ***o***. The Sun's motion through the eccentric circle models the variation in the Sun's apparent motion through the zodiac over the course of the year. The point ***r*** is the beginning of Aries, from where the position of the Sun is reckoned as projected onto the zodiac. From ***t*** to ***a*** through ***o***, the line of apsides is drawn. As viewed from ***o***, the Sun is slowest at the *aux*, ***a***, and fastest at ***t***. The mean solar anomaly is the arc from the position of the *aux*, ***a***, to the position of the body of the Sun, at ***b*** or ***c***, on the eccentric circle. The same arcs of mean anomaly appear on the ecliptic (arcs ***qe*** or ***qr***). The true Sun is defined as the projection of the Sun at ***b*** or ***c*** onto the zodiac circle at ***d*** or ***m*** from the Earth, ***o***. The arcs ***de*** and ***mr*** represent the solar equation separating the mean and true Sun. The diagram further defines two additional circles centred on ***o***: one representing the external bound of the eccentric circle with a radius ***oa***, and the other representing the internal bound of the eccentric circle with a radius ***on***.

Here, we use the term 'demonstration' (*demonstratio*) just as the historical actors did themselves, but we do so advisedly. By demonstration in this context, we do not mean a rigorously deductive argument from first principles, but instead a practical exhibition or explanation of how something works.[52] In this way, the author of the *Arabes maxime* emphasizes that diagrams can reveal the reasons for canon operations (*ratio operationis*), thereby resolving any confusion about why certain calculations should be performed to obtain a final value.[53] By the time of writing of the *Arabes maxime*, there was already a long tradition in the Latin West of understanding demonstrative reasoning in this way. According to Charles Burnett, a similar use of the term *demonstrare* can be discerned in

52 For the variety of medieval definitions applied to Euclidean 'demonstration', see Charles Burnett, 'The Latin and Arabic Influences on the Vocabulary Concerning Demonstrative Argument in the Versions of Euclid's *Elements* Associated with Adelard of Bath', in *Aux origines du lexique philosophique européen: l'influence de la 'latinitas'*, ed. by Jacqueline Hamesse (Société internationale pour l'étude de philosophie médiévale: Louvain-la-Neuve, 1997), pp. 117–35. Karine Chemla ed., *The History of Mathematical Proof in Ancient Traditions* (Cambridge: Cambridge University Press, 2012), p. 17 and pp. 41-43.

53 Appendix, p. 46, line 28. In the text the scribe appears to have inverted the common phrase *ratio operationis*, curiously referring to his description of the solar diagram as an *operatio rationis*.

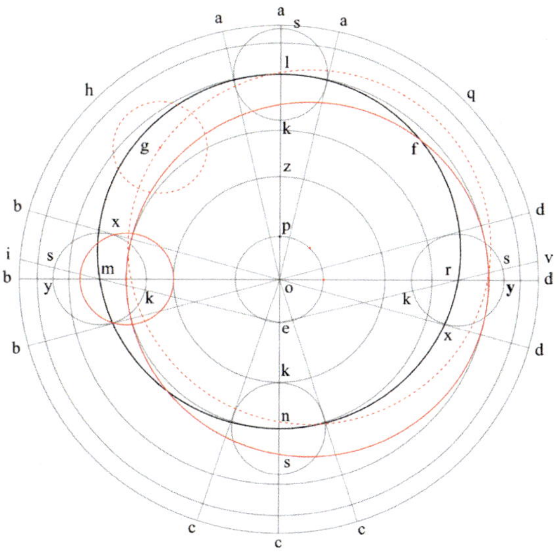

Figure 1.2. Theory of the Moon. Illustration of the diagram of the Moon in Chapter 20 (Appendix, pp. 46-48, line 130-192), where we have superimposed the moving deferent (in red) for greater clarity. The circle ***abcd*** represents the line of the ecliptic in the zodiac. The tropical period of the Moon is represented by the movement of the centre of the epicycle that travels along the circumference of the deferent circle, ***lmnr***, from west to east with uniform angular velocity. Meanwhile, the body of the Moon at ***x*** travels uniformly about the centre of its epicycle, ***sk***, from east to west. These two uniform motions account for the Moon's variable velocity in the zodiac, known as the first anomaly. The diagram also includes the second lunar anomaly, known now as evection, which Ptolemy had found to affect the position of the Moon when it was at quadratures, that is, 90° from the position of the Sun. This second anomaly manifests itself as the apparent variation in the diameter of the lunar epicycle. To model this variation, the centre of the deferent circle, ***p***, is placed eccentrically to the Earth, which is represented by the centre point ***o***. Thus, when the lunar epicycle is at the point of the *aux*, ***l***, its diameter has its least apparent length from the perspective of ***o***. When it is at the point opposite the *aux*, ***n***, its diameter has its greatest apparent size. When the epicycle centre occupies any other points on the deferent between ***l*** and ***n***, the argument of the Moon has to be adjusted by the equation of centre (by convention defined as the angle ***emo*** or angle ***ero*** where ***e*** is the point opposite the centre of the deferent). Finally, the centre of the deferent, ***p***, is set in motion uniformly about ***o*** from east to west on a circle traditionally called the *circulus brevis* (though not explicitly so named in the text). One full revolution of the deferent centre around this circle represents the Moon's synodic period.

Version 1 of Adelard of Bath's twelfth-century translation of Euclid's *Elements* from the Arabic, where ' "demonstrare" is to demonstrate in the sense of pointing to something, to show something is true by scratching the geometrical figures in the sand or (in modern terms) on the blackboard.'[54]

As we shall see with the case of the Moon, the author relies even more profoundly on his reader's power of imagination, describing three different places that the Moon's epicycle centre passes through. One position appears on the diagram as the point 'opposite the *aux*' (*oppositum augis*) at the base of the diagram (Figure 1.2). The other two positions,

54 Burnett, 'The Latin and Arabic Influences on the Vocabulary Concerning Demonstrative Argument', p. 123.

however, are indicated to be at quadratures to the line of apsides presented in the image. They are not represented as the centres of visible epicycles but rather as imagined points (*sub* ***b*** *vel* ***d***), such that the author leaves their visualization to the reader's imagination. The author refers to these as the positions where the perigee occurs 'in reality' (*realiter*).[55] They are only coherent with Ptolemy's second lunar model, moreover, if the reader imagines the deferent to be in motion with its centre rotating about the Earth in the diagram itself.

The author attempts this ambitious exercise in order to help clarify the geometrical relations lying behind certain technical terms invoked in the canon procedure. He defines the equation of centre by showing the reader how to draw a line through the point opposite the centre of the lunar deferent, ***e***, and the centre of lunar epicycle, ***m***, thereby indicating the direction of the mean *aux* (*media aux*). The arc formed between this line and the line of the true *aux* (*vera aux*), ***om***, constitutes the equation of centre, which is an important value in the tables. To this end, the author first lays out the figure. He writes:

> For greater clarity, I shall include a figure of the Moon. Let ***abcd*** be the circle of signs centred on ***o***, which is also the centre of the sphere of the Moon whose thickness ***s*** is enclosed by the two circles ***sc*** and ***zk***. The circle that is drawn as ***ac*** represents the exterior surface of the lunar sphere; the circle ***zk*** the interior surface. Let the deferent of the Moon be ***lmnr*** centred on ***p***. The diameter of the epicycle, ***sk***, is, thus, aligned with the point ***e*** in every position.[56]

After constructing the figure in this way, he goes on to say that the equation of centre will be zero when the centre of the epicycle is at the lunar apogee ***l*** and the perigee ***n***, that is, when the Moon is at mean syzygies:

> Just as ***l*** and ***p*** move to the west toward ***d***, so the centre of the epicycle, which is also signified by ***l***, moves to the east toward ***b***, with point ***e*** meeting it. When the centre of the epicycle is placed at the *aux* of the eccentric at ***l***, the Sun by [its] mean motion will, therefore, be at ***a*** or at ***c***, which amounts to the same thing at present. [This is] because the centre of the epicycle is in mean conjunction and opposition at ***l***, that is, at the *aux* of the eccentric. In such a position there are, in fact, no equations of centre because the centre is zero. Equations are also zero in the opposite location, namely, when the centre of the epicycle is at the place opposite the *aux* at point ***n***, but, in reality, that happens first under ***b*** then under ***d*** since, when the Moon is distant from the Sun by 3 or 9 signs, it is at quadratures.[57]

Thus, the text helps the reader to imagine the effects of the model's moving deferent centre, even though the diagram does not sensibly represent this motion. It serves as an explanation for why it is that, when the centre of the epicycle arrives 'under ***b***', as the text describes it, it will be at perigee and opposite the lunar *aux* lying on the line ***od***. Furthermore, it

55 The author makes use of this adverb four times to distinguish between where a point appears in a static diagram and where it would really appear as the deferent centre, ***p***, moves about ***o***. See Appendix, p. 47, lines 147, 155, 156, 161.

56 Appendix, p. 47, lines 135-40.

57 Appendix, p. 47, lines 140-48.

explains the reason why, when the epicycle centre proceeds to the point 'under ***d***', it will have returned to perigee for a second time in its synodic cycle.

In his effort to present a diagram in motion to the reader's imagination, the author of the *Arabes maxime* was participating in what Kathleen Crowther and Peter Barker have called a 'sophisticated visual and pedagogical program' to 'train the intelligent eye'.[58] Following the analysis of David Summers, Crowther and Barker draw a distinction between 'naturalistic' illustrations, which explicitly conform to an optical experience, and 'realistic' ones, which represent real objects in nature, even if they do not do so naturalistically.[59] The *Arabes maxime* features just this sort of realism in its diagrams, which bear a strong resemblance to the figures of the Latin *theorica* and Arabo-Persian *hay ʿa* traditions.

Conclusion

In all of the instances we have outlined, the teaching within the *Arabes maxime* can be said to transgress certain traditional epistemic boundaries of practical knowledge to include material typically reserved for other genres, particularly that of the *theorica*. In fact, the author did so self-consciously. In his chapter on the *kardage*, he even anticipates possible objections to the integration of practical and theoretical modes of reasoning within a canon text, and he justifies his approach, writing:

> Although in this work we mainly deal with practical knowledge, if we nevertheless insert something pertinent to theoretical knowledge for the sake of solace, it must not be considered inappropriate. In fact, the intellect is delighted all the more by the certainty of a truth that has been attained through demonstration.[60]

Of course, the *Arabes maxime* is not completely unique as a canon set that uses diagrams and geometrical arguments to reformulate and clarify tabular procedures. We find similar canons — or canon glosses, if you will — treating John of Saxony's *Tempus est mensura motus* in Oxford, Bodleian Library, Digby 97 and John of Lignères' *Priores astrologi* in Erfurt CA 4° 366.[61] Even within the *Tempus est mensura* itself, John of Saxony occasionally explores the reason or *causa* for canon procedures by alluding to diagrams in an effort to demonstrate why certain computations are performed with the tables.[62] If we cross the disciplinary divide to *theorica* texts, we find an author such as Campanus of Novara describing how table-makers (*tabularii*) relied on geometrical models or even 'movable diagrams' in the form of instruments to justify the use of their tables, and thus the line of

58 Kathleen M. Crowther and Peter Barker, 'Training the Intelligent Eye: Understanding Illustrations in Early Modern Astronomy Texts', *Isis* 104 (2013), 429-70 (pp. 435-36).

59 David Summers, *The Judgment of Sense: Renaissance Naturalism and the Rise of Aesthetics* (Cambridge: Cambridge University Press, 1987), p. 3. Crowther and Barker, 'Training the Intelligent Eye', pp. 438-39.

60 Appendix, p. 41, lines 103.

61 See *Canones in Tabulas Alphonsinas* in Oxford, Bodleian Library, Digby 97, ff. 240^{r}-90^{v}; *Canones tabularum Alfonsi* in Erfurt, Universitätsbibliothek, CA 4° 366, ff. 70^{r}-73^{r}.

62 See Emmanuel Poulle, *Les Tables Alphonsines avec les canons de Jean de Saxe* (Paris: Editions du Centre National de la Recherche Scientifique, 1984), p. 70, lines 5-9; p. 74, lines 16-29.

demarcation between practical and theoretical astronomy was always porous.[63] Even so, the canons of the *Arabes maxime* appear to be witnesses — among a growing body of similarly identified works — to a specific sort of canon text that intentionally and systematically provided geometrical reasoning to underpin practical instructions for trigonometry and finding planetary positions.

What stands out about the *Arabes maxime,* moreover, is its close attention to the Toledan canons and the Toledan Tables at a time when Alfonsine materials had become prevalent within the medieval Latin university curriculum. It is written with the clear aim to compare the Toledan canons with the Alfonsine ones. Several questions emerge from our initial investigation of this work, including the dating of composition, the significance of the sources chosen to interpret the Toledan corpus, and the relationship between Toledan and Alfonsine material following the reception of the Castilian Alfonsine Tables in Paris in the 1320s and thereafter. This contribution serves merely as a notice of these questions, which will be the subject of ongoing research. We hope that these research areas will be further addressed and expanded upon when we release the critical edition of the entire work.

63 See Benjamin and Toomer, *Campanus of Novara and Medieval Planetary Theory*, pp. 206-11.

Appendix

The aim of this Appendix is to provide a comprehensive context for the references included in the article. The Appendix includes the text of Chapters 7, 8, 9, 18, 19, and 20, excerpted from the ongoing critical edition of the work we refer to as the *Arabes maxime*. The sole witness of the work we have identified so far, namely Paris, BnF lat. 7295A, mostly lacks chapter headings and chapter initials; in the edition, the latter have been silently integrated. Both the numbering and the headings of each chapter in this Appendix are not original to the text. In fact, in the manuscript the text has no numbering and includes only the title of one chapter out of forty-nine. The headings, which in our edition consist of the *incipit* of each chapter, as well as the numbering, are therefore merely an editorial device that allows the reader to navigate the text more easily.[64]

The text in the manuscript copy presents a number of corruptions. In order to overcome this, we have had to make several conjectures, which are all detailed in the apparatus and, for diagrams, in the editorial principles. Some of these conjectures are informed by the sources of the work, while others are based on a comparison of the text to other similar texts from the same period. In some cases, the editors have relied on context to arrive at plausible conjectures. It is important for readers to be aware of the potential uncertainties and limitations involved in this critical editing process that is still ongoing. Therefore, we encourage readers to approach this edition with an attentive eye, carefully considering the evidence presented in the apparatus and in the following editorial principles.

Index of the chapters

The chapters edited in this Appendix are found in Paris, BnF lat. 7295A:
Ch. 7: *Quamvis in hoc opera de practica*, f. 183ra-83vb
Ch. 8: *Et quoniam in premissa demonstratione*, f. 183vb
Ch. 9: *Item in eadem demonstratione*, ff. 183vb-84ra
Ch. 18: *Nota in tabulis tholetanis utimur*, f. 186ra
Ch. 19: *Ad inveniendum medios motus planetarum*, f. 186ra-86rb
Ch. 20: *Solem equabis per omnem motum*, ff. 186rb-87ra

Index of the diagrams

The diagrams edited in this Appendix are found in Paris, BnF lat. 7295A:
D. 1: Kardage, Ch. 7, f. 183va
D. 2: Addition of a square, Ch. 8, f. 183vb
D. 3: Subtraction of a square, Ch. 9, f. 183vb
D. 4: Theory of the Sun, Ch. 20, f. 186rb
D. 5: Theory of the Moon, Ch. 20, f. 187ra

64 Exception made for Ch. 34, bearing the rubricated heading *De Venere*; the remaining chapters do not include any headings; see f. 190va. Chapters including chapter initials, are 6 and 34, 35, 36, 37, and 38. A few chapters (i.e. 29, 33, 40, 41) includes only guide letters.

Editorial principles

The text

In preparing this Appendix, we have attempted to reproduce the text in Paris, BnF lat. 7295A as faithfully as possible, retaining its spelling in almost all instances (e.g. *Yndia*, *dyametrum*, *seccantibus*). However, we have standardized the alterations of the consonant spelling of 'ci/ ti' (e.g. *eciam*>*etiam*, *parcium*>*partium*, *gracia*>*gratia*, *solacii*>*solatii* …), as it does not seem possible to assess the prevalence of one spelling convention over the other throughout the text. In one case, *vasis*>*basis* (Ch. 8, line 60), we have retained the original reading, and we gave the more common spelling in the apparatus to avoid ambiguities. Some readings referring to proper names did not appear to us to be acceptable spelling variants. In these cases (e.g. *Auphontii* for *Alphontii* in Ch. 19, line 87; *Tremonia* and *Tremonie* for *Cremona* and *Cremone* in Ch. 19, lines 94 and 98), we have presented the readings in the edition as conjectures and offered the original spelling in the apparatus.

To mark syntactic units, we have applied a modern system of punctuation. In a few instances, we have altered the apparent divisions of the text, and the page changes are noted in the right margin and within the text by a vertical bar.

The critical text is accompanied by an *apparatus fontium* and an *apparatus criticus*: (1) the first apparatus records the conjectures, the scribal errors, as well as the correcitons made by Heingarter to the text he wrote out, and a few necessary clarifications; (2) the second apparatus includes the sources of the text. Since the edition was conducted on the basis of a single manuscript, it was deemed appropriate to avoid a redundant and unnecessary repetition of its siglum in the apparatus.

In Chapters 7 and 8 on numerous occasions it was necessary to intervene to conjecture 'square' in the place of 'quarter' (i.e. *quadratum* and *quartum* variously declined). The abbreviation for indicating both lemmas is very similar: in fact the abbreviation for 'quarter' normally consists of the numeral 4 (ꭓ) plus the ending indicating the case at the top right of the numeral (e.g. ꭓ^{tũ}); for 'square' instead the abbreviation consists of the numeral 4, the letter 'a' surmounting the numeral and the ending at the top right (e.g. ꭓ̊^{tũ}). Since in both chapters both the word 'square' and the word 'quarter' appear, we have decided to conjecture and show in apparatus what is reported in the text, reproducing the abbreviation as faithfully as possible for the benefit of the reader.

The diagrams

The diagrams were redrawn with respect to their proportions as they were found in the manuscript. In some aspects the diagrams in the manuscript contradict what is stated in the text. For these aspects they have generally been edited with the aim to make them consistent with the indications about them given in the text (see details below). Although the manuscript diagrams were produced using a straight edge and compasses, their finish has not been done with the greatest of care. In those cases where convergence in a point or tangency of a circle is not realized in the manuscript but can be understood from the context, the redrawn diagrams have tacitly corrected those scribal imprecisions (i.e. non convergence, non-tangency).

In general, the lettering of the diagrams was taken as a reference to correct or conjecture the letters in the text passages referring to them. Substituted letters are given in italics. Reconstructed letters (that are not present in the witness) are given in bold.
Some diagrams were placed in the edited text where it appeared most suitable for the current edition, their original placement being noted in the following list.

For each diagram, specific interventions were made:

- D.1: Kardage, Ch. 7, f. 183va. The divisions of the circle were idealized to 15° sectors.
- D.2: Addition of a square, Ch. 8, f. 183vb. The hatching of one of the squares is in the manuscript.
- D.3: Subtraction of a square, Ch. 9, f. 183vb. Where the diagram was cut in the process of a rebinding, the missing lines and vertices, and corresponding letters, were reconstituted based on the context. The reconstituted line segments are dotted, and the inferred letters are in bold.
- D.4: Theory of the Sun, Ch. 20, f. 186rb. In two instances, the circles confining the orb of the Sun were said to be in red (*circuli rubri*). While in the manuscript all circles are drawn in the same ink, we restore the red colour in the redrawn diagrams. The letter at the bottom of the main circle, missing from the figure, was conjectured to be ***t***. It was difficult to harmonize the text and diagram. This is the case with regard to the points ***q*** and ***a*** and also the points where the solar equations are maximum. In the absence of any easy way of achieving full consistency, we opted to preserve the occurrence of these points and letters as in the manuscript.[65]
- D.5: Theory of the Moon, Ch. 20, f. 187ra. For the Moon the text and diagram could be harmonized more easily. A substantial change was made to the diagram regarding the position of two of the epicycles **sk** when they are at mean distance (under b or d). A close reading of the text shows that their centres ***m*** and ***r*** need to be placed on the lunar deferent to show the equation of centre at quadrature (l. 167*ss.*). The witness puts these circles at the distance of the opposition of the apogee where the equation of centre would be nil. The letters ***q***, ***f***, ***g***, and ***h*** were placed in accordance with what is stated in the text, halfway between the top and the quadratures. In the diagram, the letter ***s*** occurs four times in a script that makes it look like a capital ***B***. Therefore, the letters ***bk*** in the text (l. 139) were conjectured as ***sk*** (see also the caption to this diagram in Figure 1.2).

List of abbreviations

ante corr. = *ante correctionem* = 'before the copyist corrected'
post corr. = *post correctionem* = 'after the copyist corrected'
cf. = *confer* = 'compare/consider'

---

65 On two occasions, the ambiguity resulting from the inconsistency of the use of ***q*** and ***a*** is clarified in the *apparatus criticus* (e.g.: q] *sc.* a). Indeed, twice in the text the letter ***q*** is found in the place a modern reader would expect to find the letter ***a***. This may not have been a cause for concern for historical actors encountering the text; points labelled as ***a*** and ***q*** might have been considered somehow interchangeable due to their visual or even theoretical sense. Our approach to these two letters was to retain their occurrence as displayed in the manuscript witness, considering it unnecessary to add conjectures to the labels on the diagram.

coni. = *coniecimus* = 'we conjectured'
deest = 'this [word or letter] is missing'; *indicates that the evidence of the manuscript is unavailable for this letter or word because the manuscript is damaged or its reading is no longer possible for some other reason(s).*
del.= *delevit* = 'the copyist deleted'
fors. = *forsan* = 'perhaps'
i. m. = *in margine* = 'in the margin'
litt. = *littera-ae* = 'letter or letters'
s. l. = *supra lineam* = 'above the line'
sc. = *scilicet* = 'clearly'
suppl. = *supplevit* = 'the copyist supplied'
angular brackets (<…>) *enclose words, etc. that we have added.*

7. Quamvis in hoc opere de practica

<Q>uamvis in hoc opere de practica principaliter cogitemus, si tamen gratia solatii aliquid de pertinentibus ad theoricam inserimus, non debet inconveniens reputari, nam intellectum delectat | quam plurimum certitudo veritatis per demonstrationem accepta. 183rb

Igitur ad demonstrandum quantitatem sinus cuiuslibet portionis pono figuram. Sit circulus super ***o*** centrum descriptus ***az*** et ***gz*** dyametris super centrum se orthogonaliter seccantibus; ***ag*** quarta in sex partes sive arcus equales divisa que partes sunt kardage sicut patet. ***Dsxl*** quadratum inscribatur, ***ck*** exagoni inscripti signatur. Item ***er***, aliud latus exagoni inscripti, licet non eisdem protrahitur adhuc ***ls*** dyametri.

Reliquis vero lineis quemadmodum apparet in figura protractis, arguo sic: posita certa quantitate dyametri que in tabulis kardagarum nec non in tabulis sinus et declinationis supponitur 300 partium, nota est linea ***go***, cum sit medietas dyametri et consequenter ***ck*** et ***er*** linee sunt note, cum sint latera exagoni qui per penultimam 4^{i} Euclidis inscriptus est. Item ***dl*** que est latus quadrati inscripti, nota erit hoc modo. ***Dol*** angulus est rectus, quia cadit in quartam circuli, ergo per penultimam primi quadratum ***dl*** valet quadrata ***dol***. Sed illa duo quadrata sunt equalia, ergo si coniungantur, ut eis quadratum unum, extrahatur radix et exibit quantitas ***dl*** sive quantitas ***ds*** que eadem causa est.

Itaque preter dyametros et medietates earum notas habemus 4 lineas notas que sunt ***ck dl ds er*** et per consequens medietates earum sunt note que sunt ***cm dn dm en***. Tunc arguo sic: quadratum ***oc*** valet quadrata ***cm*** et ***mo***; ergo si a quadrato ***oc*** subtrahatur quadratum ***cm*** remanebit quadratum ***mo*** a quo extrahatur radix et exit quantitas ***mo***. Et ita per consequens nota erit ***am*** cum nota sit ***ao*** tota. Et sic demonstratum habemus tam sinum rectum quam versum duarum kardagarum.

Postea procedo sic: quadratum ***ac*** valet quadrata ***amc;*** iunctis ergo quadratis ***amc***, ut fiat ex eis quadratum unum, extrahitur radix et exibit quantitas ***ac***. Sed ***ac*** equalis ***hb*** per 28^{am} 3^{ii} Euclidis, quia equalibus arcubus subtenduntur; ergo nota erit ***hb*** et per consequens ***bi*** erit nota, cum sit eius medietas.

Tunc argumentatur sic: quadratum ***ob*** valet quadrata ***oib***; ergo si a quadrato ***ob*** tollatur quadratum ***bi*** remanebit quadratum ***io***, extrahatur radix et patebit quantitas ***io*** et sic nota erit ***ai*** que est sinus versus prime kardage, nota etiam erit ***bz*** que est sinus versus secunde kardage per hoc quod nota ***im*** que est ei equalis ratione eque distantie per 24^{a} primi et cetera. |

---

12 per penultimam 4^{i} Euclidis] Euclides, *Elementa* IV.15, 156 **14** per penultimam primi] Euclides, *Elementa* I.47, 93 **24/25** per 28^{am} 3^{ii} Euclidis] Euclides, *Elementa* III.28, 131

---

6 quarta] *coni.*; $4^{tū}$ **7** Dsxl] dlxk *ante corr.* **8** ls] *coni.*; lf **11** go] *coni.*; to **13** quadrati] *coni.*; 4ti nota] *coni.*; non **14** quadratum] *coni.*; $4^{tū}$ **15** quadrata^{2}] *coni.*; quarta quadratum] *coni.*; 4m **16** ds] *coni.*; df **19** quadratum^{1}] quadr *sed* $4a^{tū}$ *post corr.* **20** quadratum] *coni.*; quartum mo^{1}] ms *ante corr.* mo^{2}] ms *ante corr.* **23** quadrata] *coni.*; $4a^{tū}$ **27** oib] *coni.*; ozib

183va

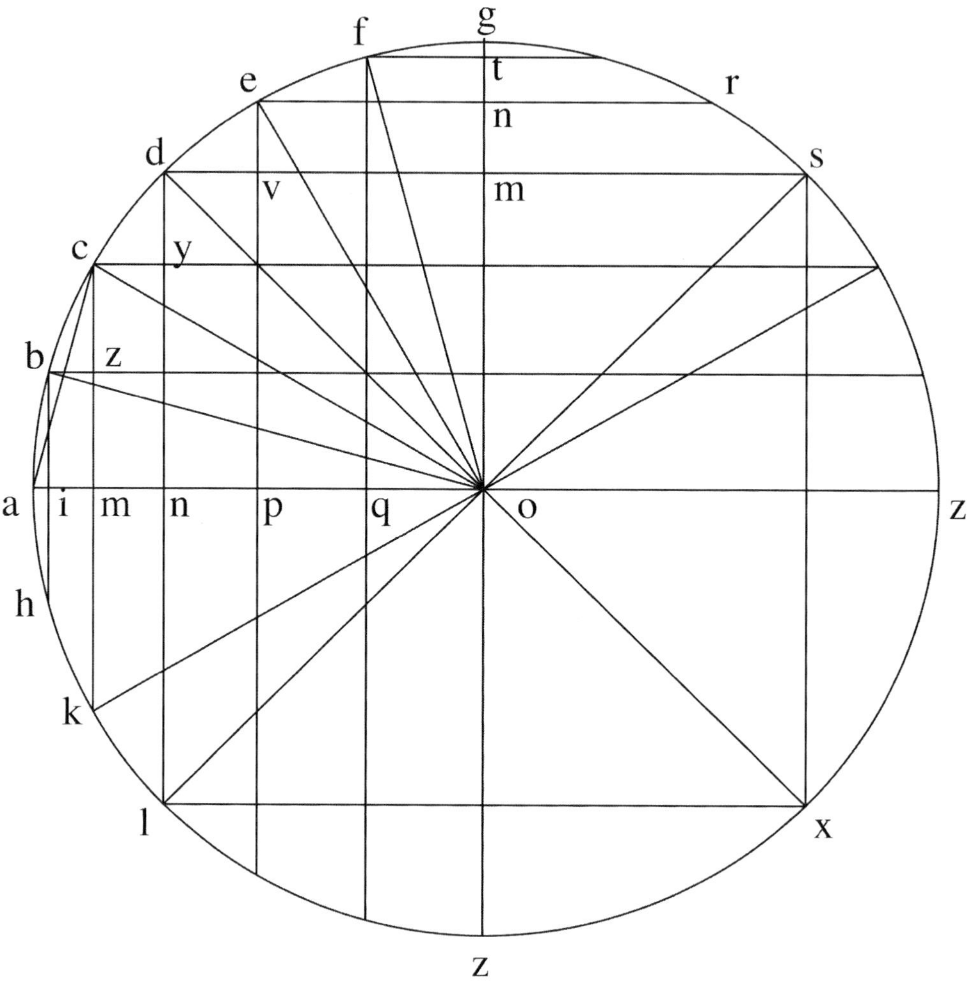

Item nota est ***bi*** que est sinus rectus prime kardage, et per consequens nota erit ***mz*** que est illi equalis ratione eque distantie. Et cum ergo ***cm*** sit nota, erit medietas eius scilicet ***cz*** nota que est sinus rectus 2^e^ kardage.

Item patet ex predictis quod ***dn*** que est sinus rectus trium kardagarum est nota, ergo eius quadratum a quadrato ***do*** auferatur, remanebit quadratum ***no***, extrahatur igitur radix et exibit quantitas ***no***; erit ergo ***an*** nota que est sinus versus trium kardagarum. Item ex dictis est ***mn*** nota; ergo et ***cy*** ei equalis ratione eque distantie erit nota que est sinus versus 3^e^ kardage. Et cum nota sit ***cm***, nota erit ***yn*** ratione eque distantie, ergo et est ***dy*** que est sinus rectus 3^e^ kardage.

Ex hiis que dicta sunt, manifesta est quantitas sinus tam recti quam versi cuiuslibet kardage et quarumlibet simul sumptarum, quia sinus rectus prime est sinus versus 6^e^ et sinus rectus 2^e^ est sinus versus 5^e^. Item sinus rectus duarum scilicet prime et 2^e^ et sinus

---

33 ergo] *coni.*; 3° **36** quadratum[1]] *coni.*; quartum quadrato] *coni.*; 4to quadratum[2]] *coni.*; quartum **37** quantitas no] no quantitas *ante corr.*

versus duarum scilicet 5^{e} et 6^{e} et versus illarum est rectus istarum. Item sinus rectus 3^{e} est sinus versus 4^{e}, et versus 3^{e} est rectus quarte. Similiter sinus rectus trium primarum est sinus versus trium ultimarum et sinus versus illarum est rectus istarum. |

Et si in ultimis tribus demonstrare volueris sicut iam dictum est in tribus primis, 183vb
eadem habebis argumenta operando in lineis ad dexteram protractis cum dyametro ***go*** que superius habuisti in lineis aliis, quia isti sunt illis equales. Sed in hoc solo est differentia, quod hic sinum versum concludis ubi sinum rectum conclusistis ibi et econverso. Et propterea lineam ***eg*** ducere oportebit loco ***ac*** linee que in parte alia est.

8. Et quoniam in premissa demonstratione

<E>t quoniam in premissa demonstratione alicui docuimus ut quadrata iungerentur in unum, qualiter hoc secundum artem geometrie fieri potest, considera, si duo quadrata fuerint equalia, tunc quadretur dyameter et habetur propositum scilicet unum quadratum equipollens duobus.

Si vero inequalia, quemadmodum in presenti <figura> sunt descripta, scilicet quadratum ***abcd*** et quadratum ***efgh*** unum alteri coniungitur, hoc modo: per 2^{am} primi Euclidis a dato puncto ***g*** linee proposite scilicet ***bcd*** ducatur equalis et sic ***gk***.

Item per 11^{am} eiusdem ***hg*** linee ex puncto ***g*** trahatur perpendicularis; et erit ***ad*** presens linea eadem scilicet ***gk***. Hoc facto subtendatur angulo ***hgk*** vasis que est linea ***hk***, quoniam ***hgk*** angulus est rectus. Patet <per> penultimam primi quod quadratum ***hk*** valet duo quadrata proposita. Quadretur ergo ***hk*** et habebitur propositum.

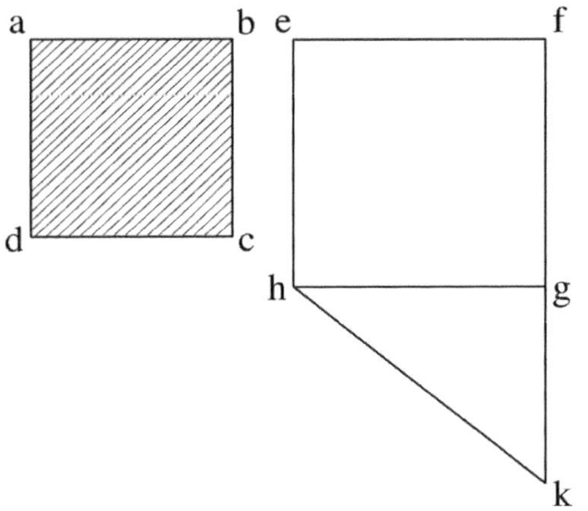

57 per 2^{am} primi Euclidis] EUCLIDES, *Elementa* I.2, 60 59 per 11^{am} eiusdem] EUCLIDES, *Elementa* I.11, 67

44 istarum] *primum scr.* Et si in ultimis tribus demonstrare volueris *post* istarum *sed del.* **45** est^{1}] *coni.*; et **52** quadrata] *coni.*; 4a **53** quadrata] *coni.*; 4ta **54** quadratum] *coni.*; quartum **60** vasis] *sc.* basis

9. Item in eadem demonstratione

<I>tem in eadem demonstratione, quoniam diximus quadratum a quadrato esse tollendum, quod qualiter per geometriam fieri debeat, patet sic. Sit quadratum ***hklm*** equale duobus quadratis suprascriptis quemadmodum ostensum est supra.

Abcd quadratum subtrahitur ab eo hoc modo: per 3^{am} primi Euclidis ex linea ***hm*** abscindatur linea ***ab*** et linee equalis sit ***nm***. Item ex linea ***ml*** abscindatur alia eidem ***nm*** equalis et sit ***mo***.

Item per 31^{am} ex puncto ***n*** trahatur eque distans ipsi ***hk*** et sit linea ***np***. Item ex puncto ***o*** per eandem 31^{am} ducatur eque distans ipsi ***kl*** et sit linea ***oe***. Ad hoc per primam petitionem Euclidis trahatur ***mk*** | dyagonaliter que transit per punctum ***r***. 184ra

Tunc argumentatur sic: cum ***nrom*** quadratum ratione compositionis figure sit equale ***abcd*** quadrato et ipsum totale quadratum ***hklm*** ex ypotesi sit equale duobus quadratis supra descriptis, patet quod gnomo residuus erit equalis ***efgh*** et sic est propositum.

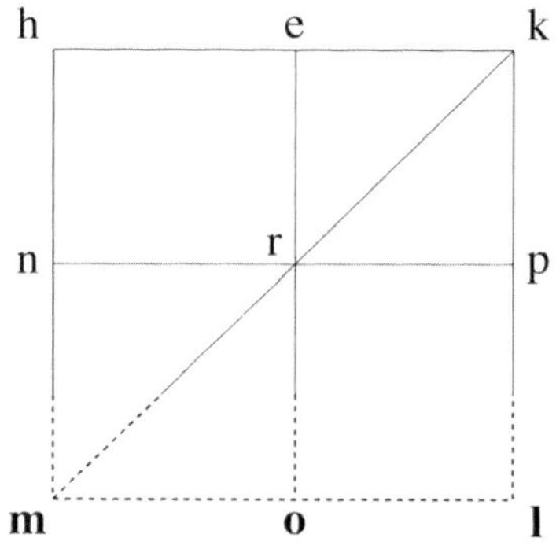

18. Nota in tabulis Tholetanis utimur

<N>ota: in tabulis Tholetanis utimur annis Arabum et ponuntur ibi radices ere Arabum. Et sunt partes signorum sive loca in quibus planete fuerint in inceptione ere Arabum et <computata> sunt <ad> meridianum Tholeti cuius longitudo est 28 graduum et dimidii ab occidente vel medio mundi, 4 horarum et 10^{e} unius hore, hoc est 61 graduum et 30 minutorum, hora enim est revolutio 15 graduum; dicitur autem medii mundi locus in Yndia magna, in civitate Arim, sub equinoxiali circulo. Dico autem orientem initium locorum habitabilium versus orientem et sic suo modo versus occidentem. Spacium autem terre habitabilis inter hos terminos secundum Alfraganum est tantum 180 graduum hoc est quantum revolvitur aliquis punctus celi in 12 horis ab oriente in occidentem; latitudo vero Tholeti est distantia eius ab equinoxiali versus septentrionem 39 graduum et 54 minutorum. 186ra

---

66 per 3^{am} primi Euclidis] EUCLIDES, *Elementa* I.3, 61 **69** per 31^{am}] EUCLIDES, *Elementa* I.31, 80 **70** per eandem 31^{am}] EUCLIDES, *Elementa* I.31, 80 **70/71** per primam petitionem Euclidis] EUCLIDES, *Elementa* I. pet. 1, 58 **77/80** Tholeti ... circulo] cf. *Canones Azarchelis*, Cb133, 430 **80/81** Dico ... occidentem] cf. ALFARGANUS, *Differentie* 6.3, 9 **84/85** latitudo ... minutorum] cf. *Canones Azarchelis*, Cb133, 430

70 oe] *coni.*; ne **72** nrom] norom *ante corr.* **75** utimur] iti *ante corr.* **82** 180] *coni.*; 12 **85** 54] *coni.*; 4

19. Ad inveniendum medios motus planetarum

<A>d inveniendum medios motus planetarum in tabulis Tholetanis eadem est operatio que in tabulis Alphontii nisi quod in tabulis Tholetanis intramus cum annis a primo collectis, postea expansis, postea mensibus perfectis, postea cum horis, si volumus ultra meridiem habere motum. Si autem volueris habere medios motus ad quamcumque aliam civitatem a Tholeto, considera in tabula per quot gradus longitudinis distet ista civitas a Tholeto et 15 gradibus horam constitues. Vide quot horas colligere possis et fractiones horarum non negligas igitur invenias medium cursum in tot horis cum fractionibus suis, et adde eum super medium cursum quem invenisti ad Tholetum si civitas ista fuerit occidentalis. Si vero fuerit orientalis subtrahatur et habebis quod queris. Cremona distat a Tholeto versus orientem 20 gradibus, | hoc est per unam horam et 3^{am} 186rb
partem hore. Invento igitur medio motu super Tholetum si ab eo subtraxeris quantum movetur planeta in una hora et 20 minutis hore, remanebit cursus eius ad medium diem Cremone.

20. Solem equabis per omnem motum sicut in Alphontio

<S>olem equabis per omnem motum sicut in Alphontio docetur. Si autem partem proportionalem, sic facies: pone 60 primum in proportione, minuta argumenti secundum, differentiam totam tertium, pars que queritur erit quartum. Itaque multiplica minuta argumenti per totam differentiam et quod exierit, divide per 60, quod tunc exierit erit pars quesita, quam adde prime equationi vel subtrahe secundum quod oportet. Habito ergo vero loco solis, computa eum a principio arietis mobilis, cui certo loco superadde motum 8^{ve} spere equatum, et habebis locum eius certissime computando a coniunctione circulorum.

89/91 Si ... constitues] cf. *Canones Azarchelis*, Cb140, 434 **93/94** et ... queris] cf. *Canones Azarchelis*, Cb140, 434 **94/96** Cremona ... hore] cf. *Canones Azarchelis*, Cb140, 434 **96/98** Invento ... Cremone] cf. *Canones Azarchelis*, Cb140, 434 **103/106** quam ... circulorum] cf. *Canones Azarchelis*, Cb142, 436

---

87 Alphontii] *coni.*; Auphontii **94** Cremona] *coni.*; Tremonia **98** Cremone] *coni.*; tremonie **105** locum] *coni.*; locus

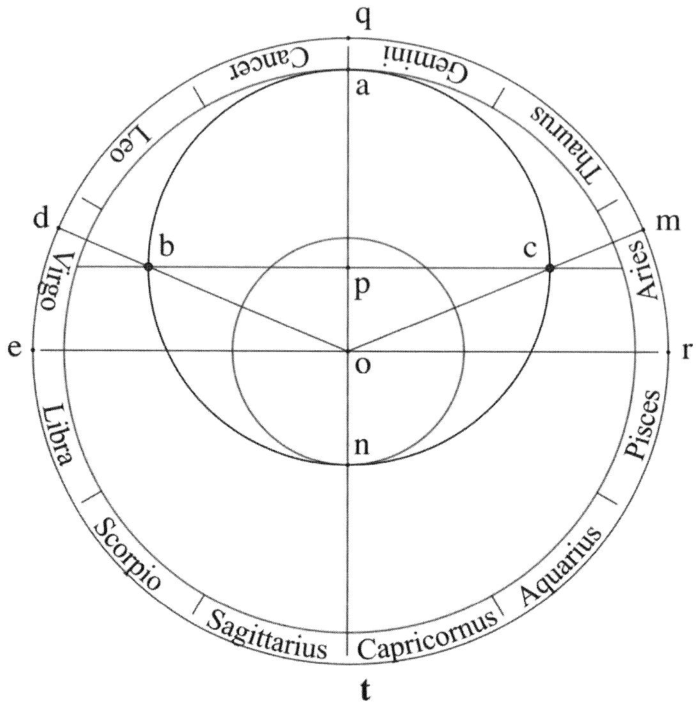

Pro maiori evidentia pono figuram: sit ***o*** centrum terre, ***p*** centrum ecentrici, et ***a*** erit punctus augis et sit ***r*** caput arietis, sol vero sit apud ***b*** item posterius sol sit in puncto ***c*** et circuli rubei comprehendant inter se spissitudinem spere.

Et dicitur exterior circulus augis scilicet puncti ***q***, interior vero circulus sit oppositus augis, scilicet puncti ***n***. Licet itaque sol in eodem percurrat zodiacum in quo pertransit ecentricum, non tamen est hec computatio in partibus circulorum proportionalibus, nisi tantum in duabus medietatibus; hec in puncto augis et eius opposito distinguuntur, quoniam in superiori medietate ecentrici que est ***cqb*** plus movetur in ecentrico quam in zodiaco quod patet quia medietas illa non correspondet medietati zodiaci proportionali ymmo parti minori. In inferiori vero medietate ecentrici que est ***bnc*** plus movetur in zodiaco quam in ecentrico. Quod patet quia medietas illa non precise correspondet medietati zodiaci ymmo arcui maiori.|

Hinc accidit quod in ista medietate ecentrici que est ex ***q*** per ***b*** in ***n*** maior est medius 186va
motus quam verus. Verbi gratia: sole existente ut positum est in ***b***, medius motus est arcus ***rqe***, sed verus motus est arcus ***rqd***, arcus equationis interiacens est ***de***. Igitur, si a medio motu subtrahatur aux id est arcus ***rq***, remanet argumentum scilicet arcus ***qe*** minor 6 signis. Ideoque equationis arcus a medio motu substrahitur et remanet arcus ***rqd*** qui est motus solis equatus et locus solis verus est punctus ***d***. Econverso accidit in alia medietate que est ex ***nt*** per ***cm*** <in> ***aq*** quod maior est verus motus quam medius. Verbi gratia: sole ut positum est posterius existente in puncto ***c*** medius locus solis est ***r*** sed verus est ***m***. Igitur cum argumentum sit plus 6 signis arcus equationis qui est ***rm*** additur medio motui et habetur motus equatus. Et sic patet istius ratio operationis.

109 circuli rubei] *circuli non rubricati in figura* **110** q] *sc.* a oppositus] *coni.*; opposıtum **113** distinguuntur] *coni.*; *fors.* distiguuntur **114** cqb] *sc.* cab **116** bnc] *coni.*; bvc **119** q] *sc.* a **125** nt] *coni.*; *post* n *deest* 1 *litt.* cm] *primum scr.* qsq *post* cm *sed del.* aq] *coni.*; qs **126** r] *coni.*; f **127** rm] *coni.*; fm **128** ratio operationis] *coni.*; rationis operatio

Exemplum autem posuimus in locis ubi maxima est diversitas inter medium locum et verum, quod patet per figuram precedentem.

Si vis equare lunam, fac totaliter sicut in Alphontio in capiendo partem proportionalem de diversitate dyametri: pone 60 pro primo numero, minuta proportionalia pro secundo, diversitatem dyametri totam pro tertio; multiplica igitur secundum per tertium vel e converso et divide per primum et habebis quartum.

Pro maiori evidentia ponam figuram lune. Sit circulus signorum ***abcd*** super centrum ***o*** quod centrum etiam est centrum spere lune, cuius spissitudo ***s*** est inclusa 2 circulis ***sc zk***. Descriptus circulus ***ac*** ostendit exteriorem superficiem spere lunaris, circulus vero ***zk*** interiorem superficiem.

Deferens lune sit ***lmnr*** super centrum ***p***; igitur ***sk*** dyametrum epicycli semper respicit punctum ***e*** in omni situ. Quia sicut ***l*** et ***p*** moventur ad occidentem versus ***d***, ita centrum epicycli, quod etiam significatur per ***l***, movetur ad orientem versus ***b***, et ei occurrit punctum ***e***. Posito ergo epicycli centro in auge ecentrici apud ***l***, sol secundum medium cursum erit apud ***a*** vel apud ***c***, quod in idem redit ad presens. Quia in coniunctione et oppositione media centrum epicycli est apud ***l***, hoc est in auge ecentrici, in tali enim situ equationes centri non sunt, quia centrum nullum | 186vb est. Equationes vero minime in situ autem opposito scilicet, cum fuerit centrum epicycli in opposito augis in puncto ***n***, quod quidem realiter contingit primo sub ***b*** deinde sub ***d***. Quoniam cum luna distitit a sole per 3 vel 9 signa, hoc est in quadraturis, quia tunc centrum epicycli est in opposito augis, centrum erit 6 signa et equationes eius nulle, quoniam eadem est linea ostendens medium locum et verum epicycli; equationes vero argumenti erunt maxime. Et hoc patet cum enim luna fuerit in puncto, ubi ***oc*** contingit epicyclum, tunc argumenti eius equatio est maxima sicut apparet ad sensum, scilicet ***a a a*** sunt minores ***c c c***, qui tunc arcus argumenti equationis vocantur. Hec autem ostensio maioritatis equationis argumenti ex necessitate descriptionis ostensa est sub ***c***, cum tamen, sicut dictum est, realiter contingat sub ***b*** vel ***d***, quod est propter revolutionem ecentrici. In situ autem intermedio, scilicet cum centrum epicycli super 3 vel 9 signa ab auge, quod realiter contingit primo in ***g*** sub ***h*** deinde in ***f*** sub ***q***, scilicet quando luna distat a sole 45 gradibus, ut sub ***h***, vel 10 signis et 15 gradibus, ut sub ***q***, tunc sunt equationes centri maxime.

Equationes vero argumenti medio modo se habent, quia sunt maiores quam in auge, minores vero quam in opposito augis. Et hec ostensio ex necessitate descriptionis facta est sub ***b*** et ***d***, licet ut dictum est, realiter sic accidat sub ***h*** et ***q***. Posito itaque epicyclo in ***m*** sub ***b*** vel in ***r*** sub ***d***, locus ipsius epicycli medius erit punctum ***i***, quem ostendit linea ***ei*** vel punctum ***v*** quem ostendit linea ***ev***; locus autem eius verus erit ***b*** quem ostendit linea ***omb*** vel ***d*** quem ostendit linea ***ord***. Centrum autem eius medium est ex puncto ***a*** usque in ***i*** vel ultra per ***c*** usque in ***v***. Centrum autem equatum sive verum est ex eodem puncto ***a*** usque in ***b*** super ***m*** vel ultra per ***c*** usque in ***d*** super ***r***.

Equatio centri in epicyclo, que etiam equat argumentum, est arcus ***sy***. Equatio vero centri in zodiaco est arcus ***ib*** vel ***vd*** predicto arcui proportionalis, quia sicut se habet ***sy*** ad totum epicyclum, ita se habet ***ib*** vel ***vd*** ad totum zodiacum. Et enim in eundem angulum cadunt, ut patet hinc est quod harum duarum equationum centri accepta una habetur reliqua. Item aux media epicycli est ***s***, aux vera est ***y*** et si corpus lune fuerit in ***x***,

---

136 est^{2}] *coni.*; et **137** zk] *coni.*; et k **138** zk] *coni.*; et k **139** sk] *coni.*; bk **142** epicycli] epicyclo *ante corr.* **157** in f] *coni.*; infra **158** 15] *coni.*; 12 **162** erit] *coni.*; et **164** puncto] *coni.*; p^{t} **165** v] *coni.*; r **166** b] *coni.*; k **167** in] *suppl. s. l.* **168** est] *coni.*; et argumentum eius medium erit arcus ***sx*** et hoc sub ***b*** vel in alio epicyclo | ***syx*** et hoc erit sub ***d***. 187ra

Argumentum autem equatum est arcus ***ysx*** sub ***b*** vel ***yx*** sub ***d***; patet igitur quod equatio argumenti est arcus ***b b*** vel arcus ***d d***. Et hec equatio equat motum lune. Hiis omnibus itaque consideratis manifestum est quare equatio centri additur argumento si centrum est minus 6 signis ut ***ys*** arcus sub ***b***, et substrahitur si est plus ut ***ys*** arcus sub ***d***. Item patet quare equatio argumenti zodiaci additur medio motui, cum ipsum argumentum fuerit plus 6 signis, verbi gratia: si esset luna in puncto contactus ***pk*** et subtrahitur ab eodem motu, si dictum argumentum fuerit minus 6 signis, verbi gratia: si luna esset aput ***x***. Hoc autem quod ex diversitate dyametri additur equationi argumenti est plus vel minus secundum quod epicyclus ab auge versus oppositum augis, plus vel minus recedit. Semicirculus autem ecentrici super quem centrum epicycli in isto recessu defertur, potest intelligi quasi per 60 partes equales divisus, quarum partium numerus minuta proportionalia dicuntur, sicut igitur partes pertransite se habuerint ad 60, sic de diversitate dyametri additur equationi argumenti.

Habito igitur motu lune equato debes eum computare, sicut dixi tibi in sole a capite arietis mobilis quod est in octava spera. Si autem eum habere volueris a capite arietis fixi, scilicet ab ista intersectione circulorum ubi caput arietis octave spere supponitur capiti arietis none spere, tunc adde super motum lune equatum motum octave spere equatum et quod pervenerit computa a tali loco, sicut dixi. Et hec omnia patent in figura.

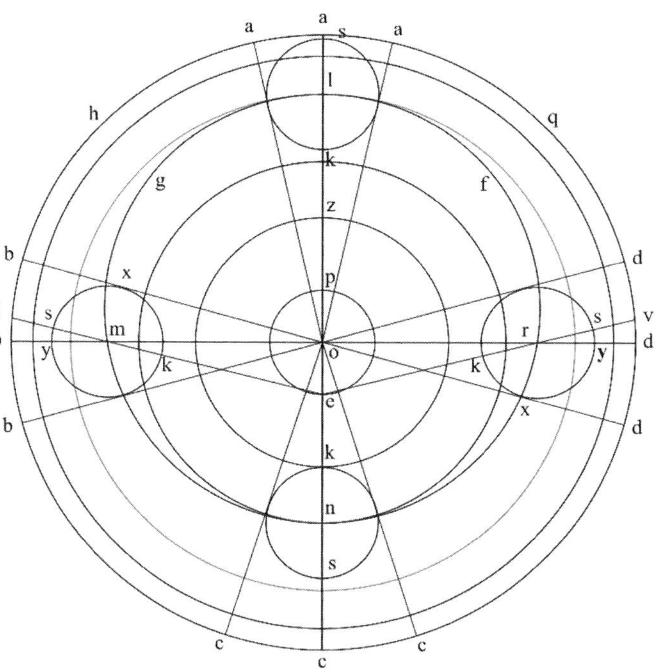

174 ysx] *coni.*; yxs **181** quod] *coni.*; quidem

List of manuscripts

Erfurt, Universitätsbibliothek, CA 4° 366
Oxford, Bodleian Library, Digby 97
Oxford, Hertford College, 4
Paris, Bibliothèque nationale de France, lat. 11232
Paris, Bibliothèque nationale de France, lat. 7197
Paris, Bibliothèque nationale de France, lat. 7273
Paris, Bibliothèque nationale de France, lat. 7295A
Paris, Bibliothèque nationale de France, lat. 7305
Paris, Bibliothèque nationale de France, lat. 7314
Paris, Bibliothèque nationale de France, lat. 7333
Paris, Bibliothèque nationale de France, lat. 7432
Paris, Bibliothèque nationale de France, lat. 7446
Paris, Bibliothèque nationale de France, lat. 7447
Paris, Bibliothèque nationale de France, lat. 7450
Toledo, Archivo y Biblioteca Capitulares (olim Biblioteca del Cabildo), 104 6
Vienna, Österreichische Nationalbibliothek, Cod. 5268

Bibliography

Almagesti minor, ed. by Henry Zepeda, *The First Latin Treatise on Ptolemy's Astronomy: The* Almagesti minor *(c. 1200)* (Turnhout: Brepols, 2018).

Canones Azarchelis, ed. and trans. by Fritz S. Pedersen, 'Part 2 Cb: Canons "Quoniam cuiusque actionis" ', in *The Toledan Tables. A Review of the Manuscripts and the Textual Versions with an Edition*, ed. by Fritz S. Pedersen (Copenhagen: C.A. Reitzels Forlag, 2002), pp. 382–499.

Theorica planetarum Gerardi, ed. by Francis J. Carmody (Berkeley: University of California, 1942).

Al-Farghānī, *Differentie in quibusdam collectis scientie astrorum*, ed. by Francis J. Carmody (Berkeley: 1943).

Aquinas, Thomas, *Commentary* on Aristotle, *On Memory and Recollection*, trans. by John Burchill, in *The Medieval Craft of Memory: An Anthology of Texts and Pictures*, ed. by Mary Carruthers and Jan M. Ziolkowski (Philadelphia: University of Pennsylvania Press, 2002), pp. 153–88.

Berschin, Walter, 'Greek Script in Latin Manuscripts', in *The Oxford Handbook of Latin Palaeography*, ed. by Frank T. Coulson and Robert G. Babcock (New York: Oxford University Press, 2020), pp. 115–17.

Burnett, Charles, 'The Latin and Arabic Influences on the Vocabulary Concerning Demonstrative Argument in the Versions of Euclid's *Elements* Associated with Adelard of Bath', in *Aux origines du lexique philosophique européen: L'influence de la 'latinitas'*, ed. by Jacqueline. Hamesse (Louvain-la-Neuve: Société internationale pour l'étude de philosophie médiévale, 1997), pp. 117–35.

Campanus de Novara, *Theorica planetarum*, ed. and transl. by Francis S. Benjamin Jr. and Gerald J. Toomer, *Campanus of Novara and Medieval Planetary Theory:* Theorica planetarum (Madison: University of Wisconsin Press, 1971).

Chabás, José and Bernard R. Goldstein, *The Alfonsine Tables of Toledo* (Dordrecht: Kluwer, 2003).

Chemla, Karine ed., *The History of Mathematical Proof in Ancient Traditions* (Cambridge: Cambridge University Press, 2012).

Crowther, Kathleen M. and Peter Barker, 'Training the Intelligent Eye: Understanding Illustrations in Early Modern Astronomy Texts', *Isis* 104 (2013), 429–70.

Dinkova-Bruun, Greti, 'Text and Gloss', in *The Oxford Handbook of Latin Palaeography*, ed. by Frank T. Coulson and Robert G. Babcock (New York: Oxford University Press, 2020), pp. 924–42.

Euclides, *Elementa*, ed. by Hubertus L. L. Busard, *Campanus of Novara and Euclid's Elements* (Stuttgart: Franz Steiner, 2005).

Goldstein, Bernard R, *Ibn al-Muthannâ's Commentary on the Astronomical Tables of al-Khwârizmî* (New Haven: Yale University Press, 1967).

Husson, Matthieu, 'Mathematical Astronomy and the Production of Multiple-Texts Manuscripts in Late Medieval Europe: A Comparison of BnF lat. 7197 and BnF lat. 7432', in *The Emergence of Multiple-Text Manuscripts*, ed. by Alessandro Bausi, Michael Friedrich, and Marilena Maniaci (Berlin: De Gruyter, 2019), pp. 247–74. <https://doi.org/10.1515/9783110645989> [accessed: 14.04.2023]

———, 'The Art of Astrological Computations: Conrad Heingarter and the Manuscript Paris, BnF latin 7295A', in *Exploring Written Artefacts: Objects, Methods, and Concepts*, ed. by Jörg B. Quenzer (Berlin: De Gruyter, 2021), pp. 513–32. <https://doi.org/10.1515/9783110753301> [accessed: 14.04.2023]

Johannes von Gmunden, *De sinibus, chordis et arcubus*, ed. by Hubertus L.L. Busard, *Der Traktat* De sinibus, chordis et arcubus *von Johannes von Gmunden* (Vienna: Springer Verlag, 1971).

Martinelli Tempesta, Stefano and Marco Petoletti, 'Il ritratto di Omero e la firma greca di Boccaccio', *Italia Medioevale e Umanistica* 54 (2013), 399–409.

Molland, Andrew G, 'Mathematics in the Thought of Albertus Magnus', in *Albertus Magnus and the Sciences: Commemorative Essays*, ed. by James A. Weisheipl (Toronto: Pontifical Institute of Mediaeval Studies, 1980), pp. 463–78.

North, John D., *Richard of Wallingford. An Edition of His Writings with Introduction, English Translation, and Commentary* (Oxford: Oxford University Press, 1976).

Pedersen, Fritz S., *The Toledan Tables. A Review of the Manuscripts and the Textual Versions with an Edition*, 4 vols (Copenhagen: C.A. Reitzels Forlag, 2002).

Poulle, Emmanuel, *Les Tables Alphonsines avec les Canons de Jean de Saxe* (Paris: Editions du Centre National de la Recherche Scientifique, 1984).

Rioust, Laure, 'Latin 7295A', <https://archivesetmanuscrits.bnf.fr/ark:/12148/cc664943> [accessed: 14.04.2023].

Summers, David, *The Judgment of Sense: Renaissance Naturalism and the Rise of Aesthetics* (Cambridge: Cambridge University Press, 1987).

Steinschneider, Moritz, 'Études sur Zarkali, astronome arabe du XIe siècle, et ses ouvrages', offprint from *Bullettino di bibliografia e di storia delle scienze matematiche e fisiche* (ed. Boncompagni), 14, 16, 17, 18, 20 (1881–87).

Sylla, Edith, 'Mathematical Physics and Imagination in the Work of the Oxford Calculators: Roger Swineshead's *On Natural Motions*', in *Mathematics and Its Applications to Science and Natural Philosophy in the Middle Ages: Essays in Honour of Marshall Clagett*, ed. by Edward Grant and John E. Murdoch (Cambridge: Cambridge University Press, 1987), pp. 69–101.

Thorndike, Lynn and Pearl Kibre, *A Catalogue of Incipits of Mediaeval Scientific Writings in Latin*, rev. ed. (London: Mediaeval Academy of America, 1963).

Van Brummelen, Glenn, *The Mathematics of the Heavens and the Earth* (Princeton: Princeton University Press, 2009).

Yates, Francis, *The Art of Memory* (Chicago: University of Chicago Press, 1966).

ELEONORA ANDRIANI, CAMILLE BUI and
NICHOLAS JACOBSON

Conrad Heingarter's Treatise on Planetary Latitude

*An Introduction and Critical Edition**

To the memory of
Helena Avelar de Carvalho (1964-2021)

Introduction

Sometime in the latter half of the fifteenth century, the Swiss astrologer and physician Conrad Heingarter composed a short treatise on planetary latitudes contained in a manuscript destined for his patron, Duke John II of Bourbon (1426-88), while in residence at Belleperche. The work is divided into two sections: a *theorica* and a *practica*. The *theorica* purports to describe the planetary models upon which the tables of planetary latitude were based. The *practica* is a set of canons, which presents instructions to be used with three different types of latitude table. The inclusion of the *practica* implies that the author had more than just a theoretical interest in the subject in mind. He intended for his readers to compute with the tables, and he offered a tutorial on how to do so.

In light of Heingarter's role at the Bourbon court and his career trajectories, his efforts in establishing a practical guide for calculating planetary latitude values may reveal a growing interest among astrologers and physicians — and not just mathematical astronomers — in this relatively advanced branch of astronomy. This shift in technical interests was itself likely

* This chapter has been written collectively. Eleonora Andriani wrote most of the pages pp. 57-67, Camille Bui wrote most of the pages pp. 68-70, and Nicholas Jacobson wrote most of the pages pp. 53-56, 71-78. The edition, translation, and glossary were established collectively.
We would like to thank to José Chabás, Matthieu Husson, and Richard L. Kremer for their reading and reactions to earlier versions of this paper. In addition, we are grateful to Helena Avelar de Carvalho, Charles Burnett, Samuel Gessner, Dorian Gieseler Greenbaum, Luís Ribeiro, Marie-Madeleine Saby, and Glen Van Brummelen for their helpful suggestions, comments, and reactions to earlier drafts. This work was supported by the ERC project ALFA: Shaping a European Scientific Scene, Alfonsine astronomy, CoG 723085, PI Matthieu Husson.

Eleonora Andriani • Centre national de la recherche scientifique, Paris
Camille Bui • Observatoire de Paris - Centre national de la recherche scientifique, Paris
Nicholas Jacobson • University of Wisconsin-Madison

Alfonsine Astronomy: Computational Practices, ed. by Matthieu HUSSON, José CHABÁS, and Richard L. KREMER, Turnhout, 2026 (*Alfonsine Astronomy: Studies and Sources*, 3), p. 53-108
© BREPOLS PUBLISHERS 10.1484/M.ALFA-EB.5.137908

spurred on, at least in part, by changes in the practice of medical astrology. Throughout the late-thirteenth and fourteenth centuries in Europe, a medical astrologer's primary tools of prognostication resided in the *figura caeli*, the tracking of the Moon through critical days, and the attribution of astral properties to different medicines. These tools were taken up, in turn, by practitioners of the late medieval and early modern periods.[1]

Paramount among these stood the *figura caeli*, or horoscope, in which the positions of the seven classical planets and the lunar nodes were depicted on a square representation of the zodiac. The nodes, and sometimes the Moon's latitude, were included to account for lunar and solar eclipses. The horoscope was then divided into 12 triangles, representing the houses, which surrounded a central square providing relevant information such as the time, place, and occasion for which the horoscope had been made. The *figura caeli* typically only included the seven planets' positions in the zodiac in terms of ecliptic longitude, neglecting the latitude values for all but the Moon.[2] With this primary tool, other crucial techniques of astrological practice could be performed, such as calculating directions and profections for the chart, determining planetary conjunctions and aspects, and establishing the date of conception for the native.[3] These latter calculations, however, relied upon the same planetary positions in longitude that a practitioner would find for the base horoscope.

By the late fifteenth century, several salient factors emerged that put pressure on physicians like Heingarter to re-evaluate the theoretical underpinnings of the practices developed for their profession. First, the influx of new texts and translations from the Greek language, especially in the decades leading up to the fall of Constantinople in 1453, had contributed to a concerted effort among scholars to uncover the original teachings of the ancients. From the perspective of the astral sciences, this humanistic impulse was directed towards the goal of finding the authentic doctrines of Ptolemy and the earliest Hellenistic practitioners of astrology.[4] Although there were some who used this search for original texts as an excuse to clear away what were considered to be accretions of later generations, there were others, like Heingarter, who instead sought to gather together and systematize the vast body of knowledge within a single framework.[5] As a consequence of the compiling efforts of this new generation of humanists in the major centres of learning in Western Europe, a greater variety of computing practices were made more widely

1 For a subtle account of the interplay of different elements in medical astrology, and how they were inherited in the Renaissance and Early Modern period by such practitioners as Girolamo Cardano, see Anthony Grafton and Nancy Siraisi, 'Between the Election and My Hopes: Girolamo Cardano and Medical Astrology', in *Secrets of Nature: Astrology and Alchemy in Early Modern Europe*, ed. by Anthony Grafton, Jed Z. Buchwald, and William Newman (Cambridge: MIT Press, 2001), pp. 69-131, particularly pp. 77-78.

2 See Alexandre Tur, 'From astronomical computation to astrological interpretation: *figurae caeli* and "astrological squares" in fifteenth-century annual prognostications' in this volume.

3 Directions and profections were both methods of forecasting by which a designated position in the horoscope was moved towards other significant points in the chart. We discuss the technique of directions in greater detail in the first section of this paper. Ptolemy concluded the *Quadripartitum* with a discussion of profections. See Ptolemy, *Tetrabiblos*, ed. and trans. by Frank E. Robbins (Cambridge, Mass.: Harvard University Press, 1940), pp. 447-57.

4 See Dag Nikolaus Hasse, *Success and Suppression: Arabic Sciences and Philosophy in the Renaissance* (Cambridge, Mass.: Harvard University Press, 2016), pp. 248-92, pp. 251-53, pp. 268-69.

5 See Nicolas Weill-Parot, *Les 'Images Astrologiques' au Moyen Âge et à la Renaissance: spéculations intellectuelles et pratiques magiques (XIIe-XVe siècle)* (Paris: H. Champion, 2002), p. 805.

available to aspiring medical and judicial astrologers that went well beyond planetary longitude and eclipse calculations.

At the same time, medical astrologers were facing a wave of sceptical criticism from those who would rid the arts and sciences of divinatory practices and fatalistic beliefs.[6] Heingarter's own fierce defence of the use of astronomy for medicine appears later in his career when such debates became more prevalent in scholarly circles.[7] In order to justify their craft, some astrologers sought a common ground with their critics by emphasizing the naturalistic foundation of their discipline. Direct causal connections between planetary motions and terrestrial effects were sought that could explain individual events on Earth in terms of specific physical changes in the heavens. Techniques involving planetary conjunctions, oppositions, and aspects were gathered together in astronomical canons, with others requiring an adept to pay close attention to the speed, direction, and nearness of a planet on its epicycle and eccentric in relation to the Earth. The latitudinal position of a planet held great significance within such a naturalistic system since it could account for the position of the physical body of the planet rather than its projection onto the ecliptic plane.[8] It also played an important role in determining the aspects between these physical bodies. Thus, authors treating techniques involving the so-called planetary accidents or passions often advised the most careful practitioners to take into account the planet's latitudinal position in their computations.[9]

Finally, the greater attention afforded to planetary passions also contributed to what was a broader movement among students, practitioners, and scholars alike to reform the curriculum available for learning astronomy in the university. During this period, the *Theorica planetarum communis*, which had served as the central text on planetary astronomy for university instruction, was routinely criticized for the inadequacy of its final three chapters that considered retrograde and direct motion, the use of planetary latitude tables and their theoretical underpinnings, and planetary aspects.[10] In addition to addressing new interests and concerns in the realm of medical astrology, Heingarter's treatise on

6 Gian Mario Cao, '*Inter alias philosophorum gentium sectas, et humani, et mites*: Gianfrancesco Pico and the Sceptics', in *Renaissance Scepticisms*, ed. by Gianni Paganini and José R. Maia Neto (Dordrecht: Springer, 2009), pp. 127-47; Cao, 'The Prehistory of Modern Scepticism: Sextus Empiricus in Fifteenth-Century Italy', *Journal of the Warburg and Courtauld Institutes* 64 (2001), 229-80 (pp. 259-60). It should be noted that recent scholarship has shown that Giovanni Pico della Mirandola's own critiques may have been motivated by an impulse to reform astrology rather than fully overturn it; see Dag Nikolaus Hasse in the note above.

7 For a discussion of Heingarter's defence, see section 1 below.

8 Understanding apparent planetary positions in the zodiac to be dependent upon the motions of real physical spheres had become a subject of lively debate among fifteenth-century astronomers, particularly at the University of Vienna. For a salient example of the relevance of latitude theory to these discussions, see Michael Shank, 'Regiomontanus *versus* George of Trebizond on Planetary Order, Distances, and Orbs (*Almagest* 9.1)', in *Ptolemy's Science of the Stars in the Middle Ages*, ed. by David Juste, Benno van Dalen, Dag Nikolaus Hasse and Charles Burnett (Turnhout: Brepols, 2020), pp. 305-86, particularly pp. 370-77.

9 This was already the case with John of Saxony in his commentary on the *Introductorius* (*Commentum super textu Alchabitii*). See section 1 below.

10 The *Theorica planetarum communis* is often referred to as the *Theorica planetarum Gerardi* in reference to Gerard of Cremona's alleged authorship of the work. Because this attribution is doubtful, we have opted to refer to this work as the *Theorica planetarum communis* throughout. See Claudia Kren 'Planetary Latitudes, the *Theorica Gerardi*, and Regiomontanus', *Isis*, 68 (1977), 194-205; Claudia Kren, 'A Medieval Objection to "Ptolemy,"' *The British Journal for the History of Science*, 4 (1969), 378-93.

latitudes also intervened in these debates over the curriculum of the Latin university system. Thus, because Heingarter's own career sat at the intersection of court politics, the medical profession, and the university, it is clear that he composed his treatise to address different audiences hailing from multiple contexts. The treatise served simultaneously as a humanistic gift of erudition to his patron at Belleperche, a boon to aspiring practitioners of the astral sciences, and was also presented as a commentary text that would aid the arts faculty of Paris in their efforts to expand their curriculum.

By way of introducing Heingarter's treatise and its significance, this article will highlight these three aspects of his career in turn. The first section will discuss his firm commitment to the integration of mathematical astronomy into his practice as a physician and counsellor in the face of sceptical forces within the profession. In addition to this, we will show that his efforts in gathering and commenting on what he considered to be the most authoritative texts in his field brought to the fore several techniques that focused on the importance of planetary accidents in medical and judicial astrology. We will limit ourselves to reviewing the features of these techniques that are relevant to planetary latitudes.

The second section will consider the humanistic impulse that spurred Heingarter to collect and preserve a wide body of textual material on planetary latitude. The many extant witnesses to his library reveal an abiding interest in the subject over the course of many years. Through the identification and analysis of several of these manuscripts, we will reconstruct features of Heingarter's research and collection practice that are relevant to understanding how he composed the treatise.

The third section discusses Heingarter's continued collaboration with masters at the University of Paris and his intention to augment the arts curriculum with his teachings on latitude theory and computation. This element of Heingarter's career has received relatively little attention in the current secondary literature, and we hope to present this text as a facet of Heingarter's larger project of guiding and training a new generation of students in the arts and medical faculties of the University of Paris. In doing so, we will provide an overview and analysis of the treatise.

A critical edition and translation of Heingarter's treatise will appear at the end as an Appendix, together with a list of relevant vocabulary terms as a glossary.

1. Heingarter's defence of medical astrology

Originally from Horgen near Zürich, Heingarter received his B.A. in 1454, his M.A. in 1455, and his licentiate in medicine at the University of Paris in March 1466. Perhaps as early as 1463 or 1464, Heingarter began his career in the service of John II, whom he served until the duke's death in 1488.[11] During his years of study in Paris, Heingarter had become

11 It is beyond the scope of this article to provide an exhaustive account of the life and the work of Conrad Heingarter. We are therefore only mentioning a selection of contributions vital to an investigation into these topics. For more on Heingarter's life and work, see Lynn Thorndike, *A History of Magic and Experimental Sciences*, 8 vols (New York: Columbia University Press, 1934-1958), IV, pp. 357-73 and pp. 374-85; Maxime Préaud, *Les méthodes de travail d'un astrologue du XV*[e] *siècle, Conrad Heingarter*, unpublished thesis, Paris, École des chartes, 1969, pp. 143–49; Maxime Préaud, *Les astrologues à la fin du Moyen Âge* (Paris: J.-C. Lattès, 1984), particularly pp. 71–94, 97-100, 141–55, 177–86, 240–43; Jean-Patrice Boudet, *Le* Recueil des plus célèbres

an avid student of the quadrivium, as has been made readily apparent in the many extant manuscripts either belonging to him or that have transmitted his own original works. From those of his codices preserved at the Bibliothèque nationale de France in Paris, we are offered a unique window into the growth and development of his competencies in mathematical astronomy.[12]

Many of these witnesses contain mathematical texts that were gathered together by Heingarter when he was still a student at the University of Paris. In one such student manuscript, BnF, lat. 7197, introductory texts on arithmetic, geometry, and astronomy filled with Heingarter's annotations sit incongruously alongside Ciceronian orations and Hugh of St Victor's *De grammatica*. Manuscripts such as BnF, lat. 7197 constitute a series of exercises amounting to a corpus of juvenilia, but, even years after writing his youthful notes on the *Sphere* of John of Sacrobosco and the *Theorica planetarum communis*, Heingarter would return to these foundational works again and again for inspiration. He wrote no less than two marginal commentaries on the full text of the *Theorica*, and the planetary latitude treatise itself should be seen as a specialized version of a third such commentary focusing on the *Theorica*'s seventh chapter. He remained committed to the application of astronomy to his profession throughout his life.

A perusal of Heingarter's practice in the first decade or so of his career shows that his methods did not deviate drastically from those of his contemporaries. As a counsellor to the duke and his household, as well as a diplomat and advisor to King Louis XI (the brother of the duchess), Heingarter availed himself of the common astrological tools of his generation. In 1469, he prepared for the king an interpretation of the Saturn-Mars conjunction, although it is no longer extant.[13] BnF, lat. 7446 contains a nativity that Heingarter prepared for Jean de la Gutte, a favourite at the Bourbon court. In BnF, lat. 7447 there appears a *iudicium anni* for the duke attributed to Heingarter.[14] Likewise Heingarter prepared a *iudicium anni* for the year 1476 addressed to King Louis that is preserved in BnF, lat. 7450.[15] These judgments conform to the typical conventions of finding the degree of the ascension and midheaven, dividing the houses, and finding planetary longitudes for a horoscope.

Something appears to have happened later in Heingarter's career that caused him to become much more self-conscious about the methods of medical astrology. Beginning in the late 1470s and continuing throughout the 1480s, he began writing medical advice and

astrologues *de Simon de Phares*, 2 vols (Paris: H.Champion 1997–1999), I, pp. 585-87, 94a-94b; Weill-Parot, *Les 'Images Astrologiques'*, pp. 710-14 and pp. 901-3; Matthieu Husson, 'The Art of Astrological Computations: Conrad Heingarter and the Manuscript Paris, BnF latin 7295A', in *Exploring Written Artefacts: Objects, Methods, and Concepts*, ed. Jörg B. Quenzer, (Berlin: De Gruyter, 2021), pp. 513–32; Steven Vanden Broecke, 'Astrological Self-Government at the Fifteenth-Century Court of Bourbon', *Early Science and Medicine* 28 (2023): pp. 34–64.

12 The Bibliothèque nationale de France preserves several manuscripts belonging to Heingarter or preserving his works, namely: lat. 7197, lat. 7295A, lat. 7305, lat. 7333, lat. 7432, lat. 7446, lat. 7450, and lat. 11232. Manuscripts of uncertain connection to Heingarter: lat. 7273, lat. 7447, and lat. 7314.

13 Though the judgment is no longer extant, Simon de Phares attests to its existence in *Le Receuil*; see Boudet, *Le Recueil*, vol. I, p. 586.

14 On the attribution of this work to Conrad Heingarter, see Ernest Wickersheimer, *Dictionnaire biographique des médicins en France au Moyen Âge*, new ed. (Geneva: Droz 1979), p. 57.

15 On the political profile of Heingarter's astrological activity, see Tur, 'From Astronomical Computation to Astrological Interpretation' in this volume.

public circulars in which he zealously defended his practice against critics. He wrote one of his most strident defences of the integration of astronomy into medicine to Duchess Jeanne of Bourbon in a work of medical advice dated to August 26, 1480. At the very beginning of this excursus, he included a long chapter, aptly entitled *On the Usefulness of the Science of the Stars and what is Profitable and Necessary to Physicians* (*De utilitate sciencie astrorum et quod sit medicis proficua et necessaria*).[16] Written three years after Marsilius Ficino's *Disputatio contra iudicium astrologorum* and only six years before Pico della Mirandola's bombshell *900 Theses* were published in Rome, thereby challenging the very epistemic foundations of astrology, Heingarter's defence entered circulation at a moment of sceptical ferment within the medical profession.[17] The text offers an eclectic mix of justifications for astronomy's role in medicine, taking turns in appealing to traditional authorities and natural causation for proof of its efficacy.

It was an inescapable truth to Heingarter that the lower realm of the earth was ruled by celestial bodies and to prove this he called to his defense a host of authorities including Aristotle, Cicero, al-Rāzī, Bernard Gordon ('a modern medical man', cit.), Arnald of Villanova, Abū Ma'shar, Hippocrates, Ptolemy, and the *Speculum astronomie*, which he attributed to Albert the Great.[18] Making heavy weather of the fact that he had tradition on his side, Heingarter was offering a response to rivals who questioned the relevance of the astral sciences to their discipline. Such physicians, he anticipated, would have much to say against what he had written, even though it contained doctrines in agreement, as he himself explains, 'with none but authoritative sources'.[19]

Not satisfied with the invocation of authorities whom he anticipated his opponents would reject out of hand, Heingarter also made an appeal to the natural connections between the heavenly bodies and terrestrial occurrences. Summarizing his opponents' position, Heingarter explained that these physicians found it rash and superstitious for a doctor to consult the stars, since medicine was a science resting on principles of its own. Conceding that, from a methodological point of view, medicine was a discipline deriving its conclusions from its own principles, Heingarter, nevertheless, denied that this implied that medical knowledge could be gleaned without recourse to the mathematical and natural sciences. Instead, he likened the active principles of medicines, natural compounds, and celestial bodies to the virtues of moral philosophy. 'Just as in ethics the virtues are connected', he wrote, 'so medicine cannot be separated from natural philosophy or astronomy'.[20] Thus, at least by the 1480s, Heingarter had made a point to explain the efficacy of the stars on the health of his patrons in terms of natural causation.

16 The medical advice to the duchess can be found in Zürich, Zentralbibliothek MS C 131, an illuminated manuscript of 49 written folios. On this medical advice, see Lynn Thorndike, 'Conrad Heingarter in Zurich Manuscripts, Especially his Medical Advice to the Duchess of Bourbon', *Bulletin of the Institute of the History of Medicine*, 4 (1936), pp. 81-87. A text for the duke closely resembling that which was addressed to his wife can be found in BnF, lat. 11232.

17 Cao, '*Inter alias philosophorum gentium sectas, et humani, et mites*', pp. 127-47.

18 See BnF, lat. 11232, ff. 3^{v}, 4^{v}-7^{r} and 29^{r}.

19 'Forsan multum contra hoc capitulum (quod nisi ex fontibus Sapientum hausimus) garrubulant …'. See BnF, lat. 11232, f. 6^{v}.

20 'Nolo negare propria medicine principia, sed volo dicere quemadmodum (ut in ethicis probatum est) quod virtutes sint connexe; pariformiter medicina nec a philosophia naturali nec ab astronomia subtrahi potest'. See BnF, lat. 11232, f. 6^{v}; see also Thorndike, *A History of Magic*, vol. IV, p. 368.

During these years, Heingarter had integrated himself into a circle of renowned medical astrologers, and he even found the time to mentor a budding new generation of scholars. While providing his services to the king, he joined the ranks of many other practitioners, such as Jean Coleman, Arnoul de La Palu, Pierre de Graville, Laurent Hutz, Jacques Loste, and Angelo Cato. More frequently, he would have interacted with his colleagues at the ducal court, including his sometime student Simon de Phares, as well as Jacques Sanglier, Antonio Chiapucin, and S. Belle.[21] In addition to teaching Simon de Phares, Heingarter also served as a mentor to Guillaume Cop and probably also Erhard Windsberger and Conrad Türst. These latter three were all physicians of Swiss origin.

It is from the young scholars whom Heingarter mentored that we find the clearest testimony of his commitment to defending astronomy's role in medicine and his interest in reforming it when necessary. In this regard, an exceptionally revealing document is a letter addressed to Heingarter by Guillaume Cop — the future physician to Louis XII and Francis I, but a *magister artium* and a baccalaureate in medicine at the University of Paris at the time of its composition.[22] The letter, which is dated to 1493, can be found in the first four folios of Vatican, Biblioteca Apostolica Vaticana, Reg. lat. 1241, a paper manuscript of 178 folios of astronomical content.[23] From the opening lines, one learns that Heingarter served as Cop's master and teacher (*dominus et preceptor*):

> To the illustrious man Conrad Heingarter of the leafy garden of Zurich, professor of the sidereal science as well as medicine, most erudite physician of the most Christian king of France and the illustrious duke of Bourbon, Guillaume Cop of Basel, Master of Arts and bachelor of medicine, wishes as much health as possible to his master and preceptor, for he venerates him as one among the first rank.[24]

Cop went on to praise Heingarter for the effectiveness of his vigorous polemics against sceptics. Either alluding to his teacher's polemical *De utilitate sciencie astrorum* or to a later work called the *Defensio astronomie*, which, in 1488, he had written in Moulins for Louis

21 See Jean-Patrice Boudet, *Astrologie et politique entre Moyen Âge et Renaissance* (Florence: SISMEL — Edizioni del Galluzzo, 2020), pp. 297-98. On S. Belle, see Helena Avelar de Carvalho, *An Astrologer at Work in Late Medieval France: The Notebooks of S. Belle* (Leiden: Brill, 2021). On Conrad Heingarter's relationship with Erhard Windsberger and Conrad Türst see: Alfred Schmid, *Conrad Türsts Iatro-mathematisches Gesundheitsbüchlein für den Berner Schultheissen Rudolf von Erlach* (Bern: P. Haupt, 1947), particularly pp. 7-13, 21. See also Bernhard Milt, 'Beitrag zur Kenntnis der mittelalterlichen Heilkunde am Bodensee und Oberrhein', *Vierteljahrsschrift der Naturforschenden Gesellschaft in Zürich* 85 (1940), pp. 298-305.

22 On the life and work of Guillaume Cop, see Wickersheimer, *Dictionnaire biographique des médecins*, 235-38; Marie-Louise Portmann, 'Der Basler Humanisten-Arzt Wilhelm Copp (um 1460-1532)', *Gesnerus* 15 (1958), 106–19; Stefania Fortuna, 'Wilhelm Kopp Possessore dei *Par. Gr.* 2254 e 2555? Ricerche sulla traduzione del *De victus ratione in morbis acutis* di Ippocrate', *Medicina nei Secoli. Arte e scienza*, 13 (2001), 47-57. Due to its rich content, a full edition, translation, and comment of this letter will be provided in a forthcoming publication by Eleonora Andriani and Nicholas Jacobson.

23 'Iterum vale. Anno 1493, sexto Idus Iunias'. BAV, Reg. lat. 1241, f. 4v. The letter covers ff. 1r-4v. See David Juste, 'MS Vatican, Biblioteca Apostolica Vaticana, Reg. lat. 1241' (update: 21.02.2022), Ptolemaeus Arabus et Latinus. Manuscripts, accessed 8 April 2023: URL = http://ptolemaeus.badw.de/ms/464.

24 'Clarissimo viro Conrado de viridariis Thuricensi, sideralis scientie medicineque professori, christianissimi regis Francie inclitique Borbonii ducis medico doctissimo. Domino et preceptori suo in primis colendo, Guilhelmus Copp Basiliensis, arcium magister et medicine baccalaureus, salutem quam plurimam optat'. BAV, Reg. lat. 1241, f. 1r.

Malet de Graville, admiral of France, he declared that Heingarter had effectively defended his abiding interest in the mathematical sciences:

> And particularly in astronomy, against the detractors of which you wrote as rigorously as you did elegantly, so that anyone who would read through your 'edition', which you composed with great labour, are forced into silence, having lowered [their] head being unable to reply to it with anything.[25]

From Cop's testimony, we also learn of Heingarter's interest in modifying or clarifying techniques deemed necessary for the practice of the medical astrologer, known as directions. Adopting the rhetorical flourish of the new humanist style, Cop declared:

> What could I say about the directions, which all the astrologers of our times have quite an aversion to? Having examined these, as if seeing the face of a Gorgon they would turn to stone and are deeply rendered speechless. You made all these things so clear and easy that to anyone even of little intelligence in a short amount of time those things would be accessible, which, formerly, even a sharp mind in a long amount of time could barely grasp.[26]

The practice of directing a planet in a horoscope had an ancient pedigree, with Ptolemy himself having described its astrological significance in the third book of the *Quadripartitum*.[27] One planet was chosen from the chart to serve as the representative of the native's physical or spiritual well-being. It was called the significator, and the technique of directing this planet involved projecting it from the line of the ecliptic onto the celestial equator, and then imagining it to move to a place where another planet or aspect from the chart had also been projected onto the equator, this latter one called a promissor.[28] Each degree of movement on the celestial equator was thought to be equivalent to one year of the native's life. When a significator and a promissor met, it was believed that a significant event would occur to affect the native's physical or spiritual health — either for good or for ill.

Cop's assessment that Heingarter had worked on directions and sought to clarify their procedures is corroborated by the fact that one of the longest marginal notes in BnF, lat. 7432, the composition of which Heingarter had supervised for John II, is a paraphrase of John of Saxony's commentary on the influential *Introductorius* of Alcabitius ('Abd al-ʿAzīz al-Qabīṣī).[29] In this passage, Heingarter preserved a detailed set of procedures

---

25 'Precipue astronomiam, in cuius emulos tam graviter tamque eleganter scripsisti, ut quicumque editionem tuam, quam pro eius defensione elucubrasti, perlegerint, demisso vultu nihil respondere potentes, tacere cogantur'. BAV, Reg. lat. 1241, f. 2^r. The *Defensio* can be found in Zürich, Zentralbibliothek, B 244, ff. 1^r-4^r.

26 'Quid autem de directionibus loquar? Quas omnes ferme nostre tempestatis astrologi abhorrent. Ipsisque inspectis, tamquam Gorgonis viso capite in saxa mutantur penitusque obmutescunt. Has tu tam claras facilesque reddidisti, ut cuique eciam mediocris ingenii brevi temporis spacio patefiant que antea per longa temporum intervalla acutissimus intellectus vix capere poterat'. BAV, Reg. lat. 1241, f. 2^v.

27 The relevant passage can be found in the tenth chapter of the third book. For a modern translation, see Ptolemy, *Tetrabiblos*, ed. and trans. by Robbins, pp. 270-305.

28 Heingarter also refers to the significator as the 'hileg' from the Persian word *haylāj*, which means 'ruler of life'. See Al-Battānī, *Opus astronomicum*, ed. by Carlo A. Nallino, 2 vols (Milan: Hoeplium, 1899–1907), I, pp. 313-14.

29 The text is found in BnF, lat. 7432, ff. 160^r-207^v. For the critical edition of Alcabitius or Al-Qabīṣī's work in both Arabic and Latin together with a modern translation, see Alcabitius, *Introductorius*, in *Al-Qabīṣī (Alcabitius):*

for calculating planetary directions.[30] It is one of several such instances in Heingarter's annotated library in which he emphasized the importance of determining planetary latitudes for astrological judgments.

Following the basic instructions for finding directions, which only required the calculation of the planetary longitudes, Heingarter went on to include a section in which he discussed the importance of taking into account planetary latitudes in more carefully considered calculations. Quoting from John of Saxony's commentary, he informed the reader that this method is more difficult.[31] In these cases, the position of the planet projected onto the celestial equator must originate from the physical body of the planet in the sky rather than from its unadjusted longitudinal position on the ecliptic.

Heingarter also detailed the use of an instrument, called a *directorium*, for directing planets with latitude.[32] Such an instrument is described in a text attributed to John of Lignères, which can be found with a crude diagram in BnF, lat. 7295.[33] The instrument is constructed in much the same manner as an astrolabe with a projection of the celestial sphere's climate circles, namely the tropics of Capricorn and Cancer, as well as the celestial equator. A projection of the ecliptic with nested circles of ecliptic latitude are then constructed in the same way that one produces the almucantars found on an astrolabe's climate plate. Furthermore, the rotating element of the *directorium* is not the ecliptic circle, but rather a moveable local horizon set to the terrestrial latitude of the user. This analogue to the rete of an astrolabe is comprised of two arms, one representing the horizon and the other the midheaven. In this way, the *directorium* can determine the points on the celestial equator that rise, set, and culminate with the true position of the planet in latitude to the north or south of the ecliptic plane.

Describing techniques of determining directions, which took into account the planetary latitude, was not the only area of medical astrology where Heingarter sought correctives to contemporary methods. In the same manuscript of BnF, lat. 7432, Heingarter included an excerpt from a work on medical astrology that he entitled the *Tractatus de pronosticis et componendis et ministrandis medicinis* taken from a work by the English Dominican friar, Robert of York (fl. 1313-1325), who was also known as *Perscrutator*.[34] In conclusion 8 of his *De impressionibus aeris*, in which the author treats the forecasting of plagues and pestilences, it is explained that planetary latitudes were to be taken into account when forecasting future onsets of plague. The text identified certain star positions in the zodiac that signified disease and mortality. In the event that a superior planet formed an aspect

The Introduction to Astrology: Editions of the Arabic and Latin Texts and an English Translation, ed. by Charles Burnett, Keiji Yamamoto, and Michio Yano (London: The Warburg Institute, 2004).

30 BnF, lat. 7432, ff. 194r-198v. This same passage can be found in the 1503 printed edition of the work, which also contains John of Saxony's commentary to it. See *Alchabitius cum comento*, ed. by Bartholomäus Alten (Venice: Per Ioannem et Gregorium de Gregoriis fratres, 1503), ff. h6r–i2r.

31 '… quando autem significatores habent latitudines difficilior est modus'. BnF, lat. 7432, f. 198r.

32 'Et propter hoc factum fuit instrumentum ad dirigendum planetas habentes latitudines et illud instrumentum vocatum fuit directorium'. BnF, lat. 7432, f. 198r.

33 BnF, lat. 7295, ff. 78r-79v.

34 See BnF, lat. 7432, f. 268r-268v. See Thorndike, *History of Magic*, vol. III, pp. 103-18 and 678-79; on the *De impressionibus aeris* see also Anne Lawrence-Mathers, *Medieval Meteorology: Forecasting the Weather from Aristotle to the Almanac* (Cambridge: Cambridge University Press, 2020), pp. 159-61. The identification was made on the basis of the text provided by Erfurt, Universitäts- und Forschungsbibliothek, Dep. Erf. CA 2° 395, ff. 104v-105r.

with one of these positions, it would portend the outbreak of pestilence. To determine these positions, moreover, the planet's latitude and its direction to the south or to the north of the ecliptic had to be found before the aspect between the star and the planet was established. Heingarter included two small tables containing the longitudinal and latitudinal positions of these stars together with their orientations to the north or south of the ecliptic.

The concern described in the *Tractatus de pronosticis* for a practitioner to find a planet's latitude before determining its aspect with certain stars in the zodiac is one case among a subset of general concerns about ensuring the coherence of astrological doctrine regarding the planetary conjunctions and their effects in the terrestrial realm. As John of Saxony described it in his commentary on the *Introductorius*, there was a difference between planetary aspects in longitude on the one hand, and planetary aspects in both longitude and latitude on the other.[35] If the vicissitudes of planetary rays were truly the cause of the various contingencies and changes on Earth, then planetary bodies surely had to be taken into account. These concerns about attending to the latitude of a planet became readily apparent to anyone who saw a planetary conjunction, for there were perceptible differences between a conjunction in longitude and a conjunction in longitude and latitude.

Finally, in his annotations on Pseudo-Ptolemy's *Centiloquium*, Heingarter duly noted the natural effects that a ruling planet could have on the bodily disposition of a native. In the 53rd aphorism, the text teaches the practitioner to calculate the ruling planet's latitude to determine the constitution of the individual for whom the birth chart had been drawn.[36] If the planet had no latitude, the native would be lean, and if the planet had a great latitude, the native would be large.

All of these careful notes that Heingarter arranged on the significance of latitudes were part of a much larger body of material that he collected on the importance of determining the velocity of the planets, identifying their stations, retrogradations, and directions, and finding the times when they were at apogee on the epicycles and eccentrics. The planetary latitudes seem to have taken pride of place in Heingarter's mind, and this may have been because ignoring them so starkly undermined any pretensions to a mathematically reliable medical astrology. Critics of the system were certainly aware of this. Deploying arguments that had been propounded earlier against practicing astrologers by such sceptics as Nicole Oresme and Henry of Langenstein, zealous opponents of medical and judicial astrology were quick to assail the insufficiencies of a practice that did not pay due attention to the latitudes of planets.[37] As Pico della Mirandola would so cuttingly observe in his posthumously published *Disputationes adversus astrologiam divinatricem*, even if most astrologers had

35 'Coniunctio autem latitudinis est cum fuerint duo planete in uno signo coniuncti et fuerint equalis latitudinis ab eclyptica …' See *Alchabitius cum comento*, ff. g8v–h1r.

36 BnF, lat. 7432, f. 140r.

37 Stefano Rapisarda, 'Doubts and Criticism on Astrology in Western Traditions from Antique to Early Modern Literature', in *Prognostication in the Medieval World*, ed. by Matthias Heiduk, Klaus Herbers, and Hans-Christian Lehner (Berlin: De Gruyter, 2021), pp. 756-58; Michael Shank, 'Academic Consulting in Fifteenth-Century Vienna: The Case of Astrology', in *Texts and Contexts in Ancient and Medieval Science: Studies on the Occasion of John E. Murdoch's Seventieth Birthday*, ed. by Edith Sylla and Michael McVaugh (Leiden: Brill, 1997), p. 252; George W. Coopland, *Nicole Oresme and the Astrologers: A Study of His* Livre de Divinacions (Cambridge, Mass.: Harvard University Press, 1952).

mastered mathematical astronomy (a fact he was reluctant to accept), they had manifestly failed to apply it correctly. 'How great is their negligence', he wrote, that:

> even among those who have an interest in calculating, which most of them lack, you will find few who do not resort to ephemerides to obtain the true position of the stars. [...] But those who do so necessarily make many false statements and omit many things, thereby invalidating the entire truthfulness of the predictions. For there is almost never a reference to the latitudes of the planets; but at what point will the astrologer not err, will he not lie, if he does not take this into account, if we suppose that the principles of this art are true?[38]

The treatise, which Heingarter finally prepared in order to establish the geometrical foundations of the latitude theory and the proper procedures for finding latitude values using the tables, appeared in the same manuscript where so many of these notes on the importance of planetary latitude for medical astrology can be found.

2. A scrupulous collector and compiler

Given the content and form of the extant manuscript in which the text was preserved, it would appear that Heingarter's treatise on planetary latitude was also written to convey the erudition of an author worthy to serve under the auspices of a great lord. The treatise can be found as a single witness in the margins of BnF, lat. 7432, which is a codex consisting of 275 folios bearing dimensions of approximately 125 × 205 mm. The manuscript's rich array of marginal notes presented alongside a superbly decorated iconographic programme make it clear that it was meant to be a presentation copy for John II.[39] It exhibits a single neat hand and large margins to include commentaries throughout. Though there may have been several individuals collectively responsible for its creation, the production of the manuscript was carried out under Heingarter's supervision at the castle of Belleperche and is an expression of his public persona. The multi-faceted nature of the manuscript's production, moreover (the three dates of 1468, 1477, and 1485, which can be found in different notes and colophons throughout the manuscript, point to a processual and multi-year composition of the texts and annotations therein), also has the effect of creating conjunctures with other works, tables, stylistic habits, and annotations that are found elsewhere in Heingarter's corpus. What we shall see in this section is that the treatise

38 'Iam vero indiligentia quanta paucos enim invenies etiam ex his qui supputandi habent rationem (qua tamen eorum maxima pars caret) qui ut veram positionem syderum assequantur ad ephemeridas non confugiant. [...] Hoc autem qui faciunt et perperam dicere et omittere multa est necessarium, unde omnis praedictionum veritas infirmetur, fere enim ibi nunquam de latitudine planetarum ulla significatio, qua ignorata ubi non errabit, ubi non mentietur astronomus, si vera istius artis decreta existimamus?' See Giovanni Pico della Mirandola, *Disputationes adversus astrologiam divinatricem, in Opera omnia Ioannis Pici Mirandulae Concordiaeque comitis* (Basel: Heinrich Petri, 1557), p. 445.

39 On this manuscript, see David Juste, *Les manuscrits astrologiques latins conservés à la Bibliothèque nationale de France* (Paris: CNRS, 2015), pp. 152-53; Matthieu Husson,'Mathematical Astronomy and the Production of Multiple-Texts Manuscripts in Late Medieval Europe: A Comparison of BnF, lat. 7197 and BnF, lat. 7432', in *The Emergence of Multiple-Text Manuscripts*, ed. by Alessandro Bausi, Michael Friedrich, and Marilena Maniaci, (Berlin: De Gruyter, 2019), pp. 247-74.

and its accompanying canons and tables cannot be read independently of other codices contained in his library. The overall effect was that the reading and use of the latitude treatise was performed at the level of a unified library in which common topics of interest found in separate manuscripts emerged in conjunction with one another.

In addition, the types of texts that appear in these manuscripts were chosen strategically to showcase their compiler's status as a consummate humanist collector. The opening works of BnF, lat. 7432, a text on the life of Ptolemy (*De vita Ptholomei*) and another on the life of Hermes Trismegistus (*De Hermete*), were not original to Heingarter, but were rather excerpted from the thirteenth-century *Liber philosophorum moralium antiquorum* attributed to John of Procida (*c.* 1210–*c.* 1298).[40] Nevertheless, their presence at the beginning of the codex signal their compiler's interest in honouring these original adepts of his craft, and perhaps drawing affinities between them and himself. In this manner, he was also mirroring the posture of his own patron who styled himself a champion of the humanist movement and a great preserver of ancient knowledge. Indeed, Heingarter credited John II with a worldwide survey project on the knowledge of various peoples and praised him for having ordered the correction of some altered works of antiquity.[41] The duke, in Heingarter's estimation, had decided to build up a library after he saw how many volumes had been lost due to the fall of Constantinople in 1453. This led him to collect worthy books from all over the word, repairing and emending them as he deemed necessary.[42]

Insofar as Heingarter contributed to the library's expansion, he did so in ways that manifest conspicuous commonalities across the manuscripts he possessed. For example, Heingarter frequently adorned the manuscripts of his collection with a patina of the humanistic aesthetic in the form of Grecisms and transliterations of Latin text into Greek script. The last four folios of BnF, lat. 7432 contain Greek letters with their orthographic variants, a list of Greek diphthongs, and Christian prayers in Greek with an interlinear Latin translation (*Gloria patri* / Δόξα τῷ πατρὶ, *In nomine patris* / εἰς τὸ ὄνομα τοῦ Πατρὸς, *Ave gratia plena Maria* / Χαῖρε κεχαριτωμένη Μαρία, *Pater noster* / Πάτερ ἡμῶν, etc.).[43]

40 Lynn Thorndike, Colette Beaune, and Jean-Patrice Boudet have mantained Heingarter's authorship of these texts. Thorndike writes, 'These [texts] are evidently by Heingarter since they contain allusions to his favourite retreat, Belle-Perche, and to the "era of John, duke of Bourbon and Auvergne"'. See *A History of Magic*, IV, p. 362n31. See also, Colette Beaune, 'La notion d'Europe dans les livres d'astrologie du XVᵉ siècle', in *La Conscience européenne au XVᵉ et au XVIᵉ siècle. Actes du Colloque international organisé à l'École Normale Supérieure de Jeunes Filles* (Paris: École normale supérieure de jeunes filles, 1982), p. 6n26; Boudet, Le *Recueil des plus celebres astrologues*, vol. I, pp. 23n16 and 80-81n26. Relevant textual evidence can be found in John of Procida, *Liber philosophorum moralium antiquorum*', ed. Ezio Franceschini: 'Il *Liber philosophorum moralium antiquorum*', in *Atti del Reale Istituto Veneto di scienze, lettere ed arti* 91/2 (1932), 393-597 (pp. 402-03, 531-48). On the mistaken attribution of this work to John of Procida, see Sonia M. Medrano, 'Atribución y restitución: hermenéutica crítica de la autoría del Liber philosophorum moralium antiquorum', in *De Falsa et Vera Historia* 2, ed. by Mikel Labiano, (Madrid: Ediciones Clásicas, 2019), pp. 203-210.

41 'Nam ipse est princeps qui, nostro tempore, fere solus in Europa scientias diligit, amat atque observat, et scientes honorat, qui, a finibus terre per mundi universa climata, omnium gentium et linguarum scientias perquirit, et libros de sapientum antiquorum sententiis perditos, inemendatos, restaurari emendarique iussit'. BnF, lat. 7432, f. 4ʳ. See also Thorndike, *A History of Magic*, vol. IV, p. 363n32.

42 '... ut enim Constantinopolitane cladi consideras qua nuper crudelis Turci sevitia tot optimarum litterarum volumina naufragio submersit, ex omnibus mundi partibus dignos quosque libros afferri iubes ut bene reparentur emendatioresque sint summopere curas'. BnF, lat. 7305, f. 4ʳ. See also Thorndike, *A History of Magic*, vol. IV, p. 385.

43 Heingarter also included, as part of the *Grammatica* by Hugh of Saint Victor, a full Hebrew and Greek alphabet in his student manuscript. See, BnF, lat. 7197, f. 119ʳ-119ᵛ.

His student Guillaume Cop characterized Heingarter as a voracious reader of Greek literature.[44] Witnesses to several works of mathematical astronomy that were Heingarter's bear the same features of transliterating or translating phrases into Greek. The final word of the fly leaf in BnF, lat. 7333 reads 'end', or τέλος,[45] to signal the termination of a contents list giving the title of the major works in the manuscript. And, again, the explicit to a treatise on arithmetic in BnF, lat. 7295A, concludes with a transliteration into Greek of the Latin phrase *Explicit seu finis per Konradum Heingarter*, thereby confirming that parts of this manuscript are indeed an autograph.[46]

These latter two sources are workaday codices, the latter of which (BnF, lat. 7295A) has recently been aptly named a 'toolbox' manuscript due to the gradually accumulated sets of relevant computational techniques preserved within it.[47] The similarity between the contents of the works found in these two manuscripts, as well as others in Heingarter's library, and that of BnF, lat. 7432, however, go well beyond the presence of Greek text. They also contain facets of common astronomical problems taken from different perspectives or different genres that clearly relate to one another when read together. For instance, the treatise on planetary latitudes appears as a marginal commentary on folios 219^{r}-222^{v} of BnF, lat. 7432, mostly framing the last two chapters of the *Theorica planetarum communis*, which are entitled 'On the Latitudes of the Six Planets and the Declination of the Sun', and 'On Finding the Aspects of the Planets'.[48] In like manner, we find two annotated texts in earlier manuscripts that were in Heingarter's possession. In BnF, lat. 7333, Heingarter composed a complete marginal commentary of the *Theorica* (ff. 54^{v}-63^{r}). Earlier than this, likely during his days as a student at Paris, Heingarter had also extensively annotated another witness to the *Theorica*, which is preserved in BnF, lat. 7197 (ff. 51^{r}-57^{v}).

In his glosses and commentaries, we can see Heingarter's interest in certain topics taking root, expanding, and developing over time. In the case of the latitudes, Heingarter's first treatment, found in BnF, lat. 7197 (ff. 56^{r}-57^{v}), is rather sparse, containing three *nota* in Heingarter's hand. In BnF, lat. 7333 (ff. 61^{v}-63^{r}), however, folios 62^{r}-63^{r} are filled with definitions and clarifications of the original text treating the latitudes. Furthermore, the introduction to this commentary, found on folio 54^{v}, is reminiscent in many important ways of the more specialized marginal commentary found in BnF, lat. 7432. They both divide their subject matter neatly into a *theorica*, treating attributes of astronomy defined as theoretical, followed by a *practica*, treating those attributes of astronomy defined as practical.

---

44 See Fortuna, 'Wilhelm Kopp Possessore', 48. 'Quantam autem in poetica, oratoria, grecaque litteratura operam consumpseris: si enarrare conabor, *ante diem clauso componet vesper olympo*'. Vatican, BAV, Reg. lat. 1241, f. 3^{r}. The passage in italics is a quotation from Virgil's *Aeneid*, book I, v. 374; see Publius Vergilius Maro, *Æneis*, rev. ed., trans. by Henry Rushton Fairclough and George P. Goold (Cambridge, Mass.: Harvard University Press, 1999), p. 288.

45 BnF, lat. 7333, f. II. In reality, the word written at the end of the table seems to be γέλως (i.e. laughter) rather than τέλος. However, the cursive tau may resemble a gamma, and the confusion between omicron and omega could be rather frequent even among the most skilled Greek scribes.

46 For more on this explicit included in BnF, lat. 7295A, see Eleonora Andriani, Samuel Gessner, and Nicholas Jacobson, 'A Note on the *Arabes maxime*: A testimony of Alfonsine astronomers' interest in the Toledan Tables', in this volume. See also BnF, lat. 7295A, f. 45^{r}.

47 See, Husson, 'The Art of Astrological Computations' in *Exploring Written Artefacts*', pp. 513-16.

48 'Capitulum VII. De latitudinibus sex planetarum et declinatione solis' (BnF, lat. 7432, ff. 219^{r}-221^{v}); 'Capitulum VIII. De planetarum aspectibus inventione' (BnF, lat. 7432, ff. 221^{v}-222^{r}).

Furthermore, the introduction to the BnF, lat. 7333 commentary already betrays Heingarter's concerns to refine planetary calculations in order to identify the precise locations of planetary positions. He emphasizes the importance of understanding all the components of a planet's position that together constitute its true motion. This includes identifying its position in both the eighth and ninth spheres, but also in attending to all of its accidental qualities or the planetary passions. And finally, Heingarter explains that calculating the planetary latitude in addition to the planetary longitude is necessary to find this true motion.[49]

In the same way, the accompanying canons, tables, and instrument texts relating to planetary latitudes that appear in BnF, lat. 7432 recur again in multiple volumes of Heingarter's library. In BnF, lat. 7333 a table of contents in the hand of Heingarter points out the treatment of the stations and retrogradations of the planets together with their latitude, as part of the *Liber super totam astrologiam* by the late-thirteenth-century Franciscan Bernard of Verdun. As Claudia Kren has noted in her seminal article on the development of planetary latitude theory and practice in the late-medieval West, Bernard of Verdun's was one of the very few pre-Alfonsine works that attempted to grapple with the confusions of latitude theory as it was inherited from the *Theorica planetarum communis* and the Toledan tradition.[50] On folio 43ʳ, where the chapter *De stationibus et retrogradationibus atque planetarum latitudinibus* begins, a marginal note in Heingarter's hand is provided, which reads *hic incipit tractatus nonus et habet distinctiones duas, quarum prima est de planetarum retrogradationibus et continet capitula tria* (*et distinctio secunda est de latitudinibus et earum annexis et continet capitula novem*). Throughout the entire *Liber super totam astrologiam*, this note in the hand of Heingarter is by far the lengthiest among a set of marginal annotations that are otherwise rather short and sporadic.

In addition to the *Liber super totam astrologiam*, in BnF, lat. 7333 Heingarter also preserved a work by Bernard of Verdun dealing with an instrument called the *torquetum* (*truquetum* in the manuscript). This is followed by another text concerning the same instrument by Franco de Polonia (ff. 49ʳ-52ᵛ), which Heingarter had earlier included in BnF, lat. 7197 (ff. 15ʳ-16ʳ). The *torquetum* may have been of some interest to Heingarter due to his personal proclivities for latitude theory, as the instrument can be used for coordinate measurement and conversion, and the determination of a heavenly body's position using either ecliptic or equatorial coordinates.[51]

The commonalities of texts and topics across manuscripts sometimes are also indicated by the scribal marks of Heingarter's editing, which may be evidence of the direct transfer of certain works from one manuscript to another. On folio 176ʳ-176ᵛ of BnF, lat. 7295A, for instance, we find the chapters of the *Priores astrologi* treating planetary latitude.[52] It is possible here to detect traces of Heingarter's careful reading of the latitude canons as

49 BnF, lat. 7333, f. 54ᵛ.

50 Claudia Kren, 'Planetary Latitudes, the *Theorica Gerardi*, and Regiomontanus', 194-205 (pp. 198-99).

51 On this instrument, see Emmanuel Poulle, 'Bernard de Verdun et le turquet', *Isis*, 55 (1964), 200-08, and Richard L. Kremer, 'How Did the *Turketum* (or *Torquetum*) Get its Name?', in *Scientific Instruments between East and West*, ed. by Neil Brown, Silke Ackermann, and Feza Günergun (Leiden: Brill, 2019), pp. 80-107.

52 These are part of a nearly complete set of the *Priores astrologi* found from ff. 174ʳ-180ᵛ. The set lacks five canons, which Heingarter does not fail to notice, writing in the lower margin of f. 180ᵛ: 'Hic deficiunt canones quinque scilicet 41us, 42us, 43us, 44us et quadragesimus quintus'. See BnF, lat. 7295A, f. 180ᵛ.

he applied a scribal technique of adding red slashes to the chapters. On the same folio there are also marginal notes written in his hand, corresponding to these same chapters.

These notes and scribal marks were either added to aid Heingarter in his reading of the chapters, or, more likely, to edit parts that would be used in his final treatise. For these same chapters on planetary latitude would be paraphrased in the *practica* of his treatise in BnF, lat. 7432, as well as being appended to his version of John of Saxony's canon set (incipit: *Tempus est mensura motus*) in the same presentation manuscript. Indeed, from folios 224^{r}-236^{v} in BnF, lat. 7432, Heingarter decided to add supplemental canons to his set of the *Tempus est mensura motus* canons. To this end, we find three canons from the set of John of Lignères (incipit: *Priores astrologi*) following John of Saxony's canons, together with four canons excerpted from the Cb canons of the Toledan Tables, and a canon for equating the houses of unknown provenance. The canons excerpted from the Toledan canons had already been preserved by Heingarter in a comparative gloss of Alfonsine and Toledan astronomy found in BnF, lat. 7295A bearing the incipit *Arabes maxime*, and those annotated canons of the *Priores astrologi* found in BnF, lat. 7295A may similarly be the source of their equivalent texts accompanying *Tempus est mensura motus* in BnF, lat. 7432.[53] Thus, earlier texts studied and annotated in his toolbox manuscripts were apparently repurposed and rewritten in his presentation copy to the duke.

That the notes contained in BnF, lat. 7295A would serve as the groundwork of significant parts of Heingarter's later *practica* is also corroborated by clear connections that one can find between his descriptions of the latitude tables associated with the treatise in BnF, lat. 7432 and the presence of these same tables and their descriptions in BnF, lat. 7295A. In BnF, lat. 7295A, f. 127^{v} and f. 164^{v}, one finds latitude tables, one of which appears to have been directly incorporated into BnF, lat. 7432. These two tables in BnF, lat. 7295A, which we shall label A and B, contain numerical values deriving from the Ptolemaic tradition, which had been adapted by John of Lignères from the Toledan Tables and the *zīj* of al-Battānī.

53 The first text on f. 237^{r} (heading: *De stacione planetarum et retrogradatione*) corresponds to Cb 152-59; see Fritz S. Pedersen, *The Toledan Tables: A Review of the Manuscripts and the Textual Versions with an Edition* (Copenhagen: C.A. Reitzels Forlag, 2002), p. 442. The second text on f. 237^{r} (heading: *De motu arumenti*) corresponds to Cb 160; see Pedersen, *The Toledan Tables*, pp. 442 and 444. The third text on f. 237^{r} (heading: *De declinatione solis*) corresponds to Cb 161-62; see *The Toledan Tables*, p. 444. The fourth text (heading: *De latitudine lune*) on f. 237^{r} corresponds to Cb pp. 163-64; see Pedersen, *The Toledan Tables*, p. 444. The fifth text covering f. 237^{r}-237^{v} (heading: *De latitudine trium superiorum*) corresponds to Cb 165; see Pedersen, *The Toledan Tables*, pp. 444 and 446. On f. 237^{v} there is also a text (heading: *De latitudine Veneris et Mercurii*) corresponding to Cb 166; see Pedersen, *The Toledan Tables*, p. 446. The following text on f. 237^{v} (heading: *Alius modus ad sciendum latitudines quinque planetarum secundum tabulas de Lineriis*) corresponds to Canon 21 of John of Lignères' *Priores astrologi*; see Marie-Madeleine Saby, *Les Canons de Jean de Lignères sur les tables astronomiques de 1321: Édition critique, traduction et étude*, 3 vols (unpublished thesis: École nationale des Chartes, 1987), I, pp. 206-07. The following text on ff. 237^{v}-238^{r} (heading: *De latitudine Veneris*) corresponds to *Priores astrologi*, Canon 22; see Saby, *Les canons de Jean de Lignères*, I, pp. 207-09. The following text on f. 238^{r} (heading: *De latitudine Mercurii*) corresponds to Canon 23; see Saby, *Les canons de Jean de Lignères*, I, pp. 210-11. We could not identify the last text on f. 238^{r} (heading: *De equatione duodecim domorum*; starting with 'Quere ascensiones tuas habitas ex parte gradus solis ex parte horarum simul iunctarum inter ascensiones circuli recti'). Interestingly, John of Lignères' *Cuiuslibet arcus* includes a canon about the equation of the houses (i.e. Canon 37: *Duodecim domos adequare*), but it does not correspond to this text on f. 238^{r}. See Saby, *Les canons de Jean de Lignères*, I, pp. 153-55.

Despite many similarities between the two tables contained in BnF, lat. 7295A, it seems that Heingarter privileged Table A. Copied in his own hand, it was addended to the eleventh quire of the manuscript alongside the equation tables of the Parisian Alfonsine Tables, and a close reading of its title and column headings show that it was adapted and integrated into BnF, lat. 7432 to be used in conjunction with the *practica* of his treatise. Table A is found on folio 255^{v} of BnF, lat. 7432 bearing the title *Tabula latitudinum quinque planetarum que est distancia eorum a via solis*, which differs from the title of Table B.[54] Furthermore, there are subtle alterations to the vocabulary of the columns containing latitude values for the inferior planets. For instance, instead of using the more common term *reflexio* for the second angle of inclination on the epicycle, which is referred to as 'slant' in modern commentaries, Heingarter opted for the slightly modified term *reflexus*. This word is repeated in the table contained on folio 255^{v} of BnF, lat. 7432.

Table A also corrects for what is most probably a copying error found in the upper part of the column for the northern latitude of Saturn in table B. Indeed, in the 24° entry, the corresponding value in table B is 2;8 instead of 2;7. 2;8 normally corresponds to the value for the next entry 30°. From 24° to 84°, the values are therefore shifted by one line. To fill the gap created by a misalignment of the values, Heingarter repeats the value of 2;30 at 84° and 90°. One final distinguishing feature of Table A is the presence of a third column for the proportional minutes (*minuta proportionalia*), which is not found in Table B, and was not common to this kind of latitude table.[55]

These differences between Tables A and B amount to only very slight alterations on the part of Heingarter. Nevertheless, he deemed them significant enough to make note of them in his treatise. In the *practica*, he associated the third method for finding latitudes discussed therein with a latitude table he claimed to have altered from the original set of John of Lignères, writing:

> I have brought to you (Michael Roscher, Master of Arts) the not so lazy load of works, for which nothing is more ready for action than when <a thirst> for acquisition has been kindled; but I have sent these, which, taken from the wealth of the wise, we have gathered for the utility of students. And because in the previous books we have thoroughly gone through the motion of the stars and planets according to longitude with however much of theory and practice as suffices, I set out to compose this treatise on the motion of the planets according to latitude, which I shall divide into two chapters: the first of these explains the theory; the second explains the practice and the method of finding [the planets'] latitude with their genzahar[56] for the era of the Illustrious John, duke of Bourbon and Auvergne.[57]

He then goes on to paraphrase the latitude canons from the *Priores astrologi* to be used with his table in BnF, lat. 7432 adapted for this purpose. Whatever reasons inspired

54 Table B is called the *Tabula latitudinis 5 planetarum quae est longitudo a via solis in circulo signorum* and is a part of the set of tables of John of Lignères. See José Chabás and Marie-Madeleine Saby, *The Tables of 1322 by John of Lignères, an Edition with Commentary*, (Turnhout: Brepols, 2022) 76–79.

55 Also referred to as the *partes latitudinis planetarum*. These numbers do not provide a higher level of precision to the original value of the proportional minutes, as it does not fit any re-computation scenario for this particular case. One may wonder if it is an independent table or an auxiliary table that could help in performing interpolations.

56 That is, the planetary nodes.

57 BnF, lat. 7432, f. 219^{r}. See Appendix, p. 84, lines 4-12.

Heingarter to copy out his own latitude table — whether it was to correct the copyist's errors in Table B or whether it was to add his own values of proportional minutes — the links between Table A in BnF, lat. 7295A and the canons of BnF, lat. 7432 are drawn fairly explicitly in Heingarter's treatise.

In addition to these tables taken from the *zīj* of al-Battānī, Heingarter would later include the tables from the *zīj* of al-Khwārizmī in his presentation copy to the duke as well. They appear in the manuscript following the treatise in BnF, lat. 7432 on f. 222^{r} and 255^{v}-258^{r}. His inclusion of these tables similarly reveals connections between BnF, lat. 7295A and BnF, lat. 7432. Following Heingarter's treatise in the latter manuscript are found the two tables of the Indian tradition. The first of these tables, the *Tabula quadripartialis ad sciendum latitudines quinque planetarum*, starts at folio 256^{v} and ends with folio 257^{r}.[58] These are followed by the *Tabula bipartialis* from folios 257^{v}-258^{r}.[59] Much of the *practica* is devoted to the explication of these tables, and the methods of calculation that one applies to use them.

Manuscript 7295A, though it does not contain the tables themselves, ends with the *Arabes maxime* (already mentioned above), which is a comparative exposition of canons to the Alfonsine Tables and the Cb canons of the Toledan Tables. This work includes five chapters on planetary latitude from folios 189rb-190rb drawn from the Toledan canons.[60] The instructions found in the *Arabes maxime* explain how to calculate a latitude using the *bipartialis* and *quadripartialis* tables. Furthermore, as we discuss more fully in the next section, just as Heingarter relied upon the *Science of the Stars* (De scientia astrorum) by Alfraganus (al-Farghānī) to explicate the canons and tables in his treatise, so too does the *Arabes maxime* invoke this text frequently in order to gloss the Toledan canons. Heingarter's decision to make use of an unusual vocabulary found in Alfraganus to describe different aspects of the planetary latitude model, which we will discuss below, becomes less surprising if one has already read the chapters on latitude in the *Arabes maxime* that developed the same terminology. Thus, even though the *bipartialis* and *quadripartialis* tables are not found in BnF, lat. 7295A, the *practica* and tables of BnF, lat. 7432 can still be considered a continuation of the same project as that one being developed by the *Arabes maxime*.

One final set of latitude tables described in the treatise, the so-called Oxford Tables, which had been computed by Englishman William Batecombe in 1348, offers the starkest example of the manner in which BnF, lat. 7432 relies upon BnF, lat. 7295A for coherency.[61] In the *practica* of his treatise, Heingarter described their format and how to use them, writing:

58 This table is called *quadripartialis* because the argument folds into four columns (from 1° to 90°, 179° to 90°, 181° to 270°, and 359° to 270°). For details on the origins of this table format and those of the *bipartialis* (see note below), see Kren, 'Planetary Latitudes, the *Theorica Gerardi*, and Regiomontanus', 194–205 (pp. 195-96n8).

59 The argument for this table is folded in two, hence the name *bipartialis*. Another table dealing with retrogradations and stations, the *Tabula retrogradationum and stationum planetarum* was placed between these two parts of the latitude tables in the manuscript. After these tables, one can find the *Tabula de exitu planetarum* (See Chabás and Saby, *The tables of 1322*, 92).

60 On this work starting with the *Arabes maxime*, see Andriani, Gessner, and Jacobson, 'A Note on the *Arabes maxime*', in this volume.

61 This third set of tables Heingarter acquired had proven successful during the century since they had been composed, as they had enjoyed wide circulation throughout the major centres of learning in Europe. See, John D. North, 'The Alfonsine Tables in England', in *Prismata, Naturwissenschaftsgesichtliche Studien*, ed. by

> The first [kind of latitude table] is that in which [the table] is entered with the mean centre in the descending line, and with the mean argument in the transversal line; and what is taken is that which is in the convergence of the [two] lines; and that will be the latitude of the planet; and this method is easy[62]

No such tables exist in BnF, lat. 7432, however, and Heingarter's only copy of them appears in the thirteenth quire of BnF, lat. 7295A.[63]

These connections between BnF, lat. 7432, BnF, lat. 7295A, and BnF, lat. 7333 reveal a fascinating practice on the part of Heingarter to not only order related works within a single manuscript, but also to arrange them across multiple manuscripts within a single library.[64] Some of the texts in question may have been composed when Heingarter was still a student; others were written at a time when Heingarter had begun his career as a medical professional following his licentiate. However, the mutually interrelated topics found across all of these texts suggest a tendency to cross-reference, augment, and edit topics of interest in such a way that no single codex was self-sufficient. The coherent application of computational techniques found in 7432, then, could only be achieved with recourse to multiple volumes of the library.

3. The planetary latitude treatise and the arts curriculum at Paris

The latitude treatise, together with the compiled texts, tables, and annotations stored in the volumes detailed above, might rightly be understood as a research agenda, and the questions they raised probably animated Heingarter's mind over many years of his life. They also suggest that Heingarter's interest in these topics belonged to a more communal practice in which multiple actors were involved. In this way, the content and concern of the treatise was of more than just a personal nature. The treatise was likely intended to serve as a touchstone for the study of latitude theory and calculation.

Yasukatsu Maeyama and Walter G. Saltzer, (Wiesbaden: Steiner, 1977), pp. 269–301 (Reprinted in J. D. North, *The Universal Frame: Historical Essays in Astronomy, Natural Philosophy, and Scientific Method* [London: The Hambledon Press, 1989], pp. 327–59).

62 BnF, lat. 7432, f. 221^{v}. See Appendix, p. 96, lines 189-91.

63 This copy of the Oxford Tables is partial, and only includes the part on latitudes. Heingarter's interest in this part of Batecombe's set is likely due to their novel composition, which applied a double argument format and was relatively straightforward, thereby allowing for latitudes to be obtained with a minimal amount of calculation. They start at folio 145^{r} and finish at folio 152^{r}. A full latitude table for each planet takes three pages (i.e. 1 ½ folios each).

64 A final manuscript worthy of comment is BnF, lat. 7314, which bears many affinities with 7432 in terms of its astrological point of view. For instance, it includes the *Introductorius maius* by Albumasar (Abū Maʿshar). In this work, Albumasar is concerned to take into account planetary latitudes when discussing 'the planets' occultation of one another and the eclipse of the two luminaries' (*Great Introduction*, I.2.3b). He also defines the 'application of the planets to each other in latitude by conjunction' to be a circumstance when 'two planets [are] conjoining', and 'their latitude is the same <and> in the same direction'. When this is the case, 'one of them eclipses the other' (*Great Introduction*, VII.5.8a). We are very grateful to Charles Burnett for pointing out these accounts. See Albumasar (Abū Maʿshar), *Liber introductorius*, in *The Great Introduction to Astrology by Abū Ma ʿšar*, ed. by Yamamoto Keiji, Charles Burnett, and David Pingree, (Leiden: Brill, 2019), pp. 53 and 769.

The general pitch of the dedicatory introduction would suggest this, as it is presented in the form of a letter to Michael Roscher, Master of Arts at the University of Paris. Roscher was procurator of the nations of England and Germany when Heingarter was still at the university, and the two seem to have developed a rapport. In 1466, Roscher had worked together with Heingarter, who, on behalf of Jean Rosée, dean of the medical faculty, had petitioned to secure funding for students pursuing the doctoral degree in medicine.[65] The opening lines of the letter reproduce, almost verbatim, the beginning of the prologue to Boethius' *De institutione arithmetica*, a reference that would hardly escape the attention of a university master. Indeed, the division of the mathematical sciences into arithmetic, music, geometry, and astronomy displayed in *De institutione arithmetica* was widely assumed to be the basis of the medieval curriculum. Equally worthy of note is the fact that the parallel construction of Boethius and Heingarter's introductions are mirrored by the circumstances under which they both appear to have been composed. For Boethius also intended his scholarly efforts to serve the purpose of a gift to a patron and friend. In his case, the treatise was addressed to the Roman senator, Symmachus, who was also his father-in-law.[66]

The relevant connection between the arts faculty and Heingarter's production of the treatise is further confirmed by his explicit statement that students were the intended recipients of the work. He explained:

> I have brought to you (Michael Roscher, Master of Arts) the not so lazy load of works, for which nothing is more ready for action than when <a thirst> for acquisition has been kindled; but I have sent these, which, taken from the wealth of the wise, we have gathered for the utility of students. And because in the previous books we have thoroughly gone through the motion of the stars and planets according to longitude with however much of theory and practice as suffices, I set out to compose this treatise on the motion of the planets according to latitude, which I shall divide into two chapters: the first of these explains the theory; the second explains the practice and the method of finding [the planets'] latitude with their genzahar[67] for the era of the Illustrious John, duke of Bourbon and Auvergne.[68]

The dedication of the text, not only to Roscher, but also to students of astronomy, reveals a third dimension to Heingarter's interest in latitude, quite different from the defence of his profession and the contributions to a humanistic library at Belleperche. He was also offering an authoritative text to be used in the context of formal instruction.[69]

---

65 '[...] supplicavit magister Conradus Heingarter, in medicina licenciatus, quod natio fide iuberet decano facultatis medicine pro summa triginta quatuor scutorum pro bursis ad doctoratum'. As quoted in *Liber procuratorum nationis anglicanae (alemanniae)*, ed. by Charles Samaran, Émile van Moé, and Susanna Vitte, 3 vols (Paris: Didier, 1935), III, p. 5.

66 See Boethius, *De arithmetica*, in *Corpus Christianorum, Series Latina 94a*, ed. by H. Oosthout and J. Schilling, (Turnhout: Brepols, 1999), pp. 3-5; for an English translation, see *Boethian Number Theory: A Translation of the 'De institutione arithmetica'*, trans. with introduction and notes by Michael Masi (Amsterdam: Rodopi, 1983), pp. 66-68.

67 That is, the planetary nodes.

68 BnF, lat. 7432, f. 219r. See Appendix, p. 84, lines 4-12.

69 Given that he intended to adjust the planetary nodes to the date of Duke John's birth, he must have dedicated his efforts to the university at a time after he had accepted patronage at Belleperche.

Heingarter continued to foster a relationship with the university even after he had departed for Belleperche. In 1471, there was something of a power struggle between the king of France and duke of Bourbon on the one hand and the corporate authorities of the University of Paris on the other. The university had appointed a certain Anthony de Busco as acting scribe of the conservator (*scriba conservationis*), who, for reasons that remain unclear, was deemed unacceptable by the king. In response to the king's displeasure, the duke sent letters to the faculty requesting that Anthony de Busco be removed from office. In his place, John II suggested Heingarter should fill the post for the rest of the former scribe's tenure.[70]

Heingarter ultimately did not replace Anthony de Busco as *scriba conservationis*, but the fact that his name was raised as a possible replacement shows that both he and his patron continued to take a great interest in the affairs of the university.[71] This included not only a desire to guide the school's administrative affairs, but also an apparent interest in determining the curriculum for lectures in astronomy. Heingarter's curriculum was memorialized in the words of Guillaume Cop, who, in his above-mentioned 1493 letter, states that it is hard to tell whether it is Heingarter who is indebted to the quadrivial curriculum, or if it is not rather the contrary that is the case. Moreover, in Cop's words, Heingarter is the one who restored these mathematical sciences to light — particularly astronomy — to which Heingarter contributed greatly with the works he wrote. [72]

Cop would go on in the letter to lay out the curriculum of astronomical texts, tables, and instruments that Heingarter introduced to him for mastery of the discipline.[73] Cop presents the *Theorica planetarum communis* as the introductory work of Heingarter's teaching regimen.[74] Neither as thorough nor as coherent as its earlier chapters treating planetary longitudes, the quality of the seventh chapter on latitude had already been challenged by Henry of Langenstein, whose *De reprobatione ecentricorum et epiciclorum* (1364) would gain currency at the University of Vienna after he moved there in the late fourteenth century.[75]

70 'Anno quo supra, die decima Marcii, congregata fuit alma Parisiensis Universitas apud Sanctum Maturinum super duobus articulis. Primus fuit ad audiendum lecturam quarumdam licterarum supremi domini nostri domini regis et domini ducis Borbonie, in quibus continebatur quod non placebat supremo domino nostro regi quod magister Anthonius de Busco amplius officium scribe conservacionis exerceret'. See, *Liber procuratorum nationis anglicanae (alemanniae)*, ed. by Samaran, van Moé, Susanna Vitte, III, p. 171.

71 'Totum negotium erat de remotione personae non gratae magistri Anthonii de Busco. Dux Borbonii commendaverat Universitati 'suum medicum, licet astrologum, videlicet magister Corardum'. Illi tamen non annuit Universitas et cum consensu domini conservatoris, scilicet episcopi Meldensis, magister Robertum Mazengarbe, iam scribam conservationis, designavit ad exercendum simul duos scribatus'. See, *Liber procuratorum nationis anglicanae (alemanniae)*, Samaran, van Moé, Susanna Vitte, III, pp. 171-72n1.

72 'In mathematicis autem discipilinis ita profecisti, ut an tu musis ipsis quadruvialibus, an ipse tibi plus debeant: haud facile coniicere possim. Cum enim iam quasi perdite extincteque essent, tu ipsas in lucem reducis. Precipue astronomiam in cuius emulos tam graviter tamque eleganter scripsisti …'. BAV, Reg. lat. 1241, f. 2^{r}. The arts of the quadrivium are here embodied by the muses, which is consistent with the humanistic and erudite tone of the letter.

73 BAV, Reg. lat. 1241, ff. 3^{v}-4^{v}. The question of Heingarter's preparatory curriculum in mathematical astronomy for arts and medical students will be treated more fully by Eleonora Andriani and Nicholas Jacobson in a future article. See note 22 on p. 59.

74 It should not be forgotten that the latitude treatise is, in fact, a commentary on the seventh chapter of the same work. See BAV, Reg. lat. 1241, f. 4^{r}.

75 BnF, lat. 16401, f. 62^{r}. See Kren, 'A Medieval Objection to "Ptolemy"', 378-93; Michael Shank, 'Regiomontanus and Astronomical Controversy in the Background of Copernicus', in *Before Copernicus: The Cultures and*

By the time of Heingarter's writing, the *Theorica planetarum* was generally agreed to be inadequate. A still more devastating attack than Langenstein's would be published at about the same time as Heingarter's latitude treatise when Johannes Regiomontanus held it up to bitter scrutiny in the *Disputationes contra deliramenta cremonensia* (1475).[76] Because the *Theorica planetarum communis* continued to play such an integral role in university instruction of astronomy even after such broadsides had been levelled, it makes sense that masters of arts at Paris would call upon someone to clarify the text and correct it where necessary.

In Heingarter's efforts to do so, we see that he offered two rather different strategies in his *theorica* and his *practica*, respectively. In the *theorica*, he sought to establish what one might call an authoritative and univocal account of planetary motion in latitude. He does not present two models, one to be associated with the *zīj* of al-Khwārizmī and the other with the *zīj* of al-Battānī. Instead, the planetary motions described in Heingarter's *theorica* are unambiguously associated with the models established by Ptolemy in the thirteenth book of the *Almagest* and transmitted by the *zīj* of al-Battānī.[77]

In the *theorica* of the treatise, Heingarter differentiates, just as Ptolemy had done, between two categories of latitudinal motion, one that pertains to the superior planets and the other to the inferior. In the case of the superior planets, both the deferent and the epicycle of the simple longitudinal model are brought out of the plane of the ecliptic to account for the planet's latitude. He explains that in the case of the inclination of the planet's deferent, the angle is fixed. The circles of the deferent and the ecliptic intersect at two points called nodes or *genzahar*, and a straight line between these two points forms the nodal line, dividing the circle of the deferent into two unequal parts, where the larger part is the north and the smaller south.[78] The second inclination is that between the epicycle and the deferent. As opposed to the inclination of the deferent, the epicycle's inclination varies over time with its axis of rotation always parallel to the nodal line.[79] When the epicycle is at the nodes, it is in the plane of the ecliptic, but the inclination reaches its maximum when the epicycle centre reaches its highest point on the deferent.[80]

As for the inferior planets, Heingarter explains that their motion in latitude has three components. Like the superior planets, their deferents incline from the ecliptic plane, but this time the angle of the tilt varies over time. When the centre of the epicycle is at the nodes, the deferent is in the ecliptic plane, but when it is at the top of the deferent,

---

Contexts of Scientific Learning in the Fifteenth Century, ed. by Rivka Feldhay and F. Jamil Ragep (Montreal: McGill-Queen's University Press, 2017), p. 84.

76 The seventh chapter of the *Theorica planetarum communis* has been edited by Francis Carmody. See *Theorica planetarum Gerardi*, ed. by Carmody (Berkeley: 1942), pp. 41-43. For an in-depth discussion of Johannes Regiomontanus' 1475 *Disputationes contra deliramenta cremonensia*, see Kren, 'Planetary Latitudes, the *Theorica Gerardi*, and Regiomontanus', 194-205.

77 Although Ptolemy revised what he had done in his later works through the *Handy Tables* and the *Planetary Hypotheses*, it was the more complex version of the *Almagest* that was to be found in the Toledan Tables. See Noel Swerdlow, 'Ptolemy's Theories of the Latitude of the Planets in the Almagest, Handy Tables, and Planetary Hypotheses', in *Wrong for the Right Reasons*, ed. by Jed Z. Buchwald and Allan Franklin (Dordrecht: Springer, 2005), pp. 41–71.

78 BnF, lat. 7432, f. 219v. See Appendix, p. 88, lines 56-62.

79 BnF, lat. 7432, f. 220r. See Appendix, pp. 88-90, lines 73-80.

80 BnF, lat. 7432, f. 220r. See Appendix, p. 90, lines 81-83.

the inclination reaches its maximum.[81] Heingarter goes on to say that, like the superior planets, the inferior planets also have an inclination of the epicycle rotating about the diameter of the epicycle that runs parallel to the nodal line, except that, in the case of the inferior planets, this inclination varies such that the epicycle is in plane when the epicycle centre is at the top of the deferent and the inclination is maximum when the epicycle centre is at the nodes.[82] In addition to this, a third inclination must be added, with an axis of rotation about the diameter of the epicycle that is perpendicular to that of the first epicyclic inclination. This inclination varies such that the epicycle is in plane when the epicycle centre is at the nodes.[83] It is at its maximum when the epicycle centre is the top of the deferent.[84]

Despite Heingarter's self-described attempt to thoroughly treat planetary latitude theory just as previous authors had done for planetary longitudes, his efforts to offer this unequivocal exposition were hindered by a general lack of standardization in the terminology, procedural methodology, and mathematical models he inherited from previous source material. The heterogeneity of latitude terminology stands in stark contrast with the terminology developed for planetary longitudes. As Olaf Pedersen observed in his seminal article, 'The Origins of the *Theorica planetarum*', the *Theorica planetarum communis* had effectively championed the models of Ptolemaic astronomy against other alternatives by consistently establishing a vocabulary that could be applied to the canonical procedures and the numerical values of the tables available to the practitioner. Pedersen wrote that the author of the *Theorica planetarum communis*

> successfully trie[d] to overcome the terminological ambiguity in Ptolemy by introducing separate Latin terms for separate concepts, for instance by creating a variety of terms for all the variables that are called 'anomaly' in the *Almagest*. This is not to say that he knew Greek, or even a Latin translation of the *Almagest*, but only that he was aware of the need for a consistent and unequivocal vocabulary in planetary theory. [...] Historically speaking there is no doubt that these five chapters were instrumental in acquainting students with the principal features of the various kinematical models of the *Almagest*, and thus in strengthening the position of Ptolemaic astronomy against attacks from teachers who, following al-Bitruji, tried to introduce other models.[85]

Heingarter did not enjoy the benefit of having inherited a similarly unified vocabulary for his *theorica* on latitudes. As if to confirm the dearth of contemporary fifteenth-century texts containing a consistent and conventional idiom, in the first chapter of the treatise Heingarter leans heavily upon the older (and largely outmoded) terminology found in the eighteenth chapter of John of Seville's translation of Alfraganus' *Science*

---

81 See Glen Van Brummelen, *Mathematical Tables in Ptolemy's Almagest*, PhD Thesis (Simon Fraser University, 1993), p. 358.
82 BnF, lat. 7432, f. 221ʳ. See Appendix, pp. 92-94, lines 126-46.
83 Again, the top of the deferent is always southerly for Mercury and always northerly for Venus. See Van Brummelen, *Mathematical Tables in Ptolemy's Almagest*, p. 358.
84 BnF, lat. 7432, f. 221ʳ. See Appendix, p. 94, lines 147-55.
85 Olaf Pedersen, 'The Origins of the "Theorica Planetarum"', *Journal for the History of Astronomy*, 12 (1981), 113-23 (p. 116).

of the Stars.[86] Written in the twelfth century, the text contains a vocabulary that is not consistent with that of the more conventional lexicon of astronomical terms that had taken root in the university in subsequent centuries. Even astronomical objects that are ubiquitous to the Ptolemaic models, like that of the eccentric circle (*circulus egresse cuspis* in Alfraganus, but more typically rendered as *circulus ecentrici*) or the epicycle (*circulus brevis* in Alfraganus, but more typically rendered as *epiciclus*), diverge from the standard vocabulary. This is to say nothing of the more specialized terminology of latitude theory like the inclination of the deferent, which is often translated as 'deviation' in modern commentaries, but that Heingarter calls the *declinatio ecentrici ab ecliptica* in the *theorica*. Heingarter's descriptive phrase expresses the spatial relations between the deferent and the ecliptic nicely, but it differs from the more common term *deviatio*, which could be found in the canons and is employed by Heingarter in his *practica*. Similarly, the second angle of inclination on the epicycles of the inferior planets, often translated as 'slant' in modern commentaries, is given the much more unusual name *involutio* in the *theorica*. This differs from most other Latin sources on the subject, including the Toledan and Alfonsine canons, which refer to this same angle as the *reflexio* — a term that Heingarter again reverts to in the *practica*.[87]

Heingarter's introduction of novel terminology in the *theorica* may have undermined his unifying efforts in some cases, but there are other instances in which he can be seen updating outdated modes of expression by either silently changing the terms in his compilation or by including the vocabulary that one would find in canon texts taught in the university alongside the older term. For example, when referring to the epicycle centre, he often calls it the *cuspis seu centrum epicicli*. He also attempted to reinterpret passages from the *Theorica planetarum communis* that did not conform with the Ptolemaic model of planetary latitude detailed by Alfraganus in the *De scientia astrorum*. For example, Heingarter quotes a passage from the *Theorica's planetarum communis*' seventh chapter verbatim when explaining how the epicycle of a superior planet inclines from its deferent plane, writing:

> The epicycle, moreover, is inclined from the eccentric such that the planet will always be between the ecliptic and the centre of the epicycle, except when the centre of the epicycle is at the head or tail; then, the epicycle is right on the eccentric.[88]

This passage is not coherent with the planetary models described by Ptolemy, however, since it is not the planet that will always be between the ecliptic and the epicycle centre due to

86 For a modern edition of work of Alfraganus, see Alfraganus (al-Farghānī), *Differentie in quibusdam collectis scientie astrorum*, trans. by John of Seville, ed. by F.J. Carmody (Berkeley: University of California Press, 1943). We follow Charles Burnett in giving John of Seville's translation of Alfraganus the title *Science of the Stars*. This is in spite of the fact that the work has been given a wide variety of different names over the years. On this topic, see the third Appendix to Charles Burnett's 'From *Astronomica* to *Exotica*: Jacob Golius' Edition of Alfraganus' *On the Science of the Stars* in Comparison with the Earlier Versions', in *Scholarship Between Europe and the Levant: Essays in Honour of Alastair Hamilton*, ed. by Jan Loop and Jill Kraye (Leiden: Brill, 2020), pp. 85-86.

87 For a more detailed account of the variations in terminology to be found in the treatise, see the appendices in this article, including the glossary.

88 BnF, lat. 7432, f. 220^r. See Appendix, p. 90, lines 86-88.

the epicycle's inclination, but rather the epicycle's apogee. Heingarter glosses this passage in a way that agrees with the description found in Alfraganus, however, by explaining that

> ... the inclination of the lower half of the epicycle from the plane of the deferent is always in that direction toward which that half of the deferent — in which is the centre of the epicycle — declines from the plane of the ecliptic. And, therefore, when the centre of the epicycle is outside of the nodes, the planet on the upper part of the epicycle is between two planes, namely that of the ecliptic and its deferent.[89]

That is to say, the epicycle's perigee is always further from the plane of the ecliptic than is the centre of the epicycle. Thus, when the planet is on the part of the epicycle closer to the apogee, it will fall between the planes of the ecliptic and the deferent. This statement is in agreement with the Ptolemaic model, though it contradicts the passage taken from the *Theorica planetarum communis*, to which it serves as a gloss.[90]

This latter example of Heingarter's efforts to bring some order and consistency to the many sources that he integrated into his treatise serves to show that, more serious than the terminological variability of the latitude theory, was the multiplicity of descriptions of the models lying behind these terms and the procedures involved in their calculation.

Such ambiguities became more apparent in Heingarter's *practica*, where multiple canons are provided for using different tables to find latitude values. As opposed to his approach in the *theorica*, Heingarter decided to leave the multiple methods alongside one another in the *practica* with no attempt to reconcile them to one another. Instead, he provides one set of instructions for using tables ultimately derived from al-Battānī's *zīj*, which require the user to find two angles of inclination for the superior planets and three angles of inclination for the inferior planets. These values are then combined to obtain the final latitudinal position of the planet. This is the method that one would find in the canons written by John of Lignères and paraphrased in Heingarter's *practica*.[91]

When drawing upon the Toledan canons, Heingarter provided a different set of instructions, which preserved the procedures designed for the Indian model as found in the *zīj* of al-Khwārizmī. These procedures required simultaneous use of the *bipartialis* and *quadripartialis* tables, the format of which is described in the previous section. For these tables, he provided two canons: one for the superior planets, the other for the inferior.[92]

That the canon instructions, the tables, and, by implication, the models discussed in the treatise's second chapter are demonstrably divergent from each other was apparently not Heingarter's chief concern in the *practica*. Instead, he appears to have cared most of all about offering workable and clear instructions for students, instructing them in how to use each table set regardless of whether the table set in question was linked to the model

89 BnF, lat. 7432, f. 220r. See Appendix, p. 90, lines 81-85.

90 This is an error in the *Theorica planetarum communis* that Regiomontanus attacked vociferously in the *Disputationes*. See Kren, 'Planetary Latitudes, the *Theorica Gerardi*, and Regiomontanus', 194-205 (p. 199).

91 For a description of the tabular procedures for planetary latitudes found in John of Lignères' canons, see José Chabás and Bernard R. Goldstein, *A Survey of European Astronomical Tables in the Late Middle Ages* (Leiden: Brill, 2012), pp. 106-10.

92 For a succinct explanation of the canon methods using the *bipartialis* and *quadripartialis* tables, see Kren, 'Planetary Latitudes, the *Theorica Gerardi*, and Regiomontanus', 194-205 (pp. 196-97n9).

described in the first chapter. He was aware that these tabular methods were different and likely knew they yielded different numerical results. Nevertheless, by giving these different methods as a *practica*, Heingarter did not necessarily wish to resolve all the ambiguities implied in the tables.

Conclusion

It remains to be seen if Heingarter ever applied the procedural instructions of his latitude treatise in any systematic way to his horoscope charts later in life. Indeed, since the extant artefacts of Heingarter's career date from the 1460s to the early 1480s and very possibly predate the writing of the treatise, we may not be able to say with any degree of certainty if he changed his practices in discernible ways later in his career.[93] Given what we know of the other arenas of Heingarter's life — gathering a manuscript collection at the court and developing a curriculum for students at the University of Paris — it is also possible that the latitude treatise could have served other purposes as well. In some instances, Heingarter's insistence on precisely calculating planetary latitudes for medical astrology may have been directed more at those who would attack a medical astrologer's practice than those who would actually do the calculations in emulation of it. Thus, there was probably at least an element of his research agenda that was much more rhetorical than practical and that intended to answer the questions of sceptics or tyros as much as to expand the repertoire of the seasoned calculator.

What is undeniable, however, is that Heingarter was a part of a new generation of practitioner in the late fifteenth century that had become profoundly interested in problems of planetary latitude, and he sought to gather as much material on the topic as possible. Furthermore, the importance of accounting for latitude in astrological procedures, as Heingarter explored in the annotations of BnF, lat. 7432, would garner more and more advocates as the discipline of medical astrology came under increasing critical scrutiny in the decades following Heingarter's composition of the treatise. This was particularly the case for the technique of finding planetary directions. Giovanni Bianchini and Johannes Regiomontanus both computed canons and tables for planetary directions with provisions that would allow a user to include planetary latitude values.[94] Later still, late-sixteenth- and early-seventeenth-century horoscopes, such as those drawn up by Girolamo Cardano and Galileo Galilei, would contain planetary latitudes for use in directing the significator of a nativity.[95] Similarly, Giovanni Antonio Magini would compute direction tables for

93 Although the horoscopes of Heingarter that we investigated do not include planetary latitude values, it is worth mentioning that he did take into account planetary latitude positions in some of his own prognostications. See for instance, BnF, lat. 7447, f. 9^{r}.

94 Giovanni Bianchini, *Tabulae astronomicae et canones in eas*, ed. by Augustin Olomoucký (Venice: Simone Bevilacqua, 1495); See Johannes de Regiomontanus, *Tabulae directionum profectionumque*, ed. by Johannes Angelus (Augsburg: Erhard Ratdolt, 1490), pp. b5^{a}-b6^{a}, c1^{a}-c3^{a}.

95 See Girolamo Cardano, *De astrorum iudiciis, tomus quintus quo continentur astronomica, astrologica, onirocritrica*, ed. by Jean-Antoine Huguetan and Marc-Antoine Ravaud (Lyon: Huguetan and Ravaud, 1663), p. 221; Galileo Galilei, *Astrologica nonnulla* in *Le opere di Galileo Galilei*, ed. by Antonio Favaro, 20 vols (Florence: G. Barbèra, 1890), XIX, pp. 205–20.

planets with different latitudes in his posthumous *De directionibus*.[96] Such calculations had become a convention by the time Jean-Baptiste Morin treated the subject in his encyclopaedic *Astrologia Gallica*.[97]

For the quotidian calculations of the practicing medical astrologer, these procedures would have been simplified by the printed ephemerides that would increasingly include planetary latitude positions as well as longitude. Evidence to this effect would include some of the most widely circulating ephemerides of eminent authors such as Regiomontanus, Cardano, Johann Stöffler, Jacob Pflaum, Philipp Imsser, Luca Gauricus, Giovanni Battista Carello, Johann Stadius, and Cyprian Leowitz.[98] In short, Heingarter's concerns about latitudes can be understood in terms of many broader trends in the practice of astronomy that were developing at the end of the fifteenth century.

96 Magini describes what he considers to be the conventional techniques for finding planetary aspects and directions, both requiring the latitude position of the planet in question. He does so in two extended passages in his work, which was published posthumously. See Giovanni Antonio Magini in Alfonso Zoboli, *De directionibus* (Venice: apud Franciscum Grossum, 1620), pp. 2-6, 13.

97 See Jean-Baptiste Morin, *Astrologia Gallica principis et rationibus propriis stabilita atque in XXVI libros distributa* (The Hague: Adriaan Vlacq, 1661), pp. 354-62 and 545-58. For an English translation of these passages, see Jean-Baptiste Morin, *Astrologia Gallica: Book Sixteen, The Rays and Aspects of the Planets* (Tempe: American Federation of Astrologers, 2008), pp. 3-54; *Astrologia Gallica: Book Twenty-two, Directions*, trans. by James Herschel Holden (Tempe: American Federation of Astrologers, 1994), pp. 31-48.

98 We thank Rich Kremer for informing us of this tendency in fifteenth- and sixteenth-century printed ephemerides. See Girolamo Cardano, *Libelli quinque* (Nuremberg: Johann Petreius, 1547); Johannes Regiomontanus, *Ephemerides* [1475-1506] (Nuremberg: Regiomontanus, 1474); Johannes Stöffler and Jacob Pflaum, Almanach nova plurimis annis venturis inservientia [1499–1531] (Ulm: Joannis Reger, 1499); Philipp Imsser, *Ephemeridum opus Ioannis Stoefleri Iustingensis mathematici praeclarissimi a capite anni Redemptoris Christi 1532 in alios 20 proxime subsequentes ad veterum imitationem accuratissimo calculo elaboratum* [1532-1551] (Tübingen: Ulrich Morhart, 1531); Luca Gaurico, *Ephemerides recognitae et ad unguem castigatae. Schemata et praedictiones ad annum usque virginei partus 1552* [1534-1551] (Venice: Lucantonio Ginti, 1533); Giovanni Battista Carello, *Le Efemeridi volgari per anni 17, al meridiano dell'inclita città di Vinegia* [1555-71] (Venice: Nicolo Bevilacqua, 1555); Johannes Stadius, *Ephemerides novae et exactae Ioannis Stadii Leonnouthesi ab anno 1554 ad annum 1570* (Cologne: Birckmann, 1556); Cyprian Leowitz, *Ephemeridum novum atque insigne opus ab A.D. 1556 usque in 1606 accuratissime supputatum* (Augsburg: Philipp Ulhard, 1557).

Appendix

Editorial principles

The text

The following critical edition represents the first edition of the treatise on planetary latitude by Conrad Heinarter found in Paris, BnF, lat. 7432 on ff. 219^{r}-222^{v}. We attempted to reproduce this text as faithfully as possible, retaining its orthography almost in all instances. However, we have standardised the alterations of the consonant spelling of 'ci/ti' (e.g. opulencia>opulentia declinacio>declinatio, inclinacio>inclination ...), as it does not seem possible to assess the prevalence of one spelling habit over the other throughout the text. Moreover, we opted for the capitalisation of proper names (e.g. iohannes>Iohannes, avernia>Avernia, alfraganus>Alfraganus) and planets (mercurius>Mercurius, mars>Mars, venus>Venus), although they are found in lower case in the manuscript witness.[99]

To mark syntactic units, we have applied a modern system of punctuation. In the process of editing the text, we have made a concerted effort to remain true to the apparent division of the text. Nonetheless, it is important to acknowledge that capital letters are frequently utilised to indicate information considered significant by the author and may not always signal the beginning of new sentences or paragraphs.

To improve the readability of the text, we have taken the liberty of creating new paragraphs where necessary, even if not explicitly indicated in the manuscript. This decision was made after careful consideration and a deep understanding of the intended meaning and sense of the text. Our primary objective is to present an accurate and comprehensible edition of the text, and we believe that our approach will enable readers to gain a more profound appreciation of the author's intent and meaning.

Within the text at hand, numerical expressions are restricted only to the sexagesimal values. Conversely, adverbs such as "primo, secundo, tertio," and so forth, are spelled out in full despite the fact that their abbreviated form employs a numeral. Moreover, ordinal numbers are also spelled out in their alphabetic form (e.g. 3^{a} latitudo > tertia latitudo).

The edited text is accompanied by an *apparatus criticus* and an *apparatus fontium*: (1) the first apparatus records few conjectures and scribal errors; (2) the second apparatus includes the sources of the text. Only on the first page of the edition a footnote is added to clarify the integration of a word in angular brackets.

In this edition, the page change is noted in the right margin and within the text by a vertical bar. Finally, the edition was conducted on the basis of a single manuscript, and therefore, to avoid unnecessary repetition, we have not included its *siglum* in the apparatus.

The translation

The translation we provide hereafter is intended to help the reader through the understanding of the treatise and offer a handy context of reference for the contribution preceding this

99 For clarity and consistency the words provided as examples in parenthesis are expressed in nominative case, although some of these words never occur in the nominative within the text.

Appendix. We attempted to translate the text as literally as possible, inserting words in square brackets only when we considered it suitable for its readability.

The very nature of Latin syntax allows for long compound sentences whereas English syntax does not. Hence, in a few instances we have altered the syntax of the text, by breaking a long sentence into two or more separate shorter sentences. In order to preserve the connection between the text and its translation, the reference to the page change is noted in the right margin and within the text by a vertical bar.

List of abbreviations

ante corr. = *ante correctionem* = 'before correction'
cf. = *confer* = 'compare/ consider'
coni. = *coniecimus* = 'we conjectured'
Angular brackets (<…>) enclose words, etc. that we have added as deemed necessary for the reconstructed text
Square brackets ([…]) enclose words that we have added in the translation to enhance clarity

Glossary of terms

Conrad Heingarter's theoretical chapter on planetary latitudes relied heavily upon John of Seville's Latin translation of Alfraganus' *Science of the Stars*, while his practical chapter drew mainly from the canons to John of Lignères' tables (bearing the incipit *Priores astrologi*) and the version of the canons to the Toledan Tables, which Fritz Pedersen has designated by the letters Cb. The technical vocabulary used by Alfraganus to describe the models for planetary latitudes differs significantly from that of the canons to the Alfonsine and Toledan Tables. Despite his occasional updating of John of Seville's terms for epicycle and eccentric, however, Heingarter chose not to standardize the lexicon of terms that he had at his disposal to describe the various angles of motion involved in finding planetary latitudes. For this reason, we have deemed a glossary of terms to be a useful addition to the edited text and translation provided above. Each entry consists of a single-word term. Phrases, which incorporate a word but signify a distinct concept, are included in bold within the definition of each entry. Each entry word is presented in its nominative and genitive form with its gender signified in italics alongside it.

abscisio, abscisionis, *f.*: intersection; in its verbal form, the term is used to describe the point where a planet's eccentric plane cuts the ecliptic plane; when used in the phrase **locus abscisionis**, the term refers to a planetary node. See also, **caput et cauda draconis, genzahar, intersectio**, and **nodus**.
arcus, arcus, *m.*: arc; the term **arcus circuli inter stellam et equinoxialem** is used to define the declination of a star or a planet from the celestial equator. See also, **declinatio**.
argumentum, argumenti, *n.*: argument; the term is used in two senses in the text: first as the elongation of the true position of the Moon from the true position of the northern lunar node, the **argumentum latitudinis lune**; second, the term is used to refer to a planet's true elongation from the apogee of its epicycle, or the **argumentum equatum planete**.
aux, augis, *f.*: aux, apogee; referred to simply as the "aux" in the translation above, this term refers to the point on the epicycle where the planet will have reached its furthest

distance from the Earth, which is the **aux epicicli**; on the eccentric circle, the term is used to designate the furthest point of the centre of the epicycle from the Earth, which is the **aux ecentrici**.

axis, axis, *m.*: axis; Heingarter uses the term to indicate the movement of a body about a fixed point of rotation. To this end, he refers to the axis of the celestial equator (**axis circuli equinoxialis**), the axis of the ecliptic (**axis circuli signorum**), and the axis of the eccentric circles (**axis ecentricorum**) of specific planets; he also describes the motion of planetary epicycles and eccentrics in terms of axial rotation (e.g. **axis qui transit per ambo puncta longitudinum mediarum epicicli**).

caput, capitis, *n.*: head; in reference to the "head and tail of the dragon" (**caput et cauda draconis**), namely the nodes of lunar latitude, in which the head signifies the point from which the Moon ascends northward above the ecliptic plane.

cauda, caudae, *f.*: tail; the "tail of the dragon" (**cauda draconis**) signifies the point from which the Moon descends southward below the ecliptic plane.

centrum, centri, *n.*: centre; the centre is the elongation of the mean planet from its apogee on the eccentric circle; because the Ptolemaic theory of planetary latitude requires an adjustment of the planet's centre to or away from the planet's eccentric apogee, Heingarter distinguishes between the planet's centre before adjusting for the nodes (**simplex centrum**) and after (**centrum cum suo addito**); the centre of an epicycle (**centrum seu cuspis epicicli**) refers to the mean motion of a planet.

circulus, circuli, *m.*: circle; the term designates several periodic and regular motions used to calculate planetary positions; Heingarter chooses to maintain certain outmoded terms for planetary motion, such as the use of the term **circulus brevis** for an epicycle (**epiciclus**), the term *circulus egresse cuspis* in reference to an eccentric (**circulus ecentrici**); the term is also deployed frequently as a synonym for the celestial equator (**circulus equinoxialis**) and the ecliptic (**circulus signorum**).

declinatio, declinationis, *f.*: declination; although generally used to describe the angle of inclination from the celestial equator to the body of a star or planet (**delinatio planete ab circulo equinoxiali**), in Heingarter's theoretical chapter it takes on a variety of meanings: it can be used to describe a planet's latitude (**declinatio planete ab ecliptica**); it also is used interchangeably with **inclinatio** (see below) to describe the tilt of a planets' epicycle at the points of apogee and perigee (**declinatio epicicli**); it can be used to describe the angle of deviation (see below) found in the canons, which is defined by the tilt of the eccentric plane from the ecliptic (**declinatio ecentrici ab ecliptica**); finally it is used interchangeably with the term **involutio** (see below) to describe the rotation of the two mean longitudes of an epicycle (**declinatio utrarum-que longitudinum epicicli mediarum**), otherwise known as "reflection" or "slant" in contemporary literature (see **reflexio** below).

deferens, deferentis[100]: deferent; a rarer term in Heingarter used as a synonym for the eccentric circle on which the centre of the epicycle rides.

deviatio, deviationis, *f.*: deviation; used only in the canons, this term is used to describe the angle of inclination formed between the plane of a planet's eccentric circle and

100 One termination participle.

the plane of the ecliptic; in the theoretical chapter, the deviation is referred to as the "declination of the eccentric from the ecliptic" (**declinatio ecentrici ab ecliptica**). See **declinatio** above.

directus, directi: direct; this term is used to describe the manner in which the epicycle circle aligns with the ecliptic plane at the planetary nodes; it is also a term used to describe values found in planetary tables at a place opposite (**in directo**) to the values of the argument of the table.

dyameter, dyametri, as *sub. m.* or *n.*: diameter; Heingarter uses the term to highlight two epicycle diameters of particular significance in the latitude models: the first of these is the diameter that passes through the apogee and perigee of the epicycle reaching the Earth's centre, and this is called the **dyameter que est ab auge vera epicicli usque in oppositum augis**; the other is the diameter that goes through the two mean longitudes of the epicycle, and is called the **dyameter que vadit super utrasque longitudines medias**.

ecliptica, eclipticae, *f.*: ecliptic; the Sun's apparent annual motion through the fixed stars of the zodiac; in Heingarter's treatise, the ecliptic is more frequently referred to as the "circle of the signs" (**circulus signorum**). See **circulus** above.

genzahar (*or* **geuzahar**; undeclinable): genzahar; from the Arabic *jawzahar* and the Persian *gawzahar*, the genzahar refer to the planetary nodes. See also, **abscissio, caput et cauda draconis, intersectio**, and **nodus**.

inclinatio, inclinationis, *f.*: inclination; the angle that cuts through the plane of eccentric circle when a planets' epicycle tilts (**inclinatio epicicli**) northward or southward at the points of apogee and perigee (aka **declinatio epicicli**). See **declinatio** above.

intersectio, intersectionis, *f.*: intersection; another word used by Heingarter to refer to the planetary nodes. See also, **abscissio, caput et cauda draconis, genzahar**, and **nodus**.

involutio, involutionis, *f.*: involution; the term in Heingarter's theoretical chapter used to describe the rotation of the two mean longitudes of an epicycle (**declinatio utrarum-que longitudinum epicicli mediarum**), otherwise known as "reflection" or "slant" in contemporary literature. See **reflexio** below.

latitudo, latitudinis, *f.*: latitude; generally the width or breadth of a thing, but specifically used to describe the angle of inclination from the ecliptic to the body of a star or planet; see also Heingarter's use of the phrase **declinatio planete ab ecliptica** under **declinatio** above.

longitudo, longitudinis, *f.*: longitude; generally the length or elongation between two bodies with astronomical significance, but specifically the position of a planet in the plane of the ecliptic as measured from the head of Aries; the terms "further longitude" (**longitudo longior**) or "nearer longitude" (**longitudo propior**) make reference to the points on an eccentric where the epicycle centre is at its furthest or nearest position with respect to the Earth (in the translation we simply render these terms as "apogee" and "perigee"); the positions on a planet's epicycle where the body of the planet is at its furthest or nearest point is also called the **longitudo longior** and the **longitudo propior** respectively; Heingarter refers to the positions of apogee and perigee together as the **longitudines diversas**, and calls the positions at right angles to these the **longitudines mediae**.

locus, loci, *m.*: place, position.

motus, moti, *m.*: motion, position.; **motus declinationis vel inclinationis epicicli**
nodus, nodi, *m.*: node; another word used by Heingarter to refer to the planetary nodes. See also, **abscissio, caput et cauda draconis**, **genzahar**, and **intersectio**.
oppositus, oppositus, *m.*: opposite, opposed; used as an adjective to describe the position on the eccentric or epicycle circle that is directly opposite the apogee (**oppositum augis**), which is, namely the perigee.
pars, partis, *f.*: part, side, direction; a term used frequently to establish the orientation of a planet in latitude, whether to the north or to the south of the ecliptic plane; the term can also be used to describe the ascending or descending position of an epicycle or eccentric circle; the word is also used to describe the fourth term in a proportional relation — the **pars proporcionalis** — that is guided by the rule of three as a kind of linear interpolation.
reflexio, reflexionis, *f.*: reflection; the term in the canons that describes the rotation of the two mean longitudes of the epicycle of one of the inferior planets (**declinatio utrarum-que longitudinum epicicli mediarum**), otherwise known as the **involutio** in Heingarter's theoretical chapter. See **involutio** above.
signum, signi, *n.*: sign; Heingarter uses the term explicitly to refer, not to the 12 zodiacal signs of 30° each, but rather the 6 "physical" signs (**signa phisica**) of 60° each.

<I>n dandis accipiendisque muneribus, ita recte officia (inter eos precipue, qui se amant colunt atque observant) estimantur, si liquido constabit nec ob hoc aliud quod liberalius offerret inventum esse, nec ab illo unquam quod iocundius benivolentia complecteretur acceptum. Hec ipse considerans attuli tibi (Michael Roscher artium magister) non ignava operum pondera quibus ad facinus nihil instructius est, cum habendi <sitis> incanduit, sed ea attuli que ex sapientum opulentia in studentium utilitatem sumpta conveximus. Et quia superioribus libris motum stellarum et planetarum secundum longitudinem quantum theorice et practice sufficit peregimus, hunc vero de planetarum motu secundum latitudinem tractatum componere intendens, quem per duo dividam capitula: quorum primum theoricam, secundum practicam et modum inventionis latitudinis cum eorum genzahar ad eram illustrissimi Iohannis Borbonii atque Arvernie ducis ostendit. 219r

<Capitulum primum>

Est autem latitudo planete declinatio eorum ab ecliptica circuli signorum in utrisque partibus septentrionalis scilicet et meridiei. Et si consideravimus circulum exeuntem per axem circuli signorum et per stellas et gradus earum ex circulo signorum, erit arcus eius circuli inter stellam et eius gradum eadem quantitas latitudinis eius a circulo. Describit autem nobis sol circulum signorum per cursum suum ab occidente in orientem, eo quod superficies eius circuli ecentrici in qua movetur sit inseperabilis a superficie circuli signorum non declinans ab ea. Ceteri autem planete preter solem sunt quemadmodum narrabo ex dictis ALFRAGANI.

Motus nanque omnium stellarum fixarum versus orientem est super utrosque polos axis circuli signorum. Et ideo quicquid ex eis fuerit in superficie circuli signorum, non declinabit ab eo, quicquid vero fuerit ex eis extra superficiem circuli signorum versus septentrionem vel meridiem, erit motus eius inseperabilis a quantitate longitudinis que est inter ipsam et circulum signorum in latitudine, nec unquam removebitur ab hoc.

1/7 In ... conveximus] Cf. SEVERINUS BOETHIUS, *De arithmetica*, prolog.; ed. SCHILLING - OOSTHOUT, p. 3. **6** sitis] The passage found in ll. 1-8 exhibits a close parallel to the first lines of Boethius' *De arithmetica*. The integration of 'sitis' in angular brackets was deemed necessary to restore the passage from corruption. Because Heingarter has followed the language of the *De arithmetica* so closely here, we have decided to use the exact word found in this source. **13/45** Est ... fere[1]] Cf. ALFRAGANUS, *Differentie*, c. 18.1-9; ed. CARMODY, p. 31-32.

In giving and receiving gifts, such kindnesses are rightly esteemed (especially among 219r those who love, honor and respect one another) if it will be clearly agreed that, on account of this, there is neither anything to be found which one should offer more freely, nor is there ever anything to be received which one should take from another more delightfully, than benevolence. Considering this, I have brought to you (Michael Roscher, master of arts) the not so lazy load of works, for which nothing is more ready for action than when <a thirst> for acquisition has been kindled; but I have sent these, which, taken from the wealth of the wise, we have gathered for the utility of students. And because in the previous books we have thoroughly gone through the motion of the stars and planets according to longitude with however much of theory and practice as suffices, I set out to compose this treatise on the motion of the planets according to latitude, which I shall divide into two chapters: The first of these explains the theory; the second explains the practice and the method of finding [the planets'] latitude with their genzahar[1] for the era of the Illustrious John[2], duke of Bourbon and Auvergne.

<The First Chapter>

Now the latitude of a planet is their declination from the ecliptic of the circle of the signs[3] in both directions, namely of the north and of the south. And if we consider a circle passing through the axis of the circle of the signs and through the stars and their degrees from the circle of the signs, the arc of that circle between the star and its degree will be the same as the distance of its latitude from the circle.[4] By its course from west to east, the Sun will describe for us the circle of the signs because the plane of that eccentric[5] circle in which it moves is inseparable from the plane of the circle of the signs, not declining from it. I shall describe the remaining planets besides the Sun from what was said by ALFRAGANUS.

Indeed the motion of all the fixed stars toward the east is [centered] on both poles of the axis of the circle of the signs. And, therefore, whichever from among these is on the plane of the circle of the signs, it will not decline from it; moreover, whichever among these is outside the plane of the circle of the signs, toward the north or south, its motion will be inseparable from the quantity of the distance which is between that [star] and the circle of the signs in latitude, nor will it ever move from this [position]. It is, there-

1 Heingarter deploys a number of different Latin terms to describe the planetary nodes, including *genzahar, nodus, abscisio*, and *intersectio*. We render Heingarter's terminology into English by keeping *genzahar* unchanged, and translating *abscisio* and *intersectio* as 'intersection'. We only ever refer to planetary nodes as such when the word *nodus* is employed.

2 John II, Duke of Bourbon and Auvergne (1426-1488).

3 With only nine exceptions, Heingarter follows Alfraganus in calling the ecliptic the 'circle of the signs'. Here in the beginning of the text he equates the two terms by introducing the phrase *ecliptica circuli signorum*. We preserve Heingarter's wording, and only refer to the ecliptic as such when it is so indicated in the text.

4 In this sentence, Heingarter invites the reader to imagine a great circle drawn through the pole of the ecliptic, passing through the body of the star and touching the ecliptic. The latitude is then defined as the arc of that circle spanning the distance between the ecliptic and the body of the star.

5 Here and throughout the treatise, Heingarter has silently changed the term for eccentric found in John of Seville's original rendering of Alfraganus, namely the *circulus egresse cuspis*, to the more familiar *circulus ecentrici*.

Patet itaque quod unaqueque stellarum fixarum vel non habet latitudinem a circulo signorum, vel eius latitudo semper sit quantitatis unius.

In luna autem vel quinque planetis erraticis erit e contrario, quia motus eorum non est semper super polos axis circuli signorum, sed sunt super ecentricorum axes, qui
abscindunt superficiem | circuli signorum super dyametrum circuli signorum et decli- 219v
nant ab eo versus septentrionem vel meridiem. Ideoque latitudo eorum fit diversa a circulo signorum.

Superficies enim circuli lune ecentrici abscindit superficiem circuli signorum super punctos utrosque capitis scilicet et caude, et declinat ab eo versus septentrionem et meridiem, eius quoque declinatio est fixa super unam quantitatem nec removetur ab hoc. Et superficies eius circuli brevis est inseperabilis a superficie circuli ecentrici, nec declinans ab ea. Ideoque fit ei una diversitas in latitudine propter declinationem ecentrici a circulo signorum.

Diversitas vero quinque planetarum erraticorum in latitudine non est una, quia circuli eorum ecentricorum declinant a circulo signorum et circuli epiciclorum declinant a circulis etiam ecentricorum. Circuli enim ecentrici abscindunt circulum signorum super dyametrum eius, et declinant ab eo versus septentrionem et meridiem. Et loca abscisionis horum quinque planetarum sunt ista: locus abscisionis Saturni est in medio inter longitudinem mediam et ustrasque longitudines diversas. In reliquis autem quatuor planetis sunt in utrisque longitudinibus mediis fere. Et dico notantur fere, quia intersectiones eorum non sunt prescise in medio longitudinis medie, ut postea patebit.

Et nota quod semper illa intersectio per quam planeta movetur ab ecliptica septentrionem versus dicitur caput draconis vel genzahar. Et alia intersectio per quam planeta ab ecliptica mutatur ad meridiem dicitur cauda draconis. Moventur autem intersectiones lune ab oriente in occidentem omni die tribus minutis fere.

fore, clear that each of the fixed stars either has no latitude from the circle of the signs, or its latitude is always of one quantity.

This will be quite the contrary for the Moon or the five wandering planets, because their motion is not always [centered] on the poles of the axis of the circle of the signs, but are on the axes of the eccentrics, which cut the plane of the circle of the signs on the 219v diameter of the circle of the signs, and decline from it toward the north or south. And therefore, their latitude appears different from the circle of the signs.

Indeed, the plane of the eccentric circle of the Moon cuts the plane of the circle of the signs on both points, that is, of the head and tail, and declines from it to the north or south. Its declination is also fixed at one quantity, and does not move from this. And the plane of its small circle is inseparable from the plane of the eccentric circle, and it does not decline from it. And therefore, it has a single difference in latitude on account of the declination of the eccentric from the circle of the signs.

In the latitude of the five wandering planets, however, there is not just one difference, because the circles of their eccentrics decline from the circle of the signs, and the circles of the epicycles also decline from the eccentric circles. In fact, the eccentric circles cut the circle of the signs on its diameter, and they decline from it to the north and south. And the places of intersection of these five planets are these[6]: the place of the intersection of Saturn is in the middle between the mean longitude and both divergent longitudes.[7] In the remaining four planets, however, they are nearly at both mean longitudes. I mean they are nearly observed [there], because their intersections are not precisely in the middle of mean longitude as shall be made clear hereafter.

And note that that intersection by which the planet moves away from the ecliptic toward the north is always called the head of the dragon or the genzahar. And the other intersection by which the planet moves away from the ecliptic to the south is called the tail of the dragon. The Moon's intersections move from the east to the west by about three minutes every day.

6 The word 'these' translates *ista* (f. 219^{v}, line 51) in the text. Apart from this word, the passage seems to follow *verbatim* John of Seville's version of Alfraganus' *On the Science of the Stars*, found on p. 32 of Carmody's edition (Note that Carmody refers to Alfraganus' work as the *Differentie*, however, the text goes by several titles according to different traditions.). On this topic, see the third appendix to Charles Burnett's 'From *Astronomica* to *Exotica*: Jacob Golius' Edition of al-Farghānī's *On the Science of the Stars* in Comparison with the Earlier Versions', in *Scholarship Between Europe and the Levant: Essays in Honour of Alastair Hamilton*, eds. Jan Loop and Jill Kraye (Leiden: Brill, 2020), pp. 85-86. Interestingly, Gerard of Cremona's translation in this place presents the same variant as Heingarter, namely *ista*, rather than John of Seville's *ita*. However, *ita* and *ista* represent equally fair variants; on the basis of this single variant, one cannot conclude that Heingarter borrowed from Gerard of Cremona rather than John of Seville. Note : we rely on BnF lat. 7400, f. 10^{r}, which is the only witness we are aware of that can be safely attributed to Gerard of Cremona.

7 Here Heingarter is referring to the apogee and the perigee.

Declinatio autem longitudinum longiorum ecentricorum defferentium trium superiorum est versus septentrionem a circulo signorum et longitudinum propiorum versus meridiem fixe sempiterneque immobilitatis quemadmodum est in luna. Sed intersectiones due ecentricorum trium superiorum cum ecliptica, non moventur, nisi motu octave spere.

Et nota quod locus genzahar Saturni, distat a loco opposito augis deferentis secundum signorum successionem per 40 gradus, alii ponunt 45. Et eius alia intersectio que cauda dicitur distat a loco augis ecentrici secundum successionem signorum per 40, alii 45, gradus similiter. Erit igitur locus partis deferentis maxime declinantis ad septentrionem distans a loco augis deferentis contra successionem signorum per 50 gradus et locus partis deferentis declinans ad meridiem distat etiam a loco oppositi augis | per 50 220r gradus; alii ponunt 45 contra signorum successionem.

Et semper locus maxime declinationis est ille qui equaliter distat ab utraque sectione. Et isti qui ponunt 45 gradus magis concordant cum ALFRAGANO. Et locus genzahar Iovis distat a loco augis sue deferentis per 70 gradus contra successionem signorum; quare locus partis deferentis sui maxime declinans ad septentrionem erit distans a loco augis secundum successionem signorum per 20 gradus. Et locus Martis distat ab auge sui ecentrici per 90 gradus contra signorum successionem. Erit igitur locus partis sui deferentis que maxime declinat ad septentrionem in loco augis sui deferentis. Et latitudines horum trium planetarum scripte in eorum tabulis posite sunt, ac si semper centrum epicicli esset in loco illo deferentis qui maxime declinat ad septentrionem vel meridiem.

Est et unus alius motus in istis tribus superioribus secundum quem movetur medietas superior epicicli et inferior declinando a superficie deferentis ex utraque parte septentrionis et meridiei. Et dicitur motus declinationis vel inclinationis epicicli. Est autem iste motus super axem qui transit per ambo puncta longitudinum mediarum epicicli, que quidem puncta distant per 90 gradus ab auge vera epicicli. Et dyameter que est ab auge vera epicicli usque in oppositum augis semper est equedistans superficiei

51/53 Declinatio ... luna] Cf. ALFRAGANUS, *Differentie*, c. 18.9; ed. CARMODY, p. 32.

51 longiorum] *coni.*; longiarum **60** 50] *coni.*; *scr.* 55 *sed* 50 *ante corr.* **61** 50] *coni.*; *scr.* 55

The declination[8] of the eccentric deferents of the three superior planets is to the north of the circle of the signs for the apogees[9], to the south for the perigees, [and is] fixed and eternally immobile, just as it is for the Moon. But the two intersections of the three superior planets' eccentrics with the ecliptic do not move except by the motion of the eighth sphere.

And note that the place of Saturn's genzahar is 40 degrees away from the place opposite of the aux of the deferent following the order of the signs; others put it at 45. And its other intersection, which is called the tail, is, in like manner, 40 (some say 45) degrees away from the place of the aux of the eccentric following the order of the signs. Therefore, the place of the part of the deferent declining most to the north is 50 degrees away from the place of the aux of the deferent against the order of the signs. And the place of the part of the deferent declining [most] to the south is also 50 degrees away from the place opposite of the aux ; others put it at 45 degrees against the order of the signs. 220r

And indeed the place of greatest declination is always that which is equally distant from either intersection. And those who posit 45 degrees agree more with ALFRAGANUS. And the place of the genzahar of Jupiter is 70 degrees away from the place of its deferent's aux against the order of the signs; therefore, the position of the side of its deferent declining most toward the north will be 20 degrees away from the place of the aux following the order of the signs. And the place of Mars' [genzahar] is 90 degrees away from the aux of its eccentric against the order of the signs. Therefore, the place of the side of its deferent that declines most to the north will be at the place of the deferent's aux. And the latitudes of these three planets written in their tables are positioned as if the center of the epicycle were always in that place of the deferent, which declines most to the north or the south.

And there is one other motion in these three superior planets according to which the upper half and lower half of the epicycle move by declining from the plane of the deferent from both the north and south's side. And this is called the motion of the epicycle's declination or inclination. This motion, moreover, is about the axis that crosses through both points of the mean longitudes of the epicycle, which points, indeed, are 90 degrees away from the true aux of the epicycle. And the diameter, which is from the true aux of the epicycle up to the point opposite of the aux, is always parallel to the plane of the

8 The 'declination … of the planets' eccentric deferents' (*declinatio … ecentricorum defferentium trium superiorum*) introduced here, is equivalent to the term 'deviation of the deferent from the ecliptic' (*deviatio deferentis ab ecliptica*) employed in the canon texts, which Heingarter used for the second chapter of the treatise below. In this case, Heingarter has chosen not to standardize the terminology of his treatise, even though he has done so for other terms elsewhere (see n. 15, 16, 17, and 18).

9 Here, and whenever this term is used, we have chosen to translate *longitudo longior* as 'apogee'. Although it is referring to the same position on an epicycle or an eccentric that we translate as 'aux' when the Latin term *aux* is found in the text, we maintain a distinction between 'apogee' and 'aux' to signal to the reader when *longitudo longior* is used, as opposed to *aux*. The same can be said for the terms *longitudo propior* and *oppositio augis* (Heingarter uses *locus oppositus augis*), which are translated as 'perigee' and 'place opposite of the aux' respectively. In considering these two pairs of terms to be equivalent, we follow chapter 12.6 of Alfraganus' *Differentie* where it reads: 'Unusquisque autem circulorum egresse cuspidis vocatur circulus egresse cuspidis, et hii sunt in quibus movetur planeta in medio cursu suo equali ab occidente in orientem. Fitque de hoc quod diximus necessario ut sint in unaquaque sperarum duo loca opposita, unus scilicet in ultima longitudine spere a terra et alius in propiori longitudine. Unus autem istorum locorum qui est longitudo longior vocatur augis [*sic*] planete et alter qui est longitudo propinquior vocatur oppositio augis'. See Carmody, *Differentie*, p. 23.

ecliptice, nisi cum fuerit centrum epicicli in nodis intersectionum. Nam tunc ipsa erit in superficie ecliptice, tunc enim nulla erit inclinatio.

Sed maxima est hec inclinatio cum centrum epicicli fuerit in locis intermediis. Et est inclinatio medietatis inferioris epicicli semper a superficie deferentis ad partem illam ad quam declinat a superficie ecliptice ipsa medietas deferentis, in qua est centrum epicicli. Et ideo quando centrum epicicli fuerit extra nodos, planeta existens in superiori parte epicicli est inter duas superficies, scilicet ecliptice et sui deferentis.

Inclinatur autem epiciclus ab ecentrico, ita quod semper erit planeta inter eclipticam et centrum epicicli, nisi cum centrum epicicli est in capite vel cauda; tunc est epiciclus directus in ecentrico.

Declinatio vero ecentricorum Veneris et Mercurii non est fixa, sed movetur super dyametrum circuli signorum et vadit super utrosque nodos motu modico versus septentrionem et meridiem, et revertitur ad locum inceptionis in anno semel quemadmodum revertitur motus longitudinis; et erit medietas longitudinis longioris ecentrici sex men|sibus versus septentrionem a circulo signorum et sex men sibus versus meridiem. Et 220v
similiter mutatur medietas propior versus septentrionem et meridiem. Supponiturque superficies ecentrici superficiei circuli signorum in anno bis et erit hoc cum fuerit cuspis circuli brevis in utrisque nodis.

Cumque fuerit cuspis circuli brevis Veneris in aliquo nodorum, erit inceptio declinationis medietatis ecentrici que succedit ipsi nodo versus septentrionem, et alia medietas

---

86/88 Inclinatur ... ecentrico] Cf. *Theorica planetarum Gerardi*; Paris, Bibliothèque nationale de France, lat. 7432, f. 219v. **89/154** Declinatio ... ecentrici] Cf. ALFRAGANUS, *Differentie*, c. 18.10-26; ed. CARMODY, p. 32-34.

ecliptic, except when the center of the epicycle is on the nodes of the intersections.[10] For then that [diameter] will be on the plane of the ecliptic[11], and then, indeed, there will be no inclination.

But this inclination is greatest when the center of the epicycle is in the intermediate places.[12] And the inclination of the lower half of the epicycle from the plane of the deferent is always in that direction toward which that half of the deferent – in which is the center of the epicycle – declines from the plane of the ecliptic.[13] And, therefore, when the center of the epicycle is outside of the nodes, the planet on the upper part of the epicycle is between two planes, namely that of the ecliptic and its deferent.

The epicycle, moreover, is inclined from the eccentric such that the planet will always be between the ecliptic and the center of the epicycle, except when the center of the epicycle is at the head or tail; then the epicycle is right on the eccentric.[14]

The declination of the eccentrics of Venus and Mercury, however, is not fixed, but moves on the diameter of the circle of the signs, and goes about both nodes with a small motion toward the north and the south; and it returns back to its starting place once in a year, just as the motion of longitude returns; and the apogee's[15] half of the eccentric will
be to the north of the circle of the signs for six months, and to the south for six months. 220v
And, similarly, the perigee's half moves toward the north and south. And the plane of the eccentric is set under the plane of the circle of the signs twice in a year, and this will be when the cusp of the small circle[16] is in both nodes.

---

10 In this sentence and the sentence before, Heingarter introduces the two diameters that Ptolemy uses to define motion in latitude on the planetary epicycles. In the first instance, Heingarter describes what is effectively the first diameter, or the line through the center of the epicycle connecting the perigee and the apogee. He does so, however, by defining the movement of the apogee and the perigee in terms of the axial rotation of the second diameter, which joins the mean points of longitude through the epicycle center (the 'motion … about the axis that crosses through both points of the mean longitudes of the epicycle …'). In the second instance, one would expect Heingarter to describe the second diameter, which could be defined as the line through the center of the epicycle connecting the mean longitudes, however, he instead defines it in terms of the first diameter ('the diameter … from the true aux of the epicycle up to the point opposite of the aux …'). We believe this may have been a slip of the pen, as the rest of the sentence clearly refers to the second, rather than the first, diameter of the epicycle, which remains always parallel to the ecliptic in the Ptolemaic model.

11 The *Theorica planetarum Gerardi* asserts that when the epicycle is at the nodes, it will be in the plane of the deferent, but Heingarter has corrected this to the plane of the ecliptic. For a discussion of the error in the *Theorica planetarum Gerardi*, see Claudia Kren, 'Planetary Latitudes, the *Theorica Gerardi*, and Regiomontanus', in *Isis* 68/2 (1977), 199-200.

12 That is, at the top and bottom of the deferent rather than the nodes.

13 Heingarter means to say that the epicycle's perigee is always further from the plane of the ecliptic than is the center of the epicycle. This statement is in agreement with the Ptolemaic model, though it contradicts the passage taken from the *Theorica planetarum Gerardi*, which is found in the following paragraph. See n. 14 below.

14 This sentence comes directly from the *Theorica planetarum Gerardi*, which Heingarter's treatise is commenting upon. It is also a quote from the work that has gained some notoriety, as Johannes Regiomontanus directly attacked this part of *Theorica planetarum* for being erroneous. It falsely claims that the planet's position must always lie somewhere between the epicycle center and the plane of the ecliptic, when it is rather the epicycle's apogee that must be found there (see n. 13) above. This statement is contradicted by Heingarter's own commentary in the passage before it. Whether this was meant as an explicit correction of the original text or not remains unsaid. For a discussion of Regiomontanus' criticism of this passage, see Kren, 'Planetary Latitudes, the *Theorica Gerardi*, and Regiomontanus', 199-200.

15 The text reads *medietas longitudinis longioris ecentrici*, which we have translated as 'apogee's half of the eccentric'. Likewise, just below, the text reads *medietas propior*, which we have translated as the 'perigee's half'.

16 Here and throughout much of the text, Heingarter preserves the terminology of Alfraganus's *Differentie*, calling the center of the epicycle the 'cusp of the small circle' (*cuspis circuli brevis*). Elsewhere Heingarter will supplement al-Alfraganus's terminology with the more conventional vocabulary of the university, such as when he refers to the *cuspis vel centrum epicicli*; see n. 17 below.

versus meridiem. In Mercurio vero est e contrario, quia cum fuerit cuspis eius circuli brevis in unoque nodorum, erit initium declinationis medietatis eius que succedit ipsi nodo versus meridiem, et alia medietas versus septentrionem.

Fit itaque necessario ut sit cuspis vel centrum epicicli unius cuiusque horum planetarum, vel in superficie circuli signorum cum fuerit apud utrosque nodos, vel in una parte circuli signorum et numquam declinat ad aliam partem. Erit enim centrum epicicli Veneris a circulo signorum versus septentrionem semper, et Mercurii versus meridiem semper. Et patet quod perfectio circulorum declinationum ecentricorum Veneris et Mercurii a circulo signorum versus septentrionem et meridiem est.

Cum fuerit cuspis seu centrum epicicli in longitudine longiori ecentrici vel propiori, declinatio autem epiciclorum quinque planetarum movetur atque mutatur; et erit reversio eius ad locum inceptionis in anno semel, ut reversio motus longitudinis. Sed motus epicicli in tribus superioribus est super dyametrum que vadit super utrasque longitudines medias. Fitque longitudo epicicli longior versus septentrionem a circulo ecentrico sex mensibus, et versus meridiem sex mensibus, et erit motus dyametri que vadit super utrasque longitudines medias in directo superficiei circuli signorum.

Eritque superficies epicicli, abscindens superficiem ecentrici et non supponitur ei, sed supponitur superficiei circuli signorum in anno bis. Hoc est cum fuerit motus cuspidis seu centri eius in utrisque locis nodorum. Et inceptio huius declinationis, que est longitudinis longioris epicicli et eius longitudinis propioris ab ecenctrico, est ex utrisque locis nodorum. Perfectio vero eius est apud longitudinem longiorem, et longitudinem propiorem.

Partes quoque declinationis sic distinguuntur: longitudines epicicli propiores declinant ab ecentricis in partem declinationis circulorum ecentricorum a circulo signorum; longitudines vero longiores e contrario. Eritque declinatio longitudinum epiciclorum propiorum, cum fuerit in medieta|tibus ecentricorum septentrionalibus, versus septen- 221r
trionem. Et cum fuerit in medietatibus meridianis, versus meridiem.

Veneri autem et Mercurio in declinatione epicicli sunt duo motus quorum unus assimilatur motui trium superiorum, qui est ex parte longitudinis longioris et longitudinis propioris epicicli ab ecentrico. Et alter est ex parte declinationis utrarumque longitudinum epicicli mediarum. Et vocatur hec declinatio involutio.

Eritque inceptio declinationis longitudinis longioris et longitudinis propioris epicicli ab ecentrico ex longitudine longiori et longitudine propiori ecentricis, et perfectio eius est apud utrosque nodos et tunc supponitur superficiei ecentrici superficies epicicli. Inceptio vero declinationis utrarumque epicicli longitudinum mediarum ab ecentrico

And when the cusp of the small circle of Venus is in one of the nodes, it will be the beginning of the declination of the half of the eccentric that is entering that node to the north; and the other half of the eccentric toward the south. But in Mercury the contrary is the case, because, when the cusp of the small circle is on one of the nodes, it will be the beginning of the declination of that half that enters that node towards the south; and the other half toward the north.

Therefore, it necessarily happens that the cusp or center of the epicycle[17] of each one of these planets is either in the plane of the circle of the signs when it is at both nodes, or to one side of the circle of the signs, and never declines to the other side. Indeed the center of the epicycle of Venus will always be to the north of the circle of the signs, and Mercury's always to the south. And it is clear that the completion of the declination circles of Venus and Mercury's eccentrics are to the north and south of the circle of the signs.

When the cusp or center of the epicycle is at the apogee or perigee of the eccentric, moreover, the declination of the five planets' epicycles moves and changes; its return to its beginning position will be once a year, just as is the return of the motion of longitude. But the motion of the epicycle in the three superior planets is on the diameter that goes through both mean longitudes. The apogee of the epicycle is away from the eccentric circle toward the north for six months, and toward the south for six months; and the motion of the diameter that goes through both mean longitudes will be right on the plane of the circle of the signs.

And this will be the plane of the epicycle, cutting the plane of the eccentric, and it is not set under it [i.e. the eccentric]; but it is set under the plane of the circle of the signs twice in a year. That is, when the motion of the cusp or its center is at both positions of the nodes. And the beginning of this declination, which is of the apogee of the epicycle and its perigee from the eccentric, happens at both positions of the nodes. Indeed its completion is at its apogee and perigee.

The directions of the declination are also distinguished thus: the perigees of the epicycle decline from the eccentrics in the direction of the eccentric circles' declination from the circle of the signs; the apogees are the opposite. And the declination of the perigees of the epicycles will be toward the north, when it is on the northern halves of 221r
the eccentrics. And it will be toward the south, when it is on southern halves.

Venus and Mercury have two motions in the declination of the epicycle, one of which is like the motion of the three superior planets, which is from the eccentric on the part of the epicycle's apogee and perigee.[18] And the other [motion] is from the direction of the declination of both of the epicycle's mean longitudes. And this declination is called involution.[19] And the beginning of the declination of the epicycle's apogee and perigee away from the eccentric will be on the apogee and perigee in the eccentrics; and its completion is at both nodes, and then the plane of the epicycle is set under on

17 Heingarter supplements John of Seville's *cuspis epicicli* with the more familiar *centrum epicicli*, and in the next instance leaves off *cuspis epicicli* entirely.

18 The 'declination of the epicycle', so defined, refers to the same angle, called by the same name, in the canons on the inferior planets in the second chapter of the treatise below.

19 The term, originally found in John of Seville's translation is *involutio*, which comes from the Arabic term *iltiwā'* used in Alfraganus' original text. It refers to the same motion in the planetary model that contemporary scholars call 'slant'. The canons below and the headings to the tables do not render *iltiwā'* into Latin as *involutio*, but rather as *reflexio*, which is another possible translation for the same Arabic word.

est ex utrisque nodis, et eius perfectio est apud longitudinem longiorem et propiorem ecentrici, et tunc erit perfectio declinationis ecentrici a circulo signorum. Patet itaque quod unaqueque illarum declinationum epicicli cum fuerit in perfectione sua, non erit ex declinatione alia aliquid omnino, quia inceptio uniuscuiusque earum est ex perfectione alterius.

Partes autem declinationis sunt ita: cum fuerit centrum epicicli in medietate que incipit a longitudine longiori ecentrici, erit declinatio longitudinis longioris epicicli Veneris ab ecentrico versus septentrionem. Eritque longitudo propior e contrario longitudini epicicli. Et cum fuerit cuspis seu centrum in alia medietate ecentrici et e contrario huic quod iam narravimus; patet igitur ex dictis in aliis planetis tribus quod plurima pars huius declinationis fit apud nodos, et tunc fuerit utreque longitudines epiciclorum horum duorum planetarum medie adherentes superficiei ecentrici et circuli signorum ab invicem suppositorum.

De parte vero declinationis involutionis, cum fuit centrum epicicli in longitudine ecentrici longiori, erit declinatio longitudinis medie epicicli orientalis ab ecentrico in Venere versus septentrionem et in Mercurio versus meridiem. Longitudo vero media occidentalis e contrario parti orientali, et cum fuerit centrum in medietate ecentrici propiori, erit e contrario huic quod diximus de medietate longiori, et patet quod plurima pars huius inclinationis fit apud longitudinem longiorem et propiorem ecentrici et tunc fiunt utreque longitudines epicicli longior, scilicet et propior coherentes superficiei ecentrici. Loca autem intersectionum in Venere et Mercurio ab ecliptica in suis deferentibus.

Tamen communitas astronomorum sic verificant genzahar planetarum. Nam ab auge Saturni subtrahunt 140 gradus, ab auge Iovis 70, a Martis 90, ab auge Veneris nihil, nam genzahar suum semper est | in loco sue augis. Ab auge Mercurii minuunt 180 gradus, 221v
quibus factis remanebunt vera loca genzahar quinque planetarum. Quantitates autem harum latitudinum in septentrionem et meridiem a circulo signorum sunt ita. Latitudo lune est quinque graduum ex gradibus circuli signorum. In planetis vero quinque narremus breviter illud quod colligitur de latitudine universorum motuum ipsorum in septentrionem et meridiem. Saturni quidem latitudo est trium graduum, Iovis duorum, Martis vero in sepentrionem quatuor graduum et 3e unius et quarte unius et in meridie septem; Veneris sex graduum et 3e unius, ut patet in *Almagesti*, et extra *Almagesti* novem, et Mercurii quatuor graduum.

Et sic peregimus universorum motuum latitudines, in hoc finiendo capitulum primum Deo auxiliante.

159/168 Quantitates ... auxiliante] Cf. ALFRAGANUS, *Differentie*, c. 18.27-28; ed. CARMODY, p. 33b. **165** ut ... Almagesti[1]] CLAUDIUS PTOLEMAEUS, *Almagestum*, XIII, c. 3.

the plane of the eccentric. The beginning of the declination of both mean longitudes of the epicycle from the eccentric is on both nodes, and its completion is at the apogee and perigee of the eccentric; and then there will be a completion of the declination of the eccentric from the circle of the signs. It is clear, therefore, that when each of these declinations of the epicycle is at its completion, there will be no effect in any way from the other declination, because the beginning of one of them happens at the completion of the other.

The directions of the declination are thus: when the center of the epicycle is on the half which begins from the apogee of the eccentric, the declination of the apogee of the epicycle of Venus will be from the eccentric to the north. And the perigee will be contrary to this for the longitude of the epicycle. And when the cusp or center is on the other half of the eccentric, it will be contrary to that which we have just described; it is, therefore, clear from what has been said about the other three planets that the greatest part of this declination happens at the nodes, and then both mean longitudes of the epicycles of these two planets are attached to the plane of the eccentric and the circle of the signs, each set under the other.

As for the direction of the declination of involution, when the center of the epicycle is at the apogee of the eccentric, the declination of the eastern mean longitude of the epicycle from the eccentric will be northerly for Venus, and southerly for Mercury; but the western mean longitude will be contrary for the eastern part. And when the center is on the perigee's half of the eccentric, it will be contrary to what we said about the apogee's half; and it is clear that the greatest part of this inclination occurs at the apogee and perigee of the eccentric; and then it happens that both the apogee and perigee of the epicycle cohere to the plane of the eccentric. The places of the intersections for Venus and Mercury, however, are away from the ecliptic in their deferents.

But the community of astronomers verify the genzahar of the planets in this way. For 140 degrees are subtracted from the aux of Saturn, 70 degrees from the aux of Jupiter, 90 degrees from Mars, nothing from the aux of Venus, for its node is always in the place of 221v
its aux. 180 degrees are subtracted from the aux of Mercury; and having done this, the true places of the genzahar of the five planets will remain. The quantities of these latitudes in the north and south from the circle of the signs are thus. The latitude of the Moon is five degrees from the degrees of the circle of the signs. But for the five planets, let us describe briefly that which is gathered from the latitude of all of those motions in the north and the south. Indeed the latitude of Saturn is three degrees; Jupiter is two; Mars is four degrees and a third and a fourth in the north and seven degrees in the south; Venus is six degrees and a third, as is made clear in the *Almagest*, and, outside the *Almagest*, nine; Mercury is four degrees.[20]

And thus we have completed the latitudes of all the motions, thereby finishing the first chapter with God's help.

20 Heingarter is relying on al-Farghānī in stating that Venus has a latitude of 6 degrees and a third, but in al-Farghānī it is actually 6 degrees and a fourth. See Alfraganus, *Differentie*, c. 18.27; ed. Carmody, p. 33b.

Capitulum secundum de practica inveniendi latitudines planetarum

<E>st autem differentia inter declinationem et latitudinem planetarum. Declinatio quidem proprie dicta sumitur a circulo equinoxiali, ita quod declinatio planete vel stelle est distantia seu quantitas arcus circuli inter stellam et equinoxialem. Circuli dico exeuntis per axem circuli equinoxialis et per stellam et gradum circuli equinoxialis.

Latitudo vero planete superius capitulo primo declarata est. Ad habendum vero latitudines lune sic procedendum est. Locum genzahar equatum de loco lune equato minue vel medium cursum genzahar loco lune equato adiunge et habebis argumentum latitudinis lune, cum quo lineas numeri equationis lune ingredere et latitudinem lune in directo illius positam summe que erit septentrionalis ascendens, si fuerit eius argumentum ab uno gradu in 90. Si autem a 90 usque 180, tunc erit septentrionalis descendens. Si autem latitudo fuerit a 180 usque ad 270, erit meridionalis descendens. Et a 270 usque ad 360, erit meridionalis ascendens.

Declinationem autem solis et stellarum reperies eodem modo, intrando cum gradu solis in tabulam declinationis, et illud quod in directo fuerit, erit vera solis declinatio, que erit ascendens in septentrionem vel descendens in meridiem, eo modo quo de luna diximus. In sex vero planetis preter solem declinatio alio modo invenitur. Nam si latitudo fuerit septentrionalis oportet eam addere ad declinationem repertam per gradum planete vel eam minuere si fuerit meridionalis, et residuum erit planete declinatio a circulo equinoxiali.

Ad inveniendum autem latitudines quinque planetarum posite sunt tabule trium manerierum: prima est in qua intratur cum centro medio in linea descendente et cum argumento medio in linea tamen transversali et capitur id quod est in con|cursu linea- 222r
rum et illud erit latitudo planete et ille modus facilis est. Est et alius modus per tabulas bipartiales et quadripartiales. Et dicuntur bipartiales, quia ostendunt latitudinem provenientem ex parte epicicli; ideo intratur in eas cum argumento equato. Sed quadripartiales ostendunt latitudinem provenientem ex parte ecentrici; ideo intratur in eas cum centro equato.

Ad habendum autem argumenta latitudinum quinque planetarum: oportet scire eorum intersectiones, id est genzahar cum ecliptica que fuerunt in principio ere Arabum, sicut patet in tabula sequenti et in principio ere illustrissimi Iohannis Borbonii atque Arvernie ducis, sicut in tabula subsequente per quam nostro tempore, hoc est

---

174/180 Locum ... ascendens] cf. *Canones Azarchelis*, Cb163-64; ed. PEDERSEN, p. 444.

175/176 argumentum latitudinis] *coni.*; *scr.* arcum

The Second Chapter On the Practice of Finding the Latitudes of the Planets

There is, moreover, a difference between the declination and latitude of planets. Indeed the declination, properly called, is taken from the equinoctial circle such that the declination of the planet or star is the distance or the quantity of the arc of the circle between the star and the equinoctial [circle]. I am speaking of the circle going through the axis of the equinoctial circle, the star and the degree of the equinoctial circle.

But the latitude of the planet is explained above in the first chapter. Indeed, one should proceed thus in order to obtain the latitudes of the Moon. Subtract the equated place of the genzahar from the equated place of the Moon, or add the mean motion of the genzahar to the equated place of the Moon, and you will have the arc of the Moon; and with this, enter the lines of number of the equation of the Moon, and take the latitude of the Moon that is placed opposite of it; this will be northern ascending if its argument is from one degree to 90. If, however, [it is] from 90 to 180, then it will be northern descending. If, moreover, it is from 180 to 270, it will be southern descending. And [if it is] from 270 to 360, it will be southern ascending.

And you will find the declination of the Sun and the stars in the same way by entering with the degree of the Sun into the table of declination; and that which is opposite it will be the true declination of the Sun, which is ascending in the north or descending in the south, in that way in which we said about the Moon. But in the six planets apart from the Sun, the declination is found in another way. For, if the latitude is northern, it is necessary to add it to the declination found for the degree of the planet, or – if it is southern – [it is necessary] to subtract it; and what remains will be the planet's declination from the equinoctial circle.

To find the latitudes of the five planets, moreover, tables of three kinds are arranged: the first [kind] is that in which [the table] is entered with the mean center in the descending line, and with the mean argument in the transversal line; and that which is taken is that which is in the convergence of the [two] lines; and that will be the latitude 222r of the planet; and that method is easy.[21] And there is another way by means of two-part number and four-part number tables.[22] They are called two-part as they show the latitude resulting from the epicycle's part; therefore, one enters into them with the equated argument. But the four-part show the latitude resulting from the part of the eccentric; therefore, one enters into them with the equated center.

Moreover, to get the arguments of the latitudes of the five planets: it is necessary to know their intersections, that is, the genzahar with the ecliptic which were at the beginning of the era of the Arabs, as is made clear in the following table and at the beginning of the era of the Illustrious John, duke of Bourbon and Auvergne, as [is is shown] in the following table by which we will operate in our own time, that is, in the year of our Lord

21 This canon instructs the reader in the use of a double argument table, likely that of the Oxford Tables, which are contained in a manuscript owned by Heingarter, namely BnF Lat. 7295A.

22 Heingarter is describing the method for finding latitudes using tables drawn from the Indian tradition. The canons detailing how to use these tables immediately follow this introduction.

anno Domini 1425 operabimur, et tempore futuro addemus supra genzahar cuiuslibet quinque planetarum, id quod ponitur in tabula continuationis motuum eorum.

De latitudine trium superiorum

Si autem per tabulam bipartialem trium superiorum planetarum invenire desideras, cum cuiusvis eorum argumento equato lineas numeri tabularum bipartialis intra, et quod in directo eius fuerit in tabula planete illius scribe ad partem et erit radix latitudinis. Depost genzahar planete eiusdem de loco eius examinato deme vel medium cursum genzahar eiusdem planete loco eius vero adde et habebis argumentum latitudinis, cum quo lineas numeri tabule quadripartialis ingredere et quod ibi inveneris in tabula planete summe et illud divide per radicem latitudinis superius inventam. Et quod provenerit ex divisione est latitudo planete. Quod si non posses dividere, frange in minutias subtiliores et postea divide sicut fit in fractionibus horarum.

Et si argumentum latitudinis cum quo intrasti fuerit ab uno gradu in 90, erit ipsa latitudo septentrionalis ascendens, eo modo quo superius de luna declaravimus. Alii tamen cum secundam latitudinem non possunt dividere per primam, faciunt e contrario et tunc latitudinem computant a limbo zodiaci versus eclipticam. |

Veneris autem et Mercurii latitudinis sic habetur scientia. Intramus cum argumento 222v eorum equato tabulas bipartialis numeri et facimus ut dictum est in aliis. Argumentum autem latitudinis sic colligimus: argumentum eorum equatum addimus medio cursui solis et quod colligitur ex eo, minuimus genzahar eorum vel addimus medium cursum genzahar eorum eodem numero collecto, et sic invenimus argumentum latitudinis eorum, cum quo intramus lineas quadripartialis numeri et facimus per omnia sicut fecimus in tribus superioribus planetis, et sic invenimus horum duorum planetarum latitudines vel ab ecliptica vel a limbo zodiaci computando; et hec de secunda manerie inventionis latitudinum planetarum sufficiant.

Est et alia maneries tertia per tabulam compositam a de Lineriis, quam tabulam faciliorem et clariorem fecimus in qua veraciter latitudines quinque planetarum per hunc modum subscriptum reperies.

202/221 Si ... latitudines] Cf. *Canones Azarchelis*, Cb165-66 ; ed. Pedersen, p. 444, 446. **224/226** Est ... reperies] Cf. *The Tables of 1322 by John of Lignères*, Table 8 ; ed. Chabás and Saby, p. 76-79.

1425; and for a future time we will add to the genzahar of any of the five planets that which is placed in the table of the continuation of their motions.[23]

On the Latitude of the Three Superior Planets

Moreover, if you wish to find [the latitude] through the two-part number table of the three upper planets, enter the lines of number of the two-part tables with the equated argument of whichever you wish, and write apart what is opposite of it in the table of that planet; and this will be the root of the latitude. Next, subtract the genzahar of that same planet from its corrected place, or add the mean motion of the same planet's genzahar to its corrected place, and you will have the argument of latitude; and with this, enter the lines of number of the four-part table, and take what you have found in table of the planet; and divide that by the root of latitude that was found above. And what results from the division is the latitude of the planet. In case you cannot divide, break it into smaller parts and then divide as happens with fractions of hours.

And if the argument of latitude, with which you entered, is from one degree to 90, this latitude will be northern ascending, in that way in which we have declared above about the Moon. Nevertheless, others, when they could not divide the second latitude by the first, do this conversely, and then compute the latitude from the limb of the zodiac toward the ecliptic.[24]

And so is had, moreover, knowledge of the latitude of Venus and Mercury. We enter 222v
the two-part number tables with their equated argument, and we operate as has been said about the others. Moreover, we compute the argument of latitude thus: we add their equated argument to the mean motion of the Sun, and, from what is added, we subtract their genzahar or we add the mean motion of their genzahar to the same summed number; and thus we find the argument of their latitude; and with this, we enter the lines of the four-part number tables and we operate for everything as we did in the case of the three upper planets; and thus we shall find the latitudes of these two planets by computing from the ecliptic or from the limb of the zodiac; and this suffices concerning the second way for finding the latitude of the planets.

And there is another, a third, way[25] in which you will truly find the latitudes of the five planets according to the following method using a table composed by [JOHN] DE LIGNÈRES, which we have made easier and clearer.

---

23 On folio 222[r] there is a table that give the values of the nodes or *genzahar* of the planets set to the *era Arabum* and which is taken from the Toledan Tables. Heingarter then includes a table of the same nodes set to the *era illustrissimi Iohannis Borbonii atque Arvernie,* although this table remains incomplete in the manuscript witness. There is also a partially complete *tabula continuationis motuum eorum* that is divided between years, months, and days.

24 Heingarter assumes an arithmetic procedure for finding the latitude, which would be a conversion of the method of division described in the Toledan canons. He borrows this line of thinking from the original text he is commenting upon where the *Theorica planetarum Gerardi* reads: 'Compositor autem tabularum ad ostensionem sui magisterii noluit ponere verso numeros praedictarum latitudinum in quibus unus subtrahitur ab altero semper, sed posuit numeros equipollentes in quibus si unus dividitur per alterum idem proveniret'. It is noteworthy that he describes this arithmetic procedure as being performed from the 'limb of the zodiac' (*a limbo zodiaci*), thereby employing a vocabulary that is most often found in instrument texts. For further discussion of this passage see Kren, 'Planetary Latitudes, the *Theorica Gerardi*, and Regiomontanus', 200-02.

25 In describing his 'third way', Heingarter has recourse to the canons of John of Lignères' *Priores astrologi,* although he presents them as a paraphrase.

De latitudine Saturni, Iovis et Martis

Primo centrum atque argumentum equatum cuiuslibet eorum scribatur per se. Secundo centro equato Saturni adde gradus 50, a centro vero Iovis gradus 20 minuuntur, de centro Martis nihil mutatur. Tertio cum centro cuiusvis eorum accipe minuta proportionalia. Quarto cum argumento vero cuiusvis planete accipe latitudinem taliter, cum enim centrum verum cum suo addito vel minuto vel invariato fuerit repertum in parte superiori tabule, tunc accipe latitudinem septentrionalem. Sin autem meridionalem. Quinto de illa latitudine accipe partem proportionalem secundum proportionem ad 60 et illa pars proportionalis erit planete latitudo.

De latitudine Veneris

Primo centrum cum argumento equato quodlibet per se scribatur. Secundo cum argumento equato cape quod est in directo tabule declinationis et reflexionis et extra scribatur. Tertio centro adde 60 gradus cum quibus summe minuta proportionalia. Quarto accipe partem proportionalem declinationis servate secundum proportionem ad 60, que pars proportionalis erit latitudo epicicli; serva eam, et si centrum cum suo addito et argumentum verum fuerit in eadem parte tabule sive superiori sive inferiori, tunc latitudo erit meridionalis. Si autem fuerit in diversis tabulis, tunc latitudo erit septentrionalis; scribe ergo nomen sue partis. Quinto cum simplici centro Veneris prius servato, accipe minuta proportionalia et scribantur ad partem et similiter cape ibidem sextam partem minutorum proportionalium que est tertia latitudo que provenit ex deviatione deferentis ab ecliptica et est semper septentrionalis, et cum minutis proportionalibus servatis ad partem, accipe partem proportionalem ex reflexione servata secundum proportionem ad 60, et hec erit reflexio secundo examinata; scribe eam.

Tunc si centrum simplex Veneris fuerit in medietate superiori et argumentum verum planete minus tribus signis phisicis, reflexio erit septentrionalis. Si argumentum fuerit plus, tunc erit meridionalis. Sed si centrum fuerit in medietate inferiori et argumentum minus tribus signis, tunc reflexio erit meridionalis. Si argumentum fuerit plus tribus signis, erit septentrionalis et est latitudo secundo examinata que provenit ex declinatione epicicli; serva eam et scribe nomen sue partis.

Ultimo illas tres latitudines collige similiter et subtrahe minus a maiori, si non fuerit eiusdem partis et a maiori denominabitur latitudo. Si vero fuerit eiusdem partis, totum in simul collige et erit latitudo Veneris.

227/234 Primo ... latitudo] Cf. Iohannes de Lineriis, *Priores astrologi*, c. 21; ed. Saby, p. 206.
235/256 Primo ... Veneris] Cf. Iohannes de Lineriis, *Priores astrologi*, c. 22, ed. Saby, p. 207-09.

On the Latitude of Saturn, Jupiter and Mars

First, the center and the equated argument of any of them should be written by themselves. Second, add 50 degrees to the equated center of Saturn; 20 degrees should be subtracted from the true center of Jupiter and from the center of Mars nothing should be changed. Third, with the center of any of these [planets] take the minutes of proportion. Fourth, with the true argument of any planet take the latitude such that, when indeed the true center, with its added, subtracted, or unchanged [value], is found on the upper part of the table, then you take a northern latitude. If not, then [you take a] southern [latitude]. Fifth, from that latitude take the proportional part according to the proportion to 60 and this proportional part will be the latitude of the planet.

On the Latitude of Venus

First, let any center be written down with the equated center by itself. Second, with the equated argument take what is opposite of it in the table of declination[26] and reflection[27], and let it be written aside. Third, add 60 degrees to the center and with those take the minutes of proportion. Fourth, take the proportional part of the saved declination according to the proportion to 60, and that proportional part will be the latitude of the epicycle; save it, and if the center, with its added [value], and the true argument are in the same part of the table, whether in the upper or the lower, then the latitude will be southern. If, however, they are in different tables, then the latitude will be northern; therefore, write down the name of its direction. Fifth, with the simple center of Venus previously saved, take the minutes of proportion and let them be written down separately and similarly take again the sixth part of the minutes of proportion, and that is the third latitude that results from the deviation[28] of the deferent from the ecliptic, and is always northern; and with the minutes of proportion that have been saved apart, take the proportional part from the reflection that has been saved according to the proportion to 60; and this will be the reflection examined for a second time; write it down.

Then, if the simple center of Venus is on the upper half and the true argument of the planet is less than 3 physical signs, the reflection will be northern. If the argument is more [than 3 physical signs], then it will be southern. But, if the center is on the lower half [of the table] and the argument [is] less than three signs, then the reflection will be southern. If the argument is more than three signs, it will be northern; and this is the second examined latitude, that results from the declination of the epicycle; save it, and write down the name of its direction.

Finally, combine together these three latitudes in like manner, and subtract the lesser from the greater if [the latitude] is not in the same direction. And the latitude will be denominated from the greater. But if all are in the same direction, add them all together at once, and this will be the latitude of Venus.

26 Heingarter's use of the term 'declination' is coherent with his terminology in the first chapter; see n. 18.

27 Again, his terminology matches the first chapter; see n. 19.

28 In the first chapter, the 'deviation' is referred to as the 'declination … of the planets' eccentric deferents'; see n. 8.

De latitudine Mercurii

Primo centrum verum et argumentum verum quodlibet per se scribatur. Secundo cum argumento vero accipe id quod in directo tabule declinationis et reflexionis reperies et quodlibet per se scribatur; cape similiter ibidem decimam partem reflexionis quam minue ex reflexione si centrum verum fuerit repertum in superiori parte tabule. Sin autem illam decimam partem reflexionis super reflexionem adde, eam serva prima deleta. Tertio cum centro vero Mercurii quatuor signa phisica et 30 gradus adde cum quibus summe minuta proportionalia et de declinatione prius servata accipe partem proportionalem secundum proportionem ad 60 et proveniet latitudo primo examinata proveniens ex declinatione epicicli; serva eam. Et si centrum cum suo addito et argumentum verum fuerint in eadem parte tabule, tunc latitudo erit meridionalis. Si in diversis, erit septentrionalis; scribe nomen partis sue. Quarto centro vero Mercurii tria signa phisica adde cum quibus accipe minuta proportionalia, similiter accipe quarta partem cum semis quarte et erit latitudo Mercurii tertia que semper est meridionalis, quam scribe ad partem, et cum minutis proportionalibus et reflexione prius servata, accipe partem proportionalem secundum proportionem ad 60, et erit reflexio examinata quam serva.

Quod si centrum verum Mercurii cum suo addito cum quo minuta proportionalia ultimo sumpsisti fuerit in medietate superiori et argumentum verum planete minus tribus signis, reflexio erit septentrionalis, et si plus, meridionalis. Sed si centrum cum suo addito fuerit in medietate inferiori et argumentum minus tribus signis, tunc reflexio erit meridionalis; si autem argumentum fuerit plus tribus signis, erit septentrionalis, et est latitudo secundo examinata que provenit ex declinatione epicicli; serva eam et scribe nomen sue partis. Demum illas latitudines tres ad invicem collige si fuerint unius partis et habebis latitudinem Mercurii. Si autem fuerint diversarum partium, tunc subtrahe minorem de maiori et remanebit vera Mercurii latitudo computata a via solis.

Hec autem, mi Michael, que de planetarum et stellarum latitudinibus breviter quo ad eorum theoricam et practicam composuimus ex dictis Sapientum. Vale.

257/280 Primo ... Mercurii] Cf. Iohannes de Lineriis, *Priores astrologi*, c. 23; ed. Saby, p. 210-11.

On the Latitude of Mercury

First, let any true center and true argument be written down by themselves. Second, with the true argument take what you find opposite to it in the table of declination and reflection and let it be written by itself; in like manner, take again the tenth part of the reflection which you will subtract from the reflection if the true center is found on the upper part of the table. If not, however, add that tenth part of the reflection to the reflection; keep it, having deleted the first value. Third, with the true center of Mercury add 4 physical signs and 30 degrees and with these take the minutes of proportion. Take the proportional part from the declination already saved according to the proportion to 60; and what results will be the latitude examined for the first time, resulting from the declination of the epicycle; keep this. And if the center with its added [value] and the true argument are in the same part of the table, then the latitude will be southern. If they are in different [parts of the table], then [the latitude] will be northern; write down the name of its direction. Fourth, add three physical signs to the true center of Mercury with which you get the minutes of proportion; in like manner, take the fourth part with half a fourth. And this will be the third latitude of Mercury which is always southern; write this apart, and with the minutes of proportion and the reflection previously saved, take the proportional part according to a proportion to 60 and this will be the examined reflection, which you will save.

Now if the true center of Mercury, with its added [value], together with its minutes of proportion that you have finally taken, is on the upper half [of the table], and the true argument of the planet is less than three signs, then the reflection will be northern. But if [the true argument] is more than three signs, then [the reflection] will be southern. But, if the center with its added [value] is on the lower half and the argument is less than three signs, then the reflexion will be southern; if the argument is more than three signs, then it will be northern; this is the latitude examined for a second time which results from the declination of the epicycle; save it, and write the name of its direction. Finally, add these three latitudes together, if they were in the same direction, and you will have the latitude of Mercury. If, however, they were in different directions, then subtract the lesser from the greater, and what remains will be the true latitude of Mercury computed from the path of the Sun.

Now this, my Michael, is what we have briefly composed from the sayings of the Wise on the latitudes of planets and stars with respect to theory and practice. Farewell.

Manuscript sources

Erfurt, Universitätsbibliothek, Dep. Erf. CA 2° 395
Paris, Bibliothèque nationale de France, lat. 7197
Paris, Bibliothèque nationale de France, lat. 7273
Paris, Bibliothèque nationale de France, lat. 7295A
Paris, Bibliothèque nationale de France, lat. 7305
Paris, Bibliothèque nationale de France, lat. 7314
Paris, Bibliothèque nationale de France, lat. 7333
Paris, Bibliothèque nationale de France, lat. 7432
Paris, Bibliothèque nationale de France, lat. 7446
Paris, Bibliothèque nationale de France, lat. 7447
Paris, Bibliothèque nationale de France, lat. 7450
Paris, Bibliothèque nationale de France, lat. 11232
Paris, Bibliothèque nationale de France, lat. 16401
Vatican, Biblioteca Apostolica Vaticana, Reg. lat. 1241
Zürich, Zentralbibliothek, B 244
Zürich, Zentralbibliothek, C 131

Bibliography

Theorica planetarum Gerardi, ed. by Francis J. Carmody (Berkeley: University of California, 1942).

Albategnius (Al-Battānī), *Opus astronomicum*, ed. by Carlo A. Nallino, 3 vols (Milan: Hoeplium, 1899–1907).

Albumasar (Abū Maʿshar), *Liber introductorius*, in *The Great Introduction to Astrology by Abū Ma ʿšar*, ed. by Yamamoto Keiji, Charles Burnett, and David Pingree, (Leiden: Brill, 2019).

Alcabitius, (Al-Qabiṣi), *Introductorius*, in *Al-Qabīṣī (Alcabitius): The Introduction to Astrology: Editions of the Arabic and Latin Texts and an English Translation*, ed. by Charles Burnett, Keili Yamamoto, and Michio Yano (London: The Warburg Institute, 2004).

Alfraganus (Al-Farghānī), *Differentie in quibusdam collectis scientie astrorum*, ed. by Francis J. Carmody (Berkeley: 1943).

Avelar de Carvalho, Helena, *An Astrologer at Work in Late Medieval France: The Notebooks of S. Belle* (Leiden: Brill, 2021).

Beaune, Colette, 'La notion d'Europe dans les livres d'astrologie du xve siècle', in *La Conscience européenne au xve et au xvie siècle. Actes du Colloque international organisé à l'École Normale Supérieure de Jeunes Filles, (30 septembre–3 octobre 1980)* (Paris: École normale supérieure de jeunes filles), pp. 1–7.

Bernhard Milt, 'Beitrag zur Kenntnis der mittelalterlichen Heilkunde am Bodensee und Oberrhein', *Vierteljahrsschrift der Naturforschenden Gesellschaft in Zürich* 85 (1940), 263–321.

Bianchini, Giovanni, *Tabulae astronomicae et canones in eas*, ed. by Augustin Olomoucký (Venice: Simone Bevilacqua, 1495).

Boethius Anicius Manlius Torquatus Severinus, *De Arithmetica*, ed. Henri Oosthout and Jean Schilling, (Turnhout: Brepols, 1999).

———, *De Arithmetica*, in *Boethian Number Theory: A Translation of the 'De institutione arithmetica'*, trans. with introduction and notes by Michael Masi (Amsterdam: Rodopi, 1983).

Boudet, Jean-Patrice, *Le* Recueil des plus célèbres astrologues *de Simon de Phares*, 2 vols (Paris: H. Champion, 1997–1999).

———, *Astrologie et politique entre Moyen Âge et Renaissance* (Florence: SISMEL –Edizioni del Galluzzo, 2020).

Burnett, Charles, 'From *Astronomica* to *Exotica*: Jacob Golius' Edition of al-Farghānī's *On the Science of the Stars* in Comparison with the Earlier Versions', in *Scholarship Between Europe and the Levant: Essays in Honour of Alastair Hamilton*, ed. by Jan Loop and Jill Kraye (Leiden: Brill, 2020), pp. 60–86.

Cao, Gian Mario, 'The Prehistory of Modern Scepticism: Sextus Empiricus in Fifteenth-Century Italy', *Journal of the Warburg and Courtauld Institutes*, 64 (2001), 229–80.

———, '*Inter alias philosophorum gentium sectas, et humani, et mites*: Gianfrancesco Pico and the Sceptics', in *Renaissance Sceptisicms*, ed. by Gianni Paganini and José R. Maia Neto (Dordrecht: Springer, 2009), pp. 127–47.

Cardano, Girolamo, *De astrorum iudiciis, tomus quintus quo continentur astronomica, astrologica, onirocritrica*, ed. by Jean-Antoine Huguetan and Marc-Antoine Ravaud (Lyon: Huguetan and Ravaud, 1663).

———, *Libelli quinque* (Nuremberg: Johann Petreius, 1547).

Carello, Giovanni Battista, *Le Efemeridi volgari per anni 17, al meridiano dell'inclita città di Vinegia* [1555-71] (Venice: Nicolo Bevilacqua, 1555).

Chabás, José and Bernard R. Goldstein, *A Survey of European Astronomical Tables in the Late Middle Ages* (Leiden: Brill, 2012).

Coopland, George W, *Nicole Oresme and the Astrologers: A Study of His* Livre de Divinacions (Cambridge, Mass.: Harvard University Press, 1952).

Fortuna, Stefania, 'Wilhelm Kopp Possessore dei *Par. Gr.* 2254 e 2555? Ricerche sulla traduzione del *De victus ratione in morbis acutis* di Ippocrate', *Medicina nei Secoli. Arte e scienza*, 13 (2001), 47–57.

Galileo Galilei, *Astrologica nonnulla* in *Le opere di Galileo Galilei*, ed. by Antonio Favaro, 20 vols (Florence: G. Barbèra, 1890), XIX, pp. 205–20.

Gaurico, Luca, *Ephemerides recognitae et ad unguem castigatae. Schemata et praedictiones ad annum usque virginei partus 1552* [1534-1551] (Venice: Lucantonio Ginti, 1533).

Grafton, Anthony and Nancy Siraisi, 'Between the Election and My Hopes: Girolamo Cardano and Medical Astrology', in *Secrets of Nature: Astrology and Alchemy in Early Modern Europe*, ed. by Anthony Grafton, Jed Z. Buchwald, and William Newman (Cambridge: MIT Press, 2001), pp. 69–131.

Hasse, Dag N, *Success and Suppression: Arabic Sciences and Philosophy in the Renaissance* (Cambridge, Mass.: Harvard University Press, 2016).

Husson, Matthieu, 'Mathematical Astronomy and the Production of Multiple-Texts Manuscripts in Late Medieval Europe: A Comparison of BnF, lat. 7197 and BnF, lat. 7432', in *The Emergence of Multiple-Text Manuscripts*, ed. by Alessandro Bausi, Michael Friedrich, and Marilena Maniaci, (Berlin: De Gruyter, 2019), pp. 247–74.

———, 'The Art of Astrological Computations: Conrad Heingarter and the Manuscript Paris, BnF latin 7295A', in *Exploring Written Artefacts: Objects, Methods, and Concepts*, ed. Jörg B. Quenzer, (Berlin: De Gruyter, 2021), pp. 513–32.

Imsser, Philipp, *Ephemeridum opus Ioannis Stoefleri Iustingensis mathematici praeclarissimi a capite anni Redemptoris Christi 1532 in alios 20 proxime subsequentes ad veterum imitationem accuratissimo calculo elaboratum* [1532-1551] (Tübingen: Ulrich Morhart, 1531).

Johannes de Regiomontanus, *Disputationem contra Cremonensis in planetarum theoricam deliramenta*, in *Spherae tractatus Ioannis de Sacro Busto*, etc. (Venice: Luce Antonii Iuntae Florentini Officina, 1531), pp. 175^b–180^a.

———, *Ephemerides* [1475-1506] (Nuremberg: Regiomontanus, 1474).

———, *Tabulae directionum profectionumque*, ed. by Johannes Angelus (Augsburg: Erhard Ratdolt, 1490).

John of Procida, *Liber philosophorum moralium antiquorum*', ed. Ezio Franceschini: 'Il *Liber philosophorum moralium antiquorum*', in *Atti del Reale Istituto Veneto di scienze, lettere ed arti* 91/2 (1932), 393-597.

John of Saxony, *Commentum super textu Alchabitii*, in *Introductorius cum commento*, ed. by Bartholomäus Alten (Venice: Per Ioannem et Gregorium de Gregoriis fratres, 1503).

Juste, David, *Les manuscrits astrologiques latins conservés à la Bibliothèque nationale de France à Paris* (Paris: CNRS Éditions, 2015).

———, MS Vatican, Biblioteca Apostolica Vaticana, Reg. lat. 1241' (update: 21.02.2022), Ptolemaeus Arabus et Latinus. Manuscripts, accessed 8 April 2023:

URL = http://ptolemaeus.badw.de/ms/464.

Kremer, Richard L., 'How Did the *Turketum* (or *Torquetum*) Get its Name?', in *Scientific Instruments between East and West*, ed. by Neil Brown, Silke Ackermann, and Feza Günergun (Leiden: Brill, 2019), pp. 80–107.

Kren, Claudia, 'A Medieval Objection to "Ptolemy"', in *The British Journal for the History of Science*, 4 (1969), 378–93.

———, 'Planetary Latitudes, the *Theorica Gerardi*, and Regiomontanus', *Isis*, 68 (1977), 194–205.

Lawrence-Mathers, Anne, *Medieval Meteorology: Forecasting the Weather from Aristotle to the Almanac* (Cambridge: Cambridge University Press, 2020).

Leowitz, Cyprian, *Ephemeridum novum atque insigne opus ab A.D. 1556 usque in 1606 accuratissime supputatum* (Augsburg: Philipp Ulhard, 1557).

Morin, Jean-Baptiste, *Astrologia Gallica principis et rationibus propriis stabilita atque in XXVI libros distributa* (The Hague: Adriaan Vlacq, 1661).

———, *Astrologia Gallica: Book Sixteen*, in *Astrologia Gallica: Book Sixteen, The Rays and Aspects of the Planets* (Tempe: American Federation of Astrologers, 2008).

———, *Astrologia Gallica: Book Twenty-two*, in *Astrologia Gallica: Book Twenty-two, Directions*, trans. by James Herschel Holden (Tempe: American Federation of Astrologers, 1994).

North, John D, 'The Alfonsine Tables in England', in *Prismata, Naturwissenschafisgeschichtliche Studien: Festschrift fir Willy Hartner*, ed. by Yasukatsu Maeyama and Walter G. Saltzer (Wiesbaden: Steiner, 1977) (Reprinted in John D. North, *The Universal Frame: Historical Essays in Astronomy, Natural Philosophy, and Scientific Method* [London: The Hambledon Press, 1989], 327–59).

Pedersen, Fritz S, *The Toledan Tables: A Review of the Manuscripts and the Textual Versions with an Edition*, 4 vols (Copenhagen: C.A. Reitzels Forlag, 2002).

Pedersen, Olaf, 'The Origins of the *Theorica Planetarum*', *Journal for the History of Astronomy* 12 (1981), 113–23.

Pico della Mirandola, Giovanni, *Disputationes adversus astrologiam divinatricem, in Opera omnia Ioannis Pici Mirandulae Concordiaeque comitis (Basel: Heinrich Petri, 1557).*

Poulle, Emmanuel, 'Bernard de Verdun et le turquet', *Isis*, 55 (1964), 200–08.

Portmann, Marie-Louise, 'Der Basler Humanisten-Arzt Wilhelm Copp (um 1460–1532)', *Gesnerus* 15 (1958), 106–19.

Préaud, Maxime, *Les Méthodes de travail d'un astrologue du XVe siècle, Conrad Heingarter* (unpublished thesis, Paris, École des chartes, 1969).

———,*Les astrologues à la fin du Moyen Âge* (Paris: éditions Jean-Claude Lattès, 1984).

Ptolomaeus, Claudius, *Tabulae manuales*, ed. Anne Tihon and Raymond Mercier, in *Ptolemaiou Procheiroi kanones: Les tables faciles de Ptolémée, Volume 1A: A1–A2* (Louvain-La-Neuve: Université catholique de Louvain, Institut orientaliste, 2011-).

———, ed. and transl. by Frank E. Robbins (Cambridge, MA: Harvard University Press, 1940).

Saby, Marie-Madeleine, *Les canons de Jean de Lignières sur les tables astronomiques de 1321: Édition critique, traduction et étude*, 3 vols (unpublished thesis: École nationale des Chartes, 1987).

Samaran, Charles, Émile van Moé, and Susanna Vitte, ed. by *Liber procuratorum nationis anglicanae (alemanniae)*, 3 vols (Paris: Didier, 1935).

Schmid, Alfred, *Conrad Türsts Iatro-mathematisches Gesundheitsbüchlein für den Berner Schultheissen Rudolf von Erlach* (Bern: P. Haupt, 1947).

Shank, Michael, 'Academic Consulting in Fifteenth-Century Vienna: The Case of Astrology', in *Texts and Contexts in Ancient and Medieval Science: Studies on the Occasion of John E. Murdoch's Seventieth Birthday*, ed. by Edith Sylla and Michael McVaugh (Leiden: Brill, 1997), pp. 245–70.

———, 'Regiomontanus and Astronomical Controversy in the Background of Copernicus', in Rivka Feldhay and F. Jamil Ragep, eds, *Before Copernicus: The Cultures and Contexts of Scientific Learning in the Fifteenth Century* (Montreal: McGill-Queen's University Press, 2017), 79–109.

———, 'Regiomontanus *versus* George of Trebizond on Planetary Order, Distances, and Orbs (*Almagest* 9.1)', in David Juste, Benno van Dalen, Dag Nikolaus Hasse and Charles Burnett, eds, *Ptolemy's Science of the Stars in the Middle Ages* (Turnhout: Brepols, 2020), pp. 305-386.

Stadius, Johannes, *Ephemerides novae et exactae Ioannis Stadii Leonnouthesi ab anno 1554 ad annum 1570* (Cologne: Birckmann, 1556)

Stefano Rapisarda, 'Doubts and Criticism on Astrology in Western Traditions from Antique to Early Modern Literature', in *Prognostication in the Medieval World*, ed. by Matthias Heiduk, Klaus Herbers, and Hans-Christian Lehner (Berlin: De Gruyter, 2021), pp. 752–59.

Stöffler, Johannes and Jacob Pflaum, Almanach nova plurimis annis venturis inservientia [1499–1531] (Ulm: Joannis Reger, 1499).

Swerdlow, Noel, 'Ptolemy's Theories of the Latitude of the Planets in the Almagest, Handy Tables, and Planetary Hypotheses', in *Wrong for the Right Reasons*, ed. by Jed Z. Buchwald and Allan Franklin (Archimedes. Berlin/Heidelberg: Springer-Verlag, 2005), 41–71.

Thorndike, Lynn, *A History of Magic and Experimental Sciences*, 8 vols (New York: Columbia University Press, 1923–1958).

———, 'Conrad Heingarter in Zurich Manuscripts, Especially his Medical Advice to the Duchess of Bourbon', *Bulletin of the Institute of the History of Medicine*, 4 (1936), 81–87.

Van Brummelen, Glen, *Mathematical Tables in Ptolemy's Almagest*, PhD Thesis (Simon Fraser University, 1993).

Vergilius Maro, Publius, *Aeneis*, rev. ed., trans. by Henry Rushton Fairclough and George P. Goold (Cambridge, Mass.:Harvard University Press, 1999).

Weill-Parot, Nicolas, *Les 'Images astrologiques' au Moyen Âge et à la Renaissance: spéculations intellectuelles et pratiques Magiques (XIIe-XVe Siècle)* (Paris: H. Champion, 2002).

Wickersheimer, Ernest, *Dictionnaire biographique des medicins en france au Moyen Âge*, new ed. (Genève: Librairie Droz, 1979).
Zoboli, Alfonso, *De directionibus* (Venice: apud Franciscum Grossum, 1620).

SAMUEL GESSNER and MARIE-MADELEINE SABY

The *Planicelium* of John Vimond

*Edition, Translation, and Interpretation of a Text Describing an Astronomical Instrument from the Early Fourteenth Century**

Important works from the major figures of Alfonsine astronomy are included in the central section of the Erfurt manuscript Universitätsbibliothek CA 2° 377. These writings from the first half of the fourteenth century have long attracted the attention of historians. The first of those texts and tables is a short work composed by John Vimond (fl. 1320) on a previously unattested astronomical instrument called *Planicelium* in the text. Vimond was a significant actor during the beginnings of the new Parisian astronomical tradition. This treatise, however, has not yet been closely studied. Indeed, the work poses significant challenges in deciphering the script and comprehending the Latin text. Nevertheless, we here present a first edition and English translation of Vimond's *Planicelium* accompanied by a brief look at the author's life and works and some explanatory comments and diagrams on this singular instrument.

The *Planicelium*'s only known copy is written on four columns and a few lines of the manuscript Erfurt, UB, CA 2° 377 (f. 21^{r-v} and top of f. 22^{r}). The treatise is attributed to John Vimond in the explicit of the copy (*Explicit tractatus Johannis Vimundi…*). The author gives a description — quite loose if compared to other treatises on astronomical instruments — of the construction and use of this previously unattested instrument: the *Planicelium*.[1] There are no figures, not even marginal ones, that would help here the reader

* Research presented in this chapter was conducted in the framework of the ERC project *ALFA, Shaping a European Scientific Scene: Alfonsine Astronomy*, (CoG 723085, PI Matthieu Husson).
We are very grateful to the members of the ALFA team for their important comments, suggestions, and careful reading of our work. Our thanks go to Jean-Patrice Boudet, Charles Burnett, José Chabás, Alena Hadravova, Petr Hadravá, Matthieu Husson, Nick Jacobson, Richard L. Kremer, Laure Miolo, Philipp Nothaft, and Alexandre Tur. We are particularly grateful to Eleonora Andriani for her rigorous verification of our edition and translation of the *Planicelium*.

1 For example, the second *Equatorium* of John of Lignères or the *Planitorbium* (1310) by G. Marchionis. See, for both, Emmanuel Poulle, *Équatoires et horlogerie planétaire du XIIe au XVIe siècle*, 2 vols (Geneva: Droz, Paris: H. Champion, 1980).

Samuel Gessner • Universidade de Lisboa
Marie-Madeleine Saby • Université Grenoble Alpes

Alfonsine Astronomy: Computational Practices, ed. by Matthieu HUSSON, José CHABÁS, and Richard L. KREMER, Turnhout, 2026 (*Alfonsine Astronomy: Studies and Sources*, 3), p. 109-148
© BREPOLS PUBLISHERS 10.1484/M.ALFA-EB.5.137909

to imagine this instrument. Moreover, some unfamiliar terms, apparently unique to this text, are used to describe certain parts of it.

At first the instrument appears to resemble an astrolabe: it consists of a double-faced circular plate, suspended from a throne when an altitude measurement is required. It also includes a mobile part, rotating around a central peg. Its principal functions seem to include two important operations known from astrology: the division of houses (domification) and the direction (measuring the equatorial arc that separates two celestial points). It might not come as a surprise that this latter function is also the explicit aim of a contemporary instrument: John of Lignères' *Directorium*.[2]

1. John Vimond and his works: '*...ad utilitatem scolarium universitatis Parisiensis et omnium aliorum*'

Little is known of Vimond's life.[3] Scarce elements can be deduced from a note in a marginal column of the last folio of the set of tables compiled by John Vimond in the manuscript BnF, lat. 7286C, f. 1^r–8^v (Figure 4.1).

This mention indicates that John Vimond is native to the diocese of Bayeux, Normandy, and that he prepared this set of tables 'for the use of students of Paris university and all others'.[4]

The registers do not show any proof that Vimond had been received as Master of Arts at the university of Paris at any time. None of the contemporaneous students or masters there appear to have quoted his name or his works. In the studies of Palémon Glorieux, the name 'Johannes Vimundus' is not found, neither the incipit of the *Planicelium* nor the name 'Johannes Parvo Sancto' (written in the upper margin of Erfurt, UB, CA 2° 377, f. 21^r).[5] Furthermore, not a single hint about Vimond is to be found in the collection on *Le travail intellectuel à la Faculté des arts de Paris: textes et maîtres (ca. 1200–1500)* by Olga Weijers.[6]

2 This work is attested by Simon de Phares as a work of John of Lignères. Simon de Phares mentioned this work by the incipit '*Accipe tabulam planam rotundam*', whereas in BnF, lat. 7295, on folio 78^r, the incipit is '*Accipe tabulam planam super cuius*'. Lynn Thorndike and Pearl Kibre (TK25) quoted this variant of the incipits. See Lynn Thorndike and Pearl Kibre, *A Catalogue of Incipits of Mediaeval Scientific Writings in Latin* (Cambridge, Massachusetts: The Mediaeval Academy of America, rev. and augm. 1963), col. 25. See also L. Thorndike, *A History of Magic and Experimental Science*, 8 vols (New York: Columbia University Press, 1923–1958), III, p. 260.

3 A recent updating of the life and works of John Vimond can be found in José Chabás, *Computational Astronomy in the Middle Ages* (Madrid: Consejo Superior de Investigaciones Científicas, 2019), pp. 133–46.

4 '*In hoc terminatur opus Johannis Vimundi Baoiceum* [sic] *dyocesis de dispositionibus planetarum et stellarum fixarum et cum istis sequitur de hiis que per ipsum ordinantur ad conversionem temporum verorum et equalium sociatorum et de dispositionibus eclipsalibus* [sic] *solis et lune sibi pertinentibus. Et de aliis dispositionibus ipsorum et aliorum corporum celestium, ad utilitatem scolarium universitatis Parisiensis et omnium aliorum*'. The first transcription and translation of this passage can be found in José Chabás and Bernard R. Goldstein, 'Early Alfonsine astronomy in Paris: the tables of John Vimond (1320)', *Suhayl*, 4 (2004), 207–94 (p. 208).

5 Palémon Glorieux, *La Faculté des arts et ses maîtres au XIIIe siècle*, Etudes de philosophie médiévale, 59 (Paris: Librairie philosophique J. Vrin, 1971). Despite the period covered in this title, one can find the *magistri* such as John of Murs, John of Lignères, and John of Saxony in this book, but there is nothing about John Vimond.

6 Olga Weijers, *Le travail intellectuel à la faculté des arts de Paris: textes et maîtres (ca.1200–1500)*, Studia artistarum (Turnhout: Brepols, 1994–2012).

A mention of John Vimond is found, though, in Simon de Phares's *Recueil des plus célèbres astrologues* from the end of the fifteenth century. Not much information is given in that text either; Simon de Phares referred to Vimond as 'a most singular and a great astrologer, very well known in his time because of his science of the stars; among other works he verified the conjunction of the luminaries, eclipses and the fixed stars for several years; he predicted great winds taking place in his time, and made several beautiful astrological judgments…'[7] This passage suggests a clear link to the set of tables of 1320 attributable to Vimond, especially to the ones concerning mean and true syzygies and to the star list.[8]

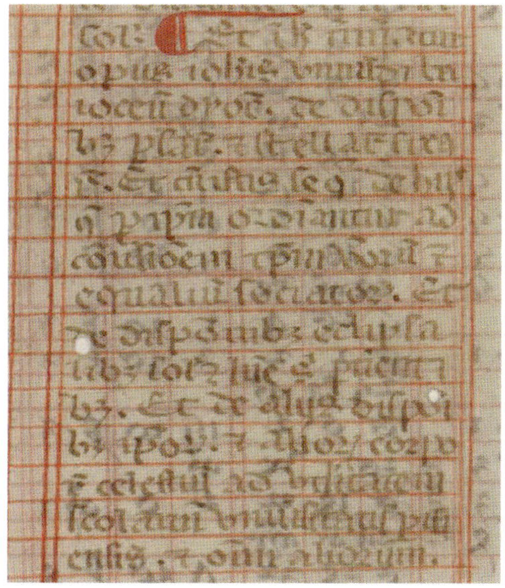

Figure 4.1. Paris, BnF, lat. 7286C, f. 8^{v}. Source gallica.bnf.fr.

John Vimond and early Alfonsine astronomy

Early Alfonsine astronomy in Paris, based on the Castilian Alfonsine Tables, is usually associated with four main astronomers: John Vimond, John of Murs, John of Lignères, and John of Saxony, who simultaneously worked on the same subjects in the same place. The last three 'Johns' knew each other, particularly John of Saxony and John of Lignères, and the former often wrote of his special respect for the latter as his teacher in his own works or in the explicits of John of Lignères' works he copied.[9]

The interactions between these astronomers go further than respectful words in their texts, however. Other connections and interdependences have been proven by José Chabás and Bernard R. Goldstein.[10] They suggest that John Vimond held a special place in the early Alfonsine astronomy Parisian 'milieu'. Their demonstration leads to the quite striking hypothesis that one can consider the existence of a kind of transmission line from

7 Jean-Patrice Boudet, *Recueil des plus célèbres astrologues de Simon de Phares*, 2 vols (Paris: H. Champion, 1997–1999), I, p. 467: *'Maistre Jehan Vymond fut à Paris, homme moult singulier et grant astrologien, lequel eut en ce temps grant cours pour la science des estoilles. Entre ses euvres, fist une veriffication de la conjonction des lu[mi] naires, aussi des eclipses et estoilles fixes pour plusieurs ans. Cestui predist les grans vens qui furent en son temps et fist plusieurs beaulx jugemens, dont il acquist grant loz et renommée en France et fut moult devost en Nostre Seigneur'.*

8 José Chabás and Bernard R. Goldstein, 'Early Alfonsine Astronomy in Paris: The Tables of John Vimond (1320)', pp. 207–94.

9 For instance, see the introduction of '*Quia plures astrologorum …*', featuring commentaries by John of Saxony on the canons of the first mobile by John of Lignères; edited by Marie-Madeleine Saby, *Les canons de Jean de Lignères sur les tables astronomiques de 1321* (Unpublished thesis: École nationale des Chartes, Paris, 1987). A summary appeared as 'Les canons de Jean de Lignères', École nationale des Chartes, Position des thèses 1987, pp. 183–90. For the explicit, see Erfurt, UB, CA 2° 377, f. 35^{r}.

10 José Chabás and Bernard R. Goldstein, 'The Medieval Moon in a Matrix: Double Argument Tables for Lunar Motion', *Archive for History of Exact Sciences*, 73 (2019), 335–59 (p. 354).

John Vimond (tables for 1320) to John of Lignères (*Tabule magne*, 1325) through John of Murs, who borrowed the table first for his tables of 1321 and later for his *Kalendarium* and his *Patefit* (*c.* 1330). We see further proof of the connections between John of Murs, John of Saxony, and John Vimond through an annotation in the top margin of the manuscript Escorial, O. II. 10, f. 204^{v} presenting a table of lunar velocities:

> *Motus Lune hic positus est ad unam horam completam in meridie post horam medie coniunctionis et vere tenet ad alias horas precise, et concordat precise cum motu Lune in primo minuto diei sicut per J[ohannem] de Vimondi primo, deinde per J[ohannem] de Saxonia.*

The hand of this annotation has been identified as that of John of Murs and transcribed by Laure Miolo.[11] Moreover, on the same folio, after his own table for velocities, John of Murs copied a part of the table for velocities by John Vimond.[12]

Finally, another hint towards a connection between John Vimond and another Parisian astronomer, John of Spira, can be read in the explicit of the Vatican manuscript Ottob. lat. 1826, folio 153^{r}. The explicit states that John of Spira has written a canon (the eleventh of the set) on a table (Table 8) composed by John Vimond.[13] This canon, beginning on folio 152^{v}, applies to the table that is copied right afterwards on the folios 153^{v}–155^{r}. This new text is attributed to John of Spira by the fifteenth-century copyist of this part of the manuscript and placed just before another canon composed by John of Murs (ff. 156^{r}–156^{v}).

John Vimond's works

The set of twenty-eight tables, studied by José Chabás and Bernard R. Goldstein (2003, 2004, 2019), is uniquely preserved in the fourteenth century manuscript BnF, lat. 7286C (ff. 1^{r}–8^{v}). Table 8 of this set, in Chabás and Goldstein numbering, is also found in the above-mentioned Ottob. lat. 1826. From a very schematic view point, one can assert that Vimond's tables were computed for Paris, around 1320, and that their parameters are mostly those of the *editio princeps* of the Alfonsine Tables (Ratdolt, 1483).[14]

Table 28, listing 225 stars, seems to have benefitted from a transmission separate to that of the other tables in the set. Found in at least eight manuscripts the various copies of the list show only small differences amongst them.[15] However, the star positions are, on the

11 We are thankful to Laure Miolo for this information on the Escorial, Real Biblioteca del Monasterio de San Lorenzo, O. II. 10. See Laure Miolo, 'A New Astronomical Text by Jean des Murs', *Cahiers de recherches médiévales et humanistes — Journal of Medieval and Humanistic Studies*, 42 (2021–2), 329–59 (p. 331, note 9).

12 For a partial reproduction of this table, see José Chabás, *Computational Astronomy in the Middle Ages*, p. 140.

13 Table 8 'Correction of the lunar position for each day between syzygies', see José Chabás and Bernard R. Goldstein, 'Early Alfonsine Astronomy in Paris: The Tables of John Vimond (1320)', pp. 228–30.

14 For a more precise characterization, see José Chabás, *Computational Astronomy in the Middle Ages*, p. 145.

15 Brussels, Bibliothèque Royale, 10117–26, ff. 50^{r}–52^{r} (the year 1320 is mentioned in the title of the star list); Cambridge, Gonville and Caius College, 141/91, pp. 377–82; Cracow BJ 551, ff. 97^{r}–98^{v}; Madrid, Biblioteca Nacional, 4238, ff. 65^{v}–66^{v}; Munich, Bayerische Staatsbibliothek, Clm 26667, ff. 46^{v}–47^{v}; Paris, BnF, lat. 7286C, ff. 8^{r}–8^{v}; Segovia, Biblioteca de la Catedral, 84, ff. 46^{r}–51^{v}; Paris, BnF, lat. 7482, ff. 61^{v}–69^{v}; Erfurt, UB, CA 2° 395, ff. 105^{v}–106^{v}. In this last manuscript, one can read the following in the left margin of 106^{r}: '*Nota istas stellas fuisse equatas ab anno Domini 1338*'. See Firmin de Beauval's treatise on astrometeorology, the *Repertorium prognosticum de mutatione aeris*, completed in 1338, where part of the star list composed by Jean Vimond in 1320 *completo*, that is, 1321 *incompleto*, is incorporated: see José Chabás and Bernard R. Goldstein, *A Survey of European Astronomical Tables in the Late Middle Ages* (Leiden-Boston: Brill, 2012), pp. 188–98.

whole, the same as those of Ptolemy's star catalogue (1028 stars) in Gerard of Cremona's version with an increment in longitude of 17° 52′. The stars are given according to the order chosen by Ptolemy in the *Tetrabiblos*.[16] The proximity of Vimond's star list to Ptolemy's is worth mentioning because Ptolemy is the authority invoked twice in the *Planicelium*. In fact, among the copies of the full star list, the name of John Vimond is only mentioned in the copy of manuscript BnF, lat. 7286C.

Finally, there is one other work attributed to John Vimond: the *Planicelium*, a singular text that shares with the 1320 table set the lack of dissemination. The *Planicelium* is, as far as we know, preserved in one single copy and is the only work of Vimond dealing with the division of celestial houses.

2. The *Planicelium*

In 'L'astronomie latine au XIVe siècle' Emmanuel Poulle noted a blossoming of new instruments at the end of the thirteenth and beginning of the fourteenth centuries.[17] They denote pronounced technical creativity, sometimes composed with a markedly pedagogical purpose. Among these were the *Turquetum* by Franco of Poland (1284); the *New Quadrant* by Jacob ben Mahir ibn Tibbon, known in Latin as Profatius Judeus (*c*. 1301); the *Saphea* described by the same ibn Tibbon and then by John of Lignères; the *Planitorbium* by the Franciscan friar G. Marchio (1310); the two types of equatoria and the *Directorium* attributed to John of Lignères; and the *Albion* (1327) by Richard of Wallingford. These works on astronomical instruments belong to a very innovative early-fourteenth-century field, along with a compilation of astronomical tables and canons and new ways of considering celestial observations and teaching astronomy. The *Planicelium*, only attested through its description by John Vimond, was part of this movement. The names themselves of all these instruments were often creative and allusive to their functions, with some resonances, for instance, between *Planitorbium* and *Planicelium*, both implying a representation of the spherical astronomy on a plane. The name of *Planicelium*, as far as we know, does not occur elsewhere. No reference could be found either in Poulle's or North's works, except with reference to Vimond.

Date matters

The single manuscript, Erfurt, UB, CA 2° 377, ff. 21^r^–22^r^ in which the *Planicelium* is preserved belongs to the Amplonian collection of the Erfurt university library. It is composed of three parts; the first and the third parts are from the fifteenth century. Only the second one is from the early fourteenth century. A study of the hands of this manuscript is available.[18] The middle section is composed of two quires (ff. 21^r^–32^v^/ff. 33^r^–47^v^).

16 See José Chabás and Bernard R. Goldstein 'Early Alfonsine Astronomy in Paris: The Tables of John Vimond (1320)', pp. 269–70.

17 Emmanuel Poulle, 'L'astronomie latine au XVIe siècle', in *Gersonide en son temps*, ed. by Gilbert Dahan (Louvain: E. Peeters, 1991), pp. 253–64.

18 Matthieu Husson and Marie-Madeleine Saby, 'Le manuscrit d'Erfurt CA F 377 et l'astronomie alphonsine', in *Miscellanées scientifiques au Moyen Age*, Micrologus, 27 (Florence: SISMEL Edizioni del Galluzzo, 2019) pp. 202–33.

This middle section is a collection of eight works from the main Parisian astronomers of that period: John Vimond, John of Lignères, John of Murs, and John of Saxony. Their works can be found in the following order in the manuscript and with dates of composition given for almost all of them:

- ff. 21^{r}–22^{r}: The *Planicelium* by John Vimond.
- ff. 22^{r}–35^{r}: Two sets of canons by John of Lignères, composed in 1322, as found in the common explicit (f. 35r).[19] Moreover, the copyist of this section identified as John of Saxony himself;[20] the same John of Saxony copied the forty-four canons *Cuiuslibet arcus* and the fourty-six canons *Priores astrologi*, composed by John of Lignères in 1322 and copied in 1323.
- ff. 35^{v}–36^{r}: The *Arbor Boecii*, by John of Murs, is dated 1324 in the explicit of the manuscript BnF, lat. 16621 (ff. 64^{r}–64^{v}).[21]
- ff. 36^{r-v}: The notes on the *Mare fusilis*, by John of Murs, are dated December 1324 in the manuscript Bruges, ms. 523 (ff. 31^{r}–32^{r}).[22]
- f. 36^{v}: The 19 *Prognosticationes*, by John of Murs cannot be dated.
- ff. 37^{r}–38^{r}: The *Tabula tabularum*, by John of Murs, dated 1322.[23]
- ff. 38^{v}–41^{r}: The *Algorismus* (including a multiplication table), by John of Lignères, can be dated around 1320.[24]
- ff. 41^{v}–47^{r}: The tables for 1322, by John of Lignères, among which some tables of John of Saxony are mixed.[25]

The very first work, in this central part, is the *Planicelium*, from folio 21^{r} to the top of folio 22^{r}, the two first folios of the quire. The hand is John of Saxony's, who wrote the *Planicelium* right before Lignères' canons, as stated in the final explicit on folio 35^{r}.

In short, this section of the codex offers texts composed by the four most representative astronomers of early Alfonsine astronomy, all written in a very short time period of four years, from 1320 to December 1324. It is a collection of diverse topics, certainly of interest to a young scholar of those times ranging from an original instrument (Vimond's *Planicelium*)

19 Erfurt, UB, CA 2° 377 f. 35^{r}: '*Expliciunt canones tabularum astronomie, ordinati per magistrum Johannem Pychardum de Lyneriis et completi Parisius anno ab incarnatione Christi filii Dei 1322. Scripte Parisius per manum Johannis de Danecowe. Anno Domini M°CCC°XXIII° in die cathedra Petri Deo Gratias*'. See Marie-Madeleine Saby, *Les canons de Jean de Lignères sur les tables astronomiques de 1321* (unpublished thesis, Paris: Ecole nationale des chartes,1987). A summary appeared as 'Les canons de Jean de Lignères sur les tables astronomiques de 1321', in *Positions des thèses* (Paris: Ecole nationale des chartes, 1987), pp. 183–90.

20 See Marie-Madeleine Saby, *Les canons de Jean de Lignères.*

21 Matthieu Husson, 'Les premiers témoins de l'*Arbor Boecii* de Jean de Murs: deux contextes distincts pour l'enseignement de l'arithmétique spéculative à Paris, au XIVème siècle', in *De l'homme, de la nature et du monde*, Mélanges d'histoire des sciences médiévales offerts à Danielle Jacquart (Genève: Droz, 2019), pp. 61–87.

22 Marie-Madeleine Saby, 'Mathématique et métrologie parisienne au début du XIVème siècle: le calcul du volume de la mer d'airain', *Archives d'histoire doctrinale et littéraire du moyen-âge*, 58 (Paris: J. Vrin, 1991), pp. 197–213.

23 Matthieu Husson, 'La *Tabula Tabularum* de Jean de Murs: nombres et opérations arithmétiques en astronomie au début du XIVème siècle', in *Cahiers de recherches médiévales et humanistes*, 27 (Auxerre: Centre d'études médiévales d'Orléans, 2014), pp. 96–122.

24 Hubertus L.L. Busard, *Het rekenen met breuken in de middleleeuwert, in het bijzonder bii Johannes de Lineriis* (Brussels: Vlaamse Academie, 1968).

25 See José Chabás and Marie-Madeleine Saby, *The Tables of 1322 by John of Lignères: An Edition with Commentary* (Turnhout: Brepols, 2022), p. 42.

with astrological purposes to the most important and canonical texts and tables of John of Saxony's teacher, John of Lignères. The amazing consistency of this collection from a temporal perspective, as well as the actual composition itself, could be considered a sound argument to put forward 1323 as the copy date of the *Planicelium* and Lignères' canons (themselves composed in 1322). Folios 21^{r}–22^{r} belong to the same quire as the one containing the beginning of the first set of those canons: *Cuiuslibet arcus*.

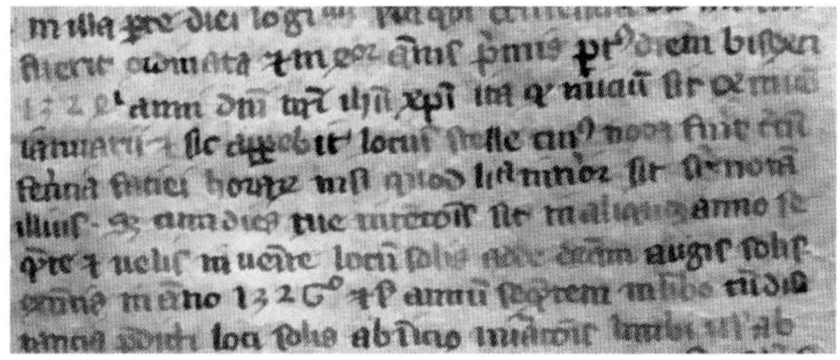

Figure 4.2. Erfurt UB, CA 2° 377, f. 21rb. Source Digitale Historische Bibliothek Erfurt/Gotha.

Further focus has to be placed on the two dates, 1324 and 1326, which are mentioned in the second column of the *Planicelium*: '... *et in quattuor annis primis post diem bisexti* 1324^{i} *anni domini nostri Jhesu Christi ita quod initium sit ex initio januarii ... adde differentiam augis solis existentis in anno* 1326°' (Figure 4.2). These two years, to which the calendar scale on the instrument was adjusted, were probably chosen by John Vimond in the nearest full bissextile cycle after the composition of the text. Thus, it can reasonably be argued that the *Planicelium* was composed before 1323, as Lignères' canons next to it were copied in that year. The composition of the *Planicelium* should not be dated prior to the bissextile cycle 1320 to 1323 because the text refers to 1324 and 1326, which are years belonging to the next bissextile cycle (1324 to 1327). Therefore, one may even presume the date of composition was nearest to this next full bissextile cycle (1324 to 1327), in around 1322 or 1323.

The text of the Planicelium and its edition

The text of the *Planicelium* is rather short, about 2700 words long. The explicit attributes the treatise to John Vimond and refers to *Ptholomeus Feludiensis*[26] (Figure 4.3).

This authority is quite consistent, but the expression used here by John of Saxony, the presumed copyist of the text of Vimond, is worth being noticed in the context of the early Alfonsine astronomy.[27] The original draft was corrected in about twelve spots (mainly on f. 21^{r}), either erasures, expunctions, or additions in margins. The same hand as for the original seems to have written these corrections. The layout of these two folios offers a text in two columns by folio parted in two main sections, with titles in a larger sized writing. The first part '*Planicelium vero componitur*', which can be considered as the title, comprises two chapters describing the various components of the instrument (limb, *balista*, face of altitude

26 Note that this adjective is used, for example, in the title of the 1515 edition of the *Almagest*: *Almagestu[m] Cl. Ptolemei: Pheludiensis Alexandrini astronomo[rum] principis*. Venice: Petrus Liechtenstein. One might speculate whether it refers to the city of Pelusium in lower Egypt. However, Ptolemy lived in or around Alexandria and died there. The adjective *Feludiensis* is probably a corruption of the Arabic *Qeludi* (for Claudius). See Gerald J. Toomer, 'Ptolemy', in *Dictionary of Scientific Biography*, ed. by Charles Gillipsie, XVI vols (New York: Charles Scribner's Sons, 1973–80, republished 2001), XI pp. 186–206.

27 '*Explicit tractatus Johannis Vimundi de floribus illorum que per instrumenta spere signorum possunt inveniri ad faciendum judicia astronomica secundum intentionem summi Ptholomei Feludiensis*'.

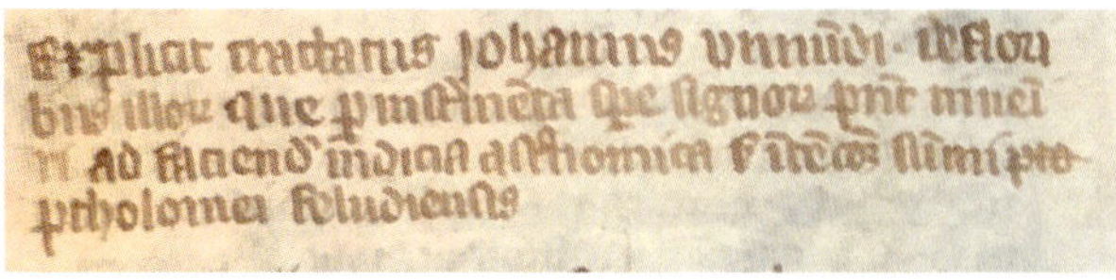

Figure 4.3. Erfurt, UB, CA 2° 377, f. 22^{r} Explicit. Source Digitale Historische Bibliothek Erfut/Gotha.

and face of hours). The second part, '*Deinde sequitur de doctrina*', with eight chapters of different length, deals with the uses. Each of the ten chapters had a guide letter as preparation for rubrication or illumination and an indentation of several lines. These decorations, however, have never been accomplished.

No figure showing the features of the instrument is included on the folios of this copy of the *Planicelium*, nor does the text itself refer to any diagram. The shape of the instrument might have appeared evident as long as the text circulated together with an actually built instrument. Here, neither the copyist nor a later reader added any diagrams, not even sketches in the margins, in contrast with what is seen on subsequent folios in the *Cuiuslibet arcus* by John of Lignères. Indeed, the copyist of these canons, John of Saxony, as he was reviewing and working on the text of his teacher, gave an illustrated interpretation of Chapters 7, 8, and 9 (ff. 22^{v}–23^{r}).

The edition of the text was difficult to establish for two reasons of different levels of significance. The first issue is typical of the edition of works where there is only one witness available. Where the scribe's hand is difficult to read, we relied more than usual on deduction about the text's content to make a transcription choice.

This process turned out to be particularly strenuous for an additional reason: the subject itself, the *Planicelium*, is not described in any other sources. The *Planicelium* being an astronomical instrument, however, it was possible to reconstitute the various graduated scales and curves inscribed on its faces by comparison with other instruments. The part on the uses of the *Planicelium*, also referring to these scales, shows a similarity with texts like those on the astrolabe or the quadrant. This similarity served as guidance.

We were always aware of the circularity of our argument when using supposed content to determine a reading, while our reading should also buttress what we understand of the content. Consequently, while the text resulting from our editorial choices is much more than a mere diplomatic transcription, we consider that the resulting text cannot constitute an actual critical edition.

3. Comment and interpretation

The treatise describes an instrument called *Planicelium*,[28] which presents an original combination of computational functions, the principles of which are known from the planispheric astrolabe. The explicit indicates that the text offers a selection of the best *flores* that can be found through instruments representing the zodiac (*instrumenta spere signorum*). These functions relate mainly to the making of astrological judgments (*ad faciendum iudicia*), although the range of problems solvable by the instrument is naturally

28 A short confrontation with *Planitorbium* by G. Marchionis, 1310, is not conclusive; *Planitorbium* is an equatorium as Emmanuel Poulle shows in *Équatoires et horlogerie planétaire*, pp. 260–77, and concerns the motions of the planets, whereas *Planicelium* refers to the motion of the *primum mobile*.

larger. Some doubts remain about the appearance of the instrument and, in particular, about the exact configuration of its scales. A careful reading of the text, however, allowed us to reconstruct the rationale behind the instructions for its various uses.

The text follows the classical *ordinatio* of instrument texts. The first part describes the structure of the instrument and its various scales. The second part provides instructions for several of the instrument's uses. In the following section we first list the elements that we can assert with some confidence. After this, we highlight some of the intriguing features that we do not fully understand.

The structure of the instrument

The heading of the first part might be loosely translated as 'The *Planicelium* consists of parts that apply to various operations…'[29] From this point of view, the *Planicelium* is not unlike a planispheric astrolabe: it presents an assemblage of a variety of scales pertaining to different uses. Also, with regard to the structure and inscribed scales and lines, we are reminded of the mater, limb, throne, rete, and alidade of a typical astrolabe. There are, however, several conspicuous differences.[30]

The *Planicelium* is described as a circular disc inscribed with different features on each face. One face is designated as the 'face of altitude' (*facies altitudinis*) and the other as the 'face of hours' (*facies horarum*; see Figures 4.4a and 4.4b). Surrounding both faces there is a limb (*limbus*; see ⟨i⟩) presenting a graduation with numbering that starts at the top. At the top, attached above the limb, a throne is fixed (*materia in exteriori parte*) so that the instrument can be suspended. This is required, as noted in the text, when measuring the altitude of a star. There is one revolving part attached to this ground plate with a central peg (*cavilla*), as is habitual with astrolabes.

The revolving part is designated as *balista* ⟨ii⟩ and described as a flat piece with several holes. Its shape is not entirely clear from the text (more on this below). A pair of pinnules are fixed on this revolving *balista*; these are aligned with a fiducial line or line of sight (*linea acuitatis*). The text also mentions that on the arc of the *balista* there is a graduated line (*ecliptica/linea longitudinis*). A straight scale described as perpendicular to the ecliptic (*linea declinationis*) is also part of the *balista*. Given that the ecliptic is represented on an arc, we infer that the two halves of it are folded onto each other on the *balista* (see Figure 4.4c).

Next, the lines inscribed on the two faces of the instrument are mentioned without, however, providing instructions on how to trace them. First, on the face of altitude ⟨iii⟩, there is the line of midday, the line of midnight, as well as the horizon, which are all familiar from the astrolabe (the horizon is traced for the latitude of 66° 27′).[31] The lines of altitude (*linee altitudinis*) and lines of direction (*linee directionis*) may be identical to what astrolabe texts call almucantars and azimuth circles. These same lines seem to be the ones

29 '*Planicelium vero componitur ex eis que sunt diversorum operum, scilicet quod apparet inferius et ideo primo notificantur*'.

30 The similarity that appears in the textual description might be the reason for the annotation, in a different hand, at the top of the page of the beginning that seems to read '*Compositio astrolabii… magistri Johannis de Parvo Sancto*'.

31 This value, 66° 27′, is a complementary angle (to 90°) of the obliquity of the ecliptic 23° 33′. Such a horizon line (a horizon parallel to the ecliptic) therefore can represent the image of the ecliptic in the stereographic projection used in astrolabes. See also the conclusion of this chapter.

that are called 'latorizal' lines (*linee latorizales*) and 'oridial' lines (*linee oridiales*) later in the text. These terms are otherwise absent in the literature and may be new words used by Vimond. Surrounding all these lines there is a double scale that coordinates days of the year and the corresponding longitude of the Sun. It is probable that this is achieved by having a uniform degree scale next to a calendar scale with uneven divisions for the days.[32]

On the face of hours ⟨iv⟩, a line of midday, a line of midnight, and a horizon are traced the same way as on the face of altitude. The horizon is numbered at two opposite points of the circumference, possibly to indicate the geographic latitude for which it stands (see Figure 4.4b). However, superposed on this framework is a bundle of lines called 'temporal lines' (*linee temporales*). These include 'house division lines' (*linee divisionis domorum*) and 'hour division lines' (*linee divisionis orarum*). The temporal lines are said to converge in two points. The lines of house division cut the circumference of the face into twelve arcs of thirty degrees. The hour division lines cut the circumference into arcs of fifteen degrees. From the astrolabe, we are familiar with the inscription of hour lines that cut the equator into parts of fifteen degrees, but there the tropic circles of both Cancer and of Capricorn are indicated. These lines are only traced on the part of the plate below the horizon.[33] On the *Planicelium*, by contrast, the equator seems to form the circumference and outer limit of this diagram (*quelibet due propinque distant ab invicem in circumferencia istius descriptionis per 15 gradus*). We can infer that, unlike on the astrolabe, the 'temporal lines' are also present above the horizon. One of the 'traditional' modes of dividing the houses would use the hour lines of the second, fourth, eighth, and tenth hours (in addition to the horizon that coincides with the twelfth, and the meridian that coincides with the sixth hour) to perform the division of the celestial houses (domification).[34] The mention of dotted lines (*de lineis continuis vel punctualibus descriptis*) suggests that every other hour line is traced as a dotted line.

Outside of all these lines on the face of hours, probably along the 360° graduations of the equator, several positions of fixed stars are marked. The text states that values of the stars' latitudes are indicated by numbers, distinguishing between positions north and south of the ecliptic.

Hereafter we propose graphic reconstructions of the *Planicelium's* parts: its two faces, and its *balista* (See Figures 4.4a, 4.4b, and 4.4c). Some parts of these diagrams are straightforward deductions from Vimond's description of the instrument or from its uses. Other parts are more speculative, such as the general shape of the *balista*, the dotted lines, and the selection of fixed stars quoted in the limb. On the whole, the text does not specify the numbering and labelling of the scales. These are our design proposals:

---

32 This way of juxtaposing concentric scales of degrees and of the calendar is often found on the reverse of Latin astrolabes.

33 See Johannes Stöffler, *Elucidatio fabricae ususque astrolabii* (Oppenheim: Jacob Köbel, 1513), f. 7v.

34 John Christopher Eade, *The forgotten Sky: A Guide to Astrology in English Literature* (Oxford: Clarendon Press, 1984), p. 44. John North called it 'the hour line method: John D. North, *Horoscopes and history* (London: Warburg Institute, 1986), p. 21. Josep Casulleras pointed out that there are astrolabes that show the counterclockwise numbering of the even numbered hours for the houses just as we have done for the reconstruction of the face of hours of the Planicelium (see Figure 4.4b). Josep Casulleras, 'The Instruments and the Exercise of Astrology in the Medieval Arabic Tradition', *Archives internationales d'histoire des sciences*, 63(170–71) (2013), 517–40, p. 530.

Figure 4.4. Hypothetical shape and design of the two faces (4.4a and 4.4b) of the *Planicelium* instrument with the rotatable part (4.4c) (*balista est tabula multipliciter perforata*) conjectured in the shape of a *balista* (a catapult or a crossbow).

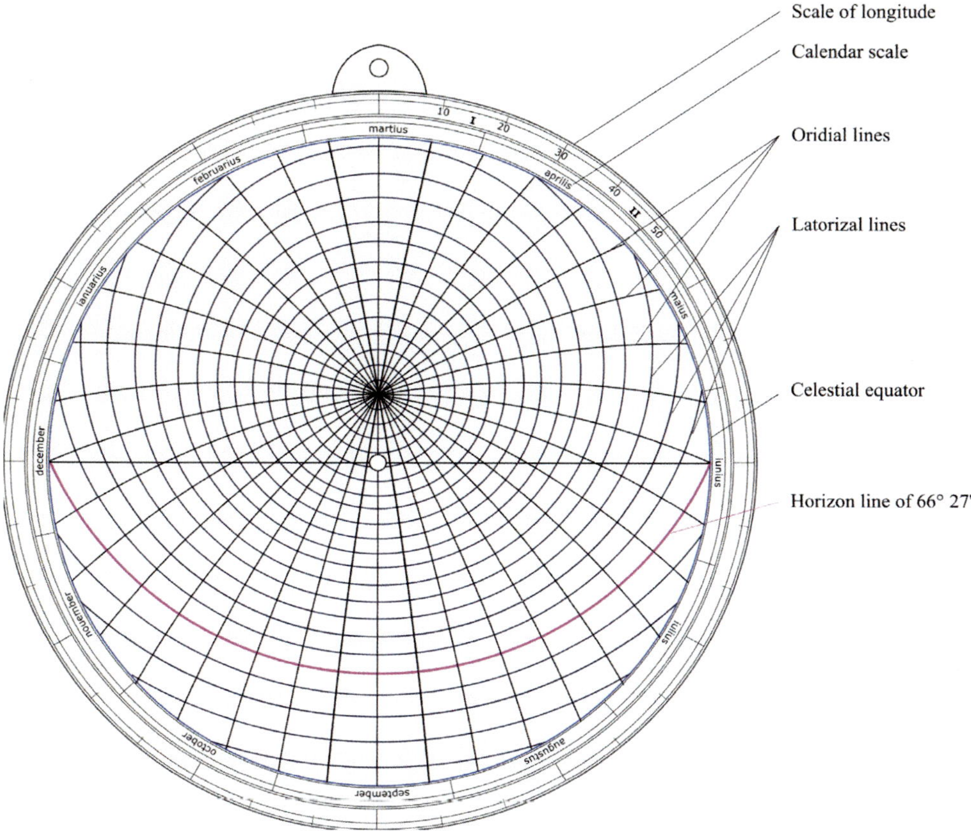

Figure 4.4a. Face of altitude. The projected lines represent ecliptic coordinates for one half of the celestial sphere, the confining exterior circle of the web being the equator (blue). The arc in bold (magenta), referred to as the 'horizon' for 66° 27', represents half of the ecliptic. The arcs and circles that do not cross it are termed 'latorizal' lines. The ones that are akin to the astrolabe's azimuth circles are termed 'oridial' lines. Hence, the point corresponding to the zenith represents the ecliptic pole, where the 'oridial' lines converge. On the limb, a degree scale is coordinated with a scale of the calendar year, the days being placed according to the corresponding solar longitude for 1324. The conjecture that the numbering starts below the throne is based on a hint in the text ('...*linea recta que procedit a cavilla versus initium numerationis divisionis limbi dicitur linea meridiei*...').

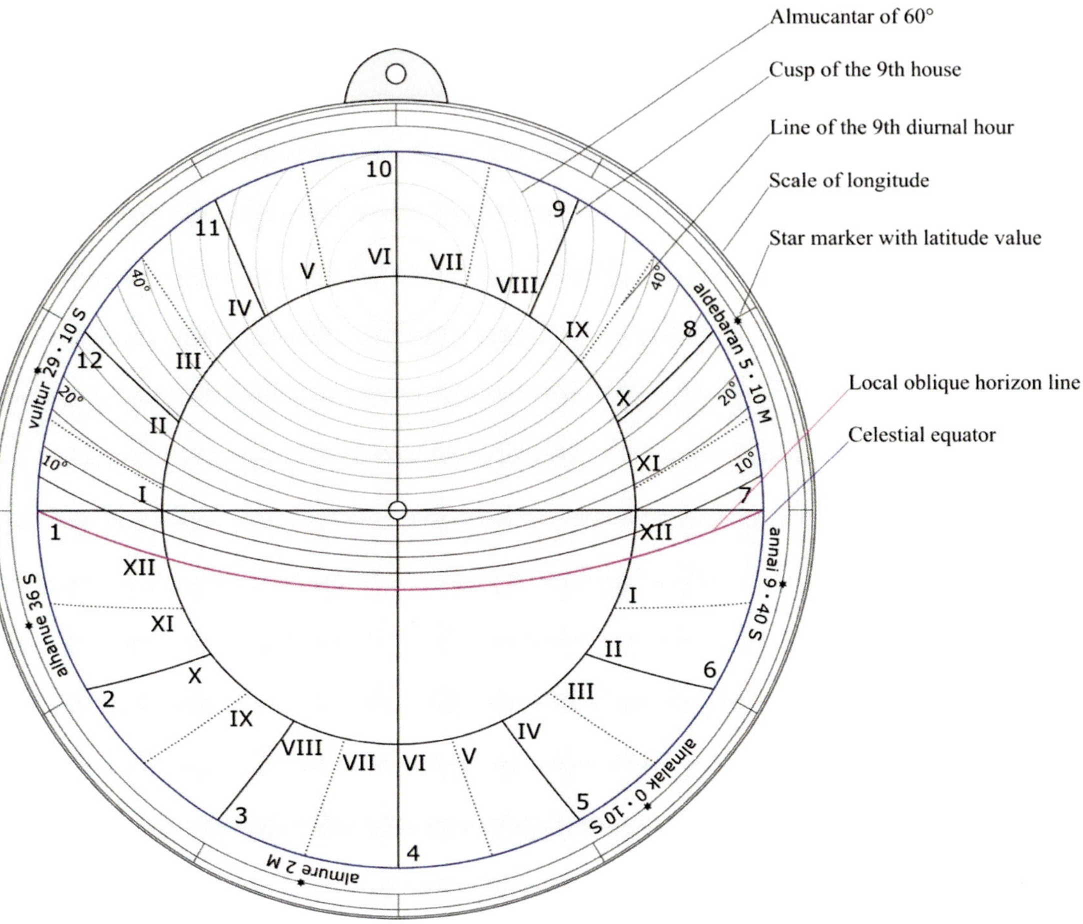

Figure 4.4b. Face of hours. Vimond does not specify the latitude for which his *Planicelium* is made (the horizon line is here arbitrarily traced for 25°). The corresponding almucantar arcs are not mentioned in Vimond's description. The use of the instrument, however, appears to presuppose their existence. The short arcs between equator (blue) and the tropic circle are termed 'temporal lines'. These include both hour lines (for the unequal hours, numbered twice from I to XII) and cusps of the celestial houses (numbered from 1 to 12). We interpret Vimond's reference to dotted lines as a means to distinguish arcs that are only hour lines. The limb is engraved with a longitude scale equipped with position marks ✶ (and possibly labels) of several fixed stars, with the indication of their latitude. As examples, we placed some stars arbitrarily picked from Vimond's star list.[35]

35 See the edition of Vimond's star list in José Chabás and Bernard R. Goldstein, 'Early Alfonsine Astronomy in Paris: The Tables of John Vimond (1320)', pp. 267–90.

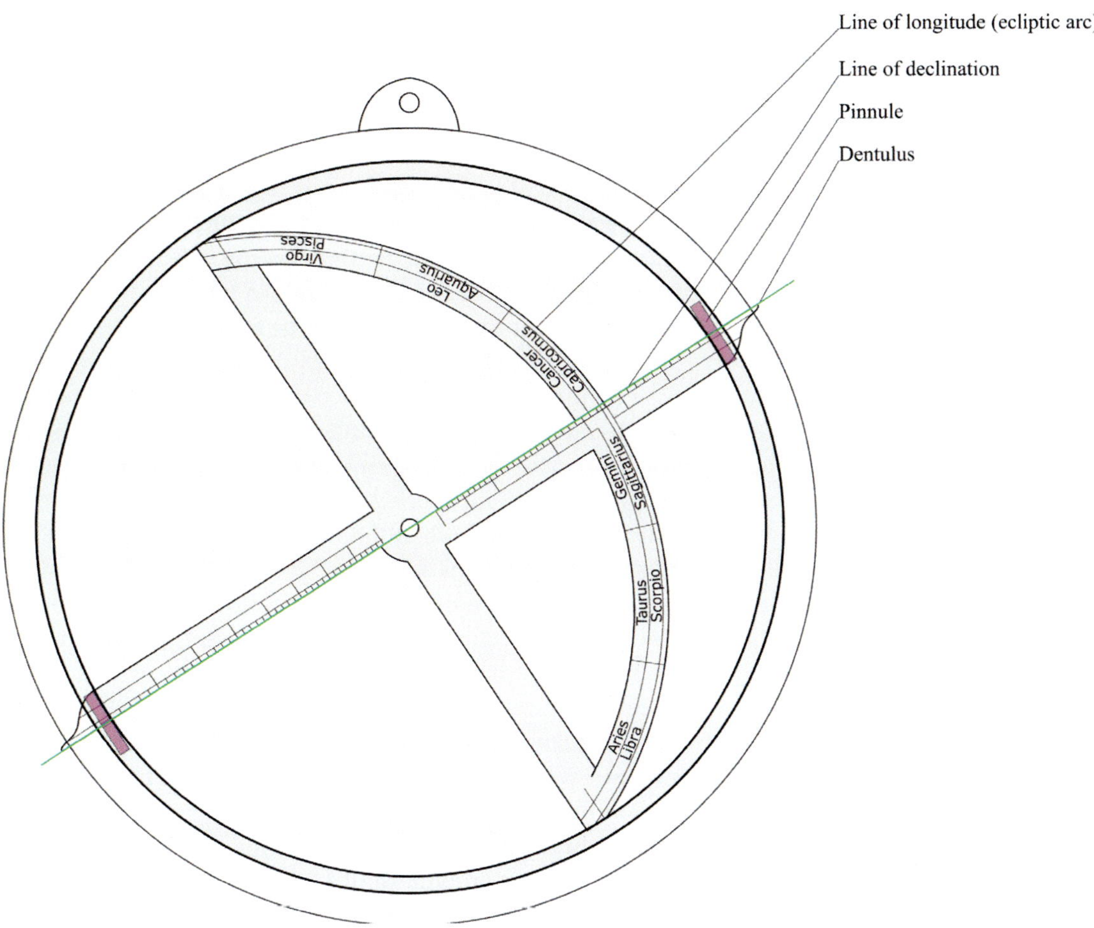

Figure 4.4c. The balista. The unevenly graduated arc of the *balista* represents the ecliptic; the pointed 'arrow' perpendicular to it, called the line of declination (green), is also graduated (unevenly) to measure declination. Pinnules (magenta) are fixed to the extremities of the line of declination. The pointed ends of the line of declination are used to read the scales on the limb of both faces. This rotating part is sometimes placed on the face of altitude, sometimes on the face of hours.

Uses of the instrument

The second part of the treatise about the use of the *Planicelium* includes instructions for operating the instrument to eight different effects. As is typical for such treatises, basic and preparatory uses are introduced first. The later uses often presuppose the results from such introductory uses. The scales of the instrument can represent only one celestial hemisphere at once as the equator forms the limit of the figure. This can be inferred from the fact that the text presents procedures for points with northern and southern declination separately, typically working with the points' opposites.

⟨1⟩ First, the text indicates the way to find the altitude above the horizon of the Sun or any other heavenly body using the pinnules. ⟨2⟩ Next, it explains how to read the solar longitude off the zodiac scale on the limb of the face of altitude for any given day of the calendar marked on an inner scale. A similar zodiac scale on the face of hours is used to determine the position of the stars that are marked alongside a number giving the value of their latitude. These explanations on recovering longitudinal positions also take into account a precessional shift. They are then followed by a general indication of how coordinates of longitude and latitude are represented on the face of altitude. ⟨3⟩ A third operation concerns the determination of local latitude using a given star. ⟨4⟩ The next operation may be understood as instructions to position the *balista* on the face of altitude such that it represents the *figura celi*. It means that the ecliptic on the *balista* is positioned with respect to the horizon to correspond to a chosen moment in time (*balista disposita super figuram celi*). Three cases are distinguished: first, taking the measured altitude of the Sun as a starting point; second, using the altitude of a known star; and, finally, just using a given day of the year and the hour of that day. ⟨5⟩ As a continuation of this operation, the text tells the reader how to transfer the resulting position of the *balista* to the face of hours, where one is then told how to read the time of day corresponding to that position. This part of the text strongly suggests that only one *balista* is present with the instrument and that it needs to be transferred from one side of the *Planicelium* to the other. ⟨6⟩ In the next passage, a string of instructions explains how to derive the division of the houses (domification) once the *balista* is adjusted to represent the *figura celi* on the face of hours by noting where the temporal lines (*linee divisionis domorum*) intersect the arc representative of the ecliptic. Furthermore, taking into account the position of the *balista*, instructions are given to determine in which celestial house a given planet or star is to be found. ⟨7⟩ The seventh operation allows us to perform what the astrologers call a direction: it measures the equatorial arc (*arcus directionis*) separating two celestial points. The arc represents the time it takes for the daily motion to take one point onto the other. These points are not just any places, but astrologically meaningful positions in the heavens (*locus alicuius puncti celestis*). In the parlance of the astrologers, a significator is 'directed' onto a promissor, where — depending on the doctrine — the ascendant, mid-heaven, a planet, or another entity can act as significator, and the promissor can be the position of one of the seven planets, one of its aspects, or a fixed star as laid out at the time of casting the horoscope. ⟨8⟩ The final exercise instructs the reader to determine the position on the face of altitude of an unknown star/planet when its altitude above horizon is first observed perpendicularly north or south (i.e. with common azimuth) of a known star. The operation amounts to obtaining the ecliptic longitude and latitude of a planet or a star that is not indicated on the instrument. This final exercise raises the question as to whether it has an astrological purpose. Arguably, it could be about both: merely demonstrating a highly technical skill or determining by observation the longitude of a planet to be placed into a horoscope.

We must stress here that the exact features of the scales and the principles underlying the functions presented are not yet fully elucidated, and further research is still required. Operation ⟨4⟩ implies taking the altitude of the Sun or a star, which means that the *Planicelium* should include almucantar arcs (lines of equal altitude) above the local horizon. The text refers to the 'face of altitude', which would conflict with the fact that there the projection is for a latitude of 66° 27'. Perhaps the intention is to use the face of hours. Such almucantar lines for a local horizon are not mentioned, however, in part ⟨iii⟩, as we would expect. The beginning of that part and the reference of diverse latitudes is difficult to interpret and may be the remnant of a fuller description that went missing during the copying of Vimond's work.

The instructions of use given by Vimond nevertheless show that the operations differ for northern or southern declinations and for the two halves of the ecliptic that are at stake. Oftentimes the opposite point (*nadir*) on the ecliptic is to be taken. The general impression suggests that the scales of the instrument are 'folded', meaning that the same lines can be interpreted in a double (northern/southern) way. For instance, in the sixth use ⟨6⟩ Vimond states that the northern houses appear directly, whereas the southern ones will be obtained by using the opposite points ('*Et in illis domibus et partibus satis apparebunt stelle septentrionales et existentes in ecliptica sed meridionales existentes in ecliptica erunt in domibus oppositis illarum in quibus erunt sua puncta opposita*').

One of our important assumptions is therefore that the outer circle of both faces represents the celestial equator, while the area enclosed in that circle sometimes signifies the northern and sometimes the southern hemisphere. The text is explicit about this feature when the hour lines are described: '*linee divisionis orarum sunt ille aliarum quelibet due propinque distant ab invicem in circumferencia istius descriptionis per 15 gradus.*' These lines are fifteen degrees apart on the outer periphery of the diagram. It is clear that this folding of the scales allows one to simplify the making of the instrument as fewer scales need to be traced; the ecliptic is thus reduced to an arc with 180 parts. In turn, the use of the instrument is rendered more complicated.

If this reading is correct, we observe that Vimond's approach to the instrument shows a tendency that is similar to what is found in John of Lignères' way of devising his second equatorium. There, the epicycle is represented by a quarter circle, and other scales are also folded onto themselves, which allows one to reduce the trouble of graduating scales. As Matthieu Husson has shown, the reduction of the instrument exploits the same 'symmetry' in the values of equation that are exploited by scribes who copy the equation tables when they double the columns of arguments. It also means, however, that the instructions need to distinguish four different cases.[36]

In the case of Vimond's *Planicelium*, the connection to the design of tables is not that direct. The author exploits the symmetries that are characteristic of the geometry of celestial circles and concepts (e.g. rising and setting). Hence, certain well-known rules for the use of ascension tables may have suggested the design of the *Planicelium*. For instance, to obtain the zodiacal degree that is setting, one searches the rising degree of the opposite point. Vimond exploits these symmetries systematically, which means that many cases need to be distinguished in using the instrument's scales (the alternative cases are introduced by the expression *vel sic*).

Subtlety of the instructions

Several of the indications about the *Planicelium*'s scales reveal the author's precise understanding and awareness of their simplified structure. These passages in the text point to the fact that the instrument is intended by Vimond not only as a 'practical aid' to computation,

36 Emmanuel Poulle, *Équatoires et horlogerie planétaire*, pp. 364–74, Fig. 75; Matthieu Husson, 'Construire un instrument astronomique au XIVème siècle: le second équatoire de Jean de Lignères', in *Science et technique au Moyen Âge (XIIe–XVe siècles)*, ed. by Joël Chandelier, Catherine Verna, and Nicolas Weill-Parot (Vincennes: Presses universitaires de Vincennes, 2017), pp. 225–45.

but first and foremost as a means to visualize and stimulate theoretical considerations. A first occurrence of such a comment on the scales can be found in the combined calendar and zodiac scales, which are mentioned at the beginning. Such a combination is commonly to be encountered on the back of planispheric astrolabes. It is called the 'theory' of the Sun. Vimond notes that the scale is used differently in the four years of a bissextile cycle. The divisions of the days, as marked on the instruments, yield positions in the first year after 29 February. In the subsequent year, the start of the day on the scale is to be taken a quarter of day early, eating into the space of the previous day. In the second year after, one shifts by two quarters, and in the third year one shifts by three quarters before the graduation of that corresponding day. This exemplifies that the instrument deliberately simplifies the scale and in change accepts a more complicated and thoughtful use.

The theory of the Sun is further elaborated on in the chapter on the uses. Here, Vimond is even more precise and adds that the calendar/zodiac scales are adjusted for the bissextile cycle starting in 1324. According to Vimond, if one searches for the longitude of the Sun in any year posterior to that cycle, one should take into account the motion of the apogee of the Sun. The rule given is to add the difference between the apogee position of 1326 and of the chosen year. Ptolemy is invoked as an authority on the motion of the apogees. Vimond reminds his readers that the motion of the apogee is very slow, involving just 'second-order minutes' and 'third-order minutes' every year.[37] Undeniably, the small amounts of change in solar longitude due to the slow motion of the apogee would hardly be visible on an engraved scale on the instrument, as they represent only about half a degree in fifty years. Again, Vimond's meticulousness in this regard highlights his understanding of the *Planicelium* as an apt instrument to anchor theoretical considerations about the precise heavenly motions.

The procedure of the correction Vimond describes is required to find the tropical longitude (as opposed to the sidereal longitude) of the Sun, that is, the one measured from the equinoctial point (*invenies distanciam loci cuiusque illorum ab initio Arietis immobilis*).

The passage is remarkable for several reasons. On the one hand, Vimond chose the year 1326 for the position of the apogee implied in the scale as if his intention was to evenly spread out the minute error that it entails over the bissextile cycle. On the other hand, adding the motion of the apogee to a value read off the scale, Vimond is assuming that solar longitude increases to the same extent as the apogee itself in terms of its longitude. In other words, this passage attributes the motion of the apogee and the eighth sphere with the same value, consistently with what is implicit in the Parisian Alfonsine Tables.[38]

37 '*Et sciendum est quod motus augis solis in quolibet anno continet minuta secunda et tertia secundum intentionem Ptholomei Feludiensis*'. In contradiction to Vimond's statement, Ptolemy's position was that the solar apogee is tropically fixed at Gemini 5° 30′, at least in the *Almagest (book III.4)*. See José Chabás and Bernard R. Goldstein, *A Survey of European Astronomical Tables in the late Middle Ages*, p. 47. At the same time, it appears that Vimond is using the *motus augis* here as the equivalent to the motion of precession. However, even with this assumption the passage does not become more intelligible. Ptolemy notoriously held the view of a uniform precession of 36″ per year (1° in 100 years), a value expressed as an integer number of seconds and not seconds and thirds. We may observe that Vimond's tables of precession in BnF, lat. 7268C contain both a linear and a periodic component quite close but not equal to the precession of the Parisian Alfonsine Tables. Its linear component is of 26″ and about 28‴, which would be in accordance with this sentence in the *Planicelium*.

38 This remark is important, as it contradicts what appears from Vimond's own tables preserved in BnF, lat. 7268C, where the solar apogee has a proper motion and differs from the motion of the eighth sphere, according to José Chabás and Bernard R. Goldstein, 'Vimond and the Alfonsine Trepidation Model', *Journal for the History*

There is a further scale on which the text provides some specifics. The rule for finding a known star's position on the face of altitude instructs us to use the horizon line drawn for 66° 27′. This particular value stands out for representing the complement of an obliquity value (of 23° 33′). This value has been assumed by several authorities (e.g. Al-Zarqālī, Ibn al-Kammād, Pierre de St-Omer).[39]

Lexical matters

The part of the *Planicelium* that rotates above the faces is designated as *balista*. In Latin dictionaries (classical or mediaeval Latin),[40] the word *ballista* belongs to a military vocabulary ranging from the topic of antiquity to that of medieval battles.[41] It refers to a kind of weapon used to launch arrows (from the Greek 'ballein'). In antiquity, the machine was rather large and was used like a catapult. It could be interesting to study more precisely the transformation into a smaller throwing weapon that was used in the Middle Ages. Given the originality of this analogy, we may wonder if Vimond wanted to appeal in particular to readers with military interests. In Spanish, Italian, or French, the words derived from *ballista* remain close to the Latin: *ballesta*/*balestra*/*balliste*. Derivations include the French *arbalète* and the Portuguese *balestilha*, designating smaller, hand-held versions of crossbow-type weapons or a measuring instrument (the cross staff or Jacob's staff). The latter was designed for angular measurements between heavenly bodies. Except for this and the *Planicelium*, the term could not be found in the indexes of any other works on astronomical subjects.

The sentence '*Balista est tabula multipliciter perforata in qua figuratur una balista et linea que dividitur in circumferentia arcus illius dicitur ecliptica*'[42] required particular attention. We take this sentence to mean that within the perforated plate the shape of a *balista* is formed. If this is true, the arc of this shape would bear the scale division of the ecliptic. The line perpendicular to it, representing the 'projectile' of the *balista*, is the scale called *linea declinationis*.

In the pertinent instrument literature there is, nevertheless, a reference to the shape of a crossbow. It is found in the interpretation given by E. Poulle about Lignères' *Equatorium*; the part of this treatise on the construction (*Fiat primo regula*) is quite clear, and only one

---

of Astronomy, vol. 34 (2003), p. 165. From this, one may conclude that by the time Vimond composed the *Planicelium*, he had abandoned an earlier version of precession and had reached the position of the Parisian Alfonsine motion of the eighth sphere.

39 José Chabás and Bernard R. Goldstein, *A Survey of European Astronomical Tables in the Late Middle Ages*, p. 23.

40 Du Cange et al., *Glossarium mediae et infimae latinitatis* (Niort: L. Favre, 1883–87). Online (16 March 2022): ducange.enc.sorbonne.fr; Aegidius Forcinelli, *Lexicon totius latinitatis*, vol. 1 (A-C). Online (16 March 2022): www.documentacatholicaomnia.eu.

41 For example, the *ballista* is known in Vitruvius, *De Architectura* (X, XI, XII), but rather refers to a catapult for launching rocks or large arrows. Such a machine can be also found in Caesar's commentaries and those of other authors. On folio 165^{v} of BnF, fr. 2643, the scene of the battle of Crécy (26 August 1346) in *Chroniques de Jehan Froissart* (illuminated *c.* 1470–75) represents what can be considered 'crossbows' or *baliste* and simple bows. We can also find a thirteenth-century image of a crossbow in Villard de Honnecourt, BnF, fr. 19093, f. 22^{v}.

42 Vimond used the term *balista* with only one 'l' in every occurrence in his text, but it is currently listed with two 'l' in dictionaries.

manuscript (Vatican, Urbin. lat. 1399) provides a figure of this part of the instrument.[43] It was E. Poulle who interpreted this picture as a shape of a 'half crossbow'. However, the word *balista* or *medietas baliste* is not used in the original text of John of Lignères according to our research.

Two adjective terms characterizing the lines inscribed on the *facies altitudinis* seem to be hapax legomena in the proper sense: *oridialis* and *latorizalis*. The text offers some hints as to what these lines are, describing the *linee oridiales* as the ones cutting the horizon, while the *linee latorizales* stay for their entire extension either above or below the horizon. As pointed out above, they have these properties in common with the azimuth circles and almucantar circles known from the astrolabe. We are inclined to identify the oridial with the former and latorizal lines with the latter — but the question remains whether Vimond used new words to avoid the Arabic terms (azimuth, almucantar) or whether they come from a different tradition that preferred them specifically when referring to a plate for 66° 27′.

4. Conclusion

The importance of the astronomer John Vimond in the history of Alfonsine astronomy is well recognized, as his set of tables constitute one of the earliest forms of the Parisian Alfonsine Tables. The edition of his *Planicelium* has allowed us to uncover some additional traits of an actor of whom otherwise only the tables and a star list are known. As a writer on this astronomical instrument, the *Planicelium*, we get to know him a bit better. That said, Vimond's terminology is original and remains occasionally obscure. Moreover, some key sentences in the instrument's description appear to be corrupted, the text still offers some resistance to interpretation. It could be of great help if more elements about the *Planicelium* were to become available. So far, however, no other manuscripts describing it have been brought to our attention. Also, it seems that no such instrument survives in any collection, and there is no text, earlier or later, referring to the *Planicelium*. Still, some of its functions and specific features are shared by various other instruments. Specifically, the procedures of domification and direction that are rendered simple by the *Planicelium* are procedures also covered by the *Directorium* instruments. The *Directorium* attributed to John of Lignères represents one type of these; another type is known from surviving instruments (Figure 4.5).[44] As the study of the *Directorium* advances, more light might be shed on the *Planicelium*.

The projection of the heavenly sphere on the *Planicelium* turns out to be of the stereographic type, the projection used for the planispheric astrolabe. Its originality, however, lies in the fact that the outer circle of the instrument represents the equator in contrast to traditional astrolabes where it is the tropic of Capricorn. This choice, however, was not unprecedented. Pierre Pèlerin (Peregrinus) of Maricourt describes astrolabic plates of this sort in the *Nova compositio astrolabii particularis* (*c.* 1263).[45] E. Poulle noted that this allowed Pierre to increase the size of the southern part of the sky for an instrument of equal size

43 Emmanuel Poulle, *Équatoires et horlogerie planétaire*, pp. 364–74, Fig. 75.

44 In Lignères' *Directorium*, the zodiac is fixed, while the horizon rotates. See note 2.

45 Emmanuel Poulle, 'L'astrolabe médiéval d'après les manuscrits de la bibliothèque nationale', in *Bibliothèque de l'Ecole des chartes* (Geneva: Librairie Droz, 1954), 112, p. 93.

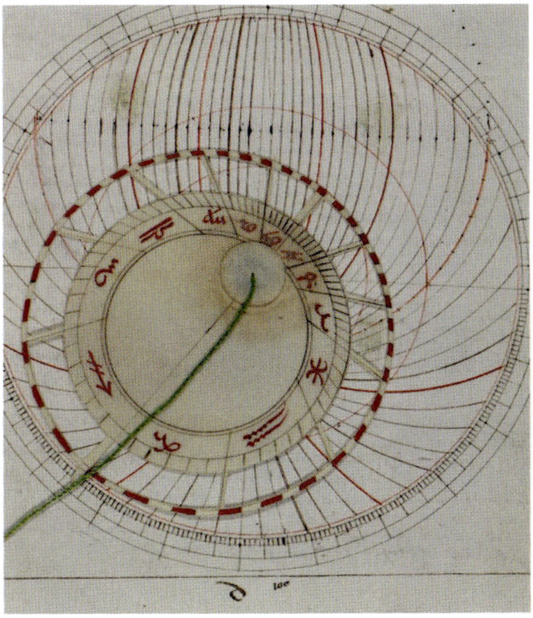

Figures 4.5a-4.5b. Contrary to the *Planicelium*, several examples of *Directorium* instruments are extant, either in metal or paper. Right: *Directorium*, 1516. Gilt copper, diameter 16.5cm. The German designation used is 'Astrologische Scheibe', Nuremberg, Germanisches Nationalmuseum, inv. nr. WI22. Reproduced with permission. Left: '*Hec est forma instrumenti directorii*', Biblioteca Apostolica Vaticana, Pal. lat. 1396, f. 36r. https://doi.org/10.11588/diglit.13038#0077.

and improve 'precision'. The further advantage of representing the entire sphere comes, however, at the cost of having to deal with the superposition of the Northern and Southern Hemispheres. This requires slightly different procedures for northern and southern points during the instrument's use. The *Planicelium* clearly uses this same device. Vimond's use of this type of projection, taking the equator as the outer limit, must therefore be seen within the context of Pierre's earlier work and is not an historical oddity.

It seems that the idea did not remain merely theoretical. The History of Science Museum in Oxford preserves a crudely made, small gothic astrolabe (HSM inv. nr. 40829) with a single plate of gilt brass (Figure 4.6).[46] It represents a projection limited to the equator. The rete includes an arc on which the zodiac signs are presented from both the northern and the southern sky. The ecliptic is thus folded onto itself. This corresponds precisely to our reconstruction of the 'line of longitude' on the *balista*. Indeed, the shape of the astrolabe's rete can be seen as representing a *balista*, as in the *Planicelium*. This reinforces our hypothetical reconstruction of the *balista* piece. However, in this Italian astrolabe (dated *c.* 1300), the piece that rotates is permanently on the front and cannot be transferred to the back of the astrolabe as in the *Planicelium*.

46 David A. King, 'A Remarkable Italian Astrolabe from ca 1300: Witness to an Ingenious Tradition of Non-standard Astrolabes', in *Musa Musaei, Studies on Scientific Instruments and Collections in Honour of Mara Miniati*, ed. by Marco Beretta, Paolo Galluzzi, Carlo Triarico (Florence: Leo S. Olschki Editore, 2003), pp. 29–52.

Figure 4.6. Oxford astrolabe, History of Science Museum, inv. nr. 40829, gilt brass, diameter 5.9 cm. Reproduced with permission.

Vimond's *Planicelium* shares another feature with Pierre's astrolabe that is even more revealing. The second part of Vimond's treatise refers to the beginning of the sign of Aries on the face of altitude as being '*in linea orizontis faciei altitudinis que fuerit descripta 66 graduum et 27 minutorum*'. A little further on in the text this face of the instrument is characterized as '*facies descriptionis altitudinis secundum latitudinem 66° 27′ que nunc deserviet pro descriptione septentrionalis medietatis spere signorum*'. The uses make it clear that this projection is used to represent either the northern or southern part of the sky. This choice allows the measuring of declinations and right ascensions of objects whose longitude and latitude are known. In modern parlance this can be described as the conversion of ecliptical to equatorial coordinates.

In some instrument texts from previous centuries, one can find this idea of astrolabe plates for such a horizon corresponding to the Arctic Circle or, in other words, a latitude of the complement of the obliquity of the ecliptic. Pierre of Maricourt's astrolabe text contains an explicit passage about the use of an astrolabe plate for 66° 24′ at the end of Chapter 21 (whose main topic is the construction of the hour lines).[47] See the edition by Ron B. Thomson, lines 57–62:

Describatur tamen cum aliis [tabulis] una tabula ad latitudinem 66 graduum et 24 minutorum in qua non sint horae neque domus et in ea erit horizon idem cum orbe signorum. Et almucantarat ostendent latitudines stellarum; azimut vero longitudines earundem. Per eam autem examinabitur volvellum et invenientur certa loca planetarum.	One shall construct with the other plates one to the latitude of 66° 24' on which there are no hour [lines] nor houses and on which the horizon will be the same as the orb of the signs [i.e. the ecliptic]. And the almuqantars show the latitudes of the stars, the azimuths their longitudes. Through this [plate] the *volvelle* will be examined and the certain places of the planets will be found.

This shows the intention behind the use of a plate for a latitude corresponding to the complement of the obliquity. Vimond's value for the obliquity is different from the value transmitted in Pierre's work, but the principle is the same. Pierre's Chapter 30 on the uses provides a striking example of how such a plate can be put to use; to know in which celestial house a planet is located, its current position (longitude and latitude) is first computed using the tables. Then, a pointer of wax or paste is fixed to the rete while using

47 Petrus Peregrinus de Maricourt, *Opera: Epistula de Magnete. Nova compositio astrolabii particularis*, ed. by Loris Sturlese e Ron B.Thomson (Pisa: Scuola Normale Superiore, 1995).

the 66° 25′ plate.[48] This 'augmented' rete is then placed on the plate with the local horizon and the house division to find the answer.

The cited passage of Pierre's *Nova compositio astrolabii particularis* is proof that such a 'horizon' to represent the ecliptic was used among astronomers concerned with instrument design. Moreover, in his catalogue of European astrolabes, King listed two other medieval European astrolabes that include plates for the latitude of the Arctic Circle.[49] Here again, we see that the idea of special horizons allowing coordinate conversion was also used in the construction of instruments. Although we were not lucky enough to find a *Planicelium* built in brass, the interpretation of Vimond's text allows us to situate the instrument, with all its originalities, among the astronomical instruments connected to early Alfonsine Astronomy.

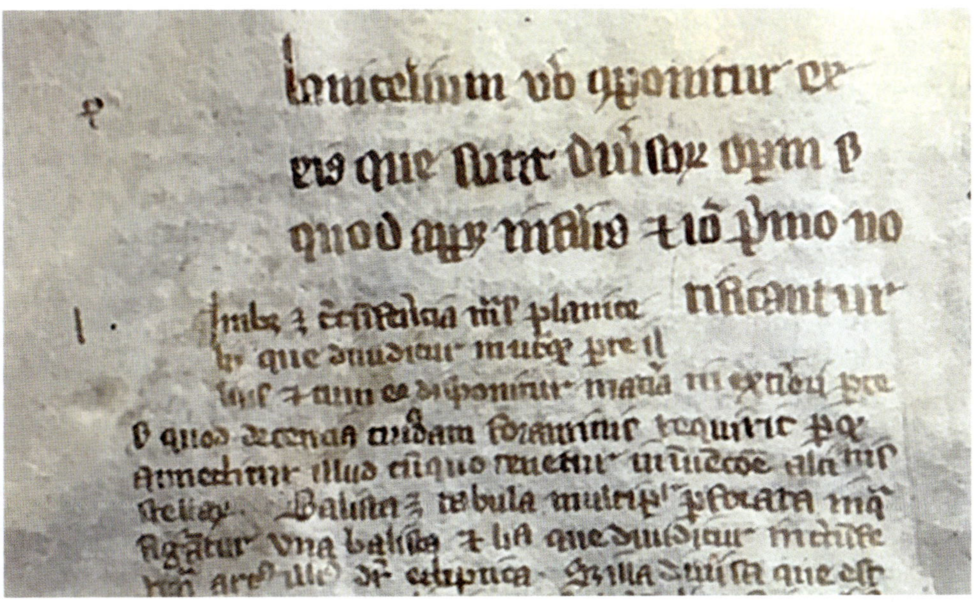

Figure 4.7. Beginning of the manuscript copy of Vimond's Planicelium. Erfurt, UB, CA 2° 377, f. 21ra. Source Digitale Historische Bibliothek Erfurt/Gotha.

48 Petrus Peregrinus de Maricourt, *Nova compositio*, Ch. 30 '*De quaerendis utilitatibus in speciali*' lines 157–73: '*Si volueris scire in qua domo sit planeta, aequa ipsum per tabulas ad quam volueris horam et scito gradum eius et latitudinem loci eius ab orbe signorum septemtrionalem vel meridionalem. Deinde pone volvellum supra tabulam descriptam ad latitudinem 66 graduum et 25 minutorum, ita quod gradus omnes arcus signorum sint in horizonte tabulae, et ab eis cooperiatur.* [...] *pone cacumen stili cerei vel caementei quem iunges cum volvello, ut moveatur cum eodem quasi sit stella fixa.* [...] *Cumque volueris scire in qua domo sit planeta, dispone volvellum supra tabulam tuae regionis in hora aequationis et aspice cacumen denticuli planetae. Cadet enim supra domum in qua fuerit in illa hora si sit septemtrionalis ab aequatore, et si sit meridionalis cadet in domo opposita ei in qua fuerit*'. Note that latitude value given in Chapter 21 is 66° 24′, which seems to be inconsistent with the value in this chapter.

49 Anglo-French astrolabe for Montpellier, Oxford, and elsewhere, dated 1326 (IC#291): London, British Museum, inv. MLA 1909, 6–17.1 (1909,0617.1); Petrus Raimundi, Barcelona, 1375 (IC#3053); Boston, Museum of Fine Arts, inv. 88.654. See David A. King's unpublished catalogue, Section 6, 'Early European Astrolabes to ca. 1500', online: https://www.academia.edu/35737805/MEDIEVAL_INSTRUMENT_CATALOGUE_EUROPEAN_6_1_6_9_pdf.

Appendix: edition and translation of Jean Vimond's *Planicelium*

Principles for the Latin edition

For the reasons exposed in the section titled 'The text of the *Planicelium* and its edition' of this chapter, the Latin text presented here cannot constitute an actual critical edition. Considering this, we used footnotes instead of a formal *apparatus criticus* to document the situation of the manuscript, our conjectures, and occasional comments. In general, we observed the following principles in editing:

When there is variation in the spelling of words (e.g. *orarum, horarum*), we opted to retain the variants in our transcription.

Several words in this copy are difficult to read due to erasures, corrections, and not least some palaeographical abbreviations. Guided by the general meaning, we ended up making conjectures. Where a word or part of a word in the Latin text is conjectured, we enclosed it in square brackets […]. We explain the specific difficulty in a footnote.

The hand of this copy used few punctuation marks (occasionally a dot slightly above the bottom line). When they do occur, they are rendered as a period mark. Our policy was to preserve the punctuation as in the manuscript, except for a few interventions in specific cases where additional periods were inserted (e.g. when a sentence was syntactically independent). These added periods are aimed at assisting the modern reader with the Latin text.

The manuscript offers the text divided in clear paragraphs. The second part of the treatise begins with a sub-title, the first paragraph of which is not indented. In transcribing, we respect the paragraph division and the subtitle, without artificially creating additional paragraphs or spacing. In this way, we seek to preserve a sense for what the scribe considered to belong together.

The manuscript text presents several underlined sentences. Our understanding is that the copyist attracts the reader's attention to these passages because he considers them important steps in the argumentation. We interpret this underlining as indicating a synthesis or conclusion of the preceding demonstration. Consequently, we made the choice to reproduce the underlining of all these passages both in the Latin edition and in the English translation.

To highlight the structure of the content, we added numbering to the text marking its different parts. First, in the description of the *Planicelium* we distinguish four parts, ⟨i⟩ to ⟨iv⟩. Then, in the uses of the instrument we divide the text into eight sections, ⟨1⟩ to ⟨8⟩.

Principles for the English translation

Our aim was to produce an easily intelligible text. We hence sometimes divided long sentences in two or solved some of the puzzles for our reader by adding a word in square brackets in English where there is no Latin counterpart. To explain these choices, or to clarify certain passages, we used footnotes.

For some rare or unknown words, we made the choice to keep the Latin terms: for example, *balista* or *planicelium*. We translated *linee oridiales* and *linee latorizales*, expressions that do not seem to occur in any other text, as 'oridial lines' and 'latorizal lines', helping us with English neologisms. For the word *tabula*, we preferred 'plate' to 'disc', as the former is typically used in astrolabe texts.

In general, we translate *descriptio* as 'drawing', in a sense of the diagram inscribed on the face of the instrument. Correspondingly, *describere* is translated as 'to draw' or 'to trace'. In the expression *in facie descriptionis altitudinis*, however, we keep the word 'description', as we think the expression refers to the side of the instrument where the scales referring to altitudes are drawn. Moreover, in the field of astrological vocabulary, we decide to keep *figura celi* in Latin and not translate it as horoscope.[50] This allows us to maintain a wider range of meanings, including both the astrological interpretation of the hour of observation and the chart of the sky at the moment of the observation.[51] For *judicia*, we use 'judgements', understood as astrological judgements based on the planetary configuration given by a particular *figura celi*. Finally, the word *intentio* is most tricky to translate. We choose to translate it by the English 'intention', although in the text of Vimond the word *intentio* seems to refer to Ptolemy's hypothesis or point of view.

The choice of punctuation was considered separately in the transcription of the Latin text and the translation. While the punctuation in the Latin text seeks to follow the manuscript as closely as possible, in the English translation we prefer to facilitate the reading of the text.

Finally, in correspondence with the Latin text, we introduced numberings in the translation, ⟨i⟩ to ⟨iv⟩ and ⟨1⟩ to ⟨8⟩, to highlight its structure. We also added the topic of each use in the margin.

50 The translation of *figura celi* as 'horoscope' is used by John D. North, *Horoscopes and history*.
51 Alexandre Tur, in this volume, 'From Astronomical Computation to Astrological Interpretation'.

Planicelium vero componitur ex eis que sunt diversorum operum, secundum quod apparet inferius et ideo primo notificantur.

⟨i⟩ Limbus est circumferentia materialis planicelii que dividitur in utraque parte illius et cum eo disponitur materia in exteriori parte, secundum quod decencia cuiusdam foraminis requirit per quod annectitur illud cum quo tenetur in inventione altitudinis stellarum. ⟨ii⟩ Balista est tabula multipliciter perforata in qua figuratur una balista et linea que dividitur in circumferentia arcus illius dicitur ecliptica. Sed illa divisa que est perpendicularis super eclipticam dicitur linea declinationis; et linea divisionis circumferencie baliste dicitur linea longitudinis. Item, linea interior est linea acuitatis extremitatis longioris dentuli exeuntis in una parte cuiuslibet sectionis circumferentie baliste. Sed linea iuxta quam incipit acuitas dicitur linea superior. Pinnule vero sunt due tabule parve que annectuntur super balistam in parte planicelii cuius partes principales simul iunguntur cum una cavilla et sibi pertinentibus. ⟨iii⟩ Facies altitudinis est illa planicies tabule continue in qua fiunt linee altitudinis atque sue divisionis et linea recta que procedit a cavilla versus initium numerationis divisionis limbi dicitur linea meridiei; sed linea medie noctis procedit e contrario. Et sciendum est quod pars orientis est illa que est versus sinistram tenentis, facie tamen existente supra, et illo puncto versus tenentem qui est oppositus illius puncti a quo incipit numeratio divisionis limbi. Et linea orizontis est illa cuius medium est in linea medie noctis et numeratur in duobus punctis oppositis circumferentie illius descriptionis et illa pars est super orizontem que est in parte concavitatis orizontis et alia sub orizonte. Item, linee dicuntur latorizales que secundum suam totam longitudinem continentur super orizontem vel sub orizonte. Alie vero que se intersecant in linea meridionali dicuntur oridiales. Et sciendum est quod extra istam descriptionem dividitur annus secundum quod requiritur ad inveniendum verum locum solis ita quod dies primi anni post diem cuiuslibet bisexti continetur inter duas divisiones; sed dies secundi anni incipit per quartam diei ante et dies tertii per duas quartas et dies quarti per tres.

Title Planicelium ... notificantur] In the upper margin: *'Compositio astrolabii ... magistri Johannis de Parvo Sancto ... magistri Johannis de Lineriis'*. This faint annotation is barely legible and was added by the same hand (however more cursive) that annotated the upper margin on folio 22r, attributed to John of Saxony. No reference to *Johannes de Parvo Sancto* could be found.

8 exeuntis] One can read *exeuntis* or *existentis* as several times infra; here, *exeuntis* makes more sense. **17** orizontis] Marginal correction *orizontis* with the same hand as *orientis*, which is omitted from the text. numeratur] The term *terminatur* would make more sense but from a strict paleographical point of view the reading is *numeratur*. **19** latorizales] *Latorizaleles* corrected by expunction as *latorizales*. **21** oridiales] One can read *oridiales* or *oridinales*. Further on in the text, the writing is clear and encourages us that the word *oridiales* is the correct one. **25** quartas] Marginal correction *quartas* with the same hand as *tertia*, which is omitted from the text.

The *Planicelium* is composed of elements that belong to various operations according to what appears below, and that is why they are first indicated[1]

⟨i⟩ The limb is the material circumference of the *Planicelium* that is divided on either side of it.[2] And together with this limb, some matter is placed at the exterior part, according to what adapting to any hole requires, through which something is attached and with which it is held while finding the altitude of the stars.[3]

⟨ii⟩ The *balista* is a plate with several perforations within which is represented a crossbow.[4] And the line that is divided on the circumference of its arc is called the ecliptic. That divided [line], however, that stands perpendicularly on the ecliptic, is called the line of declination. And the divided line of the circumference of the *balista* is called the line of longitude. Moreover, the interior line is the line of the longer extremity of the pointer that stands out on one side of any section of the *balista*'s circumference. But the line next to which the pointer begins is called the superior line. The pinnules, indeed, are two small plates that are attached on top of the *balista* on the side of the *Planicelium*, whose main parts are joined together with a pin and its parts.[5]

⟨iii⟩ The face of altitude is that surface of the full plate on which the lines of altitude are made, as well as its divisions. And the straight line that extends from the pin towards the beginning of the numbering of the limb's division is called the line of midday, but the line of midnight extends in the other direction. And one must know that the eastern part is the one that lies to the left of the person holding [the *Planicelium*], with the face still turned upwards, while the point, which is opposite the point from which the numeration of the limb's division starts, lies towards the person holding it. And the horizon line is that line whose midpoint lies on the line of midnight. And it ends at two opposite points of the circumference of that drawing. And that part is above the horizon, which is on the side of the horizon's concavity, and the other is below the horizon. Moreover, those lines are called latorizal [lines], which are contained, for their entire length, above or below the horizon. But the other lines that are mutually intersecting on the line of midday are called oridial [lines].[6] And one must know that on the outside of this drawing, the year is divided according to what is required to find the true position of the Sun, in such a way that a day of the first year after the day of any bissextile [year] is contained between two divisions. But a day of the second year begins earlier by a quarter of a day, and a day of the third by two quarters, and a day of the fourth by three quarters of a day.

---

1 The word *opera* might also refer to other works used by John Vimond to compose his own text. The explicit of the text seems to reinforce this understanding. However, the word can also be understood as 'operation' in the sense of the various uses of the *Planicelium* that are described (*notificantur*) in the second part of the text.

2 The instrument, like an astrolabe, has graduated scales on the front and the back.

3 *Stella*: a general term for 'celestial body' that includes both fixed stars, planets, and even the Sun.

4 *Balista* can be translated here with the term 'crossbow', referring to the shape of that part of this instrument.

5 *Cavilla*: the pin, as with astrolabes, is typically perforated for passing a wedge that fixes it in place.

6 See the above the section, 'Lexical matters'.

⟨iv⟩ Facies orarum est illa planicies tabule continue in qua disponitur linea meridiei, sicuti alia facie, et cursus est de linea medie noctis et de parte orientis et occidentis, et super orizontem et sub orizonte, et etiam de linea orizontis si secundum eandem latitudinem fuerint facte et licet fuerint descripte secundum diversas. Illa tamen dicitur orizon cuius medium est in linea medie noctis et numeratur in duobus punctis oppositis circumferentie illius descriptionis; et cum istis sunt linee temporales que concurrunt in duobus punctis; et linee divisionis orarum sunt ille aliarum quelibet due propinque distant ab invicem in circumferencia istius descriptionis per 15 gradus ; et ille dicuntur linee divisionis domorum quarum quelibet due propinque distant ad invicem in circumferenciam per 30 gradus. Et sciendum est quod extra istam descriptionem sunt note plurium stellarum fixarum cum numeris suarum latitudinum. Et sciendum est quod quando scribitur hoc de lineis, intelligendum est de lineis continuis vel punctualibus descriptis vel ymaginatis inter duas descriptas et tales debent judicari ymaginate nam secundum proportionem sue denominationis ad totam differentiam duarum descriptarum inter quas | debent ymaginari et haec proportio satis convenienter potest 21rb
ymaginari secundum quod apparuerit in intellectum dispositoris illarum partium.

Deinde sequitur de doctrina [inquisi]tionis alicuius in planicelio

⟨1⟩ Scito quod radius solis aut visualis directus ad stellam transeat per ambo foramina pinnularum cuiuslibet partis planicelii ita quod illud tantummodo teneatur per illud quod annectitur cum limbo et quod linea declinationis unius baliste sit in medietate superioris et alia in inferori et invenies quantitatem altitudinis illius scilicet inter caput linee declinationis superioris et initium numerationis divisionis limbi. ⟨2⟩ Et si linea interior sit super partem alicuius diei, linea superior ostendit satis sufficienter locum solis in limbo, si numeretur secundum successionem numerationis dierum, scilicet in illa parte diei longitudinis secundum quam circumferencia divisionis anni fuerit ordinata et in quattuor annis primis post diem bisexti 1324[i] anni domini nostri Jhesu Christi ita quod initium sit ex initio ianuarii et sic apparebitur locus stelle cuius nota fuerit circumferentia faciei horarum nisi quod linea interior sit super notam illius. Sed cum dies tue intentionis sit in aliquo anno sequente et velis invenire locum solis, adde differentiam augis solis existentis in anno 1326° et secundum annum sequentem in limbo cum distan-

27 cursus] Reading not certain. **30** numeratur] The term *terminatur* would make more sense but from a strict paleographical point of view the reading is *numeratur*. **31/32** concurrunt ... punctis] One part of this sentence was erased. The words '*concurrunt in duobus punctis*' written in the erased space are shorter. The ink of the new words is darker. **36** plurium] Reading not certain for *plurium*. **41** Chapter title : inquisitionis] Reading not clear for *inquisi-*.

⟨iv⟩ The face of hours is that surface of the full plate on which the line of midday is placed, as on the other face, and the course[7] [of hours] is [made] from the line of midnight, at the same time from the eastern and western sides, and above and below the horizon, and also from the horizon line, if they [the lines] are made according to the same latitude, although they may be drawn according to different ones [latitudes]. At any rate, that [line] is called the horizon, the midpoint of which lies on the line of midnight and that ends at two opposite points of the circumference of that drawing. And together with these, there are the temporal lines that run through two points. And the lines of division of the hours are those, among the others, where any two neighbouring lines are distant by fifteen degrees from each other on the circumference of that drawing. And those [lines] are called lines of division of the houses of which any two neighbouring lines are distant by thirty degrees from each other on the circumference. And one must know that on the outside of this drawing, there are marks of many fixed stars with the values of their latitudes. And one must know that when this is written about the lines, one must understand it about the continuous lines or the dotted lines[8] or imaginary lines between two drawn [lines]. And such [lines] must be judged as imaginary indeed according to the proportion of its denomination to the whole difference of two drawn [lines], between which | they must be imagined. And this proportion can be 21rb imagined in a quite convenient way according to what would appear in the mind of the designer of these parts.

Thereafter follows: On the doctrine of determining something on the *Planicelium*

Height of a celestial body ⟨1⟩ You shall know that the Sun's ray or the visual ray in direction of a star traverses both holes of the pinnules of any side of the *Planicelium*, in such a way that the latter should be held only by the piece[9] that is attached to the limb and in such a way that the line of one declination of the *balista* is in the upper half and the other [one] is in the lower [half]; and you will find the quantity of its altitude, namely between the extremity of the superior line of declination and the beginning of the numbering of the scale on the limb.

Longitude of the Sun and stars ⟨2⟩ Also, if the interior line is on a part of any day, the superior line, on the limb, indicates quite adequately the position of the Sun, if it is counted according to the succession of the numbering of the days, that is, in that part of the longitude of the day according to which the circumference of the division of the year was arranged and in the first four years after the leap day of the 1324th year of our Lord Jesus Christ such that the start be taken from the beginning of January. And in this way the position of a star whose mark is present on the circumference of the face of the hours, will appear only when the interior line is on top of its mark.

If, however, the day you consider is in some subsequent year and you intend to find the position of the Sun, add the difference of the apogee of the Sun existing in the year

7 The reading of these words is uncertain in the manuscript.

8 *'lineis continuis vel punctualibus descriptis vel ymaginatis'*: this a very rare reference to continuous or non-continuous lines (dotted lines) and even imaginary lines.

9 This part of the astrolabe is called the throne.

tia predicti loci solis ab initio numerationis limbi; vel ab ea minue si dies sit in aliquo precedente. Et similiter fac pro stella et invenies distanciam loci cuiuslibet illorum ab initio Arietis immobilis. Et sciendum est quod motus augis solis in quolibet anno continet minuta secunda et tertia secundum intentionem Ptholomei Feludiensis. Cum stella sit septentrionalis ab equinoctiali, computa distantiam illius et initii Arietis immobilis in linea orizontis faciei altitudinis que fuerit descripta 66 graduum et 27 minutorum, et incipias ab illa extremitate que fuerit in parte orientis. Et si latitudo sit ei septentrionalis, computa eam [a termino] predicte computationis, scilicet in linea oridiali transeunte per illum terminum et in parte super orizontem vel sub orizonte, si sit meridionalis. Et terminus istius computationis erit proprius locus illius stelle in facie altitudinis. Et similiter poteris invenire locum oppositi meridionalis. Sit enim linea declinationis super locum stelle septentrionalis ab equinoctiali vel oppositi puncti meridionalis in facie descriptionis altitudinis secundum latitudinem 66 graduum et 27 minutorum que nunc deserviet pro descriptione septentrionalis medietatis spere signorum. Et distancia illius loci ab initio numerationis linee declinationis erit declinatio illius septentrionalis vel oppositi meridionalis et distantia linee meridiei et linee declinationis existens in linea longitudinis, erit longitudo illius a linea declinationis. ⟨3⟩ Minue altitudinem meridionalem alicuius stelle existentis in equinoctiali de 90 vel si declinatio sit ei meridionalis, adde eam cum sua altitudine meridionali et illud quod proveniet, minue de 90. Et si declinatio sit ei septentrionalis et altidudo meridiana fuerit in linea australi, minue altitudinem de 90 et cum residuo adde illam declinationem. Sed si altitudo meridiana vel maior sit in linea septentrionali, minue declinationem de 90 et residuum de altitudine; vel si minor altitudo fuerit in linea septentrionali, minue declinationem de 90 et adde residuum cum altitudine et invenies latitudinem illius loci in quo fuerit altitudo quocumque predictorum modorum contingerit. Vel sic: minue minorem altitudinem stelle de maiori si ambe fuerint in linea australi vel septentrionali. Et si fuerint in linea septentrionali, adde medietatem illius quod remanserit cum minori altitudine. Sed si fuerint in meridionali, minue medietatem prioris remanentis de maiori altitudine et residuum de 90 et invenies sicut prius.

62 a termino] Illegible. We suggest *a termino* with quite a lot of uncertainty. oridiali] See comment to line 21: here, and further on in the text, *oridialis* is clear. **71** declinationis] There is a blank space before *minue*.

1326 and that of a following year,[10] on the limb, to the distance of said position of the Sun from the beginning of the numbering of the limb. Or, subtract it if the day is in some preceding [year].

Similarly, do this for a star and you will find the distance of the position of any of these from the beginning of the immobile Aries. Also, one must know that the motion of the Sun's apogee in any year contains second minutes and thirds[11] according to the intention[12] of Ptolemy Feludiensis.[13]

When a star is north of the equator, count the distance between it and the beginning of the immobile Aries on the line of the horizon on the face of altitude that was traced of 66 degrees and 27 minutes. And you shall start from that extremity that is on the eastern side. And if the latitude is north of it, reckon it from the endpoint of the above-said reckoning, that is, on the *oridial* line passing through that endpoint in the part above the horizon, or below the horizon if it is southern. And the endpoint of this reckoning will be the appropriate position of that star on the face of altitude. And, similarly, you could determine the position of the opposite of a southern [point].

Let the line of declination be on the position of a star that is north of the equator or [on the position] of the opposite of a southern point on the face of the description[14] of altitude according to the latitude of 66 degrees and 27 minutes, which will now serve for representing the northern half of the sphere of the signs. And the distance of this position from the beginning of the numbering of the line of declination will be its northern declination or the southern [declination] of its opposite, and the distance between the line of midday and the line of declination existing in the line of longitude will be its longitude from the line of declination.

-mination of local latitude

⟨3⟩ Subtract the meridian altitude of a given star existing on the equator from 90, or if the declination is southern, add it to its meridian altitude, and subtract the result from 90; and if the declination is northern and the meridian altitude is in the southern line, subtract the altitude from 90 and add the remainder together with the declination. But if the meridian altitude is in the northern line, and if it is a higher [altitude], subtract the declination from 90 and the remainder from the altitude; or, if a lower altitude is in the northern line, subtract the declination from 90 and add the remainder together with the altitude, and you will find the latitude of that place in which the altitude is found in whichever of the above-stated ways it occurred. Or in this way: Subtract the lower altitude of a star from the higher [altitude] if both were in the austral or the northern line. And if they are in the northern line, add half of the remainder to the lower altitude. But if they are in the southern [line], subtract half of the previous remainder from the higher altitude and [subtract] the remainder from 90 and you will find [the latitude of the place] just as before.

10 We understand *secundum annum sequentem* as being the Sun's apogee in 'some year after' 1326 and not only 1327.

11 The Latin passage could be interpreted as reading *minuta secunda et* [*minuta*] *tertia*, meaning 'second-order minutes and third-order minutes'. See the points in footnotes 37 and 38 of our chapter.

12 'Intention' translates the word *intentio* but refers here to a 'thesis' or a 'firmly grounded claim' expressed in Ptolemy's work. See also the last sentence of the *explicit*.

13 The reference to Ptolemy's theory about the solar apogee is quite difficult to understand here. See, again, comments in footnotes 37 and 38 of our chapter.

14 For this translation of *description*, see the principles of edition and translation.

⟨4⟩ Cum sol sit septentrionalis, scito distantiam illius ab initio Arietis inmobilis et computa eam ab initio Arietis | in ecliptica baliste et terminus illius computationis erit 21va locus solis in ecliptica. Deinde pone illum locum super lineam latorizalem in parte super orizontem faciei altitudinis que distet ab orizonte secundum quantitatem altitudinis solis ita quod si altitudo solis fuerit ante meridiem, pone locum eius in parte orientis; et si post meridiem, pone eum in parte occidentis. Sed si sol fuerit meridionalis, pone locum sui oppositi in ecliptica, super equidistantem in parte sub orizonte que distabit ab eodem secundum quantitatem altitudinis solis; ita tamen quod si illa altitudo fuerit ante meridiem, pone predictum locum oppositi in parte occidentis; et si fuerit post meridiem, pone eum in parte orientis. Et tunc erit balista disposita secundum figuram celi. Vel sic: quere punctum in linea declinationis qui distabit ab initio sue numerationis secundum quantitatem declinationis stelle. Et si stella sit septentrionalis, dispone illum punctum declinationis sicut dictum est de loco solis septentrionalis. Et si stella sit meridionalis, dispone predictum punctum sicut dictum est de loco solis oppositi meridionalis. Et deinde nota directum capitis linee declinationis in limbo quocumque predictorum modorum contingerit. Et si stella sit septentrionalis et in illa medietate celi que incipit ex initio Capricorni inmobilis et procedat secundum successionem signorum, nota punctum in linea longitudinis existentem in parte illorum signorum qui distet ab initio numerationis illius linee secundum quantitatem longitudinis stelle a linea declinationis. Et si stella sit in alia parte signorum, nota punctum in alia parte linee sicut dictum est. Deinde volve balistam quousque nota linee longitudinis sit in directo note que prius fuerit facta in limbo et tunc erit balista disposita. Vel sic: pone locum solis in ecliptica super lineam initii partium alicuius diei si sit septentrionalis, sed cum sit meridionalis, pone suum oppositum super lineam oppositam et quocumque istorum modorum contingerit, nota directum capitis linee declinationis in limbo. Deinde elonga predictum caput a puncto notato secundum quantitatem illarum partium diei insimul ita quod elongatio fiat secundum successionem numerationis limbi. Et tunc erit balista disposita super figuram celi. Vel sic: nota directum capitis linee declinationis unius baliste cum fuerit disposita super faciem altitudinis; et deinde pone caput alterius linee declinationis in directo similis puncti alterius partis limbi. Et tunc erit balista disposita super faciem horarum secundum modum alterius.

98 oppositi meridionalis] One *i* between *oppositi* and *meridionalis* was marked by subpunction and a misplaced *a* was also marked by subpunction in *meridiaonalis*.

Balista
ording to
figura celi

⟨4⟩ When the Sun is northern, you shall know its distance from the beginning of the immobile Aries | and reckon from the beginning of Aries on the ecliptic of the *balista*, 21va
and the endpoint of this reckoning will be the Sun's position in the ecliptic. Then put this position on a latorizal line in the part above the horizon of the face of altitude, which is distant from the horizon according to the quantity of the Sun's altitude in such a way that if the Sun's altitude is before noon, put its position on the eastern side, and if [it is] after noon, put it on the western side. But if the Sun is southern, put the position of its opposite in the ecliptic onto the parallel [circle][15] in the part below the horizon, which will be distant from it [the horizon] according the quantity of the Sun's altitude; but in such a way that if that altitude is before noon, put the aforementioned position of the opposite on the western side, and if it is after noon, put it on the eastern side. And now the *balista* will be placed according to the *figura celi*.

Or, in this way: search the point on the line of declination that will be distant from the beginning of its numbering according to the quantity of a star's declination. And if the star is septentrional, place that point of declination as it has been said about the Sun's northern position. And if the star is southern, place the aforementioned point as it has been said about the opposite of the Sun's meridional position. And then note the point[16] in front of the extremity of the line of declination in whichever of the above-stated ways it occurred. And if the star is northern and in that half of the sky that starts from the beginning of the immobile Capricorn and extends according to the order of the signs, note the point that exists in the line of longitude, in the part of those signs which is distant from the beginning of the numbering of that line according to the quantity of the star's longitude from the line of declination.[17] And if the star is in the other part of the signs, note the point in the other part of the line such as has been said. Then turn the *balista* so the mark of the line of longitude is in front of the mark previously made in the limb, and then the *balista* will be placed [according to the *figura celi*].

Or, in this way: place the Sun's position in the ecliptic on the line of the beginning of the parts of any day if it [the Sun] is northern.[18] But if it is southern, place its opposite on the opposite line, and in whichever of the above-stated ways it occurred, note the point on the limb in front of the extremity of the line of declination. Then, shift[19] the aforementioned extremity away from the noted point according to the quantity of those parts of the day at the same time, in such a way that the shift will follow the order of the numbering of the limb. And then the *balista* will be placed according to the *figura celi*.

Or, in this way: note the point directly at the extremity of one line of declination of the *balista* when it is placed on the face of altitude. Then, place the extremity of the other line of declination in front of the similar point of the other side[20] of the limb. And then the *balista* will be placed on the face of hours according to the way of the other [side].

---

15 *Equidistantem:* we understand that these are the projections of the circles parallel to the horizon that is the almucantars below the horizon (which are not usually traced on an astrolabe).

16 We understand *punctum* should be added to *directum*. The same expression occurs in the last paragraph of this section.

17 The longitude of the star is measured along the circumference of the *balista*, starting from the intersection of the line of declination with the arc of the *balista*. See supra paragraph ⟨ii⟩.

18 The line of the beginning of the parts (*linea initii partium...*) means the line of the unequal hours, as we assume.

19 The action of *elongare* is here translated as the phrasal verb 'shift away'; the noun for *elongatio* is 'shift'.

20 The face of hours.

⟨5⟩ Et erit prima diei secundum figuram celi cum balista fuerit disposita secundum illam figuram et locus solis in ecliptica fuerit super primum spatium hore in parte orientis et super orizontem. Sed si locus oppositi fuerit super illud spatium, erit prima noctis. Et si aliquis eorum sit super secundum, erit secunda, et sic de aliis suo modo in parte super orizontem. Sed si locus solis in ecliptica fuerit super primum spatium hore in parte occidentis et sub orizonte erit prima noctis vel prima diei, si sit locus oppositi solis. Et si aliquis eorum sit super, erit secunda, et sic de aliis. Et quocumque modorum contingerit pars ore denotabitur per distantiam initii hore et linee temporales supra quam locus solis vel sui oppositi erit dispositus.

⟨6⟩ Et cum balista fuerit disposita super faciem horarum secundum figuram celi, scito quod ecliptica dividetur in sex partes, secundum sex domos, quarum initia patebunt per lineas divisionis illarum domorum. Et per illa initia denotabuntur per oppositionem initia aliarum domorum. Nam prima omnium que est in parte orientis et sub orizonte opponitur septime, et secunda octave, tertia none, quarta decime, quinta undecime, sexta duodecime, quarum omnium partes apparebunt per lineas temporales que sunt inter lineas divisionis domorum.

Et in illis domibus et partibus satis apparebunt stelle septentrionales et existentes in ecliptica sed meridionales existentes in ecliptica erunt in domibus oppositis illarum in quibus erunt sua puncta opposita. Et cum stella sit extra eclipticam, nota directum capitis linee declinationis in limbo; et si stella sit in illa medietate signorum que incipit ex initio Ca|pricorni inmobilis et procedit secundum successionem signorum, nota 21vb
punctum in linea longitudinis existentem in parte aliorum signorum qui distet ab initio numerationis linee longitudinis secundum quantitatem longitudinis stelle a linea declinationis. Et si stella sit in illa medietate que incipit ex initio Cancri, nota punctum in alia parte linee longitudinis. Deinde volve balistam quousque nota linee longitudinis sit in directo note que fuerit in limbo. Et deinde quere punctum declinationis stelle in linea declinationis; nam stella disponetur in domo et in parte et cum existentibus in parte secundum quod ille punctus erit dispositus. Sed si stella sit meridionalis, illa disponetur in opposito domus et partis sui oppositi.

⟨7⟩ Et cum locus alicuius puncti celestis in aliqua divisione baliste fuerit dispositus super faciem horarum secundum figuram celi, nota directum capitis linee declinationis in limbo. Et si ille punctus celestis sit septentrionalis ab equinoctiali, volve balistam secundum motum spere signorum cum directio debeat fieri secundum successionem

122 super] After *super*, the deleted word seems to be *secundum*. **127** per oppositionem] One can read *per oppositum* or *per oppositionem*. per[2]] After *per*, one word fragment, *diui*, is deleted. **129** secunda] The word *secunda* is not clear. However, the second house should be opposed to the eighth. **147** sit] After *sit*, *sept* is deleted.

Iour of the day for a figura celi

⟨5⟩ And it will be the first [hour] of the day with respect to the *figura celi* when the *balista* is placed according to that figure, and the position of the Sun in the ecliptic is on the first space of the hour on the eastern side and above the horizon. But if the position of the opposite is on that space it will be the first [hour] of the night. And if either of these is on the second [space of the hour], it will be the second [hour], and so on also about the other in their way in the part above the horizon. But if the position of the Sun in the ecliptic is on the first space of the hour on the western side and below the horizon, it will be the first [hour] of the night, or the first of the day if it is the position of the opposite of the Sun. And if either of these is on the second, and so on about the others. And in whichever of the ways it occurred, the part of the hour will be represented by the distance of the beginning of the hour and the temporal lines on which the position of the Sun or its opposite will be placed.

nification, how to tion a star in a house

⟨6⟩ And when the *balista* is placed on the face of hours according to the *figura celi*, you shall know that the ecliptic is divided into six parts according to the six houses, the beginnings[21] of which will appear through the lines of division of those houses. And through these beginnings will be indicated, by virtue of opposition, the beginnings of the other houses. Indeed the first of all [houses] which is on the eastern side and below the horizon is opposite the seventh, and the second [is opposite] the eighth, the third [opposite] the ninth, the fourth [opposite] the tenth, the fifth [opposite] the eleventh, [and] the sixth [opposite] the twelfth, the parts of these all will appear through the temporal lines that are between the lines of division of the houses.

And in those houses and parts the northern stars located on the ecliptic will fully appear, whereas the southern [stars] located on the ecliptic will be in the opposite houses to those in which their opposite points will be. And when[22] a star is outside of the ecliptic, note the point directly at the extremity of the line of declination on the limb; and if the star is in that half of the signs that starts from the beginning of the immobile Ca|pricorn and goes on according to the order of the signs, note the point on the line of 21vb
longitude that exists on the part of the other signs, that is distant from the beginning of the numbering of the line of longitude according to the quantity[23] of the star's longitude from the line of declination. And if the star is in that half that starts from the beginning of Cancer, note the point on the other part of the line of longitude. Then turn the *balista* so the mark of the line of longitude is in front of the mark that is on the limb. And then search the point of the declination of the star on the line of declination; indeed, the star will be placed in the house and in the part [of this house]; and with those existing in the part according to which [part] that point will be placed. But if the star is southern, this [star] will be placed in the opposite of the house and of the part of its [star's] opposite.

irecting a ficator on promissor

⟨7⟩ And when the position of any celestial point in any division of the *balista* is placed on the face of hours according to the *figura celi*, note [the point] in front of the extremity of the line of declination on the limb. And if that celestial point is north of the equator, turn the *balista* according to the motion of the sphere of the signs when the

---

21 The technical term for the beginning of celestial houses is 'cusp' but we prefer to keep the non-technical word.

22 *cum* is in the non-temporal sense here, 'in the case of'.

23 Refer supra to paragraph ⟨ii⟩.

signorum, vel retrocede eam si debeat fieri contra successionem signorum quousque locus illius septentrionalis in aliqua divisione baliste sit super lineam temporalem super quam fuerit aliquis significator si sit septentrionalis ab equinoctiali; vel super lineam oppositam illi temporali super quam fuerit locus oppositi significatoris si sit meridionalis. Et distantia capitis linee declinationis et illius note que fuerit in limbo denotabit arcum directionis illius significatoris ad illum punctum celestem. Sed cum punctus celestis sit meridionalis, volve balistam sicut prius quousque locus sui oppositi in aliqua divisione baliste sit super lineam temporalem super quam fuerit locus oppositi alicuius significatoris si sit meridionalis ab equinoctiali; vel super lineam oppositam illi temporali super quam fuerit locus significatoris si sit septentrionalis. Et arcus directionis denotabitur sicut prius.

Et cum aliquis punctus celestis sit septentrionalis ab equinoctiali et locus eius in aliqua divisione baliste fuerit dispositus super faciem altitudinis, scito super quam partem linee oridialis sit et super quam latorizalem. Nam illa pars oridialis denotabit partem orizontis in qua erit ille punctus secundum figuram illius dispositionis. Et distantia orizontis et illius latorizalis denotabit altitudinem illius puncti. Sed cum punctus sit meridionalis et locus sui oppositi fuerit dispositus super faciem altitudinis, scito super quam partem linee oridialis sit et super quam latorizalem. Nam illa pars oridialis denotabit partem oppositam illius in qua erit ille punctus secundum figuram illius dispositionis, et distantia illius latorizalis et orizontis denotabit altitudinem illius puncti.

⟨8⟩ Sume altitudinem stelle ignote et illius cognite que cum ignota erit in una linea perpendiculari et hoc erit cum ille due potuerint videri equidistare lateri planicelii cum teneatur tantummodo per illud quod annectitur cum limbo. Deinde dispone balistam cum altitudine stelle cognite super faciem altitudinis. Et scito super quam lineam oridialem erit locus illius stelle cognite in aliqua divisione baliste et deinde quere locum altitudinis illius ignote, scilicet in parte super orizontem et in illa parte predicte oridialis que denotabit illam partem orizontis in qua fuerit illa ignota in tempore considerationis sue altitudinis; et si sic possit inveniri, illa stella ignota erit septentrionalis ab equinoctiali. Et si sit aliter, illa erit meridionalis; quere igitur locum sue altitudinis in parte sub orizonte et in predicta linea oridiali et quecumque predictorum modorum contingerit, pone lineam declinationis super locum altitudinis et nota punctum illius linee qui erit super illum locum; et etiam scito distanciam in limbo que fuerit inter illum locum altitudinis et locum altitudinis illius stelle cognite, et versus quam partem illa procedit a loco altitudinis illius cognite et serva eam. Et cum sic fuerit, pone lineam declinationis super locum illius stelle cognite vel sui oppositi in facie altitudinis que deserviet pro dispositione septentrionalis | medietatis spere signorum. Deinde volve 22ra balistam secundum predictam distantiam locorum altitudinis et secundum quod illa

153/154 Et … celestem] The sentence '*Et distancia capitis…celestem*' is underlined with ink in the manuscript.
162/164 Nam … latorizalis] From '*Nam illa pars…*' to '*…illius latorizalis*' the sentences are underlined with ink.
170 altitudinem] *altitudinem* was deleted.

direction must be made according to the order of the signs; or turn it backwards if it must be made in the reverse order of the signs; so that the position of the northern [celestial point] on any division of the *balista* is on the temporal line, on which was a significator, if it is north of the equator; or on the line opposite to that temporal [line], on which was the position of the opposite of the significator, if it is south [of the equator]. And the distance between the extremity of the line of declination and the mark on the limb will represent the arc of direction of that significator to that celestial point. When, however, the celestial point is southern, turn the *balista* as before so that the position of its[24] opposite on a division of the *balista* is on the temporal line, on which was the position of the opposite of some significator if it is south of the equator; or on the opposite line of that temporal [line] on which was the position of the significator if it is north [of the equator]. And the arc of direction will be represented as before.

And when any celestial point is north of the equator and its position in some division of the *balista* is placed on the face of altitude, you shall know on which part of the oridial line and on which latorizal [line] it is. Indeed, that part of the oridial [line] will represent the part of the horizon where that point will be according to the *figura* [*celi*] of that disposition. And the distance between the horizon and that latorizal [line] will represent the altitude of that point. But when the point is southern and the position of its opposite [point] is placed on the face of altitude, you shall know on which part of the oridial line and on which latorizal [line] it is. Indeed, that part of the oridial [line] will represent the opposite part of that [part] in which will be that point according to the figure of that disposition. And the distance between its latorizal [line] and horizon will represent the altitude of that point.

tic coordi- ates for an nown star

⟨8⟩ Take the altitude of an unknown star and [that] of a known [star] that will be together with the unknown in one perpendicular line. And this will be [the case] when these two can be seen at the same distance from the side of the *planicelium* when it is held only by that [part] that is attached to the limb. Then place the *balista* according to the altitude of the known star on the face of altitude. And you shall know on which oridial line the position of that known star will be in some division of the *balista,* and then search the position of the altitude of that unknown [star], that is to say in the part above the horizon and in that part of the aforementioned oridial [line] that will represent that part of the horizon in which was the unknown [star] at the moment of the observation of its altitude. And if it can be found in this way, that unknown star will be north of the equator. And if it is different, that will be south. Search, therefore, the position of its altitude in the part below the horizon and on the aforementioned oridial line. And in whichever of the aforementioned ways it occurred, put the line of declination on the position of the altitude and mark the point of that line that will be on that position. And you shall also know the distance on the limb that is between that position of altitude and the position of altitude of that known star, and towards which side it [the distance] extends from the position of the altitude of that known star, and keep it. And when such is the case, put the line of declination on the position of that known star or of its opposite on the face of altitude that will be used for the disposition of the northern | half of the sphere of the signs. Then turn the *balista* according to the aforemen- 22ra
tioned distance of the positions of altitude and according to how it [the distance]

24 'Its', that is, the celestial point's nadir or opposite point.

procedit a loco altitudinis stelle cognite. Et tunc punctus linee declinationis prenotatus erit super verum locum illius stelle prius ignote. Et si illa sit septentrionalis ab equinoctiali et ille locus sit in parte super orizontem, illa erit septentrionalis ab ecliptica, et supradictus locus sit in parte sub orizonte, illa erit meridionalis. Sed si illa sit meridionalis ab equinoctiali et predictus locus sit in parte sub orizonte, illa erit septentrionalis ab ecliptica; et si sit in parte super orizontem, illa erit meridionalis.

Explicit tractatus Johannis Vimundi de floribus illorum que per instrumenta spere signorum possunt inveniri ad faciendum judicia astronomica secundum intentionem summi Ptholomei Feludiensis.

190 meridionalis[1]] The word *meriodinalis* was first written with points under *o*, *d*, and *i* for expunction. The correct letters *d*, *i*, and *o* were inserted above the word **195** Ptholomei] *Pto.* deleted.

extends from the position of altitude of the known star. <u>And then the point of the line of declination marked before will be over the true position of the star that was previously unknown</u>. And if that star is north of the equator and that position is in the part above the horizon, that [star] will be north of the ecliptic, and [if] the aforementioned position is in the part below the horizon, that [star] will be south. If, however it is south of the equator and the aforementioned position is in the part below the horizon, that [star] will be north of the ecliptic; and if it is in the part above the horizon, that [star] will be south.

Here ends the treatise of John Vimond about the flowers of those [things] that can be found through instruments of the sphere of the signs, in order to make astronomical judgements according to the intention of the greatest Ptolemy Feludiensis.

Manuscript Sources

Bruges, Bibliothèque Publique, Ms. 523
Brussels, Bibliothèque Royale, 10117–26
Cambridge, Gonville and Caius College, 141/91
Cracow, Biblioteka Jagiellońska, 551
El Escorial, Real Biblioteca del Monasterio de San Lorenzo, O. II. 10
Erfurt, Universitätsbibliothek, CA 2° 395
Erfurt, Universitätsbibliothek, CA 2° 377
Madrid, Biblioteca Nacional, 4238
Munich, Bayerische Staatsbibliothek, Clm 26667
Paris, Bibliothèque nationale de France, fr. 2643
Paris, Bibliothèque nationale de France, fr. 19093
Paris, Bibliothèque nationale de France, lat. 7286C
Paris, Bibliothèque nationale de France, lat. 7295
Paris, Bibliothèque nationale de France, lat. 7482
Paris, Bibliothèque nationale de France, lat. 16621
Segovia, Biblioteca de la Catedral, 84
Vatican, Biblioteca Apostolica Vaticana, Pal. lat. 1396
Vatican, Biblioteca Apostolica Vaticana, Urbin. lat. 1399
Vatican, Biblioteca Apostolica Vaticana, Ottob. lat. 1826

Bibliography

Boudet, Jean-Patrice, *Recueil des plus célèbres astrologues de Simon de Phares*, 2 vols (Paris: H. Champion, 1997–1999).

Busard, Hubertus L.L., *Het rekenen met breuken in de middleleeuwert, in het bijzonder bii Johannes de Lineriis* (Brussels: Vlaamse Academie, 1968).

Casulleras, Josep, 'The Instruments and the Exercise of Astrology in the Medieval Arabic Tradition', *Archives Internationales d'Histoire des Sciences*, 63(170–71) (2013), 517–40.

Chabás, José, *Computational Astronomy in the Middle Ages* (Madrid: Consejo Superior de Investigaciones Científicas 2019).

Chabás, José and Bernard R. Goldstein, 'Early Alfonsine Astronomy in Paris: The Tables of John Vimond (1320)', *Suhayl*, 4 (2004), 207–94.

———, *A Survey of European Astronomical Tables in the Late Middle Ages* (Leiden-Boston: Brill, 2012).

———, 'The Medieval Moon in a Matrix: Double Argument Tables for Lunar Motion', *Archive for History of Exact Sciences*, 73 (2019), 335–59.

———, 'Vimond and the Alfonsine Trepidation Model', *Journal for the History of Astronomy*, 34 (2003), 163–70.

Chabás, José and Marie-Madeleine Saby, *The Tables of 1322 by John of Lignères: An Edition with Commentary* (Turnhout: Brepols, 2022).

Du Cange et al., *Glossarium mediae et infimae latinitatis* (Niort: L. Favre, 1883–87). Online (16 March 2022): ducange.enc.sorbonne.fr.

Eade, John Christopher, *The Forgotten Sky: A Guide to Astrology in English Literature* (Oxford: Clarendon Press, 1984).

Forcinelli, Aegidius, *Lexicon totius latinitatis*, vol. 1 (A-C). Online (16 March 2022): www.documentacatholicaomnia.eu.

Glorieux, Palémon, *La Faculté des arts et ses maîtres au XIIIe siècle*, Etudes de philosophie médiévale, 59 (Paris: Librairie philosophique J. Vrin, 1971).

Husson, Matthieu, 'La *Tabula Tabularum* de Jean de Murs: nombres et opérations arithmétiques en astronomie au début du XIVème siècle', in *Cahiers de recherches médiévales et humanistes*, 27 (Auxerre: Centre d'études médiévales d'Orléans, 2014), pp. 96–122.

———, 'Construire un instrument astronomique au XIVème siècle: le second équatoire de Jean de Lignères', in *Science et technique au Moyen Âge (XIIe–XVe siècles)*, ed. by Joël Chandelier, Catherine Verna, and Nicolas Weill-Parot (Vincennes: Presses universitaires de Vincennes, 2014), pp. 225–45.

Husson, Matthieu and Marie-Madeleine Saby, 'Le manuscrit d'Erfurt CA F 377 et l'astronomie alphonsine', in *Miscellanées scientifiques au Moyen Age*, Micrologus, 27 (Florence: SISMEL Edizioni del Galluzzo, 2019), pp. 202–33.

King, David A., 'A Remarkable Italian Astrolabe from ca 1300: Witness to an Ingenious Tradition of Non-standard Astrolabes', in *Musa Musaei, Studies on Scientific Instruments and Collections in honour of Mara Miniati*, ed. by Marco Beretta, Paolo Galluzzi, Carlo Triarico (Florence: Leo S. Olschki Editore, 2003), pp. 29–52.

———, *Medieval Instrument Catalogue. European 6.1–6.9*. Online (16 March 2022): https://www.academia.edu/35737805/MEDIEVAL_INSTRUMENT_CATALOGUE_EUROPEAN_6_1_6_9_pdf

Maricourt, Petrus Peregrinus de, *Opera: Epistula de Magnete. Nova compositio astrolabii particularis*, ed. by Loris Sturlese e Ron B. Thomson (Pisa: Scuola Normale Superiore, 1995).

Miolo, Laure, 'A New Astronomical Text by Jean des Murs', *Cahiers de recherches médiévales et humanistes — Journal of Medieval and Humanistic Studies*, 42 (2021–2), 329–59.

North, John D., *Horoscopes and History*, The Warburg Institute Surveys and texts, 13 (London: Warburg Institute, 1986).

Poulle, Emmanuel, *Équatoires et horlogerie planétaire du XIIe au XVIe siècle*, 2 vols (Geneva: Droz, Paris: H. Champion, 1980).

———, 'L'astronomie latine au XIVe siècle', in *Gersonide en son temps*, ed. by Gilbert Dahan (Louvain: E. Peeters, 1991), pp. 253–64.

———, 'L'astrolabe médiéval d'après les manuscrits de la bibliothèque nationale', in *Bibliothèque de l'Ecole des chartes*, 112 (Geneva: Librairie Droz, 1954), pp. 81–103.

Ptolemy, Claudius, *Almagestum Cl. Ptolemei: Pheludiensis Alexandrini astronomorum principis* (Venice: Petrus Liechtenstein, 1515).

Saby, Marie-Madeleine, 'Les canons de Jean de Lignères sur les tables astronomiques de 1321: Édition, critique, traduction et étude', unpublished thesis, Paris, École Nationale des Chartes, 1987. A summary appeared as 'Les canons de Jean de Lignères', École nationale des Chartes, Position des thèses 1987, pp. 183–90.

———, 'Mathématique et métrologie parisienne au début du XIVème siècle: le calcul du volume de la mer d'airain', *Archives d'histoire doctrinale et littéraire du moyen-âge*, 58 (1991), 197–213.

Stöffler, Johannes, *Elucidatio fabricae ususque astrolabii* (Oppenheim: Jacob Köbel, 1513).

Thorndike, Lynn, and Pearl Kibre, *A Catalogue of Incipits of Mediaeval Scientific Writings in Latin* (Cambridge, Massachusetts: The Mediaeval Academy of America, 1937; rev. and augm. 1963).

Thorndike, Lynn, *A History of Magic and Experimental Science*, 8 vols (New York: Columbia University Press, 1923–1958).

Toomer, Gerald J., 'Ptolemy', in *Dictionary of Scientific Biography*, ed. by Charles Gillipsie, 16 vols (New-York: Charles Scribner's Sons, 1973–80, republished 2001), XI pp. 186–206.

Weijers, Olga, *Le travail intellectuel à la Faculté des arts de Paris: textes et maîtres (ca.1200–1500)*, Studia artistarum (Turnhout: Brepols, 1994–2012).

NICHOLAS JACOBSON*

Exempla in figura in Alfonsine Canon Commentaries

Introduction: Diagrams in alfonsine canon texts

While surveying and analysing canon texts in the Alfonsine tradition, the ALFA research team has identified a small number of works that include illustrations of planetary motion and directions for constructing diagrams in addition to instructions for using the Alfonsine Tables. Although they represent a minority of such texts within the entire Alfonsine corpus, these canons invite our attention because of their authors' willingness to transgress certain epistemic boundaries between theoretical and practical knowledge.[1] The division between *theorica* and *practica* was frequently invoked in the introductions to thirteenth- and fourteenth-century astronomical works, and, according to this taxonomy of knowledge 'kinds', the use of diagrams to model planetary motion was understood to fall within the domain of 'theoretical' astronomy.[2] Such attitudes were aptly summarized by Campanus of Novara (c. 1220-1296) in his *Theorica planetarum* when he wrote that the theoretical part of astronomy relied upon 'infallible methods, for instance, by the elements of geometry', whereas the practical part then applied what had 'been demonstrated by suitable geometrical figures to use by clothing them in the numbers which are peculiar to

* I would like to extend my deepest thanks to members of the ALFA team for the many opportunities to discuss and develop this paper. In particular, José Chabás, Matthieu Husson, and Richard L. Kremer's reactions to earlier drafts were invaluable in encouraging me to frame and conceptualize the scope of this project. So too were the formal and informal discussions I had with Eleonora Andriani, Camille Bui, Samuel Gessner, Marie-Madeleine Saby, and Scott Trigg. This work was supported by the ERC project *ALFA: Shaping a European Scientific Scene, Alfonsine astronomy*, CoG 723085, PI Matthieu Husson.

1 See, for instance, *Canones tabularum Alfsonsi* (Incipit: *Numerus anni in ista propositione docetur modus reducendi*) in Erfurt, UB, CA 4° 366, ff. 70v–73v; *Canones in Tabulas Alphonsinas* (Incipit: *Tempus est mensura motus, ut vult Aristoteles*) in Oxford, Bodleian Library, Digby 97, ff. 240v–290v; *Arabes maxime* in Paris, BnF, lat. 7295A, ff. 182r–193v. A critical edition of the *Arabes maxime* is being prepared by Eleanora Andriani, Samuel Gessner, and Nicholas Jacobson. See the notice on this ongoing project in this volume.

2 For a discussion of the *theorica/practica* distinction in late medieval and early modern astronomy, see Richard L. Kremer, 'Incunable Almanacs and *Practica* as Practical Knowledge Produced in Trading Zones', in *The Structures of Practical Knowledge*, ed. by Matteo Valleriani (Dordrecht: Springer, 2017), pp. 340–50.

Nicholas Jacobson • University of Wisconsin-Madison

Alfonsine Astronomy: Computational Practices, ed. by Matthieu Husson, José Chabás, and Richard L. Kremer, Turnhout, 2026 (*Alfonsine Astronomy: Studies and Sources*, 3), p. 149-195

10.1484/M.ALFA-EB.5.137910

arithmetic'.[3] Thus, in Campanus's accounting of it, a shift in technique from geometrical constructions to arithmetic calculations constituted an epistemic shift as well.

This tendency for authors of medieval Latin astronomy to define practical knowledge in opposition to the theoretical has led many historians of science to follow suit, as they have drawn a line of demarcation between the content and presentation of canons vis-à-vis *theorica* texts. For example, Michael Mahoney concluded in a short article on medieval Ptolemaic astronomy for the *Dictionary of the Middle Ages* that, because they belonged 'to an independent genre, a set of tables (called a *zīj* in Arabic) had its own accompanying instructions, or canons, and could be used without reference to the models'.[4] Furthermore, when Olaf Pedersen wrote his seminal article on the establishment of Pseudo-Gerard of Cremona's *Theorica planetarum communis* as the centrepiece of the thirteenth- and fourteenth-century university curriculum, he contrasted the practical dimension of the canons with the *theorica* texts found in his *corpus astronomicum*.[5] In this way, canons have been disassociated from other works of mathematical astronomy, particularly that of the *theorica planetarum*. They have neither been characterized by an effort to demonstrate the validity of the mathematical procedures that they describe, nor have they been understood to furnish themselves with the diagrammatic visuality of their theoretical counterparts.

Given the frequent insistence on this distinction of both genre and epistemology, the presence of these canons with diagrams merits some examination, both in order to understand what the diagrams were intended to signify and in order to reconstruct how they were to function in the text. To this end, I hope to benefit from the many recent articles that have sought to historicize, and thus better understand, the multiple significations and uses of diagrams in pre-modern contexts. Such an approach takes for granted Reviel Netz's conclusion that pre-modern diagrams were not mere auxiliaries to proofs that could otherwise be purely expressed in the symbolic notation of modern mathematics.[6]

Furthermore, since the diagrams in question are found within canon texts that focused specifically on directing a user in the operation of mathematical tables to a certain goal, they will also be shown to bear a close relationship with the action-oriented language of practical knowledge that Matteo Valleriani has called its 'workflow' (i.e. a series of actions needed to achieve the desired final product).[7] In this capacity, the diagram became an image or an icon, visually representing the column headings of the tables and their numerical values as geometric quantities and relations. It was also intended to certify the workflow

3 Campanus of Novara, *Theorica planetarum Campani*, in *Campanus of Novara and Medieval Planetary Theory, Theorica planetarum*, ed. and trans. by Francis S. Benjamin, Jr. and Gerald J. Toomer (Madison: University of Wisconsin Press, 1971), p. 139. Campanus clearly distinguishes this bifurcation of theory and practice from the more common one (also discussed in his introduction) that medieval authors made between a treatment of planetary motion per se (*theorica*) and the effects of this motion on Earth (*practica*).

4 Michael S. Mahoney, 'Ptolemaic Astronomy in the Middle Ages', in *Dictionary of the Middle Ages*, ed. by Joseph R. Strayer (New York: Scribner, 1988), p. 210.

5 Olaf Pedersen, 'The *Corpus astronomicum* and the Traditions of Mediaeval Latin Astronomy', *Studia Copernicana*, 8 (1975), 57–96 (pp. 78, 81–82).

6 Reviel Netz, 'Greek Mathematical Diagrams: Their Use and Their Meaning', *For the Learning of Mathematics* 18 (1998), 33–39 (pp. 33, 37–38).

7 Matteo Valleriani, 'The Epistemology of Practical Knowledge', in *The Structures of Practical Knowledge*, p. 3

by not only representing the otherwise tabular and arithmetic steps of the procedure as an image, but also by emphasizing that the image reflected the cosmological reality that such diagrams intended to model. In this second sense, the image served not as a geometric representation of the steps of the canon, but rather as the 'cause' (*causa*) or 'reason' (*ratio*) for their operation. In other words, the status of the diagram would change, signifying the tables at certain points in the text while embodying what was perceived to be the underlying reality of planetary motion at others. By attending to the fact that, in this way, the diagram sometimes served as a sign, sometimes as the referent of a sign, we can begin to discern multiple modalities of a diagram as it functioned within a single canon text. In order to define these different functions, I will have recourse to scholarship on pre-modern diagrams and semiotic theory by Reviel Netz and, more recently, Claas Lattmann to show the kind of connections that are operative between text and image.[8]

In some of the sources that use diagrams in this way, such as a set of canons preserved in Oxford, Bodleian Library, Digby 97, the worked example is referred to as an *exemplum in figura*, and it would appear in tandem with another example that performed the steps of the canon in a purely arithmetic way. This second type was called an *exemplum in tabula*.[9] Throughout the rest of the article, I thus adopt this terminology to distinguish between these two types.[10] We see in the *exemplum in figura* a much more coherent integration of the practical and theoretical domains of medieval Latin astronomy than one normally finds in Alfonsine canon texts, as the diagrams are given as representations of tabular procedures. As these procedures are translated into geometric quantities, however, they are also revealed to be isomorphic with the models of the *theorica*. That is to say, the *exemplum in figura* proceeds in two stages that parallel the different uses the canon makes of the diagram image. The first stage constructs the diagram and 'demonstrates' that the geometry of the figure is the 'cause' of the operations of the tables. Thus, the diagram confirms that tabular methods are correct because they conform to relations constructed in the figure. The second stage, however, performs the arithmetic procedure of the *exemplum in tabula* a second time, noting which of the diagram's elements correspond to column headings in the tables in order to correlate significant tabular values with spatial positions or angular relations.

In an analytically narrow sense, the purpose of this article is to provide the documentary basis for this twofold typology of worked examples — one tabular, the other diagrammatic — while also providing a short historical account of the possible origins of the *exempla in figura*. More broadly, however, it is my hope that my flagging of these texts,

---

8 Netz, 'Greek Mathematical Diagrams: Their Use and Their Meaning', 33–39; Claas Lattmann, 'Diagrammatizing Mathematics: Some Remarks on a Revolutionary Aspect of Ancient Greek Mathematics', in *Revolutions and Continuity in Greek Mathematics*, ed. by Michalis Sialaros (Berlin: De Gruyter, 2018), pp. 107–29.

9 For the appearance of *exempla in figura* in Digby 97, see Chapter 3, lines 348–50, and Chapter 4, lines 512–14, in the Appendix below. For the appearance of *exempla in tabula*, see Chapter 3, lines 279–87 and Chapter 4, lines 453–58.

10 For the sake of brevity, I have condensed into two simple terms (*exemplum in tabula* and *exemplum in figura*, respectively) what the text of Digby 97 presents in an otherwise more descriptive manner. For instance, in introducing his tabular example for finding the true position of the Moon, the author writes: 'pono exemplum ad annum christi predictum scilicet 1355 completum … prout patet in tabula sequenti'. Likewise, in presenting the diagrammatic example, the author writes: 'pono exemplum in figura seqenti per modum theorice prout patet'. See note above.

which contain diagrammatic worked examples, will bring into sharper relief one aspect of a genre or sub-genre of canon literature that has been mentioned frequently in Alfonsine historiography, but rarely defined — namely that of the *commentum* or *expositio tabularum*.[11] For even though witnesses to *exempla in figura* in fourteenth-century manuscripts are often labelled as canons in the fly leaves and addended titles to the works that contain them, they appear to be commentary texts with diagrams and worked examples (*exempla*) serving as augmented material to be presented in conjunction with the base canon text.

In light of this broader point of analysis, the physical character of the manuscripts and the scribal conventions involved in their composition will also inform the understanding of the meaning and function of the diagrams within these canon sets, which were effectively compilations of sources drawn from multiple authors and works. Such considerations include attention to the white space between the script and the diagram, the 'seams' between different compiled works, and the integration of tabulated values of arithmetic worked examples side by side with the canon instructions. These artefacts of manuscript production, which Gregg de Young has called the 'architecture of medieval manuscripts', reveal the hybrid nature of these works, as well as the close connection between the materiality of a written text and the form of such genres as the medieval *compilatio*, *commentum*, and *expositio*.[12] By emphasizing this last point, I hope to expand our definition of canon text to include additional interpretive compilations and commentaries alongside the base text of canon instructions.

Finally, this association of these diagrammatic canon texts with the medieval genre of the *expositio* underscores the tendency of fourteenth-century Alfonsine astronomers to accommodate the language and structure of their works to the literary forms already present within the arts curriculum of their day. The status of this literature as a sort of *expositio* suggests that when practitioners of Alfonsine astronomy undertook clarifying and justifying the procedures of the canons through *exempla in figura* and *exempla in tabula*, they were proceeding in much the same way as a theologian would elucidate Peter Lombard's *Sentences* or a Master of Arts would explain Aristotle's *Metaphysics*.

Thus, in this article, I would like to (1) introduce canon texts with diagrams and explain the function of *exempla in figura* and *exempla in tabula* in the Alfonsine tradition through the lens of semiotic theory; (2) describe how the multiple modalities of *exempla in figura* performed the function of bridging the divide between canons, tables, and the models of the *theorica* in a way that enabled a user to visualize significant values in the tables on diagrams; and (3) historicize and explore the status of these texts as commentaries intended to justify the procedures of canons by analysing the 'architecture' of the parallel worked examples contained within them. An Appendix is included at the end of the article

11 I follow Charles Lohr and Daniel A. Callus in treating the *commentum* and the *expositio per modum commenti* as two titles referring to the same genre. See Lohr, 'Medieval Latin Aristotle Commentaries: Authors A–F', *Traditio* 23 (1967), 313–413 (pp. 313–14); Callus, 'Introduction of Aristotelian Learning to Oxford', *Proceedings of the British Academy* 29 (1943), 229–81 (p. 265). Lawrence D. Green expands this definition further, stating that the *commentum* was equivalent to the *expositio*, *explicatio*, *lectura*, *praelectio*, and *scholia*. See *John Rainold's Oxford Lectures on Aristotle's* Rhetoric, ed. and trans. by Lawrence D. Green (London and Toronto: Associated University Presses, 1986), p. 51.

12 Gregg de Young, 'Mathematical Diagrams from Manuscript to Print: Examples from the Arabic Euclidean Transmission *Synthese* 186 (2012), 21–54.

to showcase four such canon chapters that contain planetary diagrams: two treating the procedure for finding the true position of the Sun, and two treating the procedure for finding the true position of the Moon.

1. *Exempla in tabula* as indices of tables

Recent historical studies on mathematical diagrams in pre-modern contexts have revealed the mutually constitutive role of text and diagram in the presentation of mathematical reasoning. In his seminal article on the subject, 'Greek Mathematical Diagrams: Their Use and Their Meaning', Reviel Netz provided examples of several ways in which the content of ancient Greek mathematical texts required diagrams in order to convey a coherent argument. Netz's analysis was informed by the field of semiotics, and particularly the categories developed by Charles Sanders Peirce, and this approach has proven fruitful in a number of other papers thereafter.[13]

For instance, to show that Euclidean diagrams were included in pre-modern geometry texts to ensure that proofs were fully determined, Netz investigated (among other topics) the lettering of diagrams according to Peirce's trichotomy of signs: symbols, indices, and icons.[14] He noticed that diagrams in these texts would often be referred to by different sets of letters on the figure. If there were, say, a quadrilateral ***abcd***, it would sometimes be referred to by all its designated letters, sometimes by a pair of letters at opposite corners (say, ***ac***), and sometimes by the other pair (***bd***). This showed that the letters of pre-modern diagrams were not symbols devised as a set of conventions to refer to general mathematical objects according to a consistent notation; rather, they were indices of the specific mathematical object in question, and an author's decision to designate them according to different letters might well have indicated a desire to bring to light different aspects of the diagram when it was deemed relevant.

---

13 Netz, 'Greek Mathematical Diagrams: Their Use and Their Meaning', 33–39; Lattmann, 'Diagrammatizing Mathematics: Some Remarks on a Revolutionary Aspect of Ancient Greek Mathematics,' pp. 107–29; Bernd Mahr and Klaus Robering, 'Diagramme als Bilder, die Modelle repräsentieren: Diagrammgebrauch in der Elementargeometrie,' *Zeitschrift für Semiotik* 31 (2009), 275–309; Danielle Macbeth, 'Diagrammatic Reasoning in Euclid's Elements,' in *Philosophical Perspectives on Mathematical Practice*, ed. by Bart van Kerkhove *et al.* (London: College Publications, 2010), pp. 235–67.

14 A symbol is defined as a conventional, and hence arbitrary, sign that is related to an object in the mind through a habituated mental association (Peirce also referred to such signs as tokens). An index, on the other hand, is a sign that represents its object according to a physical or existential relation. The classic example of an index is the weathervane that points in a certain direction to signify which way the wind is blowing. It does so because the wind has physically caused it to turn. Finally, there is the icon, which represents its object through a similitude of properties, such as the relation that a painting bears to the object that it depicts. An introduction and exploration of this typology of signs can be found in two of Peirce's papers the first of which was published in the *American Journal of Mathematics*. See Peirce, 'On the Algebra of Logic: A Contribution to the Philosophy of Notation', in *Writings of Charles S. Peirce: A Chronological Edition: 1884–1886*, 8 vols, ed. by The Peirce Edition Project (Bloomington, IN: Indiana University Press, 1993), V, pp. 162–90; Peirce, 'An Elementary Account of the Logic of Relatives', in *Writings of Charles S. Peirce: A Chronological Edition, 1884–1886*, 8 vols, ed. by The Peirce Edition Project (Bloomington, IN: Indiana University Press, 1993), V, pp. 379–87.

If we were to extend these semiotic categories to worked examples in Alfonsine canons on planetary motion, we would see that values exhibited in the *exempla in tabula* and *exempla in figura* actually bear very different relationships with the tables they signify: the *exempla in tabula* serving as indices while the *exempla in figura* served as icons. In order to show this, we begin with a pair of examples, one tabular and one diagrammatic, for finding the true position of the Sun, taken from the manuscript Erfurt, UB, CA 4° 366. Here, we find a set of fourteen anonymously written canons on planetary motion.[15] At the top of folio 70^v, where the work begins, the original owner of the manuscript, Johannes de Wasia (d. 1395) wrote the title *Canones tabularum Alfonsi*, but its contents make it clear that it was a canon commentary.[16] Each chapter provides a worked example, signalled by the phrase *exempli gratia* or *verbi gratia*, that recapitulates the canon procedure in terms of a calculation established for a given date and time.[17] Before each of these *exempla* is an underlined quotation connecting the worked example to the base text, namely the first fourteen canons of the widely known *Priores astrologi* of John of Lignères (fl. 1320-1335). The thirteenth canon addresses the method for finding the true position of the Sun.[18]

The compiler does not provide the text of Chapter 13, except for the first four words that initiate the passage (*verum locum solis invenire*).[19] Instead they immediately present the *exemplum in tabula*, which introduces numerical values for steps in the original canon. The language of the worked example differs from the canon itself in terms of its specificity. That is, the Sun's mean motion is set to the day and time in Paris when it would reach six common signs, or 180°, from the head of Aries.[20] This is unlike the base text, which only gives general instructions for the tables' uses, and, rather than giving numerical values, it instead employs a lexicon of terms drawn from the titles and headings of the columns of the tables from which the numbers are sought.

These differences at the level of generality, however, belie the fact that both the canon and the *exemplum in tabula* attain their significance only in the presence of an indicated astronomical table serving as their primary referent. In this way, the canon text and the *exemplum* are tabular indices. In the following excerpt drawn from Chapter 13 of the Erfurt CA 4° 366 canons, we see that the significant values of the worked example (presented in

15 Erfurt, UB, CA 4° 366, ff. 70^v–73^v.

16 Wilhelm Schum identifies this chapter heading as being written in Johannes de Wasia's own hand. See Schum, *Beschreibendes Verzeichniss der Amplonianischen Handschriften-Sammlung zu Erfurt* (Berlin: Weidmannsche Buchhandlung, 1887), p. 613. I discuss this more fully in the final two sections of this article.

17 One exception to this is Chapter 9 on determining proportional parts, which does not signal its worked example with this conventional phrase.

18 See the equivalent titles of Chapters 1–14 in Saby, Marie-Madeleine, *Les Canons de Jean de Lignères sur les tables astronomiques de 1321. Édition critique, traduction, et étude*, 3 vols (unpublished dissertation, Paris, École Nationale des Chartes, 1987), I, pp. 174–75.

19 In our sole known witness to this commentary text, a complete version of the *Priores astrologi*, including Chapter 13 on the true Sun (f. 11^r), could be found earlier in the manuscript such that a rewriting of this original text would have been redundant. For the full text, see Saby, 'Les Canons de Jean de Lignères', I, p. 199.

20 Using the AlfonsoTab program designed by Richard L. Kremer, the Sun's mean motion would be 180° from Aries on September 12, 1321 in Paris, but this is not explicitly given in the canon. Because the compiler states that the mean motion of the Sun is six signs and also gives the solar apogee as 89° 24' 22" (i.e. two common signs, 29° 24' 22"), the reader does not need to start with a date, but can proceed directly to the tables of solar equation with the difference between these two values (i.e. the solar anomaly).

bold) are always placed in an indexical relationship with the terms found in the titles or headings of tables (presented in bold and underlined):[21]

> To find the true position of the Sun: When you would like this [value], you should know its mean motion. Let it be proposed, by way of example, that you would like to have the true position of the Sun for when its **<u>mean motion</u>** is **6 signs**. You should subtract from this mean motion the **<u>aux of the Sun</u>** in the second sense, which at this time — namely the incomplete year of 1321 — is **2 signs, 29 degrees, 24 minutes, 22 seconds**, as was seen in the tenth example [of this work]. The **<u>argument of the Sun</u>** remains, [namely] **3 signs, 0 degrees, 35 minutes, 38 seconds**. With that argument, you will enter the tables of equation. And because there are minutes, you will enter into the table of the equation of the Sun, according to the Tables of Alfonso, first with 3 signs. You will find beside this value 2 degrees, 9 minutes, 57 seconds. Then you will enter with the same argument after having added one degree, and you will find beside this value 2 degrees, 9 minutes, 59 seconds. Then take the difference of these two equations, and this is 2 seconds. From this, take the proportional part according to the proportion 36 minutes, which is the argument, in relation to 60. Do so by taking 38 seconds as one minute since there are more than 30. If, however, there had been fewer, then you would not have taken any. And note that, for all subsequent operations involving equations of planets, you can thus operate according to the method given in the equation of the motion of the eighth sphere. If you wish to operate exactly, then take the proportional part of two seconds according to the proportion of 36 minutes to 60, which is 1 second, 12 thirds as you can have by the 9th example. Therefore, add the proportional part to the first equation, because it was less than the second. **2 degrees, 9 minutes, 58 seconds, 12 thirds** will result, which is the **<u>equation of the argument of the Sun</u>**, namely 3 signs, 0 degrees, 36 minutes (taking 38 seconds as one minute). Thus subtract this equation from the mean motion of the Sun because the argument is less than 6 signs. If, however, the argument had been more than 6 signs, you would add the first equation. And the **<u>true position of the Sun</u>** after the subtraction will be **5 signs, 27 degrees, 50 minutes, 1 second, 47 thirds.**[22]

It should be noted that the compiler has proceeded with a certain degree of parsimony when presenting this worked example, choosing to forego a detailed explanation of how one would find the mean motion of the Sun or the solar apogee (called the 'aux' in the text). This is because such instructions were given in detail in Chapters 8 and 10 of the work. The example only enters into the particulars of arithmetic computation when discussing the method of linear interpolation used to find the equation values corresponding to those solar anomalies (called 'arguments of the Sun' in the text) that are not whole

21 The final value, that is, the 'true position of the Sun', is not found in the tables, but we also put it in bold and underline it, as it is what is sought using the tabular values. In preserving the 'true position of the Sun' alongside the tabular values, we follow the method of Francesco Capuano di Manfredonia in his commentary on tables where he refers to these values as the *termini* of his *exemplum in figura* for the Sun. See note 90 on Capuano in the conclusion to this article for further discussion on this.

22 See Chapter 1, lines 2–31 of the Appendix for the original Latin.

numbers. Thus, there is some discussion of how to round the value of the solar anomaly. The example also informs the reader about how to use the solar anomaly value in the argument of the table to determine whether to add or subtract the solar equation to or from the mean motion. In both of these cases, the operations described in the example and the values employed attain their significance in relation to the specific affordances of an Alfonsine solar equation table.

As we shall see, the indexical relations at play between the text of the *exemplum in tabula* and the tables clearly differ from the kind of iconic relations involved in the *exemplum in figura*. Before discussing this observation in detail, however, and for the sake of completeness, we may add that they also differ in subtle but meaningful ways from the modern symbolic notation that has been developed to define canon procedures in the commentaries of contemporary scholarship. In his 'Outline of the Ptolemaic Theory of Planetary Longitudes, as Applied in the Middle Ages,' John North provided the general procedures for calculations of planetary longitude as a programme of steps that could be applied to a known time, *t*, regardless of whether one calculated using Ptolemaic, Toledan, or Alfonsine Tables.[23] Thus, the functions that generated the numerical values found in the tables of solar motion, together with the operations for computing with them, could be signified symbolically in the following way:

1.	Mean motion of the Sun	$(\bar{\mu}_{\odot})$	$\bar{\mu}_{\odot} = \bar{\mu}_{\odot}(t)$
2.	Solar apogee	(ω)	$\omega = \omega(t)$
3.	Mean solar anomaly	$(\bar{\gamma})$	$\bar{\gamma} = \bar{\mu}_{\odot} - \omega$
4.	Equation of argument	(ξ)	$\xi = \xi(\bar{\gamma})$
5.	True position of the Sun	(λ)	$\lambda = \bar{\mu}_{\odot} \mp \xi$

North intentionally devised his notation so that he could generalize across multiple tabulated systems, writing that his second step for finding the aux or solar apogee was left under-determined in order to equally accommodate systems that localized planetary positions in a sidereal reference frame and those that localized them tropically.[24] His effort to generalize in this way implies that the referent of North's notation is not the table at all. Instead, the sign system of algebraic notation presumes the existence of mathematical functions lying behind the tables, which unify the vicissitudes of format and can also explain numerical variation across tables as the result of adjustments of key parameters in these functions.[25]

The *exemplum in tabula*, however, assumes the primacy of specific tables as objects indicated by the terms and numbers put forth in the procedure. This is in spite of the fact that the steps of the worked example proceed in an order and with reference to significant

23 John D. North, *Richard of Wallingford: An Edition of His Writings with Introductions, English Translation, and Commentary*, 3 vols (Oxford: Clarendon Press, 1976), III, pp. 175–76, 183.

24 North, *Richard of Wallingford*, III, pp. 182–83.

25 We make this point to show that North's method of representation is not dependent upon a specific table. Of course, general symbolic notation, such as the one employed above, would still bear an indexical relation to an imagined set of possible tabular functions, even if it did not stand for any table in particular. In this way, Peirce's typologies were not necessarily mutually exclusive. See the refinement of his theory in Peirce, 'On the Algebra of Logic', in *Writings of Charles S. Peirce: A Chronological Edition*, V, pp. 163–64.

values that parallel North's programme, as we can see by a summarization of the above quoted text, in the following way:

1.	Mean motion of the Sun	6s 0° 0' 0"
2.	Solar apogee	2s 29° 24' 22"
3.	Mean solar anomaly	3s 0° 35' 38"
4.	Equation of argument	0s 2° 9' 58" 12"'
5.	True position of the Sun	5s 27° 50' 1" 48"'

With a different referent, a different goal for the worked example can be inferred. By choosing the year 1321 (*incompleto*) and the city of Paris to perform the worked example, the temporal and spatial indeterminacies of the base text are resolved. Thus, the indexical nature of an *exemplum in tabula* provided a means of grounding the deictic categories of a canon to a specific time and place that could be readily identified with values found within the columns of the tables. That the writers who compiled these examples were conscious that *exempla in tabula* functioned in this way becomes apparent from the fact that they often employed the same scribal technology of tabular grids in the margins beside the canon in order to extract and showcase significant values. In the set of canons found in Digby 97, which contained worked examples for John of Saxony's (fl. 1327-1355) *Tempus est mensura motus*, the compiler has integrated tabulated numerical lists alongside the text of the *exempla in tabula* as a way of arranging and privileging the values deemed significant.[26] These numbers are labelled and invariably bear a connection to the lexical terminology found in the tables themselves.

2. *Exempla in figura* as icons

Turning to the second example found in Chapter 13 of the canon set in Erfurt CA 4° 366, we observe a different kind of semiotic relation at work. It is accomplished through the transposition of the procedure from one system of signs to another as the compiler translates the arithmetic values of the *exemplum in tabula* into geometric quantities and relations. In order to do so, they first construct a diagram that can visually represent the terms of the significant intermediate values found in the canon text.[27] Furthermore, the construction of the diagram introduces additional terminology that was not discussed in either John of Lignères' original canon or in the tabular example.

These additional terms refer to the basic elements of the solar diagram that must be constructed before the significant values of the canon procedure can be represented geometrically.[28] A circle ***acbd*** is drawn, which represents the zodiac against whose backdrop the planets are observed in their true positions. Points ***a*** and ***b*** are said to be the west and

26 These tabulated lists have been reproduced alongside the transcriptions of the Digby manuscript found in the Appendix to this article.

27 The significant intermediate values are as follows: (1) mean motion of the Sun; (2) solar apogee; (3) mean solar anomaly; (4) equation of the Sun; and the final value is (5) the true position of the Sun. These are equivalent to what Capuano would later call the *termini* of such an *exemplum* (see notes 21 and 90).

28 See the first chapter featured in the Appendix for the full text and diagram.

the east, respectively. Point ***a*** is also the beginning of the sign Aries, while ***b*** is Libra, ***c*** Cancer, and ***d*** Capricorn. An eccentric circle ***gkhi*** is constructed about point ***f*** to represent the path of the Sun on its deferent, and the physical body of the Sun (*corpus solis*) is placed at points ***g***, ***k***, ***h***, and ***i***. Insofar as none of these terms, thus far outlined, are described in the operations of the first example, we might consider them to be auxiliary constructions, as they will be necessary to define each of the five significant values geometrically. That is to say, immediately after the description of these auxiliary constructions, the compiler redefines the tabular vocabulary as parts of the diagram. The five terms, which, without the diagram, simply had the headings of the columns of the planetary tables as a referent, now have recourse to the internal relations of the geometric model to represent them. The mean motion of the Sun is said to be 'the arc taken between Aries and the line proceeding from the centre of the Earth parallel to the line proceeding from the centre of the eccentric through the body of the Sun'.[29] The aux of the Sun is similarly given a twofold definition as, first, the body of the Sun at point ***g*** and, second, 'the arc ***ac***, namely between Aries and the place of the aux' as defined in the first sense.[30] 'The aux in the second sense', so defined, then, 'is subtracted from the total arc ***acb*** which is the mean motion of the Sun. … the arc ***cb*** remains, which is called the mean argument'.[31] Next, the text reads 'the true place of the Sun is the arc taken between Aries and the line proceeding from the centre of the Earth passing through the body of the Sun, and therefore, when the body of the Sun is at point ***k***, the true position is ***acn***'.[32] Finally, 'the equation of the Sun is the arc of the zodiac located between the mean motion and the true, as is arc ***bn*** of the zodiac'.[33]

Thus, a new series of relations have been established, now between the terminology of the tables and arcs on the diagram in the following way:

1.	Mean motion of the Sun	arc ***acb***
2.	Solar apogee	arc ***ac***
3.	Mean solar anomaly	arc ***acb*** – arc ***ac*** = arc ***cb***
4.	Equation of argument	arc ***bn***
5.	True position of the Sun	arc ***acb*** – arc ***bn*** = arc ***acn***

The construction of a diagram modelling planetary motion introduces an added layer of complexity to the second worked example in the text. This is because it is at times unclear in the text if the arcs defined as significant values are intended to serve as signs representing the terms found in the table or as objects of reference for the tabular procedure. It will be shown here that the diagram functions in both roles.

It is, in fact, the diagram's dual significance, as we shall see in the next section, that enables the compiler to present it as a 'cause' for the soundness of the canon's instruction. For the formal structure of the diagram will serve as a kind of demonstration for the numerical outlay of the table. However, if there were any doubts that the diagram was not

29 See Chapter 1, lines 38–40 in the Appendix.
30 Chapter 1, lines 43–44.
31 Chapter 1, lines 44–46.
32 Chapter 1, lines 48–51.
33 Chapter 1, lines 51–52.

also intended to represent the tables as a visual sign, the *exemplum in figura* puts these to rest. The compiler systematically repeats the steps of the canon, equating the values in the diagram with the numbers of the *exemplum in tabula* in each instance. When following the instructions of the *exemplum in figura*, the referent of the worked example is still the table, as we can clearly see from the following passage:

> Therefore, the arc ***acb***, by way of example, is **six signs**, which is the **mean motion**. And the arc ***ac*** [is] **2 signs, 29 degrees, 24 minutes**, which is the **aux of the Sun** in the second sense. If it is subtracted from the arc ***acb***, which is the mean motion, the arc ***cb*** will remain, which is the **mean argument 3 signs, 0 degrees, 36 minutes.** With the argument, you will enter the table of the equation of the Sun. And you will have, according to what was seen, the **equation of argument 2 degrees, 9 minutes, 58 seconds, 12 thirds**, which is the arc ***bn***. If [the equation] is subtracted from the mean motion, namely from arc ***acb***, the **true motion** will remain, namely the arc ***acn***, since, when the argument of the Sun is fewer than 6 signs in the arc ***cbd***, the mean motion is greater than the true.[34]

The diagram, as it is described in this passage, is intended as a proxy for the table with each arc directly identified with a discrete tabular value. Following Claas Lattmann's analysis of mathematical figures as semiotic signs in 'Diagrammatizing Mathematics,' we could say that the transition from *exemplum in tabula* to *exemplum in figura* results in a transposition from a system of indexical signs to iconic ones. As Lattmann explains, an icon is

> a sign that represents its object on the basis of a similarity between its own perceivable properties and the properties of its object, such as portraits and photographs — where 'similarity,' *nota bene*, does not imply a relation of identity or isomorphism. It suffices that the relevant properties of sign and object may (at least potentially) be regarded as similar, in whatever (in principle, subjective) way someone deems appropriate.[35]

It does not matter that the visual form of the table and the diagram are not isomorphic; what is essential to this semiotic relation is that, again to quote Lattmann, the 'diagram exhibits a similarity to its object with regard to its inner structure, which is given by the direct, pairwise relations among its constitutive elements'.[36]

As we shall see in the next section, the diagram in Chapter 13 intentionally represents the mean position of the Sun at six common signs or 180° from the head of Aries, just as described in the text of the *exemplum in tabula*. The true position of the Sun as a value on the zodiac circle that is less than the mean position follows from the configuration of the two arcs on the diagram. This offers one instance in which the diagram successfully represents the internal relations of the worked example. But the diagram also demonstrates its ability to represent other possible mean motion values drawn from the tables, and especially those mean motion values that correlate with solar anomalies at 0°, 180°, or positions greater than 180°. For it is the value of the solar anomaly that crucially determines whether to add, subtract, or be free from a solar equation, and so an iconic representation

34 Chapter 1, lines 53–63.
35 Lattmann, 'Diagrammatizing Mathematics', p. 113.
36 Lattmann, 'Diagrammatizing Mathematics', p. 113.

of the Sun's mean position on the opposite side of the eccentric provides a visual cue for the operational distinctions to be found in the canon instructions. The last passage of the *exemplum in figura* reads:

> But when the argument was greater than 6 signs, the true motion is greater than the mean. And, therefore, the equation ought to be added to the mean motion to obtain the true one because, when the Sun is at point ***i***, the mean motion is the arc ***acbdm*** and the true motion is the arc ***acbdml***, which is greater on the point at arc ***ml***, the equation of the Sun, which is added to the mean motion. Therefore, when the Sun is on the half of its deferent, ***gkh***, the equation should be subtracted. But when it is on the other half, that is on the half ***hig***, it should be added. When the true [Sun] is at point ***g*** or ***h***, there is no equation because the mean motion and the true are defined in the same place.[37]

As a reader is instructed to use the diagram to equate mean motions in this way, it is understood that parts of the figure, such as the arc between the true and mean positions of the Sun, are icons no less than the figure as a whole. This conforms to Lattmann's observation that the diagrams are iconic at the level of points, lines, and arcs, as well as at the level of the total figure they compose.

The conversion of tabular values into arcs also reveals an important consequence of the translation from one sign system to another — a process known as 'intersemiotic transposition'.[38] Because the indexical sign system of the tabular worked example and the base text have been transposed from the workflow of a canon onto a diagram, certain habits that are characteristic of a written procedure are carried into the visual realm. Among these are the discrete steps of discursive speech that are particularly well-suited to the form of canon writing, but foreign to visual media that lend themselves to non-discursive, configurational reasoning. In this way, a diagram was considerably freer and more flexible than a canon. In a single instant, so to speak, a practitioner could see why the equation of the Sun is sometimes added, sometimes subtracted from the mean Sun. Insofar as the diagram was borrowed from the *theorica planetarum*, however, there were limitations set on its flexibility. Its representation of the regular course of planetary motion meant that its structure was somehow fixed and unchangeable, as it would have been understood to reflect the invariant reality of the cosmos. By anchoring the procedure to the diagram, then, it also elevated the epistemic status of the tables from objects of practical computation to that of theoretical justification.

3. *Exempla in figura* as justifications

The clearest evidence that the compilers felt their diagrams could justify tabular operations is to be found in the oft-repeated and formulaic phrase that initiated the construction of the *exempla in figura*. In Chapter 13 of the anonymous canons of Erfurt CA 4° 366 (Chapter 1

37 Chapter 1, lines 64–73.

38 Roman Jakobson is credited with having first discussed this phenomenon in his essay on aphasia in children. See Jakobson, 'Deux aspects du langage et deux types d'aphasies,' in *Essais de linguistique générale* (Paris: Les Éditions de Minuit, 1963), pp. 52–56.

of the Appendix), the compiler explains that the diagram is introduced 'so that the causes of the operations (*cause operationum*) which are in the equation of the Sun might be made apparent'.[39] At the end of the following chapter, we read, 'So that what should be said on the equation of the Moon might be made more clear, a figure shall be put forward by which the causes of the operation (*cause operationis*) that is done in the equation of the Moon might be declared'.[40]

In the chapter considering the true position of the Sun in the canons of Digby 97, the compiler introduces the *exemplum in figura* in a similar way, writing 'so that what I have said on the mean and true motion of the Sun, and on its greater or lesser equation, might be clearly shown, I present an example in the following figure (*exemplum in figura subsequenti*) in this way according to the manner of a *theorica* (*per modum theorice*) as you see'.[41] Again, in the chapter on the Moon we read, 'And so that what I have said on the mean and true motion of the Moon, and likewise its centre, argument, and equation, might be clearly shown, I present an example in the following figure (*exemplum in figura sequenti*) in the manner of a *theorica* (*per modum theorice*) as is clear'.[42]

We witness in these stock phrases a change in the diagram's function, which is made possible as a consequence of the transposition of sign systems. No longer merely standing in for the table as an icon, the figure has become, in a second and more profound sense, the reason (*ratio*) for the table's existence. The causal dependence of the tables upon the diagram are highlighted in multiple ways, of which we limit our analysis in this article to two. Most straightforwardly, the compilers of the Digby 97 and Erfurt CA 4° 366 canons both define the *cause operationum tabularum* in terms of the diagram's role in a table's construction. That is, in addition to giving diagrammatic expression to the changing numerical values in the tables, the *exemplum* also provides an explanation for how the table-makers relied upon such diagrams as the basis for the calculations they performed to arrive at these values.

For instance, in Chapter 14 of the canons of Erfurt CA 4° 366 (Chapter 2 of the Appendix), there is a passage in the *exemplum in figura* that explains why there are two columns in a lunar equation table, under the headings of *diversitas diametri* and *minuta proportionalia*, respectively, which do not appear in a solar equation table. They are necessary because Ptolemy's final lunar model introduces a second anomaly defined by a moving deferent centre, which causes a variation in the first anomaly (*equatio argumenti*) depending on the distance of the epicycle centre from the deferent's apogee. The increasing and decreasing values of the *diversitas diametri* in the fourth column of the lunar equation table are shown to be determined by differences in the same equation of argument when it is at the apogee or perigee of the lunar deferent. Or, as the language of the canon explains:

> It should be known here, according to what is in the *theorica planetarum*, that the tables of the equations of argument of the Moon are made as if the centre of the epicycle is at the aux of the deferent in which the equation is at its smallest because the centre

39 Chapter 1, lines 32–33.
40 Chapter 2, lines 89–91 in the Appendix.
41 Chapter 3, lines 348–50 in the Appendix.
42 Chapter 4, lines 512–14 in the Appendix.

> of the epicycle is furthest from the centre of the Earth. Thus, the equation is greatest when the centre of the epicycle is at the place opposite the aux of the deferent. The excess of that greater equation in relation to the least equation is called the diversity of the diameter of the small circle.[43]

The compiler goes on to explain the reason for the third column of the lunar equation tables, which contains proportional numbers under the heading of *minuta proportionalia*. These *minuta proportionalia* were used to scale the increase in the value of the lunar equation depending on the epicycle centre's distance from the apogee of the deferent, and, as the canon reads:

> The reason for this (*causa istius*) is because, the more the centre of the epicycle moves away from the aux to the place opposite the aux, the closer it gets to the centre of the Earth, and thus the equation steadily increases. And, the more it moves away from the place opposite the aux, the further it gets from the centre of the Earth, and thus the equations steadily decrease. And thus we can get the equation of the argument for the centre of the epicycle existing at any place on the deferent in the following way. First, it is necessary to know how much nearer the centre of the epicycle is to the centre of the Earth than when it is on the aux, and how much further it is away from the centre of the Earth than when it is at the place opposite the aux. And this is gotten through the proportional minutes. And the proportional minutes are the excess of the line drawn from the centre of the Earth to the aux of the deferent in relation to the line drawn from the same centre of the Earth to the place opposite the aux. That is, they are the excess of the line ***dh*** to the line ***dk*** divided by 60.[44]

When the figure functioned as an icon, as described in the previous section, quantities were translated from the table onto the diagram. Yet, in these two passages we see that this directional relationship has been reversed. The configurations of the diagram are now interpreted as determining the values in the table. The compiler invites the reader to imagine all the values of the column of the *equatio argumenti* to be derived from the various positions of the body of the Moon in relation to the epicyclic apogee when the centre of the epicycle is at point ***h*** of the diagram. The values of the column of the *diversitas diametri* are shown to be the difference between these equations and their equivalents when the centre of the epicycle is at point ***k***. Finally, the values in the column of proportional minutes are explained as the ratio of the differences between two pairs of lines on the diagram (namely, the relation of the difference between line of apogee, ***dh***, and the line of interest on the one hand, and the difference of the line of apogee, ***dh***, and the line of perigee, ***dk***, on the other).

In these particular examples, the justificatory role the diagram performs as the *causa operationum tabularum* falls within the domain of practical reasoning that scholars of medieval and early modern epistemology, following the analysis of Jaako Hintikka, have identified as 'maker's knowledge'.[45] Taking his cues from an Aristotelian framework of

43 Chapter 2, lines 133–39 in the Appendix.

44 Chapter 2, lines 143–55 in the Appendix.

45 See Jaako Hintikka, 'Practical vs. Theoretical Reason — An Ambiguous Legacy', in *Practical Reason*, ed. by Stephen Körner (New Haven: Yale University Press, 1974), pp. 83–102.

causation, as modified through the lens of Judaeo-Arabic natural philosophy, Hintikka explained that proponents of maker's knowledge assumed the epistemic priority of minds that knew how to accomplish the proper series of actions that could produce a desired final effect. Such knowledge had two aspects. First, an adept must understand the causes that enabled something to be brought into existence. Second, the object produced through the application of this knowledge was intentionally initiated, and thus the maker could evaluate whether the finished product was suitably created. Hintikka elaborated on this second observation, writing the following:

> ... one of the most characteristic features of intentional action is that the agent must be able to tell ahead of time (among other things) what counts as success and what not. An especially important element here seems to be this double requirement of awareness or knowledge: it has to comprise *both* knowledge of the causal connections *and* the knowledge needed to judge the outcome. The latter is in some cases due to the agent's knowledge of the nature of the outcome, including its mode of operation.[46]

The adaptation of Aristotelian categories of causation to table-making and canon procedures was already developed by Campanus of Novara in his thirteenth-century *Theorica planetarum*. In the devising of his diagrammatic instruments (equatoria), the author made use of the language of both final and efficient causation to distinguish between different elements of his planetary motion.[47] He also occasionally likened the arrangement of significant values found on his diagrams to the intentional causal operations of the divine will in nature.[48] Still more systematically, he included the reasoning of the tabular procedures (*ratio operis tabularum*) by means of his equatoria in the conclusion to each of his chapters on planetary motion. The diagram as the instrument, he argued, could provide 'the cause of the whole operation of the tables' (*causa totius operationis tabularum*) for each planet.[49]

As *theorica* and canon texts began to circulate more widely in the fourteenth and fifteenth centuries, together with commentaries on them, the notion of the *causa operationum* became still more closely aligned with Aristotelian causation. In one of the first commentaries to Georg Puerbach's (1423-1461) *Theoricae novae planetarum*, Francesco Capuano di Manfredonia (d. c. 1490) wrote that the concept of *causa*, as it was developed in the *theorica* tradition, was taken from the introduction of Aristotle's *Physics*.[50] In the

46 Jaako Hintikka, *Knowledge and the Known: Historical Perspectives in Epistemology* (Dordrecht: Kluwer Academic Publishers, 1974), pp. 87–88.

47 Campanus of Novara, *Theorica planetarum Campani*, in *Campanus of Novara and Medieval Planetary Theory, Theorica planetarum*, ed. by Benjamin and Toomer, 302–03. For Benjamin and Toomer's discussion on Campanus's decision to describe the elements of his planetary diagrams in terms of Aristotelian causation, see p. 425n10.

48 Campanus of Novara, *Theorica planetarum Campani*, in *Campanus of Novara and Medieval Planetary Theory, Theorica planetarum*, pp. 168–69.

49 Campanus of Novara, *Theorica planetarum Campani*, in *Campanus of Novara and Medieval Planetary Theory, Theorica planetarum*, pp. 156–59, 204–11, 268–93.

50 'Quemadmodum inquit Aristoteles in prologo physicorum tunc cognoscere arbitramur unumquodque cum causas primas et prima principia cognoscimus usque ad elementa, in quo vult quod res non possit cognosci et sciri sine noticia causae eius'. See Francesco Capuano di Manfredonia, *Theoricae novae planetarum Georgii Purbachii astronomi celeberrimi cum expositione Ioannis Baptistae Capuani de Manfredonia* in *Spherae tractatus Ioannis de Sacro Busto*, etc. (Venice: Luce Antonii Iuntae Florentini Officina, 1531), p. 200[a].

Physics, Aristotle had distinguished between two ways of arriving at the truth. The first or 'natural way' was to 'start from the things which are more knowable and obvious to us and proceed towards those which are clearer and more knowable by nature'.[51] The second, however, was to begin with what is 'more clear and knowable by nature' and to proceed from these 'principles, conditions, or elements' through which natural things arise towards the things themselves. It is to these principles, conditions, and elements that the commentator referred when he invoked the *causa*, because, as Aristotle explained, it was 'through acquaintance with these [principles] that knowledge, that is to say scientific knowledge' was attained.[52]

In a vein closer to this introductory gloss provided by Capuano, we discern a second way in which tabular operations were understood to be causally dependent upon diagrams. For, just as knowledge of the *causa*, in a narrow sense, could bring to light the intentions and starting points of the table-makers in their calculations, it was more broadly believed to reveal the often unseen first principles of nature upon which the perceivable effects in the sensible realm depended. In accordance with the conventions of the *theorica* tradition, the texts containing *exempla in figura* betray an interest in the 'realistic' qualities of the diagram, including an effort to represent the corporeal bodies of planets and the boundaries of the planetary orbs. These convex and concave boundaries are often included in the figure with conspicuously distinctive red ink.[53] As Kathleen Crowther and Peter Barker have documented in their analysis of figures found in the *Theoricae novae planetarum*, the elements of *theorica* diagrams were designed to bear hallmarks of cosmological reality, and were not 'mathematical models [intended] merely as devices to "save the phenomena"'.[54] Although we shall see, by way of a conclusion to this article, that the deployment of *exempla in figura* in *theorica* texts differed in some important ways from those of the canons, they were both committed to this kind of realism described by Crowther and Barker.

An illustrative example of the ways in which this commitment to 'realism' was upheld can be seen in the choice on the part of the compilers of the set in Erfurt CA 4° 366 to change the dates of the worked example from chapter to chapter. In doing so they ensured that the values given in the canon procedure would correspond to the positions found on the diagram, thereby faithfully maintaining a dependent relationship of the values in the worked example to the significant configurations in the image. Following instructions found in the sixth canon of the *Priores astrologi*, the compiler had stated that the date for the examples should be understood to start at noon of the last day of December in the

---

51 'Innata autem est ex notioribus nobis via et certioribus in certiora naturae et notiora'. Aristotle, *Physics* I, Ch. 1, 184^{a}16–17, in *The Basic Works of Aristotle*, trans. by R. P. Hardie and R. K. Gaye, ed. by Richard McKeon (New York: Random House, 1941), p. 218. For a discussion of the Latin text that has its origins in the translation of William of Moerbeke, see George J. McMahon, 'The *Prooemium* to the *Physics* of Aristotle', *Laval théologique et philosophique* 13 (1957), 9–57 (p. 13).

52 'Tunc enim cognoscere arbitramur unumquodque, cum causas primas et prima principia cognoscimus, et usque ad elementa', *Physics* I, Ch. 1, 184^{a}13–15. See McMahon, 'The *Prooemium* to the *Physics* of Aristotle', *Laval théologique et philosophique* 13 (1957), 13.

53 For a discussion of these conventions and their implications for critically editing diagrams found in manuscripts, see the editorial principles to the Appendix of Andriani, Gessner, and Jacobson in this volume. See also F.3 and F.4 in the Appendix below, where such rubrications are included.

54 Kathleen Crowther and Peter Barker, 'Training the Intelligent Eye: Understanding Illustrations in Early Modern Astronomy Texts', *Isis* 104 (2013), 429–70 (pp. 438–39).

complete Christian year of 1320.[55] In the case of Chapter 13, which explains how to find the true motion of the Sun, however, the compiler does not remain committed to this date. Instead, the reader is instructed to take six common signs (180°) for the value of the initial mean motion. As already noted, if a user were to calculate for this value using the tables of time, they would find that the starting point for the worked example had shifted to a time between September 12 and September 13, 1321, in Paris.

Furthermore, the compiler appears to have selected this date in order to help the user to interpret the changing values in the columns of the tables. If one looks at the strategic choice of setting the mean motion to six signs, one sees that the solar anomaly is 3s 0°; 35' 38" (i.e. 90°; 35,38). Entering the equation tables, this value is very close to the maximal value of the solar equation. What is more, if one looks at the general scheme of the diagram in the manuscript, one sees that the position of the Sun is similarly composed such that the solar anomaly is at 3 signs or 90°. Thus, the diagram has been constructed to represent the Sun at a place where its solar equation is very close to the maximum as well.[56]

In the next chapter of the canons, which sought the true position of the Moon, the compiler has altered the date of the worked example again, this time explicitly to May 2, 1321 (*incompleto*).[57] With this as the starting point, the mean centre will be 3s, 23°, 33' (113°; 33"), and, therefore, a quick perusal of the table shows that its equation (13°; 8', 33') will be very close to the maximum value (13°; 9'). The equation of argument (4°; 46', 13", 28"') is also very near, though not as close, to its maximum (4°; 56'), and, in like manner, these positions are both represented on the diagram as well.[58] The diagrams, then, were not only constructed to approximately represent the position of the given planet at the time set for the worked example; they were also constructed for strategically chosen times in order to indicate a significant maximal point in the equation tables for the given planet. By indicating these maximal values, the diagram renders the columns containing the equations of argument and centre legible in much the same way as the glosses on the tables' construction had earlier made legible the changing values in the columns of the diversity of diameter and the proportional minutes. This tight connection between numerical values and diagrammatic representation reveals a commitment on the part of the compiler to maintaining a principle of realism through a semiotic shift where the table now functions as a sign with the diagram acting as its referent.

55 See Saby, 'Les Canons de Jean de Lignères', I, pp. 191–92.

56 Another gloss with a similar diagram and set of instructions can be found in the text *Arabes maxime* in BnF lat. 7295A, f. 186rb. In this case, the intention of constructing the diagram at a position near to the solar equation's maximum is made explicit. ' 'Exemplum autem posuimus in locis ubi maxime est diversitas inter medium locum et verum …' See the Appendix to Andriani, Gessner, and Jacobson, 'A Note on Arabes maxime', p. 47, line 129.

57 Chapter 2, lines 75–77 in the Appendix.

58 In the *exemplum in figura* for the Moon, the compiler has given the value for the final equation of argument as 4°; 46', 13", 28"' where one would expect something closer to 4° 47' 29". In the Appendix all the values are kept as they are found in the original text. The compiler also included equivalent values for a reader who would use the Toledan Tables rather than the Alfonsine ones. Like the solar diagram in the previous chapter, the lunar diagram places the epicycle at four positions: one on the left side of the deferent, one on the right side, one at the eccentric apogee, and one at the eccentric perigee. Though the lettering for different parts of the epicycle on the right side of the eccentric deferent is different than those on the left as represented on the diagram, they are made uniform in the text of the *exemplum in figura*. See the editorial principles in the Appendix for a discussion of these difficulties.

4. *Exempla in tabula et figura* as commentary texts

Thus far our analysis of canons with diagrams has been largely confined to the *exempla in tabula* and *figura* as texts, graphical lists, and images. By broadening our scope to attend to the materiality of the manuscripts that contained such worked examples, however, we can glean additional insights about the genre of these canons and the role that diagrams played within them. For instance, in Erfurt CA 4° 366, the worked examples were written around the diagrams with little to no white space between text and image.[59] This suggests that a scribe or illustrator would have drawn the diagram first, the text being added later. Thus, the composition of the manuscript and its *mise-en-page* reinforce the primacy of the diagram in this particular case. Looking at the clues of scribal composition in this way, a practice that Gregg de Young usefully described as a habit of attending to the 'architecture of the manuscripts', we can begin to offer some preliminary conclusions about the genre of the works in which these *exempla* were found and the historical context from which they emerged.[60]

An examination of the composition of the chapter for finding the true Moon in Digby 97 is particularly instructive. It contains identifiable passages that the scribe has compiled from multiple sources, each of which (following the base text) signal a shift from one kind of gloss to another. Each of these discrete bodies of text contain their own authorial voice, and often bear linguistic signatures that indicate the tradition to which they belong.[61]

The chapter begins with John of Saxony's eighteenth canon from *Tempus est mensura motus* quoted at length.[62] At the end of the canon, and with a *pied-de-mouche* breaking up the flow of the text, the compiler introduces his *exemplum in tabula* set for the first day of the complete year 1355. As José Chabás has shown, the selected time used for this worked example, as well as large portions of the text of the *exemplum*, have been lifted from an earlier work called the *Practica tabularum Alfonsi* by Hermann of Saxony (f. 1350s).[63] The following significant tabular values, which were identified terminologically (but not numerically) in the procedure found in John of Saxony's canon, appear in grids in the margin of the text:

1. Mean motion of the Moon	4s	11°	22'	0''	10'''
2. Mean motion of the Sun	4s	47°	31'	59''	21'''
3. Mean centre (double elongation)	4s	47°	40'	1''	42'''
4. Mean lunar argument (mean anomaly)	4s	31°	11'	58''	0'''
5. Equated lunar argument (true anomaly)	4s	21°	3'	26''	0'''
6. Second examined equation of argument	0s	5°	42'	47''	36'''
7. True motion of the Moon	4s	17°	4'	47''	46'''

59 See Erfurt, UB, CA 4° 366, 72v–73r. For a reproduction of the diagrams in question, see F.1 and F.2 in the Appendix.

60 Gregg de Young, 'Mathematical Diagrams from Manuscript to Print: Examples from the Arabic Euclidean Transmission *Synthese* 186 (2012), 21–54.

61 Some of the conspicuous linguistic signatures that distinguish the base canon text from the commentaries that interpret it are differences in verbal tense and mood, changes in deictic language, and the level of specificity in instruction.

62 Chapter 4, lines 411–13 in the Appendix.

63 See the article by José Chabás in this volume. For an introductory bibliographic treatment of Hermann of Saxony's *Practica tabularum*, see Lynn Thorndike, *A History of Magic and Experimental Science*, 8 vols (New York: Columbia University Press, 1934), III, 256.

Another *pied-de-mouche* signals the end of the *exemplum in tabula* and initiates the *exemplum in figura*. Much like the tabular worked example, the diagrammatic one also appears to be compiled from a pre-existing source. It bears strong resemblances to the second chapter of Pseudo-Gerard of Cremona's *Theorica planetarum communis*.[64] With the exception of the mean centre, the order in which significant terms are identified and defined in the *exemplum in figura* are the same as that found in the *Theorica*. The definitions themselves are sometimes, though not always, given *verbatim*.[65] As a point of departure, however, the compiler has inserted references to the lettering of the specific diagram that appears in the canon so that the reader can easily identify the elements on the figure with the definitions drawn from the *theorica* text.

The seams and layers of compiled authorities, which can be discerned in the Digby 97 canons by looking at the scribal conventions of the manuscript, are also relevant for our purposes in identifying the status of the work as a genre. In his pioneering research into the compiling, collating, and punctuating of medieval manuscripts, Malcolm Parkes showed that medieval genres were integrally connected with the material form that they took on the pages of the manuscript.[66] In their decision to collect excerpts of multiple authorities, which were clearly relevant to explaining and interpreting the *Tempus est mensura motus*, the compiler of the Digby 97 canons was demonstrating the status of their work as a commentary.

Even the base text bears telling signs of the compiler's redactions. In the original version of the canon, John of Saxony had included quotations from the *Theorica planetarum communis* in order to justify why certain procedures should be performed. When instructing the user to double the elongation between the Sun and the Moon in order to find the mean centre, for instance, he invokes the *theorica* text.[67] He appeals to the *theorica* again at the end of the canon when instructing the reader to either add or subtract the final equation of argument from the mean motion of the Moon in order to find its true position in the ninth sphere. In both cases, the compiler has removed these references to the *theorica* found in the canon and relocated them. The definition of the mean centre is placed at the beginning of the *exemplum in tabula*, and an explanation for when to add or subtract the final equation of argument appears as part of the larger extract of the *theorica* text found in the *exemplum in figura*.[68]

All of these details gathered from the scribal conventions of Digby 97 would suggest that the compiler was intent on ordering the chapters such that the canon itself was strictly procedural, and any glosses and interpretations were to follow it as a means of clarification. The interpretive aims of these canons do not seem to have been lost on the first generation of readers. Thus, even though referred to as the *Canones in Tabulas Alphonsinas* in William Macray's catalogue of the Digby manuscripts, the Digby text was clearly an expository work, which intended to provide commentary on John of Saxony's canons. And likewise, even

64 For a modern edition of the *Theorica planetarum communis*, see *Theorica planetarum Gerardi*, ed. by Francis J. Carmody (Berkeley: University of California Press, Berkeley, 1942).

65 See, for instance, Chapter 4, lines 571–75, where the text treats the equation of argument.

66 Malcolm Parkes, *Pause and Effect: An Introduction to the History of Punctuation in the West* (London: Routledge, 2016), pp. 74–75.

67 Chapter 4, lines 459–61 in the Appendix.

68 Chapter 4, lines 561–67.

though the first owner (Johannes de Wasia) of the anonymous text in Erfurt CA 4° 366 entitled them the *Canones tabularum Alfonsi*, upon an inspection of the table of contents found on the fly leaf to the same manuscript, we see that he clarified his description of the text to read *Commentum super canones tabularum Alfonsi*.[69]

As Fabrizio Amerini and Gabriele Galluzzo have noted in their *Companion to Latin Medieval Commentaries on Aristotle's Metaphysics*, after the generic *glossa* of the eleventh and twelfth centuries gave way to more formalized expository genres in the thirteenth-century university context, there came to be two principal kinds of commentary that were favoured in the arts curriculum.[70] The first was a literal preservation of the original text, followed by an explanation, clarification, or justification of it by means of the commentator's expert opinion. This was known as a *commentum per modum scripti*, or an *expositio per modum commenti*. The second kind of commentary proceeded *per modum questionis*, and took the form of an expansion on the source material so that new questions or fields of research could be entertained.

In the case of the canon texts in Erfurt 366 and Digby 97, we are faced with commentaries *per modum scripti* rather than *per modum questionis*. The *exempla* that follow the base text are explanatory glosses on the general instructions of the canon. This form of commentary, moreover, is not unique to these two works. Whether they took the form of explicating a diagram, or were simply an arithmetic computation, such examples can be found in several instances prior to the middle of the fourteenth century. Such is John of Sicily's thirteenth-century *commentum per modum scripti* on the Toledan Tables.[71] Later, and more consistently in the form of worked examples, John of Saxony also composed his *Expositiones canonum primi mobilis* (1335) in this way, thereby treating each chapter of John of Lignères' canons on spherical astronomy (*Cuiuslibet arcus*) with computed examples and sometimes diagrams.[72]

Conclusion: *exempla in figura* and commentaries as mathematical practice

The *expositio* canons have not survived in large numbers. Despite the paucity of extant witnesses, however, they tell a compelling story of the ways in which diagrams could serve multiple functions even within a single text. The need to produce and circulate canons

69 See the fly leaf of Erfurt, UB, CA 4° 366.

70 Fabrizio Amerini and Gabriele Galluzzo, 'Introduction', in *A Companion to the Latin Medieval Commentaries on Aristotle's Metaphysics*, ed. by Fabrizio Amerini and Gabriele Galluzzo (Leiden: Brill, 2014), p. 2.

71 See John of Sicily, *Scriptum Johannis de Sicilia super canones Azarchelis*, in *Scriptum Johannis de Sicilia super canones Azarchelis de tabulis Toletanis*, ed. by Fritz Pedersen (København Paludan, 1986).

72 John of Saxony, *Expositiones canonum primi mobilis per magistrum Iohannem de Lineriis a magistro Iohanne de Saxonia*, in Saby, 'Les Canons de Jean de Lignères', II, pp. 281–411. As a point of contrast, John of Murs's *Expositio intentionis regis alfonsii circa tabulas eius* (1321) decidedly follows the form of an *expositio questionis*, and his 1339 canons unconventionally contained worked examples for the majority of its chapters in a way that blended the categories of canon and canon commentary together. He composed the latter text while he was at the Collège de Sorbonne, and the canons betray many points of borrowing from his earlier *expositio*. For a discussion of John of Murs's *Expositio* and his Canons of 1339, see Emmanuel Poulle, 'Jean de Murs et les Tables Alphonsines', *Archives d'histoire doctrinale et littéraire du Moyen Age* 47 (1980), pp. 241–71; C. Philipp Nothaft, 'Jean des Murs's *Canones Tabularum Alfonsii* of 1339', *Erudition and the Republic of Letters* 4 (2019), 98–122.

with interpretive commentaries in the form of worked examples suggests a desire in fourteenth- and fifteenth-century Alfonsine astronomy to unify various elements of the discipline. The doctrines contained in *theorica*, instrument, and canon texts of the arts curriculum were not to remain hermetically sealed. Instead, a common terminology was forged to indicate, define, and transpose disparate mathematical objects found in canons, tables, instrument texts, and the *theorica planetarum*. Worked examples would have been explored using tables and canons, but also equatoria and diagrams. The *expositiones* of the arts curriculum stand as a testament to a mathematical hermeneutic that was synthetic, flexible (multi-directional), and materially grounded. That a user of such texts could freely cross the boundaries between practical and theoretical aspects of the discipline may indicate a period of dynamism and flux in astronomical study in Latin centres of learning during the middle of the fourteenth century.

The fact that expositions and commentaries so often took the form of *exempla*, such as the ones we have detailed in this article, warrants a brief concluding remark on the term itself. The history of the highly specified and case-dependent paradigms within scientific writing that have come to be known as worked examples is yet to be written, and here we can only touch briefly on their relationship to diagrammatic canon texts. The *exemplum* seems to have become established as a convention in medieval Latin canons due to the influence of multiple disciplines from different linguistic and cultural traditions. It would neither be easy nor beneficial for our purposes to favour one of these traditions over the others as the true originator of the worked example in canon texts, as the composers of the canons likely drew upon many of these traditions at once.

In one sense, the term was used frequently within thirteenth- and fourteenth-century Latin preaching and homiletic writings to characterize the illustrative and anecdotal narratives of virtuous figures for whom a lay audience or congregation were exhorted to emulate.[73] Marion Moeser has explained that these *exempla* have their roots in antiquity where short Gospel 'example-narratives' or *paradeigmata* were adapted for the purposes of evangelization and preaching in the early Christian community.[74] In this religious context, the *exemplum* stood as an individual person or a single anecdote that embodied or represented a general moral principle.

More abstractly, the *exemplum* could be understood as taking a single instance or occurrence to stand in for the totality of a thing, and Moeser explains that ancient authors used the term *paradeigma* 'most often in the senses of pattern or model, precedent, sample or illustration, and proof from example'.[75] In Aristotle's *Rhetoric*, such *exempla* became classified as a type of argument known as 'rhetorical induction', and Cicero compared an *exemplum* to an *exornatio*, or embellishment, intended to clarify and make visible an already proven statement.[76] In this way, the *paradeigma* was not taken to be a formal and

73 See, for instance, Jacques de Vitry, *The Exempla or Illustrative Stories from the Sermones Vulgares*, ed. and trans. by Thomas Frederick Crane (New York: B. Franklin, 1971).

74 Marion Moeser, *The Anecdote in Mark, the Classical World and the Rabbis* (London: Sheffield Academic Press, 2002), 57.

75 Moeser, *The Anecdote in Mark, the Classical World and the Rabbis*, 58–59.

76 Aristotle, *Rhetoric* 1356^a36–1356^b25 in *The Basic Works of Aristotle*, trans. by W. Rhys Roberts, ed. by Richard McKeon (New York: Random House, 1941), pp. 1330–31; Marcus Tullius Cicero, *Rhetorica ad Herennium*

independent genre, nor was it presented as a syllogistic proof. Instead, it always buttressed something deemed more epistemically substantial, and was presented as a sort of rhetorical argument with its primary function being that of persuasion.

There was a third instantiation of the term *exemplum*, which emanated from translations of Hebrew, Arabic, and Persian sources into Latin, ultimately deriving from the Arabic word *mithāl* (pl. *amthāl*). The Arabic word *mithāl* means a thing with a similitude to something else.[77] Not unlike the moral overtones at play in the Greek word *paradeigma*, moreover, in Islamic mystical philosophy a *mithāl* was understood to be an illustrative parable or allegory that enabled adepts to access the invisible spiritual realm by way of the imagination.[78] The term was also taken up frequently in Arabo-Persian texts of mixed mathematics, such as astronomy, optics, or practical mechanics, where Elaheh Kheirandesh has explained that the *mithāl* served as a 'textual illustration' for mathematical demonstrations, 'textual models' for productions, and 'physical models' to be used as prototypes for reconstruction.[79]

Without discounting the first two usages of the term *exemplum* in the fourteenth century, it should be clear that the third one is particularly relevant. As a precursor to the diagrams used as worked examples in the anonymous canons to the Alfonsine Tables found in Erfurt CA 4° 366 and Digby 97, we can highlight Ibn al-Muthannā's commentary to the tables of al-Khwārizmī, which has been preserved in Hebrew in three manuscripts (in one of which the translation is attributed to Abraham ibn Ezra) and translated into Latin by Hugo of Santalla in the twelfth century.[80] Sometime between 1148 and 1154, Abraham ibn Ezra would compose a paraphrase of Ibn al-Muthannā's commentary, which has been preserved in Latin in a number of manuscripts.[81]

In these Latin witnesses, Ibn Ezra's work bears the title *De rationibus tabularum*, which is reminiscent of the original Arabic genre from which it derives.[82] For Ibn al-Muthannā's work

---

in *Rhetorica ad Herennium*, trans. by H. Caplan, Loeb Classical Library 403 (Cambridge, Mass.: Harvard University Press, 1954), pp. 274–75.

77 In the twenty-fourth surah of the Qur'ān, the word *amthāl* is used to liken the light of Allah to a lamp and a shining star, among others. See the Qur'ān 24:35, in *The Holy Qur'an: Text, Translation and Commentary*, ed. and trans. by Abdullah Yusuf Ali (Elmhurst, N.Y.: Tahrike Tarsile Qur'an, 2005), pp. 907–8.

78 Carol Bier, 'Art and *Mithāl*: Reading Geometry as Visual Commentary', *Iranian Studies* 41 (2008), 491–509 (pp. 506–7).

79 Elaheh Kheirandesh, 'Science and *Mithāl*: Demonstrations in Arabic and Persian Scientific Traditions', *Iranian Studies* 41 (2008), 465–89 (pp. 468–74).

80 Bernard R. Goldstein has identified and then edited the two Hebrew manuscripts containing Ibn al-Muthannā's commentary: Parma, Bib. Palatina 2636 (olim De Rossi 212), ff. 1^r–13^v; Oxford, Bodleian Library, Michael 400, ff. 45^r–74^r. The Parma manuscript is explicitly attributed to Abraham ibn Ezra, but the Oxford manuscript is not. This has led Goldstein to believe that the latter work was translated by a different author. The Latin translation of the work can be found in three manuscripts: Oxford, Bodleian Library, Arch. Selden B 34; Oxford, Bodleian Library, Savile 15; Cambridge, Gonville and Caius College 456. See Ibn al-Muthannā, *Commentary on the Tables of al-Khwārizmī*, in *Ibn al-Muthannâ's Commentary on the Astronomical Tables of al-Khwârizmî* ed. and trans. by Bernard R. Goldstein (New Haven: Yale University Press, 1967).

81 See Abraham ibn Ezra, *De rationibus tabularum*, in *El Libro de los Fundamentos de las Tablas Astronómicas de R. Abraham ibn 'Ezra*, ed. by José Millás Vallicrosa (Barcelona: Consejo Superior de Investigaciones Científicas, Instituto Arias Montano, 1947), p. 15.

82 Millás Vallicrosa has identified and edited Abraham ibn Ezra's *De rationibus tabularum* (which he has called the *Libro de los Fundamentos de las Tablas Astronómicas*), drawing on the following extant manuscripts: Madrid, Biblioteca Nacional de España, 10053, ff. 67ra–76ra; Oxford, Bodleian Library, Digby 40, ff. 52^r–88^v; Paris, BnF, lat. 16648, ff. 106^r–146^v; Paris, BnF, lat. 16659, ff. 72^r–101^r; Erfurt, UB, CA 4° 3811, ff. 1^r–28^r; London,

is presented as a *taʿlīl*, that is, a commentary that sets out to give the reason for or the cause (*ʿilla*) of the object it addresses. In his introduction to his *zīj* commentary, he explains that 'the author of those tables [i.e. the *zīj* of al-Khwārizmī] did not adduce proof or argument for the rules which they instructed us to follow, and they did not make known to us why they instructed us as they did'.[83] He goes on to address the reader, concluding that 'since you asked me to explain the reasons to you, I will do so in order that nothing be hidden'.[84]

From the witnesses that have survived, Ibn al-Muthannā included diagrams in his treatment of planetary motions.[85] Though he did not include specific worked examples drawn from tables as we see in the Alfonsine canons treated in this article, we can, nevertheless, see resonances between the notion of the *mithāl* presented in his commentary and the diagrammatic worked examples for planetary motion found in Erfurt CA 4° 366 and Digby 97. The fact that one of the earliest commentaries on canons (or *zījs*, to be more accurate) comes from the Arabo-Persian tradition of astronomy suggests that there is an earlier tradition from which Alfonsine canon commentaries drew. Until earlier manuscripts are analysed, either from pre-Alfonsine Latin works or Arabic works in Andalusia, however, it will remain an open question as to whether Parisian practitioners directly borrowed the convention of performing arithmetic worked examples on diagrams, or if the practice has its origins in Paris after the 1320s. It will also be important to try to survey other such commentary texts that emerged during the height of Alfonsine astronomy in the fourteenth and fifteenth centuries. For instance, this method of visualizing a worked example in terms of a diagram can also be found in a text bearing the incipit *Arabes maxime* found in Bibliothèque nationale de France Latin 7295A, whose author has collated works of *theorica* and *practica* as a means for comparing the procedures of the Alfonsine and Toledan Tables.[86]

Finally, surveying astronomical works that contain *exempla in figura* but are not canon texts also yields interesting findings. Witnesses to this specific usage of the term that I have found do not originate from canons, but rather from the mid-thirteenth-century *Theorica planetarum* of Campanus of Novara.[87] In the second chapter of the work, Campanus gives the daily values for each of the five mean motions contained within the Ptolemaic lunar model. This is followed by a description of a hypothetical moment imagined to have taken place in the city of Novara beginning when the body of the Sun and the Moon are positioned at the lunar aux. Then, the diagram is illustrated for the positions of the epicycle centre, the body of the Moon on its epicycle, and the position of the mean Sun a day later. Thus, the diagram is imagined to have been set into motion to illustrate the composite effects of principle lunar motions in a general figure. What was explained earlier in the text in terms of mean daily values has thus been translated into relationships in the diagram that Campanus had constructed.

British Library, Cotton Vespasian A. II, ff. 27ra–31vb, and Cracow , Biblioteka Jagiellońska, 573, pp. 224–53. See Millás Vallicrosa, *El Libro de los Fundamentos de las Tablas Astronómicas*, p. 20. See also C. Philipp E. Nothaft, *Graeco-Arabic Astronomy for Twelfth-Century Latin Readers* (Leiden: Brill, 2022), p. 13n30.

83 Ibn al-Muthannā, *Commentary on the Tables of al-Khwārizmī*, in *Ibn al-Muthannâ's Commentary on the Astronomical Tables of al-Khwârizmî* ed. and trans. by Goldstein, p. 15.

84 Ibn al-Muthannā, *Commentary on the Tables of al-Khwārizmī*, ed. and trans. by Goldstein, pp. 15–16.

85 Ibn al-Muthannā, *Commentary on the Tables of al-Khwārizmī*, ed. and trans. by Goldstein, pp. 30–48.

86 See Andriani, Gessner, and Jacobson's article on the *Arabes maxime* in this volume.

87 Campanus of Novara, *Theorica planetarum Campani*, in *Campanus of Novara and Medieval Planetary Theory, Theorica planetarum*, ed. and trans. by Benjamin, Toomer, p. 170.

This *exemplum figuralis*, as Campanus calls it, which was in turn an adaptation of a similar kind of dynamic diagram found in the second section of book five of the *Almagest*, differs from the *exempla in figura* treated in this article in that they do not rely on astronomical tables.[88] Instead, they take the general definitions required to model planetary motion geometrically, and they instantiate them as a specific moment represented on a diagram. If we broaden our attention to other disciplines that mathematized nature (the so-called mixed sciences), such as optics, we find the convention appearing again in such works as Theodoric of Freiberg's (c. 1250-1311) treatise on the rainbow.[89] Again, in Theodoric's case, the purpose of an *exemplum figuralis* is to offer a specific instantiation of an otherwise general physical process by localizing it with lettered points on a diagram. The transposition of arithmetic tabular values onto diagrams is not at play. Despite this notable difference, these concomitant appearances of diagrammatic worked examples demonstrate that, even if integrating them into canon texts were a novel practice in the fourteenth century, it was clearly the result of borrowing well-established conventions that were already being deployed in works of the mixed sciences long before then.

The practice of transposing arithmetic procedures onto a diagram in a way that simultaneously represented the numerical values of a table and the models of the *theorica*, as we identified in Erfurt CA 4°366 and Digby 97, continued to be employed well into the sixteenth century. They appear to have migrated from canon texts back to *theorica* commentaries once Puerbach's *Theoricae novae planetarum* became a mainstay in the university curriculum. Commentaries by Francesco Capuano di Manfredonia (publ. 1495), Pedro Nuñes (publ. 1537), Oswald Schreckenfuchs (publ. 1556), and Francesco Giuntini (publ. 1581) all contain figural worked examples that interpret the canons of the tables in light of the doctrines of the *theorica*.[90] The diagrams exhibit the same distinctive functions of both representation and justification, as well as the same semiotic relationships, that we witness in the canons described at length in this paper. Thus, even though expository texts specifically devoted to tables may have given way to *theorica* commentaries, the peculiar practices at work in Alfonsine canon commentaries persisted well into the early modern era.

---

88 For Ptolemy's worked example, which seems to have inspired Campanus's, see *Ptolemy's Almagest*, ed. and trans. by Gerald J. Toomer (London: Duckworth, 1984), pp. 220–22.

89 Theodoric of Freiberg, *De iride*, in *Theodoricus Teutonicus de Vriberg De Iride et Radialibus Impressionibus*, ed. by Joseph Würschmidt (Münster: Aschendorff, 1914), pp. 114–16.

90 These commentators provided *exempla in figura* for all the planets, as well as trepidation, but here I simply cite the appearance of these *exempla* for finding the true position of the Sun in their respective commentaries. Francesco Capuano di Manfredonia does not set his examples to a specific date, but rather transposes what he calls the 'terms' (*termini*) of the canon onto the diagram. In the case of the Sun, Capuano's terms are the mean position of the Sun, the solar apogee, the solar anomaly, the solar equation, and the true position of the Sun, which are then represented on the diagram next to his written example. Each of the other commentators included a specific date to be used with the tables. Oswald Schreckenfuchs's example is set for the birth of the Holy Roman Emperor Maximilian II (August 2, 1527). Francesco Giuntini's example is set for the birth of Gaspar Arluno of Milan (June 17, 1544). Pedro Nuñes' example is set for 1592, but, because Nuñes died in 1578, this means it must have been added by a later editor. See Francesco Capuano di Manfredonia, *Theoricae novae planetarum Georgii Purbachii astronomi celeberrimi cum expositione Ioannis Baptistae Capuani de Manfredonia* in *Spherae tractatus Ioannis de Sacro Busto*, etc., pp. 119–20; Pedro Nuñes, *In Theoricas planetarum Georgii Purbachii annotationes aliquot, per Petrum Nonium Salaciensem* in *Petri Nonii Salaciensis Opera* (Basel: Henric-Petrina, 1592), pp. 209–10; Oswald Schreckenfuchs, *Commentaria in Novas theoricas planetarum Georgii Purbachii* (Basel: Henric-Petrina, 1556), p. 33; Francesco Giuntini, *Georgii Purbachii astronomi celeberrimi theoricae planetarum liber, commentariis Francisci Iunctini theologi, mathematici peritissimi illustratus* in *Speculum Astrologiae: universam mathematicam scientiam, in certas classes digestam, complectens*, vols 2 (Lyons: Beraud, 1583), II, p. 29.

Appendix

This Appendix provides the complete text of each of the diagrammatic canons treated in this article. Each of the canons is only preserved in one witness, and all four are found in the following two manuscripts:

Erfurt, Universitätsbibliothek, CA 4° 366, henceforth abbreviated as *E*
Oxford, Bodleian Library, Digby 97, henceforth abbreviated as D

Since these canons were selected for the diagrammatic and tabular content of the worked examples contained within them, I have also included the diagrams and tables featured in each of them. Therefore, I divide my editorial principles below into considerations of the text, the diagrams, and the tables.

Index of the chapters

The numbering of the chapters is not original to these texts; it has been added for ease of use for the reader. The chapters are presented as follows:
1. *Verum locum solis invenire*, *E*, ff. 72^{r}–73^{v};
2. *Verum locum lune*, *E*, ff. 72^{r}–73^{v};
3. *Verum locum solis per tabulas invenire est opus tabule*, *D*, ff. 244^{r}–247^{r};
4. *Verum locum solis per tabulas invenire est opus tabule*, *D*, ff. 247^{r}–249^{v}

Index of the diagrams

The diagrams or figures edited in this Appendix are found in the same two manuscripts and are as follows:
F.1. Theory of the Sun, *E*, f. 72^{v}
F.2. Theory of the Moon, *E*, f. 73^{r}
F.3. Theory of the Sun, *D*, f. 256^{r}
F.4. Theory of the Moon, *D*, f. 249^{v}

Index of the tables

The tables edited in this Appendix are found in manuscript *D* and are as follows:
T.1. Table containing the intermediate values for finding the Sun's position, *D*, f. 245^{r}
T.2. Table containing the equation of the Sun, *D*, f. 246^{r}
T.3. Table containing the intermediate values for finding the Moon's position, *D*, f. 248^{r}
T.4. Table equating the mean Moon, *D*, f. 248^{v}

Editorial principles

The text

In the edition, I have attempted to faithfully reproduce the language of the original witnesses, while also offering them in a format that is accessible to a contemporary reader. Thus, while

I have left the orthography of the text unchanged as much as possible, I have normalized the consonant spelling of 'ci/ti' (e.g. *distantia* rather than *distancia*). Whenever numbers are spelled out as words (e.g. *sex, tribus*, etc.), I do likewise, but in all other cases I follow the text in presenting them symbolically. I do this even when it would be unusual to do so in modern language texts. For example, I use numerical symbols for angular and temporal seconds, thirds, and fourths (2^a, 3^a, 4^a).

To ease the reading of the text, I have applied a modern form of punctuation and organization that does not always follow that of the manuscript exactly. Perhaps the most significant example of this is the fact that I consider text three (*Verum locum solis per tabulas invenire est opus tabule*) to be a unitary chapter, whereas it is presented as two chapters by the scribe.

The edition includes an *apparatus criticus* and an *apparatus fontium*. The first primarily corrects errors in the text and offers my conjectures when there appear to be grammatical errors or problems in coherency or legibility. The second provides references to earlier sources that the authors relied upon to write their canons.

The diagrams

The diagrams have sometimes been altered to conform to their descriptions found in the text. These changes have been made primarily to the letters so that they correctly correspond to the parts of the diagram described in the text. Some letters have also been added when the text makes it clear that they should be given but the scribe has left the diagram incomplete. Finally, I have presented aspects of two of the diagrams in red because the text has called for certain lines and circles to be rubricated. I have done so even though the original diagrams were not rubricated. In what follows, I detail the alterations made to each of the diagrams.

F.1: No alterations have been made to the lettering of F.1. Two pairs of letters (***g/c*** and ***i/l***), which depict the aux and the true Sun on the eccentric circle and in the zodiac respectively, are sometimes conflated in the text itself. Therefore, I have conjectured corrections to the text that preserve the diagram's coherency.

F.2: Several additions and emendations had to be applied to F.2 in order to provide agreement between the diagram and the text. I have added ***o*** and ***q*** to the bottom left-hand corner of the diagram as points that are projections of the body of the Moon and the centre of the epicycle onto the zodiac from the centre of the Earth. The point ***o*** was written in the manuscript but has since worn away. If ***q*** had been written on the diagram, there is no remaining evidence of it. The text, however, uses these letters to indicate this part of the diagram. According to the text, the epicycle on the right-hand side should have all the same lettering as that of the left, with the exception of the epicycle centre, which is labelled ***i*** on the left and ***l*** on the right. Instead, the projections onto the zodiac are ***p, q, o*** on the left-hand side, and ***y*** (or ***v***), ***r, c*** on the right. I have changed the latter letters to ***p, q, o*** to match the former as the text demands. It seems that the ***y*** (or ***v***) and ***c*** in the zodiac are corrections to the two letters labelling the body of the Moon on the epicycle, which are labelled ***c*** and ***y***, respectively. The scribe may have noticed that these two letters were incorrectly inverted on the epicycle on the right (in comparison to the same parts of the epicycle on the left) and appears to have corrected them by re-labelling them in the zodiac rather than crossing them out. I have duly switched the letters ***c*** and ***y*** on the epicycle in the diagram so that the labelling on the left-hand and right-hand epicycles agree, but I have kept their projections as ***p*** and ***o*** in the zodiac to agree with the text. I have also switched the letters ***x*** and ***z*** indicating the east and the west on the diagram so that they match the lettering of these directions in the text.

F.3: None of the lettering in the diagram has been changed, but I have added points ***l*** and ***m*** to serve as the true and mean position of the Sun beyond the eccentric circle. I have also rubricated the eccentric circle and the line projecting the true Sun into the zodiac from the perspective of the Earth. The text calls for this, though the diagrams are not so illustrated.

F.4: I have shifted the points ***o*** and ***p*** slightly to the right to show more obviously that they indicate the tangent points of the epicycle with respect to the centre of the Earth. I have omitted the numbers of the proportional minutes because they go up to thirty in the diagram but sixty in the text. The line drawn through the centre of the epicycle from the centre of the Earth is described as rubricated in the text, so I have represented it as a red line in the diagram.

The tables

All four canons expect the reader to use tables when performing the *exempla in tabula*. The latter two found in *D*, however, actually include the worked examples in tabular form. There are two tables in Canon 3 and two tables in Canon 4 presented below. I have included the tables in the edition alongside the text. When the values in the table do not agree with the text, however, I have changed them.

T.1: The first value for the equation of the Sun drawn from the tables (0° 39' 26") is mistakenly written twice in the table. The second value should not be 0° 39' 26", as it appears in the table, but rather 0° 41' 35", as the text suggests.

T.2: The second table, which is simply the value of the equation of the Sun interpolated for the solar argument 197° 45' 50", has again given the initial value of 0° 39' 26". I have replaced this with the interpolated value 0° 41' 4", as the text indicates.

T.3: In the third table, the mean argument of the Moon is incorrectly labelled as the 'verum lune medium', when it should be referred to as the 'argumentum lune medium'. The value for the degrees is also given as 33°, which I have corrected to 31°, as the text indicates.

T.4: In the fourth table, the degrees for the equated argument are wrongly given as 31° and have been corrected to 21°.

List of abbreviations

ante corr. = *ante correctionem* = 'before correction'
cfr. = *confer* = 'compare/ consider'
coni. = *coniecimus* = 'we conjectured'
corr. = *correxit* = 'the copyist corrected'

deest = *deest* (or *desunt*), 'This [word or letter]/these [words or letters] are missing'; indicates that the evidence of the manuscript is unavailable for this letter/word/passage because the manuscript is damaged or its reading is no longer legible for some other reasons
del.= *delevit* = 'the copyist deleted'
lect. inc. = *lectio incerta* = 'the reading is uncertain'
litt. = *littera-ae* = 'letter or letters'
s. l. = *supra lineam*= 'the copyist supplied'
suppl. = *supplevit* = 'the copyist'
Angular brackets (<…>) I have added the words within the brackets

1. Verum locum solis invenire

Cum volueris hoc, scias medium eius motum. Et sit propositum verbi gratia, quod velis habere verum locum solis, existente eius medio motu 6 signa, a quo medio motu subtrahas augem solis in 2^{a} significatione, qui in isto tempore scilicet anno 1321 imperfecto est 2 signa, 29 <gradus>, 24 <minuta>, 22 <secunda>, ut visum est in exemplo decime. Et remanebit argumentum solis 3 signa, 0 <gradus>, 35 <minuta>, 38 <secunda>.

Cum quo argumento intrabis tabulas equationis. Et quia ibi sunt minuta, intrabis primo cum 3 signis in tabula equationis solis secundum tabulas alfonsii. Et invenies in directo 2 gradus, 9 <minuta>, 57 <secunda>. Deinde intrabis cum eodem argumento uno gradu addito. Et invenies in directo 2 gradus, 9 <minuta>, 59 <secunda>. Deinde accipe differentiam illarum duarum equationum, et est 2 2^{a}, de qua accipe partem proportionalem secundum proportionem 36 minuta que erant cum argumento ad 60, accipiendo 38 2^{a} pro uno minuto cum sint plurima 30; si vero essent pauciores, non acciperes aliquid. Et hoc nota in operationibus omnibus sequentibus de equationibus planetarum potes tamen facere secundum modum datum in equatione motus 8^{ve} spere.

Si volueris facere secundum omnimodam veritatem, accipe igitur partem proportionalem 2 2^{orum} secundum proportionem 36 minutorum ad 60, que est 1 2um, 12 3^{a} ut potes habere per 9. Adde igitur 1^{am} partem proportionalem cum equatione prima, quia fuit minor 2^{a}. Et provenient 2 gradus, 9 minuta, 58 2^{a}, 12 3^{a}, que est equatio argumenti solis, scilicet 3 signa, 0 in gradu, 36 minuta accipiendo 38 2^{a} pro uno minuto. Hanc igitur equationem minue a medio motu solis quia argumentum est minus sex signis. Si autem argu- f. 72v
mentum medium esset plus 6 signis, adderes 1^{am} equationem. Et sic erit verus locus solis post subtractionem 5 signa, 27 <gradus>, 50 <minuta>, 1 <secundum>, 47 <tertia>.

Ut igitur cause operationum que fiunt in equatione solis appareant, describatur circulus orbis signorum super centrum ***e*** qui sit ***acbd***. Et sit ***a*** principium arietis et occidens, ***b*** principium libre et oriens, ***c*** principium cancri, et ***d*** principium capricornis. Circulus vero ***gkih*** sit deferens solis eccentricus super centrum ***f***. Et statuatur sol ***g***, ***k***, ***h***, ***i***. Et sit punctus ***g*** aux deferentis. Et ***h*** oppositum augis.

---

1 Tit. Verum ... invenire] ; cfr. 'Les Canons de Jean de Lignères', vol. 1, canon 13, ed. by Marie-Madeleine Saby (1987), p. 199.

---

10 2] *coni.*; scr. 2 bis sed primum numerum *del.* **27** gkih] *coni.*; gklh i] *coni.*; l

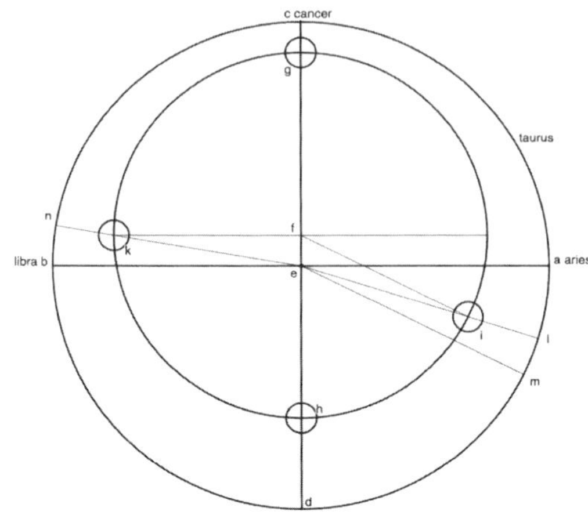

Figure 1

Tunc dico sic medius motus solis est arcus interceptus inter arietem et lineam exeuntem a centro terre equidistantem line exeunti a centro eccentrici per corpus solis. Et quia linea medii motus est semper ei equidistans, aut ille 2^{e} fiunt eadem linea, sicut cum corpus solis est in puncto ***g*** vel ***h***, hoc est in auge vel in eius opposito, quia tunc fiunt eadem linea, arcus vero ***ac***, scilicet inter arietem et locum augis, dicitur aux in 2^{a} significatione. Qui subtrahatur a toto arcu ***acb*** qui est medius motus solis, sole existente in puncto ***k*** deferentis sui ***g***. Remanet arcus ***cb*** quod dicitur argumentum medium. Vel subtrahatur a toto arcu ***acbm***, qui est medius motus, sole existente in puncto ***i***. Et remanebit ***cbm*** quod est argumentum medium. Verus locus solis est arcus interceptus inter arietem et lineam exeuntem a centro terre transeuntem per corpus solis. Et ideo corpore solis existente in puncto ***k***, verus locus est ***acn***. Equatio solis est arcus zodiaci existens inter medium motum et verum, sicut est arcus zodiaci ***bn***.

Arcus igitur ***acb*** est sex signa gratia exempli qui est medius motus, et arcus ***ac*** 2 signa, 29 gradus, 24 minuta, qui est aux solis in 2^{a} significatione. Qui si subtrahatur ab arcu ***acb*** qui est medius motus, remanebit arcus ***cb*** qui est argumentum medium 3 signa, 0 <gradus>, 36 <minuta>. Cum argumento intrabis tabulam equationis solis. Et habebis, secundum quod visum est, equationem argumenti 2 gradus, 9 minuta, 18 <secunda>, 52 <tertia>, qui est arcus ***bn***. Qui si subtrahatur a medio motu scilicet ab arcu ***acb***, remanebit verus motus scilicet arcus ***acn***.

Qui cum argumentum solis fuerit minus 6 signis in arcu ***cbd***, maior est medius motus quam verus. Et ideo equatio argumenti debet subtrahi a medio motu pro habendo vero. Sed cum argumentum fuerit plus 6 signis maior est verus motus quam medius. Et ideo equatio debet addi medio motui ad habendum verum quia, corpore solis existente in puncto ***i***, medius motus est arcus ***acbdm*** et verus motus est arcus ***acdbml***, qui est maior

29/30 medius ... solis] ; cfr. *Theorica planetarum Gerardi* in *Spherae tractatus Ioannis de Sacro Busto*, etc. (1531), f. 169^{r} (original folio misnumbered as 161^{r}). **37/38** Verus ... solis] ; cfr. *Theorica planetarum Gerardi* in *Spherae tractatus Ioannis de Sacro Busto*, etc. (1531), f. 169^{r} (original folio misnumbered as 161^{r}).

33 ac] *coni.*; ag **36** i] *coni.*; l. **51** habendum] *coni.*; *ante* ndum *desunt* 2 *litt.* **52** i] *coni.*; l

puncto in arcu ***ml***, que est equatio solis, que addatur cum medio motu. Cum igitur sol fuerit in medietate sui deferentis ***gkh***, subtrahatur equatio. Sed cum fuerit in alia medietate scilicet in medietate ***hig*** addatur. Cum verus fuerit in puncto ***g*** vel ***h*** nulla est equatio quia medius motus et verus terminantur in eodem loco.

2. Verum locum lune

Sic quod velimus habere verum locum lune in meridie 2^{e} diei maii annis christi completis 1320 queratur eius medius motus et centrum medium et argumentum medium ad illam diem. Et exemplum huius habuisti in 8vo exemplo et etiam in 5°. Et erit medius motus 3 signa, 15 <gradus>, 2 <minuta>, 1 <secundum>, et centrum medium 3 signa, 23 gradus, 33 minuta, et argumentum medium 3 signa, 5 <gradus>, 23 <minuta>.

Serva unumquodque per se. Hic est notandum quod communiter non querimus medios motus planetarum ultra 2^{a} nec argumenta media et centra media ultra minuta. Et ideo in mediis motibus si habeamus ultra 30 3^{a} accipimus pro uno 2° et si minus 30 3^{is} habeamus pro nullo computamus. Et similiter in centris mediis et mediis argumentis si habeamus ultra 30 2^{a} accipimus pro 1 minuto. Si vero minus 30 pro nullo computetur.

Ut igitur clarius pateant que dicenda sunt in equatione lune, premittatur figura per 73r
quam declarabuntur cause operationis que fit in equatione lune. Sit igitur zodiacus circulus ***fqgx*** super centrum ***d*** et deferens lune circulus ***hikl*** super centrum ***e***. Et sit ***b*** punctus oppositus centro deferentis tantum distans a centro terre quam centrum deferentis. Et sit epicyclus lune in 4^{or} locis in punctis ***h***, ***i***, ***k***, ***l*** et corpus lune sit in punctis ***c***, ***y***. Et caput arietis sit in puncto ***a***. Et punctus ***z*** sit oriens et punctus ***x*** sit occidens.

Movetur autem epicyclus in circumferentia deferentis ab occidente versus orientem supra centrum deferentis. Et corpus lune in epicyclo movetur dum est in superiore parte ab oriente versus occidentem et in inferiori e contrario. Aux deferentis est punctus deferentis maxime remotus a terre centro, sicut punctus ***h***, cuius locus terminatur in zodiaco per lineam exeuntem a centro terre transeuntem per centrum deferentis, que linea transit necessario per punctum augis et protracta totum ultra ostendit locum augis in zodiaco sicut linea ***dehf***. Est igitur ***f*** locus augis lune.

Habet etiam epicyclus augem scilicet mediam et veram. Aux media terminatur in superiori parte epicycli per lineam exeuntem a puncto ***b***, qui est punctus oppositus centro deferentis transeuntem per centrum epycicli ad superiorem partem epicycli, quia ubi abscindit superiorem partem epicycli, ibi est aux media, sicut linea ***bim*** vel ***blm***. Est igitur aux media in epicyclo in puncto ***m***. Et per respectum ad istam augem mediam movetur corpus lune in epicyclo equaliter. Aux vera terminatur per lineam exeuntem a centro terre transeuntem per centrum epicycli usque ad zodiacum sicut linea ***dinq*** centro epicycli existente in puncto ***i*** vel ***dln*** in centro epiycli existente in puncto ***l***.

---

56 Tit. Verum ... lune] ; cfr. 'Les Canons de Jean de Lignères', vol. 1, canon 13, ed. by Marie-Madeleine Saby (1987), p. 200. **74** Movetur ... orientem] ; cfr. *Theorica planetarum Gerardi* in *Spherae tractatus Ioannis de Sacro Busto*, etc. (1531), f. 169^{v} (original folio misnumbered as 161^{v}).

---

53 motu] motum *ante corr.* **60** 3^{2}] *coni.*; scr. 3 bis sed secundum numerum *del.* **67** dicenda] *lect. inc.*; *an* docenda? **68** Sit] *coni.*; hic **72** z] *coni.*; g **77** terre] terra *ante corr.* **87** dinq] *primum scr.* dlnq *sed del.* **88** i] *coni.*; l

Verus locus terminatur per lineam exeuntem a centro terre transeuntem per corpus lune in epicyclo usque ad zodiacum. Ubi terminatur in zodiaco, ibi est verus locus lune, sicut linea ***dco*** vel linea ***dyp*** et sic de aliis. Equatio centri in epicyclo est arcus epicycli contentus inter augem mediam et augem veram epicycli sicut arcus ***mn***. Equatio argumenti est arcus zodiaci contentus inter medium locum et verum locum lune sicut arcus ***qo*** vel ***qp*** centro epicycli existente in puncto ***i*** et sic de aliis. Medius motus lune et omnium planetarum incipit ab ariete et argumentum medium ab auge media epicycli.

Est igitur medius motus lune centro epicycli existente in puncto ***i*** arcus ***a*** et ***q*** zodiaci qui est motus ut perhibitum est quia est 3 signa, 15 gradus, 2 minuta, 1 2um. Et centrum medium similiter est notum et est distantia loci medii motus lune ab auge deferentis scilicet arcus ***fa*** et ***q*** qui est 3 signa, 23 <gradus>, 33 <minuta>. Est etiam argumentum lune notum copore lune existente in puncto ***c*** scilicet arcus ***mc*** qui est ut visum est prius 3 signa, 5 <gradus>, 23 <minuta>.

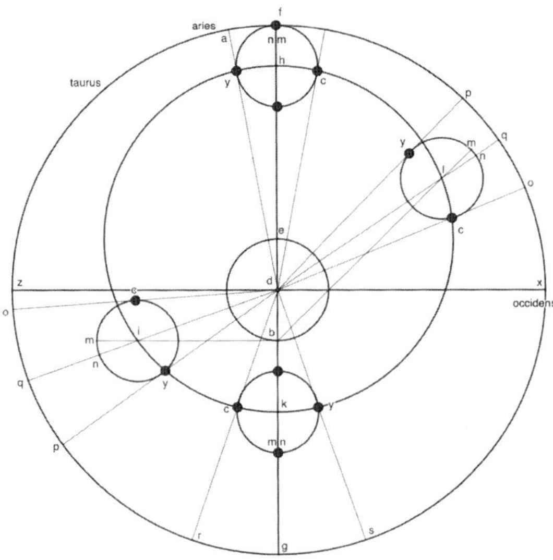

Figure 2

Notandum hic secundum quod habetur in theorica planetarum tabule equationum argumenti lune facte sunt ac si centrum epicycli esset in auge deferentis in quo loco equatio est minor que possit esse quia centrum epicycli est maxime remotum a centro terre. Sic equatio maior est centro epicycli existente in opposito augis deferentis quia excessus istius equationis maioris ad equationem minimam vocatur diversitas dyametri circuli brevis. Centra vero epicycli accedente ad oppositum augis crescunt equationes continue quousque veniat ad oppositum augis ubi sunt equationes maiores quia ab opposito augis usque ad augem decrescunt equationes quousque item centrum epicycli fuerit in auge. Et causa istius est quia dum centrum epicycli movetur ab auge ad oppositum augis dum plus movetur tanto fit propinquius centro terre et sic equatio

89/90 Verus ... lune[1]] ; cfr. *Theorica planetarum Gerardi* in *Spherae tractatus Ioannis de Sacro Busto*, etc. (1531), f. 170^{r} (original folio misnumbered as 162^{r}). **91/92** Equatio ... veram] ; cfr. *Theorica planetarum Gerardi* in *Spherae tractatus Ioannis de Sacro Busto*, etc. (1531), f. 170^{r} (original folio misnumbered as 162^{r}).

94 i] *coni.*; l **96** i] *coni.*; l **100** mc] *supp. s. l.* sed mo *ante corr.* **105** Sic] *coni.*; si

continue augmentatur. Et recedendo ab opposito augis dum plus movetur tanto fit 73v
remotius centro terre et sic continue diminuntur equationes.

Et sic ad hoc quod possimus habere equationem argumenti centro epicycli existente in quocumque loco deferentis. Oportet primo scire in quantum centrum epicycli est propinquius centro terre quam quando fuit in auge vel in quantum est remotius quam quando fuit in opposito augis. Et hoc habetur per minuta proportionalia. Et sunt minuta proportionalia excessus linee protracte a centro terre ad augem deferentis ad lineam protractam ab eodem centro terre ad oppositum augis. Scilicet excessus line ***dh*** ad lineam ***dk*** divisus in 60.

Cum igitur voluerimus invenire verum locum lune oportet primo scire medium motum, centrum medium et minuta proportionalia et argumentum medium et non solum argumentum medium sed etiam argumentum verum. Argumentum verum habetur aliquando subtrahando equationem centri ab argumento medio et aliquando addendo. Et aliquando argumentum medium et verum sunt dicta operando igitur iuxta doctrinam canonis.

Intrabimus cum centro medio ad tempus propositum tabulam equationis lune querendo similem in lineis numeri, accipiendo equationem centri et minuta proportionalia. Et equabimus pro 33 minutis que erant cum 3 signis et 23 gradibus secundum quod factam fuit in sole per 9. Et erit equatio centri – que debetur 3 signa, 23 gradus, 33 minuta – 13 gradus, 8 minuta, 33 2^{a}, et minuta proportionalia 39. Quam equationem centri adde cum medio argumento quia centrum lune est minus 6 signis, quia tunc argumentum verum est maius medio in arcu ***nm*** qui est equatio centri sicut apparet epicyclo existente in puncto ***i***. Sed si centrum medium fuisset plus 6 signis equatio centri subtraheretur quia argumentum verum minor est medio in arcu equationis centri scilicet in arcu ***mn*** ut apparet centro epicycli existente in puncto ***l***.

Adde igitur arcum ***mn*** qui est 13 gradus, 8 <minuta>, 33 <secunda> supra arcum ***mc*** qui est 3 signa, 5 <gradus>, 23 <minuta> scilicet argumentum medium. Et proveniet arcus ***nc*** 3 signa, 18 <gradus>, 32 <minuta>, accipiendo 33 secunda pro uno minuto qui est argumentum verum id est distantia lune ab auge vera epicycli.

Deinde cum argumento equato intra easdem tabulas equationis et accipe que inveneris in directo de equatione circuli brevis et de equatione argumenti. Et erit equatio circuli brevis 2 gradus, 40 minuta. Et equatio argumenti secundum ptoletanas erit 4 gradus, 52 minuta, 37 secunda, 20 tertia. Et secundum tabulas alfonsi erit equatio argumenti 4 gradus, 46 <minuta>, 13 <secunda>, 28 <tertia>. Et ista erit equatio que supponit quod centrum epicycli sit in auge deferentis.

Sed quia luna non est in auge sed alibi propinquius est centro terre ideo oportet quod equatio sit maior et hoc accipietur per equationem circuli brevis que est 2 gradus, 40 minuta et illi duo gradus et 40 minuta sunt excessus equationis argumenti ***fdc*** in opposito augis ad epicyclum ***fcyn*** in auge. Sed quia centrum epicycli non est in opposito augis ideo accipiatur pars proportionalis istius excessus secundum proportionem minutorum proportionalium que fuerunt 39 ad 60. Et est illa pars proportionalis que pars secundum quod potest sciri per 9 erit 1 gradus, 43 minuta que addantur equationi argumenti. Et erit equatio argumenti secundum tabulas tholetanas 6 gradus, 35 minuta,

---

113 remotius] *supp. s. l.* sed propinquius *ante corr.* **116** quam[1]] *coni.*; quando **123** etiam] *coni.*; et **134** i] *coni.*; l **139** nc] *primum scr.* no *sed del.* **140** verum] *primum scr.* enim *post* verum *sed del.*

37 2^a, 20 3^a. Et secundum tabulas alfonsii 6 gradus, 29 <minuta>, 13 <secunda>, 28 <tertia>.

Et sic habebis equationem argumenti examinatam scilicet arcum ***qo***. Que equatio argumenti minuatur a medio motu lune quia argumentum equatum est minus 6 signis. Et hoc est quia medius motus est maior vero in arcu ***oq***. Minue ergo eum ab arcu ***a*** et ***q*** qui est medius motus et remanebit arcus ***a*** et ***o*** qui est verus motus et hic querebatur.

Si vero argumentum equatum esset plus 6 signis equatio argumenti adderetur supra medium motum. Et hoc quia verus motus esset maior medio in arcu equationis ***qp***. Si corpus planete esset in puncto ***y***, equatio argumenti esset tunc arcus ***qp*** et medius motus est arcus ***a*** et ***q*** et arcus ***a*** et ***qp*** esset verus motus. Et sic apparet quod verus esset maior medio in arcu ***qp*** qui est equatio argumenti.

Minuatur igitur equatio argumenti a medio motu et remanebit verus: secundum ptoletanas 3 signa, 8 gradus, 26 minuta, 23 2^a, 40 3^a et secundum tabulas alfonsii erit 3 signa, 8 <gradus>, 32 <minuta>, 47 <secunda>, 32 <tertia>.

3. Verum locum solis per tabulas invenire est opus tabule

Quere primo medium motum solis et ipsum serva. Deinde scias augem solis per doctrinam traditam quam augem subtrahe a medio motu solis. Si autem medius motus solis sit minor tunc mutua 1 revolutionem medio motui et postea subtrahe augem, et quod post subtractionem provenit erit argumentum solis. Hic nota quod quotienscumque precipitur subtrahere talem numerum a tali sine conditione semper oportet subtrahere sive sit numerus a quo debet fieri subtractio minor sive maior.

Habito ergo argumento solis intra cum eo in tabulam que intitulatur tabula equationis solis et quere numerum consilem in lineis numeri. Et si totum numerum argumenti precise inveneris, id est, si sint ibi tantum signa et gradus sine minutis, illic invenies in directo equationem solis quam equationem subtrahe a medio motu si argumentum mediante quo invenisti fuerit minus 3 signis, vel adde eandem equationem 244v
cum medio motu si argumentum fuerit plus 3 signis. Et quod post additionem provenerit erit verus locus solis in 9 spera.

Qui vero precisius operari vellet, necesse est intrare easdem tabulas equationis solis per duplicem introitum prout hic declarabo de sole una pro omnibus, quia idem opus est in isto et in ceteris. Igitur intra itaque cum argumento et quere numerum signorum et vide quid sit in directo de equatione scilicet de gradibus et minutis et 2^{is}. Et quod ibi inveneris, scribe extra in duobus locis et vocetur prima equatio. Deinde si in argumento ultra signa et gradus fuerint aliqua minuta et 2^a, adde argumento unum gradum et cum illo quod provenerit ultra intra iterum in tabulam et vide equationem in directo existentem et eam scribe extra et vocetur 2^a equatio.

Quo facto subtrahe minorem a maiori et illud quod remanet est differentia duarum equationum videlicet est equatio que debetur unum gradum in illo loco tabule; de qua

169/174 Quere ... maior] ; cfr. *Les tables Alphonsines avec les canons de Jean de Saxe*, ed. by Emmanuel Poulle (1984), pp. 64, 66. **175/181** Habito ... spera] ; cfr. *Les tables Alphonsines avec les canons de Jean de Saxe*, ed. by Emmanuel Poulle (1984), p. 66. **184/200** Igitur ... spera] ; cfr. *Les tables Alphonsines avec les canons de Jean de Saxe*, ed. by Emmanuel Poulle (1984), pp. 66, 68.

160 o] *coni.*; c **163** y] *coni.*; l

differentia accipe partem proportionalem id est talem partem qualis pars sunt minuta que in argumento ultra signa et gradus reperiuntur vel dicendo per verba consueta de qua predicta differentia accipe partem proportionalem secundum proportionem minutorum ad 60.

Quam partem proportionalem adde prime equationi si secunda fuerit maior vel subtrahe eam a prima si 2^a fuerit minor. Et quod post additionem vel subtractionem provenerit erit equatio solis. Quam equationem ultimo examinatam subtrahe a medio motu solis, si argumentum fuerit minus 3 signis vel eam adde si fuerit plus 3 signis. Et quod post additionem vel subtractionem provenerit erit verus locus solis in 9 spera.

Et quia in ista sive in omnibus equationibus ubi volumus precise operari indigemus parte proportionali, ideo hic declarabo et repetam modum inveniendi partem proportionalem, nec repetam in subsequentibus sed in qualibet equatione huius tabule remittam ad casum istud sive istum canonem. Si autem in differentia duarum equationum repertarum, prout superius dictum est, fuerint minuta et 2^a, reduc ea ad idem genus scilicet ad 2^a; hoc autem facies multiplicando minuta per 60 et tunc erunt 2^a cum quibus addas 2^a quod prius habebas et quod provenit fiet in denominatione 2^{orum}. Postea minu- 245r
ta et 2^a reduc similiter ad idem genus scilicet ad 2^a.

Quo facto multiplica numerum differentie predicte per numerum residui argumenti et quod inde provenerit serva eius denominationem, quia si multiplicasti minuta per minuta provenient 2^a. Si autem multiplicasti 2^a per 2^a provenient 4^a. Qua autem multiplicatione facta, divide numerum provenientem per 60 et redibit fractio immediate grossior. Si per multiplicationem proveniunt 4^a, per divisionem redibunt 3^a in numero quotiens, et quod remanent fuerint 4^a et sic de aliis multiplicationibus. Item si in numero quotiens fuerint plura 3^a quam 60 divide iterum per 60 et provenient 2^a et quod remanent fuerint 3^a, et sic procedendo quousque fient minuta. Quo facto provenit pars proportionalis precise operando quam partem adde vel subtrahe equationi prime, sicut superius dictus est.

Et ut planius pateat, pono exemplum per hunc modum de sole ad annum supra dictum scilicet 1355 completum ad quam equationem habendam. Quere primo medium motum solis prout ante declaratum est per 4^a, 3^a, 2^a, 1^a ex predicte scilicet 4^a: 2, 3^a: 17, 2^a: 28, prima: 33. Quem motum inveni talem qualem sequens tabula denotat. Deinde a medio motu predicto subtraxi augem solis que fuit talis qualis sequens tabula demonstrat et ut superius dictum est. Et post hanc subtractionem remansit argumentum solis prout sequens tabula denotat.

201/218 Et ... est] ; cfr. *Les tables Alphonsines avec les canons de Jean de Saxe*, ed. by Emmanuel Poulle (1984), pp. 68, 70.

---

206 ad 2a] *suppl. s. l.* **214** remanent] *coni.*; remaniet **216** fuerint] *coni.*; fuerit

	Signa	Gradus	Minuta	Secunda	Tertia	Quarta
Medius motus	4	47	31	59	21	8
Aux eius	1	29	46	9	17	o
Argumentum						
eius	3	17	45	50	4	8
		Gradus	Minuta	Secunda		
Equatio inventa						
prima		o	39	26		
Equatio inventa						
secunda		o	41	35		
	Signa	Gradus	Minuta	Secunda	Tertia	Quarta
Medius motus	4	47	31	59	21	8
Aux eius	1	29	46	9	17	o
Argumentum eius	3	17	45	50	4	8

Table 1

Cum quo argumento sic equato intravi tabulam equationis solis et inveni eius equationem signis et gradibus corespondentem hanc o signis, 39 minutis, 26 2^{is} quam enim scripsi in duobus locis per hunc modum.

Et quia in argumento fuerunt ultra gradus minuta et 2^{a}, intravi iterum predictam tabulam equationum solis uno gradu addito puta cum 18 et sumpsi equationem sibi in 245v
directo corespondentem que fuit equatio secundum quod scripsi in medio duarum equationum prout patet.

Quo facto inspexi istas duas equationes scilicet primam et secundam que earum esset maior et vidi quod 2^{a} fuit maior. Subtraxi ergo ab ea primam et si 2^{a} fuisset minor prima, subtraxissem eam a prima. Si vero fuisset sibi equalis accepissem eam sicut stetissent, quia equata esset hoc modo absque labore. Post hanc subtractionem autem remansit michi differentia 2^{arum} propositarum equationum talis minuta: 2, 2^{a}: 9.

De qua differentia accepi partem proportionalem secundum proportionem minutorum et 2^{orum} que fuerunt ultra gradus in argumento ad 60, sub hac forma reduxi differentiam ad eandem denominationem scilicet ad 2^{a}, et hoc feci multiplicando minuta per 60. Et provenerunt 120 2^{a} quibus addidi 9 2^{a} que erant cum minutis in differentia predicta.

Et post additionem provenit 129 2^{a} quem numerum multiplicavi per residuum argumenti id est per 45 minuta, 50 2^{a}. Et quia in residuo argumenti predicti fuerunt 50 2^{a}, prout patet, debuisset reduxisse ad eandem denominationem scilicet ad 2^{a}, prout superius dictum est. Sed ad demonstrandum aliam regulam in hoc necessariam et satis precisam, pono pro hiis 50 2^{a} que fuerunt ultra in argumento unitatem in minutis, quia 50 2^{a} sunt plus 30. Si vero fuisset minus 30, non curasset de eis. Quam enim predictam unitatem in minutis que fuerant in argumento predicto scilicet 45 minuta et post additionem provenerunt 46 minuta per quem multiplicavi differentiam predictam et ultimo multiplicatavi 2 2^{a}, 9 3^{a}. Et provenerunt 5934 3^{a}, quia minuta multiplicata per 2^{a} denominant 3^{a} quem numerum puta 5934 3^{a}.

266 2] *coni.*; 12

Divisi per 60 et provenit in quotiente 98 2^a et in residuo remanserunt 54 3^a que servavi ad partem. Deinde istum quotientem puta 98 2^a iterum divisi per 60 et provenit in quotiente unum minutum et in residuo remanserunt 38 2^a. Quas autem fractiones omnes collegi et provenit pars proportionalis que fuit talis qualis tabula denotat sequens.

Deinde hanc partem proportionalem addidi cum equatione primo accepta, que fuit talis qualis tabula denotat, quia secunda equatio fuit maior prima equatione prout patet in tabulis. Si autem prima equatio fuisset maior secunda, tunc subtraxisset partem proportionalem a prima equatione, quia semper sciendum est in omnibus equationibus astronomie quotiens equatio 2^a fuit maior equatione prima sic faciendum est et e 246r
converso. Et sic parte proportionali addita prime equationi provenit michi pars proportionalis secunda et equata precise per hunc modum.

	Gradus	Minuta	Secunda
Equatio inventa	0	41	4

Table 2

Quam autem equationem sic equatam addi cum medio motu solis, quia argumentum fuit plus 3 signis. Si vero fuisset minus 3 signis, tunc subtraxissem equationem predictam de medio motu solis. Et facta additione medio, provenit mihi verus locus solis in 9 spera sub hac forma.

Computavi igitur ab ariete et sumpsi per 4 signa alfonsi 8 communia signa, quia in hoc opere scilicet alfonsii unum signum valet 2 signa communia. Et quia cum signa fuerunt 48 gradus, deminui gradus et addidi signis prioribus et fuerunt 9 signa communia et sic fuit verus locus solis 18 gradus, 13 minuta, 3 2^a, 21 3^a, 8 4^a signi capricorni.

Et ut plane pateat quod dixi de medio motu at vero motu solis et de eius equatione maiori vel minori pono exemplum in figura subsequenti per modum theorice sub hec forma quam vides.

---

286 medio[2]] *coni.*; media **291** 3] *coni.*; 4

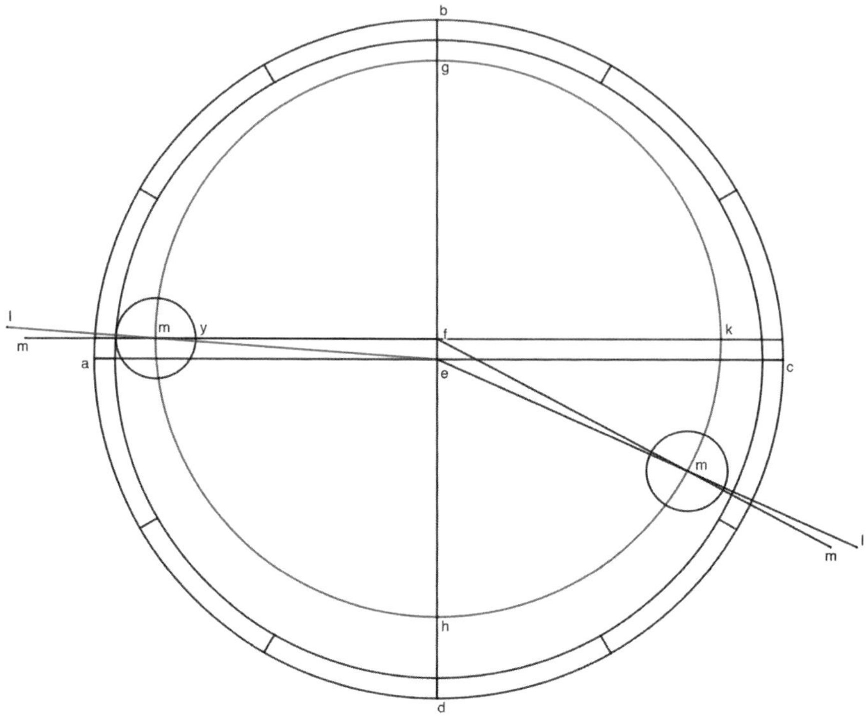

Figure 3

Breviter autem patefacere quid sit 9 spera demonstrat in precedenti figura ciclus supperius continens formam orbis signorum. Quid autem sit oriens in illa spera demonstrat punctum ***a***, quid meridies punctus ***b***, quid occidens punctus ***c***, quid sit septentrion punctus ***d***. Quid sit centrum terre punctus ***e***. Quid enim sit circulus ecentricus demonstrat quidam circulus rubeus cuius centrum elevatum est super centrum mundi sive terre quod demonstrat punctus ***f***. Quid sit aux ecentrici demonstrat punctus ***g***. Quid longitudo illius ecentrici media et prima demonstrat punctus ***m***. Quid oppositum augis ecentrici demonstrat punctus ***h***. Quid longitudo 2^{a} demonstrat punctus ***k***. f. 246v

Unde sciendum quod 9 spera sive circulus orbis signorum movetur proprio motu ab oriente in occidentem puta a puncto ***a*** per punctum ***b*** in punctum ***c***. Et circulus ecentrici movetur proprio motu e converso scilicet ab occidente in orientem a puncto ***k*** per punctum ***g*** in punctum ***y***. Et quia in capitulo precedenti scilicet in equatione solis necesse est scire quid sit medius ac verus motus solis, quid autem sit argumentum ac equatio solis, quare autem equatio addenda vel subtrahenda sit medio motui solis in brevibus demonstrat sub hac forma.

Et primo quid sit medius motus solis demonstrat quedam linea exiens a centro ***f*** ecentrici nigra transiens per centrum solis in 9 spera ***m***. Quid sit enim verus locus vel motus solis demonstrat quaedam linea exiens a centro terre et per centrum solis in 9 spera ***l***.

Unde breviter invenire medium motum solis nichil aliud est quam invenire archum zodiaci incipientem a principio arietis finientem in zodiaco ubi linea nigra exiens a centro ecentrici finitur sive demonstratur. Verum motum solis nichil aliud est quam invenire archum zodiaci incipientem a principio arietis finientem in zodiaco ubi linea

rubea exiens a centro terre procedens per centrum solis finitur in zodiaco. Quid sit enim argumentum solis demonstratur archus zodiaci incipiens ab auge ecentrici ***g*** terminans in zodiaco ubi linea rubea exiens a centro terre transiens per centrum corporis solis finitur in zodiaco. Quare enim si subtrahatur aux equata a medio motu eius scilicet a 247r
principio arietis remanet eius argumentum.

Quid autem sit equatio argumenti demonstrat ille parvulus archus zodiaci inter lineam rubeam ***m*** et lineam nigram ***l***. Quare enim sit equatio addenda vel subtrahenda. Sciendum est quod in illa parte ecentrici versus orientem scilicet a puncto ***g*** per ***y*** in punctum ***h*** medius motus precedit verum motum, quod patet in figura in illa parte ubi linea nigra designans medium motum solis scilicet punctum ***l*** precedit lineam rubeam in zodiaco scilicet punctum ***m*** que linee equationem denotant. Quare in illa parte subtrahenda est.

Item in alia parte ecentrici versus occidentem scilicet a puncto ***h*** per punctum ***k*** in punctum ***g*** verus motus precedit medium quod similiter patet in figura ubi linea rubea designans verum motum scilicet punctum ***m*** precedit lineam nigram designantem medium motum scilicet punctum ***l***. Quare in illa parte equatio addenda est. Unde hoc facto provenit verus motus vel locus in 9 spera.

Item sciendum est quotienscumque fuerit argumentum solis 0 signis, 0 gradibus, 0 minutis, scias ipsum esse in puncto ***g*** scilicet in auge sui ecentrici. Si autem argumentum fuerit minus 3 signis, scias ipsum esse inter ***g*** et ***y*** in illa parte sui ecentrici versus orientem. Si precise 3 signa fuerint scias ipsum esse in puncto ***y*** scilicet in longitudine media et prima sui ecentrici. Si plus 3 signis et minus 6 scias ipsum esse inter ***y*** et ***h***. Si autem fuerit argumentum 6 signa precise, scias ipsum esse scilicet solem esse in opposito augis sui ecentrici scilicet in puncto ***h***, et sic iterum procedendo quousque redeat ab auge sui ecentrici. Sed in hoc argumento posui signa communia.

4. Verum locum lune per tabulam invenire

Quere primo medium motum lune et ipsum serva. Quere etiam medium motum solis quem subtrahe a medio motu lune. Et quod ibi remanet dupla et proveniet centrum medium lune. Serva ipsum iuxta medium motum eius. Quere postea argumentum medium lune et ipsum serva.

Quibus quesitis et servatis intra primo cum centro medio in tabulas equationum lune. Et accipe equationem <centri> lune in directo existentem. Intra enim cum duplici introitu, si oportet, id est, si fuerint minuta ultra gradus et signa in centro predicto sicut 247v
dixi in canone solis.

Postea, vide utrum centrum medium sit minus 3 signis vel plus. Si fuerit minus, adde equationem centri cum argumento medio. Si vero fuerit plus eandem equationem centri subtrahe ab argumento medio. Et quod post additionem vel subtractionem provenerit

346/348 Quere[1] ... lune] ; cfr. *Les tables Alphonsines avec les canons de Jean de Saxe*, ed. by Emmanuel Poulle (1984), p. 70. **350/353** Quibus ... solis] ; cfr. *Les tables Alphonsines avec les canons de Jean de Saxe*, ed. by Emmanuel Poulle (1984), p. 70. **354/360** Postea ... existentem] ; cfr. *Les tables Alphonsines avec les canons de Jean de Saxe*, ed. by Emmanuel Poulle (1984), p. 70.

342 et[3]] *coni.*; o

erit argumentum lune equatum. Est autem argumentum lune equatum sive argumentum verum distantia corporis lune ab auge vera sui epicicli.

Quo habito, intra cum eo in easdem tabulas lune querendo numerum consimilem in linea numeri et accipe equationem argumenti in directo existentem. Intra semper cum duplici introitu si oportet id est si fuerint minuta ultra signa et gradus in argumento lune equato, prout dixi in canone solis. Quo facto accipe ibidem ubi intrasti in lineis numeri cum argumento equato diversitatem diametri circuli brevis intrando semper cum duplici introitu si oportet sicut dixi et cetera. Habita ergo equatione argumenti serva eam et vocatur equatio argumenti primo examinata. Postea accipe diversitatem diametri prius servatam de qua diversitate diametri accipe partem proportionalem secundum proportionem minutorum proportionabilium quam dixi prius servavi ad 60.

Quam partem proportionalem accipe hoc modo: Si in diversitate diametri fuerint gradus et minuta, reduc gradus ad minuta. Deinde multiplica per minuta proportionalia et erunt 2^{a}. Quem inventum 2^{orum} divide per 60 et proveniet in numero quotiens minuta, quia si venerunt minuta plus quam 60 iterum divide per 60 et provenient in numero quotiens gradus.

Quibus gradibus, minutis, 2^{is} habitis provenit pars proportionalis precise equate que pars proportionalis est addenda cum equatione argumenti prius servata quam nominavi primo examinata. Et post additionem provenit equatio argumenti secundo examinata. Quam equationem ultimo examinatam adde medio motui lune si argumentum equatum fuerit plus 3 signis vel subtrahe si fuerit minus.

Et post hanc additionem vel subtractionem provenit verus locus lune in 9 spera.

Et ut planius pateat sententia predicta pono exemplum ad annum christi predictum scilicet 1355 completum. Quesivi ergo medium motum lune quem inveni prout patet in tabula sequenti a quo medio motu subtraxi medium motum solis ad idem tempus qui fuit talis ut patet in tabula. Et post hanc subtractionem provenit residuum illorum 2^{orum} prout patet in tabula sequenti.

	Signa	Gradus	Minuta	2^{a}	3^{a}
Medius motus lune	4	11	22	0	10
Medius motus solis	4	47	31	59	21
Residuum istorum graduum	5	23	50	9	51
Centrum lune medium	4	47	40	1	42
Argumentum lune medium	4	31	11	58	0

Table 3

Dicit enim theorica: Si dupletur distantia inter solem et lunam habetur distantia inter 248r augem centri lune et medium motum lune. Et hec distantia vocatur centrum lune scilicet medium. Quibus sic habitis servavi per hunc modum: Quesivi enim medium argumentum posterius scriptum. Quesivi in tabula medii argumenti lune ad modum quem feci de medio motu solis vel lune per eram prius reductam prout patet in capitulo solis. Et unde provenit argumentum medium lune prout patet in tabula precedenti sub

365/367 accipe ... 60] ; cfr. *Les tables Alphonsines avec les canons de Jean de Saxe*, ed. by Emmanuel Poulle (1984), pp. 70, 72. **378** post ... spera] ; cfr. *Les tables Alphonsines avec les canons de Jean de Saxe*, ed. by Emmanuel Poulle (1984), pp. 72. **392/394** Si ... medium[1]] ; cfr. *Theorica planetarum Gerardi* in *Spherae tractatus Ioannis de Sacro Busto*, etc. (1531), f. 169v (original folio misnumbered as 161v).

suo titulo. Habui ergo ista tria principia: medium motum lune, centrum eius medium, argumentumque eius medium que tria requiruntur in omni equatione planetarum praeter solem.

Quibus habitis et servatis, intravi primo cum centro medio tabulas equationum lune puta cum 4 signis, 47 gradibus in quorum directo sumpsi equationem centri que fuit talis 10 gradus, 13 minuta. Sumpsi etiam ibidem in directo minuta proportionabilia que fuerunt 18 minuta pro quibus non intravi cum duplici introitu, quia in superiori linea fuerunt similiter minuta. Si enim fuisset plus ibi necesse fuisset intrare cum duplici introitu. Servavi igitur hoc ad partem proportionalem. Sed quia in centro predicto fuerunt minuta ultra signa et gradus scilicet 40, intravi in easdem tabulas cum duplici introitu eodem modo sicut dixi in canone solis. Et post hanc equationem provenit equatio centri talis 10 gradus, 8 minuta, 20 2^a. Que fuit equatio signis et gradibus et minutis correspondens operando recte sicut in canone solis operabar de parte proportionabili extra habenda.

Quo facto, istam equationem centri subtraxi ab argumento medio lune, quia centrum fuit plus 3 signis et provenit argumentum lune equatum tale: 4 signa, 21 gradus, 3 minuta, 26 2^a, 0 3^a. Cum quo argumento equato intravi iterum tabulas equationum lune querendo numerum consimilem in lineis numeri et sumpsi in eius directo equationem argumenti talem: 4 gradus, 55 minuta, 4 2^a. Sumpsi etiam ibidem diversitatem diametri que fuit talis: 2 gradus, 39 minuta. Quibus sic habitis servavi ad partem proportionalem. Et quia in argumento predicto fuerunt minuta ultra signa et gradus, intravi easdem tabulas cum duplici introitu prout dixi in canone solis. Et post hanc equationem provenit equatio argumenti precise equata talis: 4 gradus, 55 minuta, 5 2^a, 36 3^a que equatio nominatur primo examinata.

Deinde sumpsi de diversitate diametri partem proportionalem minutorum 248v
proportionalium prius servatorum ad 60. Que pars proportionalis fuit 18 minuta et hoc feci directe in eodem modo operando sicut feci in exemplo solis scilicet reducendo diversitatem diametri ad eandem denominationem, scilicet ad 2^a. Quo facto provenerit 159 minuta, que multiplicavi per 18 minuta scilicet per minuta proportionabilia et provenerunt 2862 2^a. Quem numerum divisi per 60 quotiens potui et provenerunt in quotiente 47 minuta et in residuo 42 2^a. Et fuit pars proportionalis de diversitate diametri que semper est addenda equationi argumenti primo examinate quod ita feci.

Et post additionem provenit equatio argumenti 2° et ultimo examinata talis: 5 gradus, 42 minuta, 37 2^a, 36 3^a. Quam secundam equationem addidi medio motui lune, quia argumentum equatum fuit plus 3 signis. Si vero fuisset minus subtraxissem. Et post subtractionem provenit mihi verus locus lune in 9 spera, prout patet in figura.

---

404 18] *coni.*; 8 **410** canone] *coni.*; centro

	Signa	Gradus	Minuta	2^a	3^a
Medius motus lune	4	11	22	0	10
Argumentum lune equatum	4	21	3	26	0
Equatio lune secunda	0	5	42	47	36
Verus locus eius in 9 spera	4	17	4	47	46

Table 4

Et ut planius pateat quod dixi de medio motu et vero lune, similiter de centro et de argumento ac de equatione eiusdem pono exemplum in figura sequenti per modum theorice prout patet. Et primo quid sit 9 spera sive orbis signorum patet in figura solis superius posita. Quid sit ecentricus lune designat circulus rubeus cuius centrum elevatum est supra centrum terre in super circumferentiam minimi circuli circumeuntis centrum terre cuius centrum ecentrici demonstrat punctus ***f***. Quid sit aux ecentrici demonstrat punctus ***g***. Quid oppositum augis demonstrat punctus ***h***. Quid sit circulus equans demonstrat circulus niger cuius centrum est sub centro terre in opposito differencie minimi circuli circumeuntis centrum terre cuius centrum demonstrat ***y*** eiusque augem demonstrat ***k***. Quid sit epiciclus demonstrat prius circulus movens in circumferentia ecentrici cuius centrum demonstrat ***l***, eiusque augem demonstrat ***m***, eiusque oppositum demonstrat ***n***. Quid sit longitudo media et prima demonstrat ***o***. Quid vero sit longitudo media et secunda demonstrat ***p***. Quid autem sit medius motus lune demonstrat linea quedam rubea exiens a centro terre exiens per centrum epicicli ***l*** in 9 spera ***q***.

Unde invenire medium motum lune nichil aliud est quam invenire quedam archum zodiaci incipientem ab ariete per successionem signorum finientem in fine line predicte 249r ***q***. Quid sit verus locus lune demonstrat quedam linea nigra exiens a centro terre, transiens per centrum corporis lune in 9 spera ***r***. Unde invenire verum motum lune nichil aliud est quam invenire archum zodiaci a principio arietis usque in finem huius linee ***r*** in zodiaco.

Quid sit aux media in epiciclo demonstrat quedam linea nigra exiens a centro equantis ***y***, transiens per centrum epicicli ***l*** in 9 spera ***s***. Quid sit aux vera epicicli demonstrat linea rubea superius dicta exiens a centro terre, transiens per centrum epicicli ***l*** in 9 spera ***q***.

Quid sit centrum lune est distantia augis ecentrici lune ab auge epicicli quod demonstrat archus zodiaci incipiens a fine rubee linee exientis a centro terre finiens per punctum ***g*** in 9 spera, et ab hoc loco procedens per succesionem signorum usque in punctum ***q*** quod demonstrat alia rubea linea exiens a centro terre, transiens per centrum epicicli ***l*** in 9 spera ***q***.

Quid sit argumentum medium lune demonstrat archus epicicli inter ***s*** et ***r*** scilicet inter augem mediam et centrum corporis planete scilicet lune constitutus secundum motum lune scilicet contra succesionem signorum, quia sic movetur luna in superiori parte epicicli scilicet a puncto ***m*** in punctum ***n*** per punctum ***o***. Quid autem sit argumentum verum dicitur archus epicicli inter ***q*** et ***r*** scilicet inter augem veram et centrum lune.

Quid sit enim equatio centri est archus epicicli ***qs*** inter augem mediam et veram que equatio nulla est centro epicicli ***l*** existente in opposito augis ecentrici vel in puncto ***h***.

Quid autem sit equatio argumenti lune dicitur archus zodiaci ***qr*** inter medium scilicet motum et verum, que equatio nulla est centro epicicli ***l*** existente in auge ecentrici ***g*** vel in opposito augis ***y***, luna tamen existente in longitudinibus longioribus ***q***.

Et cum fuerit luna in medietate epicicli que respicit occidentem scilicet in parte ***mon***, tunc medius motus eius maior est vero motu. Quare equatio argumenti subtrahenda est a medio motu et post subtrahendum provenerit verus locus lune in 9 spera. In reliqua vero medietate epicicli scilicet in parte que respicit orientem ***npm*** est e contra, quia in illa est verus motus maior medio motu. Quare tunc equatio addenda est medio motui et provenit verus locus lune equatus.

Quid sit enim diversitas diametri circuli brevis demonstrat archus zodiaci qui est inter ***m*** et ***r***, inter equationes argumenti has argumenti centro epicicli existente in auge 249v
ecentrici vel in opposito augis ecentrici. Unde equationes argumenti que scribuntur in tabulis sunt semper eodem modo ac si centrum epicicli esset semper in auge ecentrici, quia vero existente in aliis locis ecentrici crescunt equationes argumenti secundum quod centrum epicicli accedit ad centrum terre.

Et ita equaciones finiuntur per minuta proportionalia. Quid sint minuta proportionalia demonstrant quedam linee sumpte per 10 intrando usque ad 60 que vocantur minuta proportionalia. Unde sciendum est quod ista minuta proportionalia nulla sunt quando hec linea rubea exiens a centro terre procedens per punctum ***g*** idem conducens augem ecentrici dirigitur in punctum ***h*** quod est autem oppositum augis. Linea vero ***eg*** que dirigitur ad augem habet etiam omnia minuta proportionalia. Hoc enim est intelligendum aux ecentrici rursum elevata insuper. In aliis vero locis que dirigitur ad alia loca habet de minutis proportionalibus secundum accessum et recessum ab auge et ab opposito augis sicut apparet in figura sequenti subtus.

483/484 luna ... subtrahenda] ; cfr. *Theorica planetarum Gerardi* in *Spherae tractatus Ioannis de Sacro Busto*, etc. (1531), f. 170^{r} (original folio misnumbered as 162^{r}). **491/494** equationes ... terre] ; cfr. *Theorica planetarum Gerardi* in *Spherae tractatus Ioannis de Sacro Busto*, etc. (1531), f. 170^{r} (original folio misnumbered as 162^{r}). **500/503** que ... sequenti] ; cfr. *Theorica planetarum Gerardi* in *Spherae tractatus Ioannis de Sacro Busto*, etc. (1531), f. 170^{r} (original folio misnumbered as 162^{r}).

---

482 q] *coni.*; p

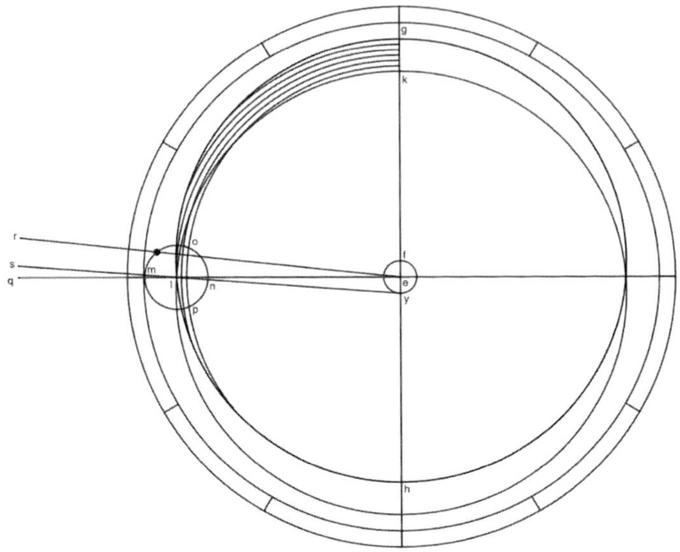

Figure 4

Manuscript sources

Cambridge, Gonville and Caius College, 456/394
Erfurt, Universitätsbibliothek, CA 4° 366
Erfurt, Univesitätsbibliothek, CA 4° 381
Cracow, Biblioteka Jagiellońska, 573
London, British Library, Cotton Vespasian A. II
Madrid, Biblioteca Nacional de España, 10053
Oxford, Bodleian Library, Arch. Selden B 34
Oxford, Bodleian Library, Digby 40
Oxford, Bodleian Library, Digby 97
Oxford, Bodleian Library, Michael 400
Oxford, Bodleian Library, Savile 15
Paris, Bibliothèque nationale de France, lat. 7295A
Paris, Bibliothèque nationale de France, lat. 16648
Paris, Bibliothèque nationale de France, lat. 16659
Parma, Biblioteca Palatina 2636 (olim De Rossi 212)

Bibliography

The Holy Qur'an: Text, Translation and Commentary, ed. and trans. by Abdullah Yusuf Ali (Elmhurst, N.Y.: Tahrike Tarsile Qur'an, 2005).

Theorica planetarum Gerardi, ed. by Francis J. Carmody (Berkeley: University of California Press, Berkeley, 1942).

Amerini, Fabrizio and Gabriele Galluzzo, 'Introduction', in *A Companion to the Latin Medieval Commentaries on Aristotle's Metaphysics*, ed. by Fabrizio Amerini and Gabriele Galluzzo (Leiden: Brill, 2014), pp. 1–18.

Aristotle, *Physics*, in *The Basic Works of Aristotle*, trans. by R. P. Hardie and R. K. Gaye, ed. by Richard McKeon (New York: Random House, 1941), pp. 218–394.

———, *Rhetoric*, in *The Basic Works of Aristotle*, trans. by W. Rhys Roberts, ed. by Richard McKeon (New York: Random House, 1941), pp. 1317–1451.

Bier, Carol, 'Art and *Mithāl*: Reading Geometry as Visual Commentary', *Iranian Studies* 41 (2008), 491–509.

Callus, Daniel A., 'Introduction of Aristotelian Learning to Oxford', *Proceedings of the British Academy* 29 (1943), 229–81.

Campanus of Novara, *Theorica planetarum Campani*, in *Campanus of Novara and Medieval Planetary Theory, Theorica planetarum*, ed. and trans. by Francis S. Benjamin, Jr. and Gerald J. Toomer (Madison: University of Wisconsin Press, 1971).

Cicero, Marcus Tullius, *Rhetorica ad Herennium* in *Rhetorica ad Herennium*, trans. by H. Caplan, Loeb Classical Library 403 (Cambridge, Mass.: Harvard University Press, 1954).

Crowther, Kathleen and Peter Barker, 'Training the Intelligent Eye: Understanding Illustrations in Early Modern Astronomy Texts', *Isis* 104 (2013), 429–70.

de Vitry, Jacques, *The Exempla or Illustrative Stories from the Sermones Vulgares*, ed. and trans. by Thomas Frederick Crane (New York: B. Franklin, 1971).

de Young, Gregg, 'Mathematical diagrams from manuscript to print: Examples from the Arabic Euclidean transmission', *Synthese* 186 (2012), 21–54.

di Manfredonia, Francesco Capuano, *Theoricae novae planetarum Georgii Purbachii astronomi celeberrimi cum expositione Ioannis Baptistae Capuani de Manfredonia* in *Spherae tractatus Ioannis de Sacro Busto*, etc. (Venice: Luce Antonii Iuntae Florentini Officina, 1531), pp. 200[a]–268[a].

Giuntini, Francesco, *Georgii Purbachii astronomi celeberrimi theoricae planetarum liber, commentariis Francisci Iunctini theologi, mathematici peritissimi illustratus* in *Speculum Astrologiae: universam mathematicam scientiam, in certas classes digestam, complectens*, vols 2 (Lyons: Beraud, 1583).

Hintikka, Jaako, 'Practical vs. Theoretical Reason — An Ambiguous Legacy', in *Practical Reason*, ed. by Stephen Körner (New Haven: Yale University Press, 1974), pp. 83–102.

———, *Knowledge and the Known: Historical Perspectives in Epistemology* (Dordrecht: Kluwer Academic Publishers, 1974).

Ibn Ezra, Abraham, *De rationibus tabularum*, in *El Libro de los Fundamentos de las Tablas Astronómicas de R. Abraham ibn 'Ezra*, ed. by José Millás Vallicrosa (Barcelona: Consejo Superior de Investigaciones Científicas, Instituto Arias Montano, 1947).

Ibn al-Muthannā, *Commentary on the Tables of al-Khwārizmī*, in *Ibn al-Muthannâ's Commentary on the Astronomical Tables of al-Khwârizmî* ed. and trans. by Bernard Goldstein (New Haven: Yale University Press, 1967).

Jakobson, Roman, 'Deux Aspects du Langage et Deux Types d'Aphasies,' in *Essais de Linguistique Générale* (Paris: Les Éditions de Minuit, 1963), pp. 43–67.

John of Saxony, *Expositiones canonum primi mobilis per magistrum Iohannem de Lineriis a magistro Iohanne de Saxonia*, in Marie-Madeleine Saby, 'Les Canons de Jean de Lignères sur les tables astronomiques de 1321. Édition critique, traduction, et étude', 2 vols (unpublished dissertation, Paris, École Nationale des Chartes, 1987), II, pp. 281–411.

John of Sicily, *Scriptum Johannis de Sicilia super canones Azarchelis*, in *Scriptum Johannis de Sicilia super canones Azarchelis de tabulis Toletanis*, ed. by Fritz Pedersen (Copenhagen: Erik Paludan, 1986).

Kheirandesh, Elaheh, 'Science and *Mithāl*: Demonstrations in Arabic and Persian Scientific Traditions', *Iranian Studies* 41 (2008), 465–89.

Kremer, Richard L., 'Incunable Almanacs and *Practica* as Practical Knowledge Produced in Trading Zones', in *The Structures of Practical Knowledge*, ed. by Matteo Valleriani (Dordrecht: Springer, 2017), pp. 333–70.

Lattmann, Claas, 'Diagrammatizing Mathematics: Some Remarks on a Revolutionary Aspect of Ancient Greek Mathematics', in *Revolutions and Continuity in Greek Mathematics*, ed. by Michalis Sialaros (Berlin: De Gruyter, 2018), pp. 107–29.

Lohr, Charles, 'Medieval Latin Aristotle Commentaries: Authors A–F', *Traditio* 23 (1967), 313–413.

Macbeth, Danielle, 'Diagrammatic Reasoning in Euclid's Elements,' in *Philosophical Perspectives on Mathematical Practice*, ed. by Bart van Kerkhove *et al.* (London: College Publications, 2010), pp. 235–67.

Mahoney, Michael S., 'Ptolemaic Astronomy in the Middle Ages', in *Dictionary of the Middle Ages*, ed. by Joseph R. Strayer (New York: Scribner, 1988), pp. 206–11.

Mahr, Bernd and Klaus Robering, 'Diagramme als Bilder, die Modelle repräsentieren: Diagrammgebrauch in der Elementargeometrie,' *Zeitschrift für Semiotik* 31 (2009), pp. 275–309.

McMahon, George J., 'The *Prooemium* to the *Physics* of Aristotle', *Laval théologique et philosophique* 13 (1957), 9–57.

Moeser, Marion, *The Anecdote in Mark, the Classical World and the Rabbis* (London: Sheffield Academic Press, 2002).

Netz, Reviel, 'Greek Mathematical Diagrams: Their Use and Their Meaning', *For the Learning of Mathematics* 18 (1998), 33–39.

North, John D., *Richard of Wallingford: An Edition of His Writings with Introductions, English Translation, and Commentary*, 3 vols (Oxford: Clarendon Press, 1976).

Nothaft, C. Philipp E., *Graeco-Arabic Astronomy for Twelfth-Century Latin Readers* (Leiden: Brill, 2022).

———, 'Jean des Murs's *Canones Tabularum Alfonsii* of 1339', *Erudition and the Republic of Letters* 4 (2019), 98–122.

Nuñes, Pedro, *In Theoricas planetarum Georgii Purbachii annotationes aliquot, per Petrum Nonium Salaciensem* in *Petri Nonii Salaciensis Opera* (Basel: Henric–Petrina, 1592), pp. 197–307.

Parkes, Malcolm, *Pause and Effect: An Introduction to the History of Punctuation in the West* (London: Routledge, 2016).

Pedersen, Olaf, 'The *Corpus astronomicum* and the Traditions of Mediaeval Latin Astronomy', *Studia Copernicana*, 8 (1975), 57–96.

Peirce, Charles S., *Writings of Charles S. Peirce: A Chronological Edition, 1884-1886.* 8 vols, ed. by the Peirce Edition Project (Bloomington: Indiana University Press, 1982-2010).

Poulle, Emmanuel, 'Jean de Murs et les Tables Alphonsines', *Archives d'histoire doctrinale et littéraire du Moyen Age* 47 (1980), 241–71.

———, *Les Tables Alphonsines avec les Canons de Jean de Saxe* (Paris: Editions du Centre National de la Recherche Scientifique, 1984).

Ptolemy, Claudius, *Almagest*, in *Ptolemy's Almagest*, ed. and trans. by Gerald J. Toomer (London: Duckworth, 1984).

Rainold, John, *Lectures on Aristotle's* Rhetoric, in *John Rainold's Oxford Lectures on Aristotle's* Rhetoric, ed. and trans. by Lawrence D. Green (London and Toronto: Associated University Presses, 1986).

Saby, Marie-Madeleine, 'Les Canons de Jean de Lignères sur les tables astronomiques de 1321. Édition critique, traduction, et étude', unpublished thesis, Paris, École Nationale des Chartes, 1987.

Schreckenfuchs, Oswald, *Commentaria in Novas theoricas planetarum Georgii Purbachii* (Basel: Henric–Petrina, 1556).

Schum, Wilhelm, *Beschreibendes Verzeichniss der Amplonianischen Handschriften-Sammlung zu Erfurt* (Berlin: Weidmannsche Buchhandlung, 1887).

Theodoric of Freiberg, *De iride*, in *Theodoricus Teutonicus de Vriberg De Iride et Radialibus Impressionibus*, ed. by Joseph Würschmidt (Münster: Aschendorff, 1914).

Thorndike, Lynn, *A History of Magic and Experimental Science*, 8 vols (New York: Columbia University Press, 1934).

Valleriani, Matteo, 'The Epistemology of Practical Knowledge', in *The Structures of Practical Knowledge*, ed. by Matteo Valleriani (Cham, Switzerland: Springer, 2017), pp. 1-19.

Part 2

Reconfiguring Tables

PETR HADRAVA

Mathematical Treatment of Alfonsine Trepidation and its Sources

Introduction

The difference between the actual precession and its model described by Ptolemy in his *Almagest*[1] reached values of several degrees at a time when Islamic astronomy was florishing, which did not escape attention of the astronomers. Moreover, various values of inclination of the ecliptic were obtained by different authors in different epochs. It was due to errors of measurements rather than to its slow real change; however, it was difficult to recognize on the basis of limited precision and the amount of data available in Middle Ages. These facts thus led to attempts to explain both effects using a geometric model of motion of the 'eighth' sphere of fixed stars.

The Aristotelian viewpoint implies that the precession, like any other eternal motion, should result from a uniform circular motion, and hence be periodic. This presumption was also satisfied by the Ptolemaic rotation of the sphere of stars around the pole of ecliptic, which, however, could not explain the apparent changes of its inclination and the non-uniformity of the increase of longitudes of stars. The assumption of a rotation around a different centre thus suggested that the increase of the longitudes of stars observed at the time is only a temporary effect due to a slow periodic 'access and recess' of the eighth sphere, later often also denoted as 'trepidation'. Despite this conception occasionally being questioned,[2] it became widespread and thus misleading until Tycho Brahe correctly

* This work has been completed in the framework of the ERC project *ALFA — Shaping a European Scientific Scene: Alfonsine Astronomy* (CoG 2016, agreement No 723085) with the support of Project RVO 67985815. I am indebted to Alena Hadravová for reading and translating the manuscripts in question. The useful discussions with Richard L. Kremer and Benno van Dalen and their help in obtaining some needed literature and data are also highly appreciated.

1 Ptolemy, *Almagest* VII, 1–3. See G. J. Toomer, *Ptolemy's Almagest* (London: Duckworth, 1984), pp. 321–38.

2 See the study by Philipp E. Nothaft, 'Criticism of Trepidation Models and Advocacy of Uniform Precession in Medieval Latin Astronomy', *Archive for History of Exact Sciences*, 71 (2017), 211–44 and references therein, to which we refer for discussion of the historical contexts of trepidation.

Petr Hadrava • Czech Academy of Sciences, Prague

Alfonsine Astronomy: Computational Practices, ed. by Matthieu Husson, José Chabás, and Richard L. Kremer, Turnhout, 2026 (*Alfonsine Astronomy: Studies and Sources*, 3), p. 199-223
© BREPOLS PUBLISHERS 10.1484/M.ALFA-EB.5.137911

explained the confusion.[3] Following Otto Neugebauer, however, it should be pointed out that in view of the contemporarily available observational evidence, the attempts to find a model of the variability of precession were 'perfectly reasonable'.[4]

The aim of this contribution is thus to study the mathematical background of the theory of trepidation in its form found in the Alfonsine Tables in detail. For this purpose, one must also consider the Pseudo-Thabitan treatise *De motu octave spere* and the Toledan Tables, from which the Alfonsinian tradition developed.

1. Pseudo-Thabitan trepidation

The most influential in medieval Europe was the model of trepidation ascribed to Thābit ibn Qurra (*c.* 830–901, active in Baghdad).[5] One may suspect that this theory, at least in the form that became known, is of a later date and also of a different (most probably al-Andalus) provenience; nevertheless, we shall maintain its traditional designation as a Pseudo-Thabitan model here.[6] It is known from the Latin treatise *De motu octave spere*, which is a translation of a supposed unpreserved Arabic original.[7] This treatise describes the model in words, depicts it in figures, and gives numerical tables for the computation of its results, but these forms of information are not mutually consistent, and none is sufficiently complete to enable a unique interpretation of details of the model and its use.

The model was based on an assumption that the ecliptic, together with planetary apogees and the whole sphere of fixed stars firmly connected to it, periodically moves around its mean position. It is obvious from the treatise and from its various later reworkings that this motion should be controlled by two opposite pivoting points, namely the head of Aries and the head of Libra, which revolve in an Aristotelian uniform circular motion, each one on a circumference of one of two opposite small circles (epicycles) circumscribed around the intersections of the mean ecliptic with the equator. To determine the positions of the movable ecliptic and all fixed stars uniquely, it is necessary to restrict the orientation of the eighth sphere around the axis defined by these heads of Aries and Libra; for example,

3 See Tycho Brahe, *Astronomiae instauratae mechanica* (Wandsbek: Philipp von Ohr, 1598), sig. F3ᵛ, translation by Alena Hadravová et al., *Tycho Brahe, Instruments of the Renewed Astronomy* (Prague: KLP, 1996), pp. 127–28. See also Petr Hadrava, 'Medieval Catalogues of Stars', in *The Stars in the Classical and Medieval Traditions*, ed. by Alena Hadravová, Petr Hadrava, and Kristen Lippincott (Praha: Scriptorium, 2019), pp. 118–19; Noel M. Swerdlow, 'Tycho, Longomontanus, and Kepler on Ptolemy's Solar Observations and Theory, Precession of the Equinoxes, and Obliquity of the Ecliptic', in *Ptolemy in Perspective*, ed. Alexander Jones (New York: Springer, 2010), pp. 151–202.

4 See Otto Neugebauer, 'Thābit ben Qurra "On the solar year" and "On the motion of the eighth sphere"', *Proceedings of the American Philosophical Society* 106 (1962), 264-99 (p. 290).

5 On Thābit ibn Qurra see, e.g. Roshdi Rashed (ed.), *Thābit ibn Qurra, Science and Philosophy in Ninth-Century Baghdad* (Berlin and New York: Walter de Gruyter, 2009).

6 A recent summary of the extended literature on the trepidation theory in al-Andalus and Maghrib and its transmission to the Christian countries is provided by Julio Samsó, *On Both Sides of the Strait of Gibraltar: Studies in the History of Medieval Astronomy in the Iberian Peninsula and the Maghrib* (Leiden and Boston: Brill, 2020), pp. 579–654.

7 The Latin text was edited by Francis J. Carmody, *The Astronomical Works of Thābit b. Qurra* (Berkley: University of California Press, 1960), pp. 84–113; its English translation has been published by Neugebauer, 'Thābit ben Qurra "On the solar year" and "On the motion of the eighth sphere"', pp. 290–99.

one can specify the inclination of the movable ecliptic, that is, the angle ε under which it intersects the equator. This orientation, however, is not clearly explained in the text.

The aim of the trepidation model was to explain, in addition to the motion of the vernal point, the apparent change of inclination of the ecliptic. For instance, sentences [13] or [16] of the treatise *De motu octave spere*[8] leave no doubts that the inclination of the movable ecliptic should be variable. On the other hand, there is a table included in the treatise named *Tabula equationis octave spere que est diversitas capitis Arietis et Libre ab equatore diei*,[9] consisting of two sub-tables, *Equatio capitis ab equatore* which gives the longitude $\lambda = \lambda(\theta)$ of the head of Aries as a function of its phase θ, and *Equatio dimidii diametri circuli parvi* of the declination $\delta(\theta)$ of the head of Aries. There is also a table *Tabula motus accessionis et recessionis*[10] of corresponding mean motion (i.e. dependence $\theta = \theta(t)$ of the phase at time t in the usual form of collected and expanded [Arabic] years). However, there is no table, $\varepsilon(\theta)$, for the inclination of the ecliptic. Moreover, at the end of the treatise (sentences [48] – [49]), an alternative method is described on how to compute λ using the table of declinations, $\delta(\theta)$, of the pivoting points and a table of declinations of degrees of the ecliptic, that is, a table of $\delta(\lambda)$, which was standardly computed for a fixed inclination of the ecliptic and is not included in the treatise. Both tables $\lambda(\theta)$ and $\delta(\theta)$ are given up to seconds of degrees, with a step of 5° from 5° to 90°, while, as we shall see, the model described in words and figures is not symmetric with respect to 90°. All this indicates that despite the tables being introduced in the treatise as a tool enabling us to numerically compute the geometric model previously described in words and figures, they are in fact computed for a different simplified model of trepidation with a constant inclination of the ecliptic. The change of the ecliptic inclination, ε, was thus probably often neglected in practical applications of the Thabitan theory of trepidation.

No precise algorithm is given in the *De motu octave spere* treatise on how to compute the values of trepidation for a chosen time or how the table was composed. It would be possible to derive such an algorithm from the geometry of the model if the mentioned uncertainty in the choice of the instantaneous orientation of the movable ecliptic is somehow overcome. An agreement of numerical results with the tables of trepidation could then indicate which specific model with which specific input parameters had been used to compute the tables. Such a calculation has been performed by Raymond Mercier[11] for two particular choices of the models denoted by him as I and II. Mercier supposed that 'the use of vector algebra rather than spherical trigonometry allows a more flexible analysis'.[12] Resulting formulae for both models are so complicated that there is no doubt that such

---

8 See Carmody, *The Astronomical Works of Thābit b. Qurra*, p. 103; Neugebauer, 'Thābit ben Qurra "On the solar year" and "On the motion of the eighth sphere"', pp. 292–93.

9 See Carmody, *The Astronomical Works of Thābit b. Qurra*, p. 94; Neugebauer, 'Thābit ben Qurra "On the solar year" and "On the motion of the eighth sphere"', pp. 297–98.

10 Carmody, *The Astronomical Works of Thābit b. Qurra*, p. 94; Neugebauer, 'Thābit ben Qurra "On the solar year" and "On the motion of the eighth sphere"', p. 296. These tables also appear in Toledan Tables — see the edition Fritz S. Pedersen, *The Toledan Tables* (Copenhagen: C. A. Reitzels Forlag, 2002), pp. 1554 and 1559. Examples of these tables can be found, for example, in MSS London, BL, Harley 13, f. 119^r; Prague, NK, XIV A 18, f. 30^{r-v}; Vatican, BAV, Pal. lat. 1398, f. 28^r; BAV, Pal. lat. 1414, f. 138^r, etc.

11 See Raymond Mercier, 'Studies in the Medieval Conception of Precession', *Archives internationales d'histoire des sciences*, 26 (1976), 197–220, and 27 (1977), 33–71.

12 Mercier, 'Studies in the Medieval Conception of Precession', 62.

an algorithm could not have been derived and used by medieval Arabic or Latin authors. We verified them using a slightly different yet quite complicated 'modern calculation'. In an attempt to find a possible method accessible to medieval astronomers, we also tried to derive an algorithm based on then available means, namely Menelaus' theorem, already known since antiquity. Somewhat surprisingly, this tool, treated nowadays as cumbersome and obsolete, proved to be more suitable to solve this problem and enabled us to easily find completely equivalent results through a combination of a few subsequent steps that would have been feasible for the medieval mathematicians. It is interesting to note that it was Thābit ibn Qurra who dealt in his treatise *On the sector-figure* with Menelaus' theorem and its applications in detail.[13] We found and published this algorithm first[14] for Mercier's Model II, which seemed to correspond better to the figures in manuscripts with the treatise *De motu octave spere*, and for which Mercier's formulae are a bit less complicated than for Model I. An alternative of this derivation is summarized here for a comparison, and there is newly added similar calculation of Mercier's Model I.

1.1 Mercier's model II

The figures of the model of trepidation included in the *De motu octave spere* treatise, as well as its later reworkings, show in several variants that the movable ecliptics always pass through the solstitial points of the fixed mean ecliptic.[15] Accepting this as a general rule, we have chosen just one from the above-mentioned infinite one-parametric set of possible orientations of the movable ecliptic around the pivoting points. This choice also determines the consequent inclination of the ecliptic to the equator at each phase of the trepidation. It corresponds to Mercier's Model II[16] and is shown in Figure 6.1. Here, Z is the northern celestial pole and $TXPY$ is the equator; the mean ecliptic XS intersects it in the vernal point X with inclination $\varepsilon_0 \equiv \angle YXS$. The movable ecliptic TVS is always passing through the solstitial points of the fixed mean ecliptic, S being its summer solstitial point, V one of the pivoting points, namely the head of Aries, and T the intersection of the movable ecliptic with the equator; hence, it is the instantaneous vernal point, and $\varepsilon \equiv \angle XTV$ is the instantaneous inclination of the movable ecliptic. As V orbits on the circumference of the circle of radius $\rho \equiv XV$, its actual longitude $\lambda \equiv TV$ periodically increases and decreases

13 This treatise has been edited, translated, and commented on by Richard Lorch, *Thābit ibn Qurra, On the Sector-Figure and Related Texts* (Frankfurt am Main: Institute for the History of Arabic-Islamic Science, 2001, reprint 2008).

14 See Hadrava, 'Medieval Catalogues of Stars', pp. 124–5.

15 The most common version of the figures also redrawn by Carmody, *The Astronomical Works of Thābit b. Qurra*, p. 86, Fig. I or by Neugebauer, 'Thābit ben Qurra "On the solar year" and "On the motion of the eighth sphere"', p. 294, Fig. 19, can be found, for example, in MSS Vatican, BAV, Pal. lat. 1398 f. 27^v, Pal. lat. 1414 f. 22^v, or Cracow, BJ 551, f. 125^v. Another form resembling a stereographic projection from the pole of the mean ecliptic, appears in MS London, BL, Harley 13, f. 118^r. Yet another figure based on the stereographic projection from the solstitial point of the fixed ecliptic can be found in the treatise *De tuplici motu octave spere* by Nicolaus Comes de Comitibus (who introduced the term 'trepidation') in MS Vatican, BAV, Vat. lat. 3379, f. 2^v; for a transcription and translation of its principal part and the redrawing of the figure, see Hadrava, 'Medieval catalogues of stars', pp. 123–4.

16 It has also been accepted, for example, by Mercè Comes, 'Ibn al-Hā'im's Trepidation Model', *Suhayl*, 2, (2001), 302, or Samsó *On Both Sides of the Strait of Gibraltar*, p. 580, as well as in my aforementioned work, Hadrava, 'Medieval Catalogues of Stars'.

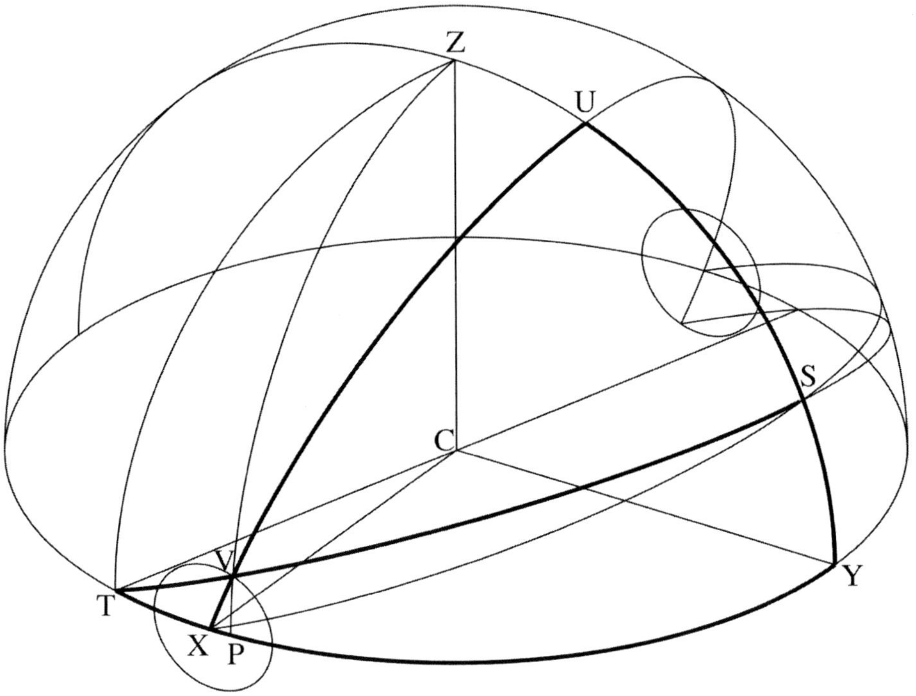

Figure 6.1. Pseudo-Thabitan model of trepidation — Mercier's Model II.

by the motion of 'access and recess'. The value of the 'equation' $\lambda = \lambda(\theta)$ of this motion (*equatio motus accessus et recessus octave spere*) is to be found as a function of the argument $\theta \equiv \angle YXV = YU$ in order to equate the longitudes of all fixed stars and the apogees of the planets. Our aim is thus to derive the formulae for both $\lambda(\theta)$ and $\varepsilon(\theta)$.

If we consider the intersection of the ecliptic TVS with the spherical triangle ΔUXY (both highlighted with bold lines in Figure 6.1, where $ZUSY$ is the solstitial colure of the fixed mean ecliptic) in the framework of the Mercier's Model II, from Menelaus' theorem we obtain the relation

$$1 = \frac{\sin(UV)\sin(XT)\sin(YS)}{\sin(US)\sin(XV)\sin(YT)} = \frac{\cos\rho\sin\alpha\sin\varepsilon_0}{\sin(\theta-\varepsilon_0)\sin\rho\cos\alpha}. \tag{1}$$

From here we can calculate the shift $\alpha \equiv XT$ in right ascension of the true instantaneous vernal point T from its mean position X,[17]

$$\tan\alpha = \tan\rho\frac{\sin(\theta-\varepsilon_0)}{\sin\varepsilon_0}. \tag{2}$$

The equation $\lambda \equiv TV$ could now be calculated using the law of cosines in the triangle ΔTXV,

$$\cos\lambda = \cos\alpha\cos\rho - \sin\alpha\sin\rho\cos\theta, \tag{3}$$

and the instantaneous inclination, ε, using the law of sines in the same triangle,

17 Here, we take α as a positive quantity, that is, an absolute value of the shift, XT, which is oriented opposite to the standard increase of right ascension from X towards Y.

$$\sin \varepsilon = \sin \rho \frac{\sin \theta}{\sin \lambda}. \tag{4}$$

These two formulae (3) and (4) thus provide the desired results using the auxiliary calculation of the right ascension α according to Eq. (2).

Admittedly, the law of cosines used in Eq. (3) and perhaps also the law of sines were not yet known in Thābit's time. However, they could be avoided again using Menelaus' theorem. If we draw the meridian passing from the pole Z through the point V to the point P on the equator, then from Menelaus' theorem for the intersection of the triangle ΔUVZ with the great circle XPY, we obtain the declination $\delta \equiv PV$ of the pivoting point,

$$\sin \delta = \sin \rho \sin \theta, \tag{5}$$

which is the law of sines in the right triangle ΔXPV. In the same way, from Menelaus' theorem for the triangle ΔXPV and circle ZUY, we reach a law of cosines in this right triangle ΔXPV,

$$\cos \rho = \cos \delta \cos \alpha_V, \tag{6}$$

where $\alpha_V \equiv XP$ is the right-ascension of the point V with respect to the vernal point X of the mean ecliptic.

Similarly to Eqs (5) and (6), the relations

$$\sin \delta = \sin \lambda \sin \varepsilon \tag{7}$$

and

$$\cos \lambda = \cos \delta \cos(\alpha_V + \alpha) = \cos \delta \cos \alpha_V \cos \alpha - \cos \delta \sin \alpha_V \sin \alpha \tag{8}$$

must be valid in the rigth triangle ΔTPV.

Finally, for the triangle ΔZPY and circle XVU, we obtain

$$1 = \frac{\sin(ZV) \sin(PX) \sin(YU)}{\sin(ZU) \sin(PV) \sin(YX)} = \frac{\cos \delta \sin \alpha_V \sin \theta}{\cos \theta \sin \delta}. \tag{9}$$

Substituting for $\sin \delta$ from Eq. (5), we obtain $\cos \delta \sin \alpha_V = \cos \theta \sin \rho$, which, after substituting into the last term in Eq. (8), changes this term to the last one of Eq. (3). Similarly, substituting the right-hand side of Eq. (6) into the previous term of Eq. (8), we fully achieve the desired law of cosines (3) in the obtuse triangle ΔTXV. The law of sines (4) can be obtained by excluding δ from Eqs (5) and (7). However, to compute the values of trepidation according to this model, it is not necessary to derive the formulae (3) and (4), but it is possible to find in successive steps the intermediate quantities, α, according to Eq. (2), δ, according to Eq. (5), α_V, according to Eq. (6),[18] λ, according to Eq. (8), and ε, according to Eq. (7).

18 Using Menelaus' theorem for the triangle ΔUXY and circle PVZ, the formula $\tan \alpha_V = \tan \rho \cos \theta$ can also be derived for a direct computation of α_V.

1.2. Mercier's model I

In addition to the above-described model, Mercier also worked on a slightly more complicated alternative, which he actually preferred and denoted as Model I. In this model, shown in Figure 6.2, the fixed and movable ecliptic mutually intersect at the point *R*, which is always at a distance of 90°, not from the vernal point *X* of the fixed mean ecliptic, but (along the movable ecliptic) from the pivoting point *V*. A support for this variant can be found in sentences [28] and [37] of Carmody's edition:

> [28]: semper caput Capricorni et caput Cancri sunt coniuncta orbi declivi fixo non recedentia ab eo, et moventur antecedentes et retrocedentes secundum quantitatem diametri circuli parvi[19]
>
> > the head of the Capricorn and the head of Cancer are always attached to the fixed inclined circle without deviating from it and move forward and backward according to the size of the radius of the small circle[20]
>
> [37]: caput Capricorni et caput Cancri removebuntur a differentiis communibus predictis et antecedent et retrocedent secundum locum capitis Arietis et Libre duorum circulorum, sed non removebuntur a coniunctione orbis declivis fixi[21]
>
> > the head of Capricorn and the head of Cancer will be removed from the above-mentioned common intersections[22] and will progress and recede according to the position of the heads of Aries and Libra on the two circles, but they will not be removed from contact with the fixed inclined circle.[23]

Actually, Mercier's interpretation of the text using Model I does not contradict the evidence provided in the figures preserved in the manuscripts, because only the extreme deviations of the movable ecliptic are always drawn from the fixed one; the solstitial point thus coincides in both versions due to symmetry. A general position of the movable ecliptic in the thus far inspected manuscripts is never drawn.[24]

In Mercier's Model I, the instantaneous ecliptic *TVR* thus intersects the mean ecliptic *XQSR* (see Figure 6.2) in a point *R*, which glides on the fixed ecliptic to both sides from its intersection *S* with the solstitial colure *ZSY* corresponding to the fixed mean ecliptic in such a way that *VR*= 90°. It means that a great circle *NVQ*, circumscribed around the point *R* as its centre, passes not only through the point *V* but also through the pole *N* of

19 Carmody, *The Astronomical Works of Thābit b. Qurra*, p. 104.

20 Modified from Neugebauer, 'Thābit ben Qurra "On the solar year" and "On the motion of the eighth sphere"', p. 295.

21 Carmody, *The Astronomical Works of Thābit b. Qurra*, p. 105.

22 That is the movable head of Cancer *R* in Figure 6.2 is removed from the summer solstitial point *S* of the fixed ecliptic (mean ecliptic *XQSR*), which is its common point (the intersection) with the meridian *YSUZN*, which has the greatest distance (the difference *YS*) from the equator *TXPYW*.

23 Modified from Neugebauer, 'Thābit ben Qurra "On the solar year" and "On the motion of the eighth sphere"', p. 296.

24 Even in the Schöner's depiction of the *Aequatorium octave sphere* is shown the situation when the heads of Aries and Capricorn coincide on the eighth and the ninth sphere. See *Ioannis Schoneri Aequatorium astronomicum*, 1551, f. a3^{r}.

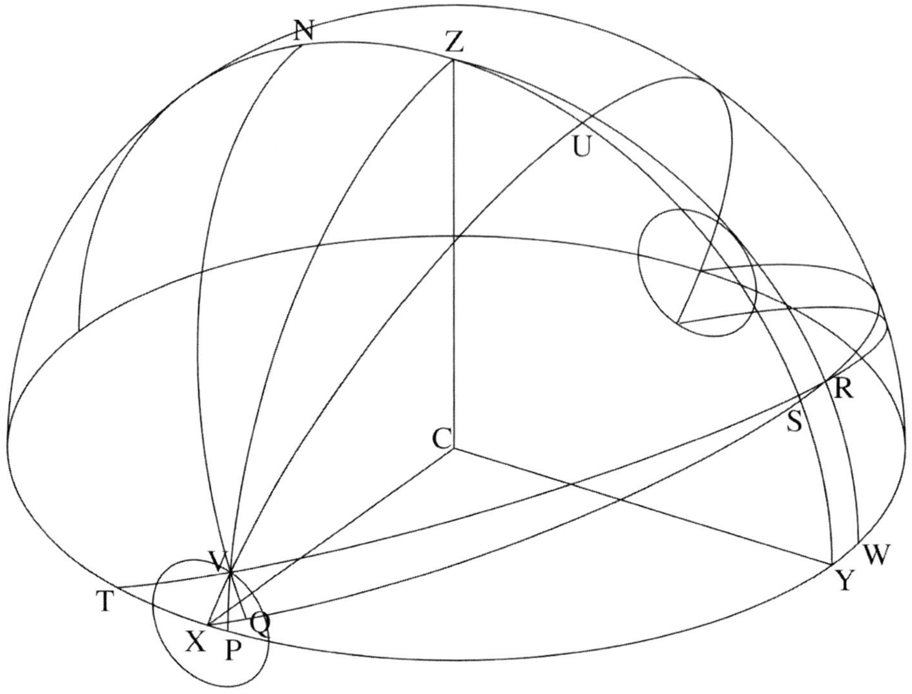

Figure 6.2. Pseudo-Thabitan model of trepidation – Mercier's Model I.

the fixed mean ecliptic. It also crosses it in the point *Q*, which is distant from *X* equally as the intersection *R* from the solstitial point *S* of the fixed mean ecliptic. In analogy with the formula in note 18, it can be found from Menelaus' theorem

$$1 = \frac{\sin(UN)\,\sin(XV)\,\sin(SQ)}{\sin(UV)\,\sin(XQ)\,\sin(SN)} = \frac{\cos(\theta - \varepsilon_o)\,\sin\rho\,\cos l_V}{\cos\rho\,\sin l_V} \qquad (10)$$

for the triangle ΔUXS and the circle *NVQ*, that the longitude of the point *V* along the mean ecliptic, that is, the side $l_V \equiv XQ$ of the rigth triangle ΔXQV with a right angle at *Q*, is given by the relation

$$\tan l_V = \tan\rho\,\cos(\theta - \varepsilon_o)\,. \qquad (11)$$

Because $SR = XQ = l_V$, we know the longitude $l_R \equiv XR = l_V + 90°$ of *R* along the fixed mean ecliptic. In analogy with Eq. (5), we can find the declination $\delta_R \equiv WR$ in the right triangle ΔXWR,

$$\sin\,\delta_R = \sin\,l_R\,\sin\,\varepsilon_o = \cos\,l_V \sin\varepsilon_o\,. \qquad (12)$$

Similarly to Eq. (7), which is valid in the triangle ΔTPV, and as $TR = TV + 90°$, we find that

$$\sin\delta_R \;=\; \sin TR\,\sin\varepsilon \;=\; \cos\lambda\,\sin\varepsilon \qquad (13)$$

in the triangle ΔTRW. Substituting here for sin ε from Eq. (7), we reach the desired formula,

$$\tan\lambda = \frac{\sin\delta}{\sin\delta_R}, \tag{14}$$

and ε can be then found either from Eq. (7) or Eq. (13). When treated in the framework of spherical trigonometry, the computation of Model I is thus as simple as for Model II; one must compute δ according to Eq. (5) and l_V according to Eq. (11), δ_R according to Eq. (12), λ according to Eq. (14), and finally ε according to Eq. (7) or (13). It should be highlighted that in both cases the declination δ of the head of Aries *V* is needed as one of the intermediate quantities and this quantity δ(θ) is also tabulated in *De motu octave spere*.

1.3. Comparison with calculation by means of vector algebra

Before we investigate numerical results, let us compare the above derivations using the classical spherical trigonometry with those obtained using tools of modern mathematics. Mercier derived his formulae (1a) or (2a) for the equation of the trepidation motion λ (denoted by him as E_I or E_{II} for the Models I or II, respectively), and similarly the formulae (1b) or (2b) for the instantaneous inclination of the movable ecliptic ε (denoted ε_I, ε_{II}).[25] He chose an orthonormal basis of vectors ***j*** in the direction of the pole of the equator (i.e. the vector from the central point *C* to the point *Z* on Figures 6.1 and 6.2), ***i*** to its cross-point (*Y*) with the solstitial colure (*ZUSY*) for the fixed (mean) ecliptic (*XS*), and ***k*** to the vernal point (*X*) on the fixed ecliptic.[26] From his conditions (4) of perpendicularity $(\boldsymbol{n}_0\boldsymbol{s}) = (\boldsymbol{n}\boldsymbol{v}) = (\boldsymbol{n}\boldsymbol{s}) = 0$, where $\boldsymbol{n}_0$ is the unit vector from the centre *C* to the pole (the point *N* in Figure 6.2) of the fixed ecliptic (*XS*), ***n*** to the pole of the moving ecliptic (*TVS*), ***v*** to the pivoting head of Aries on the eighth sphere (the point *V*), and ***s*** to the intersection of the fixed and moving ecliptic,[27] Mercier solves analytically first for directional cosines of the vector ***n***. He arrives (in his appendix) at some relatively complicated formulae (7)–(9); for example, the last one should read[28]

$$\cos\gamma \equiv (\boldsymbol{n}\boldsymbol{k}) = -\frac{\sin\rho\sin(\theta-\varepsilon_0)}{\sqrt{1-\sin^2\rho\cos^2(\theta-\varepsilon_0)+\tan^2\varphi\sin^2\rho-2\tan\varphi\sin\rho\cos\rho\cos(\theta-\varepsilon_0)}},$$

where ε_0 is the inclination of the fixed ecliptic (i.e. the angle ∠*YXS*), ρ is the radius *XV* of the small circle, θ =∠*YXV* is the argument of the trepidation, and φ is the angular distance of the intersection ***s*** of the fixed and moving ecliptic from the fixed solstitial colure *ZUSY*, i.e. a complement of its (fixed) longitude to 90°. This angle φ is actually equal to zero in Model II, but it is a varying angle *RS* in the Model I depicted here in Figure 6.2.

---

25 The computation is described in the appendix on pp. 62–5, and the resulting formulae are given on p. 212 in Mercier, 'Studies in the Medieval Conception of Precession'.

26 Mercier mistakenly wrote that 'the terminus of ***k*** is $\boldsymbol{n}_0$', which is the pole of the fixed ecliptic given by his Eq. (1) as a linear combination of ***i*** and ***j***.

27 The unit vector ***s*** is denoted $\boldsymbol{\rho}_1$ by R. Mercier. It points from the centre *C* towards *S* in Mercier's Model II or towards *R* in Model I. Similarly, the vector ***v***, which points to *V*, is denoted **ρ** while the radius ρ = *XV* of the pivoting epicycle is denoted as *r* in Mercier's calculations.

28 Due to a typo, the last term in the nominator of Mercier's formula (9) contains $\cos^2(\theta-\varepsilon_0)$ instead of $\cos(\theta-\varepsilon_0)$, but the subsequent relations derived from it are typed correctly.

An equivalent but slightly simpler and more symmetric form of Mercier's formulae (7)–(9) can be achieved by a consistent use of the vector algebra if we note that the angle $\psi = \angle VCR$ between $\boldsymbol{v}$ and $\boldsymbol{s}$ is given by the scalar product

$$\cos\psi = (\boldsymbol{vs}) = \sin\rho\cos\varphi\cos(\theta - \varepsilon_0) + \cos\rho\sin\varphi$$

(which is actually the law of cosines in the triangle ΔVXR). This means that for Mercier's Model I, $\cos\psi = 0$ and, in agreement with his formula (10),

$$\tan\varphi = -\tan\rho\cos(\theta - \varepsilon_0)\,.$$

The pole of the moving ecliptic is then given by the vector product

$$\boldsymbol{n} = \frac{\boldsymbol{v}\times\boldsymbol{s}}{\sin\psi} = \frac{\boldsymbol{m}}{\sqrt{1-(\sin\rho\cos\varphi\cos(\theta-\varepsilon_0)+\cos\rho\sin\varphi)^2}},$$

where

$$\boldsymbol{m} = \boldsymbol{i}\,(\sin\rho\sin\theta\sin\varphi - \cos\rho\sin\varepsilon_0\cos\varphi)$$
$$+\,\boldsymbol{j}\,(\cos\rho\cos\varepsilon_0\cos\varphi - \sin\rho\cos\theta\sin\varphi)$$
$$+\,\boldsymbol{k}\sin\rho\cos\varphi\sin(\varepsilon_0 - \theta)\,.$$

In Model II, where $\varphi = 0$, this expression reduces to

$$\boldsymbol{n} = \frac{-\boldsymbol{i}\cos\rho\sin\varepsilon_0 + \boldsymbol{j}\cos\rho\cos\varepsilon_0 + \boldsymbol{k}\sin\rho\sin(\varepsilon_0-\theta)}{\sqrt{1-\sin^2\rho\cos^2(\theta-\varepsilon_0)}},$$

which is equivalent to Mercier's equations (11.II). In Mercier's Model I, we can similarly find $\boldsymbol{s} = \boldsymbol{n}_0 \times \boldsymbol{v}/\sin\mu$, where $\cos\mu = (\boldsymbol{n}_0\boldsymbol{v}) = \sin\rho\sin(\theta - \varepsilon_0)$, and then

$$\boldsymbol{n} = \boldsymbol{v}\times\boldsymbol{s} \;=\; \boldsymbol{v}\times(\boldsymbol{n}_0\times\boldsymbol{v})/\sin\mu \;=\; (\boldsymbol{n}_0 - \boldsymbol{v}\cos\mu)/\sin\mu,$$

or, we can arrive at the same result

$$\boldsymbol{n} = [-\boldsymbol{i}(\sin^2\rho\sin\theta\cos(\theta-\varepsilon_0) + \cos^2\rho\sin\varepsilon_0)$$
$$+\,\boldsymbol{j}\,(\cos^2\rho\cos\varepsilon_0 + \sin^2\rho\cos\theta\cos(\theta-\varepsilon_0))$$
$$+\,\boldsymbol{k}\sin\rho\cos\rho\sin(\varepsilon_0-\theta)]\,[1-\sin^2\rho\sin^2(\theta-\varepsilon_0)]^{-1/2},$$

which is equivalent of Mercier's equations (11.I), as if we had substituted $\sin\psi = 1$, $\sin\varphi = -\sin\rho\cos(\theta - \varepsilon_0)/\sin\mu$ and $\cos\varphi = \cos\rho/\sin\mu$ into the general expression.

The vernal point on the moving ecliptic (T) is then given by a unit vector $\boldsymbol{l}$ in the direction of vector product $\boldsymbol{j}\times\boldsymbol{n}$ and the trepidation $\lambda \equiv E$ by the size of vector product $\boldsymbol{v}\times\boldsymbol{l}$:

$$\sin\lambda = |\boldsymbol{v}\times\boldsymbol{l}| = |\boldsymbol{v}.\boldsymbol{j}|\,/\,|\boldsymbol{j}\times\boldsymbol{n}|,$$

while the cosine of obliquity of the moving ecliptic is directly the j-component of $\boldsymbol{n}$. The resulting formulae (in our present notation) are

$$\sin\lambda_I = \sin\theta\tan\rho\left\{\frac{1+\tan^2\rho\cos^2(\theta-\varepsilon_0)}{\tan^2\rho\sin^2(\theta-\varepsilon_0) + [\sin\varepsilon_0 + \tan^2\rho\cos(\theta-\varepsilon_0)\sin\theta]^2}\right\}^{1/2} \qquad (15)$$

and

$$\cos \varepsilon_{I} = \frac{\cos \rho \cos \varepsilon_{o} + \tan \rho \sin \rho \cos(\theta - \varepsilon_{o})\cos \theta}{\left[1 + \tan^{2} \rho \cos^{2}(\theta - \varepsilon_{o})\right]^{1/2}} \tag{16}$$

for Model I, and

$$\sin \lambda_{II} = \sin \theta \sin \rho \left\{ \frac{1 + \tan^{2} \rho \sin^{2}(\theta - \varepsilon_{o})}{\sin^{2} \varepsilon_{o} + \tan^{2} \rho \sin^{2}(\theta - \varepsilon_{o})} \right\}^{1/2} \tag{17}$$

and

$$\cos \varepsilon_{II} = \frac{\cos \varepsilon_{o}}{\left[1 + \tan^{2} \rho \sin^{2}(\theta - \varepsilon_{o})\right]^{1/2}} \tag{18}$$

for Model II. These are quite complicated, but one can numerically check that they are fully equivalent to the formulae given in the previous subsections.

It is worth noting that Mercier[29] suggested a common approximation,

$$\sin \lambda \approx \frac{\tan \rho}{\sin \varepsilon_{o}} \sin \theta, \quad \varepsilon \approx \varepsilon_{o}, \tag{19}$$

to Eqs (15)–(18), which was also accepted and used by J. Samsó.[30] This approximation was obviously derived by neglecting the terms with tan ρ inside the curly brackets in Eq. (15), but it is inconsistent with Eq. (4), which must be satisfied for geometric reasons in both of Mercier's models and the model with a fixed inclination of ecliptic; this can be obtained by the same procedure from Eq. (17). Due to the improper tan ρ in this approximation, which should be sin ρ, its values exceed the values of Mercier's Models I and II by 17′ and 14′ at its maximum, while the approximation (4) with $\varepsilon \approx \varepsilon_{o}$ differs by only 5′ and 2′, respectively.

1.4. Iso-inclination models

It is obvious that if the movable and fixed ecliptics cross each other anywhere close to the solstitial points, then for very small radii ρ of the pivoting circle, the deviation of inclination ε of the moving ecliptic from the inclination ε_{o} of the fixed ecliptic decreases by ~ sin ρ. Regarding Eq. (4), the trepidation λ decreases with ρ nearly linearly and the correction due to variations of ε is of the second order only. It is thus reasonable to also address an iso-inclination model in which the change of inclination is completely neglected, that is, there is a fixed $\varepsilon = \varepsilon_{o}$. Regarding Eq. (4), the trepidation λ is then determined by a simple relation[31]

$$\lambda = \arcsin(\sin \lambda_{max} \sin \theta), \tag{20}$$

29 Mercier, 'Studies in the Medieval Conception of Precession', 213.
30 See Samsó, *On Both Sides of the Strait of Gibraltar*, pp. 584–85 and 645.
31 Such a relation in the form of Eq. (4) was considered by Neugebauer, 'Thābit ben Qurra "On the solar year" and "On the motion of the eighth sphere"', pp. 298–9 to interpret the meaning of the table *Equatio dimidii diametri...*, but it was then put in question because of inconsistency with the expected variations of ε. This relation can also be found in other studies; for example, José Chabás and Bernard R. Goldstein, 'Andalusian Astronomy: al-zīj al-muqtabis of Ibn al-Kammād', *Archive for History of Exact Sciences*, 48 (1994), 1–41, Comes, 'Ibn al-Hā'im's Trepidation Model', 321, José Chabás and Bernard R. Goldstein, *The Alfonsine Tables of Toledo* (Dordrecht, Boston, and London: Kluwer, 2003).

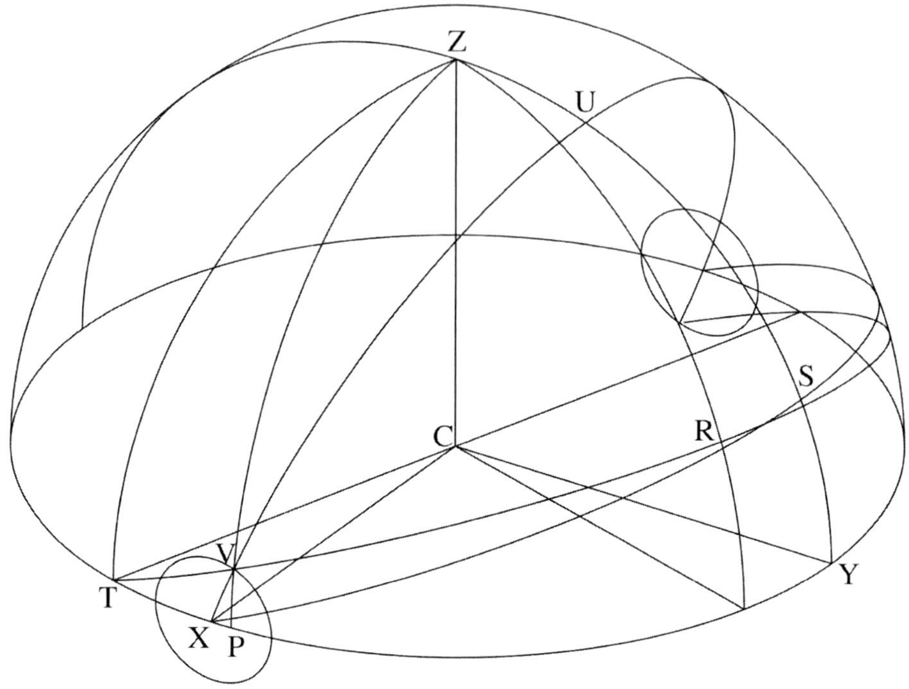

Figure 6.3. Iso-inclination model.

where

$$\lambda_{\max} = \arcsin\left(\frac{\sin \rho}{\sin \varepsilon_0}\right). \qquad (21)$$

This model is depicted in Figure 6.3, where the movable ecliptic *TVR*, with *T* being the movable equinoctial point, *R* the movable solstitial point, and *V* the head of Aries moving on the small circle, has the same constant inclination $\angle YTR = \varepsilon_0$ as the mean ecliptic *YXS*. These ecliptics intersect each other between their solstitial points *S* and *R*.

Such a model of trepidation with a constant inclination of the ecliptic is one of three models studied by Ibn al-Zarqālluh in his *Treatise on the motion of the fixed stars*.[32] As already mentioned, it also corresponds to the model on which the tables of trepidation in the treatise *De motu octave spere*, the Toledan Tables, and later the Alfonsine Tables are based. It is much simpler, but it is a good approximation to both of Mercier's models.

For an infinitesimally small ρ, the spherical trigonometry can be replaced by planar trigonometry in a plane tangential to the sphere at the point *X*. Formula (20) would then be simplified even further to

32 See Samsó, *On Both Sides of the Strait of Gibraltar*, pp. 598–9. Ibn al-Zarqālluh is also considered as one of the possible authors of the original treatise *De motu octave spere* — see Samsó, *On Both Sides of the Strait of Gibraltar*, pp. 586–7.

$$\lambda = \lambda_{max} \sin\theta, \tag{22}$$

and λ_{max} can be approximated by

$$\lambda_{max} = \frac{\rho}{\sin\varepsilon_o}. \tag{23}$$

1.5. Numerical results of the models

The treatise *De motu octave spere* gives the inclination of ecliptic equal to 23;33°. This value is first mentioned generally as an approximate one, but it is then specified as the inclination ε_o of the mean (fixed) ecliptic[33] and also the value measured in the time of caliph al-Ma'mūn.[34] It is smaller than the Indian value of 24° and Ptolemy's 23;51°, which are also mentioned in the treatise.[35] Another input parameter of the Thabitan trepidation is the diameter of the small circles (the epicycles) on which the pivoting points (the heads of Aries and Libra) are orbiting the mean vernal point *X* and the opposite autumnal point. It should be equal to 8;37,26° according to *De motu octave spere*.[36] One half of this value, that is, 4;18,43°, which is the radius ρ of the small circles, is the maximum value of the table *Equatio dimidii diametri circuli* included in the *De motu octave spere* treatise and the Toledan Tables.[37] This table gives the dependence $\delta = \delta(\theta)$ according to Eq. (5). Two more parameters needed to compute the trepidation are the radix (root) and speed of mean motion of its argument, θ. They are not mentioned in the text of the treatise, but they are given in the table attached again to both the treatise and the Toledan Tables. The radix is 1;34,2° at the beginning of Arabic era and the increment in 30 Arabic years (which is the step of the table of collected Arabic years) is 2;34,58°. This corresponds to a period of trepidation of 4181.5 Arabic lunar years, that is, approximately 4057 Julian years. This radix and rate of the argument of trepidation would equate to the phase $\theta \approx 315°$ at the time of Ptolemy and about 16° of caliph al-Ma'mūn.

The *De motu octave spere* treatise also gives the amplitude of the trepidation, which is the maximum distance of the moving vernal point *T* from the head of Aries *V*. It should be equal to 10;45°, which is really the largest value in the attached table.[38] This quantity is not an independent input parameter of the theory, but its result. It may thus test which particular model best fits this value. The same is true also for the other values given in the table of trepidation in the treatise and the Toledan Tables. These values, according to

---

33 In sentence [3] and sentences [15] and [26], respectively. See Carmody, *The Astronomical Works of Thābit b. Qurra*, pp. 102–4 and Neugebauer, 'Thābit ben Qurra "On the solar year" and "On the motion of the eighth sphere"', pp. 291, 293, and 294.

34 This means in the years 813–33; see José Chabás, *Computational Astronomy in the Middle Ages* (Madrid: Consejo superior de investigaciones científicas, 2019), p. 21.

35 In sentence [16].

36 See sentences [7] and [25] in Carmody, *The Astronomical Works of Thābit b. Qurra*, pp. 102 and 104, and Neugebauer, 'Thābit ben Qurra "On the solar year" and "On the motion of the eighth sphere"', pp. 291 and 294.

37 The values are given in the column denoted "Tol. tab." in Table 2. For sources see note 11.

38 See sentences [11] and [39] in Carmody, *The Astronomical Works of Thābit b. Qurra*, pp. 103 and 105 and the table on p. 94 and Neugebauer, 'Thābit ben Qurra "On the solar year" and "On the motion of the eighth sphere"', pp. 292, 296, and 297.

θ		Tol.tab.	Mercier I		Mercier II		Iso1	Iso2	Planar
0	180	0;00,00	0;00,00	0;00,00	0;00,00	0;00,00	0;00,00	0;00,00	0;00,00
5	175	0;55,52	0;56,31	0;56,30	0;56,18	0;56,12	0;56,23	0;55,53	0;56,13
10	170	1;50,36	1;52,36	1;52,31	1;52,16	1;51,51	1;52,21	1;51,22	1;52,00
15	165	2;45,16	2;47,46	2;47,35	2;47,27	2;46,32	2;47,30	2;46,02	2;46,56
20	160	3;39,23	3;41,34	3;41,17	3;41,24	3;39,51	3;41,25	3;39,28	3;40,36
25	155	4;31,12	4;33,36	4;33,10	4;33,41	4;31,25	4;33,41	4;31,16	4;32,35
30	150	5;22,30	5;23,24	5;22,50	5;23,52	5;20,51	5;23,56	5;21,05	5;22,30
35	145	6;09,06	6;10,37	6;09,54	6;11,33	6;07,48	6;11,47	6;08,30	6;09,57
40	140	6;53,12	6;54,53	6;54,03	6;56,21	6;51,56	6;56,51	6;53,10	6;54,36
45	135	7;36,35	7;35,52	7;34,56	7;37,53	7;32,57	7;38,48	7;34,44	7;36,05
50	130	8;14,00	8;13,17	8;12,17	8;15,49	8;10,35	8;17,18	8;12,54	8;14,06
55	125	8;47,48	8;46,51	8;45,50	8;49,52	8;44,32	8;52,02	8;47,19	8;48,21
60	120	9;17,44	9;16,22	9;15,23	9;19,47	9;14,35	9;22,45	9;17,46	9;18,35
65	115	9;43,53	9;41,37	9;40,43	9;45,20	9;40,32	9;49,11	9;43,57	9;44,34
70	110	10;05,30	10;02,28	10;01,41	10;06,20	10;02,09	10;11,07	10;05,41	10;06,06
75	105	10;22,47	10;18,47	10;18,09	10;22,39	10;19,19	10;28,22	10;22,46	10;23,01
80	100	10;35,01	10;30,28	10;30,02	10;34,11	10;31,51	10;40,47	10;35,05	10;35,12
85	95	10;42,13	10;37,28	10;37,14	10;40,52	10;39,41	10;48,17	10;42,31	10;42,33
90	90	10;45,00	10;39,43	10;39,43	10;42,42	10;42,42	10;50,47	10;45,00	10;45,00

Table 1. Values in degrees of trepidation according to the Toledan Tables and different models.

Carmody's edition, are given here in Table 1 together with values computed according to the models investigated above. They are also marked by crosses in the top panel of Figure 6.4. The upper thick sinusoidal curve denoted λ in this figure displays the Thabitan trepidation, $\lambda = \lambda(\theta)$, in degrees. This curve is a superposition of all Mercier's models and the iso-inclination models, which are not distinguishable in this scale. That is why the differences of these models are also plotted in a zoomed-in image at a higher resolution in the bottom panel of Figure 6.4.

The first obvious discrepancy between the tables in *De motu octave spere* and the values computed according to both of Mercier's models is that the table gives symmetrical values in all quadrants,[39] while Mercier's models give a maximum at phase θ less than 90°, and the decreasing part of the function in the second quadrant has smaller values than the symmetrical one, increasing part in the first quadrant. For the given ρ and ε_0, the maximum values of λ are smaller in both of Mercier's models than the given value, 10;45°. Unlike Mercier's models, the iso-inclination model (20) is symmetric, and the given input values result according to Eq. (21) just opposite in higher value of $\lambda_{max} = 10;50,47°$ (see the column Iso 1 of Tab. 1). From Eq. (21), one can see that the maximum $\lambda_{max} = 10;45°$ can be obtained for inclination of the fixed ecliptic $\varepsilon_0 = 23;46,18°$ for the given $\rho = 4;18,43°$. The trepidation of an iso-inclination model with these parameters is given in the column denoted Iso 2.

39 In some manuscripts, only values of θ up to 90° are written, but in others θ is written for two quadrants in the same form as in Tab. 1 or the table continues up to 180° and the values of trepidation are (up to scribal errors) symmetric.

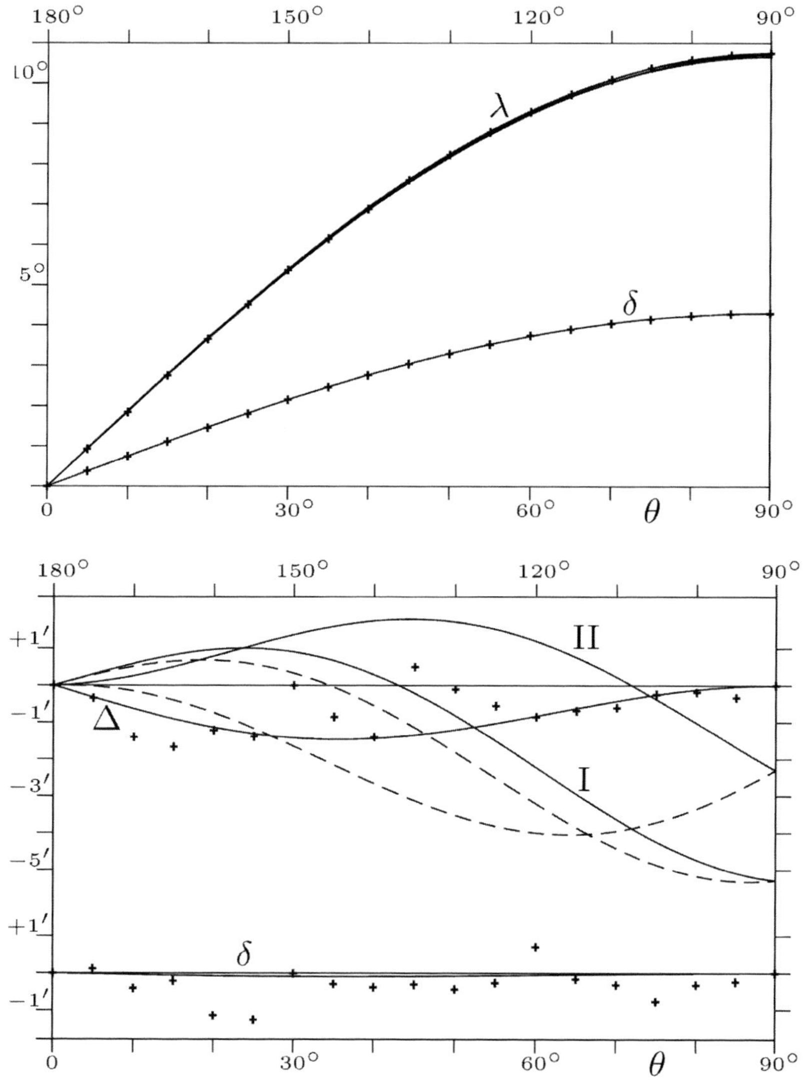

Figure 6.4. Thabitan/Toledan trepidation (top panel) and its residuals (bottom panel).

The planar model gives λ_{max} = 10;47,31° according to Eq. (23), which is the closest to the given value, 10;45°. To reach this value precisely, the inclination of the fixed ecliptic should be chosen ε_o = 23;38,52°. With the amplitude fixed at λ_{max} = 10;45°, Eq. (20) gives the simple sinusoid given in the last column of Tab. 1. This solution, previously presented for comparison by Carmody,[40] is used here as a reference solution in the bottom panel of Figure 6.4 to display deviations of both Mercier's models I and II, respectively, and of the tabulated data. The values corresponding to the first quadrant, $0 < \theta < 90°$, labelled on the bottom scale, are drawn as a solid line, while the values from the second quadrant, 90°<

40 This is denoted ψ by Carmody in *The Astronomical Works of Thābit b. Qurra*; see table on p. 94.

θ		Tol.tab.	δ_{max} sin θ	δ(θ)	ε Mercier I		ε Mercier II	
0	180	0;00,00	0;00,00	0;00,00	23;29,26	23;29,26	23;36,34	23;36,34
5	175	0;22,40	0;22,33	0;22,32	23;29,23	23;29,56	23;35,16	23;38,06
10	170	0;44,31	0;44,56	0;44,53	23;29,48	23;30,51	23;34,14	23;39,49
15	165	1;06,45	1;06,58	1;06,54	23;30,38	23;32,11	23;33,30	23;41,40
20	160	1;27,20	1;28,29	1;28,25	23;31,53	23;33,52	23;33,05	23;43,35
25	155	1;48,04	1;49,20	1;49,15	23;33,30	23;35,52	23;33,01	23;45,31
30	150	2;09,21	2;09,21	2;09,16	23;35,27	23;38,07	23;33,17	23;47,24
35	145	2;28,06	2;28,24	2;28,18	23;37,39	23;40,33	23;33,53	23;49,11
40	140	2;45,55	2;46,18	2;46,12	23;40,03	23;43,05	23;34,48	23;50,48
45	135	3;02,38	3;02,56	3;02,51	23;42,34	23;45,39	23;36,00	23;52,14
50	130	3;17,45	3;18,11	3;18,07	23;45,08	23;48,10	23;37,26	23;53,24
55	125	3;31,40	3;31,56	3;31,52	23;47,39	23;50,33	23;39,05	23;54,18
60	120	3;44,46	3;44,03	3;44,00	23;50,04	23;52,44	23;40,53	23;54,53
65	115	3;54,19	3;54,29	3;54,26	23;52,18	23;54,40	23;42,46	23;55,09
70	110	4;02,48	4;03,07	4;03,05	23;54,17	23;56,16	23;44,42	23;55,05
75	105	4;09,08	4;09,54	4;09,53	23;55,58	23;57,30	23;46,37	23;54,41
80	100	4;14,28	4;14,47	4;14,47	23;57,16	23;58,19	23;48,27	23;53,58
85	95	4;17,30	4;17,44	4;17,44	23;58,11	23;58,43	23;50,09	23;52,57
90	90	4;18,43	4;18,43	4;18,43	23;58,40	23;58,40	23;51,39	23;51,39

Table 2. Values in degrees of δ(θ) according to the Toledan Tables and ε(θ) according to Mercier's models.

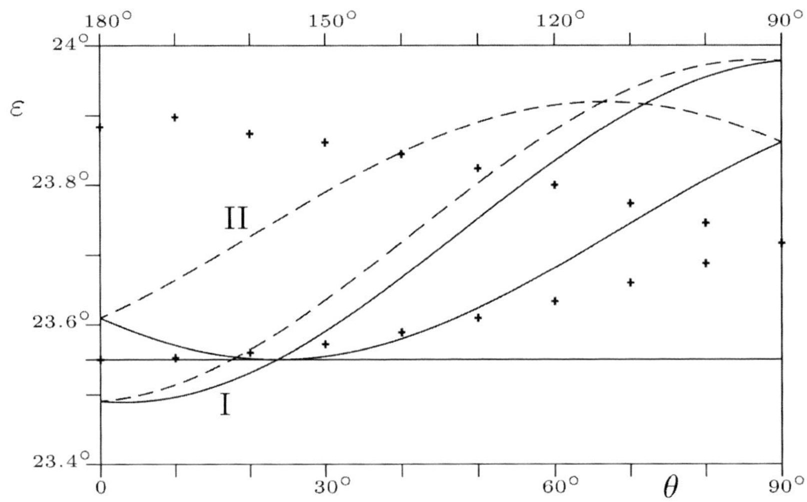

Figure 6.5. Inclination of ecliptic in Thabitan/Toledan trepidation.

θ <180°, labelled at the top of the scale, are presented by the dashed line. The symmetric iso-inclination model Iso 2 (with inclination ε_0 chosen to fit the λ_{max} = 10;45°) is also drawn, denoted as Δ. The Iso 1 model is not drawn in the figure — its curve would extend behind the upper limit of the figure. The crosses in the bottom panel of Figure 6.4 mark the deviation of tabulated values of trepidation from the sinusoidal reference model. We can see that they do not agree with Mercier's models, that is, with the model described in the *De motu octave spere* treatise. Some values (namely for $\theta = 5, 20, 25, 40, 60, 65, 75$, and $80°$) fit relatively well to the model Iso 2; however, some other values differ significantly, probably due to scribal errors in the manuscript tradition. It is interesting to note that the value for $\theta = 30°$ fits precisely with the planar sinusoidal model. This value was probably 'corrected' later in a table originally computed according to a more sophisticated model.

The declination of the pivoting points according to Eq. (5) is drawn in the top panel of Figure 6.4 by the lower curve denoted δ. Its deviation from the planar sinusoidal curve is drawn in the bottom part of the bottom panel of the figure with its own scale on the left-hand side. On the same scale, the deviations of the values given in Table 2 are also marked by crosses according to *De motu octave spere* and the Toledan Tables. Their scatter is so high that it is not possible to decide whether they were computed according to the planar or spherical trigonometry. The only exception is the conspicuous point for $\theta = 30°$, which is exactly one half of the amplitude corresponding to the planar triangle. It can again be a later 'correction', as in the case of $\lambda(\theta)$.

Table 2 also gives the inclinations $\varepsilon(\theta)$ of the movable ecliptic corresponding to Mercier's Models I and II. They are drawn in Figure 6.5 in analogy with $\lambda(\theta)$ in Figure 6.4. The straight horizontal line corresponds to the iso-inclination model, with the inclination of the ecliptic ε equal to the value ε_0 of the fixed ecliptic in Mercier's models I and II. Model II reaches this value at $\theta = \varepsilon_0$ as its minimum, while Model I can deviate from this value on both sides. Model I can get closer to the value of 24° mentioned in the *De motu octave spere* treatise. On the other hand, Model II can better fit the values of 23;51° at $\theta \approx 315°$ at the time of Ptolemy (when ε was the same as at $\theta = 135°$) and 23;33° at $\theta \approx 16°$ at the time of al-Ma'mūn, both also mentioned in the treatise. The crosses in this figure mark the values of the inclination in Ibn al-Zarqālluh's *Treatise on the motion of the fixed stars.*[41] It is obvious that they correspond to a completely different model of trepidation than Mercier's models according to the *De motu octave spere* treatise.

To summarize the results for the trepidation in the *De motu octave spere* treatise and in the Toledan Tables, we can see that the tables in these works were not computed according to the model described in words in the treatise, despite it being feasible in conjunction with the tools of contemporary mathematics. It appears that they were computed according to Eq. (20) for the iso-inclination model, but they are highly corrupted by scribal errors and later changes.

2. Alfonsine trepidation

The theory of trepidation was also included in the Alfonsine Tables, however, in a modified form. The Castilian canons to the unpreserved Alfonsine Tables of Toledo indicate that the trepidation was treated according to the third model by Ibn al-Zarqālluh, in which the 'access and recess' $\lambda(\theta)$ of the head of Aries in longitude was calculated in agreement with the iso-inclination model given by Eq. (20), and a change of the inclination was calculated separately.[42] In the following, we shall deal with the Latin Alfonsine Tables, and in particular their Parisian redaction. The canons to these tables do not explain their base

---

41 See Chabás and Goldstein, *The Alfonsine Tables of Toledo*, p. 221.

42 See Chabás and Goldstein, *The Alfonsine Tables of Toledo*, pp. 89–90 and 217–20, and Samsó, *On Both Sides of the Strait of Gibraltar*, pp. 598–610 and 643–9.

model, but it is obvious from the rules for usage of the tables that the precession, λ, was taken as a superposition, $\lambda = \lambda_1 + \lambda_2$, of a uniform increase, λ_1, of the longitude at a rate of 360° in 49000 Julian years and a periodic oscillation, λ_2, due to the 'accession and recession' motion with a period of 7000 Julian years and semi-amplitude of 9°.[43] This appears to be a compromise between the Ptolemaic uniform precession and Thabitan trepidation, but it is also similar to the model of the motion of planets taken as a superposition of mean motion, which has to be equated by a correction due to their epicyclic motion. This view is in agreement with the usual inscriptions: *Tabula medii motus augium et stellarum fixarum* for the uniform Ptolemaic motion, λ_1; *Tabula medii motus accessus et recessus octave sphere* for the argument, θ, of the trepidation; and *Tabula equationum motus accessus et recessus octave sphere* for the trepidation, λ_2.[44]

Although the algorithm by which the equation $\lambda_2(\theta)$ had been computed in the Alfonsine Tables is not explicitly described, it was mostly understood that it should correspond to the Thabitan model.[45] We can acquire further information from an analysis of numerical values given in the table of λ_2; in particular, we shall treat it here as it is given in Ratdolt's *editio princeps* of Alfonsine Tables.[46] The table (Tab. 3) gives the values up to arcseconds for arguments, θ, running from 1° to 90° with a step of 1°. In his canons *Quia ad inveniendum*, John of Lignères explicitly states that the values in other quadrants are symmetric to the first one.[47] This again contradicts the possibility that the values have been computed according to the Thabitan model either in Mercier's variant I or II, or in any other non-symmetrical model.[48] This symmetry is, however, consistent with the iso-inclination models.

θ	λ_2	Δ	θ	λ_2	Δ	θ	λ_2	Δ
1	0;09,25	−1.86	31	4;37,17	−0.27	61	7;51,49	1.09
2	0;18,49	−2.89	32	4;45,18	0.51	62	7;56,19	2.38
3	0;28,11	−2.26	33	4;53,14	1.16	63	8;00,41	2.96
4	0;37,32	−1.13	34	5;01,05	1.53	64	8;04,56	1.73
5	0;46,52	0.33	35	5;08,51	1.47	65	8;09,02	0.63

43 For an explanation of these periods, see Chabás and Goldstein, *The Alfonsine Tables of Toledo*, pp. 256–66.

44 See the edition by Erhard Ratdolt, *Tabulae astronomicae Alphonsinae*, Johannes de Saxonia: *Canones in tabulas Alphonsi* (Venice: 1483), sig. d4^v, d4^r, d3^v, resp., and Jerzy Dobrzycki, 'Teoria precesji w astronomii średniowiecznej', *Studia i materiały z Dziejów Nauki Polskiej*, Seria C, 11 (1965), 23; for English translation see 'The Theory of Precession in Medieval Astronomy', in *Jerzy Dobrzycki, Selected Papers on Medieval and Renaissance Astronomy*, ed. by Jarosław Włodarczyk and Richard L. Kremer, *Studia Copernicana* 43 (Warsaw: PAN, 2010), p. 37. For an explanation of these periods, see Chabás and Goldstein, *The Alfonsine Tables of Toledo*, pp. 256-66.

45 For instance, the author of the star catalogue in MS Prague, NK, XIV E 37 mentioned in his colophon that the positions of the stars were rectified in 1429 for the motion of access and recess in small circle as modelled by Thābit and after him by Alfonso (see Hadrava, 'Medieval catalogues of stars', p. 169).

46 Erhard Ratdolt, *Tabulae astronomicae Alphonsinae*, f. d3^v. The table can also be found in the modern edition by Emmanuel Poulle, *Les Tables Alphonsines avec les canons de Jean de Saxe* (Paris: Centre National de la Recherche Scientifique, 1984), p. 131. Up to scribal errors, the same values are in numerous manuscripts; for example, in MSS Prague, NK, X A 3, f. 6^r; NK, X B 3, f. 95^r; NK, XIV F 10, f. 94^r.

47 See Alena Hadravová and Petr Hadrava, 'John of Lignères *Quia ad inveniendum loca planetarum*: An edition and Translation', in *Alfonsine Astronomy: The written Record*, ed. by Richard L. Kremer, Matthieu Husson, and José Chabás (Turnhout: Brepols, 2021; Alfonsine Astronomy: Studies and Sources, 1), Chapter 10.

48 A supporting argument could again be the absence of a table and any mention of variations of the ecliptic inclination. Such changes would be irrelevant for computations of ecliptic coordinates of planets; however, they should influence computations of eclipses or rises and sets of zodiacal signs somewhat.

6	0;56,12	0.96	36	5;16,30	2.85	66	8;12,58	0.57
7	1;05,31	1.59	37	5;24,04	3.52	67	8;16,45	0.48
8	1;14,48	3.04	38	5;31,33	3.34	68	8;20,23	0.28
9	1;24,04	4.16	39	5;38,57	2.18	69	8;23,52	−0.11
10	1;33,20	3.78	40	5;46,16	−0.11	70	8;27,11	0.27
11	1;42,34	3.73	41	5;53,26	0.35	71	8;30,23	−1.67
12	1;51,46	3.85	42	6;00,29	1.41	72	8;33,24	−1.98
13	2;00,57	2.98	43	6;07,26	1.95	73	8;36,15	−1.72
14	2;10,06	1.94	44	6;14,17	1.84	74	8;38,56	−0.94
15	2;19,13	0.58	45	6;21,02	0.95	75	8;41,28	−0.70
16	2;28,17	−0.27	46	6;27,40	0.15	76	8;43,50	−0.06
17	2;37,16	1.23	47	6;34,10	0.32	77	8;46,02	0.96
18	2;46,11	3.92	48	6;40,33	0.34	78	8;48,05	1.30
19	2;55,02	7.63	49	6;46,49	0.09	79	8;49,59	0.91
20	3;03,49	12.21	50	6;52,58	−0.56	80	8;51,44	−0.22
21	3;12,47	2.50	51	6;59,00	−1.71	81	8;53,19	−1.15
22	3;21,36	−1.67	52	7;04,53	−1.47	82	8;54,41	1.10
23	3;30,20	−4.44	53	7;10,38	−0.98	83	8;55,55	1.51
24	3;38,57	−3.99	54	7;16,15	−0.32	84	8;57,00	1.05
25	3;47,27	−0.46	55	7;21,44	0.39	85	8;57,55	0.69
26	3;55,54	2.00	56	7;27,07	−0.95	86	8;58,40	0.42
27	4;04,17	4.22	57	7;32,21	−1.44	87	8;59,15	0.23
28	4;12,38	4.05	58	7;37,27	−2.19	88	8;59,40	0.10
29	4;20,55	3.35	59	7;42,23	−1.29	89	8;59,55	0.02
30	4;29,10	−0.04	60	7;47,10	0.17	90	9;00,00	0.00

Table 3. The Alfonsine table of trepidation from Ratdolt's edition with residuals (Δ) in arc-seconds.

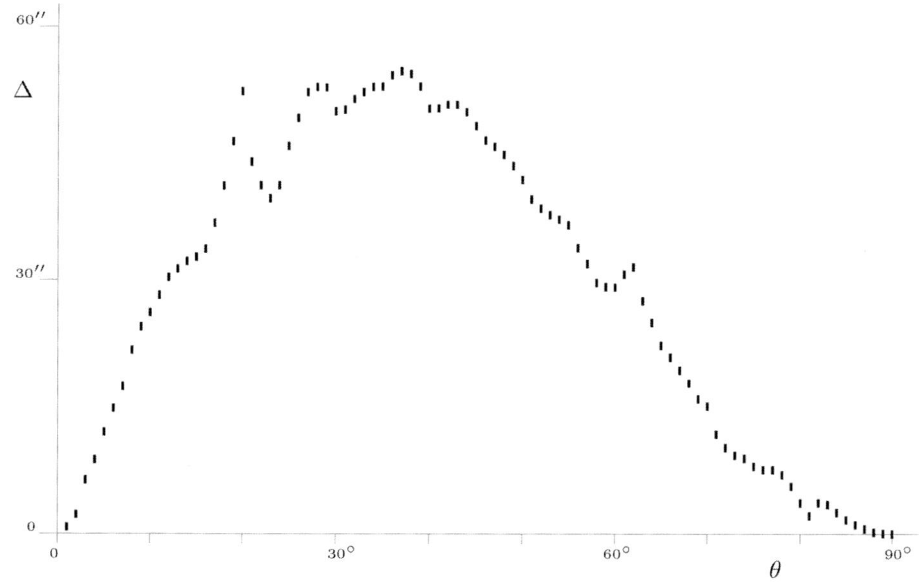

Figure 6.6. Residuals of $\lambda_2(\theta)$ with respect to planar iso-inclination model.

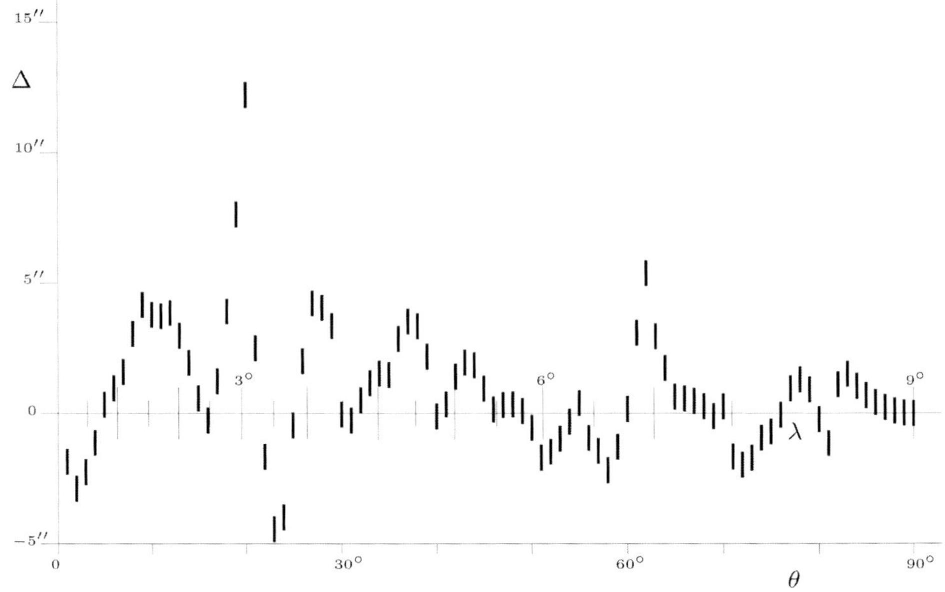

Figure 6.7. Residuals of Alfonsine trepidation with respect to the spherical iso-inclination model.

Unlike that of the Toledan Tables, the equation of trepidation in the Alfonsine Tables for argument $\theta = 30°$ is not equal to one half of the semi-amplitude, λ_{max}, as it should be according to Eq. (22) for the planar model. The residuals (in arcseconds) with respect to this formula are displayed (in absolute values) in Figure 6.6. The error bars correspond to the 1″ precision with which the values of λ_2 are given. This figure shows a systematic deflection, which reveals that this model could not be used for the computation of the Alfonsine trepidation.[49]

More appropriate appears to be the spherical iso-inclination model given by Eq. (20).[50] The residuals, Δ, with respect to this formula are given in arcseconds in Table 3 and are displayed in Figure 6.7. Their maximum is about five times smaller compared to the planar formula, and they are not systematic in the whole region as in the case of the planar model, but they are both positive and negative. This is expected from random computational errors, and it thus yields evidence that this iso-inclination model, already suspected to have been used for the Toledan Tables was even more likely used in computation of the Alfonsine trepidation.

The values of trepidation in the iso-inclination model (both planar and spherical) depend on the input parameters, ρ and ε_0, but only in terms of λ_{max} computed from them according to Eq. (21) or Eq. (23). It is thus impossible to decipher both these parameters

49 Such a formula with $\lambda_{max} = 9°$ is given for the Alfonsine trepidation by Mercier, 'Studies in the medieval conception of precession', 58. It is also sometimes used by other authors as the model of Alfonsine trepidation, but it is obvious that it is only a rough approximation.

50 This formula with $\lambda_{max} = 9°$ is presented, for example, by José Chabás and Bernard R. Goldstein in *Time, Astronomy and Calendars: Survey of European Astronomical Tables in the Late Middle Ages* (Leiden and Boston: Brill, 2012), p. 51.

φ	sin φ	fit	Δ	φ	sin φ	fit	Δ
0;30	0;31,25	0;31,23	−1.93	5;00	5;13,46	5;13,47	1.36
1;00	1;02,50	1;02,52	2.28	5;30	5;45,03	5;45,06	3.32
1;30	1;34,14	1;34,19	4.78	6;00	6;16,18	6;16,19	0.85
2;00	2;05,38	2;05,42	3.71	6;30	6;47,32	6;47,33	1.11
2;30	2;37,02	2;37,00	−1.79	7;00	7;18,44	7;18,43	−0.78
3;00	3;08,25	3;08,36	11.43	7;30	7;49,54	7;49,52	−1.66
3;30	3;39,46	3;39,40	−6.48	8;00	8;21,01	8;21,05	3.61
4;00	4;11,07	4;11,12	4.60	8;30	8;52,07	8;52,06	−0.83
4;30	4;42,27	4;42,28	0.84	9;00	9;23,10	fixed	0.16

Table 4. Table of sinus, its expected distortion, and its residuals Δ in seconds.

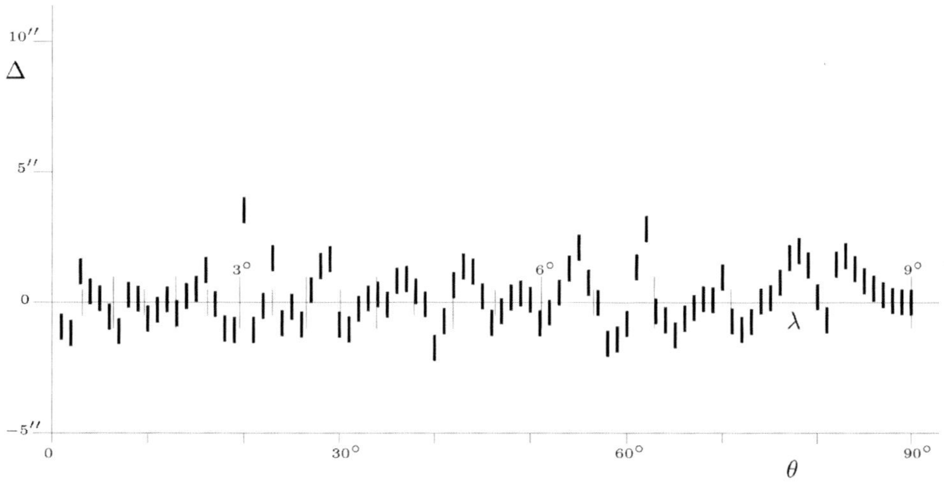

Figure 6.8. Residuals of $\lambda_2(\theta)$ with respect to spherical iso-inclination model interpolated in a corrupted table of sinus.

from the tabulated values, and the value of λ_{max} is better fit by the rounded value of 9°, even if a slightly different value would provide a slightly better fit to all data. It is nevertheless possible to make somewhat better judgements from the residual errors.

Even according to the spherical iso-inclination model, the values in Table 3 have still non-negligible errors, Δ, with a root mean square (RMS) of 2.4″. Their maximum value of 12.21″ can be found for θ =20°. The precise value of the trepidation according to this model should be $\lambda_2 = \arcsin(\sin 9° \sin 20°) = \arcsin(0.0535) = 3.067° = 3;4,1°$. The subsequent minimum value of the error −4.44″, which is also the lowest negative error, occurs at θ =23°, where the trepidation should be $\lambda_2 = \arcsin(\sin 9° \sin 23°) = \arcsin(0.0611) = 3.504° = 3;30,16°$. The preceding local minimum −0.27″ corresponds to θ =16°, where $\lambda_2 = \arcsin(\sin 9° \sin 16°) = \arcsin(0.0431) = 2.471° = 2;28,17°$. It means that the largest errors are at (or close to) the semi-integer values of λ_2. The values of the argument, θ, at which the trepidation, λ_2, should reach semi-integer value of degrees (for its semi-amplitude 9°) are highlighted in Figure 6.7 by ticks on the Δ axis. It can be seen in several cases that the trend of neighbouring residuals changes its slope in these points. It is thus evident that the errors, Δ, may be due to errors in the table of the sinus function, which was used to calculate the Table of trepidation $\lambda_2(\theta)$ and had a step of 30′.[51]

To test this hypothesis, we can replace the computations of sinus and (more importantly) its inverse arcsin in Eq. (20) by interpolation in a sine table calculated with limited precision (presumably up to seconds in sexagesimal notation) and featuring additional adjustable errors. The tabulated sine values for angles from 0;30° up to 8;30° influence not only the trepidation for these values of argument, but, because of the backward interpolation of the arcsin, also all higher arguments. If we optimize these values to yield the best fit of the residuals as it is presented in Tab. 4, we can decrease the residual noise shown in Figure 6.8, which has an RMS of only 1″.

Regarding the existence of quite precise medieval sine tables, the expected errors up to 11″ may seem to be unlikely. For instance, the sine table in al-Battānī's tables, which also have a step of 0.5°,[52] has an RMS error of 0.37″, only 32 entries (of a total of 180) have errors higher than 0.5″, of which only one exceeds 1″. Similarly, the Toledan Tables, in the form given in Pedersen's edition,[53] has an RMS error of 0.31″, only 13 entries with an error over 0.5″, and no errors over 1″. On the other hand, much larger scribal errors are also found in many manuscripts. An example of such errors can be found in the Castilian late-thirteenth-century manuscript, Paris, BA, ms. 8322,[54] in which, after the Castilian translation of al-Battānī's canons, his tables are given along with the sine table on folios 28r and 28v. This table has 85 entries with an error exceeding 0.5″, 12 of which, shockingly, have errors in whole parts/degrees; so, the RMS error of the table is 49′. In addition, 44 entries have incorrect values of the angle. The lack of precision in the numerical values is in high contrast with the perfection of the graphical layout of the table and the whole manuscript. The expected errors of the sine table used for the calculation of the Alfonsine trepidation are thus possible.

Conclusion

The above given facts show that the Alfonsine trepidation was not actually calculated according to the Pseudo-Thabitan model described in the text of the *De motu octave spere* treatise, but according to the simplified iso-inclination version used in its tables, which did not assume a change of inclination of the ecliptic. The trepidation in the Alfonsine Tables is calculated with much higher precision than that of the Toledan Tables, which were most probably computed in the framework of the same iso-inclination model. Yet, in the Alfonsine Tables, the trepidation also suffers from errors, which deserve some attention in the context of the computational practice used for the composition of the tables.

The mathematical analysis of the tabulated values of trepidation reveals that neither the Toledan Tables nor the Alfonsine Tables correspond to the Pseudo-Thabitan model

51 This step of 0.5° in tables of sinus was quite common. Its tradition may be due to a 1° step in table of chords, which are a double sinus of half an angle.

52 See the edition by Carlo Alphonso Nallino, *Al-Battānī sive Albatenii opus astronomicum, pars secunda* (Milan: Ulrichus Hoeplius, 1907), pp. 55–56.

53 See Pedersen, *The Toledan Tables*, pp. 927–59.

54 See Chabás and Goldstein, *The Alfonsine Tables of Toledo*, p. 95.

with a variable ecliptic inclination, despite it having been, in principle, feasible by means of contemporary mathematics. The *De motu octave spere* treatise was relatively often copied and known to medieval scholars, as is testified by frequent references to its basic features. These references, however, demonstrate the knowledge of the author, while the computational practice was much simpler.

Manuscript sources

Cracow, Biblioteka Jagiellońska, 551
London, British Library, Harley 13
Paris, Bibliothèque de l'Arsenal, ms. 8322
Prague, Národní knihovna, X A 3
Prague, Národní knihovna, X B 3
Prague, Národní knihovna, XIV A 18
Prague, Národní knihovna, XIV E 37
Prague, Národní knihovna, XIV F 10
Vatican, Biblioteca Apostolica Vaticana, Pal. lat. 1398
Vatican, Biblioteca Apostolica Vaticana, Pal. lat. 1414
Vatican, Biblioteca Apostolica Vaticana, Vat. lat. 3379

Bibliography

Brahe, Tycho, *Astronomiae instauratae mechanica* (Wandsbek: 1598).

Carmody, Francis J., *The Astronomical Works of Thābit b. Qurra* (Berkley: University of California Press, 1960).

Chabás, José, *Computational Astronomy in the Middle Ages* (Madrid: Consejo superior de investigaciones científicas, 2019).

———, and Bernard R. Goldstein, 'Andalusian Astronomy: al-zī al-muqtabis of Ibn al-Kammād', *Arch. Hist. Exact Sci.*, 48 (1994), 1–41.

———, and ———, *The Alfonsine Tables of Toledo* (Dordrecht, Boston, and London: Kluwer, 2003).

———, and ———, *Time, Astronomy and Calendars: Survey of European Astronomical Tables in the Late Middle Ages* (Leiden and Boston: Brill, 2012).

Comes, Mercè, 'Ibn al-Hā'im's Trepidation Model', *Suhayl*, 2, (2001), 291–408.

Dobrzycki, Jerzy, 'Teoria precesji w astronomii średniowiecznej', *Studia i materiały z Dziejów Nauki Polskiej*, Seria C, 11 (1965), 23. – English transl.: 'The Theory of Precession in Medieval Astronomy', in Jerzy Dobrzycki, *Selected Papers on Medieval and Renaissance Astronomy*, ed. by Jarosław Włodarczyk and Richard L. Kremer, Studia Copernicana 43 (Warszawa: PAN, 2010).

Hadrava, Petr, 'Medieval Catalogues of Stars', in *The Stars in the Classical and Medieval Traditions*, ed. by Alena Hadravová, Petr Hadrava, and Kristen Lippincott (Prague: Scriptorium, 2019), pp. 115–69.

Hadravová, Alena, and Petr Hadrava, 'John of Lignères *Quia ad inveniendum loca planetarum*: An Edition and Translation', in *Alfonsine Astronomy: The Written Record*, ed. by Richard L. Kremer, Matthieu Husson, and José Chabás (Turnhout: Brepols, 2021), pp. 257–302.

———, ———, and Jole R. Shackelford, *Tycho Brahe, Instruments of the Renewed Astronomy* (Prague: KLP, 1996).

Lorch, Richard, *Thābit ibn Qurra, On the Sector-Figure and Related Texts* (Frankfurt am Main: Institute for the History of Arabic-Islamic Science, 2001, reprint 2008).

Mercier, Raymond, 'Studies in the Medieval Conception of Precession', *Archives internationales d'histoire des sciences*, 26 (1976), 197–220, and 27 (1977), 33–71.

Nallino, C. A., ed., *Al-Battānī sive Albatenii opus astronomicum, pars secunda* (Milan: Ulrichus Hoeplius, 1907).

Neugebauer, Otto, 'Thābit ben Qurra "On the solar year" and "On the motion of the eighth sphere"', *Proceedings of the American Philosophical Society* 106 (1962), 264–99.

Nothaft, C. Philipp E., 'Criticism of Trepidation Models and Advocacy of Uniform Precession in Medieval Latin Astronomy', *Arch. Hist. Exact Sci.* 71 (2017), 211–44.

Pedersen, Fritz Saaby, *The Toledan Tables. A review of the manuscripts and the textual versions with an edition* (Copenhagen: C. A. Reitzels Forlag, 2002).

Poulle, Emmanuel, *Les Tables Alphonsines avec les canons de Jean de Saxe* (Paris: Centre National de la Recherche Scientifique, 1984).

Rashed, Roshdi, ed., *Thābit ibn Qurra, Science and Philosophy in Ninth-Century Baghdad* (Berlin and New York: Walter de Gruyter, 2009).

Ratdolt, Erhard, ed., *Tabulae astronomicae Alphonsinae*, Johannes de Saxonia: *Canones in tabulas Alphonsi* (Venice: 1483, ISTC ia00534000).

Samsó, Julio, *On Both Sides of the Strait of Gibraltar: Studies in the History of Medieval Astronomy in the Iberian Peninsula and the Maghrib* (Leiden and Boston: Brill, 2020).

Swerdlow, Noel M., 'Tycho, Longomontanus, and Kepler on Ptolemy's Solar Observations and Theory, Precession of the Equinoxes, and Obliquity of the Ecliptic', in *Ptolemy in Perspective*, ed. by Alexander Jones (New York: Springer, 2010), pp. 151–202.

Schöner, Johannes, 'Aequatorium astronomicum', in *Opera mathematica Ioannis Schoneri*, ed. by Andreas Schöner (Nuremberg: Berg & Neuber, 1551), ff. a1^{r}-f5^{r}.

Toomer, G. J., *Ptolemy's Almagest* (London: Duckworth, 1984).

GLEN VAN BRUMMELEN

The Emergence of Auxiliary Astronomical Tables in Europe

Introduction

Before the rise of digital computers, the need for trigonometric calculation when working with astronomical models presented either a barrier to development or a burden requiring substantial labour. In fact, some of the most significant innovations in the history of astronomy have not been observational achievements or improved planetary models, but improvements in computational methods: for instance, trigonometric tables, interpolation, prosthaphairesis,[1] and logarithms. Indeed, of the latter, Laplace is said to have exclaimed that they are 'an admirable artifice which, by reducing to a few days the labour of many months, doubles the life of an astronomer, and spares him the errors and disgust inseparable from long calculations'.[2]

Among the inventions in medieval Islam that attempted to simplify computations were tables of *auxiliary functions*, which are not primitive trigonometric quantities themselves, nor astronomical quantities, but something in between. Consider, for example, the work of fourteenth-century *muwaqqit* (astronomical timekeeper) Shams al-Dīn al-Khalīlī, from the Umayyad mosque in Damascus. One of his accomplishments was the construction of a large table of the *qibla*, the direction of the Sacred Mosque in Mecca. The table gives the *qibla* $q(\varphi, \Delta L)$ for worshippers' latitudes $\varphi = 10°, 11°, \ldots, 56°$ (as well as $\varphi = 33\frac{1}{2}°$, the latitude of Baghdad), and for the difference in longitude between Mecca and the worshipper $\Delta L = 1°, 2°, \ldots, 60°$. Converted to modern notation,[3] the formula is

* This research was supported by the ERC project ALFA: Shaping a European Scientific Scene, Alfonsine Astronomy, CoG 723085, PI Matthieu Husson, as well as the Fund for Historical Studies at the Institute for Advanced Study (Princeton). This paper benefited from conversations, advice, critiques, and other support from José Chabás, Richard L. Kremer, Matthieu Husson, and Darcy Otto.

1 This term was used several ways. Here, we refer to the use of trigonometric identities to convert multiplications/divisions to additions/subtractions.

2 Quoted in Glen Van Brummelen, *The Doctrine of Triangles: A History of Modern Trigonometry* (Princeton: Princeton University Press, 2021), p. 76.

3 Ancient and medieval trigonometric functions referred not to the unit circle, but to a circle of some other radius, often 60.

Glen Van Brummelen • Trinity Western University, Langley

Alfonsine Astronomy: Computational Practices, ed. by Matthieu Husson, José Chabás, and Richard L. Kremer, Turnhout, 2026 (*Alfonsine Astronomy: Studies and Sources*, 3), p. 225-246

10.1484/M.ALFA-EB.5.137912

$$q(\varphi, \Delta L) = \text{arc cos} \left(\frac{\sin h \tan \varphi - \frac{\sin \varphi_M}{\cos \varphi}}{\cos h} \right),$$

where φ and $\varphi_M = 21.5°$ are the latitudes of the observer and of Mecca, respectively, and h is the altitude of Mecca's zenith above the local horizon, which is calculated separately. Even with the aid of trigonometric tables, the computation of the 2,880 entries in this table by hand would have been no mean feat.[4]

Separately, al-Khalīlī computed three tables of the following functions:

$$f(x,y) = \frac{\sin x}{\cos y};\ g(x,y) = \sin x \tan y;\ \text{and}\ G(x,y) = \text{arc cos} \left(\frac{x}{\cos y} \right).$$

These three double-argument tables, comprising over 13,000 entries, have no astronomical interest on their own, but they may be combined in various ways to solve problems in spherical astronomy.[5] For instance, the *qibla* may be determined as

$$q(\varphi, \Delta L) = G(g(h, \varphi) - f(\varphi_M, \varphi), h).$$

Now, on its own, it does not seem like much of an advantage to replace the task of calculating one large table with the task of calculating three large tables. However, if one is strategic about the choice of auxiliary function, the same set of tables may be used to find many other astronomical quantities. For instance, given the Sun's altitude h and declination δ and one's terrestrial latitude φ, the hour angle t is

$$t = \text{arc cos} \left(\frac{\frac{\sin h}{\cos \varphi} - \sin \delta \tan \varphi}{\cos \delta} \right).$$

With his auxiliary tables, al-Khalīlī instructs his readers to compute t according to

$$G(f(h, \varphi) - g(\delta, \varphi), \delta).$$

He goes on to describe the uses of the same three tables to solve nine other astronomical problems.[6] The similarity in the formulas for these different problems may seem coincidental, but it is a reflection of the use of the same spherical trigonometric methods to solve each problem: in this case, a version of the spherical law of cosines.[7]

Perhaps the earliest author to compose auxiliary tables in the Islamic world was mid-ninth-century astronomer Ḥabash al-Ḥāsib. In modern notation, his functions are mathematically equivalent to

$$f_1(x) = \text{arc tan} (\tan \varepsilon \sin x)$$

4 For a description of al-Khalīlī's *qibla* table, see David King, 'Al-Khalīlī's *qibla* Table', *Journal of Near Eastern Studies* 34 (1975), 81–122.

5 For a description of al-Khalīlī's auxiliary tables, see David King, 'Al-Khalīlī's Auxiliary Tables for Solving Problems of Spherical Astronomy', *Journal for the History of Astronomy* 4 (1973), pp. 99–110; for an analysis of their methods of computation, see Glen Van Brummelen, 'The Numerical Structure of al-Khalili's Auxiliary Tables', *Physis* 28 (1991), 667–98.

6 See King, 'Al-Khalīlī's auxiliary tables for solving problems of spherical astronomy', 102–6.

7 In a spherical triangle with sides a, b, c and opposing angles A, B, C, $\cos c = \cos a \cos b + \sin a \sin b \cos C$.

$$f_2(x) = \cos(\arc\sin(\cos x \sin \varepsilon));$$

$$f_3(x) = \frac{\cos x}{f_2(x)};$$

$$f_4(x) = \frac{\cos x}{\cos \varepsilon};$$

$$f_5(x) = \tan x \sin \varepsilon; \text{ and}$$

$$f_6(x) = \frac{\sin x}{\cos x},$$

where ε is the value of the obliquity of the ecliptic, taken here to be constant.[8] The form taken by Ḥabash's functions indicates that he was using methods that are equivalent to the spherical right-angled triangle identities. Note the appearance of the tangent as $f_6(x)$.[9] While it is tempting to call this the moment when the tangent function was invented, it would be more than a century after Ḥabash that the tangent would become genuinely trigonometric. We shall return to this point.

Since they compute single-argument (rather than double-argument) functions, Ḥabash would have had a much easier time constructing his tables than al-Khalīlī's. However, a table maker might choose to sacrifice one's own time and energy by creating large tables to make an astronomer's computational life even easier. For instance, Najm al-Dīn al-Miṣrī's triple argument table (*c.* AD 1300), containing more than 400,000 entries, must have taken a significant part of his working life to construct, but it can solve any problem in astronomical timekeeping or spherical astronomy.[10] These examples serve as a brief illustration of a thriving tradition of auxiliary tables that continued for well over half a millennium in medieval Islamic science.

1. Bianchini

The same cannot be said for medieval Europe. A scan of the contents of Chabás and Goldstein's *A Survey of European Astronomical Tables in the Late Middle Ages* reveals a list of tables concerned with a variety of astronomical, chronological, geographical, and astrological topics. Chapter 2 on trigonometry and spherical astronomy discusses sines and chords, as well as a number of the foundational astronomical quantities, but no auxiliary tables are to be found among them. However, suddenly in the middle of the fifteenth century we find a fully realized set of auxiliary tables, the *Tabulae magistrales* by the Italian astrologer Giovanni Bianchini. As we shall see, this seems to have been a result of his interest in correcting mathematical errors in the practice of astrology, which had sent him into a reform of methods in spherical astronomy. At the same time, he may have felt a need to ease his readers' difficulties with his novel system of decimal arithmetic.

---

8 The obliquity of the ecliptic is the angle between the ecliptic and the celestial equator, around 23.5°. See Rida A. K. Irani, 'The "Jadwal al-Taqwīm" of Ḥabash al-Ḥāsib', unpublished MA thesis, American University of Beirut, 1956, for an analysis of the treatise containing these auxiliary tables.

9 Although the modern equivalents to f_1 and f_5 employ tangents, Ḥabash did not refer to them in his definitions.

10 See François Charette, 'A Monumental Medieval Table for Solving the Problems of Spherical Astronomy for all Latitudes', *Archives internationales d'histoire des sciences* 48 (1998), 11–64.

Bianchini (*c.* 1410–69) served much of his career as an administrator in the court of the d'Este family (Niccolò, Leonello, and Borso) in Ferrara, Italy. Beginning his working life as a merchant, he seems to have become interested in astronomy primarily in service to astrology.[11] His creativity was almost unique for his time; each of his five major astronomical treatises reveals multiple traces of a restless innovative mind. In order of composition, the five works are:

- the *Tabulae astronomiae*, a collection of tables on the motions of heavenly bodies, especially the planets;[12]
- the *Tabulae primi mobilis A*, a collection of tables and topics on spherical astronomy and astrology;[13]
- the *Flores Almagesti*, the mathematical backbone of his subsequent work – an incomplete attempt to treat the topics of the *Almagest* from scratch;[14]
- the *Tabulae primi mobilis B*, a nearly complete rewrite of the *Tabulae primi mobilis A* that repairs errors especially with respect to the conversion of star positions from ecliptic to equatorial coordinates;[15]
- the *Tabulae eclypsium*, a collection of tables for predicting eclipses.[16]

All of these works were likely written between 1440 and 1460. After this time period and near the end of his life, Bianchini also engaged in scientific correspondence (some of the earliest in Europe) with Regiomontanus, which has received scholarly attention.[17] Since the story of the development of Bianchini's notions of auxiliary functions is connected in one way or another with all but the first of his treatises, this paper shall also justify the relative chronology above.

The *Tabulae magistrales* is a set of between six and eight tables (depending on the manuscript) with no accompanying text. As we shall see below, they are strongly associated with the *Tabulae primi mobilis B*, although they were not Bianchini's first effort to compose auxiliary tables. The *Tabulae magistrales* were first analyzed by José Chabás in 'An Analysis of the *Tabulae magistrales* by Giovanni Bianchini', *Archive for History of Exact Sciences* 70 (2016), 543–52, and that paper forms part of the foundation of this study.

---

11 For biographical information on Bianchini, see, for example, José Chabás and B. R. Goldstein, *The Astronomical Tables of Giovanni Bianchini* (Leiden/Boston: Brill, 2009), p. 13.

12 These tables are the topic of Chabás and Goldstein, *The Astronomical Tables of Giovanni Bianchini*. See also José Chabás, *Computational Astronomy in the Middle Ages: Sets of Astronomical Tables in Latin* (Madrid: Consejo Superior de Investigaciones Científicas, 2019).

13 This treatise was recently re-discovered; see Glen Van Brummelen, 'Before the End of an Error: Giovanni Bianchini's Original Flawed Treatise on the Conversion of Stellar Coordinates', *Archive for History of Exact Sciences* 75 (2021), pp. 109–24.

14 The *Flores Almagesti*, beginning as an *abacus* textbook before transitioning to topics in Ptolemaic astronomy, awaits a thorough study. The earlier part of the work has been studied in several papers by Grażyna Rosińska; see the bibliography for a list.

15 See Glen Van Brummelen, 'The End of an Error: Bianchini, Regiomontanus, and the Tabulation of Stellar Coordinates', *Archive for History of Exact Sciences* 72 (2018), pp. 547–63, as well as Chabás, *Computational Astronomy in the Middle Ages: Sets of Astronomical Tables in Latin*, pp. 353–58.

16 This work is the subject of the study José Chabás and B. R. Goldstein, 'The *Tabulae eclypsium* by Giovanni Bianchini', *Aestimatio* 2 (1) (2021), pp. 1–41. See also Chabás, *Computational Astronomy in the Middle Ages: Sets of Astronomical Tables in Latin*, pp. 358–61.

17 These letters were edited in Maximilian Curtze, 'Der Briefwechsel Regiomontans mit Giovanni Bianchini, Jacob von Speier und Christian Roder', *Abhandlungen zur Geschichte der Mathematischen Wissenschaften* 12 (1902), pp. 185–336; see also Armin Gerl, *Trigonometrisch-astronomisches Rechnen kurz vor Copernicus: Der Briefwechsel Regiomontanus-Bianchini* (Stuttgart: Steiner, 1989).

2. The *Tabulae primi mobilis* A

Bianchini's first major work, the *Tabulae astronomiae*, is a collection of tables intended for the prediction of planetary positions. Dedicated to Leonello d'Este and dated around 1442,[18] the book's main innovation is its tabular structure designed to reduce the user's computational burden, especially with respect to interpolation. It is the only one of Bianchini's major treatises that does not contain calculations in what was to become his signature system of decimal arithmetic, and it mentions none of his other works. Likely the first of his books, ironically it had the longest life, at least under his name;[19] it was printed three times: in 1495, 1526, and 1553.

The *Tabulae primi mobilis A* is Bianchini's second work, dedicated to Leonello d'Este. It refers to the *Tabulae astronomicae* at the outset:

> I have come not to overturn a principle, but to illuminate those things which are in shadows. It does not seem to me to be enough to do this through that which I have written in my composition of the tables regarding the motions of the planets.[20]

None of Bianchini's other works are mentioned. Notably missing from the introduction is an extended description of the contents of the *Flores Almagesti* that appears in the later *Tabulae primi mobilis B*, an introduction that is otherwise almost identical to that of the *Tabulae primi mobilis A*. It is the first of his books to use decimal arithmetic (replacing sexagesimal arithmetic); it contains the first sine table with the radius $R = 60{,}000$, establishing a tradition of using large radii of $60 \cdot 10^n$ (or, shortly afterward, just 10^n) that would be passed on to Regiomontanus and through him to many astronomers for centuries afterward.[21] However, in this treatise the decimal arithmetic is not explicitly fractional, unlike what was to appear later in the *Tabulae primi mobilis B*.[22]

Bianchini continues with a description of the purpose of his book:

> If I calculate something about the doctrine of the prime mover and of the celestial figure, I will not yield, in order to set right a common error of the ignorant. In arranging the planets and other stars in the celestial figure [horoscope], they commit the greatest error by dismissing the latitudes of the planets and the fixed stars.[23]

18 José Chabás and B. R. Goldstein, *The Astronomical Tables of Giovanni Bianchini*, p. 14.

19 Regiomontanus's *Tabulae directionum*, which had a publishing life that stretched into the seventeenth century, may be considered at least in a large part to be a revision or second edition of Bianchini's *Tabulae primi mobilis B*.

20 *Non veni solvere legem sed illuminare his qui in tenebris sedent. Nec mihi videtur satisfecisse per id quod scripseram in compositione tabularum de motibus planetarum.* We shall reference quotations from the *Tabulae primi mobilis A* using Florence, Ashburnham 142 (216), the most complete of the five manuscripts. This passage is on folio 1r.

21 On decimal trigonometric tables after Bianchini, see Grażyna Rosińska, 'Tables of Decimal Trigonometric Functions from ca. 1450 to 1550', in *From Deferent to Equant: A Volume of Studies in the History of Science in the Ancient and Medieval Near East in Honor of E. S. Kennedy*, ed. by David A. King and George Saliba (New York: New York Academy of Sciences, 1987), pp. 419–26.

22 See Glen Van Brummelen, 'Decimal Fractional Numeration and the Decimal Point in 15th-Century Italy', *Historia Mathematica*, 66 (2024), 1-13.

23 *Si aliquid de primo mobili et figura celesti calculo doctrinae non tradam pro corrigendo communem errorem imperitorum. Qui in collocando planetas et ceteras stellas in figura celesti maximum committunt errorem dimittendo Latitudines tam planetarum que stellarum fixarum.* Florence, Ashburnham 142 (216), f. 1r.

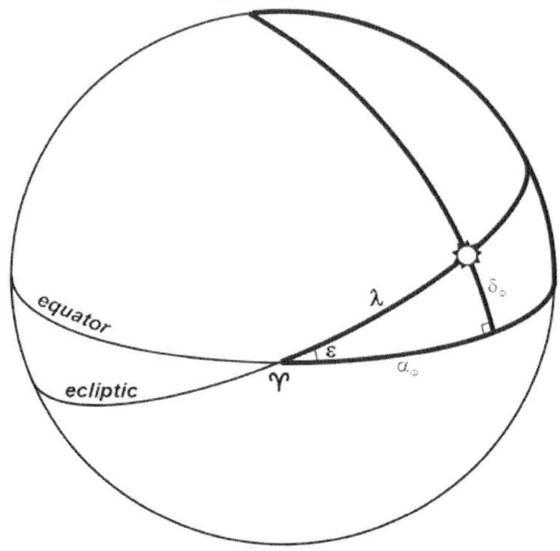

Figure 7.1. Solar declination and right ascension.

Indeed, the *Tabulae primi mobilis A* is devoted mostly to reconsidering the calculations of the positions of the planets by considering their latitudes as well as their longitudes. Bianchini demonstrates that the inclusion of latitudes can make a difference to astrological work, especially the placement of the planets in the twelve houses. However, he himself commits a significant mathematical error in the process, one that had dated from his predecessors' struggles to handle a lacuna in Ptolemy's treatment of stellar coordinates in the *Almagest*, originally transmitted to Europe through al-Battānī. Although he lavishes praise on Ptolemy in this treatise and elsewhere, in this book he occasionally seems unaware of what Ptolemy had written on topics fundamental to his work.[24]

It is in the *Tabulae primi mobilis A* that Bianchini takes the first halting steps towards what would become auxiliary tables. Chapters 7–12 of the canons contain his method of computing the declination, right ascension, and oblique ascension of any star, which he will use eventually for the astrological purposes of placing planets in the astrological houses and the computation of primary directions. He begins with a reference to his Sine table, which (as noted above) uses the radius $R = 60{,}000$; we shall call his function $\mathrm{Sin}\,\theta$ ($= R\sin\theta$). The table also gives the cosine (*sinus secundus*), which we shall call Cos. There follows a table of solar declinations $\delta_\odot(\lambda)$; in Figure 7.1, given a point on the ecliptic with longitude λ, $\delta_\odot$ is the length of the arc of the projection from that point perpendicularly onto the equator. Next, he describes a table containing values of the Sine and Cosine of the solar declination with argument λ (Figure 7.2), his *Tabula sinus declinationis*. This is the first appearance of what one might consider to be an auxiliary table. The closest precedent that one might find in Europe for such a table is the *Tabula differentie ascensionis* found in the Toledan Tables and a number of western Islamic *zījes*, which computes a constant

24 For instance, Bianchini asserts that his method of calculating the stellar right ascension was demonstrated in *Almagest* VIII.5, but this is not true, as Regiomontanus points out in a marginal note in one manuscript (Nuremberg, Cent. V. 58, f. 137v).

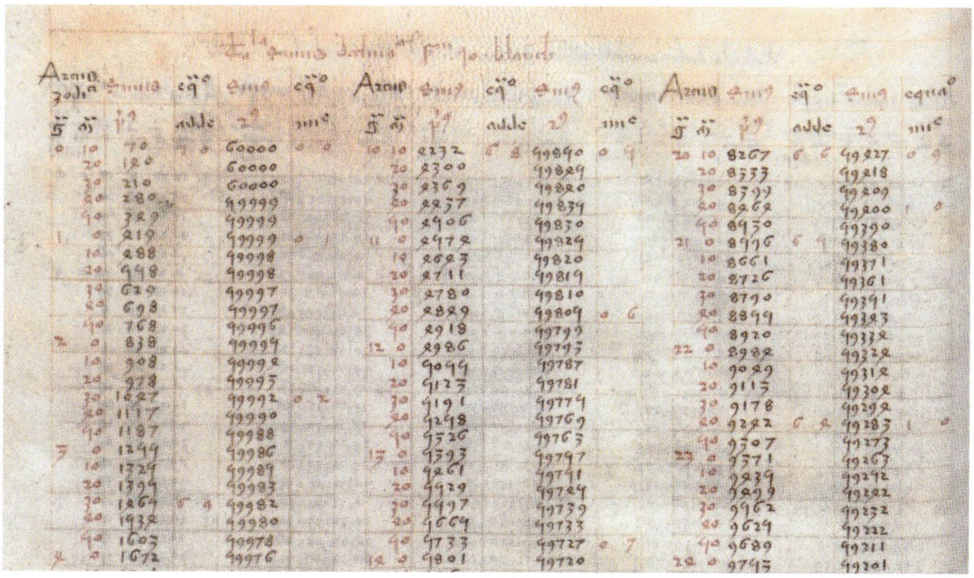

Figure 7.2. The beginning of the *Tabula sinus declinationis* in the *Tabulae primi mobilis A* (Florence, BML, Plut.29.33, f. 140^{v}), reproduced with permission.

multiple of tan $\delta_{\odot}$. This quantity had been used to find the difference between the solar right ascension and the solar oblique ascension.[25]

Bianchini's table, however, differs from these other tables both in computation and in application. Since the solar declination is

$$\text{Sin } \delta_{\odot} = \frac{\text{Sin } \varepsilon \text{ Sin } \lambda}{R}$$

(Bianchini sets the value of ε, the obliquity of the ecliptic, to 23;33,30°),[26] the values in his *Tabula sinus declinationis* are simply a constant multiple of his Sine values:

$$\frac{\text{Sin } \varepsilon}{R} \cdot \text{Sin } \lambda = \frac{23981}{60000} \text{Sin } \lambda.$$

The values of Cos $\delta_{\odot}$ may then be found from the Sines using the Pythagorean theorem. So, to calculate the entries in this table, one does not have to determine values of $\delta_{\odot}$ at all.

Unlike the quantity in the Toledan Tables, Bianchini's Sin $\delta_{\odot}$ satisfies the definition of an auxiliary function; it does not represent an astronomical quantity itself, but it may be used in different contexts to solve a number of astronomical problems. His first illustration of its use in Chapter 7 is to find the solar right ascension $a_{\odot}$, given ε and $\delta_{\odot}$. Other than the use of sines rather than chords, his method is identical to what is found in *Almagest* I.16.

25 See Julio Samsó, *On Both Sides of the Strait of Gibraltar: Studies in the History of Mathematical Astronomy in the Iberian Peninsula and the Maghrib* (Leiden/Boston: Brill, 2020), pp. 698–700, for a description.

26 We shall represent degrees, minutes, and seconds in the standard way: 23;33,30° = 23+33/60+30/60^{2}°. In our manuscripts, the numerical format was often **g** 23 **m** 33 **2** 30.

In Figure 7.1, the highlighted arcs form the configuration of Figure 7.3, the diagram of the *sector figure* (today called Menelaus's theorem), which states the following two results:[27]

$$\frac{\text{Sin } AD}{\text{Sin } BD} = \frac{\text{Sin } ER}{\text{Sin } BR} \cdot \frac{\text{Sin } AG}{\text{Sin } EG}\ (\textit{disjunction}),$$

$$\text{and } \frac{\text{Sin } AB}{\text{Sin } AD} = \frac{\text{Sin } BE}{\text{Sin } ER} \cdot \frac{\text{Sin } GR}{\text{Sin } DG}\ (\textit{conjunction}).$$

Ptolemy and Bianchini apply disjunction to the highlighted arcs in Figure 7.1, obtaining

$$\frac{\text{Sin } \varepsilon}{\text{Cos } \varepsilon} = \frac{\text{Sin } \delta_\odot}{\text{Cos } \delta_\odot} \cdot \frac{60000}{\text{Sin } \alpha_\odot}.$$

For his sample calculation with $\lambda = 54°$, he obtains

$$\text{Sin } \alpha_\odot = \frac{\text{Cos } \varepsilon \cdot \text{Sin } \delta_\odot}{\text{Sin } \varepsilon} \cdot \frac{60000}{\text{Cos } \delta_\odot} = \frac{\mathit{55000} \cdot \mathbf{19399}}{\mathit{23981}} \cdot \frac{60000}{\mathbf{56777}} = 47017,$$

and from his sine table he extracts $\alpha_\odot = 51;36°$. Our italicized numbers, which are fixed for all arguments λ, were well known to Bianchini as sines and cosines of common astronomical parameters. The numbers in bold face are obtained from the *Tabula sinus declinationis*, which thereby saves the user the bother of looking up Sin 54°, multiplying it by $\frac{23981}{60000}$ to obtain Sin $\delta_\odot(54°)$, and using Pythagoras and its troublesome square-root operation to find Cos $\delta_\odot(54°)$.

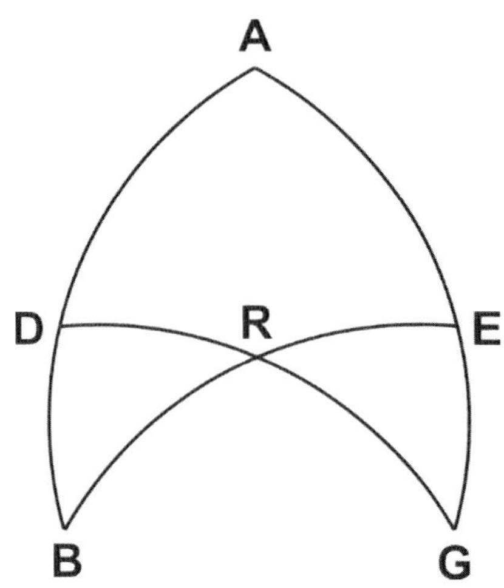

Figure 7.3. Sector figure.

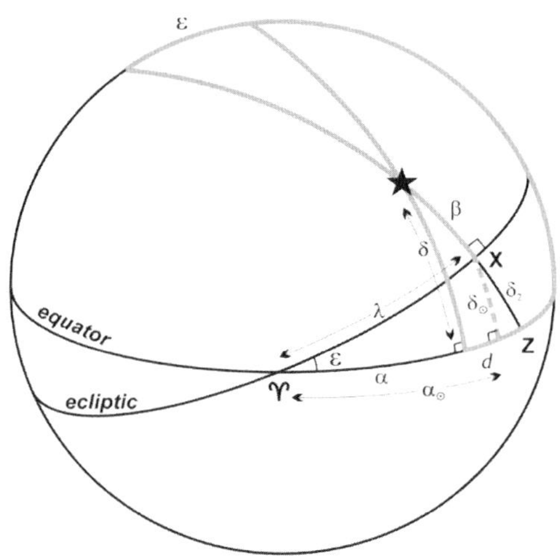

Figure 7.4. Stellar coordinates problem in the *Tabulae primi mobilis A*.

The same table is used in another more substantial example in the same chapter, Bianchini's flawed method for determining a star's declination δ given its longitude λ and latitude β (see Figure 7.4).[28] The error is to assume that the star's arc of latitude ★X is

27 We represent the results here using Bianchini's Sines. For Ptolemy's versions of the theorem, replace every reference of the Sine of an arc to the chord length of double the arc.

28 For a more detailed treatment of this, see Glen Van Brummelen, 'Before the End of an Error'.

part of the same great circle as the arc of point *X*'s declination $\delta_\odot$: in other words, that $\delta_\odot$ – which is the length of the perpendicular arc dropped from *X* onto the equator – is equal to the *second declination* $\delta_2 = XZ$, which is an arc perpendicular to the ecliptic (not the equator) at *X*. He applies the conjunction version of the sector figure to the shaded configuration in Figure 7.4 and arrives at

$$\text{Sin}\ \delta = \frac{\text{Sin}\ (\delta_\odot + \beta) \cdot \text{Cos}\ \varepsilon}{\text{Cos}\ \delta_\odot}.$$

He instructs the reader to use the *Tabula sinus declinationis* to look up Cos $\delta_\odot$ as well as Cos ε, which is equal to Cos $\delta_\odot(90°)$. He does a similar calculation in the following example, using the *Tabula sinus declinationis* twice to determine the star's right ascension.[29]

So, the *Tabula sinus declinationis* is used genuinely as an auxiliary table; it is neither a primitive trigonometric function nor an astronomical quantity, and it is used in different contexts to solve different astronomical problems. However, it is a meagre first step. It is only a single table, used only in a few contexts, and in those contexts the amount of time the user saves is not very significant. However, all ideas must start somewhere, and it would not be long before Bianchini was able to take it much further.[30]

3. The *Flores Almagesti*, the *Tabulae primi mobilis B*, and the *Tabulae magistrales*

At some point after Bianchini completed the *Tabulae primi mobilis A*, his work on spherical astronomy and other topics underwent a dramatic transformation. Although he had taken the opportunity to praise Ptolemy on numerous occasions before, the *Almagest* and the *Handy Tables* became a genuine model for his mathematical thought in later works. The *Flores Almagesti* was mature in a way that the *Tabulae primi mobilis A* was not; it expunges most (although not quite all) of the erroneous material of its predecessor and justifies its arguments adhering to a mathematical standard reminiscent of the *Almagest* itself. The first two of its ten treatises are essentially an *abacus* school textbook on arithmetic and algebra. After this, Bianchini follows the development of the *Almagest*: Treatises 3 and 4 on plane trigonometry; Treatise 5 on spherical trigonometry and astronomy; Treatises 6 and 7 on Ptolemy's models for the motion of the Sun and Moon; and Treatises 8, 9, and 10 on eclipses. The *Flores Almagesti* appears to be unfinished; indeed, of the five apparently complete manuscripts, two end after Treatise 8, while the other three end after Treatise 10.[31] While the *Almagest* would go on to treat the motions of the planets, Bianchini's book

29 Bianchini refers to the *Tabula sinus declinationis* once more in Chapter 17.

30 Later in the book, Bianchini constructs three tables containing quantities intermediary to determining the oblique ascension of a given star at a given latitude, two of which are mathematically equivalent to a tangent and an inverse sine. Glen Van Brummelen, 'Before the end of an error: Giovanni Bianchini's original flawed treatise on the conversion of stellar coordinates', p. 117. However, since they are used only in this context, they are not really auxiliary tables.

31 The manuscripts that end after the eighth treatise are Paris, BnF 10253 and Vatican, BAV, Vat. lat. 2228. The manuscripts that end after the tenth treatise are Bologna, BUB 198 (293); Cracow, BJ 558; and Perugia, Biblioteca Comunale Augusta, 1004 (M. 27). Two other manuscripts (Cracow, BJ 601 and Vatican, BAV, Reg. lat. 1904) contain only fragments of the work.

ends abruptly with eclipses. In one place there is a reference to the date 1456 as 'the present year', indicating that he was still working on the book at that time.[32]

The *Flores Almagesti* contains a complete and correct account of the problem of conversion of stellar coordinates, displaying a considerable increase in sophistication over the *Tabulae primi mobilis A*. Near the end of the introduction to the *Tabulae primi mobilis B*, otherwise almost entirely shared with the *Tabulae primi mobilis A*, we find an extended description of the contents of the *Flores Almagesti*. In all manuscripts, the first nine of the *Flores Almagesti*'s treatises are described. The introduction to the *Tabulae primi mobilis B* concludes as follows:

> Lastly, in the operations written above which are referred to in the sector figure, I have drawn out and composed five very brief tables[33] with acuteness and mastery, which I called the *Tabulae magistrales* with specific proposed quantities, which I compiled systematically in this small work. With these, it is demonstrated that by following the order in the following canons to work with them, what is sought will be brought to a conclusion. Therefore, concerning the difficulty of calculating, I have taken up nearly the whole burden – in showing that difficult necessary calculations may be avoided without the labour of calculation.[34]

This is Bianchini's full set of auxiliary tables, which eventually expanded gradually up to eight tables, not just the five mentioned here.[35] All entries except those in Table 7 (which is exceptional for other reasons; see below) are given in decimal notation. All but two of the ten known copies of the *Tabulae magistrales* are found in conjunction with a manuscript of *Tabulae primi mobilis B* and contain at least the first six tables.[36] The manuscripts of the *Tabulae primi mobilis B* containing tables (with folio numbers of the *Tabulae magistrales* indicated as **TM**) are as follows:[37]

- Bologna, Biblioteca Comunale dell'Archginnasio 1601: 1^{r}–17^{r} (canons); 31^{r}–66^{r}, 67^{r}–68^{r}, 69^{v}, 71^{v}–72^{r}, 74^{v}–85^{r}, 98^{v}–99^{r} (tables); **TM**: 66^{v}, 85^{v}–92^{v} (*Tables 1–8*);
- Cracow, BJ 556: 9^{r}–23^{v} (canons); 40^{r}, 47$^{r\text{-}v}$, 55^{r}–92^{v}, 94^{r}–105^{v} (tables); **TM**: 48^{r}–54^{v}, 93^{r} (*Tables 1–8*);

32 See Chabás and Goldstein, 'The *Tabulae eclypsium* by Giovanni Bianchini', p. 6, and for instance, in Bologna BUB 198 (293), f. 58$^{r\text{-}v}$.

33 One manuscript, Vat. lat. 2228, refers to six tables here.

34 *Ultimo, in ex omnibus superscriptis operationibus que ad figuram sectoris referuntur cum subtilitate et magisterio extraxi et composui quinque tabulas brevissimus: quas vocavi tabulas magistrales cum certus numeris ad propositum: quas in hoc opusculo ordinate compilavi: Cum quibus facillime ordinem persequendo ut in sequentibus canonibus ad ipsas operando demonstratur: ad conclusionem quaesiti pervenientur: Quare in difficultate calculandi totum fere onus suscepi: ut illi qua in indicando calculationes necessarias difficultatem obmittant absque labore calculari.* Quotations from the *Tabulae primi mobilis B* will be referenced with the manuscript Kraków, BJ 556. In this case, the passage is on folio 11^{r}.

35 The *Tabulae magistrales*, in relation to the *Tabulae primi mobilis B* and the *Flores Almagesti*, are also mentioned in Bianchini's later correspondence with Regiomontanus. See Curtze, 'Der Briefwechsel Regiomontan's mit Giovanni Bianchini, Jacob von Speier und Christian Roder', 206.

36 One manuscript of the *Tabulae primi mobilis B*, Vat. lat. 2228, includes the *Tabulae magistrales* in the list of tables provided in the canons (Chapter 27, f. 62^{r}).

37 Three manuscripts of the *Tabulae primi mobilis B* contain only canons: Innsbruck W. 3277, Paris BnF Latin 10267, and Vatican BAV Vat. lat. 2228.

- Oxford, Bodleian Can. Misc. 517: 82^r–99^r (canons); 116^r–153^r, 160^r–171^r (tables); **TM:** 172^r–177^v[38];
- Paris, BnF lat. 7270: 143^r–167^r (canons); 183^r, 186^v–227^v, 231^r–236^r (tables); **TM:** 183^v–186^r, 228^r–230^v (*Tables 1–6*);
- Paris, BnF lat. 7271: 147^r–161^r (canons); 191^r, 184^v, 190^r–216^v, 222^r–237^r, 242^r–245^v (tables); **TM:** 181^v–187^v (*Tables 1–6*);
- Paris, BnF lat. 7286: 65^r–81^v (canons); 95^r–131^v, 133^r, $136^{r\text{-}v}$, 139^r–148^v (tables); **TM:** 132^v, 149^r–154^v (*Tables 1–7*);
- Paris, BnF lat. 10265: 85^r, 88^v–91^r, 92^v, 153^v–222^r, 232^v–235^r (tables); **TM:** 223^v–232^r, 237^r–238^r (*Tables 1–6, 8*).

One copy of the *Tabulae magistrales* exists independently (Cracow, BJ 606, ff. 62^r–69^r, *Tables 1–6*), as well as one short extract (Vatican, BAV, Vat. lat. 3538, ff. 6^r–7^r, *Tables 1–4*).

The first four tables of the Tabulae magistrales[39]

Bianchini's earliest applications of the *Tabulae magistrales* are found in the canons to the *Tabulae primi mobilis B*. All but one are in Chapters 13 to 20, where he describes how to convert stellar coordinates from ecliptic to equatorial.[40] The only *Tabulae magistrales* tables mentioned are Tables 1 to 4, suggesting that these four were conceived and constructed first (see Figure 7.5). They are defined as follows:

$$\begin{aligned} T_1(x) &= R\frac{\text{Cos}\ \varepsilon}{\text{Sin}\ x} \\ T_2(x) &= \text{Tan}\ \delta_{\odot}(x) \\ T_3(x) &= \frac{\text{Tan}\ \varepsilon\ \text{Cos}\ a_2(x)}{R} \\ T_4(x) &= R\frac{\text{Sin}\ x}{\text{Cos}\ x}, \end{aligned}$$

where a_2 is Bianchini's new quantity, the *radicum ascensionum*, the arc along the equator from the vernal equinox ♈ to the intersection of the second declination arc XZ with the equator (see Figure 7.6). Note that a_2 is not the right ascension of point X on the ecliptic since arc XZ is perpendicular to the ecliptic rather than the equator.

The applications of the *Tabulae magistrales* in the canons to the *Tabulae primi mobilis B* are as follows:
- Chapter 13 (finding $a_{\odot}(\lambda)$; requires $a_2(\lambda)$ from the *radicum ascensionum* table): Tables 3 and 4;
- Chapter 15 (finding $\delta(\lambda, \beta)$): Table 1;

38 I have not had the opportunity to consult this manuscript.

39 José Chabás, 'An Analysis of the *Tabulae magistrales* by Giovanni Bianchini', *Archive for History of Exact Sciences* 70 (2016) analysed the tables themselves. Our treatment expands on it by examining how Bianchini uses the tables in the *Tabulae primi mobilis B*, its appendix, and the *Tabulae eclypsium*, revealing with some specificity for what purposes he invented them, and demonstrating that the tables were invented in two separate stages.

40 The other application is in Chapter 39 on an astrological topic, to which we shall refer later.

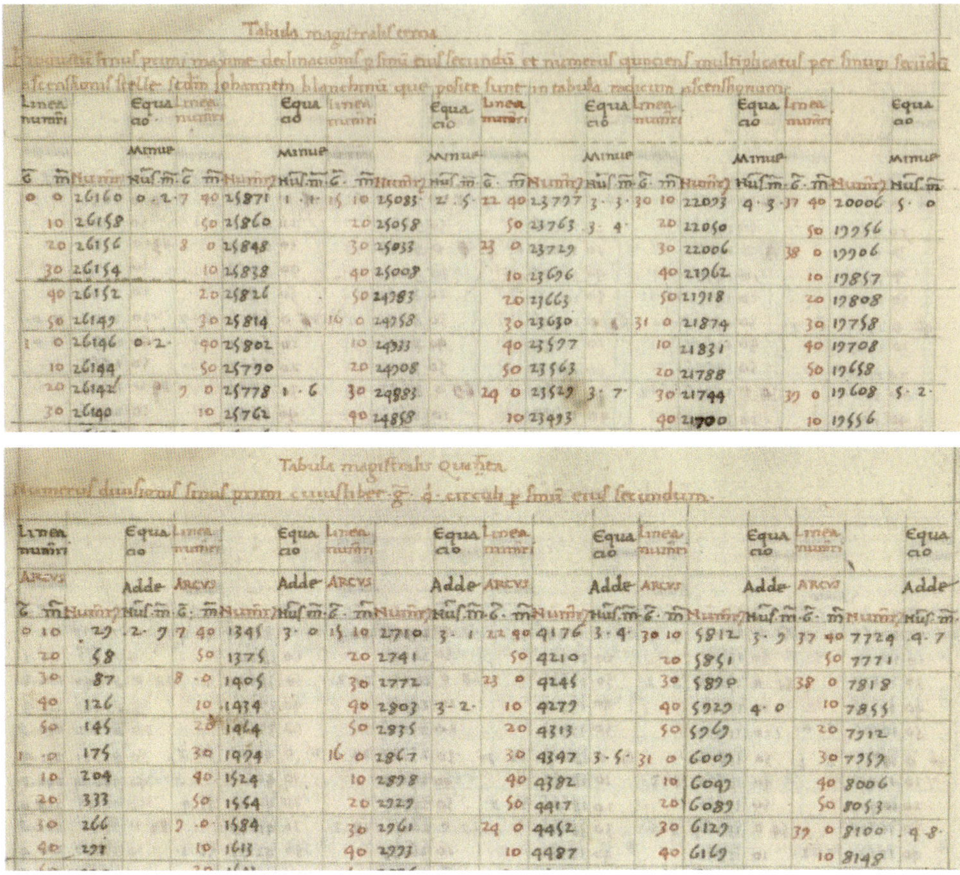

Figure 7.5. Beginnings of *Tabulae magistrales* 3 and 4 (Cracow, BJ 556, f. 51ʳ, 52ʳ).

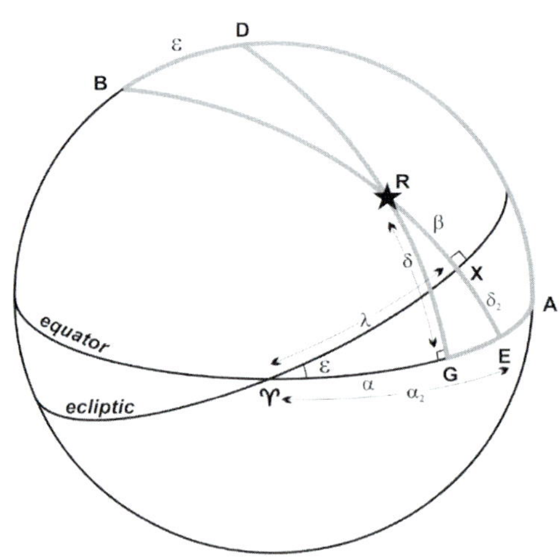

Figure 7.6. Stellar coordinates problem in the *Tabulae primi mobilis B*.

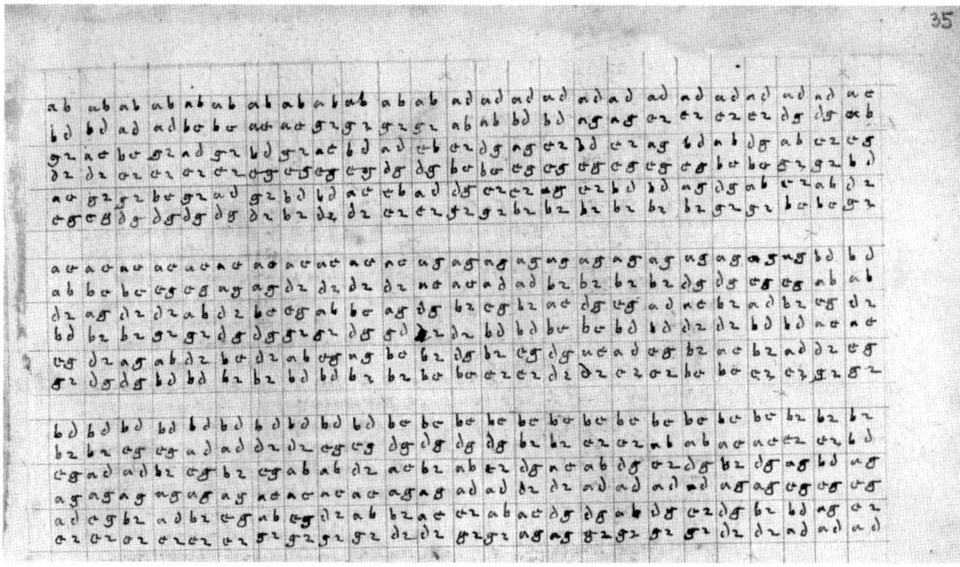

Figure 7.7. The beginning of the table of permutations of ratios of the sector figure in the *Flores Almagesti* (Cracow, BJ 558, f. 35r).

- Chapter 16 (finding $a(\lambda, \beta)$; requires $\delta(\lambda, \beta)$ from Chapter 15 and $a_2(\lambda)$ from the *radicum ascensionum* table): Tables 3 and 4;
- Chapter 18 (finding the oblique ascension $\theta(\lambda, \beta, \varphi)$ where φ is the terrestrial latitude, and the diurnal arc): Tables 2 and 4;
- Chapter 20 (finding the diurnal arc; requires $\delta(\lambda, \beta)$ from the stellar declination table): Table 4;
- Chapter 39 (an astrological application related to his new method for handling the projection of the rays): Table 4.

The chapters all use similar methods; here, we discuss the example of Chapter 16. In Figure 7.6, the goal is to find the star's declination δ given its longitude λ and latitude β; we assume that we have access to the *radicum ascensionum* table that gives $a_2(\lambda)$. As one can see by comparing Figures 7.4 and 7.6, the repair of the incorrect grey sector figure (obtained by replacing $\widehat{XZ}$ with $\widehat{XE}$) and the tables of a_2 and δ_2 provide a path to a solution. Indeed, Bianchini rarely uses anything but the sector figure in his spherical trigonometry. In every application of the sector figure in the *Flores Almagesti* and the *Tabulae primi mobilis B*, he employs it in the same distinctive way. In the *Flores Almagesti*, he enumerates and tabulates 150 ways that the ratios of line segments in the two sector figure theorems can be permuted so that one ratio may be seen to be composed (be equal to a product) of two other ratios; see Figure 7.7. He was not alone in this systematic approach; it is found as early as the work on the sector figure by ninth-century astronomer Thābit ibn Qurra, which had been translated into Latin three times.[41]

41 Thābit's treatises on the sector figure and related texts in Arabic and Latin have been edited, translated, and studied in Richard Lorch, *Thabit ibn Qurra. On the Sector-Figure and Related Texts* (Frankfurt: Institut für Geschichte der Arabisch-Islamischen Wissenschaften, 2001; reprinted Augsberg: Rauner, 2008). Bianchini's

In the *Flores Almagesti*, whenever Bianchini applies the sector figure he informs us which of his 150 permutations he applies to derive the result. In the *Tabulae primi mobilis B*, he simply states an abbreviated rule of calculation. He refers back to the relevant section in the *Flores Almagesti*, but the permutation he uses in the *Tabulae primi mobilis B* is not always the same as the one that he used in the *Flores Almagesti*. Such is the case in Chapter 16, where he implicitly uses permutation 1 in his table (see Figures 7.6 and 7.7):[42]

$$\frac{\text{Sin } AB}{\text{Sin } BD} = \frac{\text{Sin } GR}{\text{Sin } DR} \cdot \frac{\text{Sin } AE}{\text{Sin } EG}.$$

All but one of these quantities are known. For the numerical example $\lambda = 60°$ in the text:

- $AB = 90°+\varepsilon = 113;33{,}30°$; its Sine is 55000;
- $BD = \varepsilon = 23;33{,}30°$; its Sine is 23981;
- $GR = \delta(\lambda,\beta) = 42;35°$, worked out earlier; its Sine is 40604;
- $DR = 90°-\delta(\lambda,\beta) = 47;25°$; its Sine is 44178;
- $AE = 90°-a_2(\lambda) = 27;53°$ (determined from the *radicum ascensionum* table); its Sine is 28060;
- $EG = a_2(\lambda) - a(\lambda,\beta)$; since $a_2(\lambda)$ is known, we shall use EG to determine a.

Entering these values in our sector figure relation, we have

$$\frac{55000}{23981} = \frac{40604}{44178} \cdot \frac{28060}{\text{Sin } EG},$$

or

$$\text{Sin } EG = \mathit{\frac{23981}{55000}} \cdot \mathit{28060} \cdot \mathbf{\frac{40604}{44178}}.$$

However, the quantities in italics represent $T_3(\lambda)$, while the quantities in bold face represent $T_4(\delta)$. So, we simply look them up and multiply them, giving Sin $EG = 11245$. From the sine table, $EG = 10;48°$; and since $EG = a_2 - a$, we have $a = 51;19°$.

Thus, the motivation for Bianchini's auxiliary tables is straightforward. It is not to invent new trigonometric quantities or to identify astronomical structures within the calculations. Rather, he simply uses the tables to save himself and his readers the tedium of multiplying and dividing values that appear frequently in the application of the sector figure. In the section on stellar coordinates, he frequently finds himself multiplying a *radicum ascensionum* value by $\frac{23981}{55000}$, so he makes a table for it. Similarly, he frequently finds himself dividing the sine of a quantity by the cosine of the same quantity, so he makes a table for that also. The purpose of these auxiliary functions is thus computational rather than trigonometric.

approach to the sector figure is described in Henry Zepeda, 'The Medieval Latin Transmission of the Menelaus Theorem', unpublished thesis, University of Oklahoma, 2013, pp. 323–29.

42 Whenever Bianchini uses the sector figure in the *Flores Almagesti*, he matches the letters in his astronomical diagram with those in Figure 7.3. We have inserted the letters from Figure 7.3 into the diagram in Figure 7.6.

The addenda to the Tabulae primi mobilis B, *and the last four tables of the* Tabulae magistrales

Although only the first four tables of the *Tabulae magistrales* have been used, they make no further appearance in the *Tabulae primi mobilis B*. However, Bianchini seems to have revisited spherical astronomy some time later, this time exploring the amplitude of a star, the location of its rising point on the horizon. He does this at the end of Book I of Treatise 7 of the *Flores Almagesti* at the conclusion of a discussion of lunar theory, just before beginning an extensive treatment of eclipses. Separately, he converts this material to a treatment in the style of the *Tabulae primi mobilis B* and the *Tabulae magistrales*. I have found it in only two manuscripts: Cracow, BJ 556 (ff. 5^r–7^v) as a set of *Additiones* to the canons of the *Tabulae primi mobilis B*; and appended to the end of the book in Vat. lat. 2228 (ff. 73^v–78^r). In the *Additiones*, we find the only references to *Tabulae magistrales* 5, 6, and 8,[43] which are defined as follows:

$$T_5(x) = \frac{R^2}{\text{Sin}\,x}$$

$$T_6(x) = \frac{\text{Tan}\,\varepsilon\,\text{Tan}\,x}{R}$$

$$T_8(x) = R\frac{\text{Sin}\,x}{\text{Cos}\,\varepsilon}.$$

The *Additiones* tend to be briefer in their exposition than what is found in the canons to the *Tabulae primi mobilis A* and *B*; occasionally they refer to the *Flores Almagesti* for mathematical support, but often they do not. This is the case in *Additiones* Chapter 1,[44] where the amplitude a of a star is determined (Figure 7.8), given its declination δ (displayed in the figure as southward) and the altitude h of the equator above the horizon at its intersection with the meridian in the south.[45] In the figure, G is the eastern point of the horizon and B is the south pole of the equator. This problem had already been solved by Ptolemy in *Almagest* II.3,[46] in a calculation equivalent to Bianchini's. From Entry 3 in Bianchini's table of sector figure relations (Figure 7.7),

$$\frac{\text{Sin}\,AB}{\text{Sin}\,AD} = \frac{\text{Sin}\,BE}{\text{Sin}\,ER}\cdot\frac{\text{Sin}\,GR}{\text{Sin}\,DG}.$$

However, $AB = BE = DG = 90°$, $AD = h$, $ER = \delta$, and $GR = a$, so

$$\text{Sin}\,a = R\frac{\text{Sin}\,\delta}{\text{Sin}\,h}.$$

Bianchini instructs the reader to use *Tabulae magistrales* 5:

$$\text{Sin}\,a = \text{Sin}\,\delta\cdot\frac{R}{\text{Sin}\,h} = \text{Sin}\,\delta\cdot T_5(h),$$

43 Tables 1 and 4 also make a reappearance here. Table 8 is actually called Table 7 in the text.

44 The chapters are not numbered in Vat. lat. 2228; in Cracow, BJ 556, only the first six of seventeen chapters are numbered.

45 The altitude h is the complement of the observer's terrestrial latitude.

46 See G. J. Toomer, *Ptolemy's Almagest* (New York: Springer and London: Duckworth, 1984), pp. 78–79.

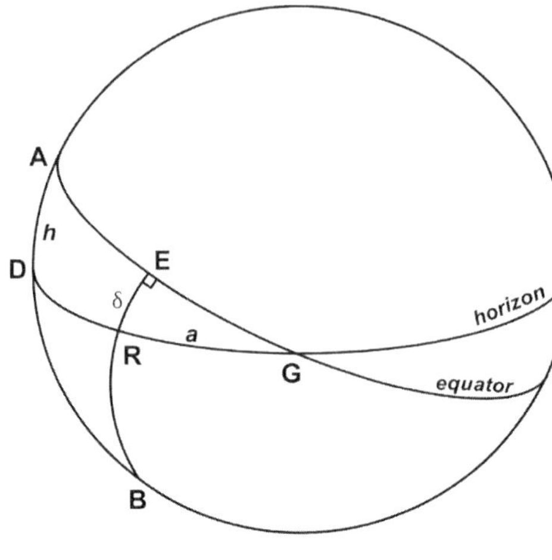

Figure 7.8. Finding a star's amplitude *a* from its declination δ and the meridian altitude of the equator *h*.

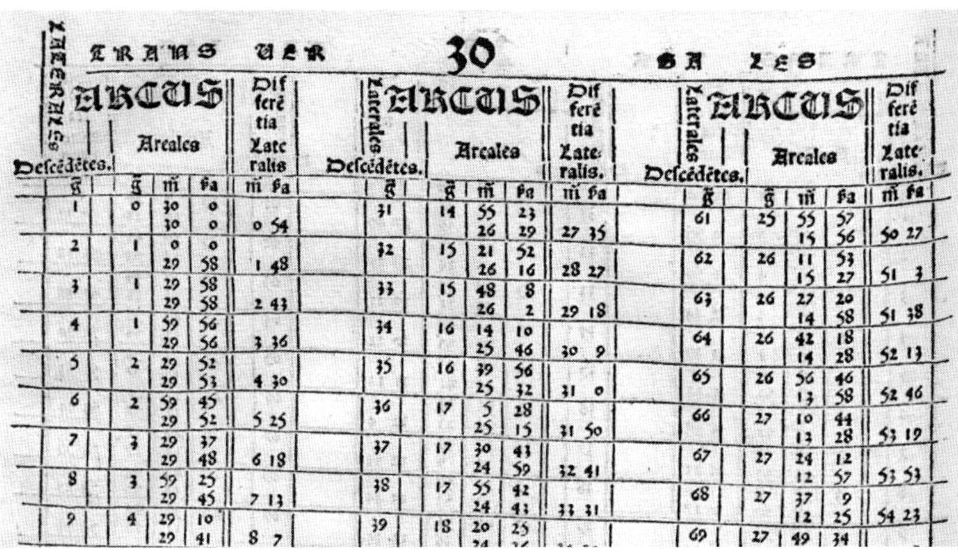

LATERALES — TRANS VER 30 SA LES

Arcus Laterales Descẽdẽtes.	Arcus Arcuales			Differẽtia Lateralis	Arcus Laterales Descẽdẽtes.	Arcus Arcuales			Differẽtia Lateralis.	Arcus Laterales Descẽdẽtes.	Arcus Arcuales			Differẽtia Lateralis.
g	g	m	s	m s	g	g	m	s	m s	g	g	m	s	m s
1	0	30 30	0 0	0 54	31	14	55 26	23 29	27 35	61	25	55 15	57 56	50 27
2	1	0 29	0 58	1 48	32	15	21 26	52 16	28 27	62	26	11 15	53 27	51 3
3	1	29 29	58 58	2 43	33	15	48 26	8 2	29 18	63	26	27 14	20 58	51 38
4	1	59 29	56 56	3 36	34	16	14 25	10 46	30 9	64	26	42 14	18 28	52 13
5	2	29 29	52 53	4 30	35	16	39 25	56 32	31 0	65	26	56 13	46 58	52 46
6	2	59 29	45 52	5 25	36	17	5 25	28 15	31 50	66	27	10 13	44 28	53 19
7	3	29 29	37 48	6 18	37	17	30 24	43 59	32 41	67	27	24 12	12 57	53 53
8	3	59 29	25 45	7 13	38	17	55 24	42 43	33 31	68	27	37 12	9 25	54 23
9	4	29 29	10 41	8 7	39	18	20 24	25 [illegible]	[illegible]	69	27	49	34	

Figure 7.9. Part of Regiomontanus's *Tabula primi mobilis* ([Peurbach 1514]).

thereby converting a division into an easier multiplication.[47]

The *Additiones* is the last time that Bianchini describes the *Tabulae magistrales* in his works of spherical astronomy. However, three copies of the *Tabulae magistrales* include another table as well.[48] It is called Table 7, but it is separated from the other tables in all three manuscripts.[49] Its nature differs from the other tables in that its values are in sexagesimal

47 Note that T^8 is essentially the reciprocal of T^1, confirming that Bianchini preferred multiplication to division enough that he was prepared to compose a new table to replace a division with a multiplication.

48 Bologna 1601, f. 66^v; Cracow, BJ 556, f. 93^r; Paris, BnF 7286, f. 132^v.

49 Table 8 appears in Bologna 1601 (f. 92^r–92^v) under the title of Table 7; it appears in Cracow, BJ 556 (f. 49^{r-v}) and Paris, BnF 10265 (ff. 237^r–238^v) under the title of Table 8. The *Additiones* calls it Table 7.

format rather than the usual decimal, and it works directly with astronomical (rather than auxiliary) quantities; it tabulates the oblique ascension of a star for terrestrial latitude of forty-five degrees (the latitude of his home town of Ferrara), and the semidiurnal arc as a function of the right ascension.

The birth and expansion of the tangent function

It may not have escaped the reader's notice that T_4 is mathematically equivalent to a tangent table. We argued earlier in this article that the appearance of an auxiliary table equivalent to a tangent (for instance, in Ḥabash al-Ḥāsib's work) should not be taken as an invention of the tangent function, assuming that the table was not used as a general tool for arbitrary arguments or defined as a primitive trigonometric tool. There are shades of meaning here; one may choose to draw the line between auxiliary and trigonometric functions as one likes. In Bianchini's case, while T_4 began like the other tables in the *Tabulae magistrales* as a computational tool for converting stellar coordinates, there is no doubt that its value as an independent entity quickly became apparent. Even within stellar coordinates, it is used in four problems in various ways (Chapter 13, 16, 18, and 20; indeed, only in Chapter 15 does he use the *Tabulae magistrales* without calling upon T_4). He extends T_4 to another context even before the end of the *Tabulae primi mobilis B*; in Chapter 39, he uses it to help solve an astrological problem.

Bianchini's last major work is his *Tabulae eclypsium*.[50] The canons begin with the first of a number of references to the *Flores Almagesti*. Later in the book he predicts future solar and lunar eclipses: for instance, in July 1460 in his home town of Ferrara.[51] He frequently uses the decimal arithmetic introduced in the *Tabulae primi mobilis B*, a more advanced system than is found in the *Flores Almagesti*. In Chapter 18 of the canons to the *Tabulae eclypsium* he briefly discusses the *Tabulae magistrales*:

> And after this, with the basis of this demonstration, I constructed 5 most succinct *Tabulae magistrales*, through which the present table can be constructed with a very easy and brief operation. Or, without the other tables, all the calculations will be performed, in *sphaera recta* as in *sphaera obliqua*, and in directions through *circulo recto* or *circulo obliquo*, through mixed ascensions.[52]

He does appeal to the auxiliary tables in the *Tabulae eclypsium*, but he only uses the tangent function T_4, employing it on four separate occasions (Chapter 11, twice in Chapter 25, and Chapter 30). T_4 was starting to take on a life of its own, finding a regular role in improving

---

50 While the *Tabulae eclypsium* refers several times to the *Tabulae magistrales*, which (as we have seen) is tightly linked to the *Tabulae primi mobilis B*, it refers only to T_4. So, we have no evidence to help us decide whether the *Tabulae eclypsium* was written before or after the *Additiones*.

51 See Chabás and Goldstein, 'The *Tabulae eclypsium* by Giovanni Bianchini', and Chapters 23 and 30 of the *Tabulae eclypsium*.

52 *Et ultra hoc cum fundamento ipsius demonstrationis construxi 5 brevissimas tabulas magistrales per quas cum facillima et brevi operatione construi presente tabule. Seu absque aliis tabuliis concludent calculos omnes, tam in spera recta, quam in obliqua, et tam in directionibus per circulum directum seu obliquum, quam per ascensiones mixtas.* Paris, BnF 7270, f. 174r. A 'mixed ascension' is a weighted average of right and oblique ascensions that arises in the calculation of primary directions in astrology.

computational work in various contexts. The other auxiliary functions, fitted more closely to particular astronomical situations, had almost no opportunities to be repurposed elsewhere.

Regiomontanus, who possessed a copy of the *Tabulae primi mobilis B*, borrowed its tabular structure for many of the tables in his *Tabulae directionum*.[53] His book contains an expanded and recalculated version of Bianchini's sine table with $R = 60{,}000$, tabulated for every minute of arc rather than every 10′. Many of the astronomical tables in Regiomontanus's book are derived from the *Tabulae primi mobilis B*, recomputed for different values of the parameters (especially ε).[54] However, of all the *Tabulae magistrales* tables, Regiomontanus borrows only T_4, calling it the *tabula fecunda* and computing it only for every degree rather than every 10′. Yet, there is no sign of it in his trigonometric text *De triangulis omnimodis*, written only a couple of years earlier.[55]

Through Regiomontanus, T_4 continued to inspire and assist astronomers in the sixteenth century. Erasmus Reinhold's 1554 *Tabularum directionum*, for instance, contains a *Canon fecundus* with $R = 10{,}000{,}000$ and entries for every minute of arc (and every 10″ for arguments between 89° and 90°).[56] Georg Rheticus's 1551 *Canon doctrinae triangulorum* and his and Valentin Otho's 1596 *Opus palatinum* brought not just the tangent but all of the six modern trigonometric functions together in a single unified system, using his own terminology, which does not refer to a 'sine', let alone a 'tangent'.[57] Rheticus's system had some influence, for instance, in François Viete's 1579 *Canon mathematicus seu ad triangula*.[58] However, the Bianchini/Regiomontanus approach and nomenclature persisted; for instance, Maurolico referred back to Regiomontanus's *tabula fecunda* when he coined his secant table, the *tabella benefica*, in 1558.[59] It was not until 1583 in the *Geometria rotundi* that Thomas Fincke gave the tangent and secant their modern names and geometric definitions, and the conversion of T_4 to the trigonometric function we recognize today was essentially completed.[60]

53 Johannes Regiomontanus, *Tabulae directionum* (Augsburg: Ratdolt, 1490).

54 See Glen Van Brummelen, 'The End of an Error', 558–62 for a comparison of the stellar coordinate conversion tables in the two books.

55 See Barnabas Hughes, *Regiomontanus* On Triangles (Madison University of Wisconsin, 1967) for a facsimile, translation, and discussion of this work.

56 Erasmus Reinhold, *Tabularum directionum* (Tubingen: Ulrici Morhardi, 1554).

57 Georg Rheticus, *Canon doctrinae triangulorum* (Leipzig: Wolfgang Gunter, 1551) and Georg Rheticus and Valentine Otho, *Opus palatinum de triangulis* (Neustadt: Matthaeus Harnisch, 1596). See Kailyn Pritchard, 'Determining the Sine Tables Underlying Early European Tangent Tables', in *Editing and Analysing Numerical Tables: Towards a Digital Information System for the History of Astral Sciences*, ed. by Matthieu Husson, Clemency Montelle, and Benno van Dalen (Turnhout: Brepols, 2021), pp. 107–42 for an analysis of the tangent tables in Rheticus's works, as well as in those of Bianchini and Regiomontanus.

58 François Viète, *Canon mathematicus seu ad triangula* (Paris: Jean Mettayer, 1579).

59 Francesco Maurolico, *Theodosii sphaericorum elementorum libri III*, etc. (Messina: Pietro Spira, 1558); see a discussion of this table in Glen Van Brummelen and James Byrne, 'Maurolico, Rheticus, and the Birth of the Secant Function', *Journal for the History of Astronomy* 52 (2021), 189–211.

60 Thomas Fincke, *Geometria rotundi* (Basel: Henric Petri, 1583). For an account of the events in these paragraphs see Glen Van Brummelen, *The Doctrine of Triangles: A History of Modern Trigonometry* (Princeton: Princeton University Press, 2021), pp. 3–16.

4. Auxiliary tables from Regiomontanus to Napier

Compared to medieval Islam, the tradition of auxiliary tables in Europe is substantially compressed, for two reasons. Firstly, Regiomontanus himself accelerated the progress of auxiliary tables, less than a couple of decades after Bianchini had begun it, with his own auxiliary table entitled *Tabula primi mobilis* and an associated text, the *Fundamenta operationum*. The *Tabula primi mobilis*[61] contains a large double-argument table, containing 8100 entries tabulating the quantity

$$\text{Sin}\, z = (\text{Sin}\, x\, \text{Sin}\, y\, /R)$$

for x and y between 1° and 90°, as well as sixty-three chapters of canons describing various astronomical problems that the table may be used to solve (see Figure 7.9). The *Fundamenta operationum* is the collection of propositions demonstrating the spherical relations upon which the canons are based.[62] From a modern point of view, this table permits the use of five of the ten fundamental identities of a right-angled spherical triangle, namely, the ones that do not require a tangent or cotangent. While this limits his available mathematical tools, he is still able to solve any astronomical problem that can be solved using the other five identities (in these two books, he never mentions his *tabula fecunda*).

The advantage of this table would have been immediately clear to any user; it takes care of two sine lookups, a multiplication, and a reverse sine lookup at once. Others followed Regiomontanus down this path, especially Antonio Magini, whose 1604 *Tabulae primi mobilis* includes a collection of four large tables, the first of which is based on the same formula as Regiomontanus's.[63]

The second reason for the compressed history of auxiliary functions in Europe is its abrupt end. Only a decade after the appearance of Magini's tables, John Napier announced the invention of logarithms.[64] This new tool rendered auxiliary functions obsolete by eliminating the need for any multiplications or divisions using identities equivalent to $\log \frac{a}{b} = \log a - \log b$ and $\log (a \cdot b) = \log a + \log b$. Indeed, Napier geared his table of logarithms specifically towards spherical astronomy by tabulating not logarithms themselves, but the logarithms of sines. The flexibility of the new tool, including applications in such diverse fields as navigation and architecture, contributed to their quick adoption. Auxiliary tables were just as quickly forgotten.

Conclusion

The parallels in the development of auxiliary tables in medieval Islam and in Europe are unmistakable. As scholars like Bianchini and Regiomontanus explored the mathematics of astronomy and astrology with more mathematical sophistication than their predecessors, they felt the need for smoother and more reliable computational methods. This gave

61 Found in Georg Peurbach, *Tabulae eclypsium* (Vienna: Johann Winterbuger, 1514).
62 Johannes Regiomontanus, *Fundamenta operationum* (Neuberg: Hans Kilian, 1557).
63 Antonio Magini, *Tabulae primi mobilis* (Venice: Damianum Zenarium, 1604).
64 John Napier, *Mirifici logarithmorum canonis descriptio* (Edinburgh: Hart, 1614).

rise to tools whose main purposes were to save users part of the computational burden, especially of spherical astronomy, allowing the user to avoid unnecessary table lookups, multiplications, and divisions. The notion of the tangent followed quickly in both cultures. However, the two developments also illustrate important differences. It took much longer for the notion to be born in Europe; prior to the fifteenth century, the needs for computational astronomy were not as varied as in Islamic regions, so auxiliary tables would not have been as useful. Once Bianchini had designed his *Tabulae magistrales,* Regiomontanus immediately saw their power, and he extended the concept very far and very quickly. Logarithms ended the tradition of auxiliary functions, providing a solution to computational burdens that rendered its predecessors obsolete almost instantly. It is a story we have seen before.

Manuscript sources

Bologna, Biblioteca Comunale dell'Archginnasio, 1601
Cracow, Biblioteka Jagiellońska, 556
Oxford, Bodleian Library, Canon. misc. 517
Paris, Bibliothèque nationale de France, lat. 7270
Paris, Bibliothèque nationale de France, lat. 7271
Paris, Bibliothèque nationale de France, lat. 7286
Paris, Bibliothèque nationale de France, lat. 10265

Bibliography

Chabás, José, 'An Analysis of the *Tabulae magistrales* by Giovanni Bianchini', *Archive for History of Exact Sciences* 70 (2016), 543–52.

Chabás, José, *Computational Astronomy in the Middle Ages: Sets of Astronomical Tables in Latin* (Madrid: Consejo Superior de Investigaciones Científicas, 2019).

———, and B. R. Goldstein, *The Astronomical Tables of Giovanni Bianchini* (Leiden/Boston: Brill, 2009).

———, and B. R. Goldstein, 'The *Tabulae eclypsium* by Giovanni Bianchini', *Aestimatio* 2 (1) (2021), pp. 1–41.

Charette, François, 'A Monumental Medieval Table for Solving the Problems of Spherical Astronomy for all Latitudes', *Archives internationales d'histoire des sciences* 48 (1998), pp. 11–64.

Curtze, Maximilian, 'Der Briefwechsel Regiomontans mit Giovanni Bianchini, Jacob von Speier und Christian Roder', *Abhandlungen zur Geschichte der Mathematischen Wissenschaften* 12 (1902), 185–336.

Fincke, Thomas, *Geometria rotundi* (Basel: Henric Petri, 1583).

Gerl, Armin, *Trigonometrisch-astronomisches Rechnen kurz vor Copernicus: Der Briefwechsel Regiomontanus-Bianchini* (Stuttgart: Steiner, 1989).

Hughes, Barnabas, *Regiomontanus On Triangles* (Madison, WI: University of Wisconsin, 1967).

Irani, Rida A. K., 'The "Jadwal al-Taqwīm" of Ḥabash al-Ḥāsib' unpublished MA thesis, American University of Beirut, 1956.

King, David, 'Al-Khalīlī's Auxiliary Tables for Solving Problems of Spherical Astronomy', *Journal for the History of Astronomy* 4 (1973), 99–110.

———, 'Al-Khalīlī's *qibla* Table', *Journal of Near Eastern Studies* 34 (1975), 81–122.

Lorch, Richard, *Thabit ibn Qurra. On the Sector-Figure and Related Texts* (Frankfurt: Institut für Geschichte der Arabisch-Islamischen Wissenschaften, 2001; reprinted Augsberg: Rauner, 2008).

Magini, Antonio, *Tabulae primi mobilis* (Venice: Damianum Zenarium, 1604).

Maurolico, Francesco, *Theodosii sphaericorum elementorum libri III*, etc. (Messina: Pietro Spira, 1558).

Napier, John, *Mirifici logarithmorum canonis descriptio* (Edinburgh: Hart, 1614).

Peurbach, Georg, *Tabulae eclypsium* (Vienna: Johann Winterburger, 1514).

Pritchard, Kailyn, 'Determining the Sine Tables Underlying Early European Tangent Tables', in *Editing and Analysing Numerical Tables: Towards a Digital Information System for the History of Astral Sciences*, ed. by Matthieu Husson, Clemency Montelle, and Benno van Dalen (Turnhout, Belgium: Brepols, 2021), pp. 107–42.

Regiomontanus, Johannes, *Fundamenta operationum* (Neuberg: Hans Kilian, 1557).

———, *Tabulae directionum* (Augsburg: Ratdolt, 1490).

Reinhold, Erasmus, *Tabularum directionum* (Tubingen: Ulrici Morhardi, 1554).

Rheticus, Georg, *Canon doctrinae triangulorum* (Leipzig: Wolfgang Gunter, 1551).

———, and Valentin Otho, *Opus palatinum de triangulis* (Neustadt: Matthaeus Harnisch, 1596).

Rosińska, Grażyna, 'Tables of Decimal Trigonometric Functions from ca. 1450 to 1550', in *From Deferent to Equant: A Volume of Studies in the History of Science in the Ancient and Medieval Near East in Honor of E. S. Kennedy*, ed. by David A. King and George Saliba (New York: New York Academy of Sciences, 1987), pp. 419–26.

———, 'A Chapter in the History of Renaissance Mathematics: Negative Numbers and the Formulation of the Law of Signs (Ferrara, Italy *c.* 1450)', *Kwartalnik Historii Nauki i Techniki* 40 (1995), pp. 3–20.

———, 'The "Fifteenth-Century Roots" of Modern Mathematics', *Kwartalnik Historii Nauki i Techniki* 41 (1996), pp. 53–70.

———, 'The "Italian Algebra" in Latin and How it Spread to Central Europe: Giovanni Bianchini's *De algebra* (ca. 1440)', *Organon* 26–27 (1997–1998), pp. 131–45.

———, 'The Euclidean *spatium* in Fifteenth-Century Mathematics', *Kwartalnik Historii Nauki i Techniki* 43 (1998), 27–42.

———, '"Mathematics for Astronomy" at Universities in Copernicus's Time: Modern Attitudes toward Ancient Problems', in *Universities and Science in the Early Modern Period*, ed. by Mordecai Feingold and V. Navarro-Brotóns (Berlin: Springer, 2006), pp. 9–28.

Samsó, Julio, *On Both Sides of the Strait of Gibraltar: Studies in the History of Mathematical Astronomy in the Iberian Peninsula and the Maghrib* (Leiden/Boston: Brill, 2020).

Toomer, G. J., *Ptolemy's Almagest* (New York: Springer and London: Duckworth, 1984).

Van Brummelen, Glen, 'The Numerical Structure of al-Khalili's Auxiliary Tables', *Physis* 28 (1991), 667–98.

———, 'The End of an Error: Bianchini, Regiomontanus, and the Tabulation of Stellar Coordinates', *Archive for History of Exact Sciences* 72 (2018), 547–63.

———, 'Before the End of an Error: Giovanni Bianchini's Original Flawed Treatise on the Conversion of Stellar Coordinates', *Archive for History of Exact Sciences* 75 (2021), 109–24.

———, *The Doctrine of Triangles: A History of Modern Trigonometry* (Princeton Princeton University Press, 2021).

———, and James Byrne, 'Maurolico, Rheticus, and the Birth of the Secant Function', *Journal for the History of Astronomy* 52 (2021), 189–211.

Viète, François, *Canon mathematicus seu ad triangula* (Paris: Jean Mettayer, 1579).

Zepeda, Henry, 'The Medieval Latin Transmission of the Menelaus Theorem', unpublished thesis, University of Oklahoma, 2013.

JOSÉ CHABÁS AND BERNARD R. GOLDSTEIN

Computing Solar and Lunar Velocities and Radii

The Table of John of Genoa and his Predecessors

In geocentric models, the Sun and the Moon progress at varying velocities, and determining their velocities at any time was a critical issue for medieval astronomers. In particular, these velocities were required for determining the times of true syzygies and eclipses, in which both luminaries are involved. This problem was addressed by compiling tables for solar and lunar velocities in much the same way as many other astronomical quantities were treated: fix a mean value and compute a variable correction. When compiling these tables, the mean value was taken to be the mean motion of the Sun and the Moon in longitude, and the correction was made depending on the solar equation in the case of the Sun and on the equation of anomaly in the case of the Moon.

Whereas the values for the mean motions of the Sun and the Moon had not changed substantially from the time of Ptolemy, there were some significant changes in the values for the equations of the luminaries and minor changes in the structure of the tables. For the Sun, the maximum value of the solar equation varied from 2;23° (Ptolemy) to 2;14° (al-Khwārizmī/Maslama),[1] 1;59,10° (al-Battānī and the Toledan Tables), and 2;10° (Alfonsine Tables), yielding different values of the solar velocity. For the Moon, the maximum value of the equation of anomaly also varied, although to a lesser extent, from 5;1° (Ptolemy) to 4;56° (al-Khwārizmī/Maslama) 5;1,0° (al-Battānī and the Toledan Tables), and again to 4;56° (Alfonsine Tables), also producing new values for the lunar velocity.

In this paper, we focus on the sets of values of the velocities computed by an early Alfonsine astronomer, John of Genoa, extant in his widely diffused table for the radii of the Sun and the Moon at syzygy and the solar and lunar velocities. We recomputed all entries in this table, as well as those by two of his distinguished predecessors, al-Khwārizmī

* Research presented in this paper by José Chabás was made in the framework of the ERC project ALFA: Shaping a European Scientific Scene, Alfonsine Astronomy, CoG 723085. The authors thank the two other editors of the series, Matthieu Husson and Richard L. Kremer, for their comments and suggestions on a previous version of this paper.

1 The zij of al-Khwārizmī only survives in the version made in al-Andalus by Maslama al-Majrīṭī (d. 1007), and here it is called the zij of al-Khwārizmī/Maslama: see Heinrich Suter, *Die astronomischen Tafeln des Muḥammad ibn Mūsā al-Khwārizmī* (Copenhagen: Andr. Fred. Høst & Søn, 1914).

José Chabás • Universitat Pompeu Fabra, Barcelona
Bernard R. Goldstein • University of Pittsburgh

Alfonsine Astronomy: Computational Practices, ed. by Matthieu HUSSON, José CHABÁS, and Richard L. KREMER, Turnhout, 2026 (*Alfonsine Astronomy: Studies and Sources*, 3), p. 247-274

10.1484/M.ALFA-EB.5.137913

(fl. 830) and al-Battānī (d. 929), and we claim that all three astronomers used the same basic algorithms, with different parameters, to compute the velocities of the Sun and the Moon (see Appendix 1).

John of Genoa was a physician and a member of the first generation of Alfonsine astronomers in the early fourteenth century; he was also the surgeon of Pope Clement VI in Avignon from 1342 to 1348. His known works are limited to astronomical matters, especially the computation of eclipses, and they have received little attention from modern scholars until recently.[2] He is the author of a table for determining the radii of the Sun, the Moon, and the shadow of the Earth during eclipses, as well as the velocities of the two luminaries. The table is associated with a short canon explaining its use and containing succinct instructions for drawing figures for solar and lunar eclipses. The table and canon were completed before January 2, 1332. This precise date is found in the explicit of another work by John of Genoa entitled 'Canon of eclipses', in which the table and the canon for the radii and velocities at syzygy are mentioned: *Deinde scias dyametrum solia et lune per tabulam quam feci de hoc secundum canonem quam super eam ordinavi* (Paris, Bibliothèque nationale de France [henceforth: BnF], MS lat. 7281, 207^{r}).

John of Genoa's short canon is entitled *Canon tabule ad sciendum motum solis et lune in una hora et semidiametros luminarium secundum tabulas Alfonstii*, and it begins *Verum motum solis et lune*. According to Miolo (2022), this canon is found in fifteen manuscripts. It provides scant information on the use of the table or on the way the entries were computed. The only information about this table is that which can be gleaned from the title. We are told, however, that the argument, given at intervals of six degrees, is the solar anomaly for the solar velocity and the solar radius and the lunar anomaly for the lunar velocity and the other radii. The canon also informs us that the equated values of the Earth's shadow are derived by subtracting a correction called the 'variation' of the Earth's shadow from the preliminary value for the Earth's shadow (see below).

John of Genoa's table is extant in at least twenty-five manuscripts and generated a few variants (some of uncertain authorship) that were also widely circulated (see Appendix 2). One copy of John of Genoa's table, preserved in Oxford, Bodleian Library, Digby 97, 130^{v}, has been used in this paper for comparison and recomputation (see Figure 8.1). It consists of seven columns: column 1 is for the argument and ranges from 0° to 180° and at intervals of six degrees; columns 2 to 5 are for the radii of the Sun, the Moon, the shadow of the Earth, and its 'variation', respectively, with entries in arcminutes, seconds, and thirds; and columns 6 and 7 display the solar and lunar velocities, respectively, in arcminutes, seconds, and thirds per hour: from 2;22,30′/h to 2;33,40′/h (Sun) and from 29;37,13′/h to 36;58,54′/h (Moon).

2 Laure Miolo, 'Retracing the Tradition of John of Genoa's *Opus astronomicum* Through Extant Manuscripts', in *Alfonsine Astronomy: The Written Record*, ed. by Richard L. Kremer, Matthieu Husson and José Chabás (Turnhout: Brepols, 2022), pp. 343–80; see also Bernard R. Goldstein 'Lunar Velocity in the Ptolemaic Tradition', in *The Investigation of Difficult Things. Essays on Newton and the History of the Exact Sciences*, ed. by Peter M. Harman and Alan E. Shapiro (Cambridge, 1992), pp. 3–17; and José Chabás and B. R. Goldstein, *The Alfonsine Tables of Toledo* (Dordrecht: Kluwer, 2003).

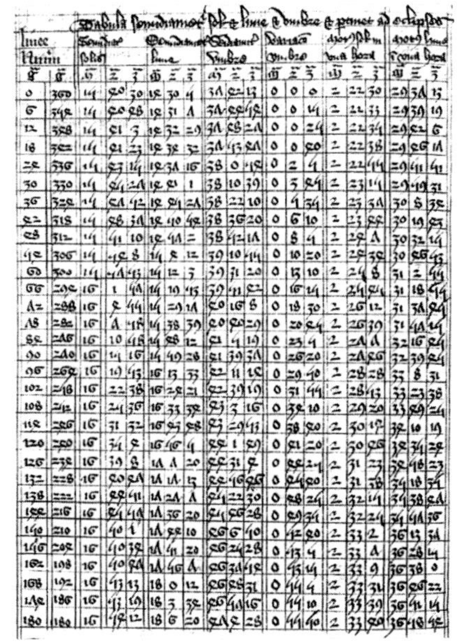

Figure 8.1. Oxford, Bodleian Library, Digby 97, 130v.

1. Column 2 for the solar radius

The argument for the solar radius is solar anomaly. The entries for the solar radius were computed from the solar velocity v_s in minutes per hour, by means of the following formula:

$$r_s = v_s\,(\,'\,/h) \cdot 11/100 = v_s \cdot 0;6,36. \qquad (1)$$

This expression agrees with the instructions provided in the Toledan Tables (Canon Cb 193) and by John of Lignères (*Priores*, Chapter 36)[3] for computing the solar diameter, where we are told to multiply the solar velocity by 2 and 1/5 and divide by 10:

$$r_s = v_s\,(\,'\,/h) \cdot ½\,(2 + 1/5)/10 = v_s\,(\,'\,/h)\,(11/100). \qquad (2)$$

In turn, for computing the diameter where the velocity is measured in degrees per day rather than in minutes per hour, expression (2) becomes

$$d_s = v_s\,(^\circ/d) \cdot 11/20 = v_s \cdot 0;33 \qquad (3)$$

for

$$d_s = v_s\,(^\circ/d)/(24/60) \cdot 11/20 = v_s\,(\,'\,/h) \cdot 0;13,12,$$

3 For the Toledan Tables, see Fritz S. Pedersen, *The Toledan Tables: A Review of the Manuscripts and the Textual Versions with an Edition* (Copenhagen: C.A. Reitzels Forlag, 2002), pp. 460–63, 1424; for the *Priores*, see Marie-Madeleine Saby, 'Les canons de Jean de Lignères sur les tables astronomiques de 1321: Édition critique, traduction et étude',unpublished thesis, Paris, École Nationale des Chartes, 1987, pp. 249–50.

or

$$r_s = v_s\,(\,'\,/h)\cdot 11/100 = v_s \cdot 0;6,36.$$

The source of Formula (3) is Brahmagupta's *The Khaṇḍakhādyaka*:[4] 'Multiply the apparent daily motion of the sun and the moon, respectively, by 11 and 10 and divide, respectively, by 20 and 247; the resulting minutes are the angular diameters of the sun and the moon'.[5]

It thus comes as no surprise that the entries in John of Genoa's table are very similar to those in the zij of al-Khwārizmī/Maslama, which were derived from Brahmagupta's formula[6] (see Table 1).

Solar Radius		
Argument	al-Khwārizmī /Maslama (′)	John of Genoa (′)
0°	15;40	15;40,30
30	15;44	15;45,27
60	15;58	15;57,53
90	16;17	16;15,16
120	16;36	16;35, 4
150	16;50	16;50, 1
180	16;54	16;54,12

Table 1. Comparison between entries for the solar radius in al-Khwārizmī/Maslama's zij and John of Genoa's table.

2. Column 3 for the lunar radius

The argument for the lunar radius is lunar anomaly. The entries for the lunar radius were computed from the lunar velocity v_m in arcminutes per hour, by means of the following formula:

$$r_m = v_m\,(\,'\,/h)\cdot 47/96 = v_m \cdot 0;29,22,30. \qquad (4)$$

This expression agrees with the instructions provided by al-Battānī, the Toledan Tables (Canon Cb 194), and John of Lignères' *Priores* (Chapter 36)[7] for computing the lunar

4 Prabodh Chandra Sengupta, *The Khaṇḍakhādyaka by Brahmagupta* (Calcutta, 1934), p. 32. For a modern explanation of these formulae for the angular diameters of the Sun and the Moon, see Bernard R. Goldstein, *Ibn al-Muthannā's Commentary on the Astronomical Tables of al-Khwārizmī. Two Hebrew Versions, Edited and Translated, with an Astronomical Commentary* (New Haven: Yale University Press, 1967), pp. 226–29.

5 Otto Neugebauer, *The Astronomical Tables of al-Khwārizmī* (Copenhagen: Munksgaard, 1962), pp. 105–6; and Fuad I. Haddad, Edward. S. Kennedy, and David Pingree, *The Book of the Reasons Behind Astronomical Tables by ʿAlī ibn Sulaymān al-Hāshimī* (Delmar, New York: Scholar Facsimiles & Reprints, 1981), pp. 316–20.

6 Suter, *Die astronomischen Tafeln*, pp. 175–80; Neugebauer, *The Astronomical Tables*, pp. 58–9.

7 For al-Battānī, see Carlo A. Nallino, *Al-Battānī sive Albatenii opus astronomicum*, 2 vols (Milan: Osservatorio di Brera, 1903–1907), I, p. 97: 21–23; for the Toledan Tables, see Pedersen, pp. 462–63; and for the *Priores*, see Saby, Les canons, p. 250.

diameter, where we are told to multiply the lunar velocity in minutes per hour, by six minus one eighth, and then divide by six:

$$r_m = v_m\,(\,'\,/h)\cdot \tfrac{1}{2}\,(6-1/8)/6 = v_m\,(\,'\,/h)\cdot 47/96. \qquad (5)$$

The analogous instruction in the Indian tradition is different.[8] In this case, the corresponding expression is

$$d = v_m\,(^\circ/d)\cdot 10/247 = v_m \cdot 0;2,25,45, \qquad (6)$$

or, equivalently,

$$r_m = v_m\,(\,'\,/h)\cdot (24\cdot 10/247\cdot 2) = v_m\,(\,'\,/h)\cdot 0;29,9, \qquad (6a)$$

in contrast to the coefficient in Formula (4), above.

Table 2 displays selected computations of the solar and lunar radii using Formulae (1) and (4).

Argument	Recomputed rs	John of Genoa	Recomputed rm	John of Genoa
	(′)	(′)	(′)	(′)
0	15;40,30	15;40,30	14;30, 6	14;30, 5
30	15;45,27	15;45,27	14;41, 1	14;41, 1
60	15;57,53	15;57,53	15;12, 3	15;12, 3
90	16;15,16	16;15,16	15;59,28	15;59,28
120	16;35, 4	16;35, 4	16;56, 5	16;56, 5
150	16;50, 1	16;50, 1	17;44,10	17;44,10
180	16;54,12	16;54,12	18; 6,20	18; 6,20

Table 2. Recomputation of a few entries of the solar (r_s) and lunar (r_m) radii.

The very good agreement between computation and text, which also applies to the other entries in the table, indicates (as above) that John of Genoa's entries were correctly computed.

3. Column 4 for the radius of the shadow at solar apogee

To determine the (equated) radius of the shadow, we are given two columns: one for the preliminary value of the radius of the shadow at solar apogee u_1 and another for its variation from solar apogee to perigee u_2. For the preliminary value of the radius of the shadow at solar apogee, the entries in the table range from 37;42,13′ to 47;4,28′, and the argument is lunar anomaly. The entries were computed using the following formula:

$$u_1 = r_m \cdot 13/5. \qquad (7)$$

8 See Goldstein, *Ibn al-Muthannā's Commentary*, pp. 104, 227.

The coefficient 13/5 already appears in *Almagest* V,14 as 2 ⅗, a value which is reproduced in the instructions given by al-Battānī, the Toledan Tables (Canon Cb 195), and John of Lignères's *Priores*\[9] We note that u_1 depends uniquely on the lunar velocity.

4. Column 5 for the 'variation of the shadow'

The entries in this column vary from 0;0,0′ to 0;55,50′ and the argument is the solar anomaly. The u_2 entries were computed by means of the following formula:

$$u_2 = \tfrac{1}{2}\,(v_S - v_O) \cdot 10, \qquad (8)$$

where v_O = 2;22,30′/h is the solar velocity at apogee and v_S is the solar velocity at a position between the solar apogee and perigee. This expression corresponds to the instructions given in the Toledan Tables (Canon Cb 197) and John of Lignères' *Priores* (Chapter 36) where, in both texts, v_O = 2;23′/h (rather than 2;22,30′/h).[10]

Table 3 displays selected recomputations of the radius of the shadow (u_1) and its variation (u_2) by means of Equations (7) and (8). We note that u_2 depends uniquely on the solar velocity.

Argument	Recomputed u1	John of Genoa	Recomputed u2	John of Genoa
	(′)	(′)	(′)	(′)
0	37;42,15	37;42,13	0; 0, 0	0; 0, 0
30	38;10,39	38;10,39	0; 3,45	0; 3,45
60	39;31,20	39;31,20	0;13,10	0;13,10
90	41;34,37	41;39,37*	0;26,20	0;26,20
120	44; 1,49	44; 1,49	0;41,20	0;41,20
150	46; 6,50	46; 6,50	0;52,40	0;52,40
180	47; 4,28	47; 4,28	0;55,50	0;55,50

* Read 41;34,37, with almost all manuscripts consulted

Table 3. Recomputation of a few entries of the uncorrected radius of the shadow (u_1) and its variation (u_2).

The total, or 'equated', radius of the shadow r_u for any given values of the arguments is found by subtracting the variation from the radius of the shadow at apogee:

$$r_u = u_1 - u_2; \qquad (9)$$

that is, the total radius of the shadow has two terms, one depending exclusively on the lunar velocity, with a maximum of about 47′, and the other exclusively on the solar velocity, with a maximum of about 1′. The variation u_2 does not seem to serve any useful

9 For the *Almagest*, see G. J. Toomer, *Ptolemy's Almagest* (London: Duckworth 1984), p. 254; for al-Battānī, see Nallino, al-Battani, I p. 97: 21–23; for the Toledan Tables, see F. S. Pedersen, pp. 462–63; and for the *Priores*, see Saby, Les canons, p. 250.

10 For the Toledan Tables, see F. S. Pedersen, pp. 462–63; for the *Priores*, see Saby, Les canons, p. 250.

purpose since the amount is too small to affect subsequent computations significantly in all cases.

For John of Genoa, the maximum value of the solar term 0;55,50 is five times the difference between solar velocities at perigee, 2;33,40′/h, and apogee, 2;22,30′/h. We have not a found an astronomically meaningful explanation for this factor of five, but it is possible that the astronomers compiling the Toledan Tables, later followed by John of Lignères and John of Genoa, were inspired by the instructions in al-Khwārizmī/Maslama's zij for computing the total radius of the shadow. To be sure, for al-Khwārizmī/Maslama, the total radius is also the difference between two terms, one depending on the lunar velocity and another depending on the solar velocity:[11]

$$\begin{aligned} r_u = s_2 - s_1 &= 8/5 \cdot v_m\,(°/h) - 5 \cdot v_s\,(°/h), \\ &= (1/5) \cdot [8\,v_m\,(°/h) - 25 \cdot v_s\,(°/h)], \\ &= (1/120) \cdot [8\,v_m\,(°/d) - 25 \cdot v_s\,(°/d)], \end{aligned} \tag{10}$$

where s_1 and s_2 are entries in the columns for the components of the radius of the shadow in al-Khwārizmī/Maslama's table.[12] In Formula (10), the factor of five applies to the solar velocity rather than to the difference between two solar velocities. Formula 10 is equivalent to

$$2 \cdot ru = (1/60) \cdot [8\,v_m\,(°/d) - 25\,v_s\,(°/d)], \tag{11}$$

which is the rule in the *Khaṇḍakhādyaka*.[13]

It follows from Formulae (9) and (10) that computations based on the table in al-Khwārizmī/Maslama's zij for the total radius differ substantially from those found following the instructions given in the Toledan Tables and slightly modified by John of Genoa. To support this claim, we display the minimum and maximum values for the total radius of the shadow computed by John of Genoa and those for the radius of the shadow derived from al-Khwārizmī/Maslama's zij:[14]

	John of Genoa	**al-Khwārizmī/Maslama**
min.	37;42,13–0;55,50 = 36;46,23′	48;19′–12;48′ = 35;31′
max.	47; 4,28–0; 0, 0 = 47; 4,28′	57; 4′–11;52′ = 45;12′

Table 4. Total radii of the shadow.

In sum, the entries for the radii of the Sun, the Moon, and the shadow of the Earth in the table by John of Genoa were computed using entries for the solar and lunar velocities found in the table itself. The values for the diameter of the shadow are previously unattested and need to be explained.

11 Neugebauer, *The Astronomical Tables*, p. 107.
12 Suter, *Die astronomischen Tafeln*, pp. 175–80.
13 Sengupta, *The Khaṇḍakhādyaka*, p. 83. For an explanation of Formula (11), see Goldstein, *Ibn al-Muthannâ*, pp. 227–30.
14 Neugebauer, *The Astronomical Tables*, p. 107; Suter, *Die astronomischen Tafeln*, pp. 175–80.

5. Column 6 for the solar velocity

The argument for the solar velocity is the solar anomaly. The entries in the table, in arcminutes per hour, were recomputed using the following formula:

$$v_s(\kappa) = 0;59,8,20\ (1 + \Delta c) \cdot 60/24, \qquad (12)$$

where 0;59,8,20°/d is the daily mean solar velocity, 60/24 = 2;30 is the coefficient to convert it into arcminutes per hour, and $\Delta c = c(\kappa + 1) - c(\kappa)$ is the increment of the equation of centre of the Sun, c, as a function of the solar anomaly κ. Formula (12) can also be presented as

$$v_s(\kappa) = (0;59,8,20 + 0;59,8,20 \cdot \Delta c) \cdot 60/24, \qquad (13)$$

which is reminiscent, but for the conversion factor, of the instructions given by Ptolemy in *Almagest* VI,4 for determining the velocity of the Moon at syzygy: add the mean lunar velocity in longitude (0;32,56°/h) to the product of the mean lunar velocity in anomaly (0;32,40°/h) times the difference of the two successive corresponding entries of the equation of anomaly, $\Delta c = c(\alpha + 1) - c(\alpha)$.[15] For the solar velocity, it would seem that John of Genoa adjusted the procedure set by Ptolemy for the simple lunar model and took both velocities in this case to be equal to the mean solar motion.

To recompute the entries, we used the solar equation in the Parisian Alfonsine Tables and the value of the Alfonsine mean solar velocity in arcminutes, seconds, and thirds of a day. The results are shown in Table 5, together with the entries displayed in the manuscript (see Figure 8.1).

The agreement in Table 5 is very good, indicating that the entries for the solar velocities were originally correctly computed, even though two outliers are found for arguments 126° and 180°. The first outlier, 2;31,23 (for argument 126°), does not seem to be a scribal error, for this precise value was used to compute the solar radius for argument 126° (see Equation 1) and 2;31,23 · 0;6,36 = 0;16,39,8′, which is exactly the entry displayed in the table. It is probably due to a Δc value that is too low that results from the difference between the entries of the equation of centre for arguments 126° (–1;47,40°) and 127° (–1;46,20°), as found in the *editio princeps* of the Parisian Alfonsine Tables. Close examination of the tabulated values of the Alfonsine solar equation around 126° shows that they were not properly computed or copied (see Table 6).

It is clear that Δc does not increase as smoothly as would be expected, indicating that the entries for the solar equation do not do so either. Perhaps John of Genoa had a better table for the solar equation, with smoother values. In any case, this table presents problems that are similar to those found elsewhere, and a systematic examination would be desirable. The second outlier in Table 5 corresponds to argument 180°. The entry in the text, 2;33,40, differs from the computation. The same entry is found in the various manuscripts consulted. This could be a computational error or an initial copyist confusing 0 and 6, because the value 2;33,46 is documented as the entry for 180° in a table for the same purpose -although with entries given at intervals of one degree — in later manuscripts (see Section 9, below).

15 Toomer, *Almagest*, p. 282; Goldstein, 'Lunar Velocity in the Ptolemaic Tradition', p. 5.

Argument	Text v_s ('/h)	Δc (°)	Recomputed v_s ('/h)	T – C (arcthirds)
0	2;22,30	−0; 2,10	2;22,30	0
6	2;22,33	−0; 2, 9	2;22,33	0
12	2;22,35	−0; 2, 8	2;22,35	0
18	2;22,38	−0; 2, 7	2;22,38	0
24	2;22,55	−0; 2, 0	2;22,55	0
30	2;23,15	−0; 1,52	2;23,15	0
36	2;23,37	−0; 1,43	2;23,37	0
42	2;23,44	−0; 1,40	2;23,44	0
48	2;24, 7	−0; 1,31	2;24, 7	0
54	2;24,34	−0; 1,20	2;24,34	0
60	2;25, 8	−0; 1, 7	2;25, 6	+2
66	2;25,45	−0; 0,51	2;25,45	0
72	2;26,12	−0; 0,40	2;26,12	0
78	2;26,39	−0; 0,29	2;26,39	0
84	2;27, 7	−0; 0,18	2;27, 7	0
90	2;27,46	−0; 0, 2	2;27,46	0
96	2;28,28	0; 0,15	2;28,28	0
102	2;28,53	0; 0,25	2;28,52	+1
108	2;29,20	0; 0,36	2;29,20	0
114	2;30,14	0; 0,58	2;30,14	0
120	2;30,46	0; 1,11	2;30,46	0
126	2;31,23	0; 1,20	2;31, 8	+15
132	2;31,38	0; 1,32	2;31,38	0
138	2;32,15	0; 1,47	2;32,14	+1
144	2;32,25	0; 1,51	2;32,24	+1
150	2;33, 2	0; 2, 6	2;33, 2	0
156	2;33, 7	0; 2, 8	2;33, 6	+1
162	2;33, 9	0; 2, 9	2;33, 9	0
168	2;33,31	0; 2,18	2;33,31	0
174	2;33,39	0; 2,21	2;33,38	+1
180	2;33,40	0; 2,24	2;33,46	−6

Table 5. Recomputation of the solar velocity.

Argument	Solar equation	Δc
(°)	(°)	(°)
121	−1;53.46	0; 1,11
122	−1;52.35	0; 1,11
123	−1;51,24	0; 1,12
124	−1;50,12	0; 1,13
125	−1;48,59	0; 1,19
126	−1;47,40	0; 1,20
127	−1;46,20	0; 1,27
128	−1;44,53	0; 1,27
129	−1:43,26	0; 1,29
130	−1;41,57	0; 1,30
131	−1;40,27	0; 1,30

Table 6. Entries of the Alfonsine solar equation and their successive differences.

6. Column 7 for the lunar velocity at Syzygy

The argument for the lunar velocity is the lunar anomaly. The entries in the table, in arcminutes per hour, were computed by means of the following equation:

$$v_m(\alpha) = 0;\,32{,}56{,}28 + 0;\,41{,}48 \cdot \Delta c, \qquad (14)$$

where α is the true lunar anomaly, 0;32,56,28°/h is the hourly lunar mean motion, 0;41,48°/h is the corrected hourly mean motion in anomaly at syzygy, and $\Delta c = c(\alpha + 1)$ — $c(\alpha)$. This formula takes into account Ptolemy's second lunar model.[16] A passage in John of Genoa's canons hints at the use of Ptolemy's second model for computing lunar velocities: 'One should understand here that what has been computed in this table for the motion of the Moon is that which affects the Moon due to the equation of centre'.[17] In the Parisian Alfonsine Tables for the lunar equation, the argument for the equation of centre is the double elongation, which affects computations with Ptolemy's second lunar model but not with Ptolemy's first or simple lunar model. Apparently, John of Genoa was aware that although the value for the equation of centre is zero when the centre is zero (i.e. at syzygy), there is still an effect on the lunar velocity. Ptolemy did not notice this effect, and it has continued to be ignored in recent times as well.[18] In terms of modern mathematics, when the value of a function is zero, in general its derivative is not zero, and the velocity depends on the derivative.[19] Since the entries in the Parisian Alfonsine Tables for the lunar equation are given only to arcseconds, the values in John of Genoa's column for arcthirds are unnecessary (see Table 7).

16 Goldstein, 'Lunar Velocity in the Ptolemaic Tradition'.

17 *Hic sciendum quod in tabula motus Lune est computatum illud quod contingit Lune propter equationem centri* (Paris, BnF, MS lat. 7282, 129v).

18 See, e.g. Toomer, *Almagest*, p. 282 n. 15.

19 For details, see Goldstein, 'Lunar Velocity in the Ptolemaic Tradition'; Olaf Pedersen, *A Survey of the Almagest* (Odense: University Press, 1974), p. 226.

Argument	Text v_m (′/h)	Δc (°)	Recomputed v_m (′/h)	T–C (arcthirds)
0	29;37,13	−0; 4,46	29;37,13	0
6	29;39,19	−0; 4,43	29;39,18	1
12	29;42, 6	−0; 4,39	29;42, 6	0
18	29;46,17	−0; 4,33	29;46,17	0
24	29;51,51	−0; 4,25	29;51,51	0
30	29;59,31	−0; 4,14	29;59,30	1
36	30; 8,34	−0; 4, 1	30; 8,34	0
42	30;19,43	−0; 3,45	30;19,43	0
48	30;32,15	−0; 3,27	30;32,15	0
54	30;46,53	−0; 3, 6	30;46,53	0
60	31; 2,55	−0; 2,43	31; 2,55	0
66	31;18,55	−0; 2,20	31;18,56	0
72	31;37,45	−0; 1,53	31;37,45	0
78	31;57,15	−0; 1,25	31;57,15	0
84	32;16,45	−0; 0,57	32;16,45	0
90	32;39,45	−0; 0,24	32;39,45	0
96	33; 8,31	0; 0,13	33; 5,31	*
102	33;23,38	0; 0,39	33;23,38	0
108	33;49,25	0; 1,16	33;49,25	0
114	34;10,19	0; 1,46	34;10,19	0
120	34;35,25	0; 2,22	34;35,24	1
126	34;58,23	0; 2,55	34;58,23	0
132	35;18,35	0; 3,24	35;18,35	0
138	35;38,47	0; 3,53	35;38,47	0
144	35;57,36	0; 4,20	35;57,36	0
150	36;13,37	0; 4;43	36;13,37	0
156	36;28,15	0; 5, 4	36;28,15	0
162	36;38, 0	0; 5,18	36;38, 0	0
168	36;46,22	0; 5,30	36;46,22	0
174	36;51,15	0; 5,37	36;51,15	0
180	36;58,54	0; 5,40	36;53,20	*

*96. All manuscripts consulted share the same reading: 33;8,31.
*180. All manuscripts consulted share the same reading: 36;58,54.

Table 7. Recomputation of John of Genoa's lunar velocity at syzygy.

The first explicit statement of Equation (14) appeared in Regiomontanus's *Epitome of the Almagest*, VI,4, published in 1496, although he invoked a slightly different parameter (0;41,49 instead of 0;41,48):

$$v_m(a) = 0;32{,}56 + 0;41{,}49 \cdot \Delta c\,. \tag{15}$$

The parameter 0;41,49, based on Ptolemy's second lunar model, represents the corrected hourly mean motion in anomaly at syzygy v_z:

$$v_z = 0;32,40 + 1;1 \cdot 0;9 = 0;41,49, \quad (16)$$

where 0;32,40°/h is the hourly mean motion in anomaly, 1;1° is the hourly mean motion in double elongation of the Moon from the Sun, and 0;9° is the difference between the entries for arguments 1° and 0° in the Parisian Alfonsine Tables' lunar correction table, column 3, 'equation of centre'. With more precise parameters,

$$v_z = 0;32,39,44,54 + 1;0,57,13 \cdot 0\ ;9 = 0;41,48,20 \approx 0;41,48,$$

which is the value underlying John of Genoa's table. Thus, John of Genoa was completely consistent in using the Parisian Alfonsine Tables for computing the lunar velocity at syzygy, as indeed was Regiomontanus (for more details, see Goldstein 1992, 'Lunar Velocity', pp. 8–9, 14).

The faulty entries for arguments 96° and 180° can be considered as markers of John of Genoa's velocity table.

As recently pointed out by Kremer, two early Alfonsine astronomers, John of Murs and Firmin of Beauval, used the lunar velocity in John of Genoa's table to compute their *Tabulae permanentes*.[20] Nicholaus de Heybech (fl. 1390s) also used John of Genoa's table for solar and lunar velocity in compiling his table for the time between mean and true syzygy; in particular, he used the faulty value 36;58,54′/h for argument 180° in John of Genoa's table.[21]

A contemporary of John of Genoa, Levi ben Gerson (Orange, d. 1344), compiled two tables for lunar velocity: one that has entries ranging from 30;18′/h to 36;4′/h, similar to those in the table of al-Battānī (see Appendix 1, Section 2.2, below), and another that has entries ranging from 29;35′/h to 36;56′/h, similar to those in the table of John of Genoa. Levi did not explain how either table was computed, but it seems that in his table based on the algorithm associated with Regomontanus and grounded in Ptolemy's second lunar model, he used al-Battānī's table for the lunar equation rather than his own table for this equation, which was based on his original lunar model.[22] Despite the similarities, it is clear that Levi and John of Genoa worked independently.

As shown above, John of Genoa computed the velocities of the two luminaries to thirds using Alfonsine mean motions and velocities. He thus went beyond his Parisian colleague John of Lignères both in precision and in consistency; in his Tables for 1322, John of Lignères included a table for the equations and the hourly velocities of the Sun and the Moon, where the equations are Alfonsine but the velocities are not; they were in fact borrowed from the Toledan Tables.[23]

---

20 Richard L. Kremer, 'Cracking the *Tabulae permanentes* of John of Murs and Firmin of Beauval with Exploratory Data Analysis', in *Editing and Analysing Numerical Tables: Towards a Digital Information System for the History of Astral Sciences*, ed. by Matthieu Husson, Clemency Montelle and Benno van Dalen (Turnhout: Brepols, 2021), pp. 363–422 (383-88).

21 José Chabás and Bernard R. Goldstein, 'Nicholaus de Heybech and His Table for Finding True Syzygy', *Historia Mathematica*, 19 (1992), 265-89 (pp. 274-75).

22 See Bernard R. Goldstein, *The Astronomical Tables of Levi ben Gerson* (New Haven: Connecticut Academy of Arts and Sciences: 1974), pp. 53–74, 111–13, 182; Goldstein, 'Lunar Velocity in the Ptolemaic Tradition', p. 10.

23 See José Chabás and Marie-Madeleine Saby, *The Tables of 1322 by John of Lignères* (Turnhout: Brepols 2022), pp. 108–13 (Table 19).

7. Variant tables

In several manuscripts, John of Genoa's seven-column table is presented in a variant form by an unknown compiler, and this is also at intervals of six degrees of the argument; the columns for the radii are excluded and new columns for the solar and lunar velocity in degrees per sixtieth of a day (mn) are included, that is, there are only columns for velocities (see Goldstein 1992, pp. 12–13: for an excerpt, see Table 8). The entries in the last two columns are related to the previous two columns by means of the following expressions:

$$v_s\,(\,'\,/mn) = v_s\,(\,'\,/h) \cdot 24/60$$

and

$$v_m\,(\,'\,/mn) = v_m\,(\,'\,/h) \cdot 24/60.$$

Argument	Solar v ('/hr)	Lunar v_m ('/h)	Solar v_s ('/mn)	Lunar v_m ('/ mn)
0	2;22,30	29;37,13	0;57, 0	11;50,53
…				
60	2;25, 8	31; 2,55	0;58, 3	12;25,10
…				
96	2;28,28	33; 8,31	0;59,23	13;14,13
…				
120	2;30,46	34;35,25	1; 0,18	13;50,10
…				
180	2;33,40	36;58,54	1; 1,28	14;47,33

Table 8. A variant of John of Genoa's table with velocities only (excerpt): Kraków, Biblioteka Jagiellońska, MS 552, 24v.

In the margin of this table in Paris, BnF, MS lat. 7283, 44v, and in Cracow, Biblioteka Jagiellońska, MS 551, 92r, we are told explicitly that it does not belong to the set by John of Lignères. The two markers in John of Genoa's lunar velocity table (faulty entries for arguments 96° and 180°: 33;8,31 and 36;58,54, respectively) are also found in this variant table. However, in the newly added column in ′/mn, the corresponding entry for argument 180° derives from John of Genoa's faulty value ($14;47,33 = 36;58,54 \cdot 24/60$), whereas the entry for 96° does not ($13;14,13 \neq 33;8,31 \cdot 24/60$). Moreover, it was derived from the 'correct' entry, 33;5,31, indicating that the unknown compiler of the newly added columns did not always derive them directly from John of Genoa's table. This variant table was integrated in the Canons to the Alfonsine Tables, written in 1365 by the Swabian astronomer Heinrich Selder, as in Erfurt, Universitätsbibliothek, CA 2° 37 and Philadelphia, University of Pennsylvania, LJS 174. Selder thus appears to be a potential candidate for assembling this table.

Another different variant table is extant in Paris, BnF, MS lat. 7282, 129r, with both the four columns for the radii and the four columns for the velocities, all at intervals of six degrees.

8. The velocity table of John of Montfort

Among the many tables for the velocity of the luminaries found in miscellaneous late medieval codices, one compiled by John of Montfort stands out. John of Montfort is only known for being the author of this table. It is extant in a number of manuscripts (see Appendix 2) and presents two main characteristics: the argument is displayed for each integer degree and the velocities are given to fourths in minutes of a day.

In two manuscripts (Paris, BnF, lat. 13014, 166v and Paris, BnF, lat. 14481, 96v), the title contains relevant information: *Tabula ad sciendum motum solis et lune in uno minuto diei, et in moto lune* ***includitur equatio centri*** *et est composita anno domini 1332 incompleto mense ianuarii per magistrum johannem de monte forti secundum equatione tabularum alfonsii.* We thus learn that the table was computed following Ptolemy's second lunar model, that is, taking into account the lunar equation of centre; this makes John of Montfort the second Alfonsine astronomer to do so. We are also told that the table was compiled in January 1332, the very same date assigned to that by John of Genoa, a coincidence that leaves room for speculation.

Upon close inspection of the table, it quickly becomes clear that the entries for both solar and lunar velocities at the beginning and end are flawed, because the columns do not show progress smoothly (see Table 9).

Argument	Solar v. v_s (′/mn)	Lunar v. v_m (′/mn)
1	0;57, 1,12	11;51, 9,11
2	0;57, 1,12	11;51,25,55
3	0;57, 1,12	11;51, 9,11
4	0;57, 2,11	11;51,25,55
5	0;57, 1,12	11;51,25,55
6	0;57, 1,12	11;51,42,38
7	0;57, 2,11	11;51,42,38
8	0;57, 1,12	11;51,42,38
9	0;57, 1,12	11;51,59,21
10	0;57, 2,11	11;52,16,55
…		
175	1; 1,28,17	14;44,30,25
176	1; 1,30,15	14;44,30,25
177	1; 1,30,15	14;44,47, 8
178	1; 1,30,15	14;44,47, 8
179	1; 1,30,15	14;44,47, 8
180	1; 1,30,15	14;44,47, 8

Table 9. Excerpts of the table of John of Montfort.

These are not copyist errors, because the same entries are found in all the manuscripts we have consulted. Thus, in particular, the extreme values are not reliable. The rest of the table displays a more homogeneous behaviour. Table 10 shows the comparison between John of Montfort's entries and recomputation, using Equation 12 for the solar velocity and Equation 14 for the lunar velocity, just as in the recomputation of John of Genoa's table (see Sections 5 and 6):

$$v_s(\kappa) = 0;59,8,20\ (1 + \Delta c) \qquad (12a)$$

and

$$v_m(a) = (0;32,56,28 + 0;41,48 \cdot \Delta c)\ 24. \qquad (14a)$$

Argument	Solar v. v_s (′/mn)	Recomputed v_s (′/mn)	Lunar v. v_m (′/mn)	Recomputed v_m (′/mn)
1	0;57, 1,12	0;57, 1,11	11;51, 9,11	11;51,10, 0
...				
15	0;57, 2,11	0;57, 2,10	11;53;22,58	11;53,23,45
...				
30	0;57,17,57	0;57,17,56	11;59,47,35	11;59,48,19
...				
45	0;57,31,45	0;57,31,44	12;10,39,45	12;10,40,24
...				
60	0;58, 3,17	0;58, 2,18	12;25, 9,18	12;25, 9,50
...				
75	0;58,31;52	0;58,31;52	12;42,42,48	12;42,43,12
...				
90	0;59, 5,23	0;59, 6,22	13; 3,53,41	13; 3,53,55
...				
96	0;59,23, 7	0;59,23, 7	13,14,12,25	13;14,12,33
...				
105	0;59,35,56	0;59,35,56	13;27,18,21	13;27,18,24
...				
120	1; 0,18,18	1; 0,18,19	13;50, 9,35	13;50, 9,26
...				
135	1; 0,40,58	1; 0,40,59	14;11,53,55	14;11,53,56
...				
150	1; 1,12,31	1; 1,12,31	14;30, 0,51	14;29,26,57
...				
165	1; 1,21,22	1; 1,21,24	14;41, 9,45	14;41, 9,12
...				
180	1; 1,30,15	1; 1,30,16	14;44,47, 8	14;45,20, 0

Table 10. Entries and recomputation of John of Montfort's table (excerpt): Paris, BnF, lat. 14481, 96v.

Comparing the entries given by John of Montfort and recomputation generally shows good agreement (see also Goldstein 1992, pp. 12–13). The entries in John of Montfort's table do not always agree with those given by John of Genoa, when converted to the same units. In particular, we note that the entry for argument 96° of the lunar velocity, 13,14,12,25′/mn, was computed from the correctly recomputed value, 33;5,31′/h, rather than from

John of Genoa's faulty entry, 33;8,31′/h. It would therefore seem that John of Montfort did not derive his table for velocities from John of Genoa's, but he used the same algorithms, expanded John of Genoa's table, and changed the units to arcminutes per minute of a day, thus providing a higher level of precision.

The use of John of Montfort's velocity table is documented in a text consisting exclusively of computed examples for 1356–61 by Hermann of Saxony.[24]

9. An extension of John of Genoa's table by John of Gmunden

About a hundred years after John of Genoa, another astronomer, John of Gmunden (*c.* 1384–1442), compiled new tables for the radii of the Sun and the Moon at syzygy and for the velocities of the two luminaries at syzygy. John of Gmunden lectured at the University of Vienna and wrote several mathematical treatises and texts on astronomical instruments.[25] He also produced three extensive sets of astronomical tables between 1430 and 1440.[26] His largest set is called *Tabulae maiores*, and among these are two extended tables, one for the radii and another for the hourly velocities of the luminaries.[27]

Although separated into two pieces, their various columns correspond very precisely to those in the table computed by John of Genoa in 1332. However, the entries in John of Gmunden's table are displayed for all integer degrees, whereas those by John of Genoa are only given at intervals of six degrees. The units of the entries in both cases are the same, and the corresponding entries generally agree, but not always. A list of manuscripts containing John of Gmunden's tables for the radii and velocities is available in Appendix 2. Tables 11 and 12 show excerpts of John of Gmunden's two tables taken from a manuscript copied in 1433 in Vienna by the author himself: Vienna, Österreichische Nationalbibliothek, MS 5268, $27^{r\text{-}v}$, a manuscript containing marginal notes in the hand of Regiomontanus (d. 1476). Comparison of the corresponding entries of the two table-makers show that John of Gmunden did not simply enlarge the number of entries by interpolating between the common entries, but that he recomputed the table and corrected a few faulty values in John of Genoa's table. For example, the lunar velocity for argument 180° is the faulty entry 36;58,54′/h in the table of John of Genoa (see Table 7, above), whereas the value in the table copied by Gmunden is 36;53,21′/h, which is very close to the recomputed value (36;53,20′/h). Analogously, for argument of 96°, John of Gmunden's table has 33;5,31′/h (John of Genoa: 33;8,31′/h; recomputation: 33;5,31′/h). The rest of the entries agree or only differ by a few thirds.

24 José Chabás, 'Computational Practices in Alfonsine Astronomy: Hermann of Saxony's *practica motuum planetarum*', in this volume.

25 Rudolf Simek and Kathrin Chlench, *Johannes von Gmunden (c. 1384–1442), Astronom und Mathematiker* (Vienna: Fassbaender, 2006).

26 Beatriz Porres, 'Les Tables astronomiques de Jean de Gmunden: Édition et étude comparative', unpublished PhD thesis, Paris, École Pratique des Hautes Études, 2003; see also José Chabás, *Computational Astronomy in the Middle Ages* (Madrid: Consejo Superior de Investigaciones Científicas, 2019), pp. 321–36.

27 For complete transcriptions, see Porres, *Les tables astronomiques*, pp. 329–37, Tables 50 and 51.

Argument	Solar v. (′/h)	Lunar v. (′/h)
0	2;22,30	29;37,11
6	2;22,34	29;39,17
12	2;22,36	29;42, 4
18	2;22,39	29;46,15
24	2;22,47	29;51,49
30	2;23,15	29;59,29
36	2;23,37	30; 8,32
42	2;23,44	30;19,41
48	2;24, 7	30;32,14
54	2;24,34	30;46,52
60	2;25, 8	31; 2,53
66	2;25,45	31;18,55
72	2;26,12	31;37,44
78	2;26,39	31;57,14
84	2;27, 6	32;16,45
90	2;27,46	32;39,44
96	2;28,28	33; 5,31
102	2;28,52	33;23,38
108	2;29,20	33;49,25
114	2;30,14	34;10,19
120	2;30,46	34;35,24
126	2;31,23	34;58,24
132	2;31,38	35;18,36
138	2;32,14	35;38,48
144	2;32,24	35;57,37
150	2;33, 1	36;13,39
156	2;33, 6	36;28,16
162	2;33, 9	36;38, 2
168	2;33,31	36;46,23
174	2;33,38	36;51,16
180	2;33,46	36;53,21

Table 11. John of Gmunden: Velocities of the luminaries at 1° intervals in Vienna, Österreichische Nationalbibliothek, MS 5268, 27ʳ (excerpt).

Interestingly, a variant of the table for velocities with the entries only given to seconds has the 'correct' readings for arguments 96° and 180° (33;5 and 36;53), rounded to seconds.[28] It is preserved in Oxford, Bodleian Library, MS Can. Misc. 499, a manuscript copied in Prague *c.* 1470. The compiler is unknown, although the table is erroneously ascribed to John of Lignères by a later hand in the margin of the table, which has generated some

28 See Goldstein, 'Lunar Velocity in the Ptolemaic Tradition', pp. 12–13; see also Kremer 2021, 'Cracking the *Tabulae permanentes*', pp. 383–84, n. 50.

confusion.[29] Aside from the odd few (about one out of ten), the rest of the 180 values were rounded correctly from John of Gmunden's table. We thus conclude that the table for hourly velocities in this manuscript copied in Prague is most likely derived from John of Gmunden's table.

Argument	Solar radius (′)	Lunar radius (′)	Radius of the shadow (′)	Variation of the shadow (′)
0	15;40	14;30	37;42	0; 0
1	15;41	14;30	37;43	0; 0
…				
6	15;41	14;31	37;45	0; 0
…				
90	16;15	15;59	41;35	0;26
…				
174	16;55	18; 3	46;55	0;56
…				
179	16;55	18; 4	46;57	0;56
180	16;55	18; 4	46;57	0;56

Table 12. John of Gmunden: Radii of the luminaries at one-degree intervals in Vienna, Österreichische Nationalbibliothek, MS 5268, 27v (excerpt).

In his *Tabulae eclypsium*, first printed in Vienna in 1514, Georg Peurbach (1423–1461) used Gmunden's table for the radii of the luminaries with entries to seconds, although restricted to values of the argument which are multiples of 5°. In addition to the printed version, Peurbach's work on eclipses is preserved in several manuscripts, including Nuremberg, Stadtbibliothek, Cent. V, 57 in the hand of Regiomontanus.[30] There is no table for the radii of the luminaries in the *editio princeps* of the Parisian Alfonsine Tables. However, the second edition (1492) filled the gap and included one that is also based on Gmunden's table, with the argument given at steps of 6° and the entries to seconds.[31]

John of Gmunden recomputed and enlarged John of Genoa's table using the same algorithm for calculating the underlying solar and lunar velocities that are not explicitly given in that table. In particular, for the lunar velocity at syzygy he used Ptolemy's second lunar model, following John of Genoa, who first introduced it in Alfonsine astronomy. As far as we can tell, the various canons to the set of tables by John of Gmunden say nothing about it. The algorithm underlying the table by John of Gmunden for the lunar velocity remained embedded in the table, and was first explicitly formulated by Regiomontanus after the death of John of Gmunden.[32]

29 We note that Kremer, in 'Cracking the *Tabulae permanentes*', p. 384, correctly found this attribution dubious.
30 See Chabás, *Computational Astronomy*, 365–75.
31 For a facsimile reproduction, see Chabás, *Computational Astronomy*, 270.
32 For further developments after John of Gmunden, see Bernard R. Goldstein and José Chabás, "Tables for the Radii of the Sun, the Moon, and the Shadow from John of Gmunden to Longomontanus", *Archive for History of Exact Sciences*, 78 (2024), 67–86.

Final remarks

John of Genoa compiled the earliest table for Alfonsine velocities of the luminaries and, in particular, he was the first to compute 2;22,30′/h as the minimum value of the solar velocity and other analogous values. In contrast to many of his colleagues, he did not merely compile a table to facilitate the task of the computer by just modifying the presentation or adding more entries to a previous table. In fact, he built on the work of his predecessors and in 1332 adapted the radii and the velocities at syzygy to the new parameters in the Alfonsine Tables that had reached Paris a few years earlier.

The table by John of Genoa presented in the same way as that of al-Khwārizmī/Maslama; although the columns are presented in a different order, the units of the various quantities are the same. The radii were partially computed following the instructions in the canons to the zij of al-Khwārizmī/Maslama that are later found in the Toledan Tables and in the *Priores* by John of Lignères. The velocities were computed following the rules outlined by Ptolemy in the *Almagest* or by means of an algorithm underlying the tables by al-Khwārizmī/Maslama and al-Battānī for the same purpose. In doing so, John of Genoa increased the precision compared to previous velocity tables.

John of Genoa was fully consistent in depending on the Alfonsine Tables, for he used Alfonsine parameters both for the mean motions and the equations of the Sun and the Moon. For the computation of lunar velocity, John of Genoa was also the first Alfonsine table-maker to use Ptolemy's second lunar model. His new table was used by other astronomers such as John of Murs, Firmin of Beauval, Nicholaus de Heybech, and John of Gmunden, but it did not find its way into the Parisian Alfonsine Tables as presented in the first edition of 1483.

John of Genoa and some other astronomers after him understood that Ptolemy's second lunar model affects lunar velocity at syzygy, even though the corresponding equation is zero. This effect was pointed out by O. Pedersen using modern mathematical tools such as partial derivatives.[33] O. Pedersen was unaware of any predecessor, medieval or modern, who called attention to this effect. However, we have demonstrated that it was already known in the late Middle Ages, judging by the tables of John of Genoa, John of Montfort, John of Gmunden, and the algorithm presented by Regiomontanus. In the absence of a synthetic mathematical language expressed by means of symbols, medieval astronomers used common language with an increasing variety of technical terms and, above all, they used arrays of numbers to convey complicated relations between mathematical quantities. Still, in some cases, such as the lunar velocity at syzygy, they did not have the tools to explain the reasoning behind their computations.

Our study of John of Genoa's table once again shows that close examination of medieval astronomical tables has proven to be a useful and efficient approach to understanding computational practices at the time and the transmission of astronomical knowledge.

33 O. Pedersen, *A Survey*, p. 226.

Appendix 1: Solar and lunar Velocities in the zijes of al-Khwārizmī/Maslama and al-Battānī

For the Sun, the velocity at any given time depends on the mean motion and the solar equation. In the case of the Moon, the velocity depends on its mean motions in longitude and anomaly as well as on the equation of anomaly. The algorithms used by John of Genoa were already used by previous astronomers — al-Khwārizmī and al-Battānī — as the sum of two terms: a constant term (mean motion) and a variable term as a function of the solar or lunar equation.

1. *Velocities of the sun and the moon in al-Khwārizmī/Maslama's zij*

1.1. The solar equation and velocity are given in steps of one degree. The maximum value of the solar equation is 2;14,0° for argument 90°, and the mean velocity is 0;59,8,10,21°/d = 2;27,50,25′/h (≈ 2;27,50′/h).[34] We have recomputed the entries in the table by means of the following formula:

$$v_s(\kappa) = 2;27,50 + 2;27,50 \cdot \Delta c, \qquad \text{(A1)}$$

where the mean velocity appears twice and $\Delta c = c(\kappa + 1) - c(\kappa)$.

Argument	Text v_s (′/h)	Δc (°)	Recomputed v_s (′/h)	T–C (arcsec)
1	2;22	−0; 2,17	2;22,12	0
10	2;22	−0; 2,15	2;22,17	0
20	2;23	−0; 2, 9	2;22,32	0
30	2;23	−0; 2, 0	2;22,54	0
40	2;24	−0; 1,48	2;23,24	+1
50	2;24	−0; 1,31	2;24, 6	0
60	2;25	−0; 1,11	2;24,55	0
70	2;26	−0; 0,50	2;25,47	0
80	2;27	−0; 0,31	2;26,34	0
90	2;28	0; 0, 1	2;27,52	0
100	2;29	0; 0,20	2;28,39	0
110	2;30	0; 0,51	2;29,56	0
120	2;30	0; 1,14	2;30,52	−1
130	2;32	0; 1,35	2;31,44	0
140	2;32	0; 1,48	2;32,16	0
150	2;33	0; 2, 1	2;32,48	0
160	2;33	0; 2, 9	2;33, 8	0
170	2;34	0; 2,15	2;33,23	+1
180	2;34	0; 2,17	2;33,28	+1

Table A1. Recomputation of the solar velocity in al-Khwārizmī/Maslama's zij.

34 See Suter, *Die astronomischen Tafeln*, pp. 175–80; José Chabás and Bernard R. Goldstein, *A Survey of European Astronomical Tables* (Leiden Brill, 2012), p. 57.

Formula (A1) yields very good agreement with the entries in the table.

1.2. The lunar equations and velocity are given in steps of one degree. The maximum value of the equation of anomaly is 4;56,0° for argument 90°. The mean motion in longitude is 13;10,34,52,46°/d = 32;56,27′/h (≈ 0;32,56°/h), and the mean motion in anomaly is 13;3,53,58,49°/d = 32;39,45′/h (≈ 0;32,40°/h).[35] We have recomputed the entries in the table using the following formula:

$$v_m(a) = 0;32{,}56 + 0;32{,}40 \cdot \Delta c, \qquad \text{(A2)}$$

where $\Delta c = c(a+1) - c(a)$. It is worth noting that al-Khwārizmī/Maslama used Ptolemy's simple model.

Argument	Text v_m (′/h)	Δc (°)	Recomputed v_m (′/h)	T–C (arcsec)
1	30;12	−0; 5, 1	30;12, 7	0
10	30;14	−0; 4,57	30;14,18	0
20	30;22	−0; 4,43	30;21,55	0
30	30;34	−0; 4,20	30;34,27	0
40	30;49	−0; 3,57	30;46,58	+2
50	31; 8	−0; 3,20	31; 7, 7	+1
60	31;29	−0; 2,39	31;29,26	0
70	31;54	−0; 1,49	31;55,20	−1
80	32;31	−0; 0,44	32;32, 3	−1
90	32;56	0; 0, 2	32;57, 5	−1
100	33;23	0; 0,56	33;26,29	−3
110	33;54	0; 1,54	33;58, 4	−4
120	34;22	0; 2,42	34;24,12	−2
130	34;48	0; 3,27	34;48,42	−1
140	35; 6	0; 4, 0	35; 6,40	−1
150	35;20	0; 4,28	35;21,54	−2
160	35;31	0; 4,46	35;31,43	−1
170	35;38	0; 4,58	35;38,14	0
180	35;40	0; 5, 1	35;39,53	0

Table A2. Recomputation of the lunar velocity in al-Khwārizmī/Maslama's zij.

In this case, the agreement with the entries in the table is not as good as in Table A1, but we consider it acceptable.

2. *Velocities of the sun and the moon in al-Battānī's zij*

2.1. The solar equation and velocity are given in steps of six degrees. The maximum value of the solar equation is 1;59,10° for argument 92°, and the mean motion is 0;59,8,20,47°/d = 2;27,50,52′/h (≈ 2;27,51°/h).[36] We have recomputed the entries in the table with the following formula:

35 See Suter, *Die astronomischen Tafeln*, pp. 175–80; Chabás and Goldstein, *A Survey*, p. 58.

36 Nallino, *Al-Battānī*, 2:88; Chabás and Goldstein, *A Survey*, p. 57.

Argument	Text v_m (′/h)	Δc (°)	Recomputed v_m (′/h)	T–C (arcsec)
1	2;23	−0; 2, 1	2;22,53	0
6	2;23	−0; 1,59	2;22,58	0
12	2;23	−0; 1,58	2;23, 0	0
18	2;23	−0; 1,55	2;23, 8	0
24	2;23	−0; 1,51	2;23,17	0
30	2;24	−0; 1,46	2;23,30	+1
36	2;24	−0; 1,39	2;23,47	0
42	2;24	−0; 1,31	2;24, 7	0
48	2;25	−0; 1,23	2;24,26	0
54	2;25	−0; 1,13	2;24,51	0
60	2;25	−0; 1, 4	2;24,51	0
66	2;26	−0; 0,53	2;25,40	0
72	2;26	−0; 0,40	2;26,12	0
78	2;27	−0; 0,30	2;26,37	0
84	2;27	−0; 0,16	2;27,12	0
90	2;28	−0; 0, 5	2;27,39	0
96	2;28	0; 0,10	2;28,16	0
102	2;29	0; 0,21	2;28,43	0
108	2;29	0; 0,36	2;29,20	0
114	2;29	0; 0,47	2;29,47	−1
120	2;30	0; 1, 3	2;30,26	0
126	2;30	0; 1,12	2;30,48	−1
132	2;31	0; 1,26	2;31,23	0
138	2;31	0; 1,35	2;31,45	−1
144	2;32	0; 1,44	2;32, 7	0
150	2;32	0; 1,55	2;32,34	−1
156	2;33	0; 1,58	2;32,42	0
162	2;33	0; 2, 1	2;32,49	0
168	2;33	0; 2, 7	2;33, 4	0
174	2;33	0; 2, 8	2;33, 7	0
180	2;33	0; 2, 9	2;33, 9	0

Table A3. Recomputation of the solar velocity in al-Battānī's zij.

Argument	Text v_m (′/h)	Δc (°)	Recomputed v_m (′/h)	T–C (arcsec)
1	30;18	−0; 4,50	30;18,36	−1
6	30;19	−0; 4,49	30;19, 1	0
12	30;21	−0; 4,45	30;21,19	0
18	30;24	−0; 5,38	31;16,18	+68 [a]
24	30;28	−0; 4,31	30;28,56	−1
30	30;35	−0; 4,18	30;36, 1	−1
36	30;43	−0; 4,25	30;32,12	+11 [b]
42	30;51	−0; 3,49	30;51,48	−1
48	31; 1	−0; 3,30	31; 2, 9	−1
54	31;12	−0; 3,10	31;13, 2	−1
60	31;25	−0; 2,27	31;36,27	−11 [c]
66	31;38	−0; 2,21	31;39,43	−2
72	31;53	−0; 0,55	32;26,22	−33 [d]
78	32; 8	−0; 1,26	32; 9,39	−2
84	32;24	−0; 0,54	32;27, 4	−3
90	32;41	−0; 0,24	32;43,24	−2
96	32;59	0; 0, 3	32;58, 6	+1
102	33;17	0; 0,44	33;20,25	−3
108	33;36	0; 1,20	33;40, 1	−4
114	33;55	0; 1,42	33;52, 0	+3
120	34;14	0; 2,24	34;14,51	−1
126	34;32	0; 3, 0	34;34,27	−2
132	34;49	0; 3,28	34;49,42	−1
138	35; 4	0; 3,54	35; 3,51	0
144	35;18	0; 4,23	35;19,38	−2
150	35;31	0; 4,47	35;32,42	−2
156	35;43	0; 5, 7	35;43,35	−1
162	35;52	0; 5,25	35;53,23	−1
168	35;58	0; 5,35	35;58,50	−1
174	36; 2	0; 5,44	36; 3,44	−2
180	36; 4	0; 5,45	36; 4,17	0

Table A4. Recomputation of the lunar velocity in al-Battānī's zij.

a. In Paris, MS Arsenal 8322 and the Toledan Tables, the entries for arguments 18° and 19° in the table for the lunar equation are, respectively, 1;25,47° and 1;30,25° (Nallino: 1;31,25°). The difference is −0;4,38°, and the recomputed value for the velocity is then 30;25,8; hence, T–C = 0.
b. In Paris, MS Arsenal 8322 and the Toledan Tables, the entries for arguments 36° and 37° in the table for the lunar equation are, respectively, 2;44,57° (Nallino: 2;45,17°!) and 2;49,2° (Nallino: 2;49,42°!). The difference is −0;4,5°, and the recomputed value for the velocity is then 30;43,5; hence, T–C = 0.
c. In Paris, MS Arsenal 8322 and the Toledan Tables, the entries for arguments 60° and 61° in the table for the lunar equation are, respectively, 4;9,6° and 4;11,53° (Nallino: 4;11;33°). The difference is −0;2,47°, and the recomputed value for the velocity is then 31;25,33; hence, T–C = −1.
d. In the Toledan Tables, the entries for arguments 72° and 73° in the table for the lunar equation are, respectively, 4;37,57° and 4;39,52° (Nallino: 4;38,52°). The difference is −0;1,55°, and the recomputed value for the velocity is then 31;53,52°; hence, T–C = −1. In the case of the Arsenal manuscript, the corresponding entries are both off by about two minutes.

$$v_s(\kappa) = 2;27,51 + 2;27,51 \cdot \Delta c, \qquad \text{(A3)}$$

which parallels al-Khwārizmī/Maslama's Formula (A1).

Again, Formula (A3) yields very good agreement with the entries in the table.

2.2. The lunar velocity is given in steps of six degrees and the equations in steps of one degree. The maximum value of the lunar equation of anomaly is 5;1,0° for argument 95°. The lunar mean motion in longitude is 13;10,35,2,7°/d = 32;56,28′/h, and the lunar mean motion in anomaly is 13;3,53,56,18°/d = 32;39,45′/h.[37] We have recomputed the entries in the table by means of the following formula:

$$v_m(a) = 0;32,56,28 + 0;32,39,45 \cdot \Delta c, \qquad \text{(A4)}$$

where $\Delta c = c(a+1) - c(a)$, using Nallino's transcription of the lunar equation of anomaly for determining Δc. The erroneous values in it produce four outliers, but these discrepancies disappear when using the entries for the lunar equation in anomaly in the Castilian version of al-Battānī's table (produced at the court of Alfonso X of Castile) in Paris, Bibliothèque de l'Arsenal, MS 8322, 52^r–53^v, or those in the Toledan Tables.[38]

It is clear that Nallino's edition (based on the only extant Arabic manuscript) of the table for the lunar equations in al-Battānī's zij is less reliable than F. S. Pedersen's edition of the corresponding entries in the Toledan Tables.[39] We consider the results obtained for the lunar velocity using the Toledan Tables acceptable.

37 Nallino, *Al-Battānī*, 2:88; Chabás and Goldstein, *A Survey*, p. 58.

38 Pedersen, *Toledan Tables*, pp. 1253–58.

39 For other examples of problematic entries in the tables transcribed by Nallino, see Benno van Dalen and Fritz S. Pedersen, 'Re-Editing the Tables in the *Ṣābi' Zīj* by al-Battānī', *Acta Historica Leopoldina*, 54 (2008), pp. 405–28.

Appendix 2: Manuscript sources[40]

The lists below are intended to shed some light on the variety of similar tables for the radii and the velocities of the luminaries at syzygy, all of them derived directly or indirectly from the table compiled by John of Genoa in 1332.

Manuscripts containing John of Genoa's table and/or the associated canon:

Basel, Universitätsbibliothek, F II 7, 81^{v}
Berlin, Staatsbibliothek – Stiftung Preußischer Kulturbesitz, MS lat. F. 246, 141^{r}
Brussels, Bibliothêque Royale, 926–40, 241^{v}
Brussels, Bibliothêque Royale, 10117–26, 49^{r}
Cambridge, University Library, Ee.III.61, 154^{r} (table and canon)
Cologne, Historicher Archiv der Stadt, W* 178, 18^{r}
Cracow, Biblioteka Jagiellońska, 459, 2^{r} (canon), 2^{v} (table)
Cracow, Biblioteka Jagiellońska, 551, 74^{r}
Cracow, Biblioteka Jagiellońska, 552, 32^{v}
Cracow, Biblioteka Jagiellońska, 613, 33^{r-v} (canon)
Douai, Bibliothèque municipale, 715, 35^{r-v} (canon), 36^{r} (table)
Florence, Biblioteca Nazionale Centrale, San Marco 192 (conv. sopp. J. V. 4) , 58^{v} (table and canon)
Florence, Biblioteca Medicea Laurenziana, San Marco 193, 65^{v} (table and canon)
Freiburg, Universitätsbibliothek, Hs. 28, 104^{r}
Leipzig, Universitätsbibliothek, 1477, 23^{r}
Madrid, Biblioteca Nacional, 4238, 60^{r}
Oxford, Bodleian Library, Can. Misc. 499, $151^{v}/141^{v}$
Oxford, Bodleian Library, Digby 97, 129^{v}–130^{r} (canon), 130^{v} (table)
Oxford, Hertford College, 4, 148^{r} 147^{r-v} (canon), 148^{r} (table)
Paris, Bibliothèque nationale de France, lat. 7281, 208^{r} (canon)
Paris, Bibliothèque nationale de France, lat. 7282, 129^{r} (table), 129^{v} (canon)
Paris, Bibliothèque nationale de France, lat. 7283, 45^{r} (table and canon)
Paris, Bibliothèque nationale de France, lat. 7284, 55^{r}
Paris, Bibliothèque nationale de France, lat. 7286C, 56^{v}
Paris, Bibliothèque nationale de France, lat. 7322, 41^{v} (canon)
Princeton, University Library, Kane 51, 47^{r}
Vatican, Biblioteca Apostolica Vaticana, Ottob. lat. 1826, 148^{r} (canon), 148^{v} (table)
Vatican, Biblioteca Apostolica Vaticana, Pal. lat. 1374, 47^{r}
Vatican, Biblioteca Apostolica Vaticana, Reg. lat. 1241, 152^{v} (table), 153^{r} (canon)
Wolfenbüttel, Herzog August Bibliothek, 2401 (Cod. Guelf. 36.21 Aug. fol), 311^{r}
Wolfenbüttel, Herzog August Bibliothek, 2637 (Cod. Guelf. 65 Aug. fol), 166^{v} (table)

40 A most useful tool for establishing these lists has been the ALFA database, which constitutes a survey of manuscripts and editions in Alfonsine astronomy, a collective endeavour that is still in progress.

Manuscripts containing variants of John of Genoa's table with velocities only:

Basel, Universitätsbibliothek, F II 7, 61^{r}
Brussels, Bibliothèque Royale, 926-40, 240^{v}
Brussels, Bibliothèque Royale, 10117-26, 49^{v}
Cracow, Biblioteka Jagiellońska, 551, 92^{r}
Cracow, Biblioteka Jagiellońska, 552, 24^{v}
Cracow, Biblioteka Jagiellońska, 553, 104^{v}
Cracow, Biblioteka Jagiellońska, 563, 93^{r}
Cracow, Biblioteka Jagiellońska, 610, 330^{v}
Erfurt, Universitätsbibliothek, CA 2° 37, 74^{v}
Munich, Bayerische Staatsbibliothek, Clm 14783, 105^{r}
Paris, Bibliothèque nationale de France, lat. 7282, 129^{r}
Paris, Bibliothèque nationale de France, lat. 7283, 44^{v}
Paris, Bibliothèque nationale de France, lat. 7295A, 137
Paris, Bibliothèque nationale de France, lat. 7432, 255^{r}
Philadelphia, University of Pennsylvania, LJS 174, 117^{v}
Prague, Národní knihovna ČR, XIII C 17, 169^{r}
Vatican, Biblioteca Apostolica Vaticana, Pal. lat. 446, 93^{r}
Vatican, Biblioteca Apostolica Vaticana, Pal. lat. 1373, 103^{v}
Vatican, Biblioteca Apostolica Vaticana, Pal. lat. 1374, 51^{r}
Vatican, Biblioteca Apostolica Vaticana, Pal. lat. 1376, 87^{v}
Vatican, Biblioteca Apostolica Vaticana, Pal. lat. 1413, 37^{r}

Manuscripts containing John of Montfort's table:

Budapest, Hungarian National Library (NSZL), 62, 45^{r}–46^{r}
Catania, Biblioteca Regionale *Universitaria*, U. 011, 199^{v}–200^{r}
Cracow, Biblioteka Jagiellońska, 551, 90^{v}–91^{v}
Cracow, Biblioteka Jagiellońska, 564, 136–38 (only to thirds)
Erfurt, Universitätsbibliothek, CA 2° 384, 24^{r}, seen (incomplete and only to thirds)
Erfurt, Universitätsbibliothek, CA 2° 395, 37^{r}–38^{v} (incomplete)
Erfurt, Universitätsbibliothek, CA 4° 364, 130^{r}–132^{r}
Oxford, Bodleian Library, Can. Misc. 499, 161^{v}–162^{v}/151^{v}–152^{v}
Paris, Bibliothèque nationale de France, lat. 7283, 43^{r}–44^{r}
Paris, Bibliothèque nationale de France, lat. 13014, 166^{v}–167^{v}
Paris, Bibliothèque nationale de France, lat. 14481, 96^{v}
Vatican, Biblioteca Apostolica Vaticana, Pal. lat. 1374, 75^{v}–76^{v}

Manuscripts containing John of Gmunden's tables and derivatives for the radii and velocities:[41]

Innsbruck, Servitenkloster, I.b. 62, pp. 61–62
Klosterneuburg, Stiftsbibliothek, 1238, 29v–32r
London, British Library, Add. 24070, 23r–v and 44r–v
Munich, Bayerische Staatsbibliothek, Clm 27, 174v
Munich, Bayerische Staatsbibliothek, Clm 14783, 109v–115r
Munich, Universitätsbibliothek, 4° 737, 36r
Nuremberg. Stadtbibliothek, Cent. VI 18, 38r
Oxford, Bodleian Library, Canon. Misc. 499, 164v–165r/154v–155r
Paris, Bibliothèque nationale de France, lat. 7286A, 117r
Prague, Národní knihovna ČR, XIV F 10, 36r
Vatican, Biblioteca Apostolica Vaticana, Pal. lat. 1376, 57v–60r
Vatican, Biblioteca Apostolica Vaticana, Pal. lat. 1411, 48r–v
Vienna, Österreichische Nationalbibliothek, 2332, 88r–90r
Vienna, Österreichische Nationalbibliothek, 5151, 13v–16r
Vienna, Österreichische Nationalbibliothek, 5226, 84v–87r
Vienna, Österreichische Nationalbibliothek, 5268, 27r–v

Bibliography

Chabás, José, *Computational Astronomy in the Middle Ages* (Madrid: Consejo Superior de Investigaciones Científicas, 2019).

Chabás, José, and Bernard R. Goldstein, 'Nicholaus de Heybech and His Table for Finding True Syzygy', *Historia Mathematica*, 19 (1992), 265–89.

———, *The Alfonsine Tables of Toledo* (Dordrecht Kluwer, 2003).

———, *A Survey of European Astronomical Tables* (Leiden and Boston: Brill, 2012).

Chabás, José, and Marie-Madeleine Saby, eds, *The Tables of 1322 by John of Lignères* (Turnhout: Brepols 2022).

Dalen, Benno van, and Fritz S. Pedersen, 'Re-Editing the Tables in the *Ṣābi' Zīj* by al-Battānī', *Acta Historica Leopoldina*, 54 (2008), pp. 405–28.

Goldstein, Bernard R., *Ibn al-Muthannā's Commentary on the Astronomical Tables of al-Khwārizmī. Two Hebrew Versions, Edited and Translated, with an Astronomical Commentary* (New Haven: Yale University Press, 1967).

———, *The Astronomical Tables of Levi ben Gerson* (New Haven: Connecticut Academy of Arts and Sciences, 1974).

———, 'Lunar Velocity in the Ptolemaic Tradition', in *The Investigation of Difficult Things. Essays on Newton and the History of the Exact Sciences*, ed. by Peter M. Harman and Alan E. Shapiro (Cambridge, 1992), pp. 3–17.

41 For detailed descriptions of most of these manuscripts, see Porres 2003, *Les tables astronomiques*, pp. 82–124.

Goldstein, Bernard R., and José Chabás, "Tables for the Radii of the Sun, the Moon, and the Shadow from John of Gmunden to Longomontanus", *Archive for History of Exact Sciences*, 78 (2024), 67–86.

Haddad, Fuad I., Edward S. Kennedy, David Pingree, *The Book of the Reasons Behind Astronomical Tables by ʿAlī ibn Sulaymān al-Hāshimī* (Delmar, NY: Scholar Facsimiles & Reprints, 1981).

Kremer, Richard L., 'Cracking the *Tabulae permanentes* of John of Murs and Firmin of Beauval with Exploratory Data Analysis', in Matthieu Husson, Clemency Montelle, and Benno van Dalen (eds), *Editing and Analysing Numerical Tables: Towards a Digital Information System for the History of Astral Sciences.* (Turnhout: Brepols, 2021), pp. 363–422.

Miolo, Laure, 'Retracing the Tradition of John of Genoa's *Opus astronomicum* Through Extant Manuscripts', in *Alfonsine Astronomy: The Written Record*, ed. by Richard L. Kremer, Matthieu Husson, and José Chabás (Turnhout: Brepols, 2022), pp. 343–80.

Nallino, Carlo A., *Al-Battānī sive Albatenii opus astronomicum*, 2 vols (Milan: Osservatorio di Brera, 1903–1907).

Neugebauer, Otto, *The Astronomical Tables of al-Khwārizmī* (Copenhagen: Munksgaard, 1962).

Pedersen, Fritz S., *The Toledan Tables: A Review of the Manuscripts and the Textual Versions with an Edition* (Copenhagen: C. A. Reitzels Forlag, 2002).

Pedersen, Olaf, *A Survey of the Almagest* (Odense: University Press, 1974).

Porres, Beatriz, 'Les tables astronomiques de Jean de Gmunden: édition et étude comparative', unpublished PhD thesis, Paris Ecole Pratique des Hautes Etudes, 2003

Regiomontanus, *Epitoma in Almagestum Ptolemaei* (Venice: Johannes Hamman, 1496).

Saby, Marie-Madeleine, 'Les canons de Jean de Lignères sur les tables astronomiques de 1321': Édition critique, traduction et étude', unpublished thesis, Paris École nationale des chartes, 1987). A summary appeared as 'Les canons de Jean de Lignères sur les tables astronomiques de 1321', *École nationale des chartes: Positions des thèses*, pp. 183–90.

Sengupta, Prabodh Chandra, ed., *The Khaṇḍakhādyaka by Brahmagupta* (Calcutta, 1934).

Simek, Rudolf, and Kathrin Chlench, *Johannes von Gmunden (c. 1384–1442), Astronom und Mathematiker* (Vienna: Fassbaender, 2006).

Suter, Heinrich, *Die astronomischen Tafeln des Muḥammad ibn Mūsā al-Khwārizmī* (Copenhagen: Andr. Fred. Høst & Søn, 1914).

Toomer, Gerald J., *Ptolemy's Almagest* (London: Duckworth, 1984).

JEAN-CLAUDE PENIN

The Trigonometric Tables of Jean Fusoris (c. 1365–1436)*

Introduction

Jean Fusoris[1] was born around 1365 in Giraumont. While his father was a craftsman, making pewter pottery in his native town, Jean studied at Paris university where he attained the Master of Arts. Heir to a craft tradition and equipped with a certain scientific background, he turned to the construction of scientific instruments, practising mathematical, astronomical, and astrological activities. He built quadrants, astrolabes, clocks, and equatoria, wrote various treatises on the astrolabe, astrology, and cosmography and composed trigonometric tables, the latter of which constitute the subject of this paper.

In the dedication of his trigonometric tables, given below, Jean Fusoris addresses King Charles VII, who reputedly asked him to proceed with observations of the movements of the sky and planets and to correct the tables for these movements. The deficiencies of planetary tables, Fusoris claimed, were due as much to the flawed manuscript tradition as to the lack of observations. This rhetoric is usual among table builders; for instance, we find the same concern in the preamble to John of Murs's *Expositio intentionis Regis Alfonsii circa tabulas ejus*.[2] However, Fusoris's claim that observations are needed to compare with the predicted motions and to correct the tables, in particular with regard to the Parisian Alfonsine Tables, is remarkable and rather rare in the Alfonsine sources we have examined

* I would like to thank Matthieu Husson for kindly having welcomed me to the ALFA group and for having entrusted me with this work on the tables of Jean Fusoris, for which he also readily provided his advice and encouragement. My thanks also go to Benno van Dalen and Samuel Gessner, who did not hesitate to offer their help when I needed it; to Richard L. Kremer and José Chabás for their careful review; and to all those who gave me judicious advice in the contexts of various seminars and workshops where I presented preliminary results of my research. This work was supported by the ERC project *ALFA: Shaping a European scientific scene, Alfonsine Astronomy* (COG 723085, PI Matthieu Husson).

1 Emmanuel Poulle, *Un constructeur d'instruments astronomiques au xve siècle: Jean Fusoris* (Paris: Champion, 1963); Léon M. Mirot, 'Le procès de Maître Jean Fusoris, Chanoine de N. D. de Paris', *Mémoires de la Société de l'Histoire de Paris et de l'Ile de France*, 27 (1900), 137–287.

2 Emmanuel Poulle, 'Jean de Murs et les tables alphonsines', *Archives d'histoire doctrinale et littéraire du moyen âge*, 55 (1980), 241–71 (p. 250).

Jean-Claude Penin • SYRTE PSL-Observatoire de Paris

Alfonsine Astronomy: Computational Practices, ed. by Matthieu Husson, José Chabás, and Richard L. Kremer, Turnhout, 2026 (*Alfonsine Astronomy: Studies and Sources*, 3), p. 275-305

10.1484/M.ALFA-EB.5.137914

in the ALFA Project.[3] On the other hand, it should be noted that he quotes the Toledan Tables and Alfonsine Tables more to underline the complexity of the calculations they require than to criticize their defectiveness.[4]

> *Illustrissime princeps Karole septime Francorum rex, cum ad aciem intellectus vestri elevati qui in rebus arduis et nobilibus scientiis delectatur pervenisset tabulas motuum celestium firmamenti, stellarum fixarum atque septem planetarum et motus eorum esse corruptos et deffectuosos, tam ex neglegencia et deffectu observationis quam ex conscriptione librorum a non intelligentibus, vos compacientes tanto casui et deffectui tam nobilis ac excellentis sciencie, michi Johanni Fusoris humili clerico vestro precepistis et mandastis supra predictos motus firmamenti et planetarum facere observationes et tabulas ipsorum rectificare*[5]

> Illustrious Prince, Charles VII, King of the French, as for the vivacity of your superior intelligence, which is attracted by difficult studies and noble sciences, you have understood that the tables of the celestial movements of the firmament, of the fixed stars, of the seven planets and of their movements are altered and defective as much by negligence and the defect of observation as by the writing of books by stupid people, you, being sympathetic to the accident and defection of so noble and excellent a science, have ordered and charged me, Jean Fusoris, your humble clerk, to make observations to compare with the predicted movements of the firmament and planets and to correct their tables precisely.

Fusoris presents the King with a trigonometric table of chords and sines that he has calculated; it is, he says, universal, safe from a defective tradition, and the basis of all astronomical calculation. The originality of Fusoris lies in the fact that he considers the establishment of reliable astronomical tables to be subordinated first to the possession of exact trigonometric tables rather than the making of new observations. Hence:

> *quod ipse sunt fundamentum et origo mensuracionum tocius astronomie motualis; unde per ipsas componuntur tabule ad habendum veros motus firmamenti, stellarum fixarum, septem planetarum, conjunctiones et oppositiones ipsorum ad invicem et alios aspectus, et eclipses solis et lune;*[6]

> they [the trigonometric tables] are precisely the foundation and the origin of the measurements of all the astronomy of movement; therefore, thanks to them, tables are composed to have the true movements of the firmament, of the fixed stars, of

3 Personal communication from Richard L. Kremer. The historiography usually attributes such claims first to Regiomontanus and Peurbach in the 1450s–60s. On the problem of observations see Bernard R. Goldstein, 'A New Set of Fourteenth Century Planetary Observations', *Proceedings of the American Philosophical Society*, 132 (1988), 371–99. One could also mention a curious fourteenth-century text entitled *Tractatus super defectibus tabularum Alfonsii*, cited by Pierre Duhem, *Le système du monde, Histoire des doctrines cosmologiques de Platon à Copernic*, 10 vols (Paris: Hermann, 1913–59), IV, p. 70, and by Poulle, 'Jean de Murs', p. 242.

4 This suggests that he may have had ideas for building tables that were easier to use than the previous ones. The rest of the text does indeed point in this direction.

5 Poulle, *Un constructeur*, p. 191.

6 Poulle, *Un constructeur*, p. 191.

the seven planets, their alternating conjunctions and oppositions, as well as other aspects and eclipses of the Sun and Moon;

Building accurate trigonometry tables is one thing, but ensuring their durability is quite another. Given the vagaries of successive manuscript copies, it is necessary for table makers to provide the means to identify possible errors and to correct them:

> *Quapropter illustrissime princeps vestre majestatis tam nobili proposito cupiens obedire primo tabulas hic presentes de sinibus et cordis cum diligencia et precisione magna rectificavi et composui, ponendo cum hoc modos conservandi veritatem ipsarum cum demonstracione in perpetuum* (…).[7]
>
> Therefore, most illustrious Prince, desirous of obeying so noble a proposal from your Majesty, I first drew up and composed tables of sines and chords, presented here with great care and precision, adding to them the means of preserving their accuracy with a true demonstration for ever (…).

In this article, I analyse the original mathematical means used by Jean Fusoris to fulfil two of his goals: (i) providing the means to check the tables and to correct them if necessary; (ii) constructing chord and sine tables to a certain level of accuracy previously unequalled in Europe. These two objectives are likely to be linked, as common means towards both objectives may have been used. My analysis will distinguish a range of mathematical tools developed by Fusoris, including specific procedures for certain types of division, geometric lemmas, the use of auxiliary tables, and a practice that we will describe as a form of 'tabular mathematics'.[8] Essentially, these 'tabular mathematics' are based on the manipulation of differences and the presentation of results in the form of a grid. They allow the construction of tables from other tables through a series of simple operations such as additions and subtraction. I pay particular attention to these tabular mathematics and more generally to the way Fusoris articulates the different mathematical tools at his disposal.

Notions of precision of numbers and calculations, which were implicit and seldom analysed in medieval tables, obviously preoccupied Jean Fusoris. He knew how to adjust the number of significant sexagesimal positions on the numbers displayed with different precisions in the table of chords. The mathematical errors are optimised to the minimum (modern rounding) both in the chords and the sines. Moreover, throughout the text we find a certain number of pertinent reflections on the issue of numerical precision: how to reach a desired precision, the 'propagation of precision' or its reduction through arithmetic operations, and the reasons for certain precisions. Fusoris was known as a sound mathematician, as attested by one of his contemporaries, Simon Phares, in his *Recueil des plus célèbres astrologues*:[9]

---

7 Poulle, *Un constructeur*, p. 191.

8 This expression was proposed to me by Karine Chemla in 2020 when I was giving a talk on this subject at Agathe Keller's Reading Mathematical Texts seminar at Laboratoire SPHere, Paris.

9 Jean-Patrice Boudet, *Le recueil des plus célèbres astrologues de Simon Phares*, 2 vols (Paris: Champion, 1997–99), I, p. 540.

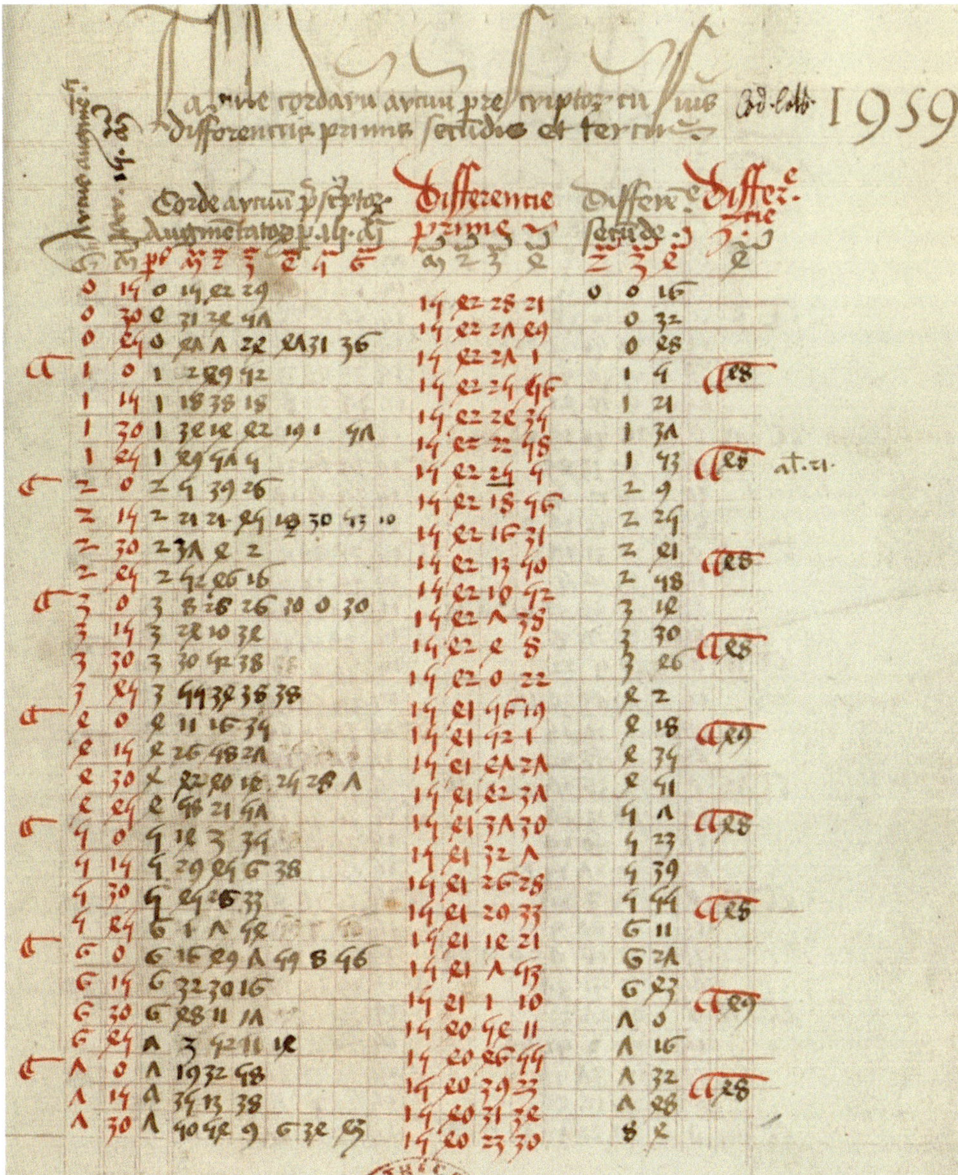

Figure 9.1. First page of the chords table. BnF lat. 7290, f. 1r.

In this time was master Jean Furois, excellent in the science of geometry and a great astrologer; he reformulated the 'Composition of the astrolabe', which is a thing of great efficiency and put there the fixed stars calculated for his time, well and truly.

The trigonometric tables of Fusoris are known only from one manuscript: Paris, BnF, lat. 7290, a beautiful and perfectly legible paper manuscript estimated to be from the fifteenth century. The scribe used two colours, red and brown, which make the tables easier to read.

Emmanuel Poulle[10] dates the construction of these tables to 1432–36 (Fusoris's death). The composition of the manuscript is as follows:

1. A table of chords from 0;15° to 180° (ff. 1^r–12^v) with a step of 15′. They are accompanied by their first, second, and third differences.
2. A table of squares of the chords from 0;15° to 180° (ff. 13^r–24^v).
3. A table of sines from 0;15° to 90° (ff. 25^r–30^v).
4. A dedication of these tables to King Charles VII followed by an explanation of them (ff. 31^v–33^r).
5. A draft table of squares and a table of sexagesimal multiplication (ff. 33^r).
6. A table of multiples of some sine verses (ff. 35^r–38^v).

This article considers Fusoris's work on tables in two sections. First, I analyze one of Fusoris's proposed methods to check and correct the tables. Second, I study the construction of certain chord table values using a method based on third differences, which, as we shall argue, is equivalent to a third-order polynomial method of interpolation.

1. Checking and correcting tables

The use of table verification and correction procedures was commonplace among table-makers. It is seen in the *Almagest*[11] and the use of differences to check values or calculate them is common among Arab mathematicians.[12] Fusoris proposes no fewer than three methods for checking and correcting the tables; although the first is common and previously attested, the following two prove his originality and ingenuity. The first refers to properties that today we would call algebraic: the Pythagorean theorem and the duplication formula. The second refers to the calculation of finite differences. Fusoris explains that since we have the differences of the entries in the chord table, it is easy to navigate from the last differences to the first differences and then to use the entries to check the latter. I will show that, in addition to being used to check or correct the tables, this method was used by Fusoris to calculate certain chord without going through the intermediary of the calculation of the 1° chord as Ptolemy and many other ancient and medieval astronomers had done. I examine this hypothesis in the second part of this chapter. The third method involves a geometrical lemma, the use of auxiliary tables and a specific procedure. It is this third method, previously unattested, that I examine first.

1.1. The third method of checking tables

> *Tertius modus et principalis per quem faciliter veritas predictarum tabularum in eternum conservabitur et per quem ille si opus esset de novo faciliter componerentur*[13]

10 Poulle, *Un constructeur*, p. 6.

11 G. J. Toomer, *Ptolemy's* Almagest (London: Duckworth, 1984), p. 48.

12 Glen Van Brummelen, *The Mathematics of the Heavens and the Earth: The Early History of Trigonometry* (Oxford: Oxford University Press, 2009), p. 163.

13 Poulle, *Un constructeur*, p. 193.

> The third and principal means by which the accuracy of the above-mentioned tables will be easily preserved in perpetuity, and by which, if this is again necessary, they will be easily composed.

Fusoris suggests that this third means of controlling tables can be also used to 'compose' the tables.[14] This possibility appears partially feasible (we can calculate certain values, but some must be pre-processed), but Fusoris said nothing more on the matter. His explanations concerning this method are provided in the form of a computed example. From two known chords of respective angles, α and β, Fusoris proposes determining the chord of the angle:

$$\frac{\alpha + \beta}{2}$$

Fusoris very clearly explains the example and method, using a geometrical diagram (Figure 9.4) and displaying a computation for a kind of division (Figure 9.2).

> *Verbi gratia sint due corde note scilicet corda 3gr. et corda sex gr., que due corde distant ad se invicem per tres gradus; volo igitur per ipsas habere cordam mediam ignitam scilicet cordam* 4^{or} *gr. 30 min.; hoc ergo sic faciam; addendo predictas duas cordas notas ad invicem et aggregatum mediando et proveniet linea gh, talis*[15]*:*
>
> > For example, let there be two obviously known chords, the three-degree chord and the six-degree chord, these two chords are mutually distant by three degrees; I therefore want, by means of these, to obtain the unknown median chord, obviously, chord of four degrees, thirty minutes; therefore, I proceed in this way; after adding the two known chords, I divide the result by two, then will come the line *gh*, such as:

p	m^a	2^a	3^a	4^a	5^a	6^a	7^a	
4	42	38	47	14	34	43	21	**linea gh**
		1	14	1	15	30	50	**a**
4	42	40	1	15	50	14	11	**b**
			12	57	13	12	54	**c**
4	42	40	14	13	3	27	5	**d**
				12	20	12	35	**e**
4	42	40	14	25	23	39	40	**f**

Figure 9.2. Algorithmic diagram for the calculation of the mean chord.

To build this algorithmic diagram, the construction of which constitutes a simple algorithm but reflects a subtle method of calculation, Fusoris uses a table of multiples of one *sinus verse,* which is at the end of the document.[16]

---

14 This method will not give the one-degree chord; it will give nothing more than Ptolemy's algebraic-geometric method.

15 Poulle, *Un constructeur,* p. 194.

16 I have included this table in Appendix 1. It is the table of multiples of *Vers* (3°/4).

cum primo igitur numero linee gh scilicet cum quatuor partibus intrabo versus sinistram tabulam predictam et in directo reperiam a quod addo predicte linee gh et consurgit b; postea cum 2º numero ipsius b scilicet cum 42 min. intrabo predictam tabulam et in directo 42 reperio c, quod addo ipsi b et consurgit d; postea cum tercio numero ipsius d scilicet cum 40 secundis intrabo predictam tabulam et in directo 40 reperio e, quod addo ipsi d et consurgit f que est corda 4 gr. 30 min. verificata usque ad 4ta. Si autem vellemus istam cordam ulterius verificare cum majori illius precisione usque ad 5ta, 6ta, etc., tunc oporteret consimiliter sicut prius intrare predictam tabulam cum 4º numero ipsius f scilicet cum 14 terciis et illud quod in directo ipsius reperitur addere ipsi f, etc. ut prius; et sic consequenter quousque fiat verificatio usque ad precisionem duorum predictarum cordarum notarum scilicet usque ad 7a.[17]

so, with the first number of the *gh* line (Figure 9.2), obviously four parts, I enter the left side of the [sine verse] table,[18] and to the right I find what I add[19] to the above line *gh* and it comes out *b*; then, with the second number of the same *b*, obviously with forty-two minutes. I enter the [versine] table, and to the right of forty-two, I find what I add precisely to *b*, and it comes out *d*; then, with the third number of *d*, obviously forty, I enter the [sine verse] table and to the right of forty I find *e*, which I add precisely to *d*, and it comes out *f*, which is the chord of four degrees, thirty minutes, true to the fourth. And if we want this chord to be exact beyond with greater precision up to the fifth, sixth, etc., then it will be necessary in the same way as before to enter the [versine] table with the fourth number of *f* evidently with fourteen to the third, and the one to the right precisely is to be added to this same *f*, etc. as before; and so consequently until it is made exact up to the precision of the two aforesaid known strings, evidently up to the seventh.

Below is a translation of this algorithm according to the text above:

p	**mª**	**2ª**	**3ª**	**4ª**	**5ª**	**6ª**	**7ª**		
<u>4</u>	42	38	47	14	34	43	21	**linea gh**	**={crd(3º)+crd(6º)}/2**
		1	14	1	15	30	50	**a**	**=4×Vers(0;45º)/60**
4	<u>42</u>	40	1	15	50	14	11	**b**	
			12	57	13	12	54	**c**	**=0;42×Vers(0;45º)/60**
4	42	<u>40</u>	14	13	3	27	5	**d**	
				12	20	12	35	**e**	**=0;0,40×Vers(0;45º)/60**
4	42	40	14	25	23	39	40	**f**	

Figure 9.3. Explanation of the algorithm.

The form of this algorithmic diagram could lead one to believe that Fusoris is doing a simple multiplication:

17 Poulle, *Un constructeur*, p. 194.

18 The table of multiples of *vers*(3º/4).

19 This number must be divided by sixty according to the previous note. The three-degree versine table is the only one which is exactly *vers*(3º/4), others were divided by sixty. I cannot explain this exception.

$$gh \times \frac{Vers(0;45)}{60} = \frac{crd(3) + crd(6)}{2} \times \frac{Vers(0;45)}{60}$$

Closer examination shows that instead of successively taking the sexagesimal positions of *gh* from the first to the seventh position, multiplying them by *vers*(0;45°), and adding this to the previous result, he successively takes these sexagesimal positions, but this is done only on the partial sums he obtained (i.e., the sexagesimal numbers *b*, *d*, and *f* of Figure 9.2 or 9.3).

1.2. The meaning of the algorithm

Further on, Fusoris gives a geometrical interpretation of this calculation:

> *Pro predictorum adhuc majori declaratione placuit ponere adhuc figuram sequentem in qua sit ac corda trium graduum nota et bd corda sex graduum etiam nota, per quas duas cordas notas volo invertigare cordam mediam ignotam 4° gr. 30 min. Componentur enim due linee note ac et bd et aggregatum medient et habebimus lineam gh quia quanto bd est major quam linea gh in tanto linea gh est major quam linea ac; habita igitur linea gh si sibi addantur due linee og et hi habebimus et quam querimus; hoc autem facimus intrando tabulas predictas sicut docebatur in exemplo predicto quia duo anguli o et i sunt recti et due linee om et ik equedistantes et linea mk equalis linee el, sic ergo per duas cordas extremas habetur corda media*[20]*;*
>
>> To better demonstrate what has been said so far, it seems apt to construct the following figure in which *ac* is the known three-degree chord and *bd* the six-degree chord; by means of these known chords, my aim is to carefully search for the unknown median chord of four degrees, thirty minutes. For this purpose, if we arrange the two known lines *ac* and *bd* and their half-sum, we obtain the *gh* line, because the greater *bd* is than the *gh* line, the greater the *gh* line is than the *ac* line; therefore, if we add the *og* and *hi* lines to the *gh* line, we will obtain what we are looking for [*el*]; and we do this by filling in the above-mentioned tables (versine table of Appendix 1, f. 36^{r}), as was taught in the previous example because the two angles, *o* and *i*, are right, and the two lines, *om* and *ik*, are of the same length, and the line *mk* equals the line *el*. So, with the two extreme chords we obtain the middle chord.

From a practical point of view, the algorithm given above is sufficient to satisfy the demand for a means of checking or correcting the tables. However, Fusoris saw it necessary to add a 'geometrical justification', which has no additional computational purpose other than to explain or illuminate this means in a different way and possibly to justify it as a precise solution to the problem. The geometrical description indeed sheds some light on Fusoris's approach and on his way of obtaining the following:

$$crd\left(\frac{3+6}{2}\right) \text{ from } gh = \frac{crd(3) + crd(6)}{2}$$

20 Poulle, *Un constructeur*, p. 195.

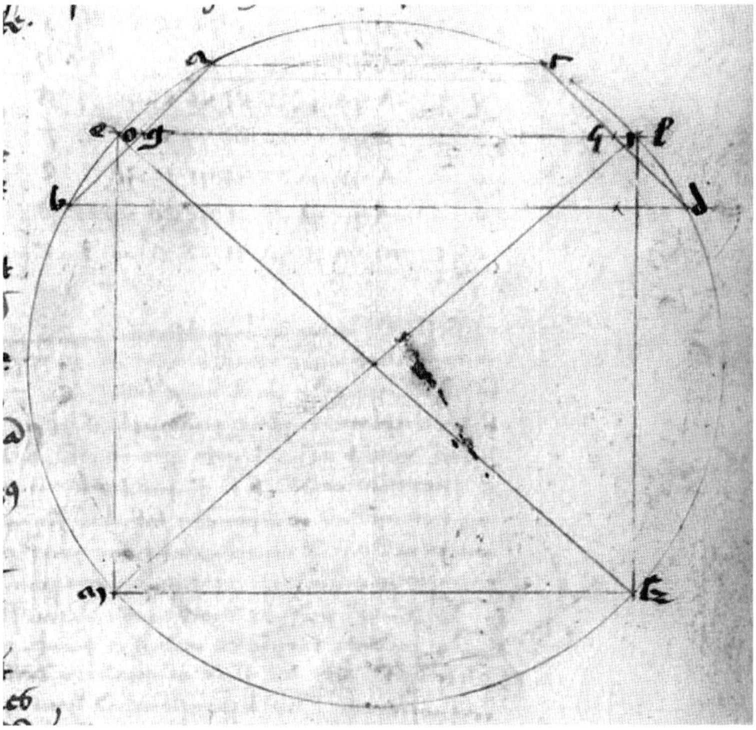

Figure 9.4. BnF lat. 7290, f. 33r.

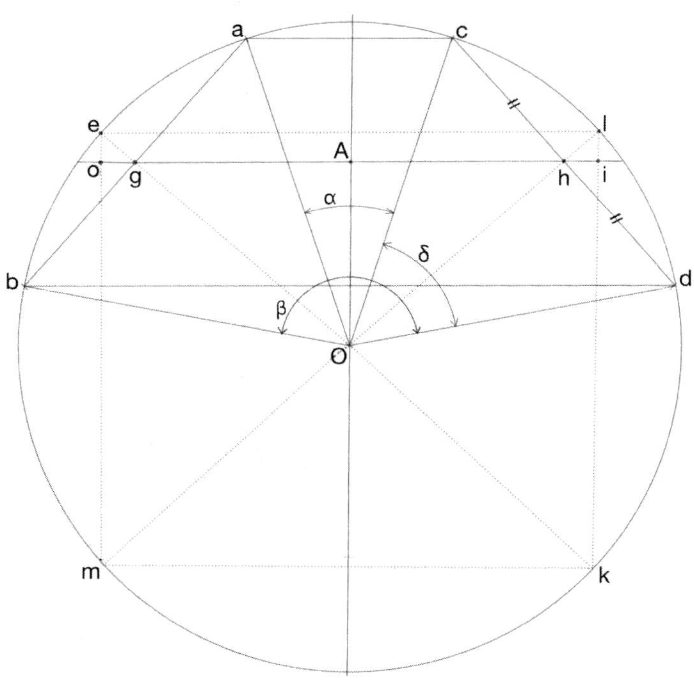

Figure 9.5. Previous figure reconstructed.

Let us reformulate Fusoris's proposition in the previous passage with algebraic formulas relying on notation defined in Figure 9.5:

$$\text{let } \beta = \alpha + 2\delta \text{ then } ac = crd(\alpha), bd = crd(\beta)$$

which becomes

$$gh = \frac{crd(\alpha) + crd(\beta)}{2}, el = crd\left(\frac{\alpha + \beta}{2}\right), el = gh + 2hi$$

The procedure then comes to an abrupt end, with the author stating that *el* 'is what we are looking for', and the explanation that follows does not tell us much more. We can reconstruct a reasonable hypothetical continuation of this procedure by comparing the procedure, the diagram and the algorithm. In Figure 9.5, similar triangles *OAh* and *hil* give

$$\frac{hi}{hl} = \frac{Ah}{Oh}$$

or

$$\frac{hi}{hl} = \frac{gh}{2 \times Oh} \text{ then } 2hi = gh \times \frac{hl}{Oh} \, [A]$$

as

$$crd\left(\frac{\alpha + \beta}{2}\right) = el \text{ and } el = oi = gh + 2hi$$

then,

$$crd\left(\frac{\alpha + \beta}{2}\right) = gh + 2hi, \text{ moreover } Oh = 60 - hl$$

comes with [A]

$$crd\left(\frac{\alpha + \beta}{2}\right) = gh + gh \times \frac{hl}{60 - hl} = gh\frac{60}{60 - hl}$$

and from there we obtain

$$(60 - hl)crd\left(\frac{\alpha + \beta}{2}\right) = 60 \times gh,$$

and, finally,

$$\left(1 - \frac{hl}{60}\right)crd\left(\frac{\alpha + \beta}{2}\right) = gh,$$

with *hl* being the sagitta of the angle δ[21] or $Vers(\delta/2)$

As stated, this is a reconstruction whose outline was probably thought up by Fusoris. Elsewhere in the text, he refers to Euclid's Elements, showing that he had some knowledge of them, including the theorem invoked. The use of the sagitta is justified by the existence of the tables of multiples of some *sinus verses* at the end of Fusoris's text. It is clear that he takes for granted

21 With trigonometry, this relation would be written as follows today: [sin(x)+sin(y)]/2 = cos[(x-y)/2]× sin[(x+y)/2].

that this sagitta depends only on the angle subtended by the arc, which a figure easily 'shows'. However, the names *sinus verse* or sagitta are never given by Fusoris to designate this length, nor is any other name for that matter; he designates this length, and its multiples, only by referring to 'a table'. Now, one may be surprised by the form of the relation on which Fusoris worked:

$$\frac{crd(\alpha) + crd(\beta)}{2} = k \times crd\left(\frac{\alpha + \beta}{2}\right)$$

It should be noted that in the Middle Ages, in addition to the classical operations (+, -, x, /), division by two and duplication (mediatio, duplicatio) were considered elementary and important operations[22]. The figure he gives in the text, because of its obvious symmetry, suggests that Fusoris first saw a link between the two averages he sought, which he then concretized with a figure and an algorithm. However, nothing in the text justifies this approach, which of course remains a hypothesis. As for the rest of the reasoning, an examination of the triangles *eOl* and *gOh* with *el* sides parallel to *gh* suggests a relationship since these two triangles differ by *eg* and *hi*, and the magnitude *hi* is precisely a sagitta.

1.3. Relationship between the algorithm of Fusoris and the formula

With the notation introduced in Figure 9.5, Fusoris's algorithm translates the following relationship:

$$crd\left(\frac{\alpha + \beta}{2}\right) = \frac{gh}{\left(1 - \frac{hl}{60}\right)}$$

This is a simple Euclidean division; the usual classical algorithm for doing such an operation requires a certain amount of trial and error to determine each 'digit'. In base 10, it is quite easy, but in sexagesimal numbering, although not very complicated, it is certainly more tedious and requires a special sexagesimal multiplication table. He gives such a division later on. However, Fusoris's algorithm is not providing such a simple division.

Fusoris does not provide any information on the origin of his calculation algorithm or why it gives the right result; he is satisfied with the example that details the manipulations. This algorithm avoids division and thus simplifies this trial-and-error procedure. Like many medieval mathematicians and astronomers, Fusoris knew arithmetic operations in decimal and sexagesimal notation;[23] others, such as John of Murs, even wrote about the sexagesimal multiplication table.[24] This familiarity with operations in sexagesimal numeration probably allowed Fusoris to make precise observations on division, in particular the division of a number by a divisor of the form $(1-\varepsilon)$, with the number ε close to 0, that is, numbers of the following form:

$D = 0;59,59,59, \dots$.

22 Sacrobosco, 'Tractatus de arte numerandi', in *Rara mathematica*, ed. by J. O. Halliwell (London: Maynard, 1839), p. 9; Matthieu Husson, 'La tabula tabularum de Jean de Murs et les modèles de l'arithmétique médiévale', *Cahiers de recherches médiévales et humanistes*, 27 (2014), 97–122 (p. 107).

23 David E. Smith, *History of Mathematics*, 2 vols (Boston: Ginn, 1925), II, pp. 128–43.

24 Husson, 'La tabula tabularum'.

Take the simple example of such a Euclidean division by a divisor of the above form:[25]

4;42,38,47 divided by 0;59,59,41,29,
where we note that the divisor is of the form 0;59,59,41,29 = 1–0;0,0,18,31, with 0;0,0,18,31 = ε.

	4;	42,	38,	47	0;	59,	59,	41, 29
-	3;	59,	58,	46				
		42,	40,	1	4;	42		
	-	41,	59,	47				
			40,	14				

Figure 9.6. Euclidian division.

I shall briefly recall the classical mechanism of a Euclidean division. We look for how many times, at most, we have the divisor 0;59,59,41,29 in the dividend 4;42,38,47; this is four, and, consequently, it is the leftmost sexagesimal position of the dividend. Then, we make the following calculation:

$$4 \times 0;59,59,41,29 \;=\; 3;59,58,46$$

which is subtracted to the left of the above division from the dividend 4;42,38,47 to obtain a remainder, which is written below: 0;42,40,1. Finally, we write '4' at the location of the divisor, and so on.

Let us go back to the first step. A calculator used to these operations will have no difficulty in noticing the following:

$$4 \times 0;59,59,41;29 \;=\; 3;59,58,46 \;=\; 4 \times (1-\varepsilon) \;=\; 4 - 4 \times \varepsilon$$

We can now perform exactly the same division by slightly changing the arrangement of the operations:

	4;	42,	38,	47	1	-	0;	0,	0,	18,	31	= 0;59,59,41,29
-	4;											
+			1	14	4;	42						
		42,	40,	1								
	-	42,										
	+			12								
			40,	14								

Figure 9.7. Modified Euclidean division.

25 I used the current scheme of the division algorithm. Although the general idea of the algorithm has changed little over time, details and patterns used have changed. See Smith, *History*, II, pp. 128–43.

This means that subtracting 3;59,58,46 from the dividend is equivalent to subtracting four and adding 4 ε = 0;0,0,1,14. The operation can then be repeated on the following lines. This remark suggests an interesting simplification of the classical division algorithm, which can be summarized in the following table:

	4	42	38	47				
+			1	14	=	4	×	ε
	4	42	40	1				
+				12	=	0;42	×	ε
	4	42	40	13				
+				0	=	0;0,40	×	ε

Figure 9.8. Gerbert's algorithm.

This is precisely the picture given by Fusoris. I explain this algorithm as a simplification of classical Euclidean division applied to this type of divisor. It should be noted that it can be faulty because of carries from one sexagesimal position to the next, but the small ε 'guarantees'[26] the accuracy of the algorithm. Among the different ways of performing and writing Euclidean division, there was an old method[27] attributed to Gerbert d'Aurillac (945–1003), whose procedure follows the scheme produced by Fusoris and uses the complement of the divisor to the higher unit, all in base 10, of course. This method was reported by Adelard of Bath, who attributed it to Gerbert. Fusoris may have been aware of this method.

1.4. Comments on accuracy

Let us return to the 'division' algorithm and analyse a passage where Fusoris comments about the accuracy of the algorithm.

> *Apparet igitur quomodo in predicto precessu majores fractiones notificant minores. Unde notanter advertendum est que semper in lineis gh est veritas in primo numero sive sint partes sive minuta; et quandoque est veritas in pluribus numeris primis sicut apparet in exemplo supra posito, in quo erat veritas in partibus et in minutis puta in 4 gr. 42 min.; deinde partes, puta 4, intrando tabulam verificabant 2ª, etc., postea minuta, puta 42, verificabant 3a et ita consequenter intelligendum est quod 2ª et 3ª etc. verificabant fractiones sequentes consimiliter intrando tabulam predictam sicut prius*[28]*;*

> It appears, therefore, from what has been said before, how the larger fractions make the smaller ones known. Hence, in particular, we must ensure that in the line *gh*, the

26 A necessary condition for the simplicity of Fusoris's algorithm is the smallness of epsilon, but this condition is not sufficient, as can be seen from some examples.

27 Smith, *History*, II, p. 134.

28 Poulle, *Un constructeur*, p. 194.

> first part of the number, the parts, and the minutes are exact. Whereas the first parts of the number are exact as it appears in the example proposed above, where the parts and minutes are exact, that is $4^{\circ}\,42^{m}$; then, entering the parts in the [versine] table, that is, four, make the second exact,[29] etc., afterwards, the minutes, that is, forty-two, make the third exact, and so, consequently, it must be understood that the second and third make the following fractions exact in the same way, entering the above [versine] table as before;

Fusoris here remarks on the evolution of the precision of a result of an operation (here a division) according to the increasing precision of one of the factors (here, the successive *dividends*: *gh*, *b*, *d*, *f*) to be taken into account. It also makes the link with the exact *n*-position of the result when the *n*-2 position of the *dividends* is used. This particular algorithm, which could have been replaced by simple division, has the merit of clearly showing this variation in the precision of the result as a function of the number of operations performed to find it. One could thus speak of the *propagation of precision* through an operation, whereas today we speak of the *propagation of errors* in calculations.

2. Hypotheses for the calculation of the chord table's short entries

The table of chords in this text by Fusoris (Figure 9.1) is its most important feature; it is made up of the table of arguments (red) and their chords and entries (brown), which run from 0;15° to 180° (ff. 1^{r}–12^{v}) with a step of 15' in the first and second columns. They are accompanied (to the right) by the difference columns. Columns 3 (red), 4 (brown), and 5 (brown) list the first, second, and third differences. Column 3 gives the differences (first differences) of successive entries in the preceding column (chords), column 4 shows the differences (second differences) of successive entries in the preceding column (first differences), and column 5 displays the differences (third differences) between the second differences in intervals of three entries.

2.1. Terminology

This table of chords proposed by Fusoris is given in the form of sexagesimal values of various levels of precision that we call entries. The first thing to note when examining the entries in this table is the regular heterogeneity of their precision:

1. *entries* with six sexagesimal positions (6th), which I call *long entry* or simply *long*;
2. *entries* with four sexagesimal positions (4th), which I call *middle entry* or simply *middle*;
3. *entries* with three sexagesimal positions (3rd), which I call *short entry* or simply *short*;
4. a few more *entries* towards the end, the precision of which is to five positions (5th).

This classification is important, and the whole work is notably subordinate to it. The different levels of accuracy of the entries in the table are not random; they correspond, in fact, to

29 *versine*(4)/60 (Appendix 1) gives 0;0,**1**,14,..., which, added to 4;42;**38**,47,..., gives 4;42;**40**,..., and **40** is the number of seconds that is exact in the final result.

different methods of calculating these numbers. Moreover, this apparent heterogeneity suggests that this table was a working table drawn up to construct the sine table of uniform precision. After some explanations concerning the differences of the entries and their errors, I briefly summarize the construction of *long entries* and *middle entries* to examine in detail the more problematic case of *short entries.*

2.2. Difference columns

One question that can be asked about the differences in the chord table concerns their origin. Were they simply calculated from the entries? The answer is linked to the calculation of the short entries. The values of the table of third differences are given only on the fourth sexagesimal position, the previous positions being zero and the following ones neglected. These differences have been grouped in threes and are placed on every third number of the second differences, perhaps for the sake of space and clarity of the table. The values of the table of second differences are given on the third and fourth sexagesimal positions, the previous positions being zero and those beyond the fourth neglected. It can be confirmed that these differences have not been calculated from the entries of this table for the following reason: the values in the table of first differences systematically have four sexagesimal positions, while the short entries have only three sexagesimal positions. When the differences are calculated from the Fusoris entries, the results diverge significantly from his differences, which can be seen in the table below.

Angles	**2d diff. given by Fusoris**	**2d diff. calculated from the chords of Fusoris**	**Angles**	**2d diff. given by Fusoris**	**2d diff. calculated from the chords of Fusoris**
0 ;45	0 ;48	0 ;35	**15 ;45**	16 ;54	16 ;47
1 ;30	1 ;37	1 ;38	**16 ;30**	17 ;42	18 ;31
2 ;15	2 ;25	2 ;37	**17 ;15**	18 ;30	18 ;36
3 ;00	3 ;14	3 ;00	**18 ;00**	19 ;18	19 ;40
3 ;45	4 ;02	4 ;16	**18 ;45**	20 ;06	20 ;20
4 ;30	4 ;51	4 ;05	**19 ;30**	20 ;53	20 ;17
5 ;15	5 ;39	5 ;16	**20 ;15**	21 ;41	21 ;10
6 ;00	6 ;27	5 ;58	**21 ;00**	22 ;29	22 ;38
6 ;45	7 ;16	7 ;28	**21 ;45**	23 ;17	23 ;28
7 ;30	8 ;04	7 ;13	**22 ;30**	24 ;04	24 ;17
8 ;15	8 ;52	8 ;44	**23 ;15**	24 ;51	24 ;54
9 ;00	9 ;41	9 ;31	**24 ;00**	25 ;39	26 ;03
9 ;45	10 ;29	10 ;04	**24 ;45**	26 ;27	26 ;30
10 ;30	11 ;17	10 ;55	**25 ;30**	27 ;14	27 ;02
11 ;15	12 ;06	11 ;34	**26 ;15**	28 ;01	28 ;28
12 ;00	12 ;54	13 ;32	**27 ;00**	28 ;48	28 ;40
12 ;45	13 ;42	14 ;24	**27 ;45**	29 ;35	29 ;26
13 ;30	14 ;30	14 ;41	**28 ;30**	30 ;22	30 ;43
14 ;15	15 ;18	15 ;58	**29 ;15**	31 ;09	31 ;22
15 ;00	16 ;06	15 ;48	**30 ;00**	31 ;56	32 ;17

Figure 9.9. Calculation of differences.

It should be noted that in this table I have only chosen certain angle arguments (step 0;45°), not all the arguments of the table of Fusoris, and that the written differences do not refer to the arguments I have written but to all the arguments of Fusoris's table (step 0;15°).[30] Moreover, Fusoris himself specifies that the short entries were calculated from the differences (see Section 3.4). It is thus clear that the differences of the entries were not directly calculated from the entries and that a more complex computation procedure was used to produce entries and their differences.

2.3. Calculation of the long and and middle entries

Fusoris explicates in detail how he uses Ptolemy's method[31] to calculate these entries, which correspond to the chords $crd(n \times 0°;45)$, and n, a natural number, to a level of precision of a sixth or a fourth.

> *Dico igitur primo quod tres sunt corde principales, scilicit 60 gr., 90 et 36 gr. que notificantur et demonstratur demonstracione geometricali et arismeticali. Secundo dico quod per has tres cordas demonstratas declaratur et notificantur omnes alie corde*[32].
>
> I say, therefore, firstly, that there are three principal chords, namely: of 60°, 90°, and 36°, which are known and demonstrated by geometrical and arithmetical proof. Secondly, I say that with these three mentioned chords, all other chords are expressed and known.

The algebraic-geometric method used to calculate the long and middle entries does not allow one to obtain the entries of all the other arcs of a table whose angular step is 15'. In particular, it cannot be used to obtain the entry of the arc of one degree, or, consequently, the entries of the arcs of 0;30° and 0;15°, which would allow us to completely fill the table presented by Fusoris.

2.4. The issue of short entry calculation

Ptolemy used a particular method of approximation for the chord of one degree[33] that Fusoris did not reproduce. Fusoris gives us only a few indications in two short passages of how he proceeded. The first passage refers to the calculation of the short entries by means of a procedure with the differences.

> *Alie vero corde intermedie notificantur et declarantur per predictarum cordarum demonstratarum differentias tam primas quam 2as quam 3as, etc., sicut tactum est superius*[34].
>
> As for the other intermediate chords, they are known and described by the differences of the aforesaid chords, both first, second, third, etc., as treated above.

30 Thus, for example, 0;48 is the second difference written opposite the entry for 0;45 in Fusoris's table (Fig. 1) and the result of subtracting the first two differences, 0;15,42,27,49–0;15,42,27,1, which are themselves the first differences of the entries for 0;30°, 0;45°, and 1°.

31 Toomer, *Ptolemy's* Almagest, p. 48 (I, 10).

32 Poulle, *Un constructeur*, p. 197.

33 Toomer, *Ptolemy's* Almagest, p. 54 (I, 10).

34 Poulle, *Un constructeur*, p. 199.

No details are given there but reference is made to a previous passage in the text where the question is treated in more detail. This second passage plays a double role: it explains how to check the chord table and how the short entries were calculated.

> *unde notanter est advertendum quod, per cordas principales demonstratas usque 6ta, fui contentus verificare differentias tam primas quam secundas quam tercias usque ad unum 4^{m} vel* ***duo 4ta*** *scilicet capiendo primo primas differentias deinde secundas deinde tercias quousque ille differentie uniformiter diminuebantur scilicet subtriplando tam primas differentias quam secundas quam tercias et consequenter procedendo a terciis differentis ad primas replendo duo spacia media pro duabus cordis mediis non demonstratis; unde isto modo reperietur quod non erit saltus inter differentias immediatas differentie igitur tercie sunt valde facilis conservationis in veritate sua, per quas verificantur secunde differentie, et per secundas verificentur prime, et per primas habentur corde*[35].
>
> Hence, in particular, it should be noted that for the principal chords demonstrated up to the sixth position, I have endeavoured to give the differences accurately, both the first, second, and third, up to one fourth position **or two at the fourth**, that is, by taking the first differences first, then the seconds, and afterwards the thirds until these differences decrease uniformly, namely by subtripling both the first differences and the seconds and the thirds and consequently by advancing from the third differences to the first ones by filling the median spaces with the two median chords not demonstrated. Hence, in this way one will find what will not constitute a jump between the immediate differences; therefore, these immediate third differences are very easy to obtain accurately, by which the second differences are verified or calculated,[36] and the first ones will be verified or calculated by the seconds, and by the first ones are obtained the chords.

This explanation is important; its general meaning is clear, yet several difficulties arise in the details. First, Fusoris calculates the differences of the long and middle entries, which are called 'principal chords'. The bold part of the text is not clear. It could mean that the third differences are exact up to one or two fourths, which at least makes mathematical sense. Fusoris calculates the first, second, and third differences until these differences are uniformly diminished. One may wonder what is meant by 'uniformly' here. The most likely explanation conveys the slow and regular decrease of the third differences: this is confirmed by the table. Then, he specifies '**subtripling**'. This is to be understood as the division of the first differences by three, of the second differences by nine, and finally of the third differences by 27. By denoting Δ^{i}_{j} the *i*-th difference of rank *j* of long and middle entries, I understand this sentence as I express it mathematically below:

$$\frac{\Delta_j^1}{3}\,\frac{\Delta_j^2}{9}\,\frac{\Delta_j^3}{27}$$

35 Poulle, *Un constructeur*, p. 193.

36 This translates the Latin word *verificantur* (verificare) in the following ways: present as true [Félix Gaffiot, *Dictionnaire illustré latin-français* (Paris: Hachette, 1934)]; or confirm the truth/authenticity of; show to be true by evidence (William Whitaker's words, online). So, the word appears to have two meanings: 1) to check 2) to give a true/accurate value.

Fusoris then says '... **by advancing from the third differences to the first ones by filling the two median spaces by the two median chords not demonstrated**'. It is easy to understand here that this is the inverse calculation of the differences; one goes back from the differences to the values of the short entries that one can thus obtain. However, here again the text is not very precise, because we are talking about specific differences, the differences of the long entries. Now, to calculate the short entries by the reverse calculation of the differences, one must use also the differences related to the middle and short entries that are distinct from the long entries differences. And the relations of these differences are complex.

Yet, on the basis of this information, we can conjecture with confidence that Fusoris exposes a method of calculation that would be described in modern mathematical term as a method of interpolation by finite differences. In other words, this last passage would suggest that Fusoris interpolated between the long, middle, and short entries by a method of third-order polynomial interpolation. Unfortunately, his unclear and clumsy explanations make it difficult to reconstruct the exact procedure. To try to emulate a method that Fusoris could plausibly have employed, I use a formal third-order polynomial interpolation that was practised and formalised from the eighteenth century onwards[37] to calculate the short entries from the long and middle entries, all while trying to adjust this method according to the scarce information given by Fusoris. The following definitions shall be used. The finite differences of all entries (*short*, *middle*, and *long*) taken together give the following:

$$\delta_j^i = \delta_{j+1}^{i-1} - \delta_j^{i-1}, x_j = \delta_j^0$$

which is a difference of the order of i (first, second, and third differences) at the rank j. The finite differences (only long and middle entries) are noted: Δ_j^i. Using this notation, I describe a hypothetical reconstruction of Fusoris's algorithm for the computation of short entries.

2.4.1. *Polynomial interpolation method by finite differences*

Let h, the step, be a real number and let f be a function whose values $f(x_0+j \times h)$ are known. Let N and M be integers with N < M, let j be an integer with $j \in \{0; M\}$, and let i be an integer with $i \in \{0; N\}$. We can then compute the differences of the order of i and rank j in the following way, which is just a rephrasing in this context of notation just given:

$$\delta_j^i = \delta_{j+1}^{i-1} - \delta_j^{i-1}, \delta_j^0 = f\left(x_0 + j \times h\right)$$

If δ_0^i are known, it is possible to obtain approximate values of $f(x)$ by simple formulae for intermediate real values of x between the following numbers: $x_j = x_0+j \times h$, $j \in \{0; M\}$. We discuss a method of the order of N[38]. For example,

Let the real z, with $0 \leq z \leq M$ and $x = x_0 + z \times h$

37 Sylvestre F. Lacroix, *Traité de calcul différentiel et de calcul intégral*, 2 vols (Paris: Duprat, 1797–98), II.

38 This is a Newtonian interpolation method. See E. Fabry, *Traité de mathématique générale*, 3d rev. ed. (Paris: A. Hermann & fils, 1916), p. 131.

$$N = 3f.(x) = f(x_0 + zh) \cong \delta_0^0 + z\delta_0^1 + \frac{z(z-1)}{2}\delta_0^2 + \frac{z(z-1)(z-2)}{6}\delta_0^3$$

gives approximate values of $f(x)$ for values of the variable x intermediate between the numbers $x_j = x_0 + j^*h$.

Generally speaking, the polynomial interpolation method of order N by finite differences amounts to approximating the function f by a polynomial of degree N, whose graphical representation passes through the following points: $(x_j; f(x_j))$ for $x_j = x_0 + j \times h$.

2.4.2. The chords table

I again present a part of the chords table with two types of finite differences:

1) the finite differences Δ^i_j corresponding to the long and middle entries X_i, which are not given by Fusoris;
2) the finite differences δ^i_j corresponding to all entries x_i (long, middle, and short), which are given by Fusoris.

$X_0=x_0$			
	δ_0^1		
x_1		δ_0^2	
	δ_1^1, Δ_0^1		δ_0^3
x_2		δ_1^2	
	δ_2^1		δ_1^3
$X_1=x_3$		δ_2^2, Δ_0^2	
	δ_3^1		δ_2^3
x_4		δ_3^2	
	δ_4^1, Δ_1^1		δ_3^3, Δ_0^3
x_5		δ_4^2	
	δ_5^1		δ_4^3
$X_2=x_6$		δ_5^2, Δ_1^2	
	δ_6^1		δ_5^3
x_7		δ_6^2	
	δ_7^1, Δ_2^1		δ_6^3
x_8		δ_7^2	
	δ_8^1		
$X_3=x_9$			

Figure 9.10. Differences of long and middle entries (Δ^i_j) and differences of all entries (δ^i_j).

2.5. My calculation hypotheses

I now give a possible pattern of the calculations for the chord table. Mathematically, the third differences of all entries truncated at the exact fourth sexagesimal position change slowly, given the step of the table (0°;15), and remain constant over large ranges. Thus, each value of the third differences remains constant on at least thirty entry values. My hypothesis is that Fusoris used this property to compute the short entries. However, it was shown above that the differences in the table were not obtained from the entries given on the left. Therefore, I assume that the differences were calculated with a procedure using only the long and middle entries which can be directly computed with the geometrical algebraic method. In order to compute the differences only from the long and middle Fusoris divided the first order difference by 3, the second order difference by 9 and the third order difference by 27 (in Fusoris words, 'subtripling'). Using the notation of Figure 9.10 we obtain:

$$\frac{\Delta_j^1}{3}\,\frac{\Delta_j^2}{9}\,\frac{\Delta_j^3}{27}$$

Now, a modern calculation with Taylor series expansion of the chord function indicates that, if h is small (here 0;15°), this method gives a fairly accurate approximation of the δ^i_j:

$$\left|\frac{\Delta_0^1}{3} - \delta_1^1\right| \leq 6h^3, \left|\frac{\Delta_0^2}{9} - \delta_2^2\right| \leq 6h^4, \left|\frac{\Delta_0^3}{27} - \delta_3^3\right| \leq 6h^3 \; [B]$$

These results are astonishing and justify dividing the first, second, and third differences of the long and middle entries described by Fusoris in the above passage by three, nine, or twenty-seven, respectively, to obtain differences δ^i_j of his table. So, the 'division by three' of the differences of the long and middle entries strongly suggests their use through the following approximations:

$$\frac{\boldsymbol{\Delta_0^1}}{\mathbf{3}} \approx \boldsymbol{\delta_1^1}, \frac{\Delta_0^2}{9} \approx \delta_2^2, \frac{\Delta_0^3}{27} \approx \delta_3^3 \; [C]$$

Mathematically speaking, it is not enough to know that the third differences are almost constant to calculate the second and first differences and, finally, the entries. In addition, it is necessary to know, for example, one second difference and one first difference. It is this necessity that led me to interpret Fusoris's not very precise text in the sense that I have given above. However, in the absence of an explanation from Fusoris, I do not know how these approximations could be found.

The end of the above passage says that by going back from the third to the first differences we will find the entries we are looking for. With these assumptions, we can now develop the calculation of the values x_1 and x_2. Let us take into account the following diagram (Figure 9.11) corresponding to the beginning of the chord table. The unknown quantities are to be determined.

$x_0=0;0$			
	δ_0^1		
x_1		δ_0^2	
	$\delta_1^1=0;15,42,28,15,50$		$\delta_0^3=16\times60^{-4}$
x_2		δ_1^2	
	δ_2^1		$\delta_1^3=16\times60^{-4}$
$x_3=0;47,7,24,47,31,36$		$\delta_2^2=48\times60^{-4}$	

Figure. 9.11. Short entry calculation diagram.

1) The third constant differences, $\delta^3_j = \delta^3_3$, with $j \in \{0, 1, \dots\}$, were obtained with the third relation of [C] above.
2) The second relation of [C] gives δ^2_2.
3) the first of relation of [C] gives δ^1_1.

Mathematically, we thus obtain a system of equations that are always solvable and, moreover, redundant, because there are more known than unknown data, taking into account the relationship between differences. For example, x_3 can be calculated from other data, but it is already known because it is a long or middle entry. The elementary solution of this problem allows us to know all the numbers in the above diagram and possibly others below them, such as x_4, x_5, and so on, as long as the δ^3_j remains 'constant'. One should note, however, that the first approximation is not accurate: $6h^3 \simeq 0;0,0,0,7^{\,p}$. The value of this is likely to influence the differences of Fusoris given at four positions.

In this case, the calculations used to fill in the above diagram are basic (Figure 9.11). Unfortunately, many mathematical errors are found in the first differences because the δ^1_i with i = 1, 2, or 3 are approximate, in particular δ^1_1. Thus mathematical errors larger than rounding errors appear. Fortunately, it is easy to empirically adjust the first differences, δ^1_j, by calculating the differences from the approximations [C]. Only additions and subtractions are needed. After calculating all the differences, we calculate the following:

$$x_0 + \delta^1_0 + \delta^1_1 + \delta^1_2 = D$$

which should give x_3 if the first differences are exact. Then, we calculate the following difference:

$$\frac{x_3 - D}{3} = d$$

Finally, we add to or subtract (according to d) from the differences δ^1_0, δ^1_1, and δ^1_2 to adjust them. Then, we can calculate x_1 and x_2. In fact, this adjustment only concerns the last three sexagesimal positions of the differences and can be done quickly. The large number of errors on the first differences may be due to this approximated adjustment.

2.6. An example

Here is an example of how Fusoris may have calculated the short entries. Let us say we want to calculate the short chords of 91;45° and 92°. The following diagram is thus considered:

x_0(91;30)=85;57,22,26,22,24,50			
	δ_0^1		
x_1		δ_0^2	
	δ_1^1=0;10,55,26,17,11		δ_0^3
x_2		δ_1^2	
	δ_2^1		δ_1^3
x_3(92;15)=86;30,8,45,14		δ_2^2=0;0,1,28,55,58	
	δ_3^1		δ_2^3
		δ_3^2	
			δ_3^3=0;0,0,0,11,6

Figure 9.12. Example calculation of short entries, known quantities and calculation grid.

The values δ^1_1, δ^2_2, and δ^3_3 were obtained with the three approximations [C] calculated from the long and middle entries. I have done the calculations up to the fifth position. Filling in the table is easily done by means of additions and subtractions to obtain the following:

x_0(91;30)=85;57,22,26,22,24,50			
	δ_0^1=0;10,56,54,50,57		
x_1		δ_0^2=0;0,1,28,33,46	
	δ_1^1=0;10,55,26,17,11		δ_0^3=0;0,0,0,11,6
x_2		δ_1^2=0;0,1,28,44,52	
	δ_2^1=0;10,53,57,32,19		δ_1^3=0;0,0,0,11,6
x_3(92;15)=86;30,8,45,14		δ_2^2=0;0,1,28,55,58	
	δ_3^1=0;10,52,28,36,21		δ_2^3=0;0,0,0,11,6
		δ_3^2=0;0,1,29,7,4	
			δ_3^3=0;0,0,0,11,6

Figure 9.13. Example calculation of short entries and filled grid.

Finally, one can adjust the first differences; for that, we calculate:

$$D = x_0 + \delta_0^1 + \delta_1^1 + \delta_2^1$$
$$D = 86;\ 30,08,45,02,51,50.$$

We can see that D is smaller than x_3 of 11,8 of the fourth position. We divide by three and find d = 3,43, which we round up to four and add to the fourths of the first three differences to adjust them. We find the following:

x_0(91;30)=85;57,22,26,22,24,50			
	δ_0^1=0;10,56,54,**55**		
x_1=**86;08,19,21**		δ_0^2	
	δ_1^1=0;10,55,26,**21**		δ_0^3
x_2=**86;19,14,48**		δ_1^2	
	δ_2^1=0;10,53,57,**36**		δ_1^3
x_3(92;15)=86;30,8,45,14		δ_2^2	
	δ_3^1		δ_2^3
		δ_3^2	
			δ_3^3

Figure 9.14. Same as Figure 9.13, but adjusted.

The results agree with those of the manuscript (see Appendix 2), with a few rounding errors. We see here that before the adjustment the errors are larger than one and are no longer rounding errors.

I must stress that all these calculations, of course, remain hypothetical,[39] even if the intervention of the approximations given above ([C]) seems probable when taking into account the scarcity of information provided by Fusoris. This method, which I

39 The method gives the numbers, but it remains possible that another method could produce the same results.

consider realistic, is close to the *false position*[40] of the mathematicians of the Middle Ages. We can imagine a grid of calculations with known numbers that can be filled in gradually. This is what the text suggests when it says *replendo duo spacia media pro duabus cordis mediis non demonstratis (…).*[41] The Latin word *replendo* indicates the filling of a grid.

2.7. Reliability of the method

As stated above, my hypothesis requires taking four long/middle entries to obtain an estimate of the third difference (Δ^3_j). This corresponds to the diagrams of Figure 9.10, with X_0, X_1, X_2, and X_3. From there, the third-order-polynomial-difference-interpolation method *gives* x_1, x_2, x_4, x_5, x_6, x_7, x_8, and x_9. These interpolations can be checked against $x_6 = X_2$, $x_9 = X_3$ values, which have previously been calculated as long/middle entries. After the proposed interpolations, x_1, x_2, x_4, x_5, x_7, and x_8, the procedure can be restarted again on X_3, X_4, X_5, X_6, and so on. However, several mathematical schemes are also possible. For example, after having worked on X_0, X_1, X_2, and X_3, next X_1, X_2, X_3, and X_4 are worked on, followed by X_2, X_3, X_4, X_5, and so on. This would have provided higher levels of accuracy, but it would also have been labour-intensive.

The only limitation of reliability of the method is the constancy of the third differences on the interval of entries $[X_j; X_{j+3}]$ considered, knowing that they are limited (in the table) to the fourth sexagesimal position with respect to the precision required for the short entries. Now, these third differences evolve on the [0; 180°] interval and this happens all the more rapidly as the angle approaches 180°. The third differences are approximately proportional to the third derivative of the chord, that is, a $\cos(x/2)$ that varies as $\sin(x/2)$ and whose maximum value is reached in 180°. Therefore, if the method is reliable with respect to the requested accuracies around 180°, it will be reliable elsewhere.

2.8. Testing the hypothesis

I applied my hypothesis to the end of the chord table for the angular values from 177°;45 to 180°. In the end of his table, the format of the entries no longer conforms to the format of the other entries in the table that Fusoris adopted in other regions of the table with alternating entries: long, short, short, middle, short, short, and so on. The chords of the angles 178°;45–179°;00–179°;30–179°;45, which should be short with three positions, have five positions; I have noted them as **L'**. In the table below (Figure 9.15), I give the entries of the angles varying from 177°;45 to 180°.

---

40 The *false position*, see Smith, *History*, 1925, II, p. 437. For example, let us solve the equation 3x–23 = 0 using a false single position. We assume x = 9, and it comes in the equation 3 × 9–23 = 4, which is different from zero because nine is a false solution. After, we calculate 9 (the guess) - 4 (the false result) /3 = 23/3, which is the correct result. Today, the false position is interpreted as a way of getting back to a simple problem of proportionality. The two equations above can be written today as 3(9–x) = 4, where 9–x is proportional to four.

41 By filling the median spaces with the two median chords not demonstrated. See Poulle, *Un constructeur*, 1963 p. 193.

Angle	My hypothesis	Table of J. Fusoris	Type	Exact values rounded
177;45	119;58,36,43,40,23,52	119;58,36,43,40,23,52	L	119;58,36,43,40,24
178;00	119;58,54,12,16	119;58,54,12	C	119;58,54,12
178;15	119;59,09,37,30	119;59;09,37	C	119;59, 9,37
178;30	119;59,22,59,22,14,35	119;59,22,59,22,14,35	L	119;59,22,59,22,14,35
178;45	119;59,34,17,53,**18**	119;59,34,17,53,**35**	L'	119;59,34,17,53,**23**
179;00	119;59,43,33, 2,**40**	119;59,43,33,2,**54**	L'	119;59,43,33, 2,**45**
179;15	119;59,50,44,50,12	119;59,50,44,50,12	M	119;59,50,44,50,12
179;30	119;59,55,53,15,**41**	119;59,55,53,15,**29**	L'	119;59,55,53,15,**37**
179;45	119;59,58,58,**19**	119;59,58,58,18,**45**	L'	119;59,58,58,18,**54**
180;00	120;00,00,00,00,00,00	120;00,00,00,00,00,00	L	120;00,00,00,00,00

Figure 9.15. Last *entries.*

One can see, first, that Fusoris's computations are accurate and match the modern values of the chord at least to the third and when provided to the fourth sexagesimal position. In other words, the finite difference method adopted by Fusoris is valid for the whole table. Second, the result of my recomputations are in good agreement with those of Fusoris. Only four discrepancies occur on the fifth sexagesimal position. Although I am not able to explain them, I note that they all correspond to the situation where Fusoris's values are also not exact with respect to the modern ones. Thus my recomputation hypothesis accounts for the values of the table and for the clues provided by the text of Fusoris.

2.9. *Comments on the construction of the chords table*

Whichever hypothesis is made, the use of a third-order-polynomial-difference-interpolation method is clear and this method of interpolation seems to have been original at that time. al-Kashi (1380–1429) developed a similar method by considering the finite differences of the second order as constant[42]. It is important to note that the expression "third-order-polynomial-difference-interpolation method" refers to a general modern mathematical concept of which Fusoris was not aware. Fusoris used an interpolation method of which I do not know the origin, but that today appears to refer to this concept. Jost Bürgi (1552–1632) used a third-order polynomial interpolation differences method to calculate the sine of small angles in his table[43], but Bürgi does not seem to have been aware of the works of Fusoris.

3. Precision of calculations

As I mentioned above, notions of precision of numbers and calculations very much preoccupied Fusoris. Much of his innovation arises from his attention to the assured precision

42 Edward S. Kennedy, 'A Medieval Interpolation Scheme Using Second Order Differences', in *A Locust's Leg: Studies in Honour of S. H. Taqizaden*, ed. by Walter Bruno Henning and Ehsan Yarshater (Lund: Humphries, 1962), pp. 117–20.

43 Denis Roegel, 'A Preliminary Mote on Bürgi's Computation of the Sine of the First Minute', unpublished paper available at: https://hal-01316358.

of tabular values. One may wonder why the long entries have six sexagesimal positions, the middle entries have four, and the short entries and the sines of the sine table have only three. I shall hypothesize here, because Fusoris gives only indirect information on this subject, that this could be a weighted choice between obtaining an assured precision for his sines at three sexagesimal positions while reducing calculation time. His tables were slightly more accurate than the other tables he might have known, as Gassendi seemed to suggest when he reported on the dedication of Fusoris in 1654.[44] This can be seen in Figure 9.16 below. There, I list some trigonometry tables that we know to be older than or contemporary to Fusoris. I have given some characteristic elements of these tables. The norm corresponds to the sine (90°).

Author[45]	**Date**	**Function**	**Norm**	**No. digits**	**Step**	**Sys. No.**
Ptolemy (Almagest)	*c.* 150	chord	**60**	**2**	0°;30	Sex
Toledan Tables (ascribed to Alkhwarizmi)	*c.* 800	sine	**150**	**2**	1°	Sex
John of Lignières (from the Toledan Tables)	*c.* 1322	chord/ sine	**60**	**2**	0°;30	Sex
Levi ben Gerson	*c.* 1342	sine	**60**	**2**	0°;15	Sex
Anonymous (John of Murs?)	*c.* 1350	sine	**60**	**2**	0°;1	Sex
Jean Fusoris	*c.* 1430	chord/ sine	60	3/4/6	0°;15	Sex
John of Gmunden	*c.* 1437	chord/ sine	**60**	**2**	0°;30	Sex

Figure 9.16. Some tables preceding or contemporary with those of Fusoris.

From the six sexagesimal positions of the long entries, he calculates the middle one,[46] of which he would keep only four exact sexagesimal positions (saving two positions), and from the long and middle entries, he would calculate the short entry, of which he would keep only three sexagesimal positions. We notice that the entries of 0°;45–1°;30–2°;15–3°;0 are all long, whereas afterwards there are alternations between the long and middle entries. This may lead us to think that after having made attempts to calculate the short entries,

44 Pierre Gassendi, 'Georgi Peurbachii et Ioannis Mulleri Regiomontani vita', in *Opera omnia*, 6 vols (Lyon: L. Anisson and J. B. Devenet, 1658), V, pp. 519-34 (520-21).

45 G. J. Toomer, 'A Survey of the Toledan Tables', *Osiris*, 15 (1968), 5–174. José Chabás and, Bernard R. Goldstein, *A Survey of European Astronomical Tables in the Late Middle Age* (Leiden: Brill, 2012); Paris, BnF, lat. 7293, f. 4^{v}; Hubertus L. L. Busard, *Der Traktat De sinibus, chordis et arcubus von Johannes von Gmunden* (New York: Springer: 1971).

46 The calculation of the middle chords is easily obtained from the long chords from the chord duplication formula known to Fusoris, $\mathrm{crd}^2(x) = 2\mathrm{x}60^2{-}60\ \mathrm{crd}(180{-}2x)$, where $\mathrm{crd}(180{-}2x)$ is a long entry. The result $\mathrm{crd}^2(x)$ is the square of a middle chord which is found at six sexagesimal positions in the table of squares and which passes to four positions after extraction of the root, in the table of strings. This confirms a light extraction of the square root.

he may have understood (I do not know how) that some entries could be calculated only at four sexagesimal positions.

All this indicates that Fusoris would have had some knowledge of the fact that in a sequence of calculations, in order to obtain a given level of precision at the end, the first calculations must initially be made with a higher level of precision since each operation introduces rounding errors. The different precisions of Fusoris tabular values thus correspond to results deftly truncated at the limit of what Fusoris perceived, in most cases rightly, as the accuracy of his computation method.

4. Conclusion

This text by Jean Fusoris on the establishment of trigonometric tables is a surprising one. It is certain that his approach emerged from a mathematical-astronomical context that was probably known to the author, although he makes few explicit references to it. As I have shown above, Fusoris created and used original and complex mathematical tools compared to those deployed by other European mathematicians of the same period.[47] He created them in a way that was adapted to his needs. In addition, he showed a definite interest in managing the precision of numbers in order to achieve the desired level of accuracy in the final calculation results and he devised algorithms that were intended to be simple so that they could be practised by non-specialists (in particular, the division algorithm). This genuine mathematical innovation would lead him to surpass his predecessors.

More generally, we see here that the history of medieval mathematics, up to the elaboration of the trigonometric table (a purely mathematical work that owes nothing to observation), was influenced by astronomical sources that, in terms of tables, have their own way of presenting and working with mathematical data.[48]

Unfortunately, these tables would remain confidential, as would the innovative means and methods he developed. They were not printed. Gassendi learned of Fusoris's work from a priest — a professor at the college in Clermont — who sent him a manuscript. This manuscript is now lost and the trigonometric work of Fusoris survives only in BnF, lat. 7290. Gassendi[49] published a large part of Fusoris's introduction while underlining the great quality of his work; moreover, the tables constituted progress in comparison with the previous ones of the same kind. Yet these tables were seldom if ever used. The reasons for this disaffection are probably multiple. Among other things, Fusoris was not able to complete his work with astronomical tables and strong competition would emerge when Giovanni Bianchini and later Regiomontanus also produced new trigonometrical tables.[50] These tables were better connected to the emerging printers-humanist milieus than were those of Fusoris.

---

47 Ghislaine L'Huillier, *Le Quadripartitum numerorum de Jean de Murs, Introduction et édition critique* (Geneva: Droz, 1990).

48 Husson, 'La tabula tabularum', p. 122.

49 Gassendi, 'Peurbachii et Regiomontani', V, p. 520.

50 Glen Van Brummelen, 'The Emergence of Auxiliary Astronomical Tables in Europe', this volume; José Chabás and Bernard R. Goldstein, *The Astronomical Tables of Giovanni Bianchini* (Leiden: Brill, 2009).

Appendix 1

Figure 9.17. Table of versine. BnF lat. 7290, f. 36r.

Appendix 2

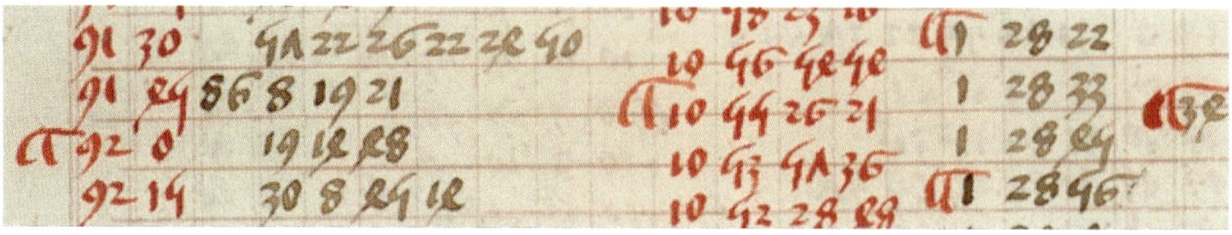

Figure 9.18. Extract of the table of chords, angle from 91; 30° until 92;15°. BnF lat. 7290, f. 7^{r}.

Manuscript sources

Paris, Bibliothèque nationale de France, lat. 7290
Paris, Bibliothèque nationale de France, lat. 7293

Bibliography

Bellhouse, David R., 'An Analysis of Errors in Mathematical Tables'. *Historia mathematica*, 42 (2015), 280–95.

Bond, John D., 'Plane Trigonometry in Richard Wallingford's Quadripartitum de sinubus demonstratis', unpublished PhD thesis, University of Michigan, 1924.

Boudet, Jean-Patrice, *Le recueil des plus célèbres astrologues de Simon Phares*, 2 vols (Paris: Champion, 1997–99).

Brummelen, Glen van, *The Mathematics of the Heavens and the Earth: The Early History of Trigonometry* (Oxford: Oxford University Press, 2009).

Busard, Hubertus L. L., *Der Traktat De sinibus, chordis et arcubus von Johannes von Gmunden* (New York: Springer, 1971).

Cardano, Girolamo, 'Liber unicus', in *Hieronymi Cardani Mediolanensis, philosophi ac medici celeberrimi, opera omnia*, 10 vols (Lyon, Huguetan & Ravaud, 1663), IV, pp. 1-220.

Chabás, José, *Computational Astronomy in the Middle Ages: Sets of Astronomical Tables* (Madrid: Consejo Superior de Investigaciones Científicas, 2019).

____, and Bernard R. Goldstein, *A Survey of European Astronomical Tables in the Middle Ages* (Leiden: Brill, 2012).

____, and ____, *The Astronomical Tables of Giovanni Bianchini* (Leiden: Brill, 2009).

Dalen, Benno van, 'A Statistical Method for Recovering Unknown Parameters from Medieval Astronomical Tables', *Centaurus*, 32, (1989), 85-145.

____, 'Ancient and Mediaeval Astronomical Tables: Mathematical Structure and Parameter Values', unpublished PhD thesis, University of Utrecht, 1993.

Debarnot, Marie-Thérèse, 'Trigonometry', in *Encyclopedia of the History of Arabic Science*, ed. by Roshdi Rashed in collaboration with Régis Morelon (London: Routledge, 1996), pp. 495–538.

Delambre, Jean Baptiste J., *Histoire de l'astronomie du moyen âge* (Paris: Courcier, 1819).

Duhem, Pierre, *Le système du monde, Histoire des doctrines cosmologiques de Platon à Copernic*, 10 vols (Paris, Hermann, 1913–59; repr. 1974).

Euclid, *Les Eléments, Traduction et commentaire par B. Vitrac*, 2 vols (Paris: Presses Universitaires de France, 1990–94).

Fabry, E., *Traité de mathématiques générales*, 3d rev. ed. (Paris: A. Hermann & fils, 1916).

Gaffiot, Félix, *Dictionnaire illustré latin-français* (Paris: Hachette, 1934).

Gassendi, Pierre, *Opera omnia*, 3 vols (Lyon: L. Anisson and J.B. Devenet, 1658).

Guenée, Bernard, '*Authentique et approuvé*. Recherches sur les principes de la critique historique au Moyen Âge', in *Politique et histoire au Moyen Âge: recueil d'articles sur l'histoire politique et l'historiographie médiévale (1956–1981)* (Paris: Publications de la Sorbonne, 1981), pp. 265–78.

Goldstein, Bernard R., 'A New Set of Fourteenth Century Planetary Observations', *Proceedings of the American Philosophical Society*, 132 (1988), 371–99.

Husson, Matthieu, 'La tabula tabularum de Jean de Murs et les modèles de l'arithmétique médiévale', *Cahiers de recherches médiévales et humanistes*, 27 (2014), 97–122.

Kennedy, Edward S., 'A Medieval Interpolation Scheme Using Second Order Differences', in *A Locust's Leg: Studies in Honour of S. H. Taqizaden*, ed. by Walter Bruno Henning and Ehsan Yarshater (Lund: Humphries, 1962), pp. 117–20.

Kremer, Richard L., 'Cracking the *Tabulae permanentes* of John of Murs and Firmin of Beauval with Exploratory Data Analysis', in *Editing and Analysing Numerical Tables: Towards a Digital Information System for the History of Astral Sciences* (Turnhout: Brepols: 2022), pp. 363–422.

Lacroix, Sylvestre F., *Traité de calcul différentiel et de calcul* intégral, 2 vols (Paris: Duprat, 1797–98).

Lalande, Joseph-Jérôme Lafrançais de, *Bibliographie astronomique avec l'histoire de l'astronomie depuis 1781 jusqu'à 1802* (Paris: Imprimerie der la République, 1803).

L'Huillier, Ghislaine, *Le Quadripartitum numerorum de Jean de Murs, Introduction et édition critique* (Geneva: Droz, 1990).

Montucla, Jean Etienne, *Histoire des mathématiques*, new edition, 4 vols (Paris: Agasse, 1798–1802).

Miolo, Laure, 'La formation d'un modèle scientifique universitaire: Le traitement du *quadrivium* au sein de la bibliothèque du collège de Sorbonne (XIIIe = XIVe siècle)', in *Apprendre, produire, se conduire: Le modèle au Moyen Âge, XLVe Congrès de la SHMESP 2014* (Paris: Éditions de la Sorbonne, 2015), pp. 275-284.

Mirot, Léon M., 'Le procès de Maître Jean Fusoris, Chanoine de N. D. de Paris', *Mémoires de la Société de l'Histoire de Paris et de l'Ile de France*, 27 (1900), 137–287.

North, John, *God's Clockmaker: Richard of Wallingford and the Invention of Time* (London: Hambledon and London, 2005).

Pedersen, Fritz S, *The Toledan Tables* (Copenhagen: C. A. Reitzels, 2002).

Pedersen, Olaf, *A Survey of the Almagest* (Odense: Odense University Press, 1974).

Poulle, Emmanuel, *Un constructeur d'instruments astronomiques au xve siècle: Jean Fusoris* (Paris: Champion, 1963).

____, 'Regiomontanus on Triangles; De triangulis omnimodis by Johann Müller, Otherwise Known as Regiomontanus', Translated by Barnabas Hughes, O. F. M., with an Introduction and Notes', *Bibliothèque de l'École des chartes*, 125 (1967), 520–22.

____, 'Jean de Murs et les tables alphonsines', *Archives d'histoire doctrinale et littéraire du moyen âge*, 56 (1980), 241–71.

____, *Les Tables Alphonsines avec les canons de Jean de Saxe:* Édition, traduction et commentaire (Paris: Éditions du centre national de la recherche scientifique, 1984).

____ and Owen Gingerich, 'Les positions des planètes au Moyen Âge: Application du calcul électronique aux tables alphonsines', *Académie des inscriptions et belles lettres, Compte rendus* (1967), 531–48.

Ptolemy, *Almagestum Cl. Ptolemei Pheludiensis Alexandrini astronomorum principis: opus ingens ac nobile omnes coelorum motus continens* (Venice: Peter Liechtenstein, 1515).

Regiomontanus, *De Trangulis omnimodis*, ed. by Johann Schöner (Nuremberg: Johann Petreius, 1533).

Roegel, Denis, 'A Preliminary Note on Bürgi's Computation of the Sine of the First Minute', unpublished paper available at https://hal.inria.fr > *hal-01316358*.

Sacrobosco, 'Tractatus de arte numerandi', in *Rara Mathematica*, ed. by J. O. Halliwell (London: Maynard, 1839).

Smith, David E, *History of Mathematics*, 2 vols (Boston: Ginn, 1925).

Thorndike, Lynn, *A History of Magic and Experimental Science*, 8 vols (New York: Macmilan, 1923–58).

Toomer, G. J., 'A Survey of the Toledan Tables', *Osiris* 15 (1968), 5–174.

____, *Ptolemy's Almagest* (London: Duckworth, 1984).

Zepeda, Henry, 'Euclidization in the *Almagestum parvum*', *Early Science and Medicine*, 20 (2015), 48–76.

____, *The First Latin Treatise on Ptolemy's Astronomy: The Almagesti minor (c. 1200)* (Turnhout: Brepols, 2018).

ELEONORA ANDRIANI AND STEFAN ZIEME

Regiomontanus on the Equation of Time

Computational Practices of an Alfonsine Astronomer[*]

Introduction

The equation of time is an essential tabular quantity in all sets of historical astronomical tables. While astronomical computations were performed in mean solar time, daily cultural life from antiquity to the early modern period was ordained by apparent solar time, which could be read on a sundial. This difference between mean and apparent time is determined by the equation of time. It is considerably small, but can amount to about half of an hour of time over the course of a year. Therefore, the difference is irrelevant when determining the positions of slow-moving celestial bodies such as Jupiter or Saturn, but it becomes pivotal for the position of the Moon and thus for the determination of solar and lunar eclipses. For astrological purposes, for example nativities and horoscopes, the equation of time may have presented a crucial component for the determination of the ascendant and the linked house divisions, as a time difference of about half an hour can amount to shifts of about eight degrees for the ascendant.[1] Notably, when a specific point in time was sought as part of an astrological inquiry, the equation of time may have been taken into account by diligent practitioners.

Regiomontanus (1436–76), one of the most renowned astronomers in early modern Europe, was a diligent practitioner and used the equation of time whenever he cast a horoscope.[2] His handwritten ephemerides that cover (with gaps) the years from 1448

* We are indebted to José Chabás, Matthieu Husson, and Richard L. Kremer for their helpful comments on our contribution. We are also grateful to Alena Hadravová for her valuable advice on the edited text included in this work. This research was supported by the ERC project *ALFA; Shaping a European scientific scene: Alfonsine Astronomy* (CoG 723085, PI Matthieu Husson). Stefan Zieme acknowledges the support of the Cluster of Excellence *Matters of Activity. Image Space Material* funded by the Deutsche Forschungsgemeinschaft (DFG, German Research Foundation) under Germany's Excellence Strategy – EXC 2025 – 390648296.

1 On horoscopes and house systems, see John D. North, *Horoscopes and History* (London: The Warburg Institute, 1986). For some historical examples, including computational practices, see, in the present volume, Alexandre Tur, 'From astronomical computation to astrological interpretation: *figurae caeli* and "astrological squares" in fifteenth-century annual prognostications'.

2 On Regiomontanus, see Ernst Zinner, *Leben und Wirken des Joh. Müller von Königsberg, genannt Regiomontanus*, 2nd ed. (Osnabrück: Zeller, 1968).

Eleonora Andriani • Centre national de la recherche scientifique, Paris.
Stefan Zieme • Humboldt-Universität zu Berlin.

Alfonsine Astronomy: Computational Practices, ed. by Matthieu HUSSON, José CHABÁS, and Richard L. KREMER, Turnhout, 2026 (*Alfonsine Astronomy: Studies and Sources*, 3), p. 307-343
© BREPOLS PUBLISHERS 10.1484/M.ALFA-EB.5.137915

to 1463 contain numerous horoscopes for various astrological configurations.[3] For the year 1448 for the entry of the Sun into Aries, the annual revolution (Figure 10.1), and for the entry of the Sun into the other signs, Regiomontanus used the equation of time that is attributed to Peter of St Omer (fl. 1293) and stated the corresponding times to the second with and without the correction from the equation of time.[4] This table for the equation of time was one of the most distributed ones of the Alfonsine period.[5] However, for the later years (1451–55) of his ephemerides, Regiomontanus switched to a different equation of time. For these years only the times for the annual revolutions — the Sun's entry into Aries — are given to seconds, which Regiomontanus corrected with the equation of time included in the *Tabulae astronomiae* of Giovanni Bianchini.[6] Bianchini's table for the equation, however, was not newly computed, but simply a conversion of al-Battānī's (d. 929) old table converted from degrees to hours.[7] For later years, starting from 1456, Regiomontanus gave the corrected times for the annual revolutions only to minutes. Thus, his correction does not allow us to specify which table for the equation of time he used, but it is in complete agreement with Bianchini's table. Since Regiomontanus is a proven user of the tables of Bianchini, it does not come as a surprise that he used the equation of time contained therein.[8] Starting with the year 1458, Regiomontanus also corrected syzygy times in his ephemerides by the equation of time. Notably, for the months of the year 1458, he first gave syzygy times without the correction by the equation of time, but he then subsequently crossed out these numbers and added corrections that agree with the equation of time contained in

3 See Vienna, Österreichische Nationalbibliothek, Cod. 4988. The ephemerides for later years, from 1464–84, bound in the same manuscript, are not in Regiomontanus's hand and were bought by Johannes Schöner in Bamberg in 1509. See Zinner, *Regiomontanus*, pp. 297–98.

4 A few values for the equation of time used by Regiomontanus are as follows: 0° Aries: 8;8min; 0° Taurus: 17;56min; 0° Gemini: 21;20min; 0° Cancer: 16;28min; 0° Leo: 11;28min. These values match the equation of time attributed to Peter of St Omer converted from degrees to hours of time. See Fritz S. Pedersen, *The Toledan Tables. A Review of the Manuscripts and the Textual Versions with an Edition* (Copenhagen: Reitzel, 2002), p. 985. For non-integer values of the longitude of the true Sun in some of the horoscopes, Regiomontanus used linear interpolation on the same table for the equation of time.

5 It is, for example, included in the manuscript Nuremberg, Stadtbibliothek, Cent. VI, 18 (ff. 46^{v}–47^{r}), which was copied by Regiomontanus around 1452 in Vienna and contains the tables of Johannes von Gmunden. On the manuscript, see Zinner, *Regiomontanus*, p. 304. See also, Ingeborg Neske, *Die Handschriften der Stadtbibliothek Nürnberg, Bd. IV: Die lateinischen mittelalterlichen Handschriften, Varia, 13.-15. und 16–18. Jahrhundert* (Wiesbaden: Harrossowitz, 1997), pp. 172–74.

6 The corresponding value reads for 0° Aries 8;40min. On Bianchini and his astronomical tables, see José Chabás and Bernard R. Goldstein, *The Astronomical Tables of Giovanni Bianchini* (Leiden: Brill, 2009), pp. 13–21.

7 Chabás, Goldstein, *Bianchini*, pp. 100–2. Al-Battānī's table was usually included in manuscripts containing the Toledan Tables. See Pedersen, *Toledan Tables*, pp. 968–77. It was reused in the Parisian Alfonsine Tables. See, for example, the first printed edition *Tabule astronomice illustrissimi Alfontij regis Castelle* (Venice: Ratdolt, 1483), sig. $k1^{r}$-$k2^{r}$. See also José Chabás, *Computational Astronomy in the Middle Ages. Sets of Astronomical Tables in Latin* (Madrid: Consejo Superior de Investigaciones Científicas, 2019), pp. 260–61.

8 In a letter dated 1456 and addressed to Johann Nihil, court astrologer to Emperor Frederick III, Georg Peurbach stated that both he and Regiomontanus used Bianchini's tables to compute planetary positions. When their results diverged, they checked the positions again. Peurbach was more used to the Alfonsine Tables. See Albin Czerny, *Aus dem Briefwechsel des Grossen Astronomen Georg von Peurbach* (Wien: Tempsky, 1888), No. 9, pp. 22–3. See also Zinner, *Regiomontanus*, p. 33. On Regiomontanus's ephemerides and his use of Bianchini's planetary tables, see, in the present volume, Richard L. Kremer, 'What Can we Learn about Computational Practices from a Study of Medieval Ephemerides?'.

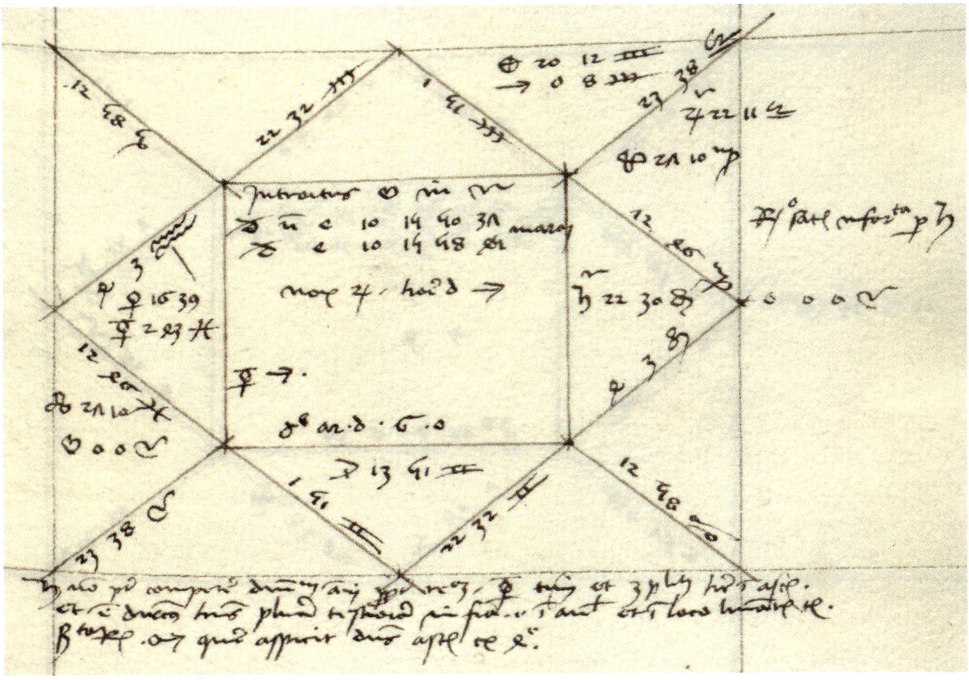

Figure 10.1. Annual revolution for 1448 from Regiomontanus's autograph ephemerides. Regiomontanus corrected the instant of time when the Sun (heart-shaped symbol) enters into Aries from non-equated to equated days by eight minutes and eight seconds (introitus solis in Arietem: diebus non equatis 10 15 50 37; diebus equatis 10 15 58 45 Marcii). Vienna, Österreichische Nationalbibliothek, Cod. 4988, f. 9r.

Bianchini's planetary tables. For later years, Regiomontanus silently corrected syzygy times by the equation of time.[9]

Regiomontanus's example shows his acute awareness of the existence of different tables for the equation of time derived in and for different periods and epochs. However, the two different tables he demonstrably used, and others that circulated in manuscript form, never indicated when or for which period they had been derived. Johannes of Gmunden's (*c.* 1384–1442) tables from the mid 1430s contain the equation of time attributed to Peter of St Omer most likely derived in and for a period around the year 1300. Although, nowhere in the different copies of this table is a date given; they are only inscribed with their usual, common, and plain title *Tabula aequacionis dierum cum noctibus suis*.[10] Giovanni Bianchini's *Tabulae astronomiae* from around 1442 contains al-Battānī's equation of time, which was also included in the Toledan Tables as well as in the Parisian Alfonsine Tables. In Bianchini's planetary tables, the equation was presented without any date, yet it was most certainly

9 For an example of crossing out and correcting syzygy times, see ÖNB, Cod. 4988, f. 113v. We are indebted to Richard L. Kremer for pointing this out to us.

10 Beatriz Porres de Mateo, 'Les Tables astronomique de Jean de Gmunden: Édition et étude comparative', unpublished PhD thesis, Paris, École Pratique des Hautes Études, 2003, pp. 539–40.

calculated for al-Battānī's own lifetime around the year 900.[11] When Regiomontanus began to use Bianchini's tables, he thus made use of an outdated equation of time that originated from a period some 400 years earlier than the one contained in Gmunden's tables.[12]

For Regiomontanus as a calculator, it may have been desirable to derive a novel equation of time valid for his own epoch, knowing that there were differing tables for the equation of time included in different sets of astronomical tables. Contained in his *Rechenbuch* (i.e. calculation book) is a unique text, written by Regiomontanus around 1456, solely dedicated to this very problem: the derivation of a new table for the equation of time. To the best of our knowledge, there is no earlier known Latin text or canon that details its tabular derivation. The *Rechenbuch*, and especially its mathematical parts, were already subject to thorough investigations; yet, the text on the equation of time has been ignored so far.[13] In this article, we provide a critical edition and translation of this text. We further include a detailed mathematical analysis of Regiomontanus's derivation within his own historical mathematical practice. This enables us to demonstrate that Regiomontanus used Bianchini's *Tabulae astronomiae* for his derivation. Unfortunately, the final table found at the bottom of Regiomontanus's text on the equation of time remained unfilled. Nevertheless, we show that the table for the equation of time contained in Georg Peurbach's *Tabulae eclypsium* is based on Regiomontanus's algorithm. Peurbach's eclipse tables were completed around 1460, just four years after Regiomontanus wrote his text in his *Rechenbuch*.[14] We therefore argue that the text in the *Rechenbuch* is some sort of note that Regiomontanus wrote for himself as a zealous student, or the draft of a letter intended for Georg Peurbach, his teacher at the University of Vienna at that time.

In Section 1, we explore the possible sources on astronomical and mathematical aspects of the equation of time that were at Regiomontanus's disposal and included among the manuscripts he owned. In Section 2, we give a short mathematical outline from a modern point of view. A codicological overview of the manuscript is given in Section 3, followed by a discussion on the genre of the text in Section 4. In Section 5, we elaborate on Regiomontanus's historical mathematical practice, and therewith demonstrate, in Section 6, that his algorithm underlies Peurbach's table. We conclude our discussion with some remarks on historical mathematical-astronomical practices and the possible pitfalls in modern analyses. The edition and translation of the text are given in the appendix.

---

11 For the equation of time, see, for example, Nuremberg, Stadtbibliothek, Cent. V, 36, f. 91^{v}. This manuscript belonged to Bernhard Walther rather than Regiomontanus. See Neske, *Handschriften Nürnberg*, pp. 66–7. For other manuscript witnesses, see Chabás, Goldstein, *Bianchini*, pp. 100–2.

12 Non-linear fits of both tables for the equation of time result in a difference of the corresponding underlying solar apogees of about $\Delta\lambda_{aux} \approx 4;30°$, which confirms the relative dating. The absolute difference between Peter of St Omer's and al-Battānī's equation of time varies between +44s and -80s of time over the whole interval of 360° solar longitude, and thus it is quite significant.

13 On mathematical aspects of the *Rechenbuch*, see Menso Folkerts, 'Die mathematischen Studien Regiomontanus in seiner Wiener Zeit', in *Regiomontanus-Studien*, ed. by Günther Hamann (Vienna: Verlag der Österreichischen Akademie der Wissenschaften, 1980), pp. 175–209. The instrument text for finding true syzygies, preceding the text on the equation of time, is discussed in Richard L. Kremer, 'Experimenting with Paper Instruments in Fifteenth- and Sixteenth-Century Astronomy: Computing Syzygies with Isotemporal Lines and Salt Dishes', *Journal for the History of Astronomy* 42 (2011), 223–58.

14 Chabás, *Computational Astronomy*, pp. 365–75.

Regiomontanus's possible sources

Among the extant manuscripts owned by Regiomontanus, there are five different Latin texts by al-Battānī, Jābir ibn Aflaḥ, and al-Farghānī, and from Ptolemy's *Almagest* and the *Almagesti minor*, that contain sections, passages, or certain details about the equation of time.[15]

The manuscript Nuremberg, Stadtbibliothek, Cent. VI, 21 contains al-Battānī's *De scientia stellarum*, which is the Latin translation of the canons to al-Battānī's *Ṣābi ʾ Zīj*: his astronomical tables translated by Plato of Tivoli. It is not an autograph copy by Regiomontanus, but he added rubrics, tables, figures, and several marginalia that attest to his close engagement with the work. The first printed edition from 1537 by Johannes Petreius in Nuremberg is based on this very manuscript and includes some of Regiomontanus's annotations.[16] Chapter 29 of al-Battānī's text is devoted to the equation of time.[17] It outlines that there are mean solar days and apparent solar days that are unequal in length. This inequality of length (i.e. the equation of time), the text continues, has two causes: one being the irregular motion of the true Sun and the other being the unequal rising of arcs of the ecliptic across the meridian, both of which add up to a maximum of 7;48°.[18] The text further states that the mean day and apparent day are about equal in length at the time when the true Sun is in 20° Aquarius and that in relation to this time al-Battānī's table for the equation of time was computed. How to find the difference between mean days and apparent days for a specific interval is also outlined in the text, yet the latter falls short in providing the details of how to derive a table.

Jābir ibn Aflaḥ's *Iṣlāḥ al-Majisṭī* or *Liber super Almagesti* in the Latin translation by Gerard of Cremona is contained as an autograph by Regiomontanus in Nuremberg, Stadtbibliothek, Cent. III, 25.[19] The last chapter of the third book of the *Iṣlāḥ* is on the equation of time.[20] In fact, Regiomontanus made a second autograph copy of the whole *Iṣlāḥ* around 1460 in Vienna, which is extant in Seitenstetten, Stiftsbibliothek, 53.[21] The

15 On Regiomontanus's library in general, see Richard L. Kremer, 'Text to Trophy: Shifting Representations of Regiomontanus's Library', in *Lost Libraries. The Destruction of Great Book Collections since Antiquity*, ed. by James Raven (Basingstoke: Palgrave Macmillan, 2004), pp. 75–90.

16 Zinner, *Regiomontanus*, pp. 321–22. Neske, *Handschriften Nürnberg*, pp. 176–77.

17 In MS Nuremberg, Stadtbibliothek, Cent. VI, 21 on ff. 39ᵛ–40ᵛ.

18 This number does not appear in al-Battānī's tables, however, where the maximum is given by 7;54°. Note that the maximum of 7;48° is identically given in the passage on the equation of time found in the Castilian Canons to the Alfonsine Tables uniquely preserved in Madrid, Biblioteca Nacional de España, 3306. See José Chabás and Bernard R. Goldstein, *The Alfonsine Tables of Toledo* (Dordrecht: Kluwer, 2003), pp. 185–88. Although the maxima in both canons — al-Battānī's and the Castilian Alfonsine Tables — are identical, the values of their two components, respectively, are not. Moreover, the two components in al-Battānī's case (Petreius, 1537) and those in the Castilian Alfonsine Tables do not add up to the given maximum. While the tables of the Castilian Alfonsine Tables have not been found to date, we conjecture that the equation of time contained therein was taken from al-Battānī as indicated by the identical estimate for the maximal value. The maximum value in the Castilian Alfonsine Tables is thus not unprecedented, as assumed in Chabás, *Computational Astronomy*, p. 130.

19 Zinner, *Regiomontanus*, p. 310. Neske, *Handschriften Nürnberg*, pp. 23–24.

20 Nuremberg, Stadtbibliothek, Cent. III, 25, f. 106ʳ.

21 Zinner, *Regiomontanus*, pp. 310–11. In Seitenstetten, SB, 53 the passage on the equation of time is on ff. 33ʳ–34ʳ, including a few marginal glosses.

Iṣlāḥ was first printed by Johannes Petreius in 1534 in Nuremberg.[22] Having copied the whole work twice, Regiomontanus might have been quite familiar with its content. Nevertheless, ibn Aflaḥ's commentary also explains what the equation of time is and does not offer a procedure on how to derive an entire table for it. It thereby follows the logic of the *Almagest*, which lacks a table for the equation of time. In fact, since the *Iṣlāḥ* is a commentary or correction to the *Almagest*, both are quite congruent in their presentation of the equation.

An autograph copy by Regiomontanus of the Latin translation of the *Almagest* by Gerard of Cremona from the Arabic is contained in the same manuscript with the *Iṣlāḥ*.[23] In both works, the estimates of the magnitudes of the two different contributions to the equation of time are identical but in turn differ from al-Battānī's explicit values. Nevertheless, al-Battānī's text is similar in content to both. Yet, al-Battānī additionally specified the true longitude of the Sun, where the apparent and mean days are equal, as 20° Aquarius, which he had used as a radix for his table. We return to this point later in our discussion of Regiomontanus's derivation in Section 5 and show that al-Battānī was one of his main sources.

The manuscript Nuremberg, Stadtbibliothek, Cent. VI, 12 is also an autograph by Regiomontanus, copied during his time in Vienna, and it holds two different works that contain passages on the equation of time: the *Almagesti minor* and al-Farghānī's *Elements of Astronomy*.[24] The first work, the *Almagesti minor*, contains various additional marginal glosses by Regiomontanus. The equation of time is covered in Book 3:18–25 and has no glosses.[25] The first seven of these eight chapters give a lengthy account of the two causes for the equation time. They address the contributions originating from the irregular motion of the Sun and unequal rising of equal arcs of the ecliptic separately. Finally, the last chapter gives a concise account of how to evaluate these contributions jointly for a specific day based on a fixed radix.[26] Most importantly, the last paragraph closes with the statement that the equation of time would always be positive when its radix is set to the beginning of the addition. The latter is the day or solar position when the apparent day starts to become longer than the mean day after both have been equal. However, there is no clear instruction on how to find this day or, consequently, the corresponding radix. The *Almagesti minor* thus contains much more detailed information than any other source, but it still falls short of giving details on how to derive an entire table.

The second work, contained in the same manuscript, al-Farghānī's *Elements of Astronomy*, also includes a very short passage related to the equation of time. The Latin

22 See Jābir ibn Aflaḥ, *Liber super Almagesti*, in Peter Apian (ed.), *Instrumentum primi mobilis* (Nuremberg: Johannes Petreius, 1534). For the passage on the equation of time, see pp. 44–45.

23 In the *Almagest*, the equation of time is covered at the end of Book 3. See Nuremberg, SB, Cent. III, 25, ff. 23^r–23^v. For an English translation, see Gerald J. Toomer (trans.), *Ptolemy's Almagest* (London: Duckworth, 1984), pp. 169–72. We note that in the *Almagest* the equation of time is used to turn apparent solar time into mean solar time, in order to utilize observational data for the determination of parameters. All other tables discussed in this article follow the opposite approach.

24 Zinner, *Regiomontanus*, p. 315. Neske, *Handschriften Nürnberg*, pp. 164–65.

25 Nuremberg, Stadtbibliothek, Cent. VI, 12, ff. 22^r–23^v.

26 For a critical edition and English translation, see Henry Zepeda, *The First Latin Treatise on Ptolemy's Astronomy: The* Almagesti minor (*c.* 1200) (Turnhout: Brepols, 2018), pp. 263–74.

text copied by Regiomontanus is a translation by Gerard of Cremona from the Arabic.[27] Chapter 11 is mainly about seasonal hours.[28] Nevertheless, al-Farghānī's text starts with a short note on the length of the day, from sunrise to sunrise, which depends on the different motions of the Sun and rising times. Essentially, this is related to the equation of time, but al-Farghānī focuses on seasonal hours in the rest of the chapter. In fact, Regiomontanus was well acquainted with al-Farghānī's work and lectured on it in Padua in the spring of 1464.[29]

What all these sources have in common is that they explain the two different causes of the equation of time: the true, irregular motion of the Sun and the unequal rising of equal arcs of the ecliptic.[30] The *Almagesti minor* is the most detailed in this respect. It includes additional subtle information about the relation between the positivity of the equation of time and its radix, but it does not explicate how to find the latter. Al-Battānī states where he found his radix but remains silent about how he determined this position. None of the sources detail a practical algorithm on how to derive an entire table for the equation of time. We argue that the lack of practical instructions on how to derive a table for the equation of time prompted Regiomontanus to write his text. As we show later, he especially relied on information provided in the *Almagesti minor* and by al-Battānī. Before doing so, we provide some further mathematical preliminaries followed by a codicological overview of the manuscript and its text on the equation of time.

Modern mathematical preliminaries

Astronomically, the equation of time is a relation between the ecliptical true Sun and an equatorial mean Sun (Figure 10.2). Mathematically, it is an intricate object that depends on several quantities and parameters. A modern mathematical analysis and discussion of the equation of time included in Ptolemy's *Handy Tables* by Benno van Dalen represents a culmination of the historiographical tradition of analysing equations of time and their underlying parameters.[31] Following van Dalen, we briefly summarize the main mathematical features of the equation of time as a starting point for our analysis.

27 For an edition of John of Seville's Latin translation, see al-Farghānī, *Differentie in quibusdam collectis scientie astrorum*, ed. by Francis J. Carmody (Berkeley: 1943). The first printed edition from 1493 (Ferrara) and the 1537 edition printed in Nuremberg by Petreius, together in one volume with al-Battānī's *De scientia stellarum*, are both based on John of Seville's translation.

28 Nuremberg, SB, Cent. VI, 12, ff. 83^{r}–83^{v}.

29 Zinner, *Regiomontanus*, pp. 110–18. The lecture notes are lost. Only the oration that Regiomontanus gave at the beginning of his lecture is extant and is contained in the 1537 printed edition of al-Farghānī's *Elements of Astronomy*. On the oration, see James Steven Byrne, 'A Humanist History of Mathematics? Regiomontanus's Padua Oration in Context', *Journal of the History of Ideas* 67 (2006), 41–61.

30 Other authors who discuss these causes, but who are not included among Regiomontanus's extant manuscripts, include John of Sicily and John of Lignères. See Fritz S. Pedersen, *Scriptum Johannis de Sicilia super canones Azarchelis de tabulis Toletanis*, 2 vols (Copenhagen: Erik Paludan, 1986), II, pp. 198–99. See Marie-Madeleine Saby, 'Les canons de Jean de Lignières sur les tables astronomiques de 1321. Edition critique, traduction et étude', unpublished thesis, Paris, Ecole nationale des Chartes, pp. 139–41.

31 See Benno van Dalen, 'On Ptolemy's Table for the Equation of Time', *Centaurus* 37 (1994), 97–153, and references therein.

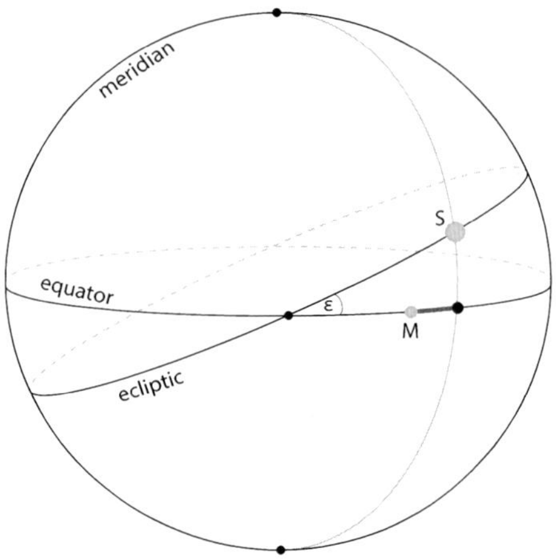

Figure 10.2. Geometrical representation of the equation of time. The equation of time, schematized here as the red line, is the difference between an imaginary mean Sun M that moves on the equator and the true Sun S moving on the ecliptic projected onto the equator along a great circle perpendicular to the equator.

The equation of time is a function of the solar equation and thus depends on the eccentricity e of the underlying solar model, as well as the ecliptic longitude λ_{aux} of the solar apogee. It is a function of the right ascension of the true Sun and thus depends on the obliquity ε of the ecliptic. Furthermore, it depends on the so-called epoch constant, c, which is related to the positivity of the equation of time. The equation of time can either be expressed as a function of true solar longitude or mean solar longitude, which is usually never indicated in historical tables. There are well-known Islamicate *zījes* that contain equations of time as a function of mean solar longitude as well as true solar longitude.[32] However, there is so far no known example for an equation of time as a function of mean solar longitude that was computed in the Latin West. All known cases in the Latin West are functions of true solar longitude. Finally, the equation of time can be expressed in degrees or hours for which a conversion factor is necessary. This conversion factor, denoted D, can take two different values depending on the intended accuracy of the table maker. In summary, every table for an equation of time is, at least, a function of the four parameters e, λ_{aux}, ε, and c. In some cases, there may be two further unknowns: the different functional dependency — either a function of mean or true solar longitude — and the conversion factor D.

All the different parameters comprise different symmetry relations originating in the underlying corresponding geometrical models that are inherited by the equation of time. Therefore, it is possible to determine fast converging estimates for the different parameters by means of modern mathematical tools of multi-parameter non-linear fitting.[33] A formula that can serve for a non-linear fit of tabular values for the equation of time E as a function of true solar longitude λ is given by Benno van Dalen:[34]

32 Edward S. Kennedy, 'Two Medieval Approaches to the Equation of Time', *Centaurus* 31 (1988), 1–8.

33 Benno van Dalen, 'Ancient and Mediaeval Astronomical Tables: Mathematical Structure and Parameter Values', unpublished PhD thesis, Utrecht University, 1993, pp. 97–152.

34 Benno van Dalen, 'al-Khwārizmī's Astronomical Tables Revisited: Analysis of the Equation of Time', in *From Baghdad to Barcelona. Studies in the Islamic Exact Sciences in Honour of Prof. Juan Vernet*, ed. by Josep Casulleras and Julio Samsó, 2 vols (Barcelona: Instituto Millás Vallicrosa de Historia de la Ciencia Árabe, 1996), I, pp. 195–252.

$$E = \frac{1}{D}\left(\lambda + q(\lambda - \lambda_{\mathrm{aux}}, e) - \alpha(\lambda, \varepsilon) + c\right). \qquad (1)$$

Here, α denotes the right ascension that depends on the obliquity ε of the ecliptic, while q is the solar equation as a function of true solar anomaly that also depends on the eccentricity e. However, the solar equation as a function of true anomaly does not appear in any historical set of astronomical tables. Historical tables only include the solar equation as a function of mean anomaly. Yet, the modern formula serves well to capture estimates for the different parameters.

Parameters of underlying geometrical models play an important role in the history of astronomy. They can appear explicitly in certain tables, such as the obliquity of the ecliptic in solar declination tables, or they may be contained implicitly in other tables, such as the eccentricity of the Sun in solar equation tables, and may be reconstructed from statistical tools. In any case, the different or identical use of parameters in astronomical models is a decisive element in mapping the dissemination of astronomical knowledge across space, time, and culture. To these historiographical tools of explicit and implicit parameters, we want to add a new category termed *historical computational practice*, which likewise circulated in space and time. Based on the example of Regiomontanus's equation of time, we wish to scrutinize the historical practice on how to derive a novel astronomical table and what tools, sources, and sub-tables have been used in the derivation. Our aim is to show that the current understanding of the historical practice of the derivation of the equation of time, which is based on compelling results of multi-parameter fits, is a modern projection rather than an actual historical practice. Current understanding overshadows historical geometrical thinking and tabular practices in favour of a modern functional character. The fact that a modern formula can capture reliable results for certain tables does not automatically allow us to draw any conclusion about how the corresponding table was derived in a given historical practice. We show this explicitly in the case of Peurbach's table for the equation of time, which was computed from the algorithm described in Regiomontanus's text. In this particular case, the modern fit does provide plausible parameters, which, however, were not used for the table. Historical practitioners of mathematical astronomy, we argue, generally used available sub-tables and geometrical thinking to lay out and solve astronomical problems at hand.[35] A modern functional approach, which in this case even includes a quantity that does not exist in tabular form (i.e. the solar equation as a function of true anomaly) may capture reliable parameter estimates, but it misrepresents any historical practices and thought.

Vienna, Österreichische Nationalbibliothek, Codex 5203

Known as Regiomontanus's *Rechenbuch*, Vienna, Österreichische Nationalbibliothek, Codex 5203 includes several works by Regiomontanus and others, mainly copied by Regiomontanus himself sometime between 1454 and 1462.[36] This manuscript not only provides clues on

35 For another example of a historical mathematical practice of deriving a table for the equation of time, see Laure Miolo and Stefan Zieme, 'Lewis Caerleon and the Equation of Time: Tabular Astronomical Practices in Late Fifteenth-Century England', *Archive for History of Exact Sciences* 78 (2024), 183–243.

36 Zinner, *Regiomontanus*, pp. 307–8.

the variety of interests characterising the Vienna years of this eminent astronomer, but it also exhibits significant evidence to assess his mathematical achievements.[37]

Codex 5203 is a paper manuscript consisting of 180 folios by one main hand, that of Regiomontanus, except for folios 67–69 and 88–92, copied by Georg Peurbach, and folios 79–86 by an unidentified third hand. The table of contents on folio 1 is the work of the Nuremberg polymath and editor Johannes Schöner, who is also responsible for assembling the volume, foliating it, and adding many titles throughout.[38]

By browsing through the table of contents of Cod. 5203, it is possible to identify various topics ranging from astronomy and astrology to music, mathematics, and physics. The table also includes a reference to a text of which a critical edition and translation is provided in this contribution; the text is entitled 'On the equations of time and how their tables should be composed'.[39] This reference is by Schöner, as is the title appearing at the top of folio 70^{r}, 'On the equations of time', as well as (likely) a marginal note on folio 71^{r} reading 'Of the ascension of the radix'.[40] This suggests that Schöner read and might have valued the work — at least he was able to classify its content.

The text in concern, covering folios 70^{r}–71^{v}, is written out by Regiomontanus in single columns. The handwriting is cursive with some slight humanistic elements (certain majuscules, such as the capitals *A* and *M*, and the angular *V*), and the script is characterized by a mostly small and consistent module.[41] Pauses in the text are usually indicated by simple *puncti*, generally signifying minor pauses, except when followed by *litterae notabiliores* (capital letters) marking the beginning of new sections. Although Regiomontanus does not seem to have taken particular care in his handwriting, the text is, overall, clearly written out, the marginal and interlinear corrections present in the text being only a few.[42] The abbreviation system used and spelling employed appear consistent throughout. The text is followed by an empty table, in the top row of which there are five draft headings. Instead of including the incomplete table in the edition of the text, we reproduce it in Figure 10.3.

The text is part of a quire (ff. 66–77) bearing a watermark resembling Piccard Waage V308 (Nuremberg 1458/59). On folios 66–69 there is a text by Georg Peurbach on an instrument to determine true syzygy; on folios 70–71 is Regiomontanus's text on the equation of time, while folios 72–77 are blank.[43]

37 Folkerts, 'Die mathematischen Studien'.

38 Zinner, *Regiomontanus*, pp. 307–08. See also Monika Maruska, 'Johannes Schöner - homo est nescio qualis: Leben und Werk eines fränkischen Wissenschaftlers an der Wende vom 15. zum 16. Jahrhundert', unpublished PhD thesis, University of Vienna, 2008, p. 97.

39 Cod. 5203, f. 1: 'De equationibus dierum et quomodo tabule earundem componi debeant'.

40 Cod. 5203, f. 70^{r}: 'De equationibus dierum', f. 71^{r}: 'Ascensionis radicis'.

41 We are very thankful to Tommaso Salvatore (Université de Genève) for his advice on the analysis of the handwritings included on ff. 70^{r}–71^{v}.

42 Most of the corrections and the redrafting by Regiomontanus, which characterize the text, is found on f. 70^{v}.

43 We are grateful to Richard L. Kremer for sharing his analysis of the quire structure and watermarks with us (personal communication, June 2022). For a general study of Regiomontanus's paper, see the digital edition of Regiomontanus's *Defensio Theonis contra Trapezuntium*, available online at https://regio.dartmouth.edu/about/about-manuscript.html, last accessed <21.03.2023>. For a discussion of the syzygy text, see Kremer, 'Experimenting with Paper Instruments'.

Text on the equation of time

In the largely technical syntax and terminology characterising these pages, it is possible to identify a few expressions that could serve as clues to assess the genre of the text included on folios 70^{r}–71^{v}. The text starts by providing 1456 as a date of reference and continues by explaining how to derive the equation of time, testing different steps for different contexts, drawing the reader's attention to the trickiest stages of the procedure, and reminding the reader when and where they must pay particular attention.

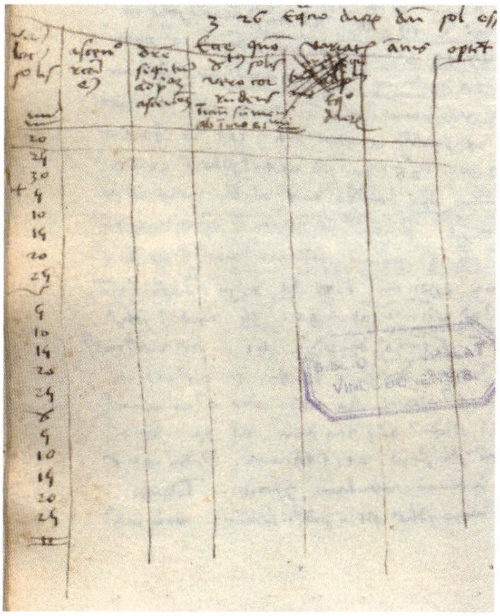

Figure 10.3. The unfilled table following the text on the equation of time in Regiomontanus Rechenbuch. The headings refer to: the true place of the Sun (verus locus solis), its true ascension (ascensio recta eius), the differences following the first ascension (differentie sequuntur ad primam ascensionem), the rising of the true Sun from 21° Aquarius (ortus solis vero correspondens inicium sum[m]ens ab inicio 21 Aquarii), and the equation of time (equatio dierum). Vienna, Österreichische Nationalbibliothek, Cod. 5203, f. 71^{v}.

After providing some initial values for his starting date, Regiomontanus describes the first steps of the procedure by adopting the first person singular ('I want to find'; 'I see how great…'; 'I look for'; 'I write'; 'I subtract'; 'I divide'). This passage is then followed by the information required for the application of the successive steps. It is after this second set of data that Regiomontanus addresses the reader and involves them in the operation by using the first-person plural ('if we will begin there').

From the following passage onwards, Regiomontanus employs not only the indicative first-person singular and the indicative and subjunctive first-person plural, but also the subjunctive second-person singular and the imperative second-person singular; indeed, he suddenly turns to the reader, pointing out a 'moment' for their 'consideration' and encourages them to pay particular attention to specific stages relevant to the understanding of the operations. In this context, a few expressions are worth noting such as 'consider this very well so that you understand', 'Learn thus briefly', 'by this do not understand', 'instead acknowledge something else', and 'produce a table anew'.

The expression that can be seen as most revealing of the text's genre is a kind of complaint found in the second half of the text, introduced by '*Hui*', a Latin exclamation of astonishment or distress. The sentence reads 'Oh, how quickly (*Hui, quam cito*) this strays from the rules: it comes from the lack of seconds in the table of the right ascensions!' Interestingly, the interjection '*hui*' + '*quam*' appears in three epistles by Cicero (*Epistulae ad Familiares XV, 19; Epistulae ad Atticum, XIII, 21a and XV, 13a)*, a classical author who represented a model of eloquence for Regiomontanus's Padua Oration, as argued by James Steven Byrne.[44] The verb used to stress the inconvenience generated by the table

44 Byrne, 'Regiomontanus's Padua Oration'.

of right ascension is '*irregulatur*', which we tentatively translate as 'strays from the rules'. This does not seem to be an attested lemma, and it is very likely a newly coined word by Regiomontanus. The production of neologisms, both in Latin and German, is indeed among his distinctive features.[45] Furthermore, a number of clauses are introduced by '*ecce*', a demonstrative adverb, the semantic function of which is to point out an object with emphasis.

These numerous calls to attention, the employment of the second-person singular (in the subjunctive or imperative), and the emphasis of certain expressions make this text likely to be some sort of note or even a draft of a letter. The correspondence Regiomontanus held with many relevant historical actors of his time is considered a great resource as it provides significant insights into the development of his work as well as into his world of ideas.[46] Similar expressions to those included in this text in Cod. 5203 on folios 70^r–71^v can be found in the letters Regiomontanus exchanged with Giovanni Bianchini, Jacob von Speier, and Christian Roder.[47]

The use of '*volo*', that is, 'I want' + an infinitive form is attested in the letters Regiomontanus wrote to Bianchini and Speier; indeed, in four instances Regiomontanus uses the expression '*volo dicere*', that is, 'I want to say'.[48] The exact same expression defining a 'moment of consideration' of the reader ('*instans considerationis tue*') is found in a letter Regiomontanus wrote to Bianchini in 1463.[49] Similarly, '*vide*' ('see'), with the connotation of 'consider', is attested in a letter Regiomontanus addressed to Speier in 1465.[50] Expressions like '*quam cito*' ('how quickly'), although the comparative and superlative forms, namely '*quam citius*' and '*quam citissime*', are found in some letters Regiomontanus addressed to Bianchini and Roder.[51] Accordingly, the demonstrative adverb '*ecce*' is found twice in a letter Regiomontanus wrote to Bianchini, and it seems to be employed with the same connotation as that found in the text in Cod. 5203, namely introducing a specific case or marking the outcome of a particular stage in the computation.[52]

In his monograph on the life and the works of Regiomontanus, Zinner describes the text we include in the appendix as a 'short work, written in 1456, dealing with the calculation of the equation of time, which, however, is unfinished, with the planned table not produced'.[53] The lack of a recipient prevents us from firmly stating that the text is a letter, and it is possible that these pages are simply home to notes that Regiomontanus

45 See Zinner, *Regiomontanus*, p. 49, p. 200.

46 See Zinner, *Regiomontanus*, p. 95.

47 Maximilian Curtze, 'Der Briefwechsel Regiomontans mit Giovanni Bianchini, Jacob von Speyer und Christian Roder', in *Urkunden zur Geschichte der Mathematik im Mittelalter und der Renaissance*, ed. by Maximillian Curtze (Leipzig: Teubner, 1902), pp. 185–336.

48 Ibid. Regiomontan to Bianchini, V, p. 246; Regiomontan to Speier, VI, p. 294, p. 297; Regiomontan to Jacob von Speier, VIII, p. 308.

49 Ibid. Regiomontanus to Bianchini, I, p. 193: 'Cumque *instans considerationis tue* fuerit horis duabus minutis 25 horologii peractis, videlicet ab occasu, additis eis ad tempus semidiurnum colliguntur hore 9 minuta 3, et hoc tempore a meridie computato fuit consideratio tua'. See also, ibid. Regiomontanus to Giovanni Bianchini, III, p. 212.

50 Ibid. Regiomontanus to Speier, VI, p. 294: 'Vide igitur hec problemata varia, [...]'.

51 Ibid. 'quam citius' (Regiomontanus to Bianchini, I, p. 194; Regiomontanus to Roder, IX, p. 336) and 'quam citissime' (Regiomontanus to Bianchini, III, p. 210).

52 Ibid. Regiomontanus to Bianchini, III, p. 220, p. 233.

53 Zinner, *Regiomontanus*, p. 65.

wrote for himself as a student. The analysis of style and comparison of some words and phrases appearing in this text and Regiomontanus's correspondence point towards the text being a letter, but there is no smoking gun. However, the question of the genre of the text seems less important when we take into consideration that, as stated above, Peurbach's table for the equation of time seems to have been computed from the algorithm described in this very text.[54] Perhaps through reading a note written by his student, or maybe in the more defined form of a letter, Georg Peurbach encountered the text in Cod. 5203. The possibility of Peurbach's awareness of the text is further strengthened by the evidence provided in Section 6.

Regiomontanus's derivation

Although not stated, Regiomontanus's derivation of the equation of time is based on Bianchini's *Tabulae astronomiae*, as we show in the following. These tables were compiled around 1442 and first appeared in print in 1495 in Venice.[55] Regiomontanus owned an autograph version that he copied around 1460 in Vienna (Nuremberg, Stadtbibliothek, Cent. V, 57).[56] However, he may not have used this version for his derivation, which he indicates to have performed four years earlier in 1456. Irrespective of the dating, there are also mathematical reasons why he could not have used his autograph version, which we explain below.[57] Because there is no other known, extant manuscript version of Bianchini's planetary tables owned by Regiomontanus, we refer to the 1495 printed edition in the following when necessary.[58] The derivation of the equation of time is structured in three parts: firstly, preliminary considerations; secondly, two successive mathematical steps; and, thirdly, concluding remarks.

The preliminary considerations concern the positivity of the equation of time. They begin with the determination of the solar apogee for the current year of 1456. It is simply read from Bianchini's tables as 90;46°, which the Sun will reach on 12 June at 21;23h past noon.[59] From the time when the mean (and true) Sun reaches its apogee, Regiomontanus simply determines the remaining time until the end of the calendar year to be 201d and 2;37h. He then notes the longitudinal distance from the apogee to the beginning of Aries to be 269;14°. From Bianchini's tables he takes the closest smaller value for the true motion of the Sun after 270d to be 268;17,28°, which differs from the distance from the apogee to

---

54 On Peurbach's table for the equation of time, see Chabás, *Computational Astronomy*, pp. 368–70.

55 Chabás, Goldstein, *Tables of Bianchini*, pp. 1–11.

56 Zinner, *Regiomontanus*, p. 311. Neske, *Handschriften Nürnberg*, pp. 90–91.

57 Regiomontanus's autograph version is not solely a copy of Bianchini's tables, as he reworked some of the them. Furthermore, he did not copy all of Bianchini's tables.

58 There is another manuscript (Nuremberg, Stadtbibliothek, Cent. V, 36) that contains Bianchini's tables at the bequest of Bernhard Walther, who had joined his library with Regiomontanus's after the death of the latter in 1476. However, there is no indication that this manuscript was previously owned by Regiomontanus.

59 See Giovanni Bianchini, *Tabulae astronomiae* (Venice: 1495), f. b2^{v}. Regiomontanus, thus, did not adjust for a different meridian but used the meridian of Ferrara implied in Bianchini's tables.

the first degree of Aries by 0;56,32°.[60] This longitudinal difference corresponds to a time difference of 22;57h, which Regiomontanus states to calculate with the mean hourly solar velocity given in Bianchini's tables.[61] The total time for the Sun to travel from its apogee to Aries is thus 270d and 22;57h. As the time from the apogee until the end of the year was 201d and 2;37h, the difference of the last two figures corresponds to the time after which the Sun will reach the equinox in 1457, that is, after 69d and 20; 20h. At the time of the 1457 equinox, Regiomontanus states, the true motion of the Sun per day would be 0;59,8°, the right ascension of which is 0;54°. Therefore, the apparent solar day would be 360;54°, but the mean solar day would be 360;59,8°. Regiomontanus then concludes that the equation of time would amount to 0;5° if he placed the beginning of the equation of time (radix) into Aries. Regiomontanus continues to explain that the equation of time would sometimes be positive and sometimes negative if the radix is set in Aries, because the mean days can also be shorter than the apparent days. Yet, it would always be positive if the radix were set to the 'beginning of the addition', which is around 19° Aquarius, according to Regiomontanus. We note that this is the position of the radix given by al-Battānī.[62] Nevertheless, Regiomontanus is inclined to find the exact point of the beginning of the addition himself, which is the first step in his derivation.

Geometrically, the beginning of the addition is the point on the sphere where the daily increase in right ascension of the true Sun is equal to the increase in daily mean motion of the Sun of 0;59,8° and from which stage the apparent days start to become longer than the mean days.[63] At this point, the apparent day is equal in length to the mean day, and the equation of time is zero by definition. There is only one such point on the sphere.[64] If this point is used as the radix for the equation of time, the latter will always be additive. Regiomontanus's intended method to find this point is by trial and error in the vicinity of the radix given by al-Battānī.

The first step begins with Regiomontanus placing the radix in 18° Aquarius.[65] At this position, the true solar motion per day is 1;0,44°. This amounts to an increase in right ascension, from 18° Aquarius to 19;0,44° Aquarius, of 0;59,44°. Hence, the difference compared to the mean solar motion of 0;59,8° is 0;0,36°. Therefore, the apparent day is

60 Ibid. f. [b5]^{r}. The true motion of the Sun after 270d corresponds to a different value in Regiomontanus's autograph copy in Nuremberg, Stadtbibliothek, Cent. V, 57, f. 23^{r} (Δ = -3s). This is because in his autograph version Regiomontanus exchanged the underlying solar equation in Bianchini's table for the true motion of the Sun with the solar equation from Peurbach's *Tabulae eclipsium*. Therefore, in the text from the *Rechenbuch* considered here, Regiomontanus did not use his autograph version Nuremberg, SB, Cent. V, 57 of Bianchini's planetary tables. He must have owned another version that is now lost.

61 The mean hourly velocity in Bianchini's tables is given as 0;2,28; see Bianchini, *Tabulae astronomiae*, f. [a7]^{v}. Although Regiomontanus explicitly reproduces this number here, he seems to have made his calculations with a more precise value of 0;59,8/24 = 0;2,27,50.

62 See, for example, Al-Battānī, *De scientia stellarum* (Nuremberg: Johannes Petreius, 1537), ff. h3^{v}–h4^{v}. See also, Al-Battānī, *Opus astronomicum*, ed. by Carlo Alfonso Nallino, 3 vols (Milan: Hoeplium, 1899–1907), I, pp. 48–49.

63 Or 1°, depending on the table-maker's preference for accuracy.

64 Strictly speaking, there is another point where both are equal, which corresponds to the beginning of the diminution. From this point, the apparent days start to become shorter than the mean days.

65 In Regiomontanus's wording, 'the beginning' of a certain degree refers to the previous degree in our modern understanding, while 'the end' of a certain degree is the same degree. For example, when he mentions 'the beginning of the 15th of Aquarius' and gives its right ascension as 316;30°, he is in fact referring to 14° Aquarius.

longer than the mean day. Unsatisfied with this difference, Regiomontanus repeats the same practice, now placing the radix in 14° Aquarius. We use this example to showcase Regiomontanus's use of Bianchini's tables.

The longitudinal difference from the solar aux 90;46° to 14° Aquarius is 223;14°. The closest value to this difference in Bianchini's table for the true motion of the Sun is after 225d, which is 223;15,28°. The difference compared to the next larger value for 226d is 1;0,44°.[66] The latter value thus corresponds to the true motion of the Sun during a day and is given precisely by Regiomontanus. Furthermore, Regiomontanus explicitly states that the right ascension of 14° Aquarius is 316;30° and the right ascension of 15;0,44° Aquarius is 317;30,44°.[67] The difference in right ascension of daily true solar motion is thus 1;0,44°, which should be compared to the daily mean solar motion of 0;59,8°. The difference is 0;1,36°, by which the apparent day exceeds the mean day. This indeed corresponds to an increase by which the apparent day exceeds the mean when moving away from 18° Aquarius to smaller longitudes. Therefore, Regiomontanus tries 19° Aquarius as the radix. For 19°, he finds the right ascension of daily true solar motion of 0;59,38°, from which he concludes that the apparent day is longer by 0;0,30° than the mean. For 21° Aquarius, he finds the right ascension 0;59,35° and a difference of 0;0,27°. Still unsatisfied with this result, he jumps forward by 3° and places the radix in 24° Aquarius, for which he finds the right ascension 0;59,30°; this still appears too large. He thus continues with 25° Aquarius and finds 0;58,30°, which is suddenly too small.

For all these calculations, Regiomontanus uses a table of right ascension with a precision to minutes only. His input values of daily true solar motion, however, are determined to seconds, and his output values for the difference in right ascension per day are also computed to seconds. The procedure is thus corrupted by an intermediate table that is given only to minutes and on which he performs linear interpolation to obtain seconds. Mathematically speaking, this is a pointless endeavour, which Regiomontanus rightly exclaims with the words '*Hui, quam cito irregulatur!*' ('Oh, how quickly this strays from the rules!'). He continues: 'It comes from the lack of seconds in the table of the right ascensions!'[68] In fact, he is unable to confirm that the point of true solar longitude where the apparent day becomes longer than the mean day is around 19° Aquarius. According to his trial-and-error calculations, the radix would be between 24° and 25° of Aquarius, which is considerably off from al-Battānī's stated value. Therefore, instead of deriving the radix, he simply submits himself to the authority of al-Battānī and places the radix in 19° Aquarius and starts to build up his table for the equation of time from there in the second step.

---

66 Bianchini, *Tabulae astronomiae*, f. [b5]^{r}. Using the closest smaller value here, as before, would lead to 1;0,54. Regiomontanus thus seems to use the closest value.

67 These values and the following ones, correspond to the use of a right ascension table found in, for example, Cracow, BJ, 556, f. 95^{r}, which is part of Bianchini's *Tabulae primi mobilis*. See Glen van Brummelen, 'Before the End of an Error: Giovanni Bianchini's Original Flawed Treatise on the Conversion of Stellar Coordinates', *Archive for History of Exact Sciences* 75 (2021), 109–24. A modern fit of the corresponding right ascension table results in a 95% confidence interval for the obliquity of <23;34,18°–23;35,11°>. This suggests that the table is based on an obliquity of 23;35°, used by al-Battānī, rather than the value of 23;33,30° apparently used by Bianchini in his solar declination table in Cracow, BJ, 556, f. 59^{r}.

68 We were unable to find any right ascension table in the intellectual surrounding of Regiomontanus (Peurbach, Gmunden, Bianchini) that is given to seconds. With regard to Regiomontanus's own interjection and the calculational importance of right ascension, this appears to be an odd fact.

A few comments are in order. We might muse that giving in to the authority of al-Battānī's text may have been an obligation for a young Regiomontanus of only twenty. In fact, both tables for the equation of time he worked and was familiar with, one attributed to Peter of St Omer and the other included in Bianchini's planetary tables, had their radices around 18° Aquarius.[69] Deviating from this value by some seven degrees may have appeared quite questionable to Regiomontanus, but, as a matter of fact, he made a mistake in his calculation. His value for the right ascension of the daily true solar motion for 24° Aquarius is wrong and too large by one minute. It must remain unanswered if this originates from a scribal error in the table of right ascension he used. Differences of one minute in tables of right ascension are very common. Or, perhaps he himself made a scribal mistake when writing down his results. A simple probe for the degrees between 21° and 24° of Aquarius may have shown this slip, but Regiomontanus did not do this. In fact, when considering the intermediate integer values skipped by Regiomontanus, the radix would lie between 21° and 22° Aquarius. In Table 1, we summarize Regiomontanus's results and our recomputation following his method exactly.[70] The mismatches of a few seconds might stem from the use of slightly different tables of right ascension or a different rounding method, and they are entirely to be expected.[71]

radix	Regiomontanus	recomputation
14°Aqu	1;0,44	1;0,44
18°Aqu	0;59,44	0;59,43
19°Aqu	0;59,38	0;59,38
21°Aqu	0;59,35	0;59,35
21;30° Aqu	*n.a.*	*0;59,05*
22;00° Aqu	*n.a.*	*0;58,35*
23;00° Aqu	*n.a.*	*0;58,35*
24° Aqu	0;59,30	0;59,33
25° Aqu	0;58,30	0;58,32

Table 1. Regiomontanus's values for the daily increase in right ascension (in degrees) of the true Sun for different radices. In the third column, we recompute the values following his algorithm exactly. The values in italics were not computed by Regiomontanus. We include them here to show that the radix sought by Regiomontanus, i.e. 0;59,08, lies in the interval [21°–21;30°] of Aquarius.

The second step starts with the assumption that the apparent and mean days are equal when the Sun is in 19° Aquarius. This corresponds to a right ascension of 321;27° that serves as the radix value for the whole table.[72] Furthermore, Regiomontanus takes

69 This can simply be estimated from the zero value of the equation of time.

70 We used the right ascension table found in Cracow, BJ, 556, f. 95^r attributed to Bianchini, which is the closest we could find compared to the values given explicitly by Regiomontanus. We note that this table of right ascension is combined with Bianchini's second equation of time.

71 On interpolation in Alfonsine astronomy, see, in the present volume, Matthieu Husson, 'Proportional Parts and Interpolation Practices in Alfonsine Astronomy: Astronomers' Works with a Mathematical Procedure'.

72 A marginal note marks this section with 'of the ascension of the radix', which is very likely the hand of Schöner. Successive values will always be compared to this radix, or as Regiomontanus puts it in the text, 'I keep these [values] continuously'.

the apogee to be in the first degree of Cancer and thus rounds the distance between the Sun and the apogee to be about 229°. Now follows the decisive element to build the whole table. The distance between the Sun and its apogee of about 229° corresponds to a time lapse of 231d (229;19,23°) according to Bianchini's tables of daily true solar motion.[73] Then, ten days later, at 241d, the Sun will be 239;24,29° from its apogee. The difference is 10;4,56°, which Regiomontanus rounds up to 10;5°. Therefore, the Sun will be in 29;5° Aquarius, which corresponds to a right ascension of 331;15°, as correctly given by Regiomontanus. Hence, the difference compared to the radix value of 321;27° is 9;48°. However, the mean motion of the Sun over ten days can also be read from Bianchini's tables as 9;51°.[74] In conclusion, the equation of time when the true Sun is in 29;5° Aquarius amounts to 0;3°. The whole table is then successively built up in the same manner. That is, twenty days later the Sun is 251d from its apogee or 249;25,56°.[75] Since the radix was 231d or 229;19,33°, the longitudinal difference of the true Sun after 20 days is 20;6,23° from the radix at 19° Aquarius, which is given exactly by Regiomontanus. This corresponds to a right ascension of 340;43°, which exceeds the radix value of 321;27° by 19;16° as given by Regiomontanus. Again, the mean motion of the Sun over twenty days can be read from Bianchini's table to be 19;43°.[76] Thus, in conclusion, when the true Sun is in 9;6° Pisces, the equation of time amounts to 0;27°. Regiomontanus gives three further examples for 30, 40, and 100 days that follow the very same approach, by which he concludes the explanation of his method.

What results but remains uncommented on by Regiomontanus is a skew table for non-integer values of the true Sun and the corresponding values for the equation of time in degrees that is based on a certain grid: ten-day intervals according to the given examples. All that remains to be done is thus to perform a linear interpolation upon the skew longitudinal grid, leaving us with a table for integer values of the true Sun for the full circle and the corresponding values for the equation of time, possibly given in minutes and seconds of time.

In his concluding remarks, Regiomontanus places the solar apogee, most likely as a test case, in the beginning of Leo, that is, at 120° longitude.[77] Again, he tries to find the radix of the equation of time by trial and error, and he now places it in 25° Aquarius. He performs one iterative step after 100 days and shows that the equation of time amounts to 3;26° when the Sun would be in 4;27° Gemini. Without any comment on how to compare this number to his previous computations, Regiomontanus concludes that the change of the solar apogee over the years necessitates a renewed derivation of the equation of time,

73 Note that the solar apogee for 1456 is 90;46° according to Bianchini's tables. To find the distance of the Sun from its apogee, however, Bianchini's tables work with days. Thus, Regiomontanus's method also requires finding a radix of days, according to the value of the apogee. Choosing an integer number of days from the table will effectively shift the solar apogee. Regiomontanus chooses 231d without stating this explicitly, but it can be inferred from his subsequent values for 20d, 30d, or 40d later that are given to seconds. This is why Regiomontanus does not bother to state the proper apogee for 1456 and only supposes that the apogee is in the first degree of Cancer.

74 Bianchini, *Tabulae astronomiae*, f. [a7]^v.

75 Ibid. f. [b5]^r.

76 Ibid. f. [a7]^v.

77 We note that this corresponds to an immense time lapse of some 4000 years.

or, as he puts it, 'Behold how having changed the years it is necessary to change the table of the equation.' However, his previous calculations for the solar aux at 90;46°, performed after 10,20,30,40, and 100 days, only resulted in values for the equation of time when the true Sun is between the end of Aquarius and the end of Taurus. At this stage, therefore, there is simply no proper value to which Regio could have compared his trial value in order to draw his conclusion. The only resolution would be that he already had a complete table for the equation of time for an aux of 90;46°. Since the table below the text is apparently empty, it appears quite likely that Regiomontanus compared his trial value to Bianchini's second equation of time, contained in his *Tabula primi mobilis*, calculated explicitly with the solar apogee at 90;46°. From Bianchini's table, the equation of time for 4;30° Gemini would amount to 20;26min.[78] Compared to Regiomontanus's value for 4;27° Gemini of 3;26°, or 13;44min when expressed in time, the difference is about seven minutes and is indeed quite considerable from a mathematical point of view. More importantly, we further conclude that Bianchini's second table for the equation of time appears to have been known to Regiomontanus when he wrote his text. Following this line of thought, we further conclude that the year 1456 (chosen for the table) is not the year in which Regiomontanus computed his table, but rather it is chosen to match the year also used by Bianchini, possibly with the aim of comparing these two tables. He might have written the text at a later date (though before 1460), possibly around 1458–59, as the dating of the paper would suggest.

The *Epitome* and Peurbach's table

The fact that Regiomontanus made a mistake in his notes did not escape his own attention. A few years later, around 1462, when he finalized and compiled the text of the *Epitome of the Almagest*, which his teacher, Peurbach, had begun and had not completed, the error of determining the radix was eventually fixed. In a neat manuscript copy of the *Epitome* given to Cardinal Bessarion, who had initially requested Peurbach to compose this work during a visit to Vienna in 1460, the radix for the equation of time was properly determined rather than being taken from al-Battānī.[79] Modelled closely after the *Almagesti minor* in that regard, the *Epitome* III: 22–30 gives a detailed account on the equation of time.[80] While the first eight chapters (22–29) parallel the content of the *Almagesti minor*, Chapter 30 is a novel addition and entirely devoted to the composition of a table for the equation of time. In fact, this part of the *Epitome* appears to have already been written by Peurbach, as Regiomontanus wrote in the dedicatory letter to Bessarion.[81] In Chapter 30 on 'composing the table of the equation of time', the radix is given as 21;15° Aquarius, if the solar apogee is

78 We obtained this value from linear interpolation in the table found in Cracow, BJ, 556, f. 95^{r}.

79 Venice, Biblioteca Nazionale Marciana, Fondo antico lat. Z. 328. This manuscript includes a dedicatory marginal note in Regiomontanus's hand. The original of Regiomontanus's *Epitome* is lost. See Zinner, *Regiomontanus*, pp. 318–19. The first printed edition appeared in 1496: Johannes Regiomontanus, *Epytoma in Almagestum Ptolomei* (Venice: 1496). It was reprinted in Basel in 1543 and in Nuremberg in 1550.

80 Venice, BNM, Fondo antico lat. Z. 328, ff. 28^{r}–31^{r}. Zepeda, *Almagesti minor*, pp. 109–14.

81 Venice, BNM, Fondo antico lat. Z. 328, f. 2^{r}. Zepeda, *Almagesti minor*, p. 109.

in the beginning of Cancer.[82] The text states that this point is generally known to be around the middle of Aquarius and that its precise location is found when the daily increase of right ascension of the true Sun is identical to the daily motion of the mean Sun. However, this point will change over time, the text explicitly remarks, due to the shifting of the solar apogee. This statement parallels and probably derives from Regiomontanus's probing of what happens with the equation of time when placing the solar apogee in the beginning of Leo (120°) in his *Rechenbuch*. We are not aware of any other earlier Latin text that explicitly states this dependence and thus suggests, at times, an update of the table for the equation of time. The text concludes that after having determined the radix, a table for the equation of time is derived into three successive steps. Firstly, readers are instructed to increase the true solar motion starting from the radix in steps of single degrees for the full circle of 360°. Secondly, they are told to determine the mean motion of the Sun and the corresponding right ascension of the true Sun. Thirdly, they are instructed to find the difference between the latter two, which in turn is the equation of time. We note that this procedure of first determining the position of the true Sun and then finding the corresponding mean motion is generally impossible. This exact sequence is only applicable when using the planetary tables of Giovanni Bianchini, which we detail in the previous section. This implies that the increase by single degrees in the first step is rather an increase by single days. Either this chapter of the *Epitome* was added by Regiomontanus or Peurbach was well aware of the method described by Regiomontanus in his *Rechenbuch*. In either case, this chapter of the *Epitome* is thus demonstrably based on three sources: the *Almagesti minor*, al-Battānī's *De scientia stellarum*, and Regiomontanus's *Rechenbuch*. Moreover, although written in a general, theoretical manner, it tacitly assumes the use of Bianchini's planetary tables.

It appears as if Regiomontanus actually returned to his *Rechenbuch* when he worked on the *Epitome*. In his *Rechenbuch*, the table that is outlined but not filled in, following his text on the derivation of the equation of time, includes headings that are somewhat incongruent with his stated examples of the text (Figure 10.3). While in the examples the true position of the Sun is determined on a grid of ten intermediate days, the first column, *verus locus solis*, suggests a grid of 5°, and thus it parallels the text of the *Epitome*, though in steps of five rather than single degrees.[83] Most importantly, the radix as the reference point for the distance to the Sun is now given as 21° Aquarius in the heading of the fourth column: *ortus solis vero correspondens inicium sum[m]ens ab inicio 21 Aquarii*. Regiomontanus struggled and failed to find this radix in the text and assumed it to be 19° Aquarius. However, this is corrected in the heading of the incomplete table according to the numbers given in the *Epitome*. We therefore conclude that Regiomontanus added the incomplete table to his

82 Venice, BNM, Fondo antico lat. Z. 328, f. 30v: '*Tabularum equationis dierum componere*'.

83 This procedure to advance the position of the true Sun by any number of integer degrees is not sensible. Using Bianchini's tables, the advancement should be in days and not in degrees. Using any other set of tables, for example the Parisian Alfonsine Tables, only the mean Sun could be advanced by integer degrees and the true Sun computed from this subsequently. The proposed 5° grid is either a scribal error or related to another procedure to derive the equation of time. The latter could be to interpolate for days that correspond to integer multiples of five degrees for the true Sun from Bianchini's table for the true motion of the Sun. This would essentially mean starting with a linear interpolation rather than ending with it. We have not checked this possibility as it is not part of Regiomontanus's text.

Rechenbuch sometime later, presumably when he worked on the *Epitome* and when he noted and corrected his error of determining the radix.

Like the *Almagest*, the *Epitome* does not present a table for the equation of time. Nevertheless, there is an equation of time attributed to Georg Peurbach that is contained in his eclipse tables. These *Tabulae eclypsium* were first printed in 1514 in Vienna, but they had already been completed and compiled in manuscript form around 1460.[84] The equation of time contained therein is reproduced in Figure 10.4. The title of the table unmistakably states that this table has been 'newly constituted'. Almost verbatim to Regiomontanus's *Rechenbuch*, it supposes the solar apogee to be in the first degree of Cancer. Quite uncommonly for almost any kind of table, the title further states that the underlying obliquity of the ecliptic used in the calculation is identical to *Declinationem Allmeonis*.[85] This means that the right ascension table used for the derivation of the equation of time was allegedly based on an obliquity of 23;33,30°. However, we show that the latter is not true and originates from an apparent misinterpretation of Bianchini's tables. Also in manuscript form, Peurbach's table for the equation of time usually circulated with the same title.[86] However, in a single manuscript, the owner and scribe, Johannes Virdung, an astrologer from Haßfurt, added a note to the common title of Peurbach's table that reads 'add four seconds and you will have the verified equations for the year 1456'.[87] These verified equations for the year 1456 unmistakably refer to Giovanni Bianchini's second, later equation of time contained in his *Tabula primi mobilis* B.[88] It was included in his table of right ascension and bears the explicit title, 'Table verified by Giovanni Bianchini on right ascensions, which <is used> universally in any region, and on the equations of time with the solar apogee in 0° 46 min of Cancer for the year 1456'.[89] Virdung's statement is of course wrong, and the difference between these two equations cannot be (and, in fact, is not) a constant. Quite obviously, he only compared the first few values of Aries from the beginning of the two corresponding tables: the first five values indeed all differ by exactly four seconds.[90] Interestingly, Virdung's note clearly shows that he compared different circulating tables for the equation of time

84 Chabás, *Computational Astronomy*, pp. 365–75. One of the earliest extant manuscripts appears to be Nuremberg, SB, Cent. V, 57, copied by Regiomontanus in Vienna around 1460 according to Zinner, *Regiomontanus*, p. 77.

85 On the attribution to al-Ma'mun, see, for example, Pedersen, *Toledan Tables*, pp. 962–63.

86 For a list of manuscripts of Peurbach's tables, see Chabás, *Computational Astronomy*, pp. 365–75. A description of Peurbach's table for the equation of time is given in ibid., pp. 368–70.

87 Vatican, Biblioteca Apostolica Vaticana, Pal. lat. 1375, f. 90r: 'Adde 4 secunda et habebis equationes verificatas pro anno 1456'. On the manuscript and its ownership, see Ludwig Schuba, *Die Quadriviums-Handschriften der Codices Palatini Latini in der Vatikanischen Bibliothek* (Wiesbaden: Reichert, 1992), pp. 88–93.

88 On versions A and B of the *Tabula primi mobilis*, see van Brummelen, 'Before the End of an Error'.

89 See, for example, Cracow, BJ, 556, f. 95r: 'Tabula verificata per Johanem de Blanchinis de ascensionibus signorum in circulo directo que universaliter <operatur> in omni regione necnon de equationibus dierum cum noctibus suis auge solis existimate in gradus 0 minuta 46 Cancri Anno Domini 1456'.

90 Virdung copied MS Vatican, BAV, Pal. lat. 1375 in Cracow (f. 1r): 'Liber magistri Joannis Virdiungi de Hasfurt scriptus manu propria Cracow'. The table with the 'verified equations for the year 1456' by Giovanni Bianchini, though, is not contained in the same manuscript. It may well be that he compared Peurbach's table with one of the manuscripts in Cracow that does contain Bianchini's second equation of time: for example, Cracow, BJ, 556.

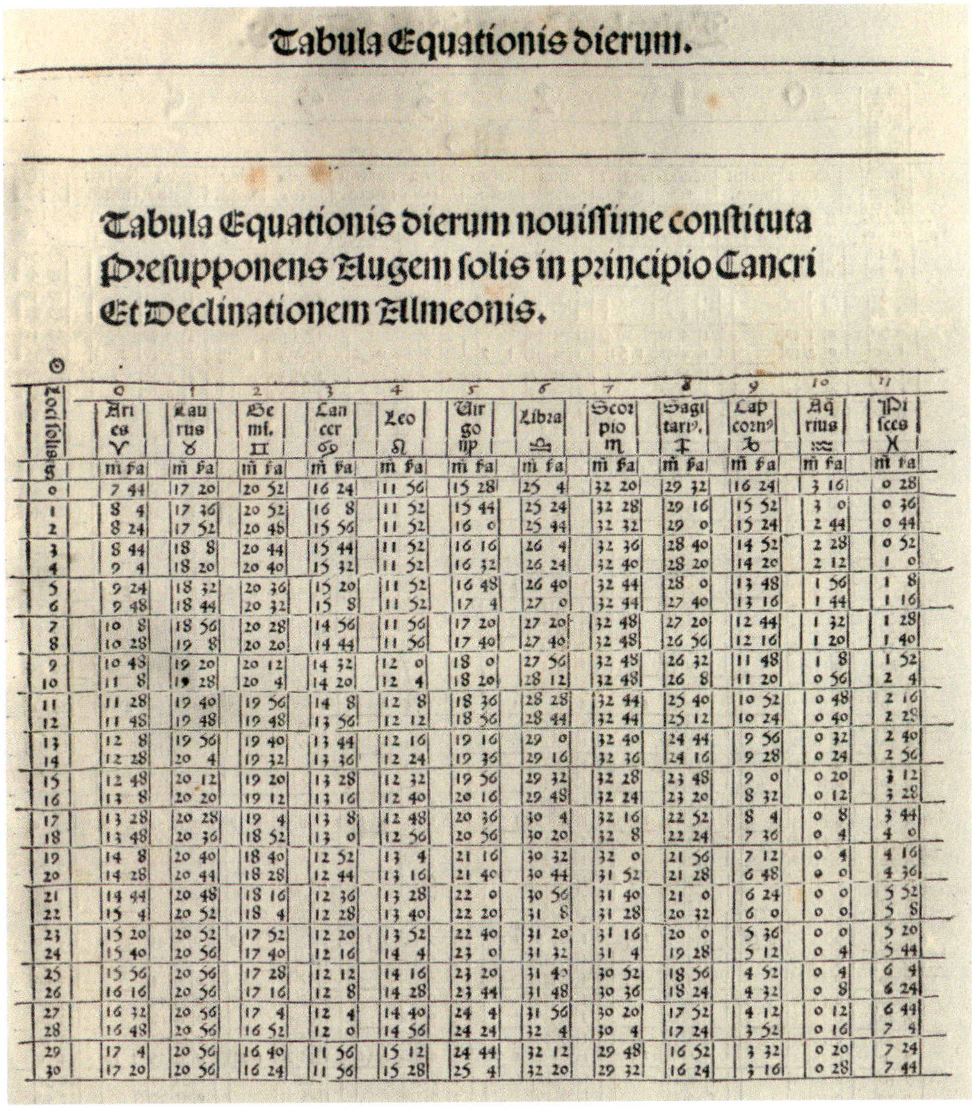

Tabula Equationis dierum.

Tabula Equationis dierum nouissime constituta Presupponens Augem solis in principio Cancri Et Declinationem Almeonis.

Loci solis G	0 Aries ♈ m sa	1 Taurus ♉ m sa	2 Gemini ♊ m sa	3 Cancer ♋ m sa	4 Leo ♌ m sa	5 Virgo ♍ m sa	6 Libra ♎ m sa	7 Scorpio ♏ m sa	8 Sagittarius ♐ m sa	9 Capricornus ♑ m sa	10 Aquarius ♒ m sa	11 Pisces ♓ m sa
0	7 44	17 20	20 52	16 24	11 56	15 28	25 4	32 20	29 32	16 24	3 16	0 28
1	8 4	17 36	20 52	16 8	11 52	15 44	25 24	32 28	29 16	15 52	3 0	0 36
2	8 24	17 52	20 48	15 56	11 52	16 0	25 44	32 32	29 0	15 24	2 44	0 44
3	8 44	18 8	20 44	15 44	11 52	16 16	26 4	32 36	28 40	14 52	2 28	0 52
4	9 4	18 20	20 40	15 32	11 52	16 32	26 24	32 40	28 20	14 20	2 12	1 0
5	9 24	18 32	20 36	15 20	11 52	16 48	26 40	32 44	28 0	13 48	1 56	1 8
6	9 48	18 44	20 32	15 8	11 52	17 4	27 0	32 44	27 40	13 16	1 44	1 16
7	10 8	18 56	20 28	14 56	11 56	17 20	27 20	32 48	27 20	12 44	1 32	1 28
8	10 28	19 8	20 20	14 44	11 56	17 40	27 40	32 48	26 56	12 16	1 20	1 40
9	10 48	19 20	20 12	14 32	12 0	18 0	27 56	32 48	26 32	11 48	1 8	1 52
10	11 8	19 28	20 4	14 20	12 4	18 20	28 12	32 48	26 8	11 20	0 56	2 4
11	11 28	19 40	19 56	14 8	12 8	18 36	28 28	32 44	25 40	10 52	0 48	2 16
12	11 48	19 48	19 48	13 56	12 12	18 56	28 44	32 44	25 12	10 24	0 40	2 28
13	12 8	19 56	19 40	13 44	12 16	19 16	29 0	32 40	24 44	9 56	0 32	2 40
14	12 28	20 4	19 32	13 36	12 24	19 36	29 16	32 36	24 16	9 28	0 24	2 56
15	12 48	20 12	19 20	13 28	12 32	19 56	29 32	32 28	23 48	9 0	0 20	3 12
16	13 8	20 20	19 12	13 16	12 40	20 16	29 48	32 24	23 20	8 32	0 12	3 28
17	13 28	20 28	19 4	13 8	12 48	20 36	30 4	32 16	22 52	8 4	0 8	3 44
18	13 48	20 36	18 52	13 0	12 56	20 56	30 20	32 8	22 24	7 36	0 4	4 0
19	14 8	20 40	18 40	12 52	13 4	21 16	30 32	32 0	21 56	7 12	0 4	4 16
20	14 28	20 44	18 28	12 44	13 16	21 40	30 44	31 52	21 28	6 48	0 0	4 36
21	14 44	20 48	18 16	12 36	13 28	22 0	30 56	31 40	21 0	6 24	0 0	5 52
22	15 4	20 52	18 4	12 28	13 40	22 20	31 8	31 28	20 32	6 0	0 0	5 8
23	15 20	20 52	17 52	12 20	13 52	22 40	31 20	31 16	20 0	5 36	0 0	5 20
24	15 40	20 56	17 40	12 16	14 4	23 0	31 32	31 4	19 28	5 12	0 4	5 44
25	15 56	20 56	17 28	12 12	14 16	23 20	31 40	30 52	18 56	4 52	0 4	6 4
26	16 16	20 56	17 16	12 8	14 28	23 44	31 48	30 36	18 24	4 32	0 8	6 24
27	16 32	20 56	17 4	12 4	14 40	24 4	31 56	30 20	17 52	4 12	0 12	6 44
28	16 48	20 56	16 52	12 0	14 56	24 24	32 4	30 4	17 24	3 52	0 16	7 4
29	17 4	20 56	16 40	11 56	15 12	24 44	32 12	29 48	16 52	3 32	0 20	7 24
30	17 20	20 56	16 24	11 56	15 28	25 4	32 20	29 32	16 24	3 16	0 28	7 44

Figure 10.4. The table for the equation of time from the printed edition of Peurbach's eclipse tables. Tabulae eclypsium magistri Georgij Peurbachij (Vienna: Winterburger, 1514) f. d5r. ETH-Bibliothek Zürich, Rar 2258.

seeking the most modern one.[91] However, Peurbach's table was in fact also derived for the year 1456, as we demonstrate.

All the values in Peurbach's table, as given in the printed 1514 edition, are divisible by four. Hence, they have been obtained in degrees, with a precision of minutes, and were

91 What is exceptional about Virdung's manuscript Vatican, BAV, Pal. lat. 1375, beyond the fact that it contains Bianchini's planetary tables, Peurbach's *Tabulae eclypsium*, and Regiomontanus's *Tabulae directionum*, is that is contains three different tables for the equation of time: the table from the *Tabulae resolutae* and Peurbach's and Bianchini's tables on ff. 40r, 90r, and 258v, respectively.

then transformed to time by dividing by 15, which in sexagesimal numbers is identical to multiplying by four and shifting the sexagesimal place. Based on Formula (1), we performed a non-linear fit for the four remaining parameters assuming an underlying one-degree grid.[92] The result is summarized in Table 2.

parameter	best-fit value	95% confidence interval
eccentricity e [°]	02;16,02	02;15,57–02;16,07
apogee λ_{aux} [°]	90;02,05	90;00,06–90;04,03
obliquity ε [°]	23;34,16	23;33,56–23;34,37
epoch constant c [°]	04;05,56	04;05,53–04;05,59

Table 2. Parameter estimates of Peurbach's equation of time based on modern analysis.

The modern fit results in parameter estimates that are quite plausible. The solar eccentricity corresponds to the value usually attributed to the Parisian Alfonsine Tables. The solar apogee is indeed in the first degree of Cancer, as stated in the table heading. The interval for the obliquity, however, does lie in between historical attested values, 23;33,30°, referred to in the heading, and 23;35°, underlying most tables of right ascension. The estimate for the epoch constant cannot be judged. However, it is related to the choice of the radix.

When we apply the algorithm from Regiomontanus's *Rechenbuch*, while correcting his calculational mistake for the radix, we find an excellent fit for Peurbach's table of the equation of time. Statistically, the fit is as good as the modern fit, but it is based on different parameters and entirely different mathematical practices.[93] To achieve this, we firstly determined the true radix using Regiomontanus's method. For the year 1456, the radix of the equation of time is determined to be 21;27° Aquarius.[94] For the rest of the method, we use Bianchini's *Tabulae astronomiae* and perform computations in sexagesimal numbers. Bianchini's planetary tables, however, do usually not contain a table for right ascension or declination. Yet, his other tables do contain these. In his *Tabulae primi mobilis*, for example, he included a table for solar declination, which he says was computed by *Almeon* and which is apparently based on an obliquity of 23;33,30°.[95] Also, Bianchini's *Tabulae magistrales* explicitly mention an obliquity of ε=23;33,30° in the canons.[96] The table of right ascension usually found in both sets, however, is based on an obliquity of 23;35°. Peurbach or Regiomontanus may thus have falsely assumed that Bianchini's preferred use of an

92 For the values given in the printed 1514 edition, we have corrected one apparent printing error for 21° Pisces: read 4;52° for 5;52°.

93 The standard deviation for the modern fit and Regiomontanus's algorithm are 2" and 3", respectively.

94 Note that this is slightly larger than the value of 21;15° Aquarius given in the *Epitome*. We obtained this value from a bisection method on interval [21°–21;30°]; see Table 1. We assume that 21;15° given in the *Epitome* is also obtained by bisection, but terminating after two steps for pedagogical purposes and, thus, is rather exemplary. Beyond this explicit number, the author of the *Epitome* states, for the final table, 'for indeed I placed the beginning in the end of 21° Aquarius'; Regiomontanus, *Epytoma in Almagestum Ptolomei* (Venice: 1496), III:30: 'Posui namque principium in fine 21 gradus Aquarii'. This is equivalent to the heading of the fourth column in the unfilled table in the *Rechenbuch*; see Figure 2.

95 van Brummelen, 'Before the End of an Error'.

96 See, for example, Cracow, BJ, 556, f. 7ra. On the *Tabulae magistrales*, see José Chabás, 'An Analysis of the *Tabulae magistrales* by Giovanni Bianchini', *Archive for History of Exact Sciences*, 70 (2016), 543–52.

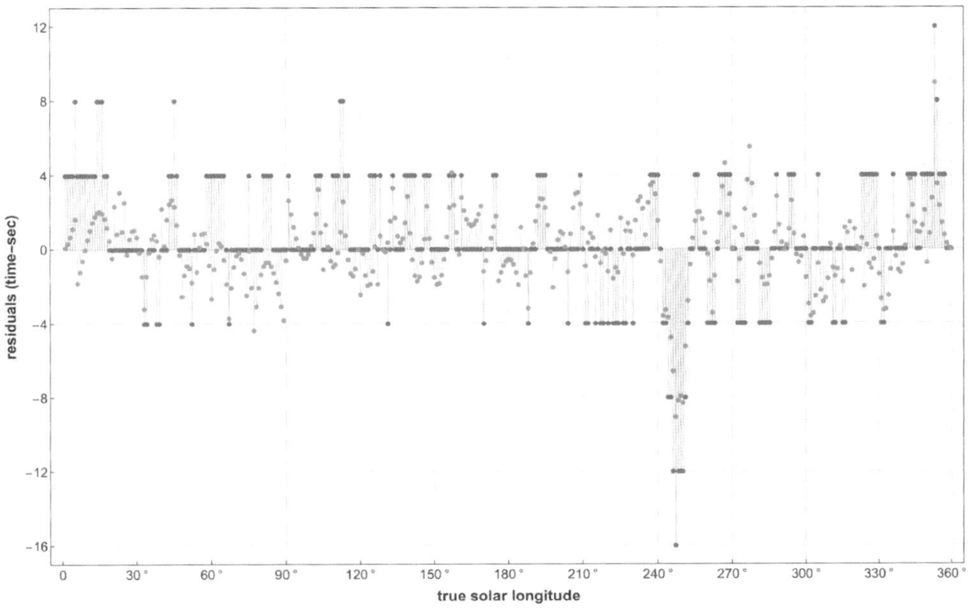

Figure 10.5. Residuals of the two recomputations in comparison to Peurbach's table for the equation of time. Blue: Regiomontanus's historical method, based on tables only and computation in sexagesimal arithmetic. Orange: Modern recomputation based on Formula (1) with best fit parameters according to Table 1.

obliquity of 23;33,30° was also applied in the table of right ascension. However, this is not true, and Bianchini's right ascension is taken from al-Battānī and based on an obliquity of 23;35°. In fact, the latter is also underlying Peurbach's equation of time.[97] With the proper radix, the *Tabulae astronomiae*, and the right ascension table from the manuscript Kraków, BJ, 556, we implemented Regiomontanus's algorithm.[98] The residuals that arise from our implementation and Peurbach's table for the equation of time are plotted in Figure 10.5, where we also give the residuals of the modern implementation for comparison.[99]

A few remarks are in order. A plain inspection of the values in Peurbach's table tentatively suggests that they have been derived on a five-degree grid with intermediate values obtained from linear interpolation. Also, the unfilled table in Regiomontanus's *Rechenbuch* suggest a five-degree grid, as can be seen from the first column in Figure 10.3. The text that details the algorithm, however, clearly indicates that a grid, if any, can only be based on days and not integer degrees of longitude. The difference is admittedly small, because daily solar mean motion is about one degree. Nevertheless, if an overall five-day grid is assumed

97 Note that this value is not contained in the 95% confidence interval for the obliquity of the modern fit given in Table 2, because the underlying formula of the modern fit is inconsistent with historical practice.

98 We have implemented the algorithm in *Mathematica* based only on looking up values from other tables. Non-integer values were obtained from linear interpolation in the corresponding tables. All computations are performed in sexagesimal arithmetic. We used the right ascension table in Cracow, BJ, 556, f. 95[r] contained in Bianchini's tables.

99 With regard to the peak in the residuals above 240°, we have collated the 1514 printed edition of Peurbach's tables with the two manuscript versions Vatican, BAV, Pal. lat. 1375, f. 90[r] and Nuremberg, SB, Cent. V, 57 f. 143[v], which, however, are identical. Hence, it is likely that the peak is the result of a computational mistake.

in the derivation then this will correspond in a few cases to four-degree intervals in the table, and in a few other cases to six-degree intervals. This is a particular consequence of the use of Bianchini's tables. The residuals given in Figure 10.5 for Regiomontanus's historical method are based on a five-day grid with intermediate values obtained from linear interpolation.[100] The fact that the table is not based on a single-day grid can be seen from the peak around 245°, where surrounding values follow the same error pattern and have thus been derived from linear interpolation of a faulty value.[101]

Beyond the fact that both the modern recomputation and the historical method fit Peurbach's table equally well, they are based on different parameters. First and foremost, the solar apogee is considerably different. The modern fit results in an apogee of 90°, while the historically attested method starts from an apogee of 90;46°. The latter is effectively shifted to a smaller longitude, by the choice of a radix of days; nevertheless, their difference amounts to forty-six minutes at the outset of the algorithm. This difference in longitude corresponds to a difference in time of some eighty years, and, should we wish to date the table from its underlying apogee, we would draw two quite different conclusions. The algorithm used can thus have a significant effect on the seemingly underlying parameters. In our example of Peurbach's table, the obliquity can also change by up to about one minute, depending on what algorithm is used for the reconstruction. Although, this difference appears to be small and will hardly be discernible from a table of right ascension itself when it is only given in minutes, yet it is of the order of magnitude discussed by Regiomontanus and Bianchini to find the proper, up-to-date obliquity for their own epoch.[102]

Conclusion

The equation of time is one of the very few tabular quantities within sets of astronomical tables that was newly computed at times due to the advancement of the solar apogee. In Johannes Virdung's autograph manuscript Vatican, BAV, Pal. lat. 1375, for example, there are three different tables for the equation of time: Peter of St Omer's table that was also used in the *Tabulae resolutae*, Peurbach's table that was part of his *Tabulae eclypsium*, and Bianchini's old table, taken from al-Battānī, that was part of his planetary tables. As we have argued, Virdung even compared Peurbach's table to a fourth, novel table for the equation of time computed by Bianchini and included in his *Tabulae primi mobilis*. All these tables for the equation of time are based on different parameters and were newly computed.

In his early works, Regiomontanus used the equation of time attributed to Peter St Omer, which was included in several sets of astronomical tables in the Alfonsine period. When he switched to using the planetary tables of Giovanni Bianchini, Regiomontanus silently

100 Statistical analysis of the residuals of Peurbach's table cannot discriminate between a single- or five-day grid used in the historical method. The standard deviation of the five-day grid is 0;0,3,20, while a single-day grid results in a standard deviation of 0;0,3,21. A ten-day grid, explicitly mentioned in Regiomontanus's text, however, can be ruled out by its standard deviation of 0;0,6,14.

101 The outlier at 353° is an apparent scribal error that we did not attempt to correct.

102 On different values for the obliquity of the ecliptic and the ensuing discussion between Regiomontanus and Bianchini, see, for example, Armin Gerl, *Trigonometrisch-astronomisches Rechnen kurz vor Copernicus: Der Briefwechsel Regiomontanus-Bianchini* (Stuttgart: Franz Steiner, 1989), footnote 201, pp. 151–55.

adopted an older equation of time contained therein, which is attributed to al-Battānī. In 1456, or shortly after, Regiomontanus wrote a short text in his *Rechenbuch* on how to derive a new table for the equation of time, which we have discussed in detail in this article. The equation of time that is included in Peurbach's eclipse tables may be based on the algorithm described in Regiomontanus's *Rechenbuch*. Hence, we ask ourselves the following question: did Regiomontanus use this equation of time derived by his own algorithm?

Around 1459, he indeed used a novel equation of time for the nativity of Maximillian I, the later Holy Roman Emperor. At the time of Maximilian's birth on 22 March 1459 at 4;49h after midday, the Sun was at 10;41° Aries.[103] To this time, Regiomontanus added eleven minutes to account for the equation of time. Neither the equation contained in Bianchini's *Tabulae astronomiae* nor the one attributed to Peter of St Omer, both used by Regiomontanus, fit this value, and both would result in a correction of twelve minutes. However, Peurbach's equation does fit the eleven-minute value. We tend to muse that once he had derived it (*c.* 1456), Regiomontanus used this equation of time. Nevertheless, Bianchini's newly derived modern equation 'verified for the year 1456' also fits this value. Without the inclusion of seconds in the nativity cast by Regiomontanus, it is impossible to differentiate between these two equations. Be that as it may, Eleanore of Portugal, Empress of the Holy Roman Empire, might have been pleased that newly derived astronomical tables came directly into use for the nativity she ordered for her royal son.

In fact, in 1451 Regiomontanus had already cast the nativity for Empress Eleonora herself, for which he had likewise used an equation of time.[104] On the date of her birth on 15 September 1436 at 15;46h after midday, the Sun was at 1;57,10° Libra. Regiomontanus corrected this time by twenty-five minutes.[105] This correction is in accordance with both the tables for the equation of time by Peter of St Omer and Bianchini's old table. However, the new table derived for 1456 by Peurbach (or Regiomontanus) and Bianchini's table 'verified for the year 1456' would both lead to a different correction of twenty-six minutes. Hence, we suppose that once derived, Regiomontanus readily used his novel equation of time.

Beyond his personal use for horoscopes, Regiomontanus's algorithm for the derivation of the equation of time, firstly laid out in his *Rechenbuch* and then later taken up again in the *Epitome of the Almagest*, did have an impact on the astronomical corpus of knowledge. In 1551, Erasmus Reinhold included a newly computed equation of time in the *Prutenic Tables*, which were based on Copernicus's novel heliocentric system presented in *De revolutionibus*. The canons start with a chapter on the equation of time and the inequality of the days. In the text, Reinhold explicitly refers to Regiomontanus's method presented in

103 Vienna, Österreichische Nationalbibliothek, Cod. 5179, f. 2^v.

104 Munich, Bayerische Staatsbibliothek, Clm 451, ff. 78^r–85^v. On this manuscript, see Zinner, *Regiomontanus*, p. 304. Note that in his edition of selected works of Regiomontanus, Schmeidler wrongly believed that the letters 'n̄ e' in the nativity referred to 'nata e[st]', and he did not seem to brood over the meaning of the single 'e' in the next line in correspondence with a slightly shifted time. Apparently, these letters indicate the use of the equation of time and refer to '[diebus] n[on] e[quatis]' and '[diebus] e[quatis]'. See Johannes Regiomontanus, *Opera collectanea*, ed. by Felix Schmeidler (Osnabrück: Zeller, 1972), p. 5.

105 Munich, BSB, Clm 451, f. 78^v.

the third book of the *Epitome*.[106] Moreover, among the tables there is one for the 'equation of time derived from the most recent notion based on the doctrine of Regiomontanus'.[107]

The example of Reinhold shows that other historical actors were aware of the principal algorithm on how to derive a novel table for the equation of time. This algorithm is based on geometrical understanding of the astronomical problem shaped by other existing tables used for the derivation. Current understanding of historical mathematical practice in terms of a functional derivation of single values from first-order principles (e.g. not based on a solar equation table) is incorrect for Regiomontanus's case.[108] We explicitly show that even parameters can be noticeably affected by the underlying algorithm used to fit the table for the equation of time. Modern fits may still be applied with caution when estimating parameters, but actual historical practice may often be embedded in an entirely different space of thought. Best-fit scenarios may not be used to argue for steps in historical calculations as they risk a false projection of modern ideas onto historical practices. To further elaborate on historical computational scenarios and thus the interdependence of astronomical knowledge, we are in dire need of more worked historical examples and texts that reveal exact procedures.

106 Erasmus Reinhold, *Prutenicae tabulae coelestium motuum* (Tübingen: Morhart, 1551), f. 16^{r-v}.

107 Reinhold, *Prutenicae tabulae*, ff. 5^{v}–6^{r}: 'Posterior canon aequationis dierum Naturalium ex recentiorem *(sic!)* sententia iuxta Regiomontani doctrinam'.

108 It is likewise incorrect for the table of the equation of time derived by Lewis Caerleon, see Miolo, Zieme, 'Lewis Caerleon and the Equation of Time'.

Appendix: A Critical Edition and Translation

Editorial Principles: The Text and its Translation

This appendix is intended to provide the edition of the text on the equation of time included in Cod. 5203 (ff. 70^{r}–71^{v}) and its translation. In preparing this edition, we have attempted to reproduce the text in Cod. 5203 as faithfully as possible, retaining its orthography in all instances. Particularly, the following orthographic features should be noted: the alteration of the consonant spelling of 'ti' into 'ci' (e.g. 'equa*ci*o' in place of the conventional spelling 'aequa*ti*o'[109]); the dropping of the diphthong 'ae' to just 'e' (e.g. '*e*quacio' in place of the conventional spelling '*ae*quatio'); the double consonant 'cc' in place of the consonant group 'ct' (e.g. 'equino*cc*ium' in place of the conventional spelling 'equino*ct*ium'); the vocalic alteration of 'i' into 'y' in the adverb '*y*mmo' (in place of the conventional spelling 'immo'). The adjective 'equinoctialis' (conventional spelling: 'aequinoctialis') never agrees with its corresponding noun in case, number, and gender; the invariability of 'aequinoctialis' is also retained in the edited text.

We have applied a modern system of punctuation to mark syntactic units. In a few instances, we have altered the original divisions of the text. This primarily helps to facilitate the understanding of the literary text, in which punctuation seems to define and distinguish the stages of the mathematical operations rather than serve a syntactic purpose. The page change is noted in the right margin and within the text by a vertical bar.

Since the edition was conducted on the basis of a single manuscript, it was deemed appropriate to avoid a redundant and unnecessary repetition of its siglum in the apparatus. In fact, the apparatus only records corrections made to the text by the main copyist (Regiomontanus) and, in two cases, additions by another hand, probably that of Johannes Schöner.

Our approach to the translation of the text has been to remain as faithful as possible to the original text, diverting from this principle only when we deemed it necessary. Square brackets in the translation enclose words that we added in order to increase the readability of the text. To preserve the connection between the text and its translation, the reference to the page change is noted in the translation in the right margin and within the text by a vertical bar.

Glossary

dies differens = apparent day
dies naturalis differens = apparent solar day
dies naturalis medius = mean solar day
dies non equati = non-equated days[110]
equatus = equated

109 The text copied by Regiomontanus presents only one instance where the abbreviation for the word 'equation' does not include its ending; in this case, this word, which is usually either found spelled out (e.g. 'equacio', line 19 of the present edition) or abbreviated as 'eq^{a}cio'(e.g. 'equacio', line 74 of the present edition), is abbreviated as 'equa^{o}'. Given that all the other occurrences highlight the spelling 'equacio', the abbreviation of this word on line 91 of the present edition is solved as 'equacio'.

110 Although non-equated days are identical to mean days, we have retained the differentiation made by Regiomontanus in order to indicate that it is the equation (of time) that turns non-equated days into equated days.

inequatus = non-equated
pars equinoctialis = equinoctial part
tempus non equatum *and* tempus inequatum = non-equated time

List of Abbreviations

add. = *addidit* = added; indicates random insertion of words or notes in another hand
ante corr. = *ante correctionem* = before correction
del. = *delevit* = deleted; indicates deletion by strikethrough
i. m. = *in margine* = in the margin
rescr. = *rescripsit idest clarius scripsit* = the copyist wrote anew to make something clearer
s. l. = *supra lineam* = above the line
scr. = *scripsit* or *scripsisse* (i.e. *with* videtur) = the copyist wrote *or* seems to have written
suppl. = *supplevit* = the copyist supplied
Angular brackets (<...>) enclose words, etc. that we have added

De equationibus dierum 70r

Anno domini 1456to currente. Sol intrabit augem suam 12^{a} die Iunii, hora 21, minuto 23^{tio} diebus non equatis et aux erit tunc 1 signum, 30 gradus, 46 minuta, 0 secunda. Dies residui de Iunio sunt 17 dies, 2 hore, 37 minuta; dies autem a fine Iunii usque in finem anni semper sunt 184. Sunt itaque ab applicatione solis in augem usque ad finem anni 201 dies, 2 hore, 37 minuta.

Volo invenire quo tempore in sequenti anno ingredietur in Arietem. Video quantum ab auge sua sit in principium Arietis et sunt 4 signa, 29 gradus, 14 minuta, 0 secunda. Hec quero in tabula motus veri solis ab auge et minorem proximo inventum in directo 270 dierum scribo, scilicet 4 28 17 28, et demo de distantia principium Arietis ab auge, remanent 0 0 56 32. Hec divido per motum solis in hora, scilicet 0 0 2 28. Exeunt 22 hore, 57 minuta.

Tempus igitur ab ingressu solis in augem usque ingressum eius in Arietem est 270 dies, 22 hore, 57 minuta sed ab ingressu eius ad augem usque in finem anni sunt 201 dies, 2 hore, 37 minuta. Quare a principio anni 1457^{mi} per 69 dies, 20 horas, 20 minuta fiet equinoccium. Sed tunc motus eius verus in die erit 0 0 59 8, huius autem ascensio recta est 54 minuta.

Est igitur dies naturalis differens tunc 360 gradus, 54 minuta equinoctialis, sed naturalis medius semper est 360 gradus, 59 minuta, 8 secunda. Quare oportet ut equacio dierum sit tunc 5 minuta equinoctialis, si faciemus ibi inicium. Quando itaque sol et principium Arietis simul sunt in meridiano, quod non dubitandum est in aliqua regione fieri, tunc citius perficitur dies differens quam medius in 5 minutis equinoctialis.

Oportet igitur ut sol transeat ultra meridianum donec 5 minuta equinoctialis recte revolvantur, si debet dies media perfici. Quando igitur instans considerationis tue pro aliquo negotio esset per diem unam mediam precise post equinoccium anno supradicto et velles scire quo tempore in die differente illud instans caderet, necesse est ut 20 secunda hore adderes super meridiem.

Nam ipsum idem instans cadit in die deferente post meridiem ad 20 secunda hore, quod quidem instans terminat diem mediam. Quare sequitur: non valet argumentum, dies medius est maior die differente. Igitur differentia inter eos est a tempore non equato subtrahenda, ymmo tunc est addenda, et vide multum bene ut hoc intelligas.

Est enim unum de raris. Breviter sic accipe. Quando dies medius est vero maior, tunc necesse est completo die medio solem transivisse ultra meridianum tanta parte equinoctialis revoluta quanta est differentia inter illos dies. Et sic quando ex inequatis volumus

1 De … dierum] *titulus alia manu inscriptus in margine superiori* **9/10** in^{2} … dierum] *suppl. i. m.* **26** illud instans] *suppl. s. l.* **28** deferente] *sic scr. pro* differente

On the Equation of Time 70r

In the current year of the Lord 1456. The Sun will enter its aux on the 12th day of June, at the 21st hour, 23rd minute, with non-equated days, and at that time the aux will be 1 sign, 30 degrees, 46 minutes, 0 seconds. The remaining days of June are 17 days, 2 hours, 37 minutes; however, the days from the end of June to the end of the year are always 184. Thus, there are 201 days, 2 hours and 37 minutes from the entry of the Sun in the aux until the end of the year.

I want to find when in the following year the Sun will enter Aries. I see how great is [the distance] from its aux to the beginning of Aries and there are 4 signs, 29 degrees, 14 minutes and 0 seconds. I look for these [values] in the table of the true motion of the Sun from the aux and I write the next smaller [value] found opposite 270 days, namely 4,28;17,28 [degrees] and I subtract the beginning of Aries from the distance from the aux, 0,0;56,32 [degrees] remain. I divide these by the motion of the Sun in one hour, namely 0,0;2,28 [degrees]. 22 hours and 57 minutes result.

Consequently, the time from the entrance of the Sun in the aux until its entrance in Aries is 270 days, 22 hours, 57 minutes, but from its entrance to the aux until the end of the year there are 201 days, 2 hours, 37 minutes. And for this reason, from the beginning of the year 1457 after 69 days, 20 hours, 20 minutes the equinox will happen. But at that time the true motion of the Sun in a day will be 0,0;59,8 [degrees], yet its right ascension is 54 minutes.

Then the apparent solar day at that time is 360 degrees, 54 equinoctial minutes, but the mean solar day is always 360 degrees, 59 minutes, 8 seconds. And for this reason, it is necessary that the equation of time is then 5 equinoctial minutes, if we will begin there. And thus when the Sun and the beginning of Aries are together in the meridian, which is not to be doubted to happen in any region, then the apparent day is sooner completed than the mean by 5 equinoctial minutes.

Consequently, it is necessary that the Sun would cross over the meridian while 5 equinoctial minutes rightly elapse, if the mean day should be completed. Then, when the moment of your consideration for some activity would be for one mean day precisely after the equinox in the above said year and you would like to know when in an apparent day that moment may occur, it is necessary that you add to the midday 20 seconds of an hour.

Indeed, that very moment occurs in the apparent day at 20 seconds of an hour after midday, and indeed this moment completes the mean day. And for this reason it follows: the argument has no influence. The mean day is longer than the apparent day. Accordingly the difference between them is to be subtracted from the non-equated time, or rather added, and consider this very well so that you understand.

Indeed, it is one of the rare cases. Learn thus briefly. When the mean day is in truth longer, then it is necessary that the Sun has crossed over the meridian in a mean complete day in as much elapsed equinoctial part as the difference between those days. And thus when from non-equated days we want to make equated days we add their difference to the non-equated time and the equated will result.

Therefore it is said that you should add to the non-equated time, but by this | do not 70v
understand that the equated days are always longer than the non-equated; instead

equatos facere, addimus differentiam eorum super tempus inequatum et proveniet equatum.

Dicitur quidem quod addere debeas super tempus inequatum, sed per hoc | non 70v intelligas quod dies equati semper sunt inequatis maiores, ymmo aliud accipe videlicet ita precipitur, ut tu scias instans temporis inequati in quo loco temporis equati ponatur. Quando autem dies medius minor est, vero tunc necesse est completo die medio solem adhuc non pervenisse ad meridianum, sed oporteat adhuc tantam partem equinoctialis revolvi quanta est dierum talium differentia.

Et sic tunc si vellemus scire pro quo instanti diei differentis terminaretur dies medius, deberemus subtrahere differentiam dictam, eo quod medius dies terminatur tunc ante meridiem. Verum quod non facimus subtractionem, quia fundamentum et radix posita sunt super principium additionis quod incipit in 19no gradu Aquarii; tunc enim verus motus solis in die medio est 1 gradus, 0 minuta, 44 secunda. Ascensio autem recta correspondens isti vero motui solis est 59 minuta, 44 secunda. Quare dies differens est maior medio in 36 secundis equinoctialis que faciunt secunda hore tantum. Ecce adhuc sequetur crementum.

Sit sol et principium 15^{mi} Aquarii in meridiano, tunc verus motus solis in die uno est 1 0 44. Ascensio autem recta ab Ariete incipiendo ad principium 15me Aquarii est 316 30, sed ascensio recta ad 15 gradus, 0 minuta, 44 secunda est 317 30 44. Differentia est 1 0 44; hec est ascensio correspondens vero motu solis in die. Sed est maior quam 59 minuta, 8 secunda, in 1 minuto et 36 secundis. Ecce dies differens et ibi maior medio in tanto.

Sit deinde sol in principio 20^{mi} gradus. Motus verus diei est 1 gradus 0 39; ascensio sibi correspondens est 59 38 que est maior 59 8 in 30 secundis; adhuc dies differens maior est medio. Item sit in principio 22^{i}. Verus motus diei est 1 0 36, ascensio sua 59 35 adhuc est maior. Item in 24to in fine. Motus diei est 1 0 34, ascensio sua 59 30.

Item in principio 26^{ti} motus in die est 1 0 33, ascensio sua 58 30. Hui, quam cito irregulatur: propter defectum igitur secundorum in tabula rectarum ascensionum provenit.|

Supponamus igitur equalibus dum fuerit sol in principio 20^{mi} / fine 19^{ni} Aquarii et 71r supponamus augem solis in principio Cancri. Tunc distantia solis ab auge erit 3 signa, 49 gradus, 0 minuta et ascensio eius recta erit 321 27. Hec continue servo.

Deinde post 10 dies medios erit sol in 10 gradibus 5 minutis distans a loco priori et sic erit 29 gradus, 5 minuta Aquarii, cuius ascensio recta est 331 15. Hoc excedit ascensionem radicis servatam in 9 48, sed medius motus solis in 10 diebus est 9 51; hic excedit differentiam ascensionum in 3 minutis. Sunt igitur isti 10 dies medii maiores 10 diebus diffe-

43 diei differentis] *suppl. i. m.* **47** 0] 4 *ante corr.* recta] *primum scr.* 19 gradus 1 minutum Aquarii *post* recta *sed del.* **48** 44 secunda] *videtur scr.* 4 secunda 44 tertia *ante corr.* maior] minor *ante corr.* **49** 36 secundis] 36 minutis *ante corr.; videtur scr.* in 36^{14} 3^{ss} *post* 36 secundis *sed del.* faciunt] *primum scr.* 19^{2} 3^{a} *post* faciunt *sed. del.* secunda] *suppl. s. l.* adhuc] *primum scr.* est in diminutione vade ultra. Sit sol et principium 19^{ni} (20^{mi} *s. l.)* gradus Aquarius in meridiano tunc verus motus solis in die media est 1 0 39 ante incip<i>et *post* adhuc *sed del.* sequetur] *suppl. s. l.* **62** fine 19^{ni}] *suppl. s. l.* **64** 321 27] *add.* ascensionis radicis *i. m.* **65** sol] *videtur scr.* mo^{i} *post* sol *sed del.* 10^{2} ... minutis] 30 gradibus 5 minutis *ante corr.*

acknowledge something else. Of course, thus is advised so that you would know to which point of equated time the moment of non-equated time is set. However, when the mean day is shorter, then in truth it is necessary that in a complete mean day the Sun has not yet arrived at the meridian, but it is necessary that still as much equinoctial part [of time] as the difference of such days revolves.

And thus then if we wish to know for which moment of the apparent day the mean day is completed, we would have to subtract the said difference, because the mean day is then completed before midday. In truth, we do not do the subtraction because the foundation and the radix are set at the beginning of the addition, which starts at the 19th degree of Aquarius; then, indeed, the true motion of the Sun in the mean day is 1 degree, 0 minutes, 44 seconds. However, the right ascension corresponding to the true motion of the sun is 59 minutes, 44 seconds. And for this reason, the apparent day is longer than the mean by 36 equinoctial seconds that make only seconds of an hour. Behold, there will follow yet the increase.

Let the Sun and the beginning of the 15th [degree of] Aquarius be in the meridian; then the true motion of the Sun in a day is 1;0,44 [degrees]. However, the right ascension starting from Aries to the beginning of the 15th [degree of] Aquarius is 316;30 [degrees], but the right ascension of 15 degrees, 0 minutes, 44 seconds [Aquarius] is 317;30,44 [degrees]. The difference is 1;0,44 [degrees]; this is the corresponding ascension in the true motion of the Sun in a day. But it is longer than 59 minutes, 8 seconds by 1 minute and 36 seconds. Behold the apparent day and there it is larger than the mean day by this much.

Now let the Sun be in the beginning of the 20th degree [of Aquarius]. The true motion of the day is 1;0,39 degrees; the corresponding ascension there is 59;38 [min], which is bigger than 59;8 [min] by 30 seconds; still the apparent day is longer than the mean. Likewise it would be in the beginning of the 22nd [degree of Aquarius]; the true motion of the day is 1;0,36 [degrees], its ascension 59;35 [min]. Still it is larger. Likewise in the end of the 24th [degree of Aquarius] the motion of the day is 1;0,34 [degrees], its ascension is 59;30 [min].

Again in the beginning of the 26th the motion in a day is 1;0,33 [degrees], its ascension is 58;30 [min]. Oh, how quickly this strays from the rules: it comes from the lack of seconds in the table of the right ascensions!|

Let us suppose they are equal, while the Sun would be in in the beginning of the 20th 71r
(end of 19th) [degree] of Aquarius and let us suppose the aux of the Sun [to be] in the beginning of Cancer. Then the distance of the Sun from the aux will be 3 signs, 49 degrees, 0 minutes and its right ascension will be 321;27 [degrees]. I keep these [values] continuously.

Then, after 10 mean days the Sun will be in 10 degrees, 5 minutes far from the previous place and thus it will be 29 degrees, 5 minutes of Aquarius, the right ascension of which is 331;15 [degrees]. This exceeds the kept ascension of the radix in 9;48 [degrees], but the mean motion of the Sun in 10 days is 9;51 [degrees]; this exceeds the difference of the ascension by 3 minutes. Then these 10 mean days are longer than 10 apparent days by 3 equinoctial minutes. Consequently, while the Sun would be in 29 degrees 5 [min] of Aquarius, the equation of time is 3 equinoctial minutes.

Again, from the beginning of the radix 20 days would pass, in which the true motion of the Sun is 20;6,23 [degrees] and thus [the Sun] will be in 9 degrees 6;23 [min] of Pisces, the right ascension of which is 340;43 [degrees]. This exceeds the ascension of

rentibus in 3 minutis equinoctialis. Dum igitur sol fuerit in 29 gradus 5 Aquarii, equacio dierum est 3 minuta equinoctialis.

Iterum a principio radicis transeant 20 dies in quibus verus motus solis est 20 6 23 et sic erit in 9 gradus 6 23 Piscium, cuius ascensio recta 340 43. Hoc excedit ascensionem radicis in 19 16, sed medius motus solis in 20 diebus est 19 43. Quare dum sol fuerit 9 gradus, 6 minuta Piscium, equacio dierum est 27 minuta equinoctialis.

Iterum a principio radicis transeant 30 dies in eis verus motus solis est 30 3 28. Erit ergo tunc in 19 gradus 3 28 Piscium et ascensio recta 349 57, que maior est ascensione radicis in 28 gradus 30, sed medius solis in 30 diebus est 29 gradus 34. Quare oportet ut hii 30 dies medii maiores sint 30 veris in 1 gradus, 4 minuta equinoctialis; et tanta est equacio dierum, dum sol fuerit in 19 gradus 3 Piscium.

Item verus motus in 40 dies est 39 gradus 57; quare esset tunc in 28 57 Piscium, cuius ascensio recta 359 2, differentia eius ad ascensionem radicis 37 35, medius solis in 40 dies 39 26. Equacio igitur dierum 1 51.

Item verus motus a loco radicis in 100 diebus 1 38 gradus 5; quare tunc esset in 27 5 Thauri, cuius ascensio 54 46, differentia eius et ascensionis radicis 93 19. Medius sol in 100 diebus 98 34. Equatio dierum 5 15, dum sol sit in 27 5 Thauri. Secundum igitur hunc modum de novo tabulam compone.

Deinde supponamus ut aux solis sit in principio Leonis. Volo videre | an adhuc sit 71v
equalitas diei naturalis medii et differentis in fine 19^{ni} Aquarii. Nam tunc sol distaret ab auge per 3 signa, 19 gradus; motus eius verus esset in die per 1 gradus 1 15, cui correspondet ascensio recta 1 0 13. Hoc excedit medium motum in die per 1 minutum, 5 secunda; venit igitur equalitas post. Volo videre in fine 24^{ti}. Verus motus in die est 1 1 14, cui tunc respondet ascensio 1 0, excedit medium motum in die.

Sit igitur in fine 25^{ti}. Verus est ut prius. Cui correspondet ascensio equalis medio motu solis in die. Ecce equalitas iam pervenisset ad finem 25^{ti}, cuius ascensio recta est 327 19. Deinde post 100 dies verus motus in eis est 1 39 27, sic esset in 4 27 Geminorum, huius ascensio recta 62 27; demo ab ea ascensionem radicis, manet 95 8. Medius solis in 100 dies est 98 34. Differentia 3 26: equacio dierum dum sol esset in 4 27 Geminorum. Ecce quomodo variatis annis oportet variare tabulam equationum.

---

72 Piscium] Capricorni *ante corr.* 43] *videtur scr.* sed verus mod *post* 43 *sed del.* **78** hii] *suppl. s. l.* in] *primum scr.* ver *post* in *sed del.* **89** gradus^{1}] *primum scr.* ut *post* gradus *sed del.* 1^{1}] *rescr.*

the radix by 19;16 [degrees], but the mean motion of the Sun in 20 days is 19;43 [degrees]. And for this reason, when the Sun is in 9 degrees, 6 minutes of Pisces, the equation of time is 27 equinoctial minutes.

Again, from the beginning of the radix 30 days would pass, in which the true motion of the Sun is 30;3,28 [degrees]. Therefore, it will then be in 19 degrees 3;28 [min] of Pisces and the right ascension will be 349;57 [degrees], which is larger than the ascension of the radix by 28 degrees 30 [min], but in 30 days the [motion of the] mean Sun is 29 degrees 34 [min]. And for this reason, it is necessary that these 30 mean days would be longer than 30 true [days] by 1 degree [and] 4 equinoctial minutes; and this amount is the equation of time when the Sun is in 19 degrees 3 [min] of Pisces.

Again, the true motion in 40 days is 39 degrees 57 [min]; and for this reason, then [the Sun] would be in 28;57 [degrees] of Pisces, the right ascension of which is 359;2 [degrees], its difference from the ascension of the radix [is] 37;35 [degrees], the mean Sun in 40 days [is] 39;26 [degrees]. Then the equation of time [is] 1;51 [degrees].

Then, the true motion from the place of the radix in 100 days [is] 1 [sign] 38 degrees 5 [min]; and for this reason then, [the Sun] would be in 27;5 [degrees] of Taurus, the ascension of which [is] 54;46 [degrees], the difference of this and the ascension of the radix [is] 93;19 [degrees]. The mean Sun in 100 days [is] 98;34 [degrees]. The equation of time [is] 5;15 [degrees], while the Sun would be in 27;5 [degrees] of Taurus. Then, according to this method, produce a table anew.

Then let us suppose that the aux of the Sun would be in the beginning of Leo. I want to check | if there is yet the equality of the mean solar day and the apparent at the end of 71v
19th [degree] of Aquarius. Indeed, then the Sun would be distant from the aux by 3 signs, 19 degrees; its true motion in a day would then be 1 degree 1;15 [min], corresponding to a right ascension of 1;0,13 [degrees]. This exceeds the mean motion in a day by 1 minute, 5 seconds; the equality then comes after [this day]. I want to check the end of the 24th [degree of Aquarius]. The true motion in a day is 1;1,14 [degrees], to which then the right ascension 1;0 [degrees] corresponds, [and this] exceeds the mean motion in one day.

Then it would be in the end of the 25th [degree of Aquarius]. It is true as before. And to this corresponds an ascension [which is] equal to the mean motion of the sun in a day. Behold the equality already came at the end of the 25th, the right ascension of which is 327;19 [degrees]. Then after 100 days the true motion is 1 [sign] 39;27 [degrees], thus [the Sun] would be in 4;27 [degrees] of Gemini, the right ascension of which [is] 62;27 [degrees]; I subtract from this [latter value] the ascension of the radix, 95;8 [degrees] remain. The mean Sun in 100 days is 98;34 [degrees]. The difference [is] 3;26 [degrees]: the equation of time when the Sun would be in 4;27 [degrees] of Gemini. Behold how having changed the years it is necessary to change the table of the equations.

Manuscript Sources

Cracow, Biblioteka Jagiellońska, 556
Madrid, Biblioteca Nacional de España, 3306
Munich, Bayerische Staatsbibliothek, Clm 451
Nuremberg, Stadtbibliothek, Cent. III, 25
Nuremberg, Stadtbibliothek, Cent. V, 36
Nuremberg, Stadtbibliothek, Cent. V, 57
Nuremberg, Stadtbibliothek, Cent. VI, 12
Nuremberg, Stadtbibliothek, Cent. VI, 18
Nuremberg, Stadtbibliothek, Cent. VI, 21
Seitenstetten, Stiftsbibliothek, 53
Vatican, Biblioteca Apostolica Vaticana, Pal. lat. 1375
Venice, Biblioteca Nazionale Marciana, Fondo antico lat. Z. 328
Vienna, Österreichische Nationalbibliothek, Cod. 4988
Vienna, Österreichische Nationalbibliothek, Cod. 5179
Vienna, Österreichische Nationalbibliothek, Cod. 5203

Bibliography

Al-Battānī, *De scientia stellarum* (Nuremberg: Johannes Petreius, 1537).
———, *Opus astronomicum*, ed. by Carlo Alfonso Nallino, 3 vols (Milan: Hoeplium, 1899–1907).
Al-Farghānī, *Compilatio astronomica* (Ferrara: Andreas Belfortis, 1493).
———, *Differentie in quibusdam collectis scientie astrorum*, ed. by Francis J. Carmody (Berkeley: 1943).
Alfonso X, *Tabulae astronomicae* (Venice: Erhard Ratdolt, 1483).
Bianchini, Giovanni, *Tabulae astronomiae* (Venice: Simon Bivilaqua, 1495).
Byrne, James Steven, 'A Humanist History of Mathematics? Regiomontanus's Padua Oration in Context', *Journal of the History of Ideas*, 67 (2006), 41–61.
Chabás, José, 'An Analysis of the *Tabulae magistrales* by Giovanni Bianchini', *Archive for History of Exact Sciences*, 70 (2016), 543–52.
———, *Computational Astronomy in the Middle Ages. Sets of Astronomical Tables in Latin* (Madrid: Consejo Superior de Investigaciones Científicas, 2019).
———, and Bernard R. Goldstein, *The Alfonsine Tables of Toledo* (Dordrecht: Kluwer, 2003).
———, and ———, *The Astronomical Tables of Giovanni Bianchini* (Leiden: Brill, 2009).
Curtze, Maximilian, 'Der Briefwechsel Regiomontan's mit Giovanni Bianchini, Jacob von Speyer und Christian Roder', in *Urkunden zur Geschichte der Mathematik im Mittelalter und der Renaissance*, ed. by Maximillian Curtze (Leipzig: Teubner, 1902), pp. 185-336.
Czerny, Albin, 'Aus dem Briefwechsel des grossen Astronomen Georg von Peurbach', *Archiv für österreichische Geschichte* 72 (1888), 281-304.
Dalen, Benno van, *Ancient and Mediaeval Astronomical Tables: Mathematical Structure and Parameter Values* (PhD dis., Utrecht University, 1993).
———, 'On Ptolemy's Table for the Equation of Time', *Centaurus*, 37 (1994), 97–153.

———, 'al-Khwārizmī's Astronomical Tables Revisited: Analysis of the Equation of Time', in *From Baghdad to Barcelona. Studies in the Islamic Exact Sciences in Honour of Prof. Juan Vernet*, ed. by Josep Casulleras and Julio Samsó, 2 vols (Barcelona: Instituto Millás Vallicrosa de Historia de la Ciencia Árabe, 1996), I, pp. 195–252.

Folkerts, Menso, 'Die mathematischen Studien Regiomontanus in seiner Wiener Zeit', in *Regiomontanus-Studien*, ed. by Günther Hamann (Vienna: Verlag der Österreichischen Akademie der Wissenschaften, 1980), pp. 175–209.

Gerl, Armin, *Trigonometrisch-astronomisches Rechnen kurz vor Copernicus: Der Briefwechsel Regiomontanus-Bianchini* (Stuttgart: Franz Steiner, 1989).

Jābir ibn Aflaḥ, *Liber super Almagesti* (Nuremberg: Johannes Petreius, 1534).

Kennedy, Edward S., 'Two Medieval Approaches to the Equation of Time', *Centaurus*, 31 (1988), 1–8.

Kremer, Richard L., 'Text to Trophy: Shifting Representations of Regiomontanus's Library', in *Lost Libraries. The Destruction of Great Book Collections since Antiquity*, ed. by James Raven (Basingstoke: Palgrave Macmillan, 2004), pp. 75–90.

———, 'Experimenting with Paper Instruments in Fifteenth- and Sixteenth-Century Astronomy: Computing Syzygies with Isotemporal Lines and Salt Dishes', *Journal for the History of Astronomy*, 42 (2011), 223–58.

Maruska, Monika, 'Johannes Schöner — homo est nescio qualis: Leben und Werk eines fränkischen Wissenschaftlers an der Wende vom 15. zum 16. Jahrhundert', unpublished PhD thesis, University of Vienna, 2008.

Miolo, Laure, and Stefan Zieme, 'Lewis Caerleon and the Equation of Time: Tabular Astronomical Practice in Late Fifteenth-Century England', *Archive for History of Exact Sciences*, 78 (2024), 183–243.

Neske, Ingeborg, *Die Handschriften der Stadtbibliothek Nürnberg, Bd. IV: Die lateinischen mittelalterlichen Handschriften, Varia, 13.-15. und 16–18. Jahrhundert* (Wiesbaden: Harrossowitz, 1997).

North, John D., *Horoscopes and History* (London: The Warburg Institute, 1986).

Pedersen, Fritz S., *Scriptum Johannis de Sicilia super canones Azarchelis de tabulis Toletanis*, 2 vols (Copenhagen: Erik Paludan, 1986).

———, *The Toledan Tables. A Review of the Manuscripts and the Textual Versions with an Edition* (Copenhagen: Reitzel, 2002).

Peurbach, Georg, *Tabulae eclypsium* (Vienna: Johann Winterburger, 1514).

Porres de Mateo, Beatriz, 'Les Tables astronomiques de Jean de Gmunden: Édition et étude comparative', unpublished PhD thesis, Paris, École Pratique des Hautes Études, 2003.

Regiomontanus, Johannes, *Epytoma in Almagestum Ptolomei* (Venice: Johann Hamann, 1496).

———, *Opera collectanea*, ed. by Felix Schmeidler (Osnabrück: Zeller, 1972).

Reinhold, Erasmus, *Prutenicae tabulae coelestium motuum* (Tübingen: Morhart, 1551).

Saby, Marie-Madeleine, '*Les Canons de Jean de Lignères sur les tables astronomiques de 1321: Édition critique, traduction et étude*', unpublished thesis, Paris, Écloe Nationale des Chartes, 1987.

Schuba, Ludwig, *Die Quadriviums-Handschriften der Codices Palatini Latini in der Vatikanischen Bibliothek* (Wiesbaden: Reichert, 1992).

Toomer, G. J., (trans.), *Ptolemy's Almagest* (London: Duckworth, 1984).

Van Brummelen, Glen, 'Before the End of an Error: Giovanni Bianchini's Original Flawed Treatise on the Conversion of Stellar Coordinates', *Archive for History of Exact Sciences*, 75 (2021), 109–24.

Zepeda, Henry, *The First Latin Treatise on Ptolemy's Astronomy: The* Almagesti minor (*c.* 1200) (Turnhout: Brepols, 2018).
Zinner, Ernst, *Leben und Wirken des Joh. Müller von Königsberg, genannt Regiomontanus*, 2d rev. ed.(Osnabrück: Zeller, 1968).

Part 3

Operating with Numbers

ALENA HADRAVOVÁ AND PETR HADRAVA

Cyclical Radices in the *Tabule resolute* and their Adaptation to the Meridian of Another City

Introduction: tables of mean motions (*tabule mediorum motuum*)

Medieval astronomy was based on the Aristotelian idea that the celestial bodies must move by an eternal uniform circular motion. The observed deviations from the uniform circular motions, such as the varying velocity of the Sun in the ecliptic or the retrograde motion of planets, was thus explained either using eccentric positions of the circular orbits of these bodies or as a superposition of several circular motions. To compute a position of each celestial body at a chosen time according to these simplified models, it was thus necessary to first calculate the so-called mean motions (*medii motus*), that is, the phases of the corresponding circular motions, followed by their equations (*equaciones*), that is, corrections for the deviations of the resulting motion from a uniformity. These corrections were derived from geometric models that were slightly different for the individual bodies.

The mean motion of each quantity at any arbitrarily chosen time can be found using the direct proportion from the initial condition, meaning from its value at a given reference instant (*radix*, the root), and from the angular velocity, which is the increment of the phase in a unit of time. To facilitate this computation, that is, to avoid the need for multiplication, the extended tables of mean motions (*tabule mediorum motuum*) were included in the sets of medieval astronomical tables. These tables give increments of individual angular quantities in various time intervals (collected years, expanded years, months, days, hours

* This work was carried out in the framework of the ERC project *ALFA — Shaping a European Scientific Scene: Alfonsine Astronomy* (CoG 2016, agreement No 723085, PI Matthieu Husson). It has also been supported by Project RVO: 68378114 (Inst. Contemporary Hist.) and RVO: 67985815 (Astr. Inst.). We are very grateful to Anna Sobańska (Cracow, Biblioteka Jagiellońska), who provided us with photographs of many folios from related manuscripts that have not yet been scanned. We also thank José Chabás for photographs of MS Oxford, Bodl., Canon. misc. 27. José Chabás, Richard L. Kremer, Matthieu Husson, Marie-Madeleine Saby, and Alexandre Tur provided us with useful comments on the draft of our contribution, which we highly appreciate.

Alena Hadravová • Czech Academy of Sciences, Prague
Petr Hadrava • Czech Academy of Sciences, Prague

Alfonsine Astronomy: Computational Practices, ed. by Matthieu HUSSON, José CHABÁS, and Richard L. KREMER, Turnhout, 2026 (*Alfonsine Astronomy: Studies and Sources*, 3), p. 347-372
© BREPOLS PUBLISHERS 10.1484/M.ALFA-EB.5.137916

and their fractions), into which the time elapsed from the reference epoch until the required date and time was decomposed.[1]

An example of the tables of mean motions can be found as early as in Ptolemy's *Almagest* (second century AD), where the increments of solar anomaly are given in periods of 18 years, single years, months, days, and hours.[2] The Toledan Tables (eleventh century) use the Arabic calendar, which has a thirty-year period giving a sequence of leap and non-leap years. The tables thus give the values of the mean motion (*medius motus* or *cursus*) of the Sun, the Moon, and so on at the beginning of each thirtieth year of Hijra (up to the years 600 or 900) inscribed as *anni collecti* or *anni coniuncti*.[3] Another table then gives increments of the mean motions from the beginning of the 30-year cycle to the beginning of each of its years (*anni expansi*), and yet another table gives the increments to the beginning of each month of the year; the next table gives them to the beginning of each day of the month, and finally there is one for the beginning of each hour and/or its minutes and smaller parts. To compute the mean motion for any required time, one must find and sum the corresponding values from each of these tables.

Applying a similar system to the Julian calendar that was used in the Christian countries required a modification. The periodicity of leap years in the Julian calendar is just four years, which would make the table of expanded years too short and that of collected years too long. A straightforward solution is to extend the table of the expanded years to some integer multiple of the four-year period. To facilitate the counting of the years, it is suitable to choose the period of twenty years, as was done, for example, in the Alfonsine Tables of Toledo; however, some other periods such as 24 or 28 years were also used.[4] Even with such a longer period, it is not really feasible to list all the cyclical radices for each period of expanded years from the beginning of the Christian era until its second millennium. The table of collected years was thus either limited to a shorter temporal period assumed to be of interest for the user of the tables, or there was just one radix given, usually for the beginning of the era. Moreover, the table of the collected years did not give the radices, only the increments of the mean motion, first in time intervals of twenty years up to hundred years, and then in steps of centuries up to 1000 years, and finally in a few more millennia.[5] The former approach can be found, for example, in MS Paris, Bibliothèque nationale de France, Latin 16208, folio 5^{r}, where the cyclical radices are given for the twelfth

1 In this contribution, we deal mainly with mean motions in the *Tabule resolute* and Parisian Alfonsine Tables. For a wider context including other sets of tables, we refer the reader to José Chabás and Bernard R. Goldstein, *A Survey of European Astronomical Tables in the Late Middle Ages* (Leiden: Brill, 2012), pp. 53–61.

2 Ptolemy, *Almagest* III, 2. See G. J. Toomer, *Ptolemy's Almagest* (London: Duckworth, 1984), pp. 142–3.

3 See Fritz S. Pedersen, *The Toledan Tables* (Copenhagen: C.A. Reitzels Forlag, 2002), p. 1147, ff. or MS Prague, NK, XIV A 18, f. 1^{r} ff. etc.

4 See, for example, José Chabás and Bernard R. Goldstein, *The Alfonsine Tables of Toledo* (Dordrecht: Springer, 2003), p. 152 or Chabás and Goldstein, *A Survey of European Astronomical Tables*, pp. 53–61, etc.

5 An early example of both approaches are al-Battānī's tables. For instance, the manuscript Paris, BA, MS 8322 contains their Castilian translation. On folio 36^{v}, we can find the cyclical radices of mean motions of the Sun and Moon in collected Arabic years from 1 to 871 with a step of 30 years; on folio 37^{r}, there are increments of the motions in expanded Arabic years; on folio 50^{r}, there are cyclical radices in collected 'Roman' years from 931 to 1631 with a step of 20 years; and on folio 50^{v}, there are increments not only in the expanded years of Julian calendar from 1 to 20, but also with steps of twenty years up to 100 years and steps of 100 years up to 600 years. See also Carlo Alfonso Nallino, *Al-Battānī sive Albatenii Opus astronomicum, Pars secunda* (Milan: Ulrichus Hoeplius, 1907), pp. 19–20 and 72–7.

century only; later, in the fifteenth century, it became a basic feature of *Tabule resolute* (TR). A variant of the latter approach is used, for example, in the Parisian Alfonsine Tables (PAT, fourteenth century), where the user has to first convert the time elapsed from the beginning of the era into the sexagesimal number of days by adding three values (for the millennia, centuries, and twenties of years) found in the table of collected years together with the number of days for expanded years, months, days, and their fractions. The increments of mean motions for each order of the sexagesimal number of days are then summed together with the radices to obtain the final mean motion. However, going through this more laborious procedure offers the reward of a practically unlimited time interval that can be treated with such tables. In the next sections, we deal with the examples of the PAT and TR in more detail.

Regarding the fact that the local time was commonly used, it was necessary to also include the increment of each quantity during the time difference between the local and basic reference meridians. The meridian of Toledo in both the Toledan and Alfonsine Tables was chosen for this reference meridian, while the geographical longitudes of local meridians were often given in marginal notes to the tables.[6] Because this correction for the geographical longitude was the same for all computations of a given quantity in a fixed observational site, it was advantageous to add it directly to the radix of the reference meridian and to compile a set of radices for the chosen site in the same way.[7] We can thus, vice versa, verify for which meridian the tables were compiled from the values of the radices. The most suitable for this purpose are the values for the Moon, the motion of which is the fastest and hence enables us to achieve the most precise results.

1. Parisian Alfonsine Tables

The Parisian edition of the Alfonsine Tables (PAT) originated in the 1320s. It is commonly accepted that its standard form is documented by Erhard Ratdolt's incunable of 1483 and the canons *Tempus est mensura motus* written in 1327 by John of Saxony, which are also included in Ratdolt's print.[8] The essential part of Ratdolt's edition has also been published and translated into French by Emmanuel Poulle.[9] However, evidence about the tables preserved in the numerous manuscript copies, as well as the other canons on the use of the tables, should also be taken into account.

All times in the PAT are consistently transformed into the sexagesimal system with a unit chosen to be one day. This choice of the unit of time is often used even today in

6 A large set of these longitudes has been compiled and discussed by Richard L. Kremer and Jerzy Dobrzycki, 'Alfonsine Meridians: Tradition versus Experience in Astronomical Practice *c.* 1500', *Journal for the History of Astronomy*, 29 (1998), 187–99.

7 It was always possible to avoid the need for radices adapted to the local meridian either by adding the zonal time difference to the local time (i.e. transforming it to a universal time), or by adding the corresponding corrections to the resulting mean motions. However, the modified radices simplify later computations with the tables.

8 *Tabule astronomice illustrissimi Alfontii, regis Castelle* (Venice: Erhard Ratdolt, 1483), GW 1257; ISTC ia00534000.

9 Emmanuel Poulle, *Les tables Alphonsines avec les canons de Jean de Saxe* (Paris: Centre national de la recherche scientifique, 1984).

modern astronomy and astrophysics, where the time expressed in Julian dates (naturally in the decimal system) facilitates searches for periodicities and computation of the motion of celestial bodies. Unlike the Toledan Tables and many others, where the mean motions must be tabulated for different time units (the collected and expanded years, months, days, hours), in the PAT the time is transformed into the days only once for all the motions, and then one table for each mean motion in the multiples and fractions of a day is sufficient. Moreover, additional tables for the conversion of dates of various calendar systems (Arabic, Egyptian, etc.) into the sexagesimal number of days make the PAT versatile and useful for processing data of different origins.[10]

As an example, let us compute positions of planets at the beginning of the year 1425, that is, at noon on the 31st December 1424, when the year 1424 was complete. Table No. 2 of Poulle's edition (*Tabula ad inveniendum omnes eras bisextiles et ad extrahendum unam eram incognitam ex altera cognita*) was included in Ratdolt's incunable[11] and naturally in a huge number of manuscripts with copies of the PAT (e.g. in MS Prague, NK X A 23, f. 2^r); from said table, we find that

1000 (Julian) years have	1, 41, 27, 30 days,
400 years have	0, 40, 35, 0 days,
20 years have	0, 2, 1, 45 days
four years have	0, 0, 24, 21 days,
the sum of which reads	2, 24, 28, 36 (i.e. in the decimal form 520116) days

from the beginning of the Christian era, which is one of the reference epochs of the PAT.

In Table No. 16 (*Tabula medii motus Lune*, MS Prague, NK X A 23, f. 7^v)[12] we can find the mean motion of the Moon in degrees for all minutes of day, that is, its sixtieths from one to 60. Using a mere shift of columns in the table, we can also find the motion in units of days and in their multiples by any power of sixty. The table gives the value of $13;10,35,1,15,11,4,35,0^\circ = 13.17639469017^\circ$ for the mean motion of the Moon in one day; so, for the above-found sexagesimal digits of 2 fourths, 24 thirds, 28 seconds, and 36 units of day, we find the following table values:

0, 26, 21, 10, 2; 30, 22, 9, 10 +
5, 16, 14, 0; 30, 4, 25, 50, 0 +
6, 8, 56; 20, 35, 5, 10, 8, 20 +
7, 54; 21, 0, 45, 6, 38, 45, 0,

the sum of which is

31, 43, 40, 53; 42, 2, 25, 16, 47, 5 degrees.

The whole circle has $360^\circ = 6 \times 60^\circ = 6,0^\circ$, that is, six seconds in the sexagesimal notation. We can thus skip the found fourths and thirds and decrease the number of $40 = 6 \times 6 + 4$ seconds to four seconds, and we then reach the result $4,53;42,2,25,16,47,5^\circ$.

10 See Julio Samsó, *On both Sides of the Strait of Gibraltar* (Leiden: Brill, 2020), p. 840 ff.

11 *Tabule astronomice*, 1483, f. 17^r; Poulle, *Les tables Alphonsines*, p. 109.

12 *Tabule astronomice*, 1483, f. 24^v; Poulle, *Les tables Alphonsines*, p. 135.

The radices of mean motions can be found in the PAT in Table No. 9 (*Tabula radicum motuum hic positorum ad eras hic positas*),[13] where they are given for different reference epochs. In particular, the radix of mean motion of the Moon to the beginning of the Christian era (the Incarnation) is 2,2;46,50,16,40°. Adding the above-calculated mean motion from the beginning of the era, we thus find the position (i.e. ecliptical longitude) of the mean Moon at the complete year 1424, which is equal to 56;28,52,41,57°.

In a tradition that dates back to the Toledan Tables, the PAT were also composed for the meridian of Toledo, as mentioned above. The authors of canons thus gave brief recommendations on how to modify the radices to the meridian of another city to enable the readers to use the tables at a different meridian. It can be documented by the following examples:

John of Lignères, *Quia ad inveniendum loca planetarum* (post 1322–ante 1327), at the end of *Capitulum octavum: De medio motu planetarum inveniendo* (*Chapter eight: On finding the mean motion of planets*):[14]

> Notandum tamen, quod omnia ista habebis ad meridianum Toleti. Si autem vis habere hoc ad aliam civitatem, que sit orientalior vel occidentalior, vide, in quanto tempore sit orientalior vel occidentalior, et in tanto tempore quere omnia ista nunc enumerata et subtrahe pro civitate orientaliori et adde pro occidentali. Notandum etiam pro capitulo sequenti, quod tempore medie coniunctionis invento per Tabulam medie elongationis Solis et Lune additur tempus illud, scilicet existens inter duos meridianos pro civitate orientali, et removetur pro occidentali. Tamen medio motui Solis et Lune et medio argumento Lune et argumento latitudinis invento hora medie coniunctionis non fit addicio nec etiam subtractio.
>
> > And note that all radices are given for the meridian of Toledo. If you want to know radices for another city, to the west or east, determine how much time it is to the east or west of Toledo, and for that time search all the radices listed here, and for an eastern city subtract and for a western city add.[15] In the next chapter on finding the time of mean conjunction using the *Table of mean elongations of the Sun and Moon*, the time difference is added for an eastern city and subtracted for a western one. However, to find the mean motion of the Sun and Moon, the mean argument of the Moon, and the argument of the latitude at the hour of the mean conjunction, nothing is added or subtracted.

The mean motions are a perpetual linear increase of longitudes of the celestial bodies or arguments of other quantities with time. This rule thus implies that for a city eastward of the reference meridian, the radices have to be decreased for the angle spun by the body during the time interval for which the local time in the eastern city precedes that in the

13 *Tabule astronomice*, 1483, f. 23r; Poulle, *Les tables Alphonsines*, pp. 124–6.

14 Alena Hadravová and Petr Hadrava, 'John of Lignères *Quia ad inveniendum loca planetarum*: An Edition and Translation', in *Alfonsine Astronomy: The Written Record*, ed. by Richard L. Kremer, Matthieu Husson, and José Chabás (Turnhout: Brepols 2021), pp. 284–5.

15 The very brief and scanty formulation of the canon can be interpreted either 'for that time search all the radices listed' (by subtracting or adding the mean motions in that time) or 'for that time find what is listed' (i.e. all mean motions) and subtract or add it to the radices.

reference meridian. Just opposite, the radices of conjunctions, which are the local times of the event of a reference conjunction, should be increased for that time. The mean motions of the Sun and Moon at the conjunction and the corresponding arguments are angular values in the event of conjunction, so they are independent of the local time of the observer.

The edition of the treatise *Quia ad inveniendum loca planetarum* is based on ten manuscripts. One of them is MS Oxford, Bodleian library, Digby 168, folios 145ra–146rb (XIV /1327–72/),[16] which is a very good copy. It often represents a different version of the work. According to the findings of Jean-Patrice Boudet and Laure Miolo, the copy of the *Quia ad inveniendum* in MS Digby 168 is of French origin, and this part of the manuscript was acquired (perhaps already in 1333) by Oxford astronomer Simon Bredon (*c.* 1300–72).[17] Instead of the words 'notandum etiam', found in the other manuscripts, this Oxford manuscript gives a wider context on folio 145va: *nota, quod meridies Toleti sequitur meridiem Parisiensem 46 minutis et quod longitudo meridianus (sic) Parisiensis est ab occidente vero 40 gradus 30 minutis et eius latitudo est 48 gradus 50 minutis et etiam.* This additional note claims the difference of 'zonal time' between Paris and Toledo to be 46 minutes, which would correspond to a difference in geographical longitudes of 11;30°. This agrees with the value given in the treatise on eclipses, *Ut annos Arabum*, attached to some copies of the Toledan Tables,[18] while in the Alfonsine tradition the time difference of 48 minutes prevails.[19] These values are exaggerated for about 5° with respect to the true longitudinal distance of these two cities, but a similar distortion was common in many medieval sources. The note also quotes geographical coordinates of Paris. It was used to list such coordinates of various cities and regions in tables, which can also be found in some sets of Alfonsine Tables, but frequently also in treatises on astronomical instruments and in other manuscripts.[20] The manuscript tradition of these tables of geographical coordinates is very rich but full of scribal errors. The longitude was given either in degrees or in temporal difference from the reference point. The most common choice of the prime meridian was the supposed westernmost edge of an inhabited region of the world assumed in the Fortunate Islands in the Atlantic Ocean (probably the Canary Islands). The longitude of Toledo with respect to this zero point was taken as 11°, and for Paris values around 23;50° were often given. The longitude 40;30° of Paris given here is, however, most probably counted in the coordinate system introduced in canons to the Toledan Tables ascribed to al-Zarquali. This system defines a 'true west' shifted for 17;30° westward from the Fortunate Islands and *medium mundi* at the longitude of 90° and identified by the mythical city of Arim, which was either 61;30° or 62°

16 See a description of the manuscript: William D. Macray, *Catalogi codicum manuscriptorum Bibliothecae Bodleianae, Pars nona: Codices a viro clarissimo Kenelm Digby anno 1634 donatos, complectens* (Oxford: Clarendon Press 1883), pp. 172–77.

17 See Jean-Patrice Boudet and Laure Miolo, 'Alfonsine Astronomy and Astrology in Fourteenth-Century Oxford: The Case of MS Bodleian Library Digby 176', in *Alfonsine Astronomy: The Written Record*, ed. by Richard L. Kremer, Matthieu Husson, and José Chabás (Turnhout: Brepols 2021), pp. 72–3.

18 See Pedersen, *The Toledan Tables*, p. 555.

19 See Kremer and Dobrzycki, 'Alfonsine Meridians', 194.

20 See e.g. Alena Hadravová and Petr Hadrava, *Stavba a užití astrolábu* (Prague: Filosofia 2001), pp. 420–32.

to the east of Toledo.[21] It is interesting to note that the same value (40;30°) is also given in John of Lignères' canons to the year 1321, edited by Marie-Madeleine Saby[22] (see below).

John of Saxony (a disciple of John of Lignères), writes the following in Chapter 12 of his above-mentioned canons *Tempus est mensura motus*:[23]

> Medium motum Solis et Lune et aliorum planetarum... per tabulas invenire. ... Intelligendum, quod radices posite in istis tabulis sunt posite ad meridianum Toleti, et ideo per istam operationem non haberes motus nisi ad meridianum Toletanum. Si autem ad meridianum tuum velis motus, radices et alios prenotatos invenire, scias, utrum tua civitas fuerit orientalior vel occidentalior [quam Toletum, et scias, per quantum tempus est orientalior vel occidentalior], et in tanto tempore quere motum, quem volueris, et illum subtrahe a motu invento, si civitas fuerit orientalior, et adde, si civitas fuerit occidentalior, et quod post additionem vel subtractionem provenerit, est motus, quem queris ad meridianum civitatis tue. Nota, quod illud opus potes multum abbreviare, videlicet reduc omnes radices ad meridianum tuum una vice et deserviet tibi quamdiu vives.
>
> > To find the mean motion of the Sun, Moon, and other planets... using the tables... It should be known that the radices given in these tables are done for the meridian of Toledo, and hence by this procedure you will get motion to the Toledan meridian only. If you would like to find the motion, radices, and other above-mentioned matters to your meridian, you must know whether your city is eastward or westward [of Toledo, and you must know for what time it is eastward or westward], find the motion that you want in that time, and subtract this motion from that found if the city is more to the east, or add it if the city is to the west, and what you will get after the addition or subtraction is the motion for your city that you seek. Note that you can shorten this work a lot if you reduce all radices to your meridian at once, and it will serve you for your whole life.

It is interesting to note that this canon, which is also included in Ratdolt's edition, advises first finding the time difference between the two meridians, and then finding the mean motion of the treated quantity spanned over this time difference, and, finally, changing the radix for the found mean motion, just as the canon of John of Lignères quoted above. On the other hand, the tables in Ratdolt's (and Poulle's) edition also contain Table 10 (*Radices motuum ad eram Incarnacionis ad loca diversa*),[24] which directly gives the radices (for the Christian era only) of all quantities for meridians with integer values of geographical longitudes between 11 degrees (what was the traditionally accepted longitude of Toledo) and 40 degrees. This table, however, does not appear (to our knowledge) in the sets of

---

21 See Pedersen, *The Toledan Tables*, pp. 255, 1192, etc. See also Dana B. Durand, *The Vienna–Klosterneuburg Map Corpus of the Fifteenth Century: A Study in the Transition from Medieval to Modern Science* (Leiden: 1952), p. 101. This western base meridian is also called 'meridian of water'; for example, by Mercè Comes, 'The Meridian of Water in the Tables of Geographical Coordinates of al-Andalus and North Africa', *Journal for the History of Arabic Science*, 10 (1994), 41–51.

22 Marie-Madeleine Saby, 'Les Canons de Jean de Lignères sur les tables astronomiques de 1321. Édition critique, traduction et étude', unpublished dissertation, Paris, Ecole Nationale des Chartes, 1987.

23 Poulle, *Les tables Alphonsines*, pp. 50–5. The text contained in the square brackets ([quam Toletum... occidentalior]) is not found in Ratdolt's edition, *Tabule astronomice*, 1483, ff. 4^{v}–5^{r}.

24 *Tabule astronomice*, 1483, f. 24^{v}; Poulle, *Les tables Alphonsines*, pp. 127–9.

PAT in manuscripts. It also has not been included in the prints of the PAT by Johannes Lucilius Santritter[25] and Petrus Liechtenstein.[26] As opposed to the radices in Table 9, those in Table 10 are rounded to a smaller number of sexagesimal digits.

Another remarkable point is that John of Saxony recommends saving the radices computed for the user's meridian to simplify future work, which is a practice followed by the TR, as we shall see below.

A slightly different procedure is presented in Canon 5 of *Priores astrologi* by John of Lignères:[27]

> Et hoc notandum, quod radices hic posite sunt ad meridianum Parisiensem, cuius longitudo est ab occidente vero 40 graduum 30 minutorum et ipsius latitudo est 48 graduum 50 minutorum. Si autem velles hec omnia invenire ad aliam civitatem, vide longitudinem inter illam civitatem et Parisius et gradus differentie multiplica per 4 et habebis minuta hore, que si fuerint plura 60, per 60 divide, et numerus quotiens erunt hore, et residuum, si sit post divisionem, erunt minuta; et sic habebis horas et minuta hore, que sunt inter illam civitatem et Parisius, vel minuta solum, si non sunt hore. Quere ergo medios motus et alia proposita in tot horis et minutis, quot habuisti, si habuisti horas et minuta,[28] et subtrahe medium motum in illo tempore inventum a medio prius invento ad Parisius et hoc, si civitas illa sit orientalior quam Parisius, vel adde, si sit occidentalior; et hoc faciendum est pro omnibus contentorum in ista propositione.
>
> > And it should be noted that the radices are given here for the meridian of Paris, the longitude of which is 40 degrees and 30 minutes from the west, and its latitude is 48 degrees and 50 minutes. If you thus wish to find them all for another city, have a look at the longitude between that city and Paris, and multiply the difference in degrees by four and you will have the minutes of the hour. If it is more than 60, divide it by 60, and the quotient are hours and the remainder, if there is any left after the division, they are the minutes. And you have thus the hours and minutes of the hour, which are between that city and Paris, or only the minutes if there are no hours. Search thus for the mean motions and other wanted quantities in so many hours and minutes that you had if you had hours and minutes and subtract the mean motion found in this time from the mean [motion] previously found for Paris if the city is eastward of Paris, or add it if it is westward. And in this way is done everything contained in this rule.

According to this text, the reference meridian of the tables is directly that of Paris, and corrections to other cities are to be computed according to their longitudinal distance from Paris. The geographical longitude 40°30′ of Paris is identical to that given in the manuscript Oxford, Bodleian library, Digby 168, as described above (Arim is also mentioned in some

25 *Tabulae astronomicae Alfonsi Regis*, with additions by Augustinus Moravus, ed. by Johannes Lucilius Santritter (Venice: Johannes Hamman, 1492), GW 1258; ISTC ia00535000.

26 *Tabule astronomice divi Alfonsi, regis Romanorum et Castille* (Venice: Petrus Liechtenstein, 1518).

27 Saby, *Les Canons de Jean de Lignères*, pp. 189–90.

28 The rest of the paragraph is modified according to the more precise formulation in the manuscript Paris, BnF, lat. 7281, f. 188r.

manuscripts with *Priores astrologi*). It is assumed in this canon that the user knows the difference of his meridian from Paris in degrees, and thus an algorithm is given to convert it to the difference in time. The correction for the longitude should then be applied to the resulting mean motion instead of to the radices as recommended by John of Saxony.

2. Tabule resolute

The *Tabule resolute* (TR) are a variant of the Alfonsine Tables, which originated in the 1420s and were disseminated in the Central Europe, especially at universities, throughout the fifteenth century. Compared to the PAT, the TR somewhat simplify the computation of mean motions by usage of the so-called cyclical radices computed directly for the city in which the tables were to be used.[29] As stated above, these cyclical radices are values of mean motions for a set of selected leap years with a step of twenty years. This set covers the whole range of time for which the tables are intended to be used. The corresponding tables are again entitled *anni collecti*, but they give the values of all quantities of interest including the radices directly, not only the number of days as in the PAT. Similarly, the subsequent tables give the increments of mean motions in single years (from 1 to 20, *anni expansi*), in all months in the leap and non-leap years, in single days (1 to 31), and finally in hours and their sexagesimal parts (1 to 60 or to 30). The TR thus provide a consistent usage of the sexagesimal system in the counting of time, which is treated simultaneously in units of hours (which are equal to 2.5 minutes of a day), in days, months, and years. The TR also often violate the sexagesimal system in counting the angles using the degrees and the common zodiacal signs (30° each) simultaneously instead of the physical signs (60° wide).[30] These changes make the TR more user-friendly for laymen accustomed to this variety of units, but at the expense of the mathematical self-consistency and beauty.[31]

A prototype of the TR was composed by Petrus Cruciferus for the meridian of Wrocław (Breslau) and it has the basic cyclical radix just for the complete year of 1424, which we gave as an example above (see Figure 11.1). In the only known copy of Petrus Cruciferus's tables preserved in the manuscript Toruń, BU 74/III, we find the radix of the mean motion of the Moon for the year 1424 to be equal to 1^{s} 25;40,33,54° on folio 251^{r}. This value is for 0;48,18,48° = 0.805222° smaller than the above-given value to the reference meridian of Toledo computed from the PAT. According to the given speed of the Moon's mean motion, this difference corresponds to the difference of local times 0.061111 day = 0;3,40 day = 1 hour 28 minutes.[32] This corresponds to angular distance of the meridians of 22°.[33] And

29 With regard to the TR, see José Chabás, *Computational Astronomy in the Middle Ages. Sets of Astronomical Tables in Latin* (Madrid: Consejo superior de investigaciones científicas, 2019), pp. 311–20.

30 An exception is, for example, the manuscript Vatican, BAV, Pal. lat. 1381, ff. 113^{r}–120^{v}, which gives the mean motions in the TR format but uses physical signs instead of zodiacal ones.

31 See, for example, Matthieu Husson, 'Physical Signs and Minutes of Days: Mean Motion Computations in the Parisian Alfonsine Tables', in *Le Moyen Âge et les sciences*, ed. by Danielle Jacquart and Agostino Paravicini Bagliani, Micrologus 100 (2021), pp. 223–50.

32 To verify this result by means of the PAT, we can find the mean motion of the Moon in 0;3 days equal to 0;39,31,45° and in 0;0,40 days equal to 0;8,47,3°, the sum of which is really 0;48,18,48°.

33 The true difference between the geographical longitudes of Toledo and Wrocław is approximately 21°.

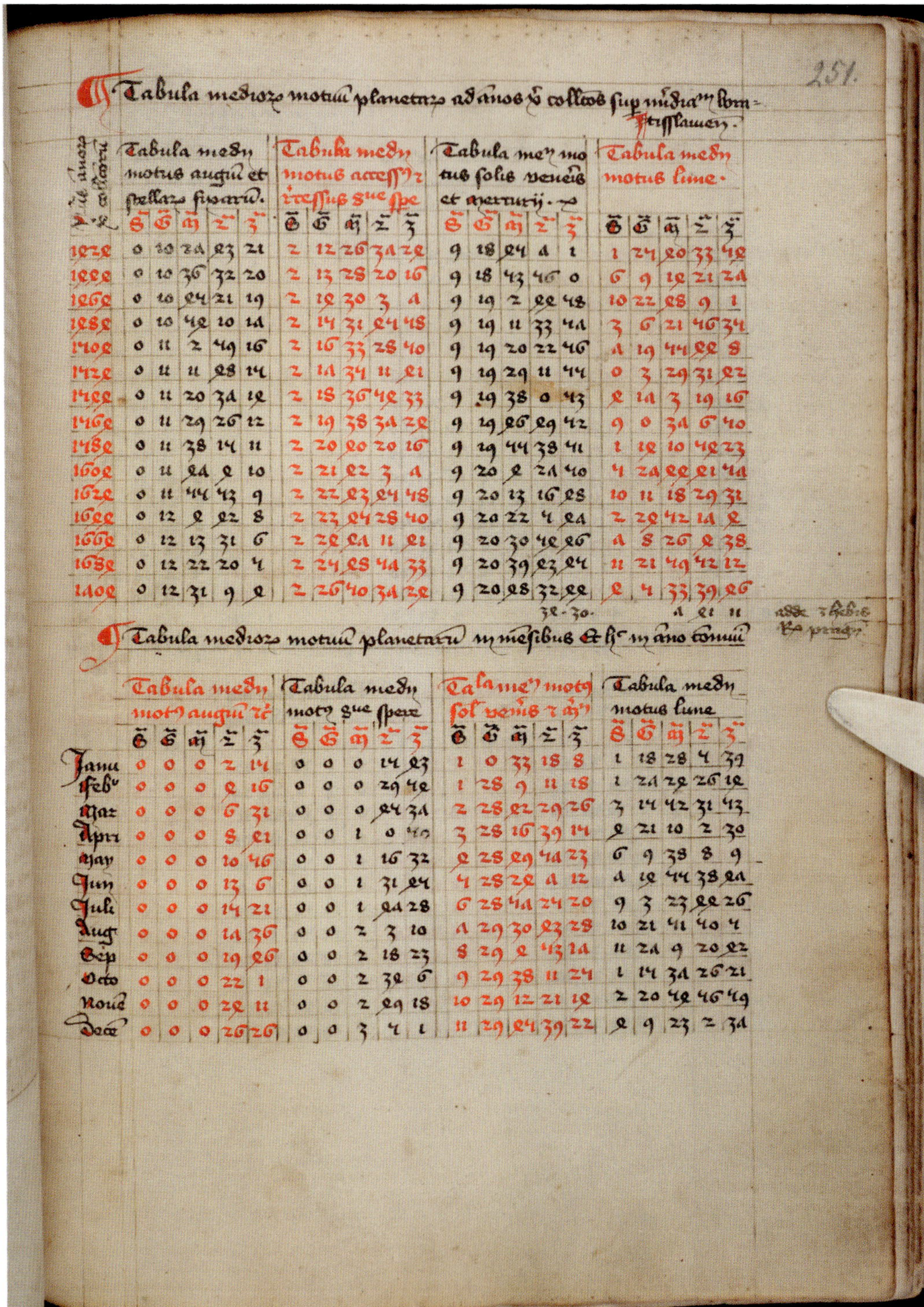

Figure 11.1. *Tabule resolute*, the table with cyclical radices to the year 1424. Toruń, BU, 74/III, f. 251r.

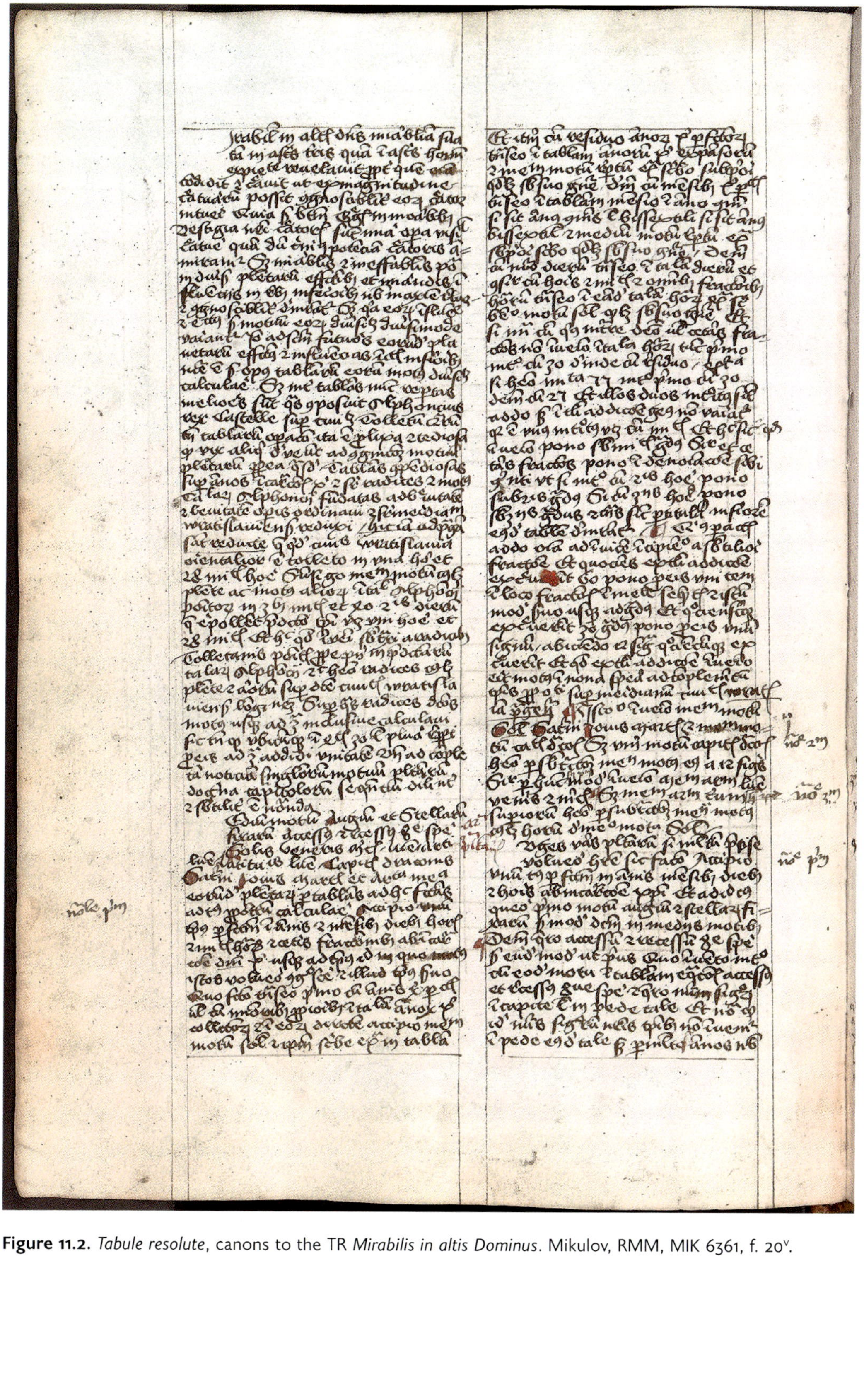

Figure 11.2. *Tabule resolute*, canons to the TR *Mirabilis in altis Dominus*. Mikulov, RMM, MIK 6361, f. 20v.

it is this value of time difference between the meridians of Toledo and Wrocław which is given in canons *Mirabilis in altis Dominus* by Petrus Cruciferus.

In the introduction to his canons, Petrus Cruciferus himself explains the reason why and how he composed his tables (see Figure 11.2):[34]

> Sed inter tabulas nunc repertas meliores sunt, quas composuit Alfoncius, rex Castelle, super civitatem Tholetum; quarum tamen tabularum operacio ita est prolixa et tediosa, quod vix aliqui deveniunt ad cognicionem motuum planetarum. Propterea quasdam tabulas compendiosas super annos Incarnacionis Christi et super radices et motus tabularum Alfoncii fundatas ad brevitatem et levitatem operis ordinavi et super meridianum civitatis Wratisslaviensis reduxi,[35] que quidem civitas Wratisslavia[36] est orientalior Tholeto in una hora et 28 minutis hore. Sumpsi ergo medium motum cuiuslibet planete ac motus aliorum in tabulis Alfoncii positorum in tribus minutis et 40 secundis dierum, que equipollent predicto tempori, videlicet uni hore et 28 minutis, et hoc, quod inveni, subtraxi a radicibus Tholetanis positis prope principium predictarum tabularum Alfoncii, et sic habeo radices cuiuslibet planete et aliorum supradicte civitatis Wratislaviensis[37] longitudinem. Super quas radices dictos medios motus usque ad tercia inclusive calculavi, sic tamen, quod ubicumque in quartis 30 vel plura reperi, pro eis ad tercia addidi unitatem. Unde ad completam noticiam singulorum motuum planetarum doctrina capitulorum sequencium diligenter et subtiliter est notanda.
>
> > Among the tables that have been found up to now, the best are those composed by the King of Castile for the city of Toledo. However, the operation with these tables is lengthy and tedious, so some scarcely arrive at cognition of planetary motions. For this reason, I have composed brief tables to the years of the Incarnation of Christ based on radices and motions of the Alfonsine Tables to shorten and to facilitate the work, and I reduced them to the meridian of the city of Wrocław. This city of Wrocław is one hour and 28 minutes of an hour to the east of Toledo. I thus took mean motions of each planet and other motions given in the Alfonsine tables in three minutes and 40 seconds of a day, which is equivalent to the above-mentioned time, namely one hour and 28 minutes. And what I had found, I subtracted from the Toledan radices given at the beginning of the above-mentioned Alfonsine Tables, and I have thus radices of any planet, and the other for the longitude of the above-mentioned city of Wrocław. Using these radices, I calculated the mentioned mean motions up to thirds included in such a way that whenever I found, in fourths, 30 or more, I added, instead of them, a unit to thirds. For a complete cognition of individual motions of planets, the teaching contained in the subsequent chapters is thus to be carefully and precisely noted.

---

34 Extracted from the critical edition in preparation, based on the following: MSS *A*: Toruń, BU 74/III, ff. 268va–273ra (Prague, 1425); *C*: Cracow, BJ 610, ff. 352ra–357vb (Cracow or Prague, 1440); *O*: Oxford, Bodl., Canon. Misc. 499, ff. 122ra–125vb (upper pagination), 112ra–115vb (lower pagination) (Prague, 1451); *P*: Prague, NK, X B 3, ff. 73ra–75va (Prague, 1435–1465); *R*: Mikulov, RMM, MIK 6361, ff. 20^{v}–22^{v} (Prague, 1470–1480); *S*: Mikulov, RMM, MIK 6361, ff. 91^{v}–92^{v} (Prague, 1470–1480).

35 Wratisslaviensis reduxi *ACO*] Pragensis reduxi *PS* : Wratisslaviensis reduxi, hic iam ad Pragam satis reducte *R*.

36 civitas Wratisslavia *ACOR*] civitas *PS*.

37 Wratislaviensis *ACOR*] Pragensis *PS*.

In the canons *Mirabilis in altis Dominus*, there is no other mention of how to modify the calculation for another city. This fits with the author's aim to simplify the usage of the tables as much as possible. In some copies of these canons (Prague, NK, X B 3 and Mikulov, RMM, MIK 6361), the name of Wrocław is skipped or replaced by the name of Prague without changing the numerical values valid for Wrocław. In the Table of mean motions of planets in collected years of Christ for the meridian of Wrocław (*Tabula mediorum motuum planetarum ad annos Christi collectos super meridianum Wratisslaviensem*), at the end of the sub-table entitled Table of mean motion of the Moon (*Tabula medii motus Lune*) in the manuscript Toruń, BU 74/III, f. 251^{r}, there is inscribed an additional note: '*7 <minuta>, 41 <secunda>, 11 <tercia> adde et habebis radicem Pragensem*', which translates as 'add 0;7,41,11 (=0.128106°), and you will have the radix of Prague'. This correction corresponds to the time difference of 14 minutes, that is, the difference of geographical longitudes 3.5°. For this value, Prague should be westward of Wrocław, that is, 18.5° eastward of Toledo. This is in good agreement with the reality, as well as with the values of longitudes of Toledo 11;00° and Prague 29;30°, given in many medieval tables of geographical coordinates.[38] In some other copies of the TR, corrections of radices for adaptation to other cities are given.

Later sets of the TR use the cyclical radices for the year 1428 and its shifts for multiples of twenty (i.e. 1448, 1468, etc.) instead of the year 1424 (see Figure 11.3). The values of the cyclical radices are thus naturally also different in these manuscripts. For instance, repeating the same computation of the mean motion of the Moon for the complete year 1428, as done above for the year 1424, or simply adding the motion in 4 Julian years = 1461 days, which is 2,50;42,45,30,45°,[39] to the previous result, we find the value for the meridian of Toledo: 3,47;11,38,12,42°.

The most frequent value of the mean motion of Moon for the year 1428 that we find in manuscripts with the TR is 7^{s}16;24,12,7° (or 6 thirds).[40] In the vast majority of them, the inscription of the tables declares that they are composed for the meridian of Cracow. This value is for 0;47,26,6° = 0.7906° smaller than the value for Toledo according to the PAT. This corresponds to the mean motion of the Moon in 0.06 day = 86.4 minutes of an hour, that is, at a location 21;36° east of Toledo.[41] In the commonly used coordinate system in which the longitude of Toledo is 11°, it would mean the longitude of Cracow equal to 32;36°. This value can be found in some other canons to the TR, as well as in tables of

---

38 For example: MSS Cracow, BJ 1915, f. 7^{r} (*c.* 1338); Cracow, BJ 551, f. 31^{v} (*c.* 1388); Kalocsa, FK 326, f. 13^{r} and 13^{v}–14^{r} (after 1434); Cracow, BJ 550, back of the cover (*c.* 1440); Cracow, BJ 1927, f. 154^{r} (*c.* 1445); Cracow, BJ 560, pp. (*c.* 1448); Vienna, ÖNB, Cod. 5145, f. 67vb (1440–60); Cracow, BJ 257, f. 135ra (after 1460); Cracow, BJ 1852, f. 170^{r} (*c.* 1464); Cracow, BJ 1857, pp. 163–74 (15th century); Cracow, BJ 1858, p. 71 (15th century); Prague, NK III C 2, f. 41^{r} (15th century). However, slightly different values can be found elsewhere.

39 The same value, 5^{s}20;42,45,31°, but rounded to thirds can also be found in *Tabula mediorum motum planetarum in annis expansis*; for example, MS Prague, NK X B 3, f. 81^{v}.

40 It is given, for example, in the following manuscripts: Cracow, BJ 560, f. 9^{v}; BJ 600, f. 86^{r}; BJ 605, f. 16^{v}; BJ 611, f. 1^{v}; BJ 614, f. 1^{v}; BJ 616, f. 1^{r}; BJ 617, f. 13^{v}; BJ 1846, f. 3^{r}; BJ 1858, f. 121^{v}; BJ 1864, f. 36^{v}; BJ 1865, f. 72^{v}; BJ 1916, p. 58; BJ 1917, f. 1^{r}; BJ 1920, f. 13^{v} (p. 26); BJ 1966, f. 7^{v} (p. 14); BJ 2252, f. 47^{v}; BJ 2478, f. 1^{v}; BJ 2480, f. 1^{v} (p. 2); BJ 2650, f. 37^{v} (p. 76); BJ 2664, f. 1^{r}; Munich, BSB, Clm 51, f. 124^{r}; BSB, Clm 214, f. 18^{r}; Vatican, BAV, Pal. lat. 1375, f. 21^{r}; Vienna, ÖNB, Cod. 5245, f. 62^{v}.

41 See appendix to Kremer and Dobrzycki, 'Alfonsine Meridians', 187, where the value of 21;36° is the most frequent for the longitude of Cracow.

Figure 11.3. *Tabule resolute*, the table with cyclical radices to the year 1428. Prague, NK, X B 3, f. 79[r].

geographical coordinates of cities: for example, in the manuscripts Cracow, BJ 560, p. 6 or Cracow, BJ 1858, p. 71.[42]

The next frequently found value in the TR for the cyclical radix of the mean motion of the Moon in the year 1428 is $7^{s}16;31,0,35^{\circ}$ (or 34 thirds), which is in the manuscripts Cracow, BJ 553, f. 68^{v}; Cracow, BJ 617, f. 13^{r}; Cracow, BJ 1852, f. 7^{v}; Mikulov, RMM, MIK 6361, f. 47^{v} and 135^{v} (note that there are two copies of the TR in this manuscript); Oxford, Bodl., Canon. misc. 499, f. 72^{r}; Prague, NK VIII G 24, f. 2^{v}; Prague, NK X B 3, f. 79^{v}; Vatican, BAV, Pal. lat. 1381, f. 113^{v} (with a value of $3,46;31,0,34^{\circ}$ in physical signs). The difference $0;40,37,38^{\circ} = 0.6771^{\circ}$ corresponds to the mean motion of the Moon in 0.0514 days = 1 hour 14 minutes, and hence to a longitude of $29;30^{\circ}$, which is 18.5° eastward of Toledo at 11°. Most of the tables in these manuscripts are inscribed to be composed for the meridian of Prague.[43] This agrees with the above-mentioned value of longitude of Prague commonly found in tables of geographical coordinates.[44]

Another value, namely $7^{s}17;17,38,26^{\circ}$, of the mean motion of the Moon in the year 1428 appears in the manuscript Oxford, Bodl., Canon. misc. 27, f. 34^{v}.[45] This value is higher for $0;6,0,13^{\circ}$ than that of Toledo, which indicates that this table should correspond to a city westward of Toledo. These TR are really computed for the meridian of Salamanca, as it is also written on folio 33^{r}, and they had been composed by Nicolaus Polonius, who also wrote brief canons on their use.[46] If we divide the difference between the radices for Toledo and Salamanca by the angular velocity of the mean Moon in longitude, we find that it corresponds to a difference of zonal time equal to 0.007593897 day = 10.93521 minutes of an hour, which is a difference in geographical longitude of $2;44,1,41^{\circ}$. Nicolaus's canons give the longitude of Salamanca $8;16^{\circ}$ in the coordinate system with Toledo at 11° or $25^{\circ}46'$ in the aforementioned system of al-Zarquali.[47] These coordinates are thus equivalent to the distance from Toledo $2;44^{\circ}$, which should give a difference of radices $0;6,0,9^{\circ}$. If we inspect the table of cyclical radices again, which are given for the years 1348–1628,[48]

42 Some other values, for example $33;0^{\circ}$, are also frequently found in the tables of coordinates (e.g. MSS Cracow, BJ 1915, f. 7^{r}; Cracow, BJ 1927, f. 154^{r}; Cracow, BJ 1865, f. 155^{r}).

43 The exceptions are the manuscripts Cracow, BJ 617, ff. 13^{r}–37^{r} and Vatican, BAV, Pal. lat. 1381, ff. 113^{r}–120^{v}, which do not specify their meridian. The manuscript Cracow, BJ 617 gives values for Prague on folio 13^{r} and values for Cracow on folio 13^{v}. The manuscript Vatican, BAV, Pal. lat. 1381, f. 113^{r} uses physical signs that are explained in the margin: 'Signum phisicum valet duo communia, videlicet 60 gradus. Unus gradus <valet> 60 minuta, unum minutum 60 secunda, unum secundum 60 tercia, unum tercium 60 quarta et sic de ceteris'.

44 See the value $18;30^{\circ}$ for Prague in Kremer and Dobrzycki, 'Alfonsine Meridians', 195.

45 While the TR for Wrocław, Prague, and Cracow are usually organized in subsequent tables that always give mean motions for all quantities first in collected years and then in the shorter intervals of time, the TR for Salamanca in the manuscript Oxford, Bodl., Canon. misc. 27, ff. 33^{r}–88^{r} give all input data for one quantity on a separate page.

46 José Chabás, 'Astronomy in Salamanca in the Mid-Fifteenth Century: The Tabulae resolutae', *Journal for the History of Astronomy*, 29 (1998), 167–75; Beatriz Porres and José Chabás, 'Los cánones de las Tabulae Resolutae para Salamanca: origen y transmisión', *Cronos* 1 (1998), 51–83.

47 Porres and Chabás, 'Los cánones', 60: 'alie tabule utilissime que vocantur resolute eis (= studentibus) proponuntur, in quibus se exerceant, ad civitatem Salamantinam in regno Castelle constitutam ordinate. Cuius longitudo a Gadibus Herculis, a quibus communiter longitudines civitatum accipiuntur, est octo graduum et 16 minutorum, a vero autem occidente 25 graduum 46 minutorum distat' (Oxford, Bodl., Canon. misc. 27, f. 122^{v}).

48 It is one of many examples demonstrating that the year of the first cyclical radix in the TR does not indicate the date of origin of the tables. Users often also needed to calculate positions of planets in the past, and thus the tables were extended to earlier dates.

we find their subsequent differences[49] to be $4^{s}13;33,47,34°$, $4^{s}13;33,47,34°$, $4^{s}13;33,47,33°$, $4^{s}13;33,47,38°$, $4^{s}13;33,47,30°$, $4^{s}13;33,47,33°$, and so on. It means that, just for the year 1428, there is a scribal error and the radix should read $7^{s}17;17,38,22°$.

The TR were later also published several times in early modern prints. In the first print edited by Ambrosius Lacher in 1511,[50] we can find the value of mean motion of the Moon — $7^{s}16;32,59,10°$ — for the year 1428 on folio F3^{v}. This is for 0;38,39,3° = 0.6442° less than the value for Toledo, which corresponds to the mean motion of the Moon in 0.0489 days, which is approximately in 1 hour 10 minutes, and hence to a city 17;36° eastward of Toledo. In his canons, Lacher wrote about the meridian of his tables and its possible modification to another city:

> |f. A6va:| Quod si forte loco habitacionis consideracionisve exigente ad alium meridianum motus supercelestium corporum infestigare[51] proposueris, necesse est, ut tempus distantie meridianorum tue consideracionis locum et Lypsigk[52] interiacens. Addas tempori proposito, ad quod motus invenire desideras, si tue consideracionis locus fuerit occidentalior Lypsigk,[53] aut subtrahas, si fuerit orientalior, prout tabula huius rei gracia facta et abunde informabit. Et tempus illud, quod facta addicione conflatur vel post subtractionem relinquitur, inserviet exacte tuo proposito, ad quod si motum alicuius planete vel mobilis inveneris, non aliter eveniet, quam si tabule nostre ad meridianum consideracionis tue vel habitacionis essent composite.
>
> > If, by chance, you would like to investigate the motion of celestial bodies at another meridian of the place of your habitation or consideration, it is necessary to determine the temporal distance between the meridians of your consideration and of Leipzig. Add this time to the time for which you want to find the motion if your place of consideration is westward of Leipzig, or subtract it if it is eastward, as the table prepared for that provides sufficient information about it. And the time that will result after the addition or remain after the subtraction will serve exactly your purpose. And if you find for it the motion of any planet or something else movable, you will find nothing else than if our tables were composed for the meridian of your consideration or dwelling.

In his tables, Lacher did not state explicitly for which meridian they were composed; however, the canons clearly indicate that they should correspond to Leipzig. The time difference computed from Lacher's and the usual Prague TR would correspond approximately to 4 minutes of an hour westward of Prague, which agrees with the 1° difference between

49 These differences should be equal to the increment in 20 years, which is given in the table of *anni expansi*. Its more accurate value computed according to the PAT to a higher accuracy is $4^{s}13;33,47,33,43,33°$. It is truncated to $4^{s}13;33,47,33°$ in the manuscript Oxford, Bodl., Canon. misc. 27, f. 34^{v}, while it is rounded to $4^{s}13;33,47,34°$, for example, in the manuscripts Toruń, BU 74/III, f. 252^{v}; Prague, NK VIII G 24, f. 4^{v}; and Cracow, BJ 605, f. 18^{v}.

50 Ambrosius Lacher, *Tabule resolute de motibus planetarum aliorumque supercelestium mobilium* (Frankfurt/Oder: Johannes Hanau, 1511), VD16 L 12.

51 *Schöner's editions:* investigare.

52 *Schöner's editions:* Norimbergam.

53 *Schöner's editions:* Norimbergae.

the usual Alfonsine meridians of Prague and Leipzig (while the modern value of the difference of their geographical longitudes is about 2°).[54] The very unreliable table of time differences of regions and cities with respect to Leipzig printed on folio O5^{v} at the end of Lacher's TR gives the distance of Prague 14 minutes to the east, but for Vienna it is only 5 minutes to the east. On the same folio is the following explicit: *Finiunt Tabule resolute… Franckfordie ad Oderam*, that is, Frankfurt an der Oder, where Lacher published his tables. He moved there from Leipzig in 1506 to teach at the then newly founded university.[55] Frankfurt (Oder) is in reality only 8′ (about 10 km) to the east of the Prague meridian, but Lacher's table gives its time difference as 20 minutes from Leipzig. To use his tables for a different meridian, Lacher gives the simple instruction to convert the desired local time to the time at the reference meridian of Leipzig.

Subsequent editions of the TR published in 1536, 1551, 1587, and 1588[56] were compiled by Johannes Schöner for the meridian of Nuremberg. The canons included in these prints are based on Lacher's canons with only slight modifications: first of all with Leipzig being replaced by Nuremberg (as indicated in footnotes to the citation from Lacher's edition above). In one of Lacher's examples of computation of the position of the Sun on April 4th, 1502, Schöner modified the numerical values according to his tables. Such a borrowing is not exceptional in Schöner's work; for instance, in his *Opera mathematica* published in 1551, he included, in a similar way, the treatise on eclipse instruments by Johannes Šindel.[57]

For the year 1428, in Schöner's TR the mean motion of the Moon equal to $7^{s}16;41,16°$ is given,[58] which is for $0;30,22° = 0.6442°$ smaller than the value for Toledo. Regarding the speed of mean motion of the Moon, this implies a time shift of 0.0384 days ≅ 55 minutes, and hence a distance in longitude 13;50° eastward of Toledo. However, a table of time differences of European cities from Nuremberg, which is included in these prints, gives a longitudinal distance of Toledo corresponding to 1 hour 11 minutes.

Another print of the TR was prepared by Johannes Virdung in Heidelberg and printed in Nuremberg in 1542.[59] This set of tables is organized in a manner similar to that of Nicolaus Polonius for Salamanca; it also gives all tables of mean motions at various time intervals, always on the same folio and for each quantity separately. Moreover, there are only five cyclical radices starting from (the non-leap) year 1510, again with the step of 20 years, but the sub-tables of mean motions in years continue after the first twenty years, with larger

---

54 See Kremer and Dobrzycki, 'Alfonsine Meridians', 196.

55 See Heinrich Grimm, 'Lacher, Ambrosius', in *Neue Deutsche Biographie, Bd. 13* (Berlin: Duncker et Humblot, 1982), pp. 367–68.

56 Johannes Schöner, *Tabulae astronomicae, quas vulgo… resolutas vocant* (Nuremberg: Johannes Petreius, 1536), VD16 S 3505; VD16 M 6553; Johannes Schöner, *Opera mathematica Ioannis Schoneri… in unum volumen congesta* (Nuremberg: Johannes Montanus and Ulricus Neuber, 1551), VD16 S 3465; Johannes Schöner, *Tabulae resolutae astronomicae* (Wittenberg: Mattheus Welack, 1587), VD16 S 3506; VD16 M 6554; Johannes Schöner, *Tabulae resolutae astronomicae* (Wittenberg: Mattheus Welack, 1588), VD16 S 3507; VD16 M 6555.

57 See Alena Hadravová and Petr Hadrava, 'The Eclipse Instrument by John Šindel', in *Studying the Arts in Late Medieval Bohemia, Reception and Transmission of Knowledge*, ed. by Ota Pavlíček (Turnhout: Brepols 2021), pp. 305–40.

58 See folio G 3^{v} in the print from 1536, folio 5 4^{v} in the print of 1551, and folio I 3^{r} in the prints of 1587 and 1588.

59 Johannes Virdung, *Tabulae resolutae de supputandis siderum motibus* (Nuremberg: Johannes Petreius, 1542), VD16 V 1320.

steps up to 1000 years, similarly to al-Battānī's tables[60] or the table of *anni collecti* in the PAT.[61] In his explanations (on f. 19^r), Virdung gives radices for Heidelberg (in the table *Radices nativitatis Christi ad meridianum Edelbergensem reductae*) of all mean motions at the beginning of the Christian era. For the mean motion of the Moon it reads $4^s2;22,8,21°$, which is $0;24,41,56°$ smaller than the value for Toledo in the PAT. This corresponds to the mean motion of the Moon in 45^m and hence to a difference in longitude of $11;15°$ from Toledo. These values are closer to the true difference of $12;43°$ than the time shift of Toledo of 1 hour 10 minutes given in the table of cities on folio 43^v, which corresponds to the longitudinal distance of $17;30°$, as well as the longitude of Heidelberg *ab occidente habitabili* 26° (i.e. 15° from Toledo) mentioned on folio 19^r.

Although the original aim of the TR was to simplify their use by reducing them directly to the meridian for which they should be used, after they were disseminated, the need to use them for other meridians arose again. Later canons thus included instructions concerning this procedure. An example of a brief description can be found in the canons *Girum recensendo* by Andreas Grzymała de Posnania in the manuscript Cracow, BJ 573, ff. 227^r–233^r:[62]

> |f. 229^r:| Radices ad quemcumque meridianum placuerit reducere. Considera, in quot horis distat meridianus Cracoviensis a meridiano tuo et cum his intra Tabulam medii motus illius, cuius radicem vis reducere, et inventum adde radici Cracoviensi, si locus tuus fuerit occidentalior, vel subtrahe, si locus fuerit orientalior. In Tabulis vero coniunccionum reducendo radices sit per oppositum, videlicet si locus tuus fuerit occidentalior, subtrahe, si orientalior, adde et patebit propositum. Distanciam autem meridiani unius ab altero: scias longitudinem subtrahendo unius a longitudine alterius et residuum divides per 15, si longitudines in tabulis certis invenias, vel queras secundum doctrinam canonum astrolabii.

>> Consider by how many hours the meridian of Cracow is distant from your meridian and with this value enter into the *Table of mean motion*, the radix of which you want to reduce. What you find, add it to the radix for Cracow if your place is westward, or subtract it if the place is eastward. In the *Tables of conjunctions*, the reducing of radices is done in the opposite way: subtract if your place is to the west, add if to the east, and you will get what you need. The distance of one meridian from another: find the (difference of) longitude by subtracting one from the other longitude and divide the remainder by 15 if you have found the longitudes in certain tables or search for it according to the rules for astrolabe.

Let us note that this canon is practically identical to canon 4 by the above-mentioned Nicolaus Polonius,[63] which again confirms a strong dependence of Nicolaus Polonius on Andreas Grzymała.

Extended canons to the TR with the incipit *Pro introductione generali* were written by Iohannes de Glogovia (i.e. Glogow in Lower Silesia). The canons were obviously

60 See the manuscript Paris, BA, MS 8322, f. 50^v and Nallino, *Al-Battānī sive Albatenii Opus astronomicum*, p. 77.
61 *Tabule astronomice*, 1483, f. 17^r; Poulle, *Les tables Alphonsines*, p. 109.
62 Extracted from an edition in preparation.
63 Porres and Chabás, 'Los cánones', 66.

intended for his teaching at Cracow' university and hence they are mostly very detailed and often accompanied by examples, which enable us to follow the contemporary practice of computations. We can find them, for example, in the manuscript Vatican, BAV, Pal. lat. 1439, on folios 165^{r}–194^{r}. On folios 188^{v}–189^{v}, the modification of radices to other cities is also described:[64]

> |f. 188^{v}:| [C]anon 36 de reduccione radicum ad quemcumque alium meridianum. Experiencia docet civitates notabili distancia a se inficem (sic) esse constructas. Quedam enim sunt propinquiores orienti, quedam vero propinquires occidenti. Orientales itaque prius habent ortum Solis, meridiem et occasum quam occidentales. Necesse itaque est modum ostendere reducendo tabulas pro certo meridiano factas ad alium meridianum. Cum itaque hoc facere volueris, scilicet radices unius civitatis reducere ad radices alterius civitatis, primo scias longitudines illarum civitatum, quas poteris scire ex Tabula longitudinum civitatum, que communis est. Est autem longitudo civitatis distancia zenith civitatis ab oriente vel occidente. In civitatibus enim orientalibus longitudo computatur vero ab oriente, vel a Gadibus Alleyri in civitatibus vero occidentalibus, sicut apud nos ab occidente, sive a Gadibus Herculis, longitudo civitatis computatur. Secundo: habitis longitudinibus duarum subtrahe minorem a maiori et manebit differencia longitudinum, ita scilicet una civitas plus distat ab oriente vel occidente quam alia. Tercio: eandem distanciam vel differenciam reduc ad fracciones tali modo. Si fuerint in differencia plures gradus quam 15, tunc semper pro 15 gradibus accipe unam horam eo, quod in una hora 15 gradus equinocciales oriuntur. Et pro quolibet gradu residuo infra 15 accipe minuta 4 hore. Hoc est gradus infra 15 multiplica per 4 et habebis minuta horarum. Quarto: si fuerint minuta graduum, in tali distancia pro quolibet minuto graduum accipe 4 secunda hore, multiplicando illa per 4. Quinto: si in minutis vel secundis plura excreverint quam 60, per 60 divide et in quociente excrescit fraccio grossior precedens. Si autem non est excessus in minutis vel secundis ultra 60, in forma propria dimitte. Sexto: cum fraccionibus tali distancie correspondentibus queratur medius motus planete, huius, cuius radicem verificare volueris. Si habes horas, intra cum horis, si non, similiter intra |f. 189^{r}:| cum minutis et secundis. Septimo: motum inventum tali distancie correspondentem subtrahe a radice illius planete, in cuius tabula eum invenisti, si civitas tua fuit orientalior, vel adde, si fuerit occidentalior.
>
> Verbi gracia: Si vis reducere Pragenses radices ad meridianum Cracoviensem, scias primo longitudines ambarum civitatum. Longitudo civitatis Cracoviensis est 32 gradum, 36 minuta, longitudo civitatis Pragensis est 29 gradus, 30 minuta. Secundo: subtrahe minorem a maiori et manebit differencia unius civitatis ab alia, hoc est, quod gradus equinocciales intercipiuntur inter Cracoviam et Pragam, et manebit in differencia 3 gradus, 6 minuta. Tercio: quia isti gradus non excedunt 15, tunc nulla hora erit distancia inter meridianum Cracoviensem et Pragensem. Quarto: gradus 3 multiplicabo per 4 et veniunt 12 minuta hore. Multiplicabo eciam minuta graduum per 4 et veniunt 24

64 Extracted from an edition in preparation. See https://digi.ub.uni-heidelberg.de/diglit/bav_pal_lat_1439/0386/image,info, consulted June 2023 for a digital reproduction of the manuscript folio.

secunda hore. Pertingitur distancia inter Cracoviam et Pragam in tempore o hora, 12 minuta, 24 secunda: licet alii resolucius loquentes dicunt, quod sit nulla hora, 12 minuta, 23 secunda, 59 tercia, 35 quarta. Quod si addas ad radicem Pragensem in coniunccionibus, exibit radix Cracoviensis in eisdem. Verum radix Pragensis est iste: 5 in diebus, 2 hore, 35 minuta, 57 secunda, 46 tercia, 30 quarta. Addita itaque predicta distancia, scilicet 12 minuta, 23 secunda cum aliis fraccionibus, exibit radix Cracoviensis in mediis coniunccionibus ista: 5 dies, 2 hore, 48 minuta, 21 secunda, 46 tercia, 5 quarta. Pro radice autem habenda Cracovie in motibus annorum collectorum intra cum 12 minuta et 24 secunda et non cum aliis fraccionibus in tabulam planete, cuius motum vis reducere, ut in Solis tabula invenies: o minuta, 30 secunda, 33 tercia. Et illa subtrahe a radice Pragensi, que est 9 signa, 18 gradus, 47 minuta, 27 secunda, 18 tercia, et veniet radix Cracoviensis ista: 9 signa, 18 gradus, 46 minuta, 56 secunda, 45 tercia.

Est eciam alter modus querendi radices Cracovienses per distanciam a Praga. Considera, quot sunt miliaria inter Cracoviam et Pragam modo precisiori, quo potes. Et quia communiter dicitur, quod sunt 60 miliaria, tum quasi vie sepius obliquantur propter valles et colles, recipio ergo 50 miliaria |f. 189^{v}:| communia. Si ergo esset transitus per directum a Cracovia ad Pragam, tot essent miliaria pro quibuslibet ergo miliaribus 16 capias 1 gradum et habebis 3 gradus in quociente dividendo per 16 et in residuo manebunt 2 miliaria. Et quia cuilibet miliari correspondent 3 minuta graduum et 45 secunda, tunc capio pro quolibet miliari 3 minuta et habebis 6 minuta. De secundis non est curandum. Vel si vis operacionem precisiorem facere, accipe 7 in minutis, sed non est necessarie, quia ergo in gradibus non est excessus ultra 15, nec in minuta ultra 16. Illud utrumque multiplica per 4 iuxta predicta et habebis 12 minuta hore et 24 secunda hore, quemadmodum autem invenisti.

> Rule 36 on a reducing of radices to some other meridian. The experience teaches us that cities have been built in some significant mutual distance. Some are thus nearer to the east, others are nearer to the west. Therefore, the eastern cities have the rise of the Sun, noon, and sunset earlier than the western ones. It is thus necessary to show the way to reduce the tables made for a certain meridian to another meridian. If you thus want to do it, that is, to reduce the radices of one city to radices of another city, find first longitudes of both cities, which you can obtain from a *Table of longitudes of cities*, which is common. The longitude of a city is the distance of zenith of the city from the east or west. In the eastern cities, the longitude is computed from the east, and in the western cities it is computed from pillars of Hercules equally as we compute the longitude of city from the west. Secondly, if you have longitudes of two cities, subtract the smaller from the greater one and there will remain difference of longitudes, so one city will be more distant to the east or west than the other. Thirdly: reduce this distance or difference to smaller fractions in this way. If the difference is greater than 15 degrees, then for each 15 degrees take one hour, because in one hour it rises 15 degrees from the equator. And for each from the fewer than 15 remaining degrees take four minutes of an hour. This means multiply the degrees below 15 by four and you will have minutes of an hour. Fourthly, if there are minutes of degrees, take at that distance four seconds of hour for each minute of a degree, multiplying them by four. Fifthly, if there are more than 60 minutes or seconds,

divide them by 60 and increase the preceding greater fraction for the quotient. If, however, there is no excess of minutes or seconds over 60, leave it in the previous form. Sixthly, with the fractions that correspond to that distance is questioned the mean motion of the planet whose radix you want to correct. If you have hours, enter with the hours, if not enter similarly with minutes and seconds. For the seventh step, if your city is to the east, subtract the found motion corresponding to the distance from the radix of that planet in whose table you found it, or add it if it is to the west.

An example: If you want to reduce Prague radices to the meridian of Cracow, find first longitudes of both cities. The longitude of the city of Cracow is 32 degrees 36 minutes, and the longitude of the city of Prague is 29 degrees 30 minutes. Secondly, subtract the smaller from the greater one and there will remain the difference of one city from the other, meaning how many degrees of equator are contained between Cracow and Prague, and there will remain a difference of three degrees, six minutes. Thirdly, because these degrees do not exceed 15, there are zero hours in the distance between the meridian of Cracow and Prague. Fourthly, multiply three degrees by four and the result will be 12 minutes of an hour. I shall also multiply the minutes of a degree by four, and there will be 24 seconds of hour. Concerning the distance in time of zero hours, 12 minutes, 24 seconds between Cracow and Prague: although some who speak more precisely say that it is zero hours, 12 minutes, 23 seconds, 59 thirds, 35 fourths. If it is added to the Prague radix of conjunctions, it shows the Cracow radix of the conjunctions. Really, the Prague radix is this: 5 days, 2 hours, 35 minutes, 57 seconds, 46 thirds, 30 fourths.[65] So adding the above mentioned distance, namely 12 minutes, 23 seconds with the other fractions, it will show this Cracow radix in mean conjunction: 5 days, 2 hours, 48 minutes, 21 seconds, 46 thirds, 5 fourths.[66] However, to obtain Cracow radices of motion in the collected years, enter with 12 minutes and 24 seconds but not with the other fractions into table of the planet whose motion you wish to reduce. So in the *Table of the Sun*, you find 0 minutes, 30 seconds, 33 thirds.[67] And subtract it from the Prague radix, which is 9 signs, 18 degrees, 47 minutes, 27 seconds, 18 thirds,[68] and this Cracow radix will be the result: 9 signs, 18 degrees, 46 minutes, 56 seconds, 45 thirds.[69]

There is also another way to find Cracow radices using distance from Prague. Consider how many miles are between Cracow and Prague as precisely as possible.

65 This value can be found for the year 1429; for example, in the manuscript Prague, NK VIII G 24, f. 11ᵛ.

66 This value can be found for the year 1429; for example, in the manuscript Cracow, BJ 611, f. 11ʳ.

67 It means in a *Table of mean motion in hours*; for example, in the manuscript Cracow, BJ 611, f. 9ᵛ, we find the mean motion of Sun in 12 hours to be equal to 0;29,34,10° and in 24 hours equal to 0;59,8,20°, so by shifting the orders we find the sum for 12 minutes 24 seconds to be 0;0,30,33,18,20°.

68 This radix value can be found for the year 1428 in the third sub-table entitled *Tabula medii motus Solis, Veneris et Mercurii* (Table of mean motion of Sun, Venus and Mercury); for example, in the manuscript Prague, NK X B 3, f. 79ʳ or Prague, NK VIII G 24, f. 2ʳ. A marginal note in the latter manuscript on the same folio explains, 'Iste sunt thabule facte et redacte ad merid(ianum) Pragensem' — 'Here are the tables made and reduced for Prague meridian', and the explicit on folio 37ʳ says, 'Expliciunt thabule de mediis et veris motibus planetarum per P(etrum) Cruc(iferum) compilate' — 'End the tables of mean and true motions of planets compiled by Petrus Cruciferus'.

69 See value for the year 1428 in the manuscript Cracow, BJ 611, f. 1ʳ.

> And because it is commonly claimed that it is 60 miles, then, as the road often bends for the valleys and hills, I shall take 50 common miles. If there were a direct way from Cracow to Prague, then to each 16 miles would correspond to 1 degree.[70] After dividing by 16, you would have three degrees in the quotient and two would remain. And because to each mile corresponds three minutes of a degree and 45 seconds, then I take three minutes for each mile, and you will have six minutes. There is no need to take care of the seconds. Or, if you wish to execute the operation more precisely, accept seven in minutes, but it is not necessary, because there is no excess over 15 in degrees, nor over 16 in minutes. Multiply these both by four as said before and you will have 12 minutes of hour and 24 seconds as you already found.

This canon follows an explanation of the radices of conjunctions, so it is not surprising that in the example it first gives the adaptation of the radix of conjunction. It is, moreover, given by the time difference between the two cities, which is an intermediate step for finding the corrections of radices of the mean motion. However, regarding the otherwise great effort of Iohannes de Glogovia to present the explanations on a highly didactic level, it is strange that after the very detailed description of the derivation this time difference, he does not mention the radices of conjunctions in the first general exposition of this canon at all and does not explain that these radices have to be increased for the eastern cities, while the radices of mean motions are to be decreased for them. The presented alternative method for deriving the corrections of radices from the mutual distance of the cities is completely misleading, because it does not mention the role of geographical latitude nor of the azimuth of direction from one city to the other on the relation between increment of geographical longitude and increment of distance along the path.

Nevertheless, it is characteristic, that the author derives the radices for Cracow from those for Prague. This is how the TR spread from Wrocław through Prague and to Cracow. The radices for Cracow were then used to derive radices for other cities where the TR were disseminated. It can be documented, for example, by marginal note on f. 189^r in the above-quoted copy of the canons by Iohannes de Glogovia, which reads:[71]

> Longitudo Cracovie 32 gradus, 36 minuta, longitudo Heydelberge 27 gradus, 40 minuta. Distancia 4 gradus, 56 minuta, que cum multiplicatur per 4, veniet 0 hora, 19 minuta, 44 secunda. Distancia inter Cracoviam et Heydelbergam in horis et minutis subtrahenda a radicibus Cracoviensibus venit ad Heidelbergam et hoc in gradus et minutis. In motibus autem planetarum adde ad Cracovienses radices ipsius, quod correspondet in motibus 0 horas, 19 minuta, 44 secunda. Alii ponunt longitudinem Heidelberge 25 gradus, 38 minuta. Differencia inter Cracoviam et Heidelbergam sunt 6 gradus, 58 minuta, que faciunt 0 horas, 27 minuta, 52 secunda.

> The longitude of Cracow is 32 degrees, 36 minutes; the longitude of Heidelberg is 27 degrees, 40 minutes. Distance is 4 degrees, 56 minutes, which multiplied by 4, results in 0 hours, 19 minutes, 44 seconds. The distance between Cracow and

70 The author of the text completely ignores the influence of geographical latitudes of both cities.
71 Extracted from an edition in preparation.

Heidelberg in hours and minutes is to be subtracted from the Cracow radices it order to get Heidelberg radices and that in degrees and minutes. In motions of planets add to Cracow radices that correspond in motions to 0 hours, 19 minutes, 44 seconds. Others lay on the longitude of Heidelberg at 25 degrees, 38 minutes. The difference between Cracow and Heidelberg is 6 degrees, 58 minutes, which makes 0 hours, 27 minutes, 52 seconds.

Conclusion

The basic parameters used in the TR are the same as in the PAT, from which they were derived. The TR thus do not bring any improvement of precision with respect to the PAT. Nevertheless, they became quite successful, as can be seen by their relatively wide dissemination. They were a particular favourite in Cracow, where different canons on their use were written. The popularity of the TR is mainly due to the fact that they facilitated the usual computations, and the procedure was more understandable for a less specialized community. The beginning of the Christian era typically precedes the target date for 1400 years $\cong 5 \times 10^5$ days $\cong 2 \times 60^3$ days and thus requires four sexagesimal digits. The replacement of radices for the Christian era by a set of cyclical radices preceding the target date by no more than 20 years (i.e. about $7 \times 10^3 \cong 2 \times 60^2$ days) enabled users to decrease the number as well as the length in sexagesimal digits of items to be summed in computation of the mean motion. In a calculation limited to whole days, in the PAT it is necessary in to sum five items (the radix and contributions of 4 sexagesimal orders of time), while in the TR, only four are needed (cyclical radix, *anni expansi*, months, and days). In the TR, these items must be found in different tables (including a table of hours). They were, however, more intuitive for common users accustomed to counting time in these units than the table of days and their multiples by powers of 60 used in the PAT. It was mostly sufficient for common purposes to find the mean motions with levels of precision up to thirds of a degree, and hence the TR provided the mean motions in five sexagesimal digits (signs, degrees, minutes, seconds, and thirds), while the PAT usually gave the mean motions in nine digits. This is needed to bridge the vast interval of time between the epoch of the radix and the target date and to count the number of completed circles circumscribed by the planet, which, however, are subsequently skipped. The smaller number of columns in the TR thus compensates the higher overall number of lines in all the aforementioned tables and renders a reproduction of the TR feasible. It can be concluded that the TR represent a step in increasing the user-friendliness of the Alfonsine Tables. The TR are designed for the 'working' astrologer, while the PAT are designed for the 'theoretical' mathematician who delights in consistent sexagesimal arithmetic but rarely computes ephemerides or horoscopes.

Manuscript Sources

Cracow, Biblioteka Jagiellońska, 257
Cracow, Biblioteka Jagiellońska, 550
Cracow, Biblioteka Jagiellońska, 551
Cracow, Biblioteka Jagiellońska, 553
Cracow, Biblioteka Jagiellońska, 560
Cracow, Biblioteka Jagiellońska, 573
Cracow, Biblioteka Jagiellońska, 600
Cracow, Biblioteka Jagiellońska, 605
Cracow, Biblioteka Jagiellońska, 611
Cracow, Biblioteka Jagiellońska, 614
Cracow, Biblioteka Jagiellońska, 616
Cracow, Biblioteka Jagiellońska, 617
Cracow, Biblioteka Jagiellońska, 1846
Cracow, Biblioteka Jagiellońska, 1852
Cracow, Biblioteka Jagiellońska, 1858
Cracow, Biblioteka Jagiellońska, 1864
Cracow, Biblioteka Jagiellońska, 1865
Cracow, Biblioteka Jagiellońska, 1916
Cracow, Biblioteka Jagiellońska, 1917
Cracow, Biblioteka Jagiellońska, 1920
Cracow, Biblioteka Jagiellońska, 1966
Cracow, Biblioteka Jagiellońska, 2252
Cracow, Biblioteka Jagiellońska, 2478
Cracow, Biblioteka Jagiellońska, 2480
Cracow, Biblioteka Jagiellońska, 2650
Cracow, Biblioteka Jagiellońska, 2664
Kalocsa, Föszékesegyházi Könyvtár, 326
Mikulov, Regional Museum in Mikulov, MIK 6361
Munich, Bayerische Staatsbibliothek, Clm 51
Munich, Bayerische Staatsbibliothek, Clm 214
Oxford, Bodleian Library, Canon. misc. 27
Oxford, Bodleian Library, Canon. misc. 499
Oxford, Bodleian Library, Digby 168
Paris, Bibliothèque de l'Arsenal, MS 8322
Paris, Bibliothèque nationale de France, lat. 7281
Paris, Bibliothèque nationale de France, lat. 16208
Prague, Národní knihovna, III C 2
Prague, Národní knihovna, VIII G 24
Prague, Národní knihovna, X A 23
Prague, Národní knihovna, X B 3
Prague, Národní knihovna, XIV A 18
Toruń, Biblioteka Uniwersytecka, 74/III
Vatican, Biblioteca Apostolica Vaticana, Pal. lat. 1375

Vatican, Biblioteca Apostolica Vaticana, Pal. lat. 1381
Vatican, Biblioteca Apostolica Vaticana, Pal. lat. 1439
Vienna, Österreichische Nationalbibliothek, Cod. 5145
Vienna, Österreichische Nationalbibliothek, Cod. 5245

Bibliography

Printed Sources

Lacher, Ambrosius, *Tabule resolute de motibus planetarum aliorumque supercelestium mobilium* (Frankfurt/Oder: Johannes Hanau, 1511).

Schöner, Johannes, *Tabulae astronomicae, quas vulgo… resolutas vocant* (Nuremberg: Johannes Petreius, 1536).

———, *Opera mathematica Ioannis Schoneri… in unum volumen congesta* (Nuremberg: Johannes Montanus and Ulricus Neuber, 1551).

———, *Tabulae resolutae astronomicae* (Wittenberg: Mattheus Welack, 1587).

———, *Tabulae resolutae astronomicae* (Wittenberg: Mattheus Welack, 1588).

Tabulae astronomicae Alfonsi Regis. With additions by Augustinus Moravus, ed. Johannes Lucilius Santritter (Venice: Johannes Hamman, 1492).

Tabule astronomice divi Alfonsi, regis Romanorum et Castille (Venice: Petrus Liechtenstein, 1518).

Tabule astronomice illustrissimi Alfontii, regis Castelle (Venice: Erhard Ratdolt, 1483).

Virdung, Johannes, *Tabulae resolutae de supputandis siderum motibus* (Nuremberg: Johannes Petreius, 1542).

Secondary Sources

Boudet, Jean-Patrice, and Laure Miolo, 'Alfonsine Astronomy and Astrology in Fourteenth-Century Oxford: The Case of MS Bodleian Library Digby 176', in *Alfonsine Astronomy: The Written Record*, ed. by Richard L. Kremer, Matthieu Husson, and José Chabás (Turnhout: Brepols, 2021), pp. 57–105.

Catalogus codicum manuscriptorum medii aevi Latinorum, qui in Bibliotheca Jagellonica Cracoviae asservantur, Vol. I–XI, Wrocław, etc. 1980–.

Chabás, José, 'Astronomy in Salamanca in the Mid-Fifteenth Century: The Tabulae resolutae', *Journal for the History of Astronomy*, 29 (1998), 167–75.

———, *Computational Astronomy in the Middle Ages. Sets of Astronomical Tables in Latin* (Madrid: Consejo superior de investigaciones científicas, 2019).

———, and Bernard R. Goldstein, *The Alfonsine Tables of Toledo* (Dordrecht: Springer, 2003).

———, and ———, *A Survey of European Astronomical Tables in the Late Middle Ages* (Leiden: Brill, 2012).

———, and ———, *Essays on Medieval Computational Astronomy* (Leiden: Brill, 2015).

Comes, Mercè, 'The "Meridian of Water" in the Tables of Geographical Coordinates of al-Andalus and North Africa', *Journal for the History of Arabic Science*, 10 (1994), 41–51.

Durand, Dana Bennett, *The Vienna–Klosterneuburg Map Corpus of the Fifteenth Century: A Study in the Transition from Medieval to Modern Science* (Leiden: Brill, 1952).

Grimm, Heinrich, 'Lacher, Ambrosius', in *Neue Deutsche Biographie, Bd. 13* (Berlin: Duncker et Humblot, 1982), pp. 367–68.

Hadravová, Alena, and Petr Hadrava, *Stavba a užití astrolábu* (Prague: Filosofia 2001).

———, and ———, 'The Eclipse Instrument by John Šindel', in *Studying the Arts in Late Medieval Bohemia, Reception and Transmission of Knowledge*, ed. by Ota Pavlíček (Turnhout: Brepols 2021), pp. 305–40.

———, and ———, 'John of Lignères *Quia ad inveniendum loca planetarum*: An Edition and Translation', in *Alfonsine Astronomy: The Written Record*, ed. by Richard L. Kremer, Matthieu Husson, and José Chabás (Turnhout: Brepols, 2021) (Alfonsine Astronomy: Studies and Sources, 1), pp. 257–302.

Husson, Matthieu, 'Physical Signs and Minutes of Days: Mean Motion Computations in the Parisian Alfonsine Tables', in *Le Moyen Âge et les sciences*, ed. by Danielle Jacquart and Agostino Paravicini Bagliani, Micrologus 100 (2021) pp. 223–50.

Kremer, Richard L., and Jerzy Dobrzycki, 'Alfonsine Meridians: Tradition versus Experience in Astronomical Practice c. 1500', *Journal for the History of Astronomy*, 29 (1998), 187–99.

Macray, William D., *Catalogi codicum manuscriptorum Bibliothecae Bodleianae, Pars nona: Codices a viro clarissimo Kenelm Digby anno 1634 donatos, complectens* (Oxford: Clarendon Press 1883).

Nallino, Carlo Alfonso, *Al-Battānī sive Albatenii Opus astronomicum, Pars secunda* (Milan: Ulrichus Hoeplius, 1907).

Pedersen, Fritz Saaby, *The Toledan Tables* (Copenhagen: C.A. Reitzels Forlag 2002).

Porres de Mateo, Beatriz, 'Les tables astronomiques de Jean de Gmunden: édition et étude comparative' (unpublished thesis, Paris, École pratique des hautes études, 2003).

———, and José Chabás, 'Los cánones de las Tabulae Resolutae para Salamanca: origen y transmisión', *Cronos* 1 (1998), 51–83.

Poulle, Emmanuel, *Les Tables Alphonsines avec les canons de Jean de Saxe* (Paris: Centre national de la recherche scientifique, 1984).

Rosińska, Grażyna, *Scientific Writings and Astronomical Tables in Cracow. A Census of Manuscript Sources (XIVth–XVIth Centuries)* (Wrocław, etc.: Ossolineum, 1984).

Saby, Marie-Madeleine, 'Les Canons de Jean de Lignères sur les tables astronomiques de 1321. Édition critique, traduction et étude' (unpublished thesis, Paris, École nationale de chartes, 1987).

Samsó, Julio, *On Both Sides of the Strait of Gibraltar* (Leiden and Boston: Brill, 2020).

Tabulae codicum manu scriptorum praeter Graecos et orientales in Bibliotheca Palatina Vindobonensi asservatorum, ed. Academia Caesarea Vindobonensis, 10 vols (Graz: Akademische Druck — und Verlagsanstalt, 1965).

Toomer, Gerald James, *Ptolemy´s Almagest* (London: Duckworth, 1984).

Truhlář, Josef, *Catalogus codicum manu scriptorum Latinorum, qui in bibliotheca publica… asservantur* (Prague: Regia Societatis Scientiarum Bohemicae, 1905).

JOSÉ CHABÁS †

Computational Practices in Alfonsine Astronomy

Hermann of Saxony's Practica motuum planetarum*

Alfonso X of Castile and León (reigned: 1252–84) gathered a multicultural group of scholars at his court to translate texts from Arabic into Castilian and Latin and to pursue new knowledge in a wide range of disciplines. Among these scholars, Isaac ben Sid and Judah ben Moses ha-Cohen were responsible for the compilation of a new set of astronomical tables, which are only known through their Castilian canons and now called Castilian Alfonsine Tables. When this set reached Paris in the early fourteenth century, it was first used by John Vimond and John of Murs and, together with other Parisian astronomers such as John of Lignères, they adapted it and wrote texts and canons explaining their use.[1] The most influential canons were those written in 1327 by John of Saxony.[2] In subsequent years, canons and tables were both presented in a variety of ways, although not challenging the basic parameters and geometrical models on which they were built.

How the tables were actually used and how the instructions given in the canons were followed in computational practices in Alfonsine astronomy are topics addressed in this paper. For this purpose, we present two texts exclusively consisting of worked examples, showing mathematical practices developed by early Alfonsine astronomers. Together with other similar texts, they can be considered a new category or genre of astronomical

* I am most grateful to Bernard R. Goldstein for his helpful advice with regard to the matters in this paper and for the continuous and stimulating conversation over the years. This paper was written in the framework of the ERC project *ALFA, Shaping a European Scientific Scene: Alfonsine Astronomy* (CoG 723085). A draft of this paper was presented at the annual meeting of Project ALFA and received comments from the participants, which were much appreciated. Matthieu Husson and Richard L. Kremer provided insightful comments and suggestions and corrected the final proofs.

1 For the context and a transcription of the Castilian canons, see José Chabás and Bernard R. Goldstein, *The Alfonsine Tables of Toledo* (Dordrecht: Springer, 2003). For the Parisian Alfonsine Tables, see José Chabás, *Computational Astronomy in the Middle Ages: Sets of Astronomical Tables* (Madrid: Consejo Superior de Investigaciones Científicas, 2019), pp. 237–76.

2 The canons are extant in a great number of manuscripts and were printed by E. Ratdolt in Venice in 1483 together with the first edition of the Parisian Alfonsine Tables. For a transcription of the text and a translation into French, see Emmanuel Poulle, *Les Tables Alphonsines avec les canons de Jean de Saxe* (Paris: Éditions du CNRS, 1984).

José Chabás • Universitat Pompeu Fabra, Barcelona

Alfonsine Astronomy: Computational Practices, ed. by Matthieu Husson, José Chabás, and Richard L. Kremer, Turnhout, 2026 (*Alfonsine Astronomy: Studies and Sources*, 3), p. 373-412
© BREPOLS PUBLISHERS 10.1484/M.ALFA-EB.5.137917

literature; indeed, the specific computational procedures associated with the Alfonsine corpus of tables and canons is an indissoluble part of Alfonsine astronomy.

In several of his publications, Lynn Thorndike mentioned Hermann of Saxony, an otherwise unknown scholar, as the author of a set of tables for the meridian of Paris accompanied by a text.[3] These tables and canons, extant in three manuscripts, are addressed in part II. In addition to these, Hermann of Saxony is also to be credited with a similar and longer text, with no tables, which is preserved in another five manuscripts and examined in detail in part I. Both works deal with the many computational problems faced by early Alfonsine astronomers, from counting days to computing the circumstances of eclipses, and consist entirely of numerical examples computed for Paris for the years 1356, 1361, and 1362. These two texts, together with the *Quia plures astrologorum* by John of Saxony, which is devoted to numerical examples of John of Lignères' canons beginning *Cuiuslibet*,[4] can be considered as the first elements of a new category (or genre) of Alfonsine texts consisting exclusively or mostly of worked examples showing the computational practices in Alfonsine astronomy and complementary to the standard canons explaining the use of tables.

In various manuscripts containing these two texts, we are told that Hermann Stilus worked in Paris and was originally from Nochem or Northem, Saxony, a locality probably to be identified with Northeim (Lower Saxony). The dates of his numerical examples make Hermann of Saxony a member of the generation following that of the earliest Alfonsine astronomers working in Paris, the 'Johns', which includes John Vimond, John of Murs, John of Lignères, John of Saxony, John of Genoa, and John of Montfort.

As will be shown below, the two texts by Hermann of Saxony reflect the computational practices in Alfonsine astronomy developed in the early fourteenth-century milieu in Paris and closely follow the instructions given in John of Saxony's canons to the Parisian Alfonsine Tables beginning *Tempus est mensura motus*, dated 1327.

Part I: Worked examples for 1356

1. *Copies of the 1356 worked examples*

We have located the text by Hermann of Saxony for 1356 in six manuscripts:
E: Erfurt, Universitätsbibliothek, CA 2° 386, 34^r–48^r;
O: Oxford, Bodleian Library, Digby 97, 240^v–290^v;
P: Paris, Bibliothèque nationale de France, latin 7407, 1^r–26^v;

3 Lynn Thorndike, 'Astronomy at Paris around 1485 and 1360', *Humanisme et Renaissance* 3,2 (1936), pp. 165–68; Lynn Thorndike, 'Pre-Copernican Atronomical Activity', *Proceedings of the American Philosophical Society*', 94:4 (1950), 321–26, see spec. p. 323; Lynn Thorndike, 'Notes on some Astronomical, Astrological, and Mathematical Manuscripts of the Bibliothèque Nationale, Paris', *Journal of the Warburg and Courtauld Institutes* 20, 1/2 (1957), pp. 112–72; L. Thorndike and P. Kibre, *A Catalogue of Incipits of Medieval Scientific Writings in Latin* (The Medieval Academy of America: London, 1963), 102(3), 102(13), and 688(4).

4 For a transcription of the text, see Marie-Madeleine Saby, *Les canons de Jean de Lignères sur les tables astronomiques de 1321*. Unpublished thesis: École Nationale des Chartes (Paris, 1987). A summary appeared as: 'Les canons de Jean de Lignères sur les tables astronomiques de 1321', *École Nationale des Chartes: Positions des thèses*, pp. 183–90. It is worth noting that John of Saxony and Hermann of Saxony's texts are complementary in that they provide numerical examples for different topics of computational astronomy.

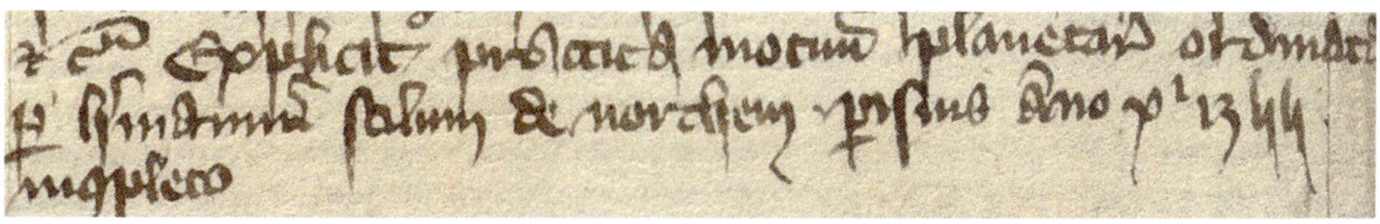

Figure 12.1. Explicit in MS Erfurt, CA 2° 386, 47^{v}.

Q: Paris, Bibliothèque nationale de France, latin 15104, 122^{v}–137^{v};
K: Prague, Knihovna metropolitní kapituly, M.CI (1460), 27^{r}–34^{v};
W: Wolfenbüttel, Herzog August Bibliothek, 2637 (Guelf 65 Aug 2°), 311^{r}–323^{v}.

MS E is taken here as the base manuscript. Its title, *Canonum Alfonsii practica per Hermannum* (34^{r}), is written in a different hand than that of the text, and the explicit provides critical information on the author: *Explicit practica motuum planetarum ordinata per Hermannum Stilum de Norchem Parisius anno Christi 1355 incompleto* (47^{v}); see Figure 12.1. We note that the text written in 1355 is incomplete, and the computed examples are for the first day of January 1356 or later dates in the same year.

In most manuscripts, Hermann's work is accompanied by texts due to early fourteenth-century astronomers.

MS E is a mid-fourteenth-century manuscript containing various astronomical items, all of them dated in the first half of the century: the commentary to John of Lignères' *Cuiuslibet arcus* by John of Saxony, beginning *Quia plures astrologorum* (26^{v}–32^{r}); a text on the comet of 1337 by Geoffrey of Meaux (59^{r-v}), two prognostications on the great conjunction of 1345 by Firmin of Beauval (59^{v}–60^{r}) and John of Murs (60^{r}–61^{v}); and John of Saxony's ephemerides consisting of canons (62^{r}–63^{r}) and tables (63^{v}–107^{v}) and beginning *Cum animadverterem*.[5] The work by Hermann of Saxony under consideration here (34^{r}–48^{r}) consists of an introductory chapter beginning *Non fuit mortuus* and a long text with worked examples for the year 1356.

MS P opens with Hermann of Saxony's work (1^{r}–26^{v}), followed by John of Saxony's *Quia plures astrologorum* (26^{v}–40^{r}). Among the texts of astronomical interest is an incomplete version of John of Lignères' *Priores astrologi* (40^{r}–66^{v}), after which a few additional canons were inserted.

MS Q is a manuscript of French origin composed of various parts written at different times. The texts on astronomy date from the fourteenth century. Among them, one finds the prognostication on the great conjunction of 1345 by Firmin of Beauval (112^{r-v}), together with that by John of Murs (112^{v}–114^{r}) and the letter written by both of them to the Pope on the reform of the Christian calendar (114^{v}–121^{r}). Then follows Herman of Saxony's work (122^{r}–137^{v}). On 143^{r}–146^{r} there is a copy of the canons to the Oxford Tables of 1348, but the associated tables are missing.

MS K is dated to the fifteenth century and has only 34 folios. It contains several texts on geometry and proportions by the German scholar Johannes de Cubito, and ends with a truncated version of the Hermann of Saxony's *practica*.

5 For a detailed description of this manuscript, see Wilhelm Schum, *Beschreibendes Verzeichnis der Amplonianischen Handschriften-Sammlung zu Erfurt* (Berlin: Weidmann, 1887), where a few items are incorrectly identified.

MS O is an early fifteenth-century manuscript of English origin containing several astronomical items. Among them are an incomplete copy of the Parisian Alfonsine Tables (5^r–13^v); the tables of William Rede for Oxford, with 1340 as epoch (14^r–32^v), and their associated canons (64^v–71^r); planetary tables belonging to the Oxford Tables of 1348 (33^v–39^v); canons by John of Murs to the Tables of Alfonso dated 1339 (122^r–125^r); and canons for the computation of eclipses (125^v–128^v) and a table for the hourly velocities and semidiameters of the luminaries (130^v), both by John of Genoa. The manuscript closes with Hermann of Saxony's work (240^v–290^v), also beginning *Tempus est mensura motus*, a fact that has misled scholars into considering this text as a copy of John of Saxony's canons to the Parisian Alfonsine Tables.

MS W is dated to the second half of the fifteenth century and has copies of the table for the hourly velocities and semidiameters of the luminaries by John of Genoa (166^v), the table and canons for determining the time from mean to true syzygy by Nicholaus de Heybech (307^v–309^r), and the text by John of Saxony beginning *Quia plures astrologorum* (339^r–348^r). Of special interest is the copy of the Parisian Alfonsine Tables (185^r–229^v), consisting of the usual tables for radices, mean motions, and equations plus a number of tables taken from the set by John of Lignères for 1322. Just after this collection of tables, in the explicit dated 1422 (348^r), the copyist, Wilhelm Getzenstorff, explains that there end the Alfonsine Tables, indicating that he considered the Parisian Alfonsine Tables as a set that included many tables in John of Lignères`s set for 1322. The text under review here is on 311^r–323^v, with blank pages in between, and some intermediary chapters are missing.

In four cases (MSS E, P, Q, and W), Hermann's text begins with the sentence *Non fuit mortuus qui scientiam vivificavit nec est pauper qui intellectui dominatus est.* In MS K, we instead read *Non est mortuus*… An explanation follows: *Hanc propositione scribit Ptholomeus in prologo prime dictionis sui magni voluminis Almagesti.* The first sentence, although attributed to Ptolemy, was not his. Rather, it is an aphorism added by Gerard of Cremona to the preface of his Latin translation of Ptolemy's *Almagest* (*c.* 1175) and is frequently used as a *dictum* by medieval scholars writing on science.[6] An example of it is shown in Figures 12.2 and 12.3, taken from a thirteenth-century manuscript now in Madrid. As mentioned above, in contrast to these five manuscripts, MS O begins *Tempus est mensura motus*, followed by the first three chapters of John of Saxony's canons to the Parisian Alfonsine Tables. This fact illustrates from the very beginning that the text in the Oxford manuscript differs in many ways from that in MSS E, P, Q, K, and W, as is repeatedly demonstrated below.

The introductory text contains general considerations on astronomy and judicial astrology. MSS P, Q, and K share the same text, with literary references to the tables of Alfonso and the Toledans, as well as a reference to the book on judgements by the Maghribi astrologer Ibn Abī l-Rijāl (d. *c.* 1038). MS E opens with the same text, but the second half is replaced by another on the movements of the Sun and the Moon, including a few diagrams. The Alfonsine tables are mentioned, and several Alfonsine parameters are explicitly given. Another author mentioned is Thadeus of Parma (fl. 1318–21). MS O has no such texts, and we instead find the first three chapters of John of Saxony's *Tempus est mensura motus*. In summary, the introduction preceding the numerical examples is

6 On the introduction of this and other *dicta* in the *Almagest*, see Charles Burnett, *'Ptolemaeus in Almagesto dixit': The Transformation of Ptolemy's* Almagest *in its Transmission via Arabic into Latin*, in G. Toepher and H. Böhme, eds., Transformationem antiker Wissenschaft (De Gruyter: Berlin, 2010), pp. 115-40

presented in five manuscripts as two different versions of one text opening *Non fuit (est) mortuus*, as well as a highly enlarged text in the sixth manuscript, which shows evidence of the intervention by a third party.

The computed examples following the introduction are much more homogenous. They are presented according to a logical structure of increasing difficulty, as in John of Saxony's canons, *Tempus est mensura motus*. Moreover, in most cases the chapter titles given by Hermann are the same as those in John's canons, although with different texts under them. To facilitate referencing, we have numbered the sections in Hermann's work, although this convention is not found in any manuscript. Hermann departs occasionally from the order in *Tempus es mensura motus*, and sometimes he inserts sections absent from John's canons or provides different procedures.

Hermann of Saxony uses signs of 60°, as did John of Saxony, although zodiacal signs of 30° are often used when presenting a final result. For time, the preferred unit is the day, especially when referring to differences between two dates; these are always presented as a sexagesimal number rather than as a number of years, months, etc. Sexagesimalization is extended to other units of time, and the basic submultiple of the day is a minute of a day (1/60d), not an hour (1/24d). In contrast, final results are given in months, days, hours, etc. It is quite clear that Hermann considered enhanced sexagesimalization as a useful computing tool and avoided mixing different bases, keeping to base 60 in his computations for time, angles, and angular distances as much as possible.

2. *Determining the sexagesimal number of days since a given era without using tables (E, 38^{r-v}; P, 2^{v}–4^{r}; Q, 123^{r}–124^{r}; K, 27^{v}–28^{v}; W, 311^{v}–312^{r})*

The title of this section is the same as in Chapter 2 of John of Saxony's canons, *Tempus est mensura motus*, beginning *Numerum annorum… sine tabulis*, and differs slightly in MS K. Hermann of Saxony gives an example for the first day of 1356. We are told to add the results of multiplying 1355 by 365d (that is, 494575 days) and multiplying 1355 by a quarter of a day (that is, 338d). The resulting value, 494913d, is converted into a sexagesimal number, 2,17,28,33d, counted from the Incarnation (Dec. 31, 0 AD), by taking the remainders of successive divisions by 60.

MSS P, Q, and K display a list of the names of the months in a year according to calendars of the different eras: Christian (beginning in October), Alexander, Diocletian, Nabonassar, Yazdegird, and Arabic. This list, incomplete in MS K, is preceded by a short text on the beginning of the year in each of the calendars and on the use of the Alfonsine tables for collected and expanded years. MS O does not include Section 2. It does, however, reproduce Chapter 2 of John of Saxony's canons in its entirety (240^{v}–241^{r}).

3. *Determining the sexagesimal number of days for a given era using tables (E, 38^{v}–39^{r}; P, 4^{r}–6^{v}; Q, 124^{r}–125^{v}; K, 28^{v}–30^{v}; W, 312^{r}–313^{r})*

This section has the same title as Chapter 3 of John of Saxony's canons, beginning *Numerum annorum… per tabulas ad hoc factas*. The numerical example given here is the same one

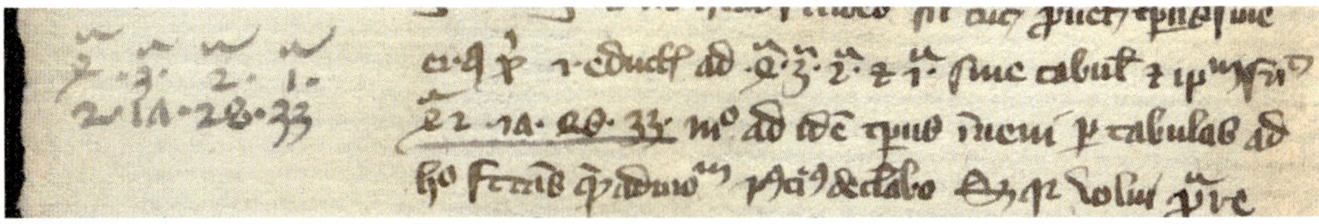

Figure 12.2. Excerpt of the example in Section 3 (MS E, 38^{v}). note how the incorrect 2, 17, 48, 33 is underligned and corrected to 2, 17, 28, 33 in the margin

as above, the first day of year 1356. We are asked to enter the table to convert collected years in days, expressed as sexagesimal numbers, and find the corresponding entries in the columns for days expressed as fourths, thirds, seconds, and primes:

1000	1	41	27	30
300	0	30	26	15
40	0	4	3	30
15	0	1	31	18
Total	2	17	28	33

Deciphering numbers in manuscripts was not always, and still it is not, an easy task. Sometimes, fortunately, the scribe decides to rewrite it in the margin or give the correct value (MS E, 38^{v}; see Figure 12.4).

As was the case above, MS O does not contain this section. It reproduces Chapter 3 of John of Saxony's canons almost entirely (241^{r-v}), including John of Saxony's example for July 2, 1327.

4. *Finding the sexagesimal number of days for a given era from the corresponding number in another era (E, 39^{r}; P, 6^{v}–7^{v}; Q, 125^{v}–126^{r}; K, 30^{v}–31^{r}; W, 313^{v}, unfinished)*

The title corresponds to Chapter 4 of John of Saxony's canons, beginning *Cognitis quartis, tertiis, secundis et primis… a principio alicuius ere*. Paralleling John, Hermann gives 5,14,42,39d as the time between the Flood and the Incarnation. By simple addition, the time between the Flood and the beginning of 1356 is said to be 7,32,11,12d.

In Hermann's text, there is a second example referring to the time from Alfonso's era (2,6,57,59d after the Incarnation), which confirms this result. It is worth noting that, although not explicitly said, the time for the era of Alfonso employed here is that used by the Parisian astronomers (June 1, 1252), which differs from that in the canons to the Castilian Alfonsine Tables (January 1, 1252).[7]

7 Chabás and Goldstein, *The Alfonsine Tables of Toledo*, p. 144.

5. *Transforming a sexagesimal number of days since a given era into years, months, and days (E, 39^{r-v}; P, 7^{v}–8^{r}; Q, 126^{v}; K, 31^{r-v})*

Once again, the title of this section is the same as in John of Saxony's canons (Chapter 5), which begins *Cognitis… in eius contentorum invenire*. Starting from the numbers of days between the Flood and the first day of January 1356, 7,32,11,12d, the example requires successive subtractions of the number of days in groups of collected years, expanded years, and months found in the appropriate table (6,45,50,0d for 4000 years, 0,40,35,0d for 400 years, etc.). The conclusion is that the beginning of 1356 corresponds to November 15, 4456 years after the Flood.

6. *Determining the day of the week of a given date (E, 39^{v}; O, 241^{v}–242^{r}; P, 8^{v}–9^{r}; Q, 126^{v}–127^{r}; K, 31^{v}–32^{r})*

The text parallels Chapter 6 of John of Saxony's canons, beginning *Notam anni, mensis vel diei*. To find the day of the week of January 1, 1356, one enters the table for finding the days of the week with each of the parts of the sexagesimal number 7,32,11,12, and adds the corresponding four entries to the day of the week of Incarnation, which is 7, to obtain 27. The remainder after dividing this number by 7 is 6. With the standard rule according to which Sunday is the first day of the week, day 6 corresponds to Friday.

7. *Correcting the tables of Alfonsine radices (E, 39^{v}; P, 9^{r-v}; Q, 127^{r-v}; K, 32^{r}; W, 314^{v})*

There is no parallel text in *Tempus est mensura motus*. The example given here is intended for checking the consistency of the radices for the mean motion of the Sun at the time of the Flood, 5,1;23,32,58,49° (MS P reads 5,1;**33**,32,**28**,49°), and the Incarnation, 4,38;21,0,30,28° (MSS E, Q, K, and W add 54 fifths and MS P adds 24 fifths); see the *editio princeps* of the Parisian Alfonsine Tables (Ratdolt 1483, c8r, with no fifths). We note that these two values for the radices are valid for the meridian of Toledo and expressed in signs of 60°, as is the case throughout the text. The time between the Flood and the Incarnation is said to be 5,14,42,39d, as indicated previously (see Section 4). One enters the table for the solar mean motion with each of the parts of this sexagesimal number and adds the corresponding four parts, to obtain 5,36;57,28,31,37°. Adding this time to the radix for the Flood should give the radix for the Incarnation. And it almost does, but the result, 4,38;21,1,30,26°, differs by one second and two thirds from that displayed by Hermann. This is an error in the text: the entry for 14 thirds in the time between the Flood and Incarnation is given as 47,56;34,**43**,29,15, whereas the table for solar mean motions reads 47,56;34,**42**,29,15 (Ratdolt 1483, d5r). In any case, the relevant issue here is the use of a full sexagesimal mean motion table, that is, one with 60 successive multiples of the basic parameter (0;59,8,19,37,19,13,56°/d for the Sun).

It is also worth noting that the *editio princeps* displays an incorrect value for the radix of the motion of the Sun at the Flood, given as **3**,1;**25**,32,58,49°. The correct value, **5**,1;**23**,32,58,49°, is found in most manuscripts. The second edition of the Parisian Alfonsine Tables (Santritter 1492) corrected the number of signs to five, but not the number of minutes. One wonders how medieval and more contemporary computers coped with such faulty values in their reference copies. Hermann of Saxony, probably well aware of this difficulty, included sections to correct the values of some other critical quantities.

8. Correcting the tables of mean motions (E, 39v–40r; P, 9v–11r; Q, 127v–128r; K, 32v–33r; W, 314v–315r)

The title of this section is the same as that of Chapter 15 of John of Saxony's canons, beginning *Modum corrigendi tabulas motuum equalium*. In Hermann's text in MSS E and K, the example chosen refers to the table for the mean motion of the Sun and uses the entries for arguments 29° and 1° to obtain that for argument 30°. The values in the text are 28;35,1,19,2,17,44,4° and 0;59,8,19,37,19,13,56°, respectively, and the result is given as 29;34,9,38,39,36,57,60° (*sic*) in MS E and 29;34,9,38,39,36,56,0° (!) in MS K. We note that the entry for argument 29° has a faulty value in the thirds (19 instead of 29), implying an error of ten thirds in the result.

Manuscripts P and Q present a different numerical example. They refer to the table for the mean motion of the eighth sphere: the entry for argument 3° results from adding the entries for arguments 1° and 2°. And this is indeed the case. MS W has, instead, an example for the mean motion of access and recess. As in Section 7, full sexagesimal mean motion tables are used in both examples.

Then follow two short texts on the way to correct the tables for the equations and an explanation of the meaning of the term 'radix', which are given titles in MSS P, Q, and K, but not in MS E. The second text follows almost verbatim Chapter 9 of John of Saxony's canons on the meaning of the term radix, beginning *Quid intelligatur per nomen*.

9. Transforming a time given in hours and its fractions into days and minutes of a day by means of tables (E, 40r–v; O, 242r–v; P, 11r; Q, 128v; K, 33r–v)

The title of this section is the same as that of Chapter 7 of *Tempus est mensura motus*, beginning *Numerum horarum ac minutorum hore*. We are here asked to convert 9;45,20h into days and minutes of a day. For that purpose, one enters the conversion table from hours into days and minutes and finds 0;22,30d, 0;1,52,30d, and 0;0,0,50d for 9h, 0;45h, and 0;0,20h, respectively. The result is 0;24,23,20d, which MS Q gives erroneously as 24;23,20 hours. The same example is used to convert, reversely, days and minutes of a day into hours and fractions of an hour.

10. Determining any mean motion by means of tables (E, 40v; O, 242v–243v; P, 11r–v; Q, 128v–129r; K, 33v; W, 316v)

As was the case in previous sections, the title reproduces almost verbatim that of a text in John of Saxony's canons: Chapter 12, beginning *Medium motum solis et lune*. The computed example is for the Sun, for the beginning of the year 1356. The radix for the Incarnation is given as 4,38;19,2,13,49° (MS W reads 4,38;19,2,**3**,49°), which differs from the value 4,38;21,0,30,28° presented above (see Section 7). The reason is clear: the previous value corresponds to Toledo, while the new one is for Paris, as explicitly stated in MSS Q, K, and W. We note that the difference (0;1,58,16,39°) between the two radices is equivalent to a time difference of 0;48h between Toledo and Paris, which is the standard value used by Alfonsine astronomers.

Since the Incarnation, 2,17,28,33 days have elapsed, and for each of the four parts of this sexagesimal number we are given the corresponding mean motions taken from a full sexagesimal table of mean motions. The result is 4,47;31,59,21,6° in MS E, 4,47;31,59,21,10° in MSS P, Q, and W, and 4,47;31,59,19,8° in MS K, whereas MS O does not give it explicitly. This divergence is probably due to the different rounding in each of the intermediary entries.

11. Determining the longitude of the apogee (E, 40^{v}; O, 243^{r}–244^{r}; P, 11^{v}–12^{r}; Q, 129^{r}; K, 33^{v}–34^{r}; W, 316^{v})

Hermann deviates here from the order of the chapters in *Tempus est mensura motus* and refers at that point to its Chapter 16, on the determination of the longitude of the apogees of the planets by means of tables. The example is, again, for the beginning of 1356. We are first given the mean motion of the apogees and the fixed stars (0,9;57,18,16°) and the mean motion of access and recess (1,8;53,48,8°), and we are told that they derive from the corresponding mean motions tables, following the procedure explained in the previous section. Then, one enters the table for the equation of access and recess with the mean motion as argument. Hermann gives the correct result, 8;23,30° (MS O reads 8;23,38°), without specifying the details of the interpolation. This value is added to the mean motion of the apogees and the result is given as 0,18;20,48,19° (MS W reads 0,18;20,48,**13**°). By adding it to the radix for the apogee of the Sun and Venus at Incarnation (not specified in the text, but 1,11;25,23° in all tables consulted), the resulting longitude of the apogee at the beginning of 1356 is given as 1,29;46,11° (MS W reads 1,29;46,5° and MS O and K read 1,29;46,9,17°, perhaps having been copied from an erroneous source).

The procedure outlined here corresponds very precisely to the two-component Alfonsine model of precession/trepidation, with one linear term for uniform precession (mean motion of the apogees and the fixed stars) and a periodic term for trepidation (equated motion of access and recess).[8]

Then follows a list of the apogees of the other planets:

Saturn	4,11;44,30,21°
Jupiter	2,51;57,49,21°
Mars	2,13;33, 1,31°
Mercury	3,29; 0,21,21°

In MSS O and K these four values differ systematically in the seconds, but they do not display the same readings in all cases.

12. Determining the true solar position by means of tables (E, 40^{v}–41^{r}; O, 244^{r}–246^{r}; P, 12^{r}–13^{r}; Q, 129^{r-v}; K, 34^{r-v}; W, 316^{v}–317^{r})

As expected, the issue addressed in the subsequent section is the determination of the true solar position by means of tables, called *astrologicas* in MSS E, P, Q, and K. The worked

8 For further details, see José Chabás and Bernard R. Goldstein, *A Survey of European Astronomical Tables in the Middle Ages* (Leiden-London: Brill, 2012), pp. 48–52.

example given here is also for the beginning of the year 1356, and it parallels Chapter 17 of John of Saxony's canons.

The solar anomaly at that time results from two previously computed values, the longitude of the solar apogee, 1,29;46,9,17° (see Section 11), and the mean motion of the Sun, 4,47;31,59,21° (see Section 10). Subtraction yields 3,17;45,48°. We are then told to enter the table for the solar equation with this particular value of solar anomaly. It is worth giving a detailed explanation of the interpolation procedure in place here. One finds a 'first equation' (0;39,26°), corresponding to argument 3,17°, and a 'second equation' (0;41,35°), corresponding to 3,18°. The difference between the two equations is 0;2,9°, which is equivalent to 129 seconds. We are then told that, since the number of seconds in the solar anomaly found above (3,17;45,48°) exceeds 30, we are supposed to round off the minutes to 46. Multiplying this number by 129 yields 5934 thirds, which is to be divided by 60, the result being 98 seconds and a remainder (*residuus*) of 54. The result is then transformed into minutes, 1;38,54°, which is added to the first equation, 0;39,26°, so that 0;41,4,54° is the value sought for the solar equation. We note the addition of two values with different degrees of precision, a usual practice in most of the worked examples and the approximation in the procedure for interpolation. The text indeed indicates that the resulting true position of the Sun is 4,48;13,3,54°; however, when adding the previously obtained mean solar position (4,47;31,59,21,6°) to the derived solar equation (0;41,4,54°) one obtains a different value (4,48;13,4,15°). This is due to the fact that a second part of the interpolation was made silently. Complete interpolation, using the value 3,17;45,48° for solar anomaly instead of 3,17;46° as indicated in the text, yields a solar equation of 0;41,4,28° and thus a value of the true position of the Sun, 4,48;13,3,49°, which is much closer to the actual result in the text. In short, interpolation was carried out in two phases, the second of which was not explicit. Modern computation for Paris yields 4,48;13,4°.

In modern notation, the procedure followed for determining the true solar position, λ_s, can be written as follows:

$$\lambda_s = \bar{\lambda}_s \pm c(\bar{\kappa}),$$

where $\bar{\lambda}_s$ is the mean solar position and $c(\bar{\kappa})$ is the solar equation, a function of the mean solar anomaly, $\bar{\kappa}$.[9]

In MS O, 246^r–247^r, a diagram and an explanation follow, showing the position of the Sun on the ninth sphere; these are not extant in the other manuscripts.

13. Determining the true lunar position by means of tables (E, 41^{r-v}; O, 247^r–249^v; P, 13^r–14^r; Q, 129^v–130^r; K, 34^v; W, 317^{r-v})

This section not only has the same title as Chapter 18 of John of Saxony's canons, *Verum locum lune per tabulas invenire*; entire sentences of it are copied verbatim. The worked example displayed by Hermann, also for January 1, 1356, follows the algorithm explained in *Tempus est mensura motus*. The true longitude of the Moon, λ_m, is obtained by adding

9 For further details, see Chabás and Goldstein, *A Survey*, pp. 63–7.

or subtracting the lunar equation, c, to its mean longitude, $\bar{\lambda}_m$. Using modern notation, this can be written as follows:

$$\lambda_m = \bar{\lambda}_m \pm c.$$

The lunar equation depends on both the true anomaly of the Moon, α, and the double mean elongation between the Moon and the Sun, $2\tilde{\eta}$, where

$$\tilde{\eta} = |\bar{\lambda}_m - \bar{\lambda}_s|,$$

and it can be expressed by

$$c = c_6(\alpha) + c_5(\alpha) \cdot c_4(2\tilde{\eta}),$$

with

$$\alpha = \bar{\alpha} + c_3(2\tilde{\eta}).$$

The mean anomaly of the Moon is $\bar{\alpha}$, the mean longitude of the Sun is $\bar{\lambda}_s$, and c_3, c_4, c_5, and c_6 are the equation of centre, the minutes of proportion, the increment (*diversitas diametri*), and the equation of anomaly, respectively, found in columns 3, 4, 5, and 6 in the Alfonsine table for the lunar equations.[10]

We are first given the three mean motions involved:

$$\begin{aligned} \bar{\lambda}_m &= 4{,}11;22,\ 0{,}10^\circ \\ \bar{\lambda}_s &= 4{,}47;31{,}59{,}21^\circ \\ \bar{\alpha} &= 4{,}31;11{,}55{,}35^\circ \end{aligned}$$

MS K ends abruptly here.

Double elongation is then computed and serves as the argument in the table for the following lunar equations:

$$\begin{aligned} 2\tilde{\eta} &= 4{,}47;40{,}1{,}46^\circ \\ c_3(2\tilde{\eta}) &= 10;8{,}20^\circ \\ c_4(2\tilde{\eta}) &= 18' \end{aligned}$$

The true lunar anomaly is then derived and used as the argument in the table for the following lunar equations:

$$\begin{aligned} \alpha &= 4{,}21;3{,}35^\circ \\ c_5(\alpha) &= 2;39^\circ \\ c_6(\alpha) &= 4;55{,}5^\circ \end{aligned}$$

Not all manuscripts agree on the values above, due to scribal errors or faulty computation at the source. For example, for the equation of anomaly, $c_6(\alpha)$, MSS E and W read 4;56,5°, MS P reads 4;56,35°, and MS O reads 4;55,5,36°, whereas MS Q gives the correct result

10 For further details, see Chabás and Goldstein, *A Survey*, pp. 67–73.

shown above. It is clear that some copyists or students were not aware that the equation of anomaly in Alfonsine astronomy does not exceed 4;56,0°. The total lunar equation and thus the true longitude of the Moon can then be easily derived:

$$c = 5;42,47^\circ$$
$$\lambda_m = 4,17;4,47^\circ$$

As was the case for the Sun, MS O includes a diagram and a short text describing the situation and showing the position of the Moon on the ninth sphere (249^v), which are not extant in the other manuscripts.[11]

14. Determining the true position of the lunar node (E, 41^v; O, 250^{r-v}; P, 14^r; Q, 130^r; W, 317^v)

This section follows Chapter 19 of *Tempus est mensura motus* very closely. For the first day of 1356, the mean position of the ascending node is given as 19;40,57,28° (MS O: 19;40,55,57°), and its 'true position', that is its complement in 360°, is given as 5,40;19,2,32° (MS O: 5,40;19,4,3°).

MS O also gives the value of the argument of lunar latitude for that date, 4,36;45,43,43°, which is obtained by subtracting the true position of the node from the true position of the Moon. A new chapter follows determining lunar latitude by means of tables (O, 250^v–252^r), which parallels Chapter A4 associated with John of Saxony's canons (Ratdolt 1483, b7^v), beginning *Latitudinem lune per tabulas invenire*. MS O includes a diagram and a worked example for January 1, 1356. After interpolation between the corresponding entries in the table for lunar latitude with the argument of lunar latitude previously found, the resulting latitude is given as 4;57,53,24°, which is an unusually precise value, since the table is only given to seconds. It is called *latitudo equata*, where *equata* means 'interpolated', rather than 'corrected', which is the usual meaning of the term.

15. Determining the true position of any planet (E, 41^v; O, 252^r–253^v; P, 14^r–15^r; Q, 130^{r-v}; W, 317^v–318^v)

In contrast to MSS E, P, and Q, in MSS O and W there are two separate sections devoted to the computation of planetary positions: one for the superior planets and another for the inferior planets. They correspond to Chapters 20 and 21 of John of Saxony's canons, respectively. In the text found in MSS E, P, and Q, only the superior planets are addressed and only one example is given (Saturn), and this is based on the instructions explained in Chapter 20 of *Tempus est mensura motus*. Using modern notation, the procedure involved can be presented as follows.

The true longitude of a superior planet, λ, is obtained by adding or subtracting a total correction, c, to or from its mean longitude, $\bar{\lambda}$:

11 For a transcription of Chapter 13 and the additional short text in MS O, see Nicholas Jacobson's contribution in this volume.

$$\lambda = \bar{\lambda} \pm c(\bar{\kappa}, \bar{\alpha}).$$

This correction depends on both the mean centre, $\bar{\kappa}$, and the mean anomaly, $\bar{\alpha}$, of the planet. The mean centre of the planet is defined as

$$\bar{\kappa} = \bar{\lambda} - \lambda_A,$$

where λ_A is its apogee, and its mean anomaly is defined as

$$\bar{\alpha} = \bar{\lambda}_s - \bar{\lambda},$$

where $\bar{\lambda}_s$ is the mean solar longitude. Then, the total correction, $c(\bar{\kappa}, \bar{\alpha})$, results from adding the corresponding equation of centre, q_1, and the equation of anomaly, q_2, both derived from the tables:

$$c(\bar{\kappa}, \bar{\alpha}) = q_1 + q_2.$$

The equation of centre, q_1, is simply $c_3(\bar{\kappa})$, which can be positive or negative, and it is used to determine the true anomaly of the planet:

$$\alpha = \bar{\alpha} \pm c_3(\bar{\kappa}).$$

The term, q_2, involving the equation of anomaly, is computed differently depending on the sign of $c_4(\bar{\kappa})$:

$$q_2 = c_6(\alpha) + c_5(\alpha) \cdot c_4(\bar{\kappa})/60 \text{ if } c_4 \leq 0$$

or

$$q_2 = c_6(\alpha) + c_7(\alpha) \cdot c_4(\bar{\kappa})/60 \text{ if } c_4 \geq 0.$$

The quantities c_4, c_5, c_6, and c_7 correspond to the minutes of proportion, the subtractive difference, the equation of anomaly, and the additive difference, respectively, and are labelled according to the numbering of the columns in the table of the Alfonsine planetary equations.[12]

In Hermann of Saxony's text, the example concerns the true position of Saturn at the beginning of 1356. First of all, one needs to know the mean positions of Saturn and the Sun (see Section 10), as well as the longitude of the apogee of Saturn (see Section 11):

$$\bar{\lambda} = 1{,}32;\,3{,}23^\circ$$
$$\bar{\lambda}_A = 4{,}11;44{,}30^\circ$$
$$\bar{\lambda}_s = 4{,}47;31{,}59^\circ$$

We note that in MS O the example is recalculated with one additional sexagesimal place, but an error is already introduced in the seconds of the apogee (4,11;44,28,21°), which affects the rest of the computation. The mean centre and the mean anomaly of Saturn can then be computed:

---

12 For further details, see Chabás and Goldstein, *A Survey*, pp. 73–81.

$$\bar{\kappa} = 3,20;18,53^\circ$$
$$\bar{\alpha} = 3,15;28,36^\circ$$

The mean centre serves as the argument in the table for the equations of Saturn to find its equation of centre and the minutes of proportion:

$$c_3(\bar{\kappa}) = +2;22,54^\circ$$
$$c_4(\bar{\kappa}) = +56' \ (\textit{propior})$$

The true centre and the true anomaly of Saturn are then derived:

$$\kappa = 3,22;40^\circ$$
$$\alpha = 3,13;\ 4^\circ$$

In passing, we note that MS O displays an uncommon precision to thirds: 3,13;5,41,19°. The values of the equation of anomaly and the additive difference are found in the table for the equations of Saturn with true anomaly as argument:

$$c_6(\alpha) = 1;34,28^\circ$$
$$c_7(\alpha) = 0;7^\circ$$

The two terms of the total correction and the true longitude of Saturn follow:

$$q_1 = +2;22,54^\circ$$
$$q_2 = 1;41,\ 0^\circ$$
$$\lambda = 1,32;52,55^\circ$$

The resulting true longitude of Saturn in MSS E, P, Q, and W (1,32;52,55°) and in MS O (1,30;44,4°) differ by about two degrees. In both cases, all intermediate calculations in the procedure were performed correctly. However, mistakes are found in all manuscripts at the very end of the computation. In MSS E, P, and Q, the difference $q_1 - q_2$ between the equation of centre (+2;22,54°) and the term associated with anomaly (1;41,0°) is given as 0;49,32°, instead of 0;41,54°. In MS O, the difference $q_1 - q_2$ between the equation of centre (+2;22,54°) and the term associated with anomaly (1;41,14°) is said to be 1;19,20° (= 1;41,14°–0;22,54°); that is, the computer or the scribe misread the number of degrees of the equation of centre, 2°, and replaced it by 0°, resulting in too low a value for the true longitude of Saturn. For the sake of comparison, the true longitude of Saturn according to the Alfonsine Tables, computed with modern tools for the same date for Paris, is 1,32;39,50°. It is not surprising to confirm that the length of the computation increases the possibility of arithmetical errors.

MSS P, Q, and W also give values for the length of the tropical year, 365d 5;49,15,43,42h. Then follows a short text for converting the lunar longitude into time using the table for mean motion. The starting longitude is 4,17;56,10° and the result is 19;30,44d, which is said to be 19d 12;16,48h (correctly, 19d 12;17,36h). In MS O, a diagram follows showing the position of Saturn together with an explanation of the situation and a text on retrogradation beginning *Utrum sit stationarius, directus vel retrogradus per istas tabulas invenire*. This text is reminiscent of Chapter 18 of John of Lignères' *Priores astrologi*, beginning *Utrum planeta sit stationarius, directus aut retrogradus per tabulas invenire*. The same text is also found in MS Q (122^{r}) before Hermann of Saxony's text. In MS O, 258^{r}–259^{v}, there is a section on the true positions of Venus and Mercury, which closely follows Chapter 21 of John of Saxony's canons to the Parisian Alfonsine Tables, and, on folios 260^{v}–261^{r}, another text on conjunctions of the planets and the stars taken from Chapter 27 of these canons.

16. Determining the entry of the Sun into the signs (E, 42^{r-v}; O, 261^{r}–262^{r}; P, 16^{r}–18^{r}; Q, 131^{r}–132^{v}; W, 318^{v}–319^{v})

A worked example is given for the incomplete year of 1356. The computation parallels the procedure in Chapter 23 of John of Saxony's canons to the Parisian Alfonsine Tables. The starting point is the mean motion of the Sun at the beginning of the year 1356, 4,47;31,59,21° (see Section 10). For a date when the Sun is thought to be close to Ari 0°, such as March 11, one computes the mean motion of the Sun, 5,57;30,50,29° (MS W has 40 instead of 30), and its mean anomaly, 4,27;50,39° (MS O: 4,47;44,41,17°). The corresponding solar equation is found in the appropriate table, 2;10,0° (MS O: 2;10,1,0°, which is a value exceeding the Alfonsine maximum); hence, the true solar motion is 5,59;40,50,29°. The distance between the Sun and Ari 0° is said to be 0;19,9,31° (MS O: 0;19,9,36° and MS W: 0;19,9,37°).

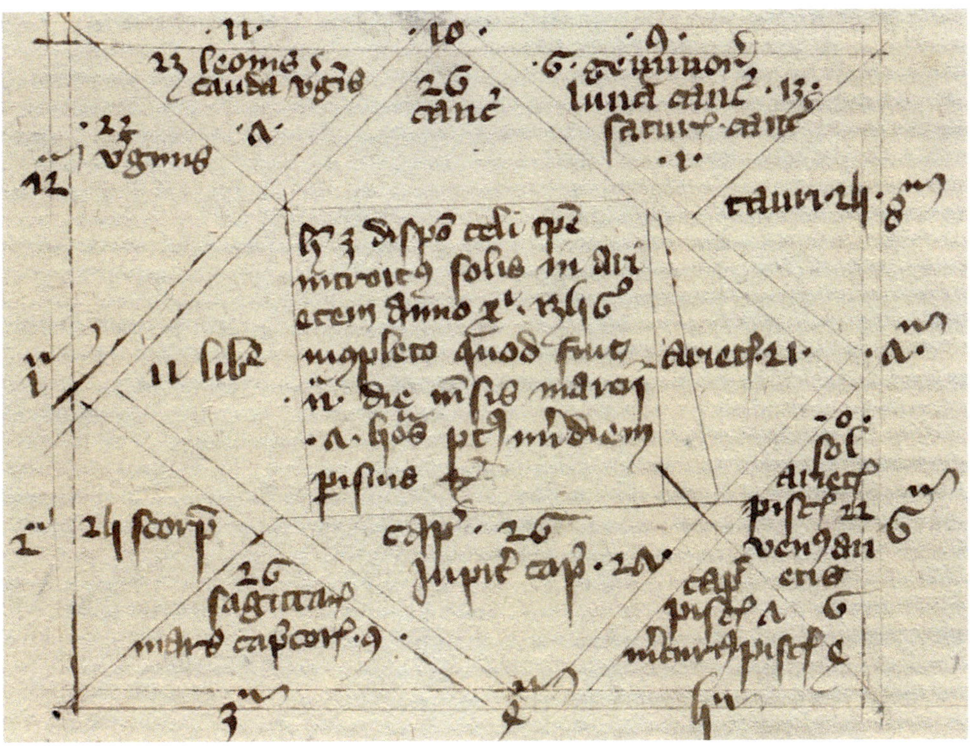

Figure 12.3. Astrological chart for the entry of the Sun in Aries in 1356, MS E, 42^{v}.

We are then asked to find the corresponding solar velocity in a minute of a day in the table for that purpose, 0;59,8°/d, and to divide the distance between the Sun and Ari 0° by it, yielding 19;26,22 minutes of a day. Hermann of Saxony's procedure differs here from that of John of Saxony, where no table for solar velocity is required, and the reader is asked to compute the true longitude of the Sun one minute of a day after the value previously derived and compute the difference between them.

According to Hermann, the time needed for the Sun to reach Ari 0° is then added to the date initially considered. The result is 11;19,26,22 days of March, and the true position of the Sun is computed anew for that date, yielding 0;0,0,5°. Note the strictly sexagesimal way to express a day and a time for computational purposes, avoiding the simultaneous use of two different units. This time is said to be equivalent to 11d 7;46,32,48h. The equation of time corresponding to Psc 30° (2;2° or 0;8,8h) is then applied, and the final result for the entry of the Sun in Aries is March 11, at 7;54,40,48h. Not all manuscripts display the same number of seconds and thirds.

Conversion of this time in degrees gives 118;45°, which results in 208;45° (= 118;45° + 90°) for the equatorial position of the ascendant. In the table of oblique ascensions for the seventh climate, valid for the latitude of Paris ($\varphi = 48°$), the argument corresponding to 208;45° is Lib 21°. This is the (ecliptic) longitude of the ascendant that was sought.

The computation of the other houses for casting the astrological chart at the time of the entry of the Sun in Aries at the latitude of Paris follows. The chart is only found in MS E, 42^{v} (see Figure 12.5, where the longitude of the ascendant is given as Lib 11° rather than Lib 21°).

17. Finding the revolution of the years of the world and the ascendant (E, 42^{v}; O, 262^{r}–263^{r}; P, 18^{r-v}; Q, 132^{v}–133^{r}; W, 319^{v}–320^{r})

This section parallels Chapters 10 and 11 of John of Saxony's canons that concern the time and the ascendant, respectively, for the entry of the Sun in Aries after a certain number of years. The numerical example given here is based on the data derived in the previous section for the beginning of 1356 and considers a date five years later, computing the time and the ascendant at the beginning of 1361.

In MSS E, P, and Q, the entry of the Sun in Aries is given as 7;54,42,48h, a time differing slightly from that found above. We are then told to add the values for five years in the corresponding columns of the table for the excess of revolution to the previously found quantities (time and longitude of the ascendant). This table provides multiples of the difference between a tropical year and a year of 365 days, both in time and in arc. The values mentioned in the text for five years, 5;6,19h and 76;35,30°, do not fully agree with those in the standard version of the Parisian Alfonsine Tables, 5;6,18,38h and 76;34,39,37°, but the results obtained, 11d 13;1,1,48h and 285;20,30°, are correctly computed. The equatorial position of the ascendant has then to be converted into an ecliptic position by means of a table for oblique ascensions valid for the meridian of Paris (7th climate). The result is Sag 18°.

The example in MS O differs from that in MSS E, P, Q, and W, for it involves a date that is four years rather than five years later.

18. Determining the time of mean conjunctions and oppositions by means of tables (E, 42^{v}–43^{r}; O, 263^{r}–264^{v}; P, 18^{v}–19^{v}; Q, 133^{r-v}; W, 320^{r-v})

The procedure described here follows that in Chapter 13 of *Tempus est mensura motus*. First, a date close to the conjunction or opposition under consideration is chosen and the mean elongation between the Sun and the Moon is computed. In this case, since the purpose is to determine the first conjunction in the year 1356, the selected date is the beginning of that year. The corresponding mean elongation is given as 5,23;50,0,47°, and thus the distance between the two luminaries at that time is its complement in 360°, 36;9,59,13° (MS O: 35;9,59,12°). From the table for the mean motion in elongation, one derives the time needed to reduce this distance to zero by a series of computations equivalent to the division of two sexagesimal numbers to thirds. The result given in the text is 2;58,0,11d (MS O: 2;58,11d). Hence, the first mean conjunction of the year 1356 corresponds to January 2 at 23;12,4h.

The text also specifies the length of the Alfonsine mean synodic month, 29;31,50,7,36d; half of it, 14;45,55,3,48d; and a quarter of it, 7;22,57,31,45d, so that the times of previous or later mean syzygies can be determined. We note that John of Saxony's Chapter 13 has the same values associated with the length of the mean synodic month, except for the quarter of it, which John's canons correctly give as 7;22,57,31,54d. It is also noteworthy that the value of the length of the mean synodic given by both John and Hermann (29;31,50,7,36d = 29d 12;44,3,2,24h) is slightly lower than the standard Alfonsine parameter (29;31,50,7,37,27d = 29d 12;44,3,2,48h, or more precisely 29;31,50,7,37,27,8,25d = 29d 12;44,3,2,59h) used in Alfonsine astronomy.[13]

19. Determining a true conjunction of the Sun and the Moon by means of tables (E, 43^{r-v}; O, 264^{v}–266^{v}; P, 20^{r}–21^{v}; Q, 133^{v}–135^{r}; W, 320^{v}–321^{r})

The example given here starts at the first mean conjunction in 1356, occurring on January 2;58,0,11d at the meridian of Paris, as found in the previous section. With no further explanation, we are given the following quantities at mean conjunction:

Mean solar and lunar longitudes	4,50;27,26°
Mean lunar anomaly	5, 9;57,32°
Mean solar anomaly	3,20;41,15°
Solar equation	0;47,20°
True solar longitude	4,51;14,46°
Equation of lunar anomaly	3;34,30°
True lunar longitude	4,54; 1,56°

We note that the longitude of the solar apogee resulting from the mean solar longitude and anomaly is 1,29;43,11°, which is the value already found in Section 11. We also note

13 Chabás, *Computational Astronomy*, pp. 154 and 315.

that the solar equation is obtained *prescise operando* (by computing exactly) according to the text, that is, by interpolation, but it should be 0;47,21° if accurately interpolated. The rest of the values are correctly computed.

The next step is to compute the time from mean to true conjunction, which is historically one of the most challenging problems in ancient and medieval astronomy. For this purpose, Hermann follows the procedure set in *Tempus est mensura motus* and first computes the difference in longitude between the true positions of the Moon and Sun. This quantity, true elongation η, is called *longitudo* and in this particular case, *longitudo lune*, for the true longitude of the Moon exceeds that of the Sun. Hermann correctly finds η = 2;47,10°. Then he determines the velocities of both luminaries in this interval by entering the tables for velocities with the solar and lunar anomalies. Before that, in the case of the Moon we are told to correct its mean anomaly. This correction is found by adding to the elongation a twelfth of it and halving the result, that is, $13\eta/24$, which he finds equal to 1;30,33°. Hence, the equated lunar anomaly should be 5,8;26,59° (= 5,9;57,32°–1;30,33°), but all manuscripts read 5,8;27,7°. This particular procedure is described in Chapter 22 of John of Saxony's canons, where no justification is offered. The following explanation has been suggested.[14] In the interval between the true longitudes of the two luminaries, the solar velocity can be considered constant, but not the lunar velocity. To take into account that the lunar velocity is about 13 times greater than the solar velocity, a coefficient 13/24 (= 0;32,30) is introduced as an approximation of the Ptolemaic mean velocity in anomaly (0;30,40°/h).

One then finds the lunar and solar velocities in the corresponding tables. Hermann explicitly states that the entries in the velocity table he used are displayed in minutes of a day. We note that in his canons, John of Saxony mentions the possibility of using velocity tables in degrees per hour or in degrees per minutes of a day, and he adds that the best such table he had ever seen is that by John of Lignères, given in hours.[15] Division of the elongation by the difference of the hourly velocities of the Moon and the Sun provides a first approximation to the time between mean and true conjunction. Before proceeding, a new correction is introduced; one must add to or subtract from the difference of the hourly velocities a number of seconds equal to the number of degrees of elongation minus one, that is, ± 0;0,1 ($|\text{Int}(\eta)-1|$). This correction is used in the Alfonsine procedure described in John of Saxony's canons, again with no justification.[16] A few years earlier, John of Lignères, master of John of Saxony, used a simpler algorithm for the computation of the time between mean and true syzygy when addressing this issue in Chapter 32 of his *Priores astrologi*.[17] John of Lignères followed the approximation procedure described in Ptolemy's *Almagest*, where the lunar velocity is taken to be 13 times the solar velocity.[18]

14 José Chabás and Bernard R. Goldstein, 'Nicholaus de Heybech and His Table for Finding True Syzygy', *Historia Mathematica*, 19 (1992), pp. 265–89; see espec. 269–71.

15 José Chabás and Marie-Madeleine Saby, *The Tables of 1322 by John of Lignères* (Brepols: Turnhout, 2022), Table 19.

16 See Chabás and Goldstein, 'Nicholaus de Heybech', pp. 269–71.

17 Saby, *Les canons de Jean de Lignères*, pp. 222–25.

18 Gerald J. Toomer, *Ptolemy's Almagest* (New York-Berlin: Springer, 1984), p. 281; Chabás and Goldstein, 'Nicholaus de Heybech', p. 268.

Back to Hermann's text, we are told that the result of the division is 14;52 minutes of a day, a first approximation to the time between mean and true conjunction. The process is iterated for this new time and all quantities involved are recomputed. Hermann does not provide the details of the computation, but in a small side table we are told that the resulting true longitudes of the Sun and the Moon are 4;50;59,40° and 4;51;2,36°, respectively. Since in this case the elongation η is 0;2,56°, the time between mean and true conjunction still needs to be corrected. According to the text, this correction amounts to 14 seconds of a day, and thus the time from mean to most true (*verissima*) conjunction is 14;38 minutes of a day (= 14;52–0;14). The final time for the true conjunction is said to be January 2 at 14;18,32h after applying the equation of time. Hermann adds that the ascendant at that time is found to be Sco 12°, its oblique ascension is 237;23°, and the *partes horarum* (read: seasonal hours) are 11;59°, given in degrees. Except for a variation in the minutes of the oblique ascension, these are indeed the values that apply in the seventh climate for the latitude of Paris. MS O just provides the time of true conjunction, 2;35,22d, that is, January 2 at 14;8,48h, without including the equation of time.

The computational method above, described by John of Saxony and applied by Hermann of Saxony, has at least two specific and crucial characteristics: the correction of mean lunar anomaly and the correction of the difference between the hourly velocities. This practice was followed by Alfonsine astronomers and is to be considered as another characteristic feature of Alfonsine astronomy, as are the tables and the canons accompanying them.

20. *Determining the time at quadrature of the Sun and the Moon by means of tables (E, 43v–44r; O, 266v–269v; P, 21v–23v; Q, 135r–136v; W, 321r–322v)*

This example refers to the first quadrature after the first conjunction of 1356 and parallels Chapter 24 of *Tempus est mensura motus*. We note that this section on quadrature is located after the sections on mean and true syzygies; this is in contrast to the printed version of *Tempus est mensura motus* (Ratdolt, 1483) and a few manuscript copies of it in which the subject of the entry of the Sun into the zodiacal signs is quite illogically inserted between the chapters on syzygies and quadratures.

First, the time of mean quadrature is found by adding a quarter of the length of the mean synodic month, 7;22,57,32d, to the time of the first mean conjunction, 2;58,0,10d, of January. For slightly different values, see Section 18, above. The resulting time is 10;20,57,42d. We are then asked to compute the mean motions of the two luminaries at that time; for the Moon, one must simply add *unum signum 30 gradus Alfonsi sive 90 gradus* to that of the Sun. This means that Hermann of Saxony considered an 'Alfonsine sign' to be 60°. The true motions are then computed from the mean motions. The results are given in a small table, but detailed computations are not displayed.

Mean solar longitude	4,57;44, 2°
Mean solar anomaly	3,27;57,51°
Mean lunar longitude	27;44, 2°
Mean lunar anomaly	46;24,47°
True solar longitude	4,58;46,52° (MS O: 4,58;48,52°)
True lunar longitude	22;39,18°

We are then asked to enter the table for normed right ascensions and find the values corresponding to the true positions of the Sun and Moon at the time of mean quadrature, α_s and α_m, by interpolating in the intervals Cap 28°–29° and Ari 22°–23°, respectively. The results are 30;56,21° (correctly, 30;56,12°) and 110;55,24° (correctly, 110;55,41°) for the Sun and the Moon, respectively. Again, we note the computer's difficulty in handling interpolation. The difference between the right ascensions is $\alpha_m - \alpha_s$ = 79;59,3°. Since this value is less than 90°, true quadrature is still to come. We are later told to compute 90°–79;59,3° and add α_m, yielding 120;56,21°, and then to find the longitude corresponding to the derived right ascension in the table. The result is Tau 3;10,20°. Subtracting the true longitude of the Moon from it, one obtains the 'longitude of the Moon at the future quadrature', 10;31,2°.

By contrast, in MS O no interpolation is used in the example, and the right ascensions found, 30;7° and 110;19°, correspond exactly to the solar and lunar longitudes, Cap 28° and Ari 22°, respectively. The partial result is given as Tau 3° and the lunar longitude at the future quadrature simply as 10°.

The next step is to determine the time from mean to true quadrature. To achieve that goal, the solar and lunar velocities in a minute of a day have to be computed. Their difference, 11;10° (= 12;11°–1;1°), is called *superatio* here, although this term usually applies to a difference of velocities in an hour, not in a minute of a day. We are not told, however, whether the velocities are found in a table for that purpose, as in previous sections, or computed following John of Saxony's canons, that is, as the differences in the longitudes of the luminaries at times separated by one minute of a day.

The time from mean to true quadrature is finally derived by dividing the longitude of the Moon at the future quadrature by the difference between the velocities of the two luminaries; the result is 56;30 minutes of a day. All quantities involved are then computed for this new time:

Mean solar longitude	4,58;39,43°
Mean lunar longitude	40; 8,30°
Mean lunar anomaly	58;42,56°
Mean centre	3,22;57,34°
Mean solar anomaly	3,28;53,32°
True solar longitude	4,59;44,29°
True lunar anomaly	51;12°
True lunar longitude	34;46,18°

Since the distance between the two luminaries differs significantly from 90°, we are asked to iterate the procedure and find the right ascensions corresponding to the new true positions of the Sun and Moon, 31;56,12° and 122;28,28°, respectively. The difference between the two values is 90;32,16°. In the table for mean elongation between both luminaries, interpolation yields the time needed to overcome the extra bit, 0;32,16°, which is correctly computed as 2;38 minutes of a day. When subtracted from the time between mean and first true quadrature computed above, 56;30 minutes of a day, one finds the time interval, 53;52 minutes of a day. This is to be added to the time of mean quadrature to obtain the final time of true quadrature. The text does not display this time explicitly, but we are told that the equation of time has to be applied to it. The true positions of the Sun and the Moon at true quadrature are given as 4,59;12,59,34,20° and 23;53,48,58,48°, respectively. Note, again, the meaningless precision, to fourths.[19]

Clearly, the emphasis in this example is on the computing method rather than on the results. The method followed in this numerical example departs from that outlined in Chapter 24 of John of Saxony's canons, where no right ascensions are involved and the whole process is iterated to determine the instant of true quadrature and the true positions of the luminaries.

21. Determining planetary conjunctions by means of tables (E, 44^{r-v}; O, 269^{v}–270^{r}; P, 23^{v}–24^{v}; Q, 136^{v}–137^{r}; W, 322^{v})

The procedure described by Hermann requires the use of a table for the daily unequal motion of the planets and differs from that in Chapter 25 of John of Saxony's canons, where we are asked to compute the true positions of the planets one day after the initial date. The numerical example given here concerns two superior planets, Jupiter and Mars, and the starting date for the computation is the last day of March 1356. In a small table integrated in the text, we are given the following values, in addition to the mean longitude of the Sun, 0,17;13,37°, with no further explanation:

	Jupiter	**Mars**
Mean longitude	4,48;51,46°	4,17; 7, 4°
Mean centre	1,56;53,58°	2, 3;34, 3°
Mean anomaly	1,28;21,51°	2, 0; 6,33°
True centre	1,51;26,46°	1,53;31,27°
True anomaly	1,33;49, 3°	2,10; 9, 9°
True longitude	4,54;34,34°	4,50;24,17°

We are then told to compute the difference between the two true longitudes of Jupiter and Mars, 4;10,17° (= 4,54;34,34°–4,50;24,17°), and transform it into seconds, given as 15000^2 in all MSS examined (correctly computed, 15017^2). We are also asked to enter the table for the daily unequal motion of the planets with the true centre and the true anomaly of each planet. This table is found, for example, among the Toledan Tables and the Tables of 1322 by John of Lignères.[20] For each planet, the velocity in a day has two components displayed in two columns, one for each of the two variables on which planetary velocities depend: centre and anomaly. The given entries for Jupiter are 5;8° and 2;11°, and those for Mars are 33;30° and 0;44°. Addition yields 7;19° for Jupiter and 34;14° for Mars. The positive difference between them, 26;55°, or equivalently 1615^2, is called *superatio*, here taken as the relative daily velocity. Dividing the difference between the true longitudes of the planets

19 All manuscript give the reading 23;53,48,58,48 but a correct one would be 33;53,48,58,48. This noted was added by the volume editors.

20 Pedersen, Fritz S., *The Toledan Tables: A review of the manuscripts and the textual versions with an edition* (Copenhagen: C.A. Reitzels Forlag, 2002), pp. 1433–36. See also Chabás and Goldstein, *The Alfonsine Tables*, pp. 170–82; Chabás and Saby, *The Tables of 1322*, Table 10.

by their relative velocity, one finds the time of the conjunction of Jupiter and Mars after the initial moment considered. The given result is 9;17,9 days, already in April 1356.

We note that the entry for the velocity in anomaly of Jupiter should be 2;11° rather than 1;30° when correctly interpolated, and that the division of 15017 or 15000 by 1615 does not yield the given result. Again, both interpolation and conversion of a decimal number into a sexagesimal one were frequent sources of errors.

22. Determining the position of the fixed stars at any time (E, 44^v; O, 259^v–260^v; P, 24^v–25^r; Q, 137^r; W, 322^v–323^r)

In this section, Hermann follows the instructions given in Chapter 26 of John of Saxony's canons describing the two-component model for Alfonsine precession/trepidation. Here again, the instant chosen is the beginning of the year 1356. He indicates that the equation of access and recess corresponding to Alfonso's time is 8;4,1° and that it is 8;23,30° at the beginning of 1356. The value for Alfonso's time is found in the table for the radices of the motion of the eighth sphere. As for the second value, the text specifies that the difference in days between Alfonso's time and the beginning of the year 1356 is 10,30,34d and that in this time interval the mean motion of access and recess has increased in 0;45,39,39°. Although not explained in the text, this number and the radix at Alfonso's time serve as arguments to enter the table for the equation of access and recess in order to find the equation above (8;23,30°).

Finally, we are told to add the difference between the two equations of access and recess, 0;19,29°, to the increase in the mean motion, 0;45,39,39°. The result, 1;5,8,39°, is the progress in the true motion of access and recess since Alfonso's era. However, MSS E, P, Q, and W read 1;58,39°, indicating that this faulty value was produced at an early stage, perhaps by the author himself. On the other hand, MS O, where many worked examples are recomputed, as noted above, reads 1;5,8,42°.

23. Determining planetary latitudes by means of tables (E, 44^v; P, 25^{r–v}; Q, 137^{r–v}, where the text ends abruptly; W, 323^{r–v})

The example given by Hermann refers to the latitude of Jupiter and it is computed for the beginning of year 1356. The true anomaly of Jupiter at that time is given as 1,33;39,3°, despite the value found in Section 21, above, 1,33;49,3°, and its true longitude as 4,54;34,34°. From this last value and the true longitude of the ascending node, 1;22;1°, not displayed anywhere above, one derives the true argument of latitude of 3,32;33° (P reads 3,31;33°). We are then told to enter the tables called *tabula bipartialis* (using true anomaly as argument) and *tabula quadripartialis* (with true argument of latitude as argument) to find the first and second latitudes of Jupiter: 1;32° (erroneously for 1;17°) and 1;20°, respectively. The faulty entry derived from the *tabula bipertialis*, 1;32°, was probably taken for the same argument from the column for Mars, rather than from that of Jupiter, next to it. Using the wrong column in a table is of course another common source of error. The entries and the titles mentioned above correspond to the tables for latitude in the Toledan Tables in the Indian tradition.[21]

21 Pedersen, *The Toledan Tables*, pp. 1311–21; Chabás and Goldstein, *A Survey*, pp. 110–12.

By contrast, Chapter A5 on planetary latitudes of the superior planets associated with John of Saxony's canons to the Parisian Alfonsine Tables in the *editio princeps* and canon 21 in John of Lignères' *Priores atrologi* refer to a different table for latitudes, at intervals of 6° in the tradition of the *Almagest*.[22] The computer probably thought that tables at intervals of 1° of the argument, as is the case of the tables *bipartialis* and *quadripartialis*, were more user-friendly than those at steps of 6°. The total latitude of the planet is then found by dividing the second latitude by the first: 1;20° / 1;32° = 0;52,10° (W reads 0;50°).

MS O (256^r–258^r) provides a different text and a different example for the latitude of the planets. It concerns Saturn and the same date, the first day of the year 1356. A diagram of the situation is also given, and it is worth noting that, as was the case in the example of Jupiter, the tables used here are also the *bipartialis* and *quadripartialis*.

MSS P (26^{r-v}) and Q (122^r) also have a short text on planetary stations. The text begins *Utrum planeta sit retrogradus, stationarius vel directus sine tabulis invenire*. This incipit is reminiscent of the title of canon 18 in John of Lignères' *Priores astrologi*, where, in contrast, tables have to be used to address the same topic. In MS P, Hermann's text ends here and it is immediately followed by a different text beginning *Quia plures astrologorum*. This is precisely the text by John of Saxony consisting of numerical examples to the canons *Cuiuslibet arcus* by John of Lignères.[23] It should be noted that in MS Q, these short texts are found on folio 122^v, that is, at the very beginning of Hermann's text. MS P adds a small table for the maximum values of the planetary latitudes. The values of the extreme northern and southern latitudes are 3;2° and 3;5° (Saturn), 2;8° and 2;8° (Jupiter), 4;0° and 6;0° (Mars), 4;56° and 4;56° (Venus), and 4;28° (Mercury, both for north and south). The values for Saturn and Jupiter are in the tradition of al-Battānī; we have no explanation for the rest.

MS O adds texts concerning the ascendant (270^r–273^v) and the planetary aspects (273^v–275^v), with a diagram, both with worked examples for early 1356. On folio 275^v, there is an astrological chart for the date and time of the first true conjunction of the Sun and the Moon, a topic already addressed above, with full computation (see Section 19).

MS W ends here after the section on planetary latitudes, and a date is given "1474° fluente". Another text, attributed to John of Saxony, follows. It begins *Modum inveniendi partem proportionalem subiungere*. It is the first canon in John of Saxony's canons for the Alfonsine Tables.

As mentioned above, Hermann's text in MSS P and Q ends after the section on planetary latitudes, with no explicit. The last two sections of the text concern the computation of eclipses and are only found in MSS E and O. To make things worse, the beginning is missing in the latter.

As for the rest of the text, it should be stressed that eclipse computation is the most complex topic a medieval computer faced, especially that of a solar eclipse, where parallax is also involved. In Hermann's text, first comes the computation of a lunar eclipse, followed by that of a solar eclipse, in contrast to the ordering in other major texts dealing with eclipses, such as John of Lignères' *Priores* and the canons on eclipses associated with John of Saxony's *Tempus est mensura motus*.

22 Pedersen, *The Toledan Tables*, pp. 1325–26; Chabás and Goldstein, *A Survey*, pp. 106–10.

23 Saby, *Les canons de Jean de Lignères*, pp. 281–411.

24. Computing a lunar eclipse (E, 44v–46r; O, 277r–280v)

Hermann's text closely follows the procedure explained in canon A3 on lunar eclipses associated with John of Saxony's *Tempus est mensura motus* reproduced in the *editio princeps*.

We are first told that the conditions for possibility (Sun and Moon in the proper range) and visibility (the event to take place at night rather than in daytime) are met at the opposition of the Sun and the Moon of February 1356. Hermann takes advantage of the fact that the first conjunction of the year was already computed (see Section 18). To the date and time of this first mean conjunction, he adds the length of one and a half synodic months (29;31,50,8d and 14;45,55,4d) and finds February 16;15,45,22 as the date and time of mean opposition. With no further explanation, we are given the following quantities at mean opposition:

Mean solar longitude	5,34; 7, 2°
Mean lunar longitude	2,34; 7, 2°
Mean lunar anomaly	2,48;41, 3°
Mean solar anomaly	4, 4;20,51°
True solar longitude	5,36; 6,17°
True lunar longitude	2,33; 3,41°

The true Sun is 3;2,36° in advance with respect to the Moon, but for three signs (i.e., 180°). The procedure to compute the time from mean to true opposition is the same as in Section 19. One first adds to the elongation a twelfth of it and halves the result, to obtain 1;38,59° (correctly, 1;38,55°). Then, the equated lunar anomaly is found by adding this amount to the mean lunar argument; the result is given as 2,50;19,57°. Taking this value as the argument, the lunar velocity found in the corresponding table is said to be 14;43 minutes per minute of a day. Likewise, for the mean solar anomaly, the corresponding entry in the table for solar velocity is said to be 1;0 minutes per minute of a day. Among the very few known tables for the motion of the luminaries in a minute of the day, only one displays the values reported by Hermann: the table computed by John of Montfort and dated 1332.[24] The entries in other tables for the same purpose do not agree with Hermann's values.

The difference between the two velocities is given as 13;43 minutes per minute of a day, which has to be increased by 0;2 (that is, the number of seconds equal to the degrees of elongation, 3;2,36°, minus 1) according to the procedure described in Section 19. The result, 13;45°, is called *superatio equata*. Dividing the elongation by this corrected relative velocity, Hermann finds 13;16,4 minutes of a day (correctly, 13;16,48) for the time interval between mean and true opposition, which yields 16;29,1,26d for the time of the true opposition in February. Again, this is a new example of the difficulty that Hermann of Saxony, and medieval computers in general, encountered in dividing two sexagesimal numbers.

24 Chabás and Goldstein, *The Alfonsine Tables of Toledo*, p. 290; see also Chabás and Saby, *The Tables of 1322*, Table 20.

The process is iterated for this new date, and the following quantities are given:

Mean solar longitude	5,34;20, 6°
Mean lunar longitude	2,37; 1,50°
Mean solar anomaly	4, 4;33,54°
Mean lunar anomaly	2,51;34,23°
Lunar centre	5;23,28°
True solar longitude	5,36;19,34°
True lunar anomaly	2,52;22,27°
True lunar longitude	2,36;19,36°

The true Moon is said to be 0;0,46° in advance with respect to the point (*nadyr*) opposite the true Sun. However, according to the values reported above, the difference is only 0;0,2°. It is very unfortunate that of the two manuscripts containing the sections on eclipses, MS O lacks the first part of Section 24, so that we are left with MS E.

Nevertheless, there are grounds to believe that Hermann considered the difference to be 0;0,46°, since he indicates that he does not find that value among the entries in the table for mean elongation and that the closest is 0;0,48°. This entry corresponds to an argument of 4 seconds of a day, which he adds to the time of the true opposition previously found, resulting in February 16;29,5,26. After conversion into days and hours, this time corresponds to 16d 11;38,10,24h, which is the *verissima oppositio*, a term already met in Section 19. Thus, the date and time of the middle of the lunar eclipse is February 16 at 11;39,50h, after applying the equation of time. These data enable the identification of the table for the equation of time used by Hermann of Saxony. To be sure, the equation of time resulting from the difference of these two values, 0;1,40h, or its equivalent in time degrees, 0;25°, should correspond to a true solar anomaly of about 5,36° (Psc 6°), as previously computed. To our knowledge, the only table for the equation of time where this occurs is the one in the *editio princeps* of the Parisian Alfonsine Tables, already found among the Toledan Tables and in the zij of al-Battānī;[25] this means that in this case he did not use the corresponding table included in John of Lignères' Tables of 1322.

The second part of the computation of the eclipse consists of determining its duration and magnitude. For this purpose, one needs to compute the lunar argument of latitude at the time of true opposition. This is achieved by adding the mean motion of the lunar node (22;11,50°) to the previously determined true lunar longitude. The result, 2,58;31,26°, serves as argument to enter the tables for lunar eclipses at greatest and least distances on the epicycle (*longitudo longioris* and *longitudo propioris*). We note that Hermann converts this value to 178;31,26°, indicating that in the eclipse tables he used the argument was not displayed in signs and degrees, but was given in degrees in the vicinity of the node. The corresponding table in the Tables of 1322 by John of Lignères, taken from the Toledan Tables shares this characteristic.[26] Three entries are taken from the sub-table for greatest distance and three others from that for least distance: magnitude (*puncta*, that is, the number of twelfths of the diameter or the area

25 See e.g., Chabás, *Computational Astronomy*, p. 261.
26 Pedersen, *The Toledan Tables*, pp. 1467–70; Chabás and Saby, *The Tables of 1322*, Table 26.

of the eclipsed body), the arc of immersion (*minuta casus*, the distance between first and second contacts), and the arc of half-totality (*minuta more*, the distance between second contact and mid-eclipse). In all six cases, interpolation is needed. The results reported by Hermann are shown in Table 1.

Table 1. Computations for the lunar eclipse of February 1356

	Greatest distance	Least distance
Digits of eclipse	17;59,40 p.	19; 7,20 p.
Arc of immersion	29; 5,20' *	35; 4,15'
Arc of half-totality	20;13,55'	26;33,26'

* Correctly computed: 30;2,28'

New interpolations are needed in all three cases between the greatest and least distances. For the given lunar argument of latitude, we are asked to find the corresponding minutes of proportion in the table for proportions at steps of 2°. The result is said to be 59;37 (correctly computed, 59;14), and the three interpolated quantities are given as 19;7,29,29 p. (digits of eclipse), 35;2,3' (arc of immersion), and 26;30,30' (arc of half-totality). When recomputed, all these values fail to render the seconds correctly. Moreover, the reported digits of eclipse fall outside the interpolation interval.

The passage regarding the computation of the relative velocity of the luminaries is of special interest. We are first told to transform the solar and lunar anomalies from values expressed in signs of 60° (here called *signa Alfonsii*) to signs of 30° (*signa communia*), and to enter a table providing the motions in one hour. The purpose of this change is to use a table using zodiacal signs and degrees as argument. We also note that, when computing the time from mean to true opposition in this section, Hermann appealed to a table for the velocities of the luminaries in minutes of a day. In this case, he appeals to a different table: both the entries given in the text (35;59,41′ for a lunar argument of 5s 22;23° and 2;29′ for a solar argument of 8s 4;34°) and the use of signs of 30° are compatible with the velocity table by John of Lignères in his tables for 1322.[27]

Dividing the arcs of immersion and half-totality by the difference in the velocities of the Sun and the Moon, Hermann obtains 1;2,43h and 0;47,27h, respectively. The sum of these two times is the half duration of the eclipse. All dates for the beginning and the end of the various phases of the eclipse are then computed, before and after the time of mid-eclipse, which is February 16 at 11;39,50h. The duration of the eclipse is given as 3;40,20h and the duration of totality as 1;34,54h.

Finally, to draw the figure of the eclipse, Hermann derives the true longitude of the Moon at the beginning and the end of the eclipse by subtracting and adding to the true longitude at mid-eclipse the product of the semi-duration of the eclipse by the hourly velocity of the Moon. The results are 2,34;13,0° and 2,37;26,12°, respectively. Likewise, the argument of lunar anomaly is computed at the beginning and the end of the eclipse. We are then told to find the values of the latitude of the Moon corresponding to the beginning, the middle, and the end of the eclipse in a table converting the argument of lunar latitude into lunar latitude: 0;13,26°,

27 Chabás and Saby, *The Tables of 1322*, Table 19.

0;7,45, and 0;1,44°, respectively. As is frequently the case where interpolation is concerned, Hermann's results are off by seconds. It is likely that they were computed using the table for that purpose found among the Tables of 1322 by John of Lignères.[28] In addition to latitudes, one needs to know the semi-diameters of the Moon and the Earth's shadow. The values corresponding to a lunar anomaly of 172° are to be found in a table for that purpose: 18;2′ and 46;54′, respectively. It is plausible that these values resulted from interpolation between entries in John of Genoa's table for semi-diameters and velocities given in steps of 6°.[29]

This long section closes with an astrological chart for the time of the lunar eclipse. In MS E, 46^r, the houses are given in degrees and minutes, and in MS O, 280^v, they are only given in degrees.

25. Computing a solar eclipse (E, 46^v–47^v; O, 281^r–286^r)

In this case, the numerical example refers to the solar eclipse of May 1361 and parallels canon A2, on solar eclipses, associated with John of Saxony's canons. This example culminates the computational flow followed since the beginning of the text; many of the specific procedures required were already described in previous sections, and so Hermann refrains from describing them in detail.

We are first told that in the table for mean syzygies, the conjunction of May 1361 meets the requirement of a solar eclipse, because the argument of lunar latitude, 2,55;18,58°, falls within the range of eclipse possibility. The associated time for mean conjunction is May 4;59,8,33d. According to the procedure described in Section 18, Hermann computes all values for that time, without displaying any intermediate steps. The final values are listed in Table 2 in the column "Mean conjunction". Then, following the Alfonsine procedure explained in Section 19, Hermann finds a first approximation for the time between mean and true conjunction by determining the equated anomaly and the equated relative velocity. The result is 7;22 minutes of a day, which is to be subtracted from the time of mean conjunction to obtain the time of the 'first equated conjunction'. All quantities involved are recomputed again for this new time. The results are displayed in Table 2 in the column labelled 'first equated conjunction'.

Table 2. Computed distances for the solar eclipse of May 1361

	Mean conjunction (May 4;59,8,33d)	**First equated conj. (0;7,22d before)**
Mean solar longitude	0,51;30, 2,54°	0,51;22,47°
Mean lunar longitude	0,51;30, 2,54°	0,49;52,59°
Mean lunar anomaly	3,33;52,18,26°	3,31;15,55°
Apogee	1,29;49,11°	
Solar anomaly	5,21;40,52°	5,21;33,36°
True solar longitude	0,52;48, 0°	0,52;40,56°
True lunar longitude	0,54;27,18°	0,52;41, 4°
Elongation	1;39,18°	0; 0,28°

28 Chabás and Saby, *The Tables of 1322*, Table 29.

29 Goldstein 1992; Chabás and Goldstein, 'Computing solar and lunar velocities and radii: The table of John of Genoa and his predecessors', in the present volume.

The procedure is then iterated for the new values of the true positions of the luminaries and a new time correction is found. The resulting time is that of the *verissima coniunctio*, May 4;51,48,53d. After adding the equation of time, Hermann obtains 4d 21;4,56h, that is, at 2;55,4h before noon. This is said to be the 'time of conjunction at first distance before noon'.

The next step is to determine the time of the middle of the eclipse. In order to do so, one has to compute the lunar parallax, a challenging problem that Hermann addresses by referring to a table for that purpose. Parallax has two components and it is a function of three variables: the observer's position (specifically, the zenith distance of the Moon), the position of the Moon on the epicycle (that is, lunar anomaly), and the position of the centre of the epicycle on the deferent (that is, the lunar centre or double elongation). This implies that several consecutive interpolations between the entries in the table are needed. As can be deduced from the instructions in the text, the argument of the table is time, given in integer hours before and after noon, and the entries are displayed as two columns for the components of parallax, in longitude and latitude. The information provided indicates that a table for parallax of the type in the *Handy Tables* or al-Battānī's zij is used. It is worth noting that, as the argument is given in integer hours and not in minutes of a day, the computations after this point will have to be carried out in hours. We are told to select the table that corresponds to the observer (7th climate, corresponding to Paris),[30] to enter the sub-table for the sign where the true Moon is (Taurus), and to interpolate between 2h and 3h in order to find the parallax in longitude at the time of true conjunction 'at first distance', 2;55,4h before noon. An analogous interpolation has to be made for the following sign (Gemini) as well. These two values serve as the extremes to compute the component of parallax in longitude for the true lunar longitude, Tau 22;41,24°, by interpolating between Tau 0° and Gem 0°. This double interpolation leads to a first value of parallax in longitudes of 16;0°, which then has to be equated with a correction that depends on the lunar anomaly and is found in the table for correcting the parallax (*tabula attacium*).[31] The final longitudinal parallax value is 18;56min. When divided by 0;33,3°/h, that is the velocity in elongation (here called *superatio lune*), a time correction due to parallax in longitude is obtained (0;34,22h), resulting in a new time for the true conjunction (20;30,34h = (4d) 21;4,56h−0;34,22h), that is, 3;29,36h before noon. This time is called 'time of conjunction at second distance before noon'.

This new time is used to enter the parallax table once again and determine a new (third) parallax in longitude (21;33'). The iteration of the whole process yields a 'time of conjunction at third distance before noon', 3;34,4h, and a new (fourth) parallax in longitude is then derived (21;57'). A new iteration yields the 'time of conjunction at fourth distance before noon', 3;4,4h. The difference between the third and fourth parallaxes in longitude is used to determine a *superatio reducta vel equata*, here given as 0;28,15°/h and resulting in the time at the middle of the eclipse: 4d 20;25,5h. Table 3 shows the successive times that were computed in this long process.

30 Chabás and Saby, *The Tables of 1322*, Table 22.
31 Chabás and Saby, *The Tables of 1322*, Table 23.

Table 3. Computed times for the solar eclipse of May 1361

	May 1361 (days)	May 4 (hours before noon)
Mean conjunction	4;59, 8,33	
First equated conjunction	4;51,46,33	
True conjunction	4;51,48,53	
Conjunction at first distance before noon		2;55, 4
Conjunction at second distance before noon		3;29,36
Conjunction at third distance before noon		3;34, 4
Conjunction at fourth distance before noon		3;24, 4
Time at mid-eclipse	4d 20;25,5h	3;34,55

Once the time of mid-eclipse has been determined, one computes the solar and lunar longitudes, the lunar anomaly, and the argument of lunar latitude. The latter is converted into latitude by means of the specific table for latitude in the vicinity of an eclipse already used in Section 24 for lunar eclipses.[32] The latitudinal component of parallax, to be added to latitude, is deduced from the parallax table mentioned above; this is done following the same procedure for determining parallax in longitude. Again, the process includes multiple interpolations.

The last part of the computation is to determine the magnitude and the duration of the eclipse. For that purpose, we are told to refer to the tables for solar eclipses at the greatest and least distances. Incidentally, we note that the scribe of MS O erroneously has *denominatio* instead of *duratio*. Either he did not master the specific terminology of eclipses or did not pay too much attention to it. He was also not familiar with the quantities involved, for he transforms '5 puncta 2 minuta' into an impossible 'puncta 52 minuta' when dealing with the magnitude of the eclipse.

After multiple corrections and interpolations, as was the case for the lunar eclipse in Section 24, the resulting arc of immersion (*minuta casus*) is given as 28;20min. Dividing this number by the relative hourly velocity, one obtains 0;51,26h as the time between the middle of the eclipse and its beginning. The results given in the text are displayed in Table 4; we note that they do not always agree with recomputation.

	Beginning of eclipse	Middle of eclipse	End of eclipse
Time	4d 19;33,46h	4d 20;25,5h	4d 21;16,32h
True solar longitude	0,52;37,38°		0,52;41, 2°
True lunar longitude	0,51;44,14°		0,52;50,34°
Argument of lunar latitude	3, 3; 8°	3, 3; 47°	3, 4;14°
Latitude	16;20°	19;14°	22; 3°

Table 4. Final results for the solar eclipse of May 1361.

32 Chabás and Saby, *The Tables of 1322*, Table 29.

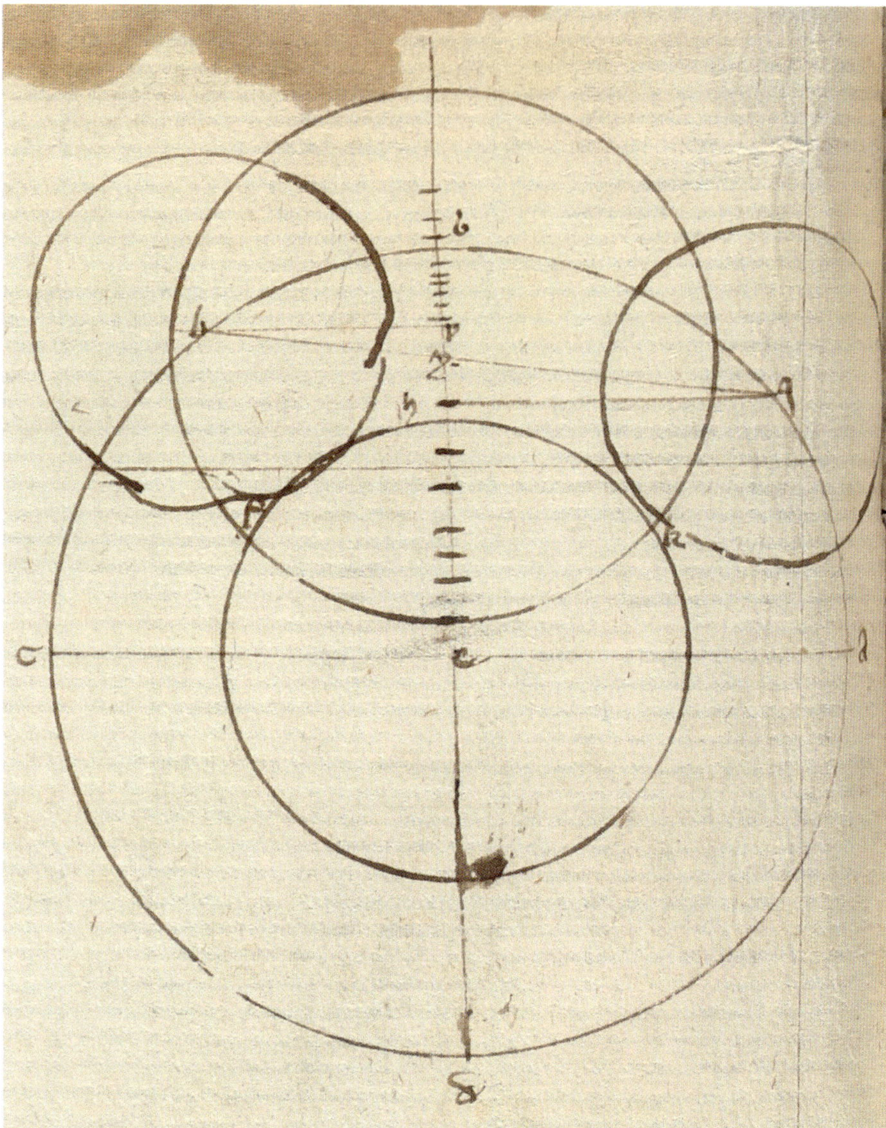

Figure 12.4. Solar eclipse of May 1361 in MS E, 47ᵛ.

Finally, the corresponding semidiameters of the Sun and the Moon are given as 15;47′ and 17;45′, respectively. All these values are then used to draw the diagram of the solar eclipse; see Figure 12.6 (MS E, 47ᵛ).

In conclusion, we call attention to some characteristic features of Hermann's text. The worked examples for 1356 and 1361 closely follow the structure and content of John of Saxony's canons beginning *Tempus est mensura motus*, indicating that, by that time, these canons were a standard text used by astronomers. However, Hermann's worked examples go further and are an expanded version of the *Tempus est mensura motus*, for they include several additional canons and omit only a few. Among the new canons are those usually known as *additio* to John of Saxony's canons, labelled here with a capital A; these are a natural continuation of John of Saxony's standard canons (A2 andA3 on eclipses) or fill a gap in them (A4–A7 on latitude). As seen above, Hermann sometimes departs from John of Saxony's algorithms and appeals directly to existing tables, such as that for the unequal velocities of the planets. In particular, he seems to have used several tables compiled by his Alfonsine predecessors in Paris, John of Montfort, John of Genoa, and John of Lignères.

The numerical examples also provide valuable information on the algorithms set up in early Alfonsine astronomy to determine specific quantities. These procedures were usually followed by others Alfonsine astronomers and have to be considered as characteristic elements defining Alfonsine astronomy in their own right.

Following the pattern set in John of Saxony's canons, the text by Herman of Saxony does not include diagrams to support the explanations and computations. These drawings are only found in the enlarged version in MS O, to be attributed to a later scholar.[33]

Hermann's text also tells us much about the computational effort, the achievements, and the difficulties Alfonsine astronomers encountered in their practice of astronomy: the use of base 60 in counting days, hours, and fractions of them; extensive use of tables where intermediate values could be found; and confusion between precision and accuracy. In the case of Hermann's calculations, sources of error are frequently found in the interpolation procedure, or the lack of it, as well as in the conversion of decimal numbers into sexagesimals and in the division of sexagesimal numbers.

Part II: Tables and worked examples for 1362

We now turn to a shorter work by Hermann of Saxony, consisting of a set of tables and computed examples for 1362. This work, mentioned by Thorndike, is examined here in detail for the first time. We have located three manuscripts containing both the tables and the text:

B: Bern, Burgerbibliothek, Cod. 454: tables, 66^{r}–88^{v}; text, 89^{r}–93^{v};[34]

L: Lübeck, Stadtbibliothek, Math. 2° 1: text, 122^{r}–125^{r}; tables, 125^{v}–132^{r} and 133^{v}–144^{r};

M: Metz, Bibliothèque municipale, 287: text, 49^{r}–51^{v}; tables, 52^{r}–68^{v}, used here as the base manuscript;[35]

33 On the diagrams in MS O, see Nicholas Jacobson's contribution to this volume: '*Exempla in figura* in Alfonsine Canon Commentaries'.

34 I am grateful to Dr Florian Mittenhuber (Bern Library) for his detailed description of the manuscript and for providing me a copy of it. See also Hermann Hagen 1875. *Catalogus codicum bernensium (Bibliotheca Bongarsiana)*, Bern: B. F. Haller, p. 396. The catalogue reports the following item on ff. 65a–93b: *Tabulae de motibus stellarum M. Hermanni Stili Saxonis de Northem*. This miscellaneous manuscript combines parchment (last third of the fourteenth century) and paper (last third of the fifteenth century). The text and tables by Hermann of Saxony are in the oldest part of the manuscript. Other Alfonsine items also extant in the manuscript are partial sets of tables by Nicholas de Heybech, John of Lignères, and William Batecombe.

35 I thank Jean-Patrice Boudet for information on this manuscript, which is accessible at https://bvmm.irht.cnrs.fr/mirador/index.php?manifest = https://bvmm.irht.cnrs.fr/iiif/22969/manifest.

R: Paris, Bibliothèque nationale de France, lat. 11250: tables, 2^{r}–42^{v}, 43^{v}–44^{r}; text, 46^{r}–57^{r}.[36]

Apparently, there was a fourth manuscript containing this work by Hermann of Saxony. It was also extant in the public library in Metz, shelf mark 286, according to the old catalogue.[37] Unfortunately, this seems to be one of the manuscripts destroyed during World War II. The notice in the catalogue of 1889 reads *Tabulae et canones tabularum magistri Hermanni de Saxonia. –Incipit: Anno Christi 1361 completo, ad meridiem 10 diei.* There is little doubt that this refers to the work discussed here.

The incipits to the tables indicate that Hermann of Saxony is the author.

MS M, 52^{r}: *Incipiunt tabule de motibus stellarum, puta tam planetarum quam etiam stellarum fixarum, quas compusit magister Hermannus Stilus Saxo de Nochem pro alleviacione operationum tabularum Alphoncii et aliorum astronomorum seu philosophorum. Et hoc ad meridanum Parisiensem anno Christi 1359. Et primo tabula differentiarum unius regni ad alliud et nomina regum atque ere differentie comprehenduntur in tabula subscripta.* MS B, 65^{r}, has the same incipit, but reads 'Northem' instead of 'Nochem'. MS R has no incipit.

We note that the incipit reproduces the recurrent idea *pro alleviacione operationum*, so frequently used by table-makers to introduce their works and said to facilitate the computational tasks of table users.

1. Tables

The last sentence of the long incipit refers to the table for the differences between eras, which is found in MSS B, 65^{r}; M, 52^{r}; and R, 43^{v}–44^{r}. There are 45 items listed, as is the case in the *editio princeps* of the Parisian Alfonsine Tables (Ratdolt 1483). They are numbered successively and display the names of two eras, the sexagesimal number of days between them, and the corresponding number of years, months, and days. In MS R, the table for the differences between eras is not the first table in the collection; it was added after the general explicit of the tables and a blank page. Moreover, the table lacks two columns, intended to be written in red ink, as is the case with many previous tables in this collection. MS R also has a table for the radices of the mean motion of the eighth sphere at the Incarnation and the radices of the apogees of the Sun and the planets for Alfonso's era (4^{v}), all of them in agreement with those in the *editio princeps* of the Parisian Alfonsine Tables. Before the proper set, MS B has tables for the entry of the Sun into Aries for Paris in 1361 and for the revolution of the years (65^{v}). MS M features the standard table for the equation of access and recess, with a maximum value of 9;0,0° (52^{v}).

Hermann's own tables then follow. The first ones are for mean motions and equations (B, 66^{r}–88^{v}; M, 52^{r}–58^{v}; R, 2^{r}–42^{v}). The mean motions are presented in collected years from 1300 to 3040, at intervals of 60 years, expanded years from 1 to 60, months for both leap and non-leap years, and days from 1 to 30 (see Figure 12.7). For the Sun, Moon, and the planets, we are also given sub-tables for the minutes of an hour, from 1 to 60. There are tables for the mean motions of the apogees and the fixed stars; access and recess; the Sun; lunar anomaly; lunar longitude; lunar node; longitudes of Saturn, Jupiter, and Mars;

36 This fifteenth-century manuscript of 58 folios only contains Hermann of Saxony's text and tables.

37 *Catalogue Général des manuscrits des Bibliothèques Publiques des Départements, Tome V (Metz, Verdun, Charleville)* (Paris: Imprimerie Royale, 1889), pp. 122–23.

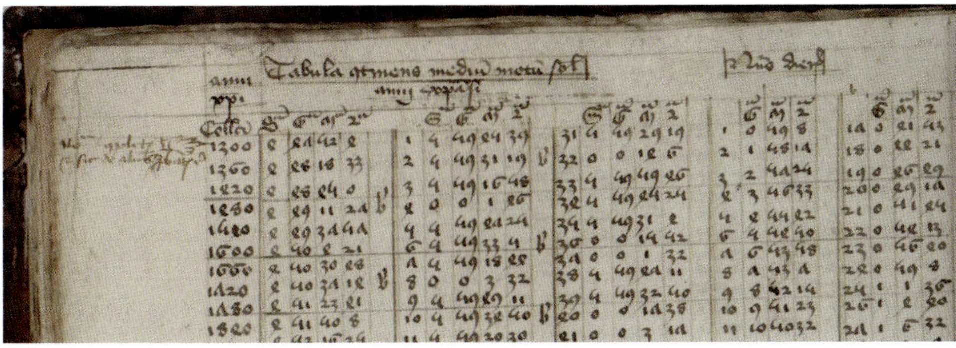

Figure 12.5. Mean motion of the Sun in Hermann of Saxony's tables in MS M, 53ᵛ.

anomalies of Venus and Mercury; argument of lunar latitude; and lunar elongation from the Sun. All entries are displayed to seconds, in signs of 60°, and the year begins in January. All mean motions conform strictly to Alfonsine parameters.

In MSS B and R, the tables for the equations are displayed between those for the mean motions of the corresponding planets. In contrast, in MS M the tables for the equations of the Sun, the Moon, and the five planets appear separately. As was the case for the mean motions, all equations are in the framework of Alfonsine astronomy.

The interval for collected years is 60, an uncommon number for that purpose. However, what is worth noting here is the use of collected and expanded years for mean motions, etc., rather than the use of a list of 60 successive multiples of a basic unit. This is yet another example undermining the claim that full sexagesimal tables for the mean motions are an essential feature of Alfonsine astronomy.

The explicits in the three manuscripts are quite similar: MS B, 88ᵛ, *Expliciunt tabule magistri Hermanni de Saxonia*; MS M, 68ᵛ, *Expliciunt tabule alfoncij illustris regis Castelle, Deo gratias*; MS R, 42ᵛ, *Expliciunt tabule magistri* (the next three words have been erased, but the remaining traces could easily be interpreted as Hermanni de Saxonia).

2. *Worked examples for 1362*

The text accompanying the tables consists essentially of worked examples computed from the tables above. The goal is to determine the true positions of the luminaries and the five planets at noon on January 10, 1362 (1361 completed) in Paris.

The incipits of the texts in the three manuscripts show slight variations: *Anno domini Christi 1361 ad meridiem…* (MS B); *Anno domini nostri Ihesu Christi 1361 completo ad meridiem…* (MS M); and *Anno Christi 1361 completo ad meridiem…* (MS R).

The first worked example is for the motion of the apogees and the fixed stars on that date. The following data are found in the corresponding table (MS M, 53ʳ):

1360	9;59,31°
1y	0; 0,26°
10d	0; 0, 1°

The entries correctly add up to 9;59,58°. The next step refers to the motion of access and recess of the eighth sphere. The data for mean motion are taken from the corresponding table (MS M, 53^{r}):

1360	1, 9; 9, 8°
1y	0; 3, 5°
10d	0; 0, 5°

correctly adding up to 1,9;12,18° (= 69;12,18°). As explained by Hermann, 'one sign in these tables amounts to two common signs', that is, 1s = 60°. With the value so obtained, we are told to enter the table of the equation of access and recess, where the entry for argument 69° is 8;23,52° and that for argument 70° is 8;27,11°. These are indeed the values given in the text. Here, Hermann of Saxony is following the specific Alfonsine procedure for determining precession/trepidation. By linear interpolation between these two entries, one should find 8;24,33° for the equation, which is to be added to the mean motion of the apogees and fixed stars previously found (9;59,58°). However, the author refrains from interpolating and, with no further explanation, simply adds 8;23,52° to 9;59,58°, thus obtaining 18;23,50°. This value is 0;41° less than when computed using interpolation.

The value so determined for the motion of the apogees since the Incarnation, 18;23,50°, has to be added to each of the apogees at the Incarnation to obtain 1,29;49,13° (Sun and Venus), 4,11;47,32° (Saturn), 2,52;0,50° (Jupiter), 2,13;36,3° (Mars), and 3,29;3,23° (Mercury). We note that all values for the apogees at the Incarnation are given in the text, but not in the tables associated with it, and they are all rounded values from those in the standard version of the Parisian Alfonsine Tables.

The next topic is the determination of the true position of the Sun at noon on January 10, 1362. For that purpose, we are told to find the mean longitude of the Sun by adding the following entries in the corresponding table (see Figure 12.7):

1360	4,48;18,33°
1y	5,59;45,39°
10d	9;51,23°

The result is correctly computed as 4,57;55,35°. We also note that Hermann reckoned the mean solar longitude for 10 days completed, that is, for the beginning (noon) of January 11. The mean solar longitude yields the mean solar anomaly by subtracting the longitude of the solar apogee at the following time: 4,57;55,35°−1,29;49,13° = 3,28;6,22°.

The mean anomaly obtained, which is off by 0;41°, as previously noted, is used as the argument to determine the solar equation in the corresponding table (MS M, 59^{r-v}). The relevant entries in this table are 1;2,54° for argument 3,28° and 1;5,1° for argument 3,29°. Again, Hermann avoids interpolation and indicates that the true position of the Sun is found by adding its mean longitude, 4s 57;55,35°, to the solar equation, 1;2,54°. The result is given as 4s 58;58,29°. Had he interpolated between the two entries of the mean anomaly, he would have found that the value of the solar equation was 0;0,13° higher. In sum, by avoiding the two interpolation procedures, he introduces two errors (0;0,41° and 0;0,13°) fortunately for him in opposite directions.

After computing the true position, Hermann indicates that the Sun is located at Cap 28;58,29°, clearly giving the impression that signs of 60° are just used for computational purposes, whereas zodiacal signs reflect what any practitioner would recognize as a celestial position.

The text then turns to the determination of the appropriate lunar quantities. The first step requires the computation of its mean longitude. The relevant data are found in the corresponding table: MS M, 53^v: 3s 24;38,23° (1360 years); 2s 9;23,3° (1 year); and 2s 11;45,50° (10 days). The result is indeed 1s 45;47,16°. The mean elongation is computed by subtracting the mean solar longitude, 4s 57;55,35° (computed above), from the mean lunar longitude, 1s 45;47,16°, yielding 2s 47;51,41°. The double elongation is twice this value: 5s 35;43,22°.

The next quantity computed by Hermann is the mean lunar anomaly. This is done by referring to the corresponding table (MS M, 54^r). The following values are given in the text:

1360	0s 20;55,58°
1y	1s 28;43,15°
10d	2s 10;39, 0°

The result is 4s 0;18,13°. The value for the year 1360 is erroneously given as 0s 20;45,58° in the table in MS M, instead of 0s 20;55,58°, which is the correct value used in the text. This is possibly a sign that Hermann's calculations reflected in the text were not based on this particular faulty table but on an original, correct table.

In order to compute the true anomaly of the Moon, one has to determine the equation of centre. This is done by entering the table for the lunar equations (MS M, 60^r–61^r) with the double elongation 5s 35;43,22° (computed previously) as the argument. The following entries are found for the equation of centre:

5s 35°	3;40°
5s 36°	3;31°

Hermann uses the rounded value of 5s 36° rather than 5s 35;43,22° to enter the table, and once again he avoids interpolation, thus rendering the seconds in his calculations meaningless. He keeps 3;31° and subtracts it from the mean lunar anomaly computed above (4s 0;18,13°) to obtain 3s 56;47,13° for the true anomaly. It is worth noting that in this case Hermann has a different interpolation practice than in the first text. The author's arithmetic at this point is certainly striking to a modern reader, in the sense that he combines values displaying different orders of magnitude and keeps the highest order in the result. At this stage, the seconds are of little use and the minutes are dubious.

The true anomaly of the Moon is needed to determine other quantities on which the total lunar equation depends: the equation of anomaly and the increment (*diversitas diametri*). To do so, Hermann enters the table for the lunar equations (MS M, 60^r–61^r) with the rounded value 3s 57° and finds 4;19,38° and 2;27°, respectively. We are then told to multiply 2;27° by 2, without any further explanation. Indeed, 2 is the value of the minutes of proportion found in the table for a double elongation of 5s 36°. To perform this simple multiplication, with the obvious result of 0;4,54°, the author details the following procedure: first convert 2;27° to minutes, then multiply 147 minutes by 2, divide the result by 60, and, finally, keep the quotient as the number of minutes and the remainder as the number of seconds. The result, 0;4,54°, is then added to the equation of anomaly, 4;19,38°, to obtain the total lunar equation: 4s 24,32°.

Finally, the true longitude of the Moon, 1s 50;11,48°, results from adding its mean longitude, 1s 45;47,16° (computed above), to the total equation, 4s 24,32°. In fact, MS M gives 1s 50;11,45° instead, but the seconds –totally meaningless by now, because they have not been taken into account in all steps — are corrected to 48 in the margin. As was the case for the Sun, the author displays the lunar position in zodiacal signs: Cnc 20;11,48°.

The lunar node is addressed in a way similar to that used to find the previous quantities. We are told to add the relevant entries in the table for the mean motion of the lunar node (MS M, 55^{r}): 1s 56;25,48° (1360 years), 0s 19;19,42° (1 year), and 0s 0;31,46° (10 days), resulting in a mean motion of 2s 16;17,16°. Its complement in 360°, 3s 43;42,44°, is called the 'true' longitude of the lunar ascending node, and it is said to be at Sco 13;42,44° or Sco 14°, whereas the lunar descending node is at Tau 14°.

The text then turns to the three superior planets, for which the same procedure applies. The mean longitude of Saturn is derived by adding the following entries in the corresponding table (MS M, 55^{v}):

1360	2s 33;15,20°
1y	0s 12;13,35°
10d	0s 0;20, 6°

The result is 2s 45;49,1°. By subtracting the longitude of the apogee found above (4s 11;47,32°) from this value, one obtains the mean centre of Saturn: 4s 34;1,29°. We are then asked to compute the difference between the mean positions of the planet and the Sun to find its mean anomaly: 4s 57;55,35°–2s 45;49,1° = 2s 12;6,34°.

Using the rounded value of the mean centre (4s 34°) as the argument, we are told to enter the table of the equations of Saturn (MS M, 61^{v}–62^{v}), and find the corresponding entry for the equation of centre: 6;28°. It is said that this amount must be added to the mean centre, 4s 34;1,29°, because it is greater than 3s (= 180°). The resulting true centre is 4s 40;29,29°. Note again the addition of terms with different orders of magnitude, making the seconds meaningless. The text proceeds with the computation of true anomaly; one has to subtract the equation of centre, 6;28°, from the mean anomaly, 2s 12;6,34°, to obtain 2s 5;38,34°.

This is just the first step for determining the true longitude of Saturn. We are then told to enter the table for the equations. For the true centre, 4s 40;29,29°, one finds the corresponding entry for the minutes of proportion, 8; for the true anomaly, 2s 5;38,34°, the given value for the equation of anomaly is 5;21°. This means that, again, interpolation is neglected, since 5;21° corresponds to an anomaly of 2s 6° (the equation of anomaly corresponding to 2s 5° is 5;24°). Going forward, we see that 2s 6° is the value actually used, as the corresponding subtractive difference (*longitudo longior*) taken from the table is 0;18°. This value is to be multiplied by the 8 minutes of proportion found above, resulting in 0;2,24°, following the procedure already explained in the derivation of the total lunar equation. The corrected equation of anomaly (5;18,36°) results from applying this negative correction to the value for the equation of anomaly previously found, 5;21°. Note again the odd combination of numbers, some given to minutes and others to seconds. The total correction for Saturn amounts to 11;46,36°, which results from adding both equations, 6;28° and 5;18,36°. Finally, the true longitude of the planet is obtained by adding the total correction to the mean longitude of the planet, 2s 45;49,1° (found above). The result is 2s 57;36,31°, which the author renders as Vir 27;35,37°, using zodiacal signs as in previous cases.

There are no examples for Jupiter and Mars. In fact, no other calculations are displayed in the rest of the text. A short explanation of planetary stations follows the computation for Saturn as well as information on the apogees and mean motions of Venus and Mars. The text closes with succinct explanations of the meaning of conjunction, quadrature, and argument of lunar latitude. It would thus seem that Hermann of Saxony's second collection of numerical examples for 1362 was left unfinished and that it was intended for a less advanced readership.

This text also differs in many ways from the previous one, with examples for 1356 and 1361 (see Part I). However, both share the same pedagogical purpose of facilitating the learning of computational astronomy and explaining and transmitting the complex and often obscure procedures described in the canons through worked examples. Moreover, in both collections the emphasis is on the algorithms rather than the specific results.

To sum up, we present two new texts and a set of tables composed by an almost unknown author, Hermann of Saxony, thus widening the cast of Alfonsine astronomers in the fourteenth century.

It is also worth stressing that the two texts by Hermann of Saxony examined here consist of ordered collections of worked examples and are not limited to one or another specific problem. They form a new category, or genre, of astronomical texts. There is another relevant text fully belonging to this category, the *Quia plures astrologorum* by John of Saxony, featuring numerical examples to illustrate John of Lignères' *Cuiuslibet arcus*. There are also other manuscripts containing texts exclusively devoted to numerical examples, although not as systematically developed as in the texts by the two scholars from Saxony, John and Hermann. Among these texts are an anonymous work with numerical examples for Vienna for the year 1406 extant in Vatican, MS Pal. lat. 1354; a text by Johannes Schwab with a computed example of an eclipse for 1413 for Vienna; and a text by an anonymous author, preserved in Naples, Biblioteca Nazionale, MS VIII.C.41, containing a number of worked examples for 1438 and 1439. Of course, there are also texts that include only a few worked examples to illustrate their contents, a rising trend in the second half of the fifteenth century.

Several of the algorithms developed by John of Saxony and applied here by Hermann of Saxony are specific to Alfonsine astronomy and have thus to be considered as one of its characteristic elements, as are tables and canons explaining their use.

Computing the true positions of the Sun, the Moon, and the planets was not an easy task in the absence of modern algebraic notation and formulas, and medieval astronomers depended exclusively on complex and repetitive written instructions found in canons. As attested in his texts, Hermann of Saxony mastered the mathematical algorithms to compute celestial quantities. However, in his mathematical practice he applied different levels of precision in his calculations and often avoided interpolation. Nevertheless, the detailed calculations presented in his practical canon were no doubt a very useful guide to understanding many of the subtleties of mathematical astronomy; indeed, it is likely that they were used for pedagogical purposes and contributed to widening access to this discipline.

Manuscript sources

Bern, Burgerbibliothek, Cod. 454
Erfurt, Universitätsbibliothek, CA 2° 386
Lübeck, Stadtbibliothek, Math. 2° 1
Metz, Bibliothèque municipale, 287
Oxford, Bodleian Library, Digby 97
Paris, Bibliothèque nationale de France, lat. 7407
Paris, Bibliothèque nationale de France, lat. 11250
Paris, Bibliothèque nationale de France, lat. 15104
Prague, Knihovna metropolitní kapituly, M.CI (1460)
Wolfenbüttel, Herzog August Bibliothek, 2637 (Guelf 65 Aug 2°)

Bibliography

Burnett, Charles, '"Ptolemaeus in Almagesto dixit": The Transformation of Ptolemy's Almagest in its Transmission via Arabic into Latin', in *Transformation antiker Wissenschaften*, ed. by Georg Toepher and Hartmut Böhme (Berlin: de Gruyter, 2010), pp. 115-40.

Catalogue Général des manuscrits des bibliothèques publiques des départments, tome V (Metz, Verdun, Charleville) (Paris: Imprimerie Royale, 1889).

Chabás, José, *Computational Astronomy in the Middle Ages: Sets of Astronomical Tables* (Madrid: Consejo Superior de Investigaciones Científicas, 2019).

______, and Bernard R. Goldstein, 'Nicholaus de Heybech and His Table for Finding True Syzygy', *Historia Mathematica*, 19 (1992), pp. 265–89.

______, *The Alfonsine Tables of Toledo* (Dordrecht: Springer, 2003).

______, *A Survey of European Astronomical Tables in the Middle Ages* (Brill: Leiden, 2012).

______ and Marie-Madeleine Saby, *The Tables of 1322 by John of Lignères* (Brepols: Turnhout, 2022).

Goldstein, Bernard R., 'Lunar Velocity in the Ptolemaic Tradition', in *The Investigation of Difficult Things: Essays on Newton and the History of the Exact Sciences*, ed. by P. M. Harman and A. E. Shapiro (Cambridge, 1992), pp. 3–17.

Hagen, Hermann, *Catalogus codicum bernensium (Bibliotheca Bongarsiana)* (Bern: B. F. Haller, 1875).

Pedersen, Fritz S., *The Toledan Tables: A Review of the Manuscripts and the Textual Versions with an Edition* (Copenhagen: C.A. Reitzels Forlag, 2002).

Poulle, Emmanuel, *Les Tables Alphonsines avec les canons de Jean de Saxe* (Paris: Éditions du CNRS, 1984).

Ratdolt, Erhard, *Tabulae astronomicae* (Venice, 1483).

Saby, Marie-Madeleine, 'Les canons de Jean de Lignères sur les tables astronomiques de 1321', unpublished thesis, Paris, École Nationale des Chartes, 1987. A summary appeared as: 'Les canons de Jean de Lignères sur les tables astronomiques de 1321', *École Nationale des Chartes: Positions des thèses*, pp. 183–90.

Schum, Wilheim, *Beschreibendes Verzeichnis der Amplonianischen Handschriften-Sammlung zu Erfurt* (Berlin: Weidmann, 1887).

Thorndike, Lynn, 'Astronomy at Paris around 1485 and 1360', *Humanisme et Renaissance*, 3 (1936), 165-68.

______, 'Pre-Copernican Atronomical Activity', *Proceedings of the American Philosophical Society*, 94 (1950), 321-26.

______, 'Notes on some Astronomical, Astrological, and Mathematical Manuscripts of the Bibliothèque Nationale, Paris', *Journal of the Warburg and Courtauld Institutes*, 20 (1957), pp. 112–72.

______, and Paul Kibre, *A Catalogue of Incipits of Medieval Scientific Writings in Latin*, rev. and augmented ed. (Cambridge: Mediaeval Academy of America, 1963).

Toomer, G. J., *Ptolemy's Almagest* (New York: Springer, 1984).

MATTHIEU HUSSON*

Proportional Parts and Interpolation Practices in Alfonsine Astronomy

Astronomers' Work with a Mathematical Procedure

Introduction

In a paper published in 2012, I studied the interpolation procedures of four double-argument tables from the works of John of Murs and John of Lignères from a computational and epistemological perspective.[1] I concluded that variations in those procedures depended on the astronomical and mathematical context of application as well as on the audiences targeted by the text and the intellectual profile of the scribes copying them. Relying on modern algebraic expressions of these procedures, I also analysed the numerical discrepancies produced among these different procedures and showed that they can be numerically significant. Following on from these results, this contribution proposes a different analysis aimed at achieving a more refined understanding of the means used by Alfonsine astronomers to create these different procedures and address the central issue of interpolation in mathematical astronomy.

In fourteenth- and fifteenth-century Latin, *interpolation* refers to a piece of text added to a main text after its initial composition. One of the first identified mathematical uses of the word in Latin is in John Wallis's arithmetic, *Arithmetica Infinitorum* (1656), and is related to function theory.[2] Alfonsine astronomers did not have a generic word referring to these mathematical practices that we can retrospectively designate as interpolation. One may pose the following question: if interpolation is a set of operations not identified in the actors' terminology, is there anything pertinent its analysis could teach us about astronomer categories? I argue that the response would be affirmative if efforts were made to slightly alter the analytical tools used to

* This contribution benefited from many insights provided during informal discussions, seminars, and workshops on draft versions from colleagues of the ALFA team: Richard L. Kremer, José Chabás, Samuel Gessner, Nicholas Jacobson, Stefan Zieme, Eleonora Andriani, Camille Bui, Sophie Serra, Petr Hadrava, Alena Hadravová, and Marie-Madeleine Saby. The research was conducted in the framework of the ERC project ALFA: Shaping a European Scientific Scene (2017–23, CoG 723085, P.I. M. Husson).

1 Matthieu Husson, 'Ways to Read a Table: Reading and Interpolation Techniques in Canons of Early Fourteenth-Century Double-Argument Tables', *Journal for the History of Astronomy* 43 (2012), 299–319.

2 The first mention is found in John Wallis's dedication to Oughtred, Jacqueline A. Stedall, *The Arithmetic of Infinitesimals, John Wallis 1656* (New York: Springer, 2004), p. 4.

Matthieu Husson • SYRTE, Observatoire de Paris – Université PSL, CNRS, Sorbonne Université

Alfonsine Astronomy: Computational Practices, ed. by Matthieu Husson, José Chabás, and Richard L. Kremer, Turnhout, 2026 (*Alfonsine Astronomy: Studies and Sources*, 3), p. 413-433
© BREPOLS PUBLISHERS 10.1484/M.ALFA-EB.5.137918

study these practices. I make a proposition to this end in the following paragraphs and introduce what I term 'proportional-part procedures' and a few other related analytical notions such as the 'layered properties' of procedures and procedural texts or the 'extension' of procedures.

Firstly, the lack of an overarching word does not imply that the practices retrospectively assembled under the name interpolation appeared disconnected to mediaeval astronomers. Their connection, I argue, was not set by a word, but by a relation to a core computation: the computation of proportional parts. This proposition is fruitful because it allows us to capture a connection between procedures embedded in practices rather than in terminology. It also leads to the study of an aspect of mathematical procedures different from their often-addressed sequential feature: their layered construction (i.e. the fact that sub-procedures are embedded in larger procedures and on several levels). One primary aim of this chapter is to understand how astronomers extended the core computation of proportional parts embedding it in larger (new) procedures. They did so by working with the terminology, the operations and their sequences, numbers, different numerical tables, and even diagrams. This specific procedural work innervates a large portion of their mathematical practices; consequently, it constitutes and maintains a set of proportional-part procedures defined by their common essential relation to the computation of proportional parts.

Proportional-part procedures are largely overlapping with, but not exactly similar to, the set of practices retrospectively assembled under the word interpolation. Analysing these discrepancies is one of the secondary aims of this chapter as it can contribute to a deeper understanding of how the word *interpolatio* acquired a mathematical meaning by describing the situation of mediaeval Latin astronomical sources.

To analyse different facets of proportional-part procedures, I have relied primarily on the corpus of Alfonsine works preserved in fourteenth- and fifteenth-century Latin manuscripts. I also consider the corpus attached to the earlier Toledan Tables when pertinent to the discussion, especially because it is likely that they continued to be read by Alfonsine astronomers in the fourteenth and fifteenth centuries. This chapter unfolds in three sections. I begin by establishing the fundamental component of the proportional-part procedure in the context of single-argument tables. The following section moves on to a first series of extensions of proportional-part procedures related to double-argument tables. The last section considers proportional-part procedures in relation to the design of complete table sets. Although the progression of the chapter essentially follows the different tabular contexts in which proportional-part procedures are developed, I include their connection to diagrams whenever pertinent.

1. Single-argument tables and proportional-part procedures

In Chapter 17 of the *Tempus est mensura motus* (1327) on the Sun's true position, John of Saxony develops the procedure to compute proportional parts with a wealth of detail. This description is an excellent starting point towards the aims of this chapter. It is a rather long passage, which I have separated into several quotes for the sake of the commentary.[3]

3 Here (and after), quotes from John of Saxony's *Tempus est mensura motus* are taken from Emmanuel Poulle, *Les Tables Alphonsines avec les canons de Jean de Saxe: Édition, traduction et commentaire* (Paris: Éditions du Centre

> *Et quia in omnibus equationibus seu operationibus ubi volumus prescise operari indigeamus parte proportionali, idea repetam modu inveniendi partem proportionalem, [nec] repetam in sequentibus sed in qualibet operatione remittam te ad istum canonem.*[4]
>
> And because in all equations or operations where one wishes to operate with precision the proportional part is needed, I will repeat the method to find the proportional part so as not to repeat it again but simply refer you to this rule for this operation.

Two different claims can be distinguished in this first quotation. First, John of Saxony indicates that the computation of proportional parts is linked to a search for precision. From the outset, we are pointed towards this very general question, which will be one of the threads of discussion among historical actors. The claim of John of Saxony implies that a practice of computation without recourse to the computation of proportional parts (and thus to interpolation) is, at least in principle, something he could have conceived, even at the cost of certain sacrifice. The second claim of the quote indicates that the following presentation of proportional-part computation is not the first in the canons. As a matter of fact, the explanation is the third in the canons, and one more follows in the next chapter. All of them differ slightly in interesting ways. The second part of the claim, indicating that cross-references to this particular passage are to be expected in the following chapters of the canons, is also true. The redundancy of *Tempus est mensura motus* on the question of proportional parts is not specific to John of Saxony's canons. It is a rather common feature of the genre, one extending, for instance, to the canons of the Toledan Tables. Proportional parts are not the only recurring topic; descriptions of how to handle sexagesimal units or rules to manipulate additive or subtractive operands in computations are also very frequent. The study of this redundancy could be the subject of an independent analysis. I simply point out here that recurring topics all seem to be linked to mathematical practices, that they might suggest difficulties in these practices, and that the frequent repetitions of these rules also provides clues concerning the consumption of canons as texts that scribes or readers could cut into parts or decompose or recompose in various ways.

Before reading the next passage, it is important to introduce a terminology related to numerical tables, upon which I constantly rely in this and other contributions. I consider a numerical table as a mathematical object constituted by an ordered set of relations between argument and entry. This set of mathematical relations between argument and entry is most often presented in a grid format, but other representations are also possible and sometimes found (e.g. uneven scales). Arguments are the quantities used to enter the table. Most tables are single-argument ones, but double — or even triple-argument tables do exist. Entries are the quantities extracted from the table and carried to the next steps of the computation. For instance, in the solar equation table considered here by John of Saxony, the table argument is the Sun's mean argument and the table entry is its equation.

The following quote describes the proportional-part computation. It considers two key quantities: the *differentia*, that is, the difference between the two successive entries

national de la recherche scientifique, 1984). A critical edition of this fundamental text is lacking. Emmanuel Poulle's text is a silently emended transcription from the 1483 princeps edition of the text.

4 Poulle, *Les Tables Alphonsines*, p. 68.

of the tables from which the user is to interpolate; and that which is named the *residuus* in the next quote, that is, the portion of the argument with which it was not possible to enter in the table. Some ways to transform the numerical format of these two quantities, preparing them for multiplication, are described. It is important to know that the particular table concerned by this text have entries given for argument with a step of one degree.

> *Si in differentia duarum equationum fuerit minuta et secunda, reduc ea ad secunda, scilicet multiplica minuta per 60 et habebis secunda, cum quibus adde secunda que habes, et illud quod proveniet serva, et non obliviscaris cujus denominationis sint, quia sunt secunda. Postea vide minuta et secunda, que sunt in argumento ultra signa et gradus, si multum prescise operari velis, et reduc similiter ea ad secunda; si autem non curas habere tantam prescisionem, accipe minuta solum et tunc si fuerint plura secunda quam 30 pone pro eis unum minutum.*[5]
>
> If in the difference of the two equations there are minutes and seconds, reduce them to seconds; that is, multiply the minutes by 60 and you have seconds, which you add to the seconds you already have, and what you find you keep, and do not forget which denomination it is, that is, they are seconds. After, see the minutes and seconds that are in the argument after the signs and the degrees. If you want to operate with great precision, reduce them in the same way to seconds. If, however, you do not care for such precision, take only the minutes, and if there are more than 30 seconds, take one minute for them.

The proposed transformation of the *differentia* and *residuus* consists of reducing them to an integer number of the smallest sexagesimal unit of which they are composed: arcseconds in this case. I highlight that the text specifically asks the user to take note of this unit. This transformation allows the application of the rules of integer algorisms (as described, for instance, in Sacrobosco) to compute arithmetical operations. This is a general option for astronomical computation that is usually adopted in canons aimed at an advanced audience. Canons aimed at non-expert audiences, for instance John of Lignères' Canons to the *Tabule magne*, generally provide tabular tools to avoid this kind of transformation.[6] The issue of computation precision is also directly addressed here. A choice is given to the users to take into account (or ignore) the number of seconds of the *residuus*. Perhaps surprisingly, the same option is not given with respect to the *differentia*. This choice is given with the idea that there are different degrees of precision with which one can choose to operate. In 2012, I also noted that a choice was given to the users, sometimes implicitly, by the procedural texts and used the term 'open' to denote such a procedural feature. This is also relevant here, and I consider it important to keep track of how different procedures are 'open' to varying degrees. The following quote then proceeds to the multiplication of the *differentia* and the *residuus*. On this occasion, the two words, which are fundamental to the terminology of proportional-part computations, are used together, establishing their technical meaning in this context:

---

5 Poulle, *Les Tables Alphonsines*, p. 68.

6 Matthieu Husson, 'Astronomers' Elementary Computations in Paris between 1321 and 1327: The Circle around John of Lignères and John of Saxony', in *Mathematical Practices in the Astral Sciences*, ed. by Matthieu Husson and Karine Chemla (Springer, forthcoming).

> *Quo facto, multiplica numerum differentie per numerum residui argumenti et serva bene denominationem ejus quod provenerit, qui si multiplicasti minuta per minuta provenient secunda, et si minuta per secunda provenient tertia, et si scunda per secunda provenient quarta.*[7]
>
> This done, multiply the number of the difference by the number of the rest of the argument and carefully keep the denomination of what is obtained; that is, if you multiply minutes by minutes seconds are obtained, and if minutes by seconds thirds are obtained, and if seconds by seconds fourths are obtained.

John of Saxony is open about the way the multiplication is to be done. I surmise that this multiplication of two integers is to be executed according to the rule described in common algorism texts, like that of Sacrobosco. In multiplication procedures for integers in algorism texts, there is no issue of making the computation with different accuracies. All digits are taken into account. The second part of the quote attracts the attention of the user to the *denominatio* of the result of the multiplication and recalls the rules applying, in this respect, to sexagesimal parts. This point of attention is first a means to prepare the reverse transformation of the large integer resulting from the multiplication to a sexagesimal number. It is also an operation of precision management, that is, keeping track of the level of precision (minutes, seconds, thirds, etc.) to which the computation is conducted, even at a stage of the computation where the transformation of the numbers into a large integer has somehow erased this information from the way the number itself is expressed. It is interesting to note that, while the only possible outcome if strictly applying the previous instruction is that of multiplying seconds by seconds, all possibilities are described here. This might be because these kinds of rules are fixed formulas that can simply be repeated and are not always adapted to their context. It could also be that restating all options is a way to show the logic of the rule at the same time as prescribing it.

The following and final quote describes the reconstruction of the result as a sexagesimal number. In the context of the Parisian Alfonsine Tables, the step of the tables in which interpolation is necessary is always one degree; so, the division by the step of the table (i.e. one degree here), which is generally necessary, is simply omitted by John of Saxony. The choice of a step of one degree makes it possible to avoid the division in this particular context. Given the difficulty linked to division, this situation is, in all probability, a deliberate choice of the table-makers.

> *Qua multiplicatione facta, divide numerum provenientem per 60 et redibit immediate fractio grossior, ut si per multiplcationem provenerint quarta, tunc per divisionem tertia redibunt in numero denotante quotiens et quod remanet erunt quarta. Item, si in numero denotante quotiens fuerint plura tertia quam 60, divide per 60 et provenient secunda et quod remanebit erunt tertia; que secunda si fuerint plura quam 60 divide per 60 et provenient minuta et quod remanet erunt secunda. Quo facto, habes partem proportionalem ita precise sicut vis*[8]
>
> Once this multiplication is done, divide the obtained number by 60 and you obtain the fraction immediately greater, so if from the multiplication fourths are obtained,

7 Poulle, *Les Tables Alphonsines*, p. 68.
8 Poulle, *Les Tables Alphonsines*, pp. 68–70.

> then from the division thirds are given in the number named 'quotient', and what remain are fourths. Again, if in the number named quotient there are more than 60 thirds, then divide by 60 and find seconds, and what remain are thirds; these seconds, if they are more than 60, divide them by 60 and find minutes, and what remain are seconds. This done, you have the proportional part at the precision you want.

Just as the initial transformation of the *differentia* and *residuus* in large integers of the smallest sexagesimal unit was done via a series of multiplications, the reverse transformation of the result into a sexagesimal number is done via a series of Euclidean divisions. The freedom given to the user with respect to the precision is the final note of the description of the procedure.

To sum up, this description of a proportional-part computation introduces the basic terminology of the *differentia* and the *residuus*. It also (albeit less clearly) introduces the fundamental sequence of operations of the procedure: multiply the *differentia* and the *residuus*; then, divide the result by the step of the table. With these two essential aspects of the proportional-part procedure, two supplementary features of the description appear: the manipulation of number format between sexagesimals and integers and the issue of precision in the computation.

In other words, even the core computation of proportional parts appears as a procedure with different intertwined layers, some of which are central, while others depend on the way computations are to be performed, the precision desired by the user, the types of number used (divisors for the fractional parts, precision), or even the particular tables that are read. This layered property of the procedure is both the result of the astronomers' work and the means they used to adapt it to different practical contexts. In this contribution, I wish to unveil how this works for proportional-part procedures.

Before moving on to more complex proportional-part procedures, where the core computation of proportional parts is embedded in different astronomical contexts, I would like to point out a small diagram showing this computation in a different light. It can be found in Bibliothèque nationale de France, lat. 7197, a fifteenth-century manuscript that the astrologer Conrad Heingarter assembled at the Paris arts faculty as a student (Figure 13.1). This diagram is not directly discussed by the texts found in the neighbouring folios. It does, however, address the same topic: computation with natural and sexagesimal fractions.[9] The diagram, quite unique in the Alfonsine corpus as far as I know, maps simple natural fractions to a sexagesimal scale so that, for instance, $\frac{2}{3}$ is mapped to 40, $\frac{2}{4}$ to 30, $\frac{1}{4}$ to 15, and so on. This little device, especially once memorized by the astronomer, can be very useful to make quick estimations of proportional parts that in many cases might be accurate enough for the computation or to serve as a way of controlling a result that is computed more accurately with a longer arithmetical procedure. Thus, in addition to being the layered procedure attested by textual evidence, proportional-part computation is also linked to diagrammatic tools allowing a different representation and different approaches to the questions of approximation and precision.

---

9 On folios 30^r–31^r, there is an *Algorismus de minutiis phisicis* with the incipit *Quoniam opus tabularum sine scientia phisicarum minutiarum*, and on 31^r–36^v there is John of Lignères' *Algorismus de minutiis* with the incipit *Modum representationis vulgarium et phisicarum*; see H. Busard, *Het rekenen met breuken in de middleleeuwen, in het bijzonder bii Johannes de Lineriis* (Brussels: Vlaamse Academie, 1968); both texts address sexagesimal and natural fractions.

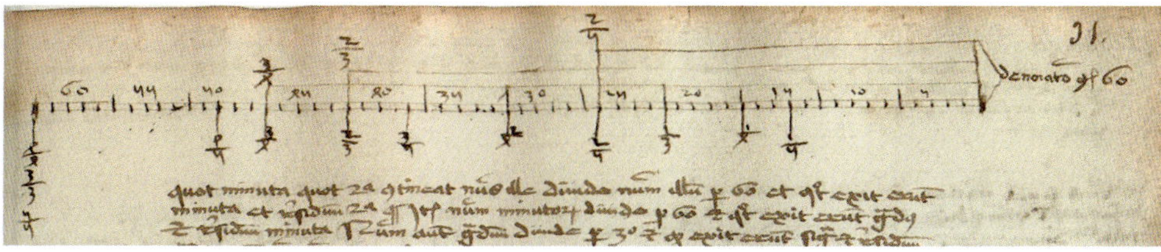

Figure 13.1. Diagram with natural fractions on sexagesimal scale (BnF, lat. 7197 f. 31r).

The first and most common situation in which the computation of proportional parts occurs is that of interpolation in single-argument tables. The following quote shows how John of Saxony integrates the proportional-part procedure in the context of the computation of the equation of the eighth sphere. Since we have just analysed one description of a proportional-part computation, I have not included the description of the same procedure here, but its position in the text is clearly indicated by the square brackets.

> *Quere ergo numerum graduum in lineis numeri, quo invento videndum est quid sit de gradibus, minutis et secundis in directo, et scribe illud extra in duobus locis. Quo facto, si tu habes aliqua minuta ultra gradus cum numero graduum cum quo intrasti, intra adhuc semel uno gradu addito; hoc est dicere: si intrasti primo cum 67 gradibus et si fuerient ultra 67 gradus aliqua minuta, intra postea cum 68 gradibus et quod sit in directo vide de equatione, et scribe illud extra. Quo facto, vide utrum prima equatio, scilicet illa quam accepisti primo, sit major quam secunda vel econtra, et subtrahe minorem de majori et istud quod remanet vocatur differentia inter duas equationes; de qua differentia accipe partem proportionalem secundum proportionem minutorum que fuerunt ultra gradus ad 60. [...] Quam partem proportionalem adde equationi primo accepte si secunda fuerit major vel subtrahe a prima si secunda fuerit minor, et quod post additionem vel subtractionem provenerit erit equatio motus octave spere.*[10]

> Seek first the number of degrees in the line of numbers, in front of which we see the degree, minutes, and seconds, which we write outside in two places. This done, if you have any minutes after the degrees, with the number of degrees you entered; then, enter again with one more degree. That is to say, if you entered first 67 degrees and if there are, after 67 degrees, any minutes, enter after 68 degrees, and what you see in front of the equation write it outside. This done, see if the first equation, that is, the one you took first, is greater than the second, or the opposite, and subtract the smallest from the largest and what remains is called the difference between the two equations; From this difference, take a proportional part according to the proportion of minutes after the degrees to 60 [minutes]. [...] This proportional part is added to the first evaluation of the equation if the second is greater or subtracted from the first if the second is smaller, and what comes from the addition or subtraction is the equation of the motion of the eighth sphere.

10 Poulle, *Les Tables Alphonsines*, pp. 62–64.

The reference to the proportional-part procedure is made just before the omitted passage. This reference includes the mention of the *differentia* and of the two other quantities, which are to be used, respectively, for a multiplication (the *residuus* here mentioned as *minutorum que fuerunt ultra gradus*) and the step of the table (*ad 60*). The way the *residuus* is mentioned depends on the layout of the table. Since the table step is one degree, the *residuus* is identified very easily as the sexagesimal parts of the number with which it was not possible to enter the table: that is, the minutes and seconds composing that number. The rest of the passage explains different steps that allow us to extend the core proportional computation to a more extended procedure allowing the use of the proportional-part computation in order to provide an estimation of the eighth-sphere equation for a non-tabulated argument. Ahead of the reference to the proportional-part procedure, instructions are given that allow us to produce the necessary input for it: the *differentia*, *residuus*, and step of the table. The last two are not explicitly defined but implied in the description of the way to obtain the *differentia*. The way to compute the *differentia* is precisely described, even with a small numerical example. The prescription of the subtraction between the *prima equatio* and the *secunda* [*equatio*] is also detailed with the reminder to subtract the lesser from the greater figure. The orientation of the subtraction is important for the final step of the procedure because it indicates how the result of the proportional part should be used: it should be added to the *prima equatio* if the *secunda equatio* is the larger, and it should be subtracted from the *prima equatio* if the *secunda* is the smaller.

The text thus extends from the core procedure by articulating the tabular context of application to the proportional-part computation. This is done by using and adapting the basic terminology of the proportional-part procedure. This is also done by introducing new terminology (e.g. *prima equatio*, *secunda equatio*) or by extending the series of operation; here, two subtractions — one being implicit — before the reference to the proportional-part procedure and one addition or subtraction after. As is the case for the description of the proportional-part procedure, the tabular context and the specificities of the numbers manipulated (esp. divisors used in the fractional parts, precision) are fully integrated in the description. This is especially clear here in the way the *residuus* is defined. The issue of precision is not addressed explicitly. In other words, the management of precision seems to be a concern only for the core procedure and for multiplication and division.

I hope that it can clearly be seen, from these first examples, how the layered features of the procedure are a means to extend a core procedure in different directions and how new layers can be added and existing layers can be adapted. The next case study is conveniently taken from a satellite text of the Canons Cb to the Toledan Tables describing how to compute an ephemeris. This choice also shows that, in working with proportional-part procedures, Alfonsine astronomers were participating in much longer discussions of mathematical astronomy.[11] It describes a different context of adaptation of the core procedure, where the *differentia* needs to be distributed in several parts instead of one. The ephemeris-maker is instructed to compute a true solar position every fifteenth day and then determine

11 See, in this volume, Richard L. Kremer, 'What Can We Learn About Computational Practices from a Study of Medieval Ephemerides?'.

intermediary positions according to a type of proportional-part procedure. The beginning of the procedure is now familiar.

> *Ponamus locum solis in prima die anni esse 10 si 23 gr et 48 mi, et in 15 die 11 si 7 gr et 45 mi. Considerata igitur differentia inter utraque loca, invenimus 13 gra et 57 m'a. Reduximus igitur gradus ad inferius genus, scilicet ad minuta, et provenerunt 780 m'a, quibus aggregavimus 57, et fuerunt 837.*[12]
>
> Let us assume that the solar position on the first day of the year is 10s 23° and 48' and on the 15th day 11s 7° and 45'. Thus, with this difference between these two places being considered, we find 13° and 57'. Consequently, we reduce the degrees to the inferior genre, namely minutes, 780' come, to which we add 57, and they are 837'.

This is all about finding the *differentia* and transforming the value into an integer number of the smallest unit, here minutes. The following instructions present an interesting variation to the procedure described just above. Fourteen different proportional parts are required instead of just one to match the 14 gaps and the 13 empty day positions of the ephemerides table. The principle of the adaptation relies on a different use of the division and a change in the order of the operation.

> *Quae cum per 14 diviserimus, exiverunt 59 m'a et remanserunt 11. Scimus utique quod unusquisque horum 14 pro parte reciperet 59 m'a; sed 11 quae remanserant, si incipientes a principio usque ad completionem eorum distribueremus aut, praetermittentes primum et secundum et tertium, a quarto inciperemus, nescivimus.*[13]
>
> And when we divide these by 14, 59 come out and 11 remain. Certainly, we know that each of these 14 parts would receive 59', but of the 11 that remain, we do not know if we would distribute starting from the beginning until their completion, or, if neglecting the first, second, and third, we would start from the fourth.

Division comes first, and instead of dividing by the step of the table, the divisor here is the number of empty gaps in the ephemerides table. Moreover, the division is an exact Euclidean one that gives us a quotient and a remainder. This move from the usual core implies that the computation will be exact to the precision of minutes and will not consider smaller divisions. The quotient and the remainder are then treated separately. The quotient is distributed evenly in each of the 14 empty gaps. This operation of distribution is a kind of iterated addition, which corresponds in a way to the multiplication in the simple core procedure. However, the procedure is not complete because the reminder also needs to be taken into account. Since the reminder is smaller than the divisor and since it is out of question to divide it by fourteen using a rounded number of seconds, a choice needs to be made in order to distribute it.

---

12 Fritz S. Pedersen, *The Toledan Tables: A Review of the Manuscripts and the Textual Versions with an Edition* (Copenhagen: C. A. Reitzels Forlag, 2002), p. 543.

13 Pedersen, *The Toledan Tables*, p. 543.

Consideravimus igitur differentiam inter 15 diem a primo et 15'um a 15'o, hoc est 30, et invenimus maiorem eam, quae est inter primum et 15'mum. Scivimus itaque per hoc quod 11, quae remanserant, deberemus incipere distribuere a 4'o, praetermittendo scilicet primum et secundum et tertium. Provenerunt itaque unicuique horum 11, qui sunt praeter primum et secundum et tertium, unus gradus, et primo et secundo et tertio 59 m'a solummodo.[14]

> Then, we consider the difference between the 15th day from the first and [between] the 15th day from the 15th, which is 30th, and we find that it is bigger than that which is between the first and the 15th day. And thus, we know by this that we would have to distribute [these] 11, which remain, from the fourth, by neglecting the first, as well as the second and the third. Therefore of these 11, which are after the first, second, and third, come one degree and only 59' for the first, the second, and the third.

From a strictly combinatorial perspective, there are many ways to distribute eleven units into fourteen places. The text considers only two of them: the units are distributed one by one in each gap, either in ascending order from the first or backwards from the last. The fact that the distribution of the remainder is done one by one reflects a fundamental concern for smoothness in the way astronomical values vary. This concern is the very basis on which the whole family of procedural part procedures rely. It is because medieval and, in fact, ancient astronomers held this belief that the variations of the quantities they considered ought to be smooth that they felt authorized to use proportional-part computations and extended the core procedure to all kinds of situations and astronomical contexts.

Back to our case study, and the choice to be made is one of directionality. It is settled by considering further differences, which are a key tool for this kind of issue. The differences between the solar positions of the first fifteen days and the following fifteen days are compared. If the difference between day 15 and 30 is smaller than the first (between day 1 and day 15), then the unit must be distributed from the first value. If the difference between days 15 and 30 is larger than the first (between day 1 and day 15), then the unit must be distributed from the last value. There are different ways one can understand this procedure. In kinematic terms, the procedure considers the acceleration of the Sun and adapts the distribution of the remainder so that it is not in conflict with it. In mathematical terms, the procedure considers the concavity of the solar position curve and adapts the distribution of the remainder so that it is not conflicting. These explanations might be useful to a modern reader, but I believe they do not reflect the actors' understanding at the time. In actors' terms, this is a work with tabular differences aimed at the smooth variation of the table differences. In the first case, the distributed difference between day 15 and 30 will be smaller than the one between day 1 and day 15; thus, it is better to have a decreasing day difference between day 1 and day 15 to ensure a smooth transition to the next series. These two ideas of working with tabular differences and considering smoothness both in the difference values themselves and in their increasing or decreasing orientation are deeply embedded in the mathematical work with tables.[15]

14 Pedersen, *The Toledan Tables*, p. 543.

15 See, in this volume, Jean-Claude Penin, 'The Trigonometric Tables of Jean Fusoris'.

Figure 13.2. Solar equation from the Parisian Alfonsine Tables, with tabulated differences in red (BnF, lat. 14481 f. 81r).

In this case study, the adaptations of the core procedure are more significant than in the preceding one: the order and nature of the operation are changed. Division comes first and is Euclidean, and the multiplication is replaced by an iterated addition, complex in nature, because the quotient and the remainder of the division are treated separately. The issue of precision is also treated differently here, as no options are proposed to the user. However, especially because of the beginning of the procedure that clearly employs the familiar vocabulary and operations of setting the application condition of a proportional-part computation, I believe that (even though this could be a limit case) this procedure can still be considered a sort of proportional-part procedure.

These examples also show that tabular layouts are deeply intertwined with the ways proportional-part procedures are expressed. Before moving to a more complex tabular environment with double-argument tables, I would like to briefly consider three more non-textual examples of the various ways the tabular layout itself is dependent on the use of a proportional procedure. One simple way this appears is the direct tabulation of *differentia* so that there is no need to compute them. This is the case of the solar equation tables of the Parisian Alfonsine Tables, as copied in the manuscript Bibliothèque nationale de France, lat. 14481 (Figure 13.2).

In this table, the scribe added the value of the difference in red in between each value of the solar equation. It is important to consider that John of Saxony's canons, which were the most common ones associated with the Parisian Alfonsine Tables, make no mention of these differences as tabulated values, but instead give instructions to compute them. As a matter of fact, many if not most of the copies of this solar equation table do not include the computed difference. It is a scribal choice to include or omit them. This is hence a scribal work on the proportional-part procedure in a tabular context. The inclusion of these differences by the scribe can clearly be helpful for a user and the scribe for computations or as means to check the tabular values, but more fundamentally it transfers an element of the procedure described in a text to the table with which the procedure is to be used. In short, a procedural feature described in a text also becomes a tabular feature of the

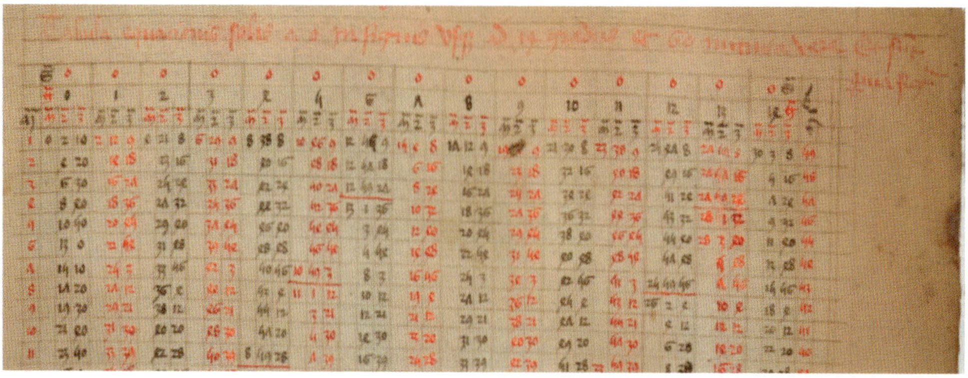

Figure 13.3. Solar equation from the Parisian Alfonsine Tables with tabular step of one arcminute (BnF, Colbert 60 f. 109^{r}).

procedure. This kind of transfer phenomenon between textual descriptions of procedures and the tables is interesting to trace and shows that although they have their autonomy, these different tools are deeply rooted in the astronomer's practices.

A second way in which procedural-part procedures can influence table layout is the choice of the tabular step for the argument. A spectacular example is provided again by a witness of the solar equation of the Parisian Alfonsine Tables, this time found in the fifteenth-century manuscript Bibliothèque nationale de France, Colbert 60 (Figure 13.3).

In this table, the chosen tabular step is one arcminute. The values of the argument are distributed both horizontally for the signs and degree and vertically for the arcminute. The values of the entries are given to the third sexagesimal unit but are taken from the Parisian Alfonsine Tables for integer degrees. Preliminary examinations indicate that intermediate values were linearly interpolated. Further analysis is needed to confirm this and specify which exact proportional-part procedure was used.[16] For my purposes, it is sufficient to remark here that this particular table layout pushed the logic of transferring a procedural feature to an element of the table layout even further by computing all the possible proportional parts of the solar equation of the Parisian Alfonsine Tables with a step of one arcminute. To my knowledge, there is no other such instance in the Alfonsine corpus of the solar equation; so, again this is a scribal choice, even if here, as elsewhere, the scribe is certainly knowledgeable in mathematical astronomy. However, it must be noted that the tendency to reduce the tabular step (and simultaneously increase the precision of their entries, as is the case here) is a general trend that started during the second half of the fifteenth century, if not earlier (e.g. Gmunden, Peuerbach, etc.).

Finally, the very frequent recourse to proportional-part procedures in the manipulation of tables and the associated computational burden led astronomers to develop dedicated proportion tables that would help them perform the necessary multiplication and division and avoid the transformation of the sexagesimal numbers into large integers of a smaller

16 It is also important to note that a similar table is proposed in the same manuscript for the sine along with a wealth of different proportions and arithmetical tables. I intend to publish a separate analysis of these different elements of the manuscript.

unit. There are many such tables essentially varying according to the type of number considered (divisors for the fractional parts, precision) or to the table step with which they can be used. The most basic and simple one is the sexagesimal multiplication table. The existence of these proportion tables maintains some sort of arithmetical interest for certain astronomers who were at the time redacting dedicated canons that can go far beyond their simple use in proportional-part procedures. Such is the case, for instance, of John of Murs with his *Tabula tabularum* text, which considers all sorts of arithmetical operations, including roots, in light of a sexagesimal multiplication table.[17] In some ways, this last instance shows an even more complete interaction between procedural texts and tables, as tables induced by a specific class of procedures end up fostering such new texts.

It would be simple to extend small case studies of the extension of the core procedure of proportional-part computation to other situations concerned with single-argument tables. Such a survey would certainly further our understanding of how astronomers treated the formatting of their numbers, the issue of precision, the astronomical meaning of their computations, and how they adapted these to different kinds of audiences. However, the textual and tabular examples considered so far indicate enough about the means astronomers used to adapt and extend the core proportional-part procedure to different contexts by working on the terminology, the order and nature of the prescribed operations, the numbers they manipulated (even with recourse to some heuristic graphical devices), and the layout of the table they used. All these different interventions constitute a work on the procedure that we are now going to pursue in the context of double-argument tables.

2. Double-argument tables and proportional-part procedures

The situation of double-argument tables presents a specific challenge for proportional-part procedures, because in principle there are at least four different table entries that can be taken into account. For double-argument tables, there are many options for different successive applications and combinations of the core proportional-part computation. Table 1 summarizes this situation with a little collection that will also conveniently allow me to describe the different options adopted in historical texts.[18] The user seeks the value of entry e for the argument (a,b). The only entries they can read directly from the table are $E1$, $E2$, $E3$, and $E4$.

	A1	*a*	*A2*
B1	*E1*		*E2*
b		*e*	
B2	*E3*		*E4*

Table 1. Schematic interpolation situation of a double-argument table.

17 Matthieu Husson, 'La *Tabula tabularum* de Jean de Murs et les modèles de l'arithmétique médiévale', *Cahiers de recherches médiévales et humanites*, 27 (2014), 97–122.

18 A similar notation is used in Husson, 'Ways to Read a Table'.

The first text is extracted from John of Murs's canons to the *Tabule permanentes*.[19] This short canon text, probably written in the late 1340s according to Richard L. Kremer,[20] explains the use of a double-argument table providing the time between mean and true syzygies. The horizontal and vertical arguments of the table are angular quantities expressed in signs and degrees, with a step of six degrees, and standing, respectively, for the solar and lunar mean arguments. The layout of these two arguments is symmetrical; half of the vertical arguments are read from left to right on the first column of the tabular grid, while the other half of the horizontal arguments are read in the reverse direction on the last column of the tabular grid. Similarly, the horizontal arguments are read from top to bottom in the top line of the tabular grid and in the reverse direction in the bottom line of the tabular grid. The top line and left column are to be used in conjunction to define an initial reading direction for the tabular grid. The bottom line and right column are also to be used together to define a second reading direction for the tabular grid. The entries of the table are temporal quantities, expressed in hours and minutes, that can be either additive or subtractive. The additive and subtractive values of the entries depend on the reading direction in which the table was entered. Finally, alongside the actual entries of the table, the canons also refer to horizontal and vertical differences of the entries that are also to be read from the table rather than computed through table entries. Thus, beyond double arguments, the tabular context presents other elements of complexity by mixing angular and temporal quantities, having arguments linked to different astronomical objects, featuring additive and subtractive entries and multiple reading directions, and by noting entries and their differences.

Four rules out of the six composing the text are dedicated to the description of proportional-part procedures allowing us to interpolate in the table depending on the different levels of complexity. The central concern of John of Murs seems to be the additive and subtractive value of the entries. He gives one proportional-part procedure for the case where all considered entries are additive, another symmetrical one for the case where all considered entries are subtractive, and a final, quite different procedure for the mixed case where some considered entries are additive and others are subtractive.[21]

In this work, I only study the first of these procedures. It considers three out of the four potential entries that could be taken into account. Considering the first reading direction of the tabular grid, it can be schematically described in the following way. *E1* is the base value. The horizontal difference is considered with *E2*, and a proportional part of this difference is taken with respect to the rest (*a-A1*). This proportional part is added to or subtracted from *E1*. Similarly, the vertical difference with respect to *E3* is taken, as well as

19 Beatriz Porres and José Chabás, 'John of Murs's *Tabulae permanentes* for Finding True Syzygies', *Journal for the History of Astronomy*, 32 (2001), 63–72; Richard L. Kremer, 'Cracking the *Tabulae permanentes* of John of Murs', in *Editing and Analysing Numerical Tables*, ed. by Matthieu Husson, Clemency Montelle, and Benno van Dalen (Turnhout: Brepols, 2021), pp. 363–424; Matthieu Husson, 'Tables as Tools to Manage Information Flow in Astronomical Computations: Determining Syzygies in Paris 1320–1340', in *Mathematical Practices in the Astral Sciences*, ed. by Matthieu Husson and Karine Chemla (Springer, forthcoming).

20 Kremer, 'Cracking the *Tabulae permanentes*', 368.

21 For a different arrangement of tables and computation avoiding the need for positive or additive quantities, see José Chabás and Bernard R. Goldstein, 'Displaced Tables in Latin: The Tables for the Seven Planets for 1340', *Archive for History of Exact Sciences*, 67 (2013), 1–42.

the proportional part of this difference with respect to the rest (b-$B1$). This proportional part is also added to or subtracted from the result of the previous correction to $E1$. This gives an evaluation of e.

> *Si precise argumenta predicta non invenire, intra cum minoribus propinquioribus a dextris vel a sinistris prout debes, et quod in angula communi invenies scribe cum signo addicionis vel minucionis cum differencia solis, que in lineis transversalibus invenitur, et cum utraque differencia fac signum addicionis si numerus anguli sequentis fuerit maior, vel minucionis si minor. Et postea de differencia solis summe partem proporcionalem secudum proporcionem residui argumenti solis cum quo intrare non potuisti ad 6 gradus, quam partem proportionalem tempori in angulo communi invento adde si cum differencia solis fuerit signum addicionis, vel minue si minucionis. Similiter fac de differencia lune secundum proporcionem residui argumenti lune, et quod provenit adde vel minue tempori preequato, et proveniet tempus inter mediam et veram coniunctionem vel opposicionem.*
>
> If the above-mentioned precise argument is not found, enter the next smallest to the right or to the left as you must, and write what is found in the common angle with the addition or subtraction symbol with the difference of the Sun found in the horizontal line, and with this difference add an addition symbol if the number of the following angle is greater, or a subtraction symbol if smaller. And after, add the proportional part of the difference of the Sun according to the rest of the argument of the Sun with which you could not enter to 6 degrees; add this proportional part to the time in the common angle if, with the difference of the Sun, there as a sign of addition, or subtract it if there is one of subtraction. In the same way, take a proportional part with the difference of the Moon according to the rest of the argument of the Moon, and what is obtained you add to or subtract from the time you have just equated and the time between the mean and true conjunction or opposition is found.[22]

John of Murs's first concern in the description of the procedure is the reading orientation of the tabular grid. One of these reading directions is described as *a dextris* (to the right), while the other is described as *a sinnistris* (to the left). Then, a second instruction concerns the additive or subtractive value of the entry, which also depends on the reading direction. The signs of addition and subtraction are, respectively, *a* and *m*, which are abbreviations of *adde* and *minue*. All of these additional instructions to the core proportional-part procedure are direct elements of management of the tabular grid. The next instruction introduces a first element directly related to proportional-part computation and defines the *differencia solis*. A little later, the text also defines the *residui argumenti solis*. Similarly, the text introduces a *differencia lune* and a *residuus argumenti lune*. The two differences of the Sun and the Moon are the first to be associated with additive or subtractive signs following the exact same rule as the one adopted in the case of a single-argument table. These differences and remainders are then the initial values from which two proportional-part computations are made. Each of these proportional parts are added to or subtracted from

22 Porres and Chabás, 'John of Murs's *Tabulae permanentes*', 69.

the initial entry, in each case following the same rule as the one adopted in the case of a single-argument table.

Concerning terminology, the core procedure is thus adapted by introducing specifiers to the basic terms: *differentia* and *residuus*. Moreover, concerning the sequence of operations, two core proportional-part computation procedures are simply juxtaposed. Around this, a set of instructions is provided to manage the element of tabular layout. These are fundamental mechanisms of extension of the core procedure that were also in use in the simpler situations analysed in the first section of this chapter. I point out that, in this particular proportional-part procedure description, the actual computation of the proportional part itself is not described. Also absent are specific rules related to the precise management of the additive and subtractive signs. This second element is described with more detail in the other proportional-part procedures of the same canons, probably because John of Murs considered it was more important for these procedures to strictly constrain the choices of the users.

Thus, the different layers composing a procedure can be explicitly unfolded or implicitly referred to by expressions such as *adde vel minue* (add or subtract) or even *partem proportionalem* (proportional part). Just as the possibility to extend a core procedure relies on a specific use of the terminology (here forming new expressions by adding specifiers to the terminology of the core procedure), the possibility to implicitly refer to a full procedure also relies on a specific use of the technical terminology: key words or expressions stand, in a metonymic fashion, for a full procedure. These two mechanisms of extension and contraction work in pairs, because were it not possible to contract some internal layers of a procedure through these metonymic means, it would become very difficult to create complex procedures with many layers and long sequences of operations.

A second description of a proportional-part procedure from a double-argument context will give us a more refined view of the way the core procedure can be extended. It comes from John of Lignères' *Tabule magne* (*c.* 1325).[23] More precisely, it comes from a canon describing the use of double-argument tables to obtain the complete equations of the planets, that is, the algebraic combination of their equations of centre and anomaly. The tabular context here is quite similar to the one of the *Tabule permanentes*, except that the values of the entries are, like the arguments, angular quantities. The horizontal argument stands for the mean centre, the vertical argument for the mean argument of the planet. Both are given with a step of six degrees, except for Mars and Venus, which are given a more precise step of three degrees for a few values near the maximum and minimum. Since the procedure proposed by John of Lignères is more involved than the preceding one from John of Murs, I have separated the description into several quotes for the sake of clarity.

This procedure relies on the same three entries as the preceding one: *E1*, *E2*, and *E3*. It also considers the differences *E2-E1* and *E3-E1*, as the preceding one did. However, in addition to this it considers the difference of those two differences, that is, *(E2-E1)-(E3-E1)*. Once these three differences are defined, a set of instructions is given to manage their additive or subtractive properties.

23 Eleonora Andriani and I are currently preparing a critical edition of this work along with the two other instrument texts dedicated to the *Tabule magne* by John of Lignères for Robert Bardis. The Latin texts and translation of the *Tabule magne* I quote here were established by Eleonora Andriani.

Si vero argumentum medium et centrum medium planete precise non inveneris, tunc in eisdem locis quere numerum minorem et propinquiorem et equationem, quam in angulo communi invenies, extra scribe, et scribe etiam differentiam suppositam que est differentia argumenti et hoc si argumentum precise non inveneris et similiter differentiam versus dextram positam extra scribe, que est differentia centri et hoc etiam si centrum precise non inveneris. Et similiter extra scribe differentiam istius differentie ad differentiam sibi recte suppositam, quam habebis per substractionem minoris a maiore, que est differentia argumenti et vocetur differentia differentie et, in accipiendo istas tres differentias, nota bene utrum numerus sequens cuius sunt differentie, quem scilicet non extra scripsisti, sit maior vel minor illo quam accepisti; et si sit maior, pone iuxta differentiam sibi correspondentem 'a', si minor 'm'

> In truth, if you cannot find the mean argument and the mean centre of the planet precisely, then in these places seek the smaller and closer number, and write outside the equatio, that you will find the common angle, and also write the difference set out, which is the difference of the argument and this is if you cannot find the argument precisely, and similarly write outside the difference set towards the right, which is the difference of the centre and this also if you cannot find the centre precisely. And, similarly, write outside the difference of this difference to that set out vertically with respect to it, which you will have by subtraction of the smaller from the greater, which is the difference of the argument, and it is called difference of the difference, and, in taking these three differences, note well if the following number, of which the differences are, namely that you did not write outside, larger or smaller than that you took; and if this is greater, place, near the difference corresponding to it, 'a', and, if smaller, 'm'.[24]

In a manner very similar to John of Murs, John of Lignères defines these three differences by adding specifiers to the basic terminology of the core computation of proportional-part procedure. In this case, this corresponds to the expressions *differentia argumenti* (difference of the argument), *differentia centri* (difference of the centre), and *differentia differentie* (difference of the differences). The last expression has a connotation that is different from the expressions encountered thus far. Until this one, they have used a qualifier with an astronomical meaning: difference *of the Sun*, that is different from the expressions. Here, the conation induces a recursive use of the proportional-part computation, which is distinct in nature, even though this difference of difference is mathematically equivalent to $E2$-$E3$. When giving the rules to assign the additive and subtractive value of these differences, it is interesting to note that the given rule only applies to the *differentia argumenti* and *differentia centri*. The way to define an additive or subtractive value of the *differentia differentie* is not clear. This problem will cause a procedural gap in the next section of this passage. This also fits with John of Lignères' apparent claim that the *differentia argumenti* will always be greater than the *differentia centri*.[25] There, an astronomical insight concerning the relative

24 Edition and translation by Eleonora Andriani.

25 This claim seems to be only statistically true; it is easy to spot situations where the *differentia argumenti* is smaller than the *differentia centri*.

numerical importance of the different anomalies of a planet comes into play in the way the rules are presented.

The next section of this description provides instructions to deal with the differences. First, the vertical dimension is addressed. The user is asked to compute proportional parts of the *differentia differentie* (*E3-E2*) and the *differentia argumenti* (*E3-E1*) according to the portion of the mean argument with which it was not possible to enter the table, that is, the *residui argumenti* (*b-B1*). The proportional part of the *differentia argumenti* (*E3-E1*) is kept for later use. The proportional part of the *differentia differentie* (*E3-E2*) is used to correct the *differentie centri* (*E2-E1*). This corrected (or 'equated', in the language of the text) *differentie centri* (*E2-E1*) is then further transformed by taking its proportional part with respect to the part of the mean centre with which it was not possible to enter the table, that is, the *residui centri* (*b-B1*).

> *et tunc accipe partem proportionalem secundum proportionem residui argumenti ad numerum per quem crescit argumentum in illo loco, partem proportionalem differentie argumenti secundum proportionem residui argumenti et eciam partem proportionalem de differentia differentie secundum proportionem residui argumenti; quo facto partem proportionalem que proveniet ex differentia differentie adde cum differentia centri si iuxta differentia differentie scriptum sit 'a' vel minue si m et proveniet tibi tunc differentia centri equata de qua differentia centri sic equata accipe partem proportionalem, secundum proportionem residui centri ad numerum per quem crescit centrum, et hoc per doctrinam precedentis;*
>
> and then take the proportional part according to the proportion of the rest of the argument to the number by which the argument grows in this place: the proportional part of the difference of the argument according to the residual of the argument and, similarly, the proportional part of the difference of the differences according to the rest of the argument. This done, add the proportional part coming from the difference of the differences to the difference of centres if with the difference of the differences 'a' is written, or subtract if 'm' is written, and you find the equated difference of centre. From this difference of centre thus equated, take the proportional part according to the proportion of rest of the centre to the number by which the centre grows, and this by the preceding instructions.[26]

This passage introduces the *residui argumenti* (rest of the argument) and the *residui centri* (rest of the centre). The recursive connotation of the expression *differentia differentie* is here confirmed by the sequence of operations prescribed. This third difference is used to correct, via the computation of a proportional part, the difference of centre. This corrected difference of centre is then corrected a second time via the computation of a proportional part. This sequence of operations, using the core procedure in a recursive way, is also something that had not been encountered before. In other words, the work on the core procedure of proportional-part computation is done concurrently and cohesively with regard to the terminology and the sequence of operation levels. In this process of correcting the *differentia centri*, the proportional part of the *differentia differentie* is to be

26 Edition and translation by Eleonora Andriani.

either subtracted or added depending on the sign that was assigned to it. However, the rule for assigning this sign is not exactly clear. This creates a choice for the users; I analysed the numerical implications of this in 2012.[27] This procedural gap is a consequence of the layered aspect of how mathematical astronomical procedures are shaped in this and in many other mathematical traditions.[28]

The final section of this description simply takes the two corrections that were computed with their additive and subtractive signs and aggregates them with the initial value of the equation to obtain the final value. The rules for managing additive and subtractive signs are again not very clearly stated. For instance, we do not know what to do depending on whether the initial equation itself comes with either a positive or negative sign.

It is also worth commenting on the astronomical component of this proportional-part procedure. It reverses the order in which the influence of the argument and centre are taken into account in a standard Ptolemaic set of equations to correct planetary longitude. In said equations, a correction depending only on the centre is computed first. Then it is used to correct the argument, and from this corrected argument a further correction is computed depending mainly on the argument and secondarily on the centre. In the procedure described here, a correction depending only on the argument is first computed, and then a composed quantity relying on both the centre and the argument is computed. I cannot help but think that this inversion is a means to reflect the fact that the difference of the argument is often larger than the difference of the centre in the proportional-part procedure; in other words, the argument is numerically dominant in this computation in the eyes of John of Lignères.

In terms of mechanisms of extension of the core procedure, the essential teaching of this second case study is that the terminological level and sequence operations level work concomitantly; a change in the sequence of operations will often be reflected by a change in the way names of the different quantities manipulated in the operations are constructed. Additionally, there are different ways of combining several instances of the core computation of the proportional part. They can be juxtaposed, as in the first case, or they can be composed or recursively used, as in the second case.

In ways similar to the first extension series of the core procedure in the context of a single argument table, the proportional-part procedures described here are essentially dependent on their tabular context. Thus, it is also possible to look at the impact of the proportional-part procedures on the table layout, as was done in the previous section. The different instances encountered in the corpus are the same as in the preceding section: tables showing differences, proportion tables dedicated to the computation of proportional parts, and tables with a reduced tabular step to avoid the need to use proportional-part procedures. They lead to the same conclusion as in the preceding section; that is, elements of the proportional-part procedure tend to be expressed directly in the table, these new tables' layouts might change the procedural text as they take into account the changes of the tables, and many of these table layout adaptations are dependent on scribal choices.

---

27 Husson, 'Ways to Read a Table'.

28 On the notion of procedural gap, see Matthieu Husson, 'Work Cohesion as a Test of Manuscript Transmission: The Case of John of Lignères *Tabule magne*', in *Alfonsine Astronomy: The Written Record*, ed. by Richard L. Kremer, Matthieu Husson, and José Chabás (Turnhout: Brepols, 2022), pp. 317–42.

3. Table set design and proportional-part procedures

Among the different passages where John of Saxony refers to the core computation of proportional parts he described in the chapter on the Sun, with which I started this study, one in particular is interesting. It is found in the chapter about the computation of the true lunar position. At one point in this computation, one must interpolate between different possible values for the equation of anomaly taking into account the various distances between the Earth and the centre of the Moon's epicycle. This non-linear interpolation, sometimes called Ptolemaic interpolation in the literature, is a procedure frequently used in the *Almagest* when a quantity depends on two variables, one of which is stronger than the other. Ptolemy selects the 'strong' variable and computes the target quantity for fixed values of the 'weak' variable. Then he computes interpolation coefficients in order to take into account the variation of the weak variable.[29] My intention here is not to discuss this procedure in itself, but rather to point out how (in the way it is executed as a tabular procedure) a connection is made to proportional-part procedures. The Ptolemaic interpolation is done with the assistance of specific sub-tables usually presented under the heading of *minuta proportionalia* and *diversitas diametri*. It is in this context that a direct reference is made to the computation of proportional parts.

> *Postea accipe de diversitate dyametri partem proporionalem secundum proportionem minutorum proportionalium que dixi prius servari ad 60, quam scies accipere per canonem solis, quam partem proportionalem adde equation argumenti primo examinate et erit equatio argumenti secundo examinata*
>
> After, take the proportional part of the diversity of the diameter according to the minutes of proportion that were before kept to 60, which you know how to find by the chapter on the Sun. To this proportional part, add the equation of the argument you determined first, and you obtain the equation of the argument second determination.[30]

The key is that the reference is made to the interpolation procedure of the chapter on the true position of the Sun. This is a new mechanism through which the core procedure is extended and adapted to a new context. This mechanism does not rely on an extension of the terminology or on a transformation of the nature or sequence of operations. It builds more directly on the layered structure of the procedures of mathematical astronomy and functions by cross-referencing to a previous chapter. In a way, it is close to the previously seen method where the reference to the proportional-part core computation was made simply by using the expression *partem proportionalem*, but it is somewhat more abstract because it refers to the procedure through the designation of the textual part in which

29 Olaf Pedersen, *A Survey of the Almagest, with Annotation and New Commentary by Alexander Jones* (Springer, 2010), pp. 85–90 and pp. 292–94; John D. North, *Richard of Wallingford: An Edition of his Writings*, 3 vols (Clarendon Press, 1976), III, pp. 178–92; Glen van Brummelen, 'Lunar and Planetary Interpolation Tables in Ptolemy's Almagest', *Journal for the History of Astronomy*, 25 (1994), 297–311.

30 Poulle, *Les Tables Alphonsines*, p. 72.

the procedure is described. Another interesting aspect of this extension of the core procedure is that it produces a proportional procedure that we would classify today as a rather complex kind of non-linear interpolation. This is, however, very simply done; the coefficient that enters the computation of the proportional part is not produced on the spot by the computation of tabular differences, but it comes from specific tables — the *diversita diametri* and the *minuta proportionalia* — that are computed beforehand. One interesting aspect is that the tables providing the coefficients to be used in the proportional-part computations become an integral part of the proportional procedure itself, and they provide a new example of the general tendency of proportional-part procedures to induce the creation of tabular objects. Another consequence of this is that the table providing the coefficients used in the proportional-part computation can be transferred from one astronomical topic to another, thus creating profound connections between them. This is the case, for instance, in the theories of declination and latitudes following the Ptolemaic tradition. The proportion tables that are used in the computation of planetary latitude are a rescaling (to a maximum of 60) of lunar latitude tables, which themselves are rescaled from solar declination tables. Realizing this kind of connection, deeply embedded in the proportional-part procedures used across various aspects of mathematical astronomy can perhaps help to understand some terminological difficulties in declination and planetary latitude theory as well as why, at least in Europe until the time of Bianchini, it was so simple to approximate a declination with a latitude or a latitude with a declination in certain spherical astronomy computations.[31]

A quick glimpse at a final diagram from the *theorica planetarum* tradition may enrich our view of this particular extension of the proportional-part core computations. Diagrams in this text present the geometrical models of the planets, with a focus on explaining how the tables compute what they are intended to compute.

The diagram in Figure 13.4 comes from the manuscript Escorial O II 10, which belonged to and was intensely used by John of Murs. It represents the theory of the superior planets. The planet itself is represented six times in three pairs of symmetrical positions of the epicycle on the deferent circle. A pair covers the case when the epicycle is at apogee (at the top of the figure) or perigee (at the bottom). In these situations, the equation of centre is null. Another pair shows the epicycle with a mean argument equal to 90° or 270°, that is, near to the position where the equation of centre is maximal. The third and final pair of positions shows the epicycle in an intermediary position between the two others. In each of these three pairs of epicycle positions, the body of the planet on the epicycle is placed so that the equation of anomaly is close to its maximum value. These different positions of the planet and epicycle allow us to define the different arcs and quantities relevant for the computation of planetary longitude and to understand their relations in a graphical and qualitative way.

31 Glen Van Brummelen, 'The End of an Error: Bianchini, Regiomontanus, and the Tabulation of Stellar Coordinates', *Archives for the History of Exact Sciences*, 72 (2018), 547–53; Glen Van Brummelen, 'Before the End of an Error: Giovanni Bianchini's Original Flawed Treatise on the Conversion of Stellar Coordinates', *Archives for the History of Exact Sciences*, 75 (2021), 109–24.

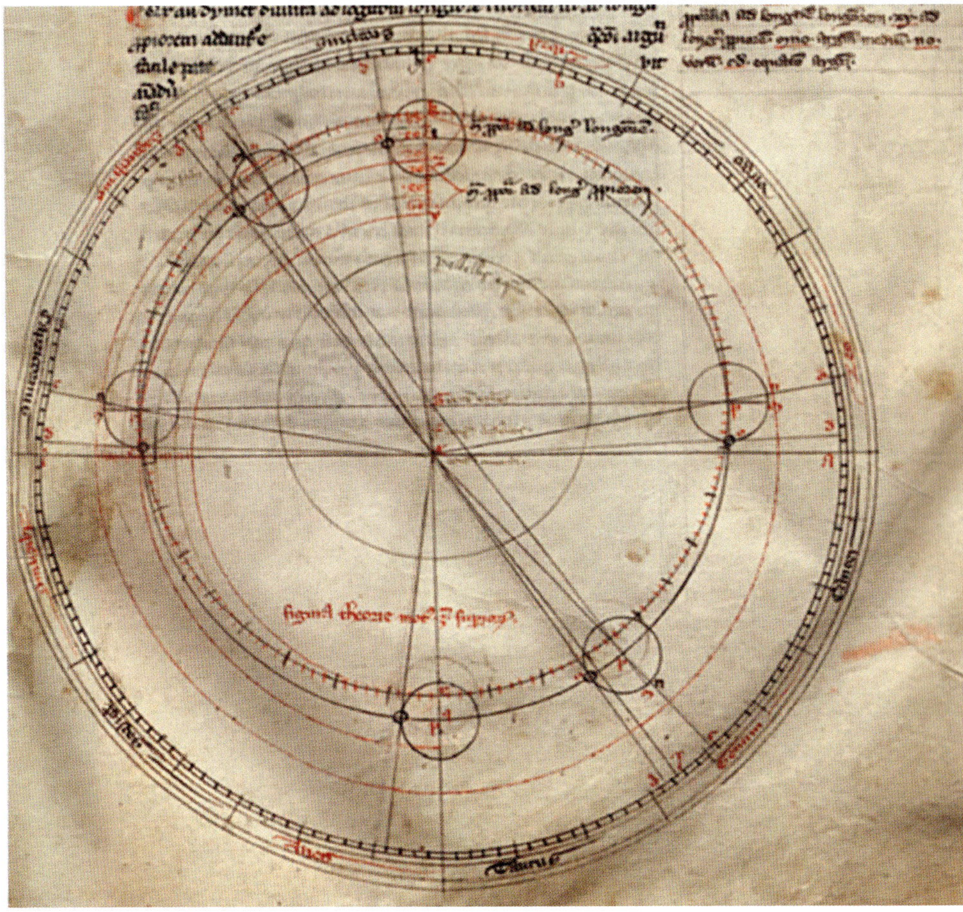

Figure 13.4. *Theorica planetarum* diagram for the superior planet (Escorial O II 10 f. 75r).

A zoomed-in view of the epicycle position at the apogee reveals a series of lines, numbers, and labels that have no direct links to the geometrical models and are in fact only necessary in connection with the tabular procedure (fig. 13.5). The two labels read *minuta proportionalia ad longitudo longior* and *minuta proportionalia ad longitudo propior.* They correspond directly to the two sub-tables that are found in planetary equations set following the Ptolemaic interpolation method and are to be used in the interpolation procedure described just above. Attached to these labels, we find a double graduation (20, 40, 60) extending above (for the *longitudo longior*) and below (for the *longitudo propior*) a central line. This graduation directly represents the values of the *minuta proportionalia* found in the tables. Finally, each of these numerical labels is attached to a given circular line representing the different possible distances between the Earth and the centre of the epicycle. These changing distances need to be considered when evaluating the effect of the equation of anomaly on the true position of the planet; the closer the epicycle is, the greater the importance of the equation of anomaly. Thus, these diagrams propose a geometrical interpretation of the tabular device attached to Ptolemaic interpolation.

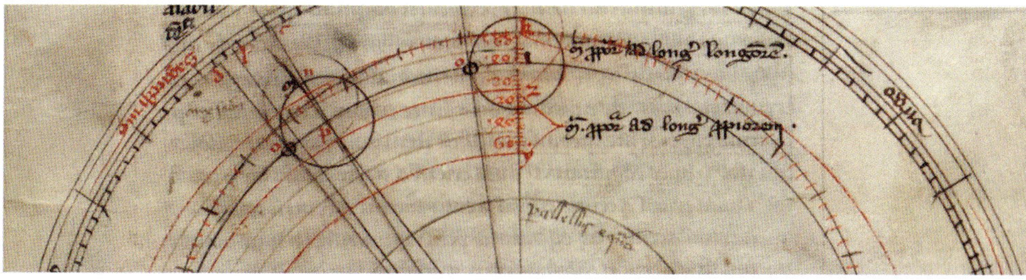

Figure 13.5. *Zoomed-in view of* the *minuta proportionalia* lines on the *theorica planetarum* diagram for the superior planet (Escorial O II 10 f. 75r).

The point of interest here is that those diagrams include lines for the *minuta proportionalia*, thus providing a geometrical interpretation of a tabular quantity of which the initial motive is computational. This is a final instance of this general tendency that procedures (and especially proportional-part procedures in this case) tend to induce specific tabular and diagrammatic objects that are directly connected to them.

Conclusion

Proportional-part procedures form a family of mathematical procedures linked by a common use of a core procedure: the computation of proportional parts. The core procedure is characterized by a specific terminology and a sequence of operations. The core procedure is expressed textually in a layered way with different sets of rules to handle the numbers, the tabular context, precision, and so on. Astronomers built on these to extend the core procedure to various proportional-part procedures adapted to different contexts. The mechanisms they use to extend the core procedure include a work with the terminology (formation of new expression with the specific terminology of the core procedure), a work with the operation sequence (changing the order, juxtaposition, iteration), a work with the nature of the operations (different divisions, iterative addition instead of multiplication), and a work on the tabular and diagrammatic expressions of the procedures. In all these extensions, the layered nature of the textual description of the procedure is important because new layers can be added or developed, while others can be referred to in very concise ways or even implicitly using keywords from technical terminology or more abstract textual references. Identifying these very general mechanisms concerning the ways astronomers worked with their procedural material is the first result of this contribution.

Analysing proportional-part procedures is also a way of better understanding the synergies between the textual, tabular, and diagrammatic modes of expressions of mathematical astronomy. In particular, several instances of procedural objects inducing new features of tabular and diagrammatic objects were shown. Some of these new diagrams and tables in turn induce the production of new procedures.

Proportional-part procedures animate a large portion of the mathematical practices of astronomers. They are important in the manipulation and conception of tabular expression of relation between astronomical quantities from the basic level of single argument tabular manipulations up to the conception of full table sets, thus building deep connections between different astronomical topics. One reason for this ubiquity, I surmise, is that proportional-part procedures operatively express a fundamental mathematical intuition of astronomers concerning the regularity of the relation they are dealing with; tabular differences should not have sudden breaks in their values or in the way they increase or decrease.

I have analysed proportional-part procedures mostly relying on the corpus of Alfonsine astronomy. The study of this corpus in particular could be systematized, but it would be also very rewarding to build a global and long-term analysis of this family of procedures. Comparing different astronomical traditions could give new insights into astronomers' conceptions of quantities (discrete and continuous), approximations, and precision, or ways to shape mathematical understanding of astronomical phenomena. Finally, it could also contribute to a deeper understanding of how the word 'interpolation' took on a mathematical meaning.

Manuscript sources

Paris, Bibliothèque nationale de France, lat. 7197
Paris, Bibliothèque nationale de France, lat. 14481
Paris, Bibliothèque nationale de France, Col. 60
El Escorial, Biblioteca del Monasterio de El Escorial, O-II-10

Bibliography

Brummelen, Glen van, 'Lunar and Planetary Interpolation Tables in Ptolemy's Almagest', *Journal for the History of Astronomy*, 25 (1994), 297–311.

______, 'The End of an Error: Bianchini, Regiomontanus, and the Tabulation of Stellar Coordinates', *Archives for the History of Exact Sciences*, 72 (2018), 547–63.

______, 'Before the End of an Error: Giovanni Bianchini's Original Flawed Treatise on the Conversation of Stellar Coordinates', *Archives for the History of Exact Sciences*, 75 (2021), 109–24.

Busard, Hubertus L., *Het rekenen met breuken in de middleleeuwen, in het bijzonder bii Johannes de Lineriis* (Brussels: Vlaamse Academie, 1968).

Chabás, José and Goldstein, Bernard R., 'Displaced Tables in Latin: The Tables for the Seven Planets for 1340', *Archives for the History of Exact Sciences*, 67 (2013), 1–42.

Husson, Matthieu, 'Ways to Read a Table: Reading and Interpolation Techniques in Canons of Early Fourteenth-Century Double-Argument Tables', *Journal for the History of Astronomy*, 43 (2012), 299–319.

______, 'La *Tabula tabularum* de Jean de Murs et les modèles de l'arithmétique médiévale', *Cahiers de recherches médiévales et humanités*, 27 (2014), 97–122.

______, 'Computing with manuscripts: Time between Mean and True Syzygies in John of Lignères *Tabule magne*', in *Editing and Analysing Numerical Tables*, ed. by Matthieu Husson, Clemency Montelle, and Benno van Dalen (Turnhout: Brepols, 2021), pp. 425–69.

______, 'Astronomers' Elementary Computations in Paris between 1321 and 1327: The Circle Around John of Lignères and John of Saxony', in *Mathematical Practices in the Astral Sciences*, ed. by Matthieu Husson and Karine Chemla (Springer, forthcoming).

______, 'Tables as Tools to Manage Information Flow in Astronomical Computations: Determining Syzygies in Paris 1320–1340', in *Mathematical Practices in the Astral Sciences*, ed. by Matthieu Husson and Karine Chemla (Springer, forthcoming).

______, 'Work Cohesion as a Test of Manuscript Transmission: The Case of John of Lignères *Tabule magne*', in *Alfonsine Astronomy: The Written Record*, ed. by Richard L. Kremer, Matthieu Husson, and José Chabás (Turnhout: Brepols, 2022), pp. 317-42.

Kremer, Richard L., 'Exploring a Late-Fifteenth-Century Astrologer's Toolbox: British Library Add MS 34603', in *Alfonsine Astronomy: The Written Record*, edited by Richard L. Kremer, Matthieu Husson, and José Chabás (Turnhout: Brepols, 2022), pp. 107–41.

______, 'Cracking the *Tabulae permanentes* of John of Murs', in *Editing and Analysing Numerical Tables*, ed. by Matthieu Husson, Clemency Montelle, and Benno van Dalen (Turnhout: Brepols, 2021), pp. 363–424.

North, John D., *Richard of Wallingford: An Edition of his Writings* 3 vols (Oxford: Clarendon Press, 1976).

Pedersen, Fritz S., *The Toledan Tables: A Review of the Manuscripts and the Textual Versions with an Edition* (Copenhagen: C. A. Reitzels Forlag, 2002).

Pedersen, Olaf, *A Survey of the Almagest, with Annotation and New Commentary by Alexander Jones* (New York: Springer, 2010).

Porres, Beatriz and Chabás, José, 'John of Murs's *Tabulae permanentes* for Finding True Syzygies', *Journal for the History of Astronomy*, 32 (2001), 63–72.

Poulle, Emmanuel, *Les Tables Alphonsines avec les canons de Jean de Saxe: Édition, traduction et commentaire* (Paris: Éditions du Centre national de la recherche scientifique, 1984).

Stedall, Jacqueline A., *The Arithmetic of Infinitesimals, John Wallis 1656* (New York: Springer, 2004).

Part 4

On the Border of Mathematical Astronomy

RICHARD L. KREMER

What can We Learn about Computational Practices from a Study of Medieval Ephemerides?*

The recomputation of historical astronomical ephemerides originated about fifty years ago when Emmanuel Poulle and Owen Gingerich began to convert medieval astronomical tables into a FORTRAN 77 code able to run on electronic computers. Rather than interpolating the equations (i.e., corrections added to or subtracted from the mean motions) directly from tables, which in manuscript transmission may exhibit scribal errors and inconsistencies in the parameters, their algorithms recomputed the equations from the underlying geometrical models and the medieval parameters and thus replicated the astronomical theories rather than the extant tables.[1] Poulle and Gingerich first implemented

* This work was supported by the ERC project ALFA: Shaping a European Scientific Scene, Alfonsine astronomy, CoG 723085, PI Matthieu Husson. I thank José Chabás, Matthieu Husson, and Stefan Zieme for critiquing earlier drafts of this chapter and members of the ALFA Project, with whom earlier versions were workshopped, for their suggestions. I am also indebted to Benno van Dalen for a discussion of massive computation in Islamic astronomy; to Lars Gislén, with whom I worked to build digital spreadsheets to recompute ephemerides positions; and to the ongoing project, funded by the German Research Council, to digitize and make freely available manuscripts from the Bibliotheca Amploniana at the University of Erfurt. All images of manuscripts or printed editions in this chapter are reproduced with permission of the holding institutions or Creative Commons licenses.

1 Emmanuel Poulle and Owen Gingerich, 'Les Positions des planètes au Moyen Âge: Application du calcul électronique aux tables alphonsines', *Académie des inscriptions & belles-lettres, Comptes rendus* (1968), 531–48. Somewhat earlier, E. S. Kennedy had used FORTRAN II to programme individual astronomical tables (e.g. oblique ascensions, solar equation, latitudes); with Honorino Mielgo, he also developed a program to compute mean and true planetary positions for given dates (not ephemerides) with parameters extracted from Arabic zijes, including the Toledan Tables. See E. S. Kennedy, 'The Digital Computer and the History of the Exact Sciences', *Centaurus*, 12 (1968), 107–13; *idem*, 'Computer Applications in the History of Medieval Exact Sciences: Suggestions for Further Research', in *Science and Technology in the Islamic World: Proceedings of the XX*th *International Congress of History of Science, vol. 21*, ed. by S. M. Razaullah Ansari (Turnhout: Brepols, 2002), pp. 65–70; Julio Samsó, *On Both Sides of the Strait of Gibraltar: Studies in the History of Medieval Astronomy in the Iberian Peninsula and the Maghrib* (Leiden: Brill, 2020), p. 175, n. 91. Slightly earlier, historical ephemerides of planetary positions, following modern theories of celestial mechanics, had been computed with electronic computers. See Bryant Tuckerman, *Planetary, Lunar, and Solar Positions, 601 B.C. to A.D. 1649*, 2 vols (Philadelphia: American Philosophical Society, 1962–64); W. D. Stahlman and Owen Gingerich, *Solar and Planetary Longitudes for Years -2500 to +2000* (Madison: University of Wisconsin Press, 1963); Owen Gingerich, 'Applications of High-Speed Computers to

Richard L. Kremer • Dartmouth College

Alfonsine Astronomy: Computational Practices, ed. by Matthieu Husson, José Chabás, and Richard L. Kremer, Turnhout, 2026 (*Alfonsine Astronomy: Studies and Sources*, 3), p. 441-486
© BREPOLS PUBLISHERS 10.1484/M.ALFA-EB.5.137919

the Parisian Alfonsine Tables (henceforth PAT), computing a daily ephemeris of planetary longitudes for a period stretching from the last quarter of the thirteenth century to the mid-sixteenth century. They also computed daily positions according to modern theory for those dates (the 'actual' locations of the planets in the sky, as observed from Earth) and thereby evaluated the 'accuracy' or 'errors' of the PAT, finding that the Alfonsine predictions generally would have matched the sky to about ± 3 degrees of longitude or less, except for Mercury, where the errors can reach fourteen degrees. Although earlier astronomers such as Regiomontanus, Copernicus, and Tycho Brahe had compared isolated observed positions against Alfonsine predictions, Poulle and Gingerich were the first to systematically explore patterns of error that would have been generated by the most widely circulating set of astronomical tables in medieval Latin Europe.

Gingerich and his collaborator, Barbara Welther, would go on to implement computer subroutines for other sets of astronomical tables (Ptolemy's *Almagest*, the eleventh-century Toledan Tables, Copernicus's *De revolutionibus*, Rheinhold's Copernicus-based Prutenic Tables, and Kepler's Rudolphine Tables). In each case, they produced ephemerides, stretching over dozens of years in the relevant centuries; matched the tabular results against modern positions; and then compared the error patterns of the tables. With such data, they were able to show, for example, that Copernicus's planetary models did not noticeably improve the accuracy of the Ptolemaic or the Alfonsine Tables, correcting a long-standing positivist narrative that had praised Copernican astronomy for its increase in 'accuracy'. In their final work, Gingerich and Welther compared more than a dozen ephemerides, published between the 1650s and the 1780s, against modern positions, finding 'appreciable improvement', as errors, say, for the superior planets, decreased from forty arcminutes of longitude to only a few arcminutes.[2] As is now well known, data from large state-supported observatories starting in the 1660s, new theories of celestial mechanics based on Newton's laws, and computational tools of calculus and logarithms had, not surprisingly, upended the practices of Ptolemaic mathematical astronomy and, by the eighteenth century, had introduced new regimes of precision into positional European astronomy.[3]

In this paper, however, I am not interested in analysing how well earlier computations of planetary motions matched the sky. Rather, I want to examine how earlier computers,

---

the History of Astronomy', *Vistas in Astronomy*, 9 (1967), 229–36. For a recent example of computing historical planetary positions with parameters rather than tables, see Johannes Thomann, 'Reverse-Engineering Applied to Ephemerides: Analysis and Edition of the Arabic Ephemeris of 1327/7 CE (Ms Cairo, Dār Al-Kutub, Mīqāt 817)', in *Editing and Analyzing Numerical Tables: Towards a Digital Information System for the History of Astral Sciences*, ed. by Matthieu Husson, Clemency Montelle, and Benno van Dalen (Turnhout: Brepols, 2022), pp. 459–510. For a cursory comparison of Regiomontanus's lunar ephemerides against modern positions, see Wolfgang Kokott, 'Regiomontans Ephemeriden für die Jahre 1475 bis 1506', *Sterne und Weltraum*, 36 (1997), 446–51.

2 Owen Gingerich, 'The Role of Erasmus Reinhold and the Prutenic Tables in the Dissemination of Copernican Theory', *Studia Copernicana*, 6 (1973), 43–62, 123–25; *idem*, '"Crisis" Versus Aesthetic in the Copernican Revolution', *Vistas in Astronomy*, 17 (1975), 89–95; Owen Gingerich and Barbara L. Welther, 'The Accuracy of the Toledan Tables', in *Prismata: Festschrift für Willy Hartner*, ed. by Y. Maeyama and W. G. Saltzer (Wiesbaden: Steiner, 1977), pp. 151–63; Owen Gingerich and Barbara L. Welther, *Planetary, Lunar, and Solar Positions, New and Full Moons, A.D. 1650–1805, Memoirs of the American Philosophical Society*, 59S (Philadelphia: American Philosophical Society, 1983), pp. xi–xxiv, at xvii.

3 For a very useful survey of seventeenth-century mathematical astronomy, see Curtis Wilson, 'Predictive Astronomy in the Century after Kepler', in *The General History of Astronomy*, ed. by René Taton and Curtis Wilson (Cambridge: Cambridge University Press, 1989), pp. 161–206.

human rather than electronic, generated daily ephemerides for periods extending over many years. My interest lies in analysing these computers' mathematical practices, as expressed in their surviving ephemerides. These historical sources reveal, with differing degrees of visibility, two levels of agency.

Most visible are what we might call the content and formatting decisions of the ephemeris-maker. Does the work present only planetary longitudes or are planetary latitudes, syzygies, eclipses of the luminaries, horoscopes for those times, astrological aspects, or other content also included? Are the data given for intervals of one day (usually local noon) or a multiple of days (e.g. every five or ten days)? To what levels of precision are the data computed (to signs, degrees, arcminutes, arcseconds)? Is the local meridian explicitly named in the text or must it be extracted by recomputation of the syzygy or lunar data? Are the signs used to define positions in the ecliptic 'physical' (sixty degrees, as in the early versions of the PAT) or 'natural' (thirty degrees, matching the twelve zodiacal signs)? Does a canon accompany the ephemerides, instructing readers on the meanings of the symbols employed or the procedures required to 'read' positions from the work? Taken together, these visible decisions characterize what we might call the maker's 'signature' for an ephemeris. By examining various signatures in ephemerides produced during the era of Alfonsine astronomy (late-thirteenth to mid-sixteenth centuries), we can learn about the circulation of mathematical practices through late-medieval Latin Europe.

Less visible agency appears in the computational practices required to generate the quantitative data of the ephemerides. Computing planetary positions with astronomical tables, a practice that can be traced back to Babylonian goal-year texts and, of course, to the later *Almagest*, requires a series of steps accompanied by decisions, especially about rounding (or truncating) computed values and about interpolating in tables. For example, computing a planetary longitude for a given date using the PAT requires seven steps:

1) For the given date, find the number of days since the epoch of the table.
2) For the given date, find the motion of the eighth sphere since the epoch of the table.
3) For the given local meridian, find the radices of motions.
4) For the given date, find the mean motions and apogees.
5) For those mean motions, find the equation of center, the true center, and the first equation of the argument.
6) Find the minutes of proportion, the diversity of the diameter, and the corrected equation of the argument
7) Find the true longitude by combining, algebraically, the mean longitude, equation of the center, and corrected equation of the argument.

Completing these steps requires the maker to extract values from tables of calendrical data, mean motions, and equations.[4] Poulle once estimated that it would have taken a medieval computer using the PAT perhaps three hours to compute a set of planetary positions for a

4 In his edition of the *editio princeps*, Poulle distinguished twenty tables as 'alphonsines proprement dites', most of which are needed for any given computation of a longitude. See Emmanuel Poulle, ed., *Les Tables alphonsines avec les canons de Jean de Saxe: Édition, traduction et commentaire* (Paris: Éditions du Centre national de la recherche scientifique, 1984), p. 23.

single time (i.e. to cast one horoscope).[5] However, medieval ephemeris-makers (potentially) computed 365 sets of positions for a year and, in many cases, extended their ephemerides to cover many years. Following Poulle's reckoning, if an astronomer computed nine hours per day, it would have taken him four months of steady work to prepare one year of daily ephemerides. Surely, we might guess, he would have used tricks, shortcuts or approximations to reduce the computational labor. My laptop computer can generate a year of Alfonsine planetary positions in less than a tenth of a second, and I can easily explore alterations in the seven steps described above. Can we, therefore, discover the labor-saving moves of medieval computers by recomputing their ephemerides, following the above seven steps and looking for patterns in the deviations (residuals) between their positions and ours?

The algorithms in my spreadsheet compute true planetary positions by means of the tables in the *editio princeps* (1483) of the PAT. They convert Julian calendar dates to days since epoch (Incarnation) and sum the mean motions by multiplication, rather than using the sexagesimal tables of the PAT. Yet all the equations, including for trepidation, are interpolated linearly from the PAT, however inconsistently constructed. I do not round or truncate intermediate values, carrying through the ten-digit, floating-point precision of my spreadsheet. My algorithms seek to replicate the seven computational steps that late-medieval users of the 1483 PAT (canons and tables) would have followed, assuming that we both understand the canons similarly. When my results diverge from theirs, I shall seek to explain how the residuals could have arisen (e.g. internal rounding, scribal errors in their or my tables, use of different tables, computational errors, sexagesimal versus decimal arithmetic, etc.).[6]

Clearly, a search for the less visible practices of medieval ephemeris-makers will be more challenging than the search for their more visible practices. The goal of this chapter, therefore, is to explore what we can learn about mathematical practices by analysing several Alfonsine ephemerides. We limit our scope to sources produced in that era and will not consider whether pre- or post-Alfonsine ephemerides might also reveal interesting insights into earlier or later computational practices in Ptolemaic astronomy.[7]

In the Alfonsine corpus, as defined by the ALFA Project, the most frequently copied work (more than 180 known manuscript copies) is the PAT. Many of these copies show signs of use (heavily thumbed leaves, additional radices or glosses added in the margins, underlinings, corrected tabular values). Presumably many, if not most, of the early owners of these manuscripts used their copy to compute planetary positions. Unfortunately, however, extant manuscripts produced during the Alfonsine era record few examples of complete computation (i.e. intermediate steps) of planetary positions.[8] Rather, we have

5 Emmanuel Poulle, *Astronomie théorique et astronomie practique au Moyen Âge* (Paris: Palais de la découverte, 1967), p. 16. J. D. North, *God's Clockmaker: Richard of Wallingford and the Invention of Time* (London: Hambledon and London, 2005), p. 306, estimated that a late-medieval 'true expert' may have required 1 to 2 hours to compute a set of eight positions.

6 The basic structure of my Alfonsine algorithms can be seen in the spreadsheets for various historical tables that I developed, over the years, with Lars Gislén. These spreadsheets can be found in DISHAS: https://dishas.obspm.fr/resources.

7 In a project currently underway, I am examining these earlier and later ephemerides.

8 However, see José Chabás, in this volume, 'Computational Practices in Alfonsine Astronomy: Hermann of Saxony's *Practica motuum planetarum*'.

only the final results of their computations, the ephemerides. Nonetheless, the extant ephemerides offer the most extensive examples of large-scale (thousands of individual calculations) computational astronomy in late-medieval Latin Europe.

This chapter will thus consider the two best known, large-scale ephemerides produced during the Alfonsine era, that is to say, John of Saxony's ephemerides for the years 1336–80 (preserved in about twenty manuscripts) and Regiomontanus's ephemerides for the years 1448–84 (manuscript, autograph through 1463) and 1475–1506 (printed in Nuremberg in 1474). By identifying the tables they used to compute daily planetary positions and then using those tables myself to compute positions for the same dates, I try to uncover aspects of their computational practices that are not generally discussed in the canons or instructional texts accompanying the tables. These comparisons will shed light on their intermediate steps, their choices, their ability to manage tens of thousands of sexagesimal numbers, and their search for shortcuts to reduce the computational labor of these massive projects.

Occasionally, however, rhetoric about computational labor does surface as table makers sought to convince readers of the value of new tables. For example, in the canon to the Tables of 1321, one of the first new sets of tables to be based parameters from the PAT, we read: "I, John of Murs, relying on *thousands of numbers*, wanted to impose many painful efforts on myself so that the results of those who conjecture and calculate exactly are based not on approximations but are established over certain durations and regular returns for years to come and whose validity is therefore perpetual."[9] Table makers, according to John of Murs, must endure pain so that table users can calculate more exactly.

1. John of Saxony's ephemerides for 1336–80

We begin with the earliest substantial ephemerides produced by computation with the PAT, a set of daily longitudes presented to a precision of arcminutes for the years 1336–80 authored by John of Saxony (fl. first half of the fourteenth century).[10] These ephemerides

9 Transcribed and translated in Emmanuel Poulle, 'Les Astronomes parisiens au XIVe siècle et l'astronomie alphonsine', *Histoire littéraire de la France*, 43 (2005), 1–54 (p. 26) (my italics). See José Chabás and B. R. Goldstein, 'John of Murs's Tables of 1321', *Journal for the History of Astronomy*, 40 (2009), 297–320.

10 I shall ignore a nearly contemporary ephemeris prepared by John of Lignères, which starts for the year 1341 and covers cycles extending as long as eighty-three years (Jupiter) or as short as eight years (Venus), in the format of Jacob ben Makhir's (also known as Profatius Judaeus) perpetual almanac (*c.* 1300). Also computed from the PAT, John of Lignères' ephemeris gives positions to a precision of degrees and thus does not allow a close analysis of his computational procedures. For an overview of these two first-generation Alfonsine ephemerides, see José Chabás and B. R. Goldstein, 'The Master and the Disciple: The Almanac of John of Lignères and the Ephemerides of John of Saxony', *Journal for the History of Astronomy*, 50 (2019), 82–96. For this study, I took John of Saxony's ephemerides from Erfurt CA 2° 386, ff. 63[v]–107[v], an early copy dated *c.* 1359, according to Wilhelm Schum, *Beschreibendes Verzeichniss der Amplonianischen Handschriftensammlung zu Erfurt* (Berlin: Weidmannsche Buchhandlung, 1887), p. 271. Erfurt CA 2° 386, f. 97[r] contains positions for Saturn for three additional years, from 1381–83. An analysis of those longitudes suggests, however, that they were computed by a different astronomer. The standard deviation of their residuals from my recomputation (0.62 arcminutes) is somewhat higher than that I find for the year 1380 (0.38 arcmin), and for each year, a larger residual (from two-to-four arcminutes) appears near the first stationary point, suggesting that the calculator of the Saturn positions for 1381–83 used a table of Saturn equations with peculiar scribal errors for several entries. John of Saxony's Saturn positions probably extend only through 1380.

follow the format of Jacob ben Makhir's widely circulated perpetual almanac of 1300, computed from the eleventh-century Toledan Tables, in presenting positions for each planet separately rather than listing the eight positions (luminaries, five planets, lunar nodes) for each date on a single line.[11] The slightly earlier ephemerides of William of Saint-Cloud, produced in Paris for the years 1292–1311 and also based on the Toledan Tables but with some parameters altered, follows the latter, all-bodies-per-line format. We might wonder, therefore, how much the early Alfonsine astronomers — John of Saxony, his teacher John of Lignères, or John of Murs — knew of William of Saint-Cloud's computations since all four of them worked in Paris.[12]

John of Saxony's external signature was analysed by José Chabás and B. R. Goldstein in 2019. They note that his ephemerides use thirty-degree signs and, by calculating at intervals of differing numbers of days, significantly reduce the computational labor required to generate forty+ years of positions. The luminaries are presented at one-day intervals, the inferior planets at four-day intervals, Mars at six-day intervals, Jupiter at eight-day intervals, and Saturn and the nodes at ten-day intervals. Rather than computing 2900 positions for a year of daily intervals, John of Saxony's ephemerides required only 1090 positions, reducing the labor by nearly two-thirds. Although John correlated his interval lengths with the mean speeds of the celestial bodies, linear interpolation across some of his intervals would generate errors of some arcminutes in the resulting true longitudes, as I note below.[13] In another labor-saving move, John of Saxony did not include syzygies, planetary aspects (as William of Saint-Cloud did), eclipses, or planetary latitudes. The earliest Alfonsine ephemeris, prepared by a Parisian astronomer who in the 1320s had played a key role in shaping the PAT, thus presents a visible signature that i) offers Alfonsine positions, to a precision of arcminutes, for an extended span of years; ii) significantly reduces the computational labor by foregoing daily positions; and iii) provides no auxiliary information for astrologers casting horoscopes.

We turn next to the less visible features of John of Saxony's computational practices, beginning with the four years of true solar longitudes. The PAT, like the earlier zij of al-Battānī and the Toledan Tables, give the solar equations to a precision of arcseconds (all other equations are given to the arcminute). John of Saxony's ephemerides, however, list solar positions only to arcminutes, so we might expect our recomputation (using equations to the arcsecond) to closely match his if he computed his solar longitudes to a

11 See J. Boffito and C. Melzi d'Eril eds, *Almanach dantis aligherii, sive Profhacii Judaei Montispessulani Almanach perpetuum ad annum 1300 inchoatum* (Florence: Olschki, 1908); José Chabás and B. R. Goldstein, 'The Almanac of Jacob Ben Makhir', in *Editing and Analyzing Numerical Tables: Towards a Digital Information System for the History of the Astral Sciences*, ed. by Matthieu Husson, Clemency Montelle, and Benno van Dalen (Turnhout: Brepols, 2022), pp. 53–78.

12 John of Murs's *Expositio intentionis regis Alfonsii circa tabulas eius*, *c.* 1321, draws heavily on the prologue to William's *Almanach planetarum* of 1292, as noted by José Chabás and B. R. Goldstein in *The Alfonsine Tables of Toledo* (Dordrecht: Kluwer, 2003), pp. 245, 257, 278–80. For a transcription of the *Expositio*, see Emmanuel Poulle, 'Jean de Murs et les tables alphonsines', *Archives d'histoire doctrinale et littéraire du moyen âge*, 55 (1980), 241–71.

13 Canons circulating with the earlier Toledan Tables advise ephemerides makers to make exactly such a correlation, albeit at slightly different intervals (luminaries daily, inferior planets at three days, Mars at five days, Jupiter and Saturn at eight days). See Fritz S. Pedersen, *The Toledan Tables: A Review of the Manuscripts and the Textual Versions with an Edition* (Copenhagen: C. A. Reitzels Forlag, 2002), pp. 542–46.

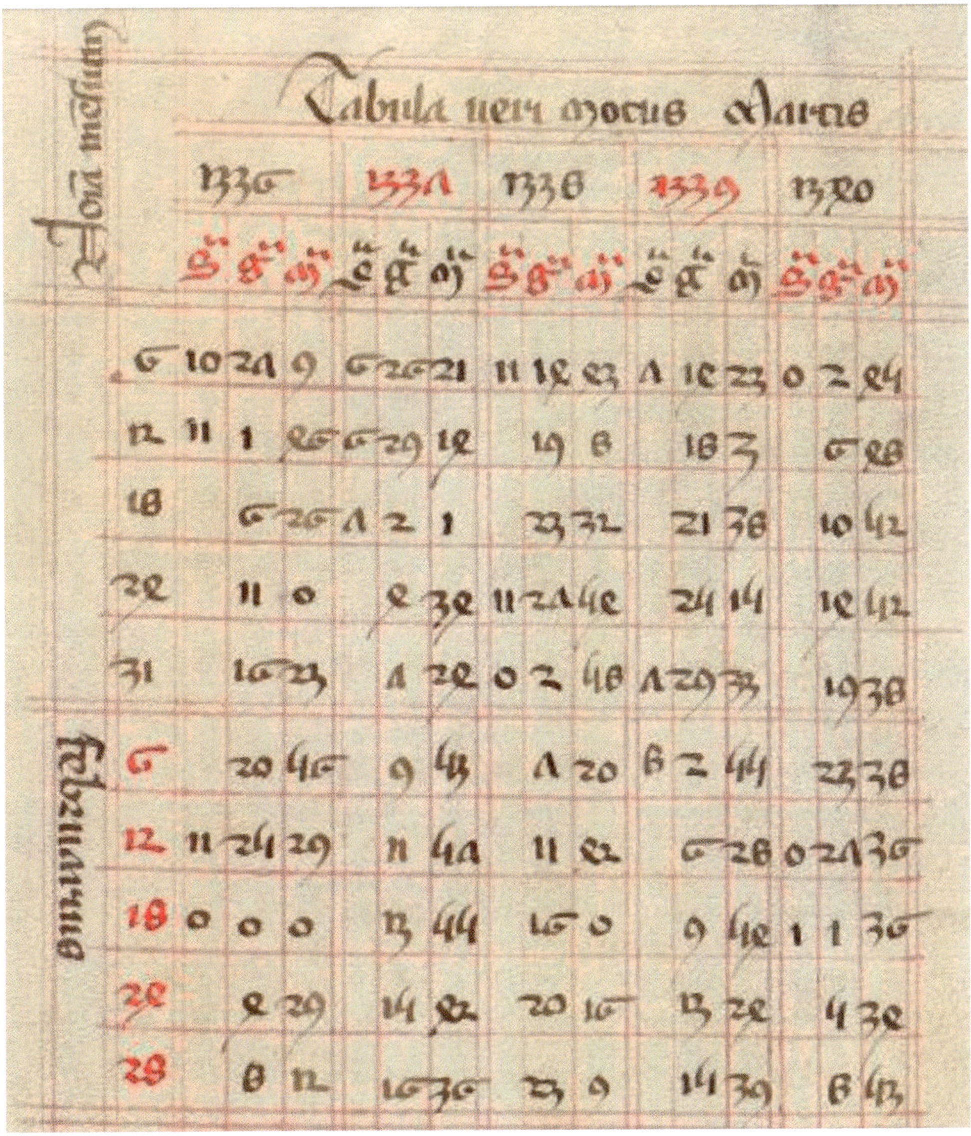

Figure 14.1. John of Saxony's ephemerides for Mars, January-February 1336–40, at six-day intervals, longitudes to signs, degrees and arcminutes. Erfurt CA 2° 386, folio 94v (detail).

precision of arcseconds and then rounded to arcminutes.[14] We will also need to consider the number of sexagesimal digits John may have used when incrementing the mean motions

14 The next known Alfonsine solar ephemeris, prepared in Oxford by William Reed for the years 1341–44, gives longitudes to the arcsecond. For an edition, see Jean-Patrice Boudet and Laure Miolo, 'Alfonsine Astronomy and Astrology in Fourteenth-Century Oxford: The Case of MS Bodleian Library Digby 176', in *Alfonsine Astronomy: The Written Record*, ed. by Richard L. Kremer, Matthieu Husson, and José Chabás (Turnhout: Brepols, 2022), pp. 57–105 (95–98).

over the course of a year (as opposed to incrementing them over hundreds of years since the epoch of the PAT).

Figure 14.2 shows the residuals in the daily solar longitudes for the years 1336–39 (residuals are always defined as historical ephemeris longitude — Kremer longitude). Three features of John's computational procedures immediately emerge. We notice, first, that for all four years, the average residual is very close to zero, which confirms the Chabás-Goldstein finding that John of Saxony set his local meridian consistently to forty-eight time minutes east of Toledo (the meridian of the PAT) and suggests that John and I used the PAT similarly.[15] Likewise, the scatter in the residuals, considered by year, is essentially unchanged over this period (standard deviations in the residuals of 18, 19, 19, and 19 arcseconds, by year). John of Saxony's solar positions, as preserved in Erfurt CA 2° 386, have been very carefully copied. In these four years, I find only one solar longitude with a residual greater than one arcminute (three standard deviations) as the scribe wrote a '12' for what should be a '13' in the minutes place.[16] 107 of the solar residuals for the four years (about seven percent) exceed an absolute value of half an arcminute. Computing positions for the Sun (one equation, requiring a single interpolation) requires fewer steps than for any other body except the lunar nodes (no equation required); the solar positions thus reveal the upper limits of John of Saxony's computational and data management skills and of the abilities of fourteenth-century scribes to control copying errors in large sets of sexagesimal numbers.[17]

The standard deviation in John's solar residuals for each of the four years is about nineteen arcseconds (no outliers dropped). If he had computed the true solar longitudes to the precision of the PAT solar equations (arcseconds) and then rounded them consistently to arcminutes, the rounding errors would be approximately uniformly distributed among ± 30 arcseconds. In this case, it is known that the expected standard deviation (uniform probability distribution) should be about seventeen arcseconds. In each year, however, we find a small number of errors greater than ± 30 arcseconds, which explains why my annual standard deviations are slightly greater than seventeen arcseconds. This analysis also suggests that John did not independently round the intermediate values (mean longitude and solar equation) before adding them to produce the true longitudes in arcminutes; it can be shown that such double rounding (both with uniformly distributed errors) would produce, in the true longitudes, a triangular probability distribution, and the expected standard deviation in the true longitudes would be about 24 ½ arcseconds, considerably larger than that found in my data.[18]

Second, we notice that the linear trend lines (least-squares, for a first-order polynomial fit, i.e. a line) through the residuals differ by years. For the first year, 1336, the trend line is essentially flat (zero slope), which suggests that John's and my computational procedures for the mean motions in longitude are similar. For the years 1337–39, however, the trend lines

15 To extract a meridian from solar longitudes in an ephemeris, I shift the local meridian of my computed positions until the sum of the squares of the residuals reaches a minimum (after discarding outliers).

16 For 27 March 1336. The digits '2' and '3' are often interchanged in fourteenth-century copies of astronomical tables.

17 By 'data management' I mean the practices required to transfer computed results, generated over months of labor, to a medium used by the scribes who prepared the manuscript copies now extant. We have no evidence showing how Alfonsine ephemeris-makers organized, preserved, and copied their results.

18 For the statistics of rounding, see Benno van Dalen, 'A Statistical Method for Recovering Unknown Parameters from Medieval Astronomical Tables', *Centaurus*, 32 (1989), 85–145. I thank Benno van Dalen for helpful conversations on these issues.

show positive slopes: roughly ten arcseconds per year in 1337, seven arcseconds in 1338, and seventeen arcseconds in 1339. In each case, the zero point of the trend line occurs roughly mid-year, when the Sun's mean center is near zero.[19] This pattern suggests that John of Saxony computed his mean motion 'exactly' for a mid-year date, that is, using all the sexagesimal digits offered by the PAT mean motion, and then rounded up the PAT daily solar motion, which he added or subtracted to determine mean longitudes for the remaining days of the year. The *editio princeps* of the PAT sets the Sun's daily motion at 0;59,08,19,37,19,13,56°/d. If John of Saxony had rounded this value to 0;59,08,20°/d, this would, over the course of one year, increase the mean longitude (i.e. tilt the trend line of the residuals) by about two arcsec. If, however, John (for whatever reason) had used al-Battānī's daily motion of 0;59,08,20,46,56,14 °/d rounded up to 0;59,08,21, this would increase the mean longitudes by nearly nine arcseconds per year, a value closer to what we find for the years 1337–39 (the red trend lines in Figs 2b-d). More evidence is needed before we conclude that John of Saxony's ephemerides were computed from hybrid sources. Unfortunately, the Canons to the Castilian Alfonsine Tables do not provide a value for the mean solar motion.[20] However, the trend lines in the residuals for 1337–39 suggest that after computing with the full PAT values an initial true longitude for the date when the Sun was at its apogee, John then constructed his remaining mean solar longitudes by adding a daily increment of 0;59,08,21 rather than consistently using the PAT mean motion tables with eight sexagesimal places.

Third, we notice in Figures 14.2a-d (red ovals) that for each year a series of -1 arcminute residuals cluster around day ninety, roughly the first week of April, and, although less prominent, a series of +1 arcminute residuals cluster around day 225 (third week of August). In Figure 14.3, we combine all the residuals for the twelve-day period from 1–12 April over the four years, when the mean center of the PAT is between 288 and 299 degrees. The red oval in Figure 14.3 marks thirteen residuals that, if shifted by one minute, would match my PAT longitudes to the nearest arcminute. Such a pattern of residuals would result if the solar equation used by John of Saxony differed from my PAT solar equation by one minute for mean centers of 294–98 degrees. Since the solar equation is symmetrical around 180 degrees, we might expect to find a similar one-minute shift (with reversed signs) in John's longitudes for mean centers of sixty-two to sixty-six degrees. As can be seen in Figure 14.2, however, only four such residuals appear in these latter data.

A search of the extant copies of the PAT finds exactly this deviation in the solar equation in several manuscripts (see Figure 14.4). Each contains five consecutive deviant entries in the minutes column of the solar equation for arguments of sixty-three to sixty-seven degrees (or 293–97 degrees).[21] If I compute true solar positions using these five deviant entries, my

19 The trend lines indicate a zero residual, i.e., unrounded PAT mean longitude, for solar mean centers of about 10° in 1337, 337° in 1338, and 353° in 1339. Noise in John's rounding to minutes of longitude could, I hypothesize, slightly affect the slopes and zero intercepts of the trend lines I extract by least-squares. Note that Alfonsine motion of the eighth sphere in the 1330s increases the position of the apsidal lines by about thirty-eight arcseconds/year. However, this shift is not added directly to mean longitudes; it only affects mean centers and therefore the equations of center. A shift of 38 arcseconds in mean center around 1330 can generate changes of about two arcseconds in solar longitude over a year, which are smaller than those seen in Figures 2b-d.

20 Chabás and Goldstein, *The Alfonsine Tables of Toledo*, pp. 35–36, 151–52.

21 See Vatican, Pal.lat. 1403, f. 17r; Munich, Cgm 737, f. 152v; Brussels, 10117-26, 39r, identified as variant S_1 in Richard L. Kremer, 'The Parisian Alfonsine Tables in Manuscript, 1351-1525', in *Alfonsine Astronomy: Expanding the Scene*, ed. by Matthieu Husson and Richard L. Kremer (Turnhout: Brepols, forthcoming), n. 78. Note

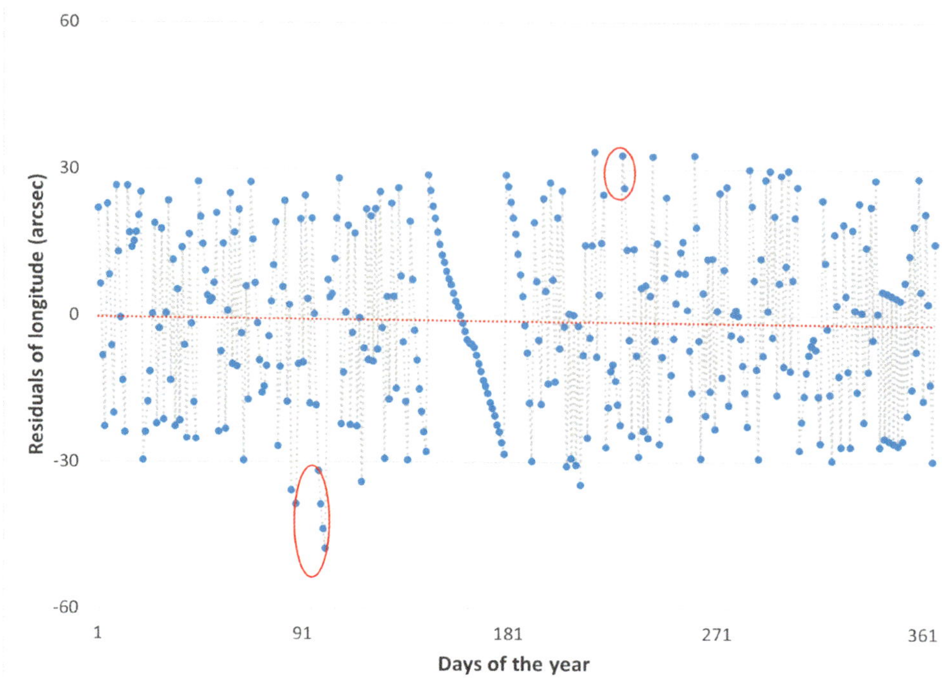

Figure 14.2a. Sun in 1336 at daily intervals, John of Saxony–PAT (forty-eight time minutes east), from Erfurt CA 2° 386, folio 64[r]. One outlier (-68 arcseconds) dropped.

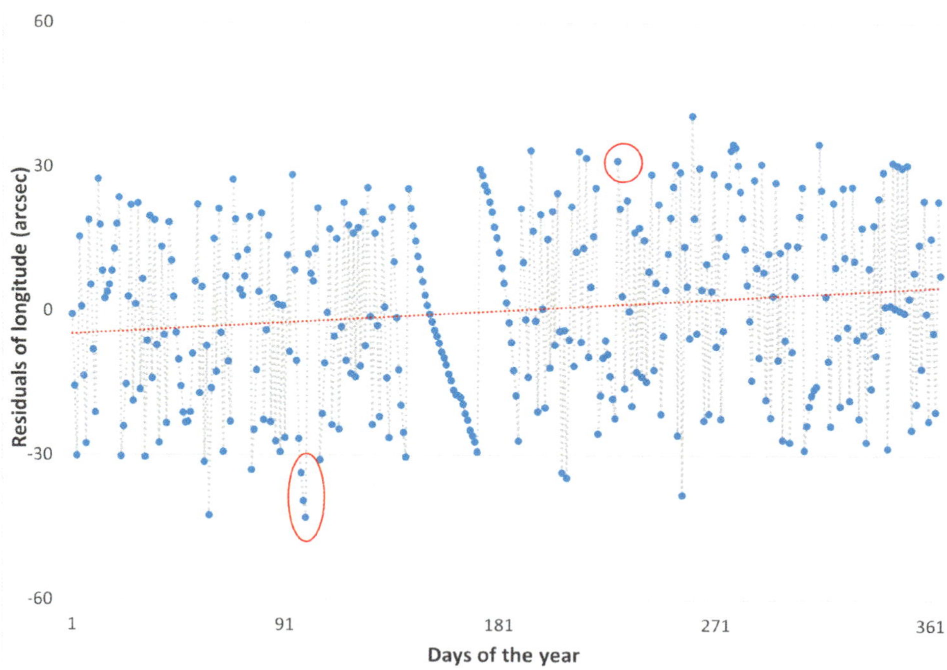

Figure 14.2b. Sun in 1337 at daily intervals, John of Saxony–PAT (forty-eight time minutes east), from Erfurt CA 2° 386, folio 64[v].

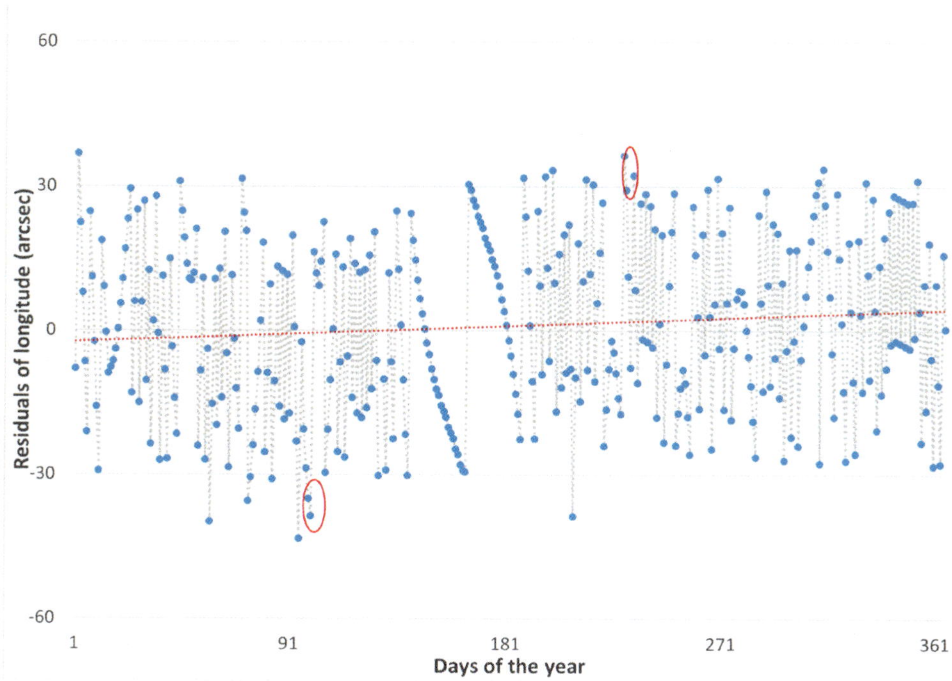

Figure 14.2c. Sun in 1338 at daily intervals, John of Saxony–PAT (forty-eight time minutes east), from Erfurt CA 2° 386, folio 65r. One outlier (69 arcseconds) dropped.

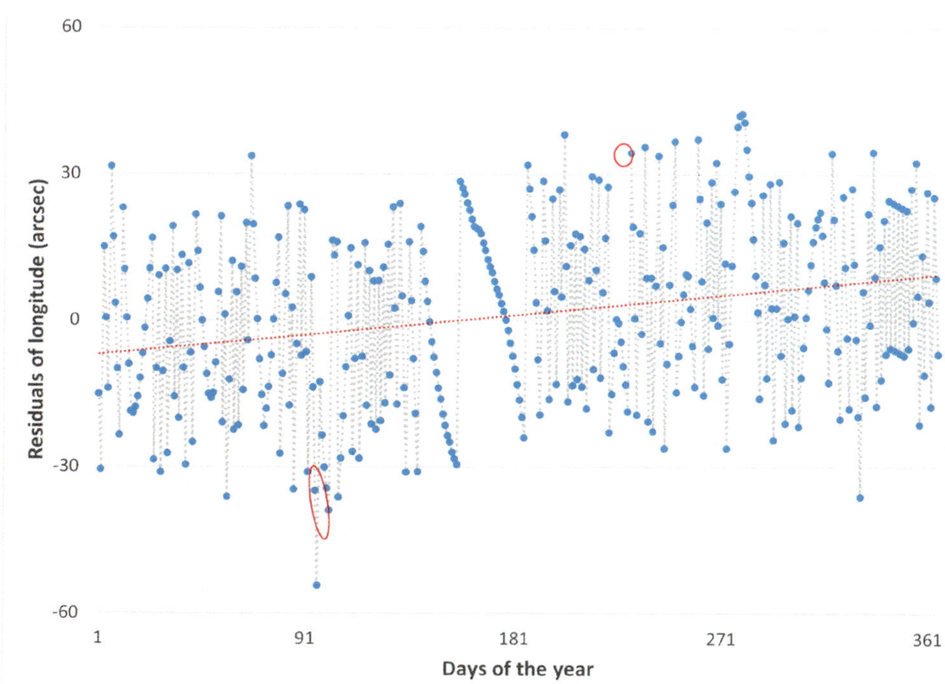

Figure 14.2d. Sun in 1339 at daily intervals, John of Saxony–PAT (forty-eight time minutes east), from Erfurt CA 2° 386, folio 65v.

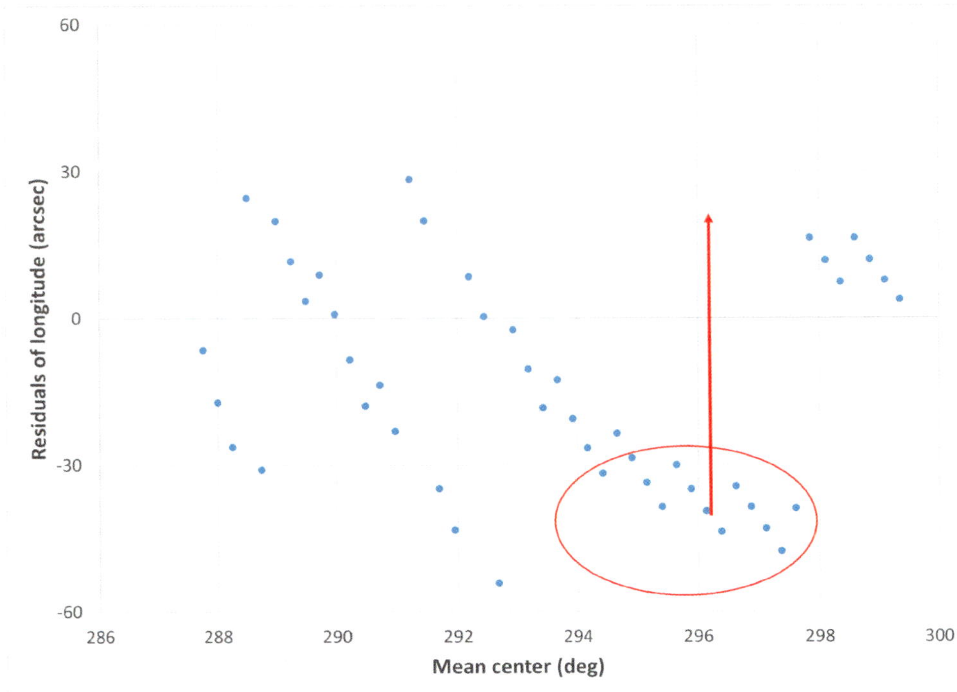

Figure 14.3. Sun in 1336–39, scatter plot of residuals for 1–12 April, from Figs 14.2a-d. Residuals in the red oval could be produced by a deviant solar equation for mean centers of 293–97 degrees.

values match those in John of Saxony's ephemerides to within ± 1 arcmin. It seems quite likely that John of Saxony used these deviant solar equations (they are incorrect, according to modern recomputation of the PAT solar equation) as he prepared his ephemerides.

Finally, we might ask whether John of Saxony's solar positions show any evidence of an interpolation grid, that is, of computing only every n^{th} entry and then linearly interpolating to find the intermediate entries. Might he have, for example, only computed every tenth or fifth value in an effort to save labor? Evidence for such a practice might appear in the first differences of consecutive entries in the solar longitudes. As can be seen in Figure 14.5, no regular, stepwise pattern is present in the differences for 1336; a similar lack of stepwise patterns appears in first differences for the years 1337–39. John of Saxony, at least for the Sun, must have equated his positions for each day.

We turn next to John's lunar positions. Since the PAT provide lunar equations to a precision of minutes, we might expect the residuals in John's lunar positions to be larger than in his solar positions (see Figure 14.6b). If we plot the residuals for the lunar positions,

---

that the PAT solar equations more generally exhibit a non-random pattern of errors that extends up to thirty arcseconds. First noticed by Ludwik A. Birkenmajer, *Nicolaus Copernicus, Part One: Studies on Copernicus's Works and Biographical Materials* [1900], transl. André Goddu (Toruń: Wydawnictwo naukowe Uniwersytetu Mikołaja Kopernika, 2023), p. 181 and independently by Julio Samsó, 'Algunas notas sobre el modelo solar y la teoría de la precesión de los equinoccios en la obra astronómica de Alfonso X', *Dynamis*, 4 (1984), 81–114 (p. 111), this pattern, to the best of my knowledge, remains unexplained.

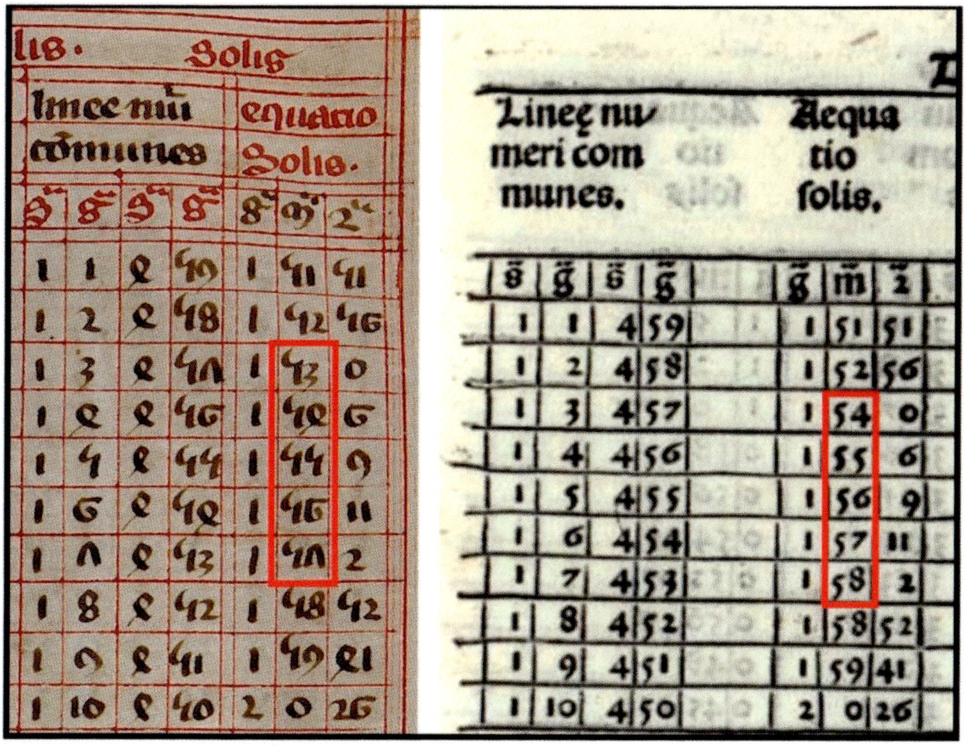

Figure 14.4. Vatican Pal. lat. 1403, folio 17r (left), PAT 1483, sig. e3r (right), showing a scribal attempt to smooth values in the minutes column of the solar equation, which, however, produced one-arcminute errors for five entries in the manuscript; the 1483 edition gives the correct entries.

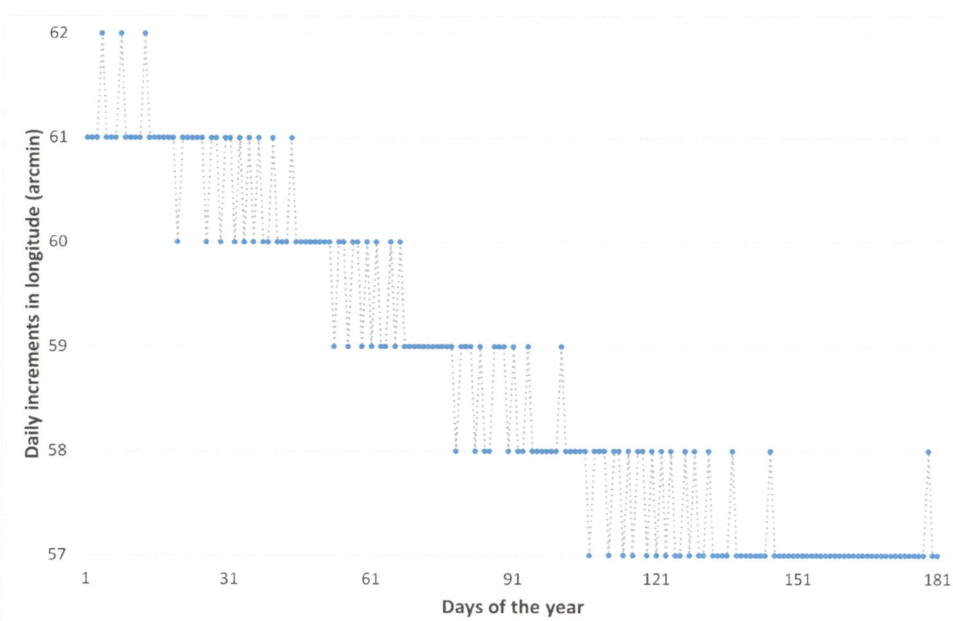

Figure 14.5. First differences (arcminutes/day) in John of Saxony's solar positions for 1336.

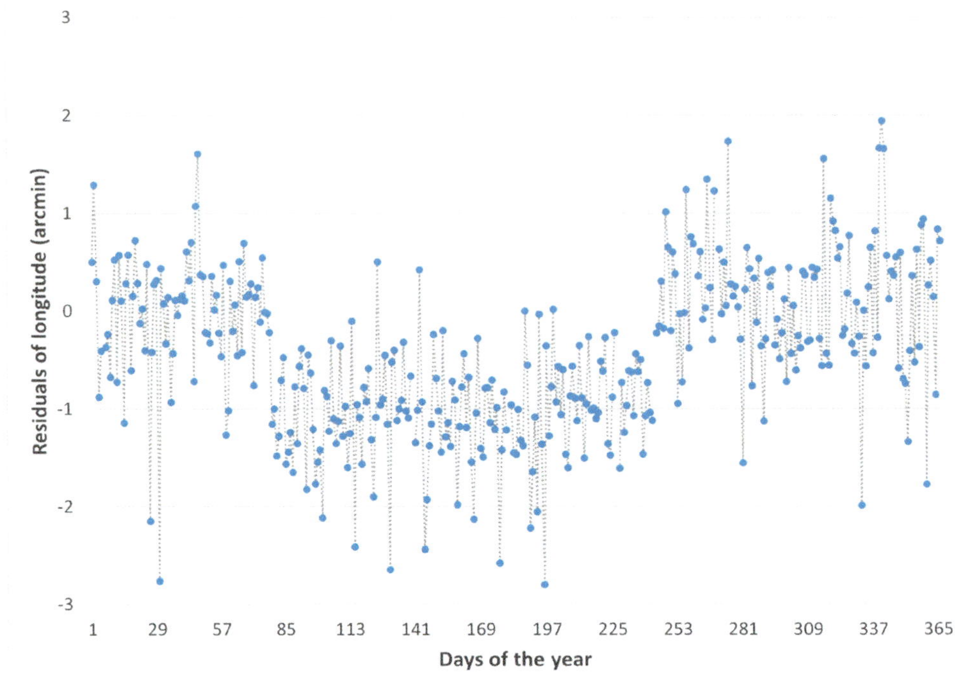

Figure 14.6a. Moon in 1336 at daily intervals, John of Saxony–PAT (forty-eight time minutes east), from Erfurt CA 2° 386, folio 66ʳ. Seven outliers dropped (-5, -9, -3, -5, 4, -9, and 7 arcminutes), one twenty-arcminute and two ten-arcminute scribal errors corrected.

we do indeed find a larger scatter, with standard deviations of the residuals reaching about two-thirds of an arcminute for 1337, after dropping three outliers.

As seen in Figure 14.6a, the residuals for the 1336 lunar positions fall into two samples. From January through 19 March and from 1 September through December, John's meridians match those computed for forty-eight time minutes east. From 20 March through August, however, it appears that John's meridian is fifty time minutes east, producing a shift of one arcminute in the computed true longitudes. John's lunar positions for the years 1337–39 consistently match a meridian of forty-eight minutes (Figure 14.6b). The manuscript copies of John of Saxony's lunar positions for 1336 offer no clues that might explain this slight shift of meridian for the middle six months of the year. Perhaps someone other than John computed the fifty-time-minute lunar longitudes, using a manuscript copy of the PAT bearing a different glossed lunar radix for Paris.[22] Perhaps John divided the work into two batches and inadvertently set the radix to different meridians for each batch (note that the breaks between the two batches occur when the mean lunar argument is at 180 degrees). In any case, the scatter (standard deviation) in the lunar residuals for

22 Note that lunar radices 'for Paris' glossed into fourteenth- and fifteenth-century manuscript copies of the PAT can vary considerably; e.g. from forty-four time minutes east to sixty time minutes east in the glosses described in Richard L. Kremer and Jerzy Dobrzycki, 'Alfonsine Meridians: Tradition Versus Experience in Astronomical Practice *c.* 1500', *Journal for the History of Astronomy*, 29 (1998), 187–99 (p. 194).

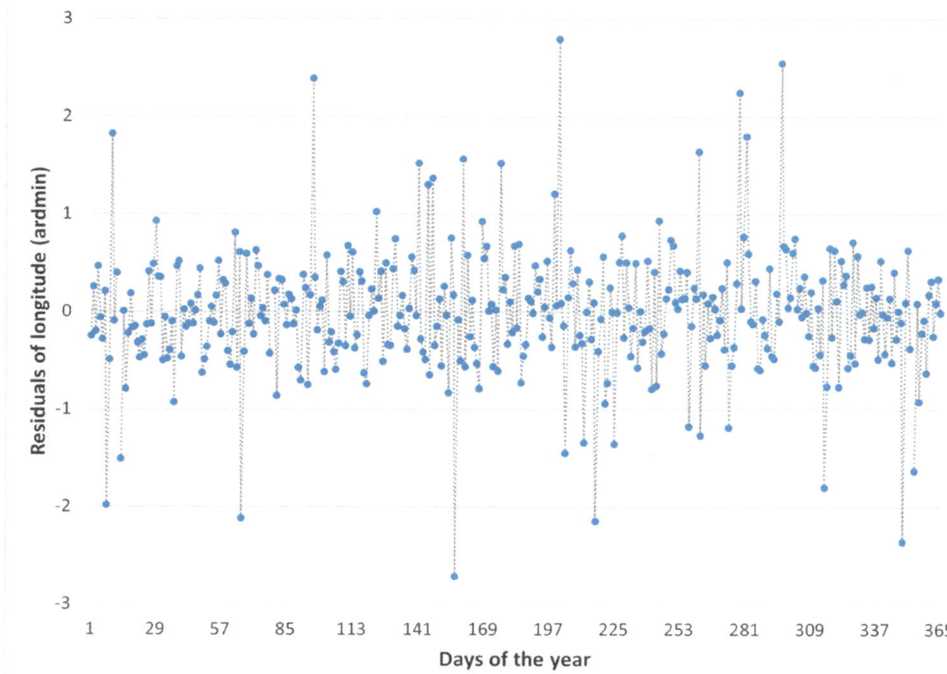

Figure 14.6b. Moon in 1337 at daily intervals, John of Saxony–PAT (forty-eight time minutes east), from Erfurt CA 2° 386, folio 66v. Three outliers (7, -5, and 6 arcminutes) dropped, one ten-arcminute scribal error corrected.

the following year, 1337, is small (2/3 arcminute) and random and strongly suggests that John of Saxony computed each day's lunar longitude and did so with a precision that will exceed, considerably, that found in Regiomontanus's lunar ephemerides computed a century later, as we shall see below.

Finally, we consider John of Saxony's Venus computations, data he provided at four-day intervals, starting on the fourth day of each month. The PAT equations for this planet are to arcminutes, and John's ephemeris provides that precision. As can be seen in Figure 14.7, the residuals are generally less than ± 1 arcminute, except for three values in July 1336 when they approach -4 arcminutes (red points in Figure 14.7). We might suspect scribal errors in the Erfurt manuscript; however, closer inspection reveals that John of Saxony used a deviant set of equations, this time in the values for that planet's *longitudo longior* for mean arguments between sixty-seven and ninety degrees. In 1968, Poulle and Gingerich noticed that 'very many' manuscript copies of the PAT consistently display incorrect values for that column, resulting from an initial scribal error of repeating the same entry three times going down a column, when the correct entry should only have been repeated twice.[23] This 'column slippage' produces a shift of the *longitudo longior* of up to four arcminutes. The *editio princeps* did not reproduce this column slippage; hence, the large deviations of the Venus positions, visible in Figure 14.7, occurring when the

23 Poulle and Gingerich, 'Application du calcul électronique', p. 542. I have found 103 (of 181) manuscripts with this deviation in the Venus equations.

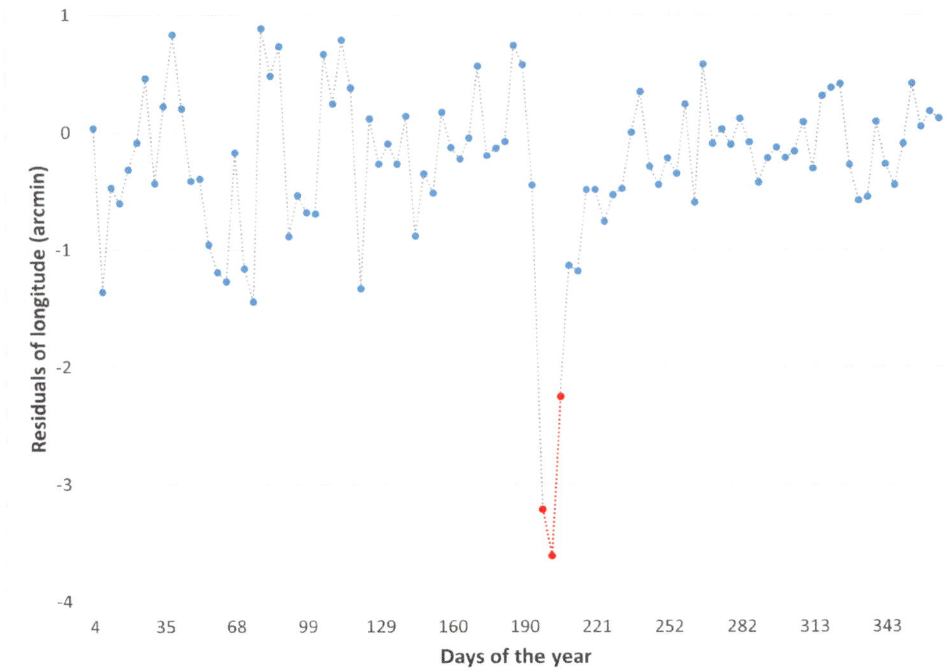

Figure 14.7. Venus in 1336 at four-day intervals, John of Saxony–PAT (forty-eight time minutes east), from Erfurt CA 2° 386, folio 99[r]. Red points mark places where the mean argument is between eighty and ninety degrees.

mean arguments (motion in the epicycle) are between eighty and ninety and 270 and 280 degrees. In 1337, the Venus residuals similarly show large deviations when the mean arguments are between eighty and ninety degrees. In 1338, those mean arguments for Venus occur twice, but the residuals show no large deviations (Figure 14.8). This suggests that for 1338, John of Saxony had corrected the scribal errors in his Venus equations for *longitudo longior*.

Further analysis of the Venus positions reveals a labor-saving computational practice employed by John of Saxony: rounding the daily increments of mean argument in the planet's motion around the epicycle. This becomes visible if we plot the Venus residuals against the mean argument for one anomalistic period (Figure 14.9). Ptolemy's model employs a large epicycle for Venus (radius of 43;10 parts where the deferent radius is sixty parts). When viewed from the Earth, any shift in the mean argument is thus magnified on the 'lower' side of the epicycle. Figure 14.9 shows that the residuals increase significantly for a mean argument ranging from about 150 to 210 degrees, that is, the lower side of the epicycle where the planet is closest to the Earth. This, plus the fact that the residuals trend toward more negative values as the mean argument increases over time (red line in Figure 14.9), suggests that John of Saxony slightly reduced the daily amount by which he incremented the mean anomaly, in contrast to the PAT value of 0;36,59,27,23,59,31°/d. Indeed, I find that I can flatten the peak in the residuals from 150 to 210 degrees and eliminate their downward trend simply by reducing the daily increment

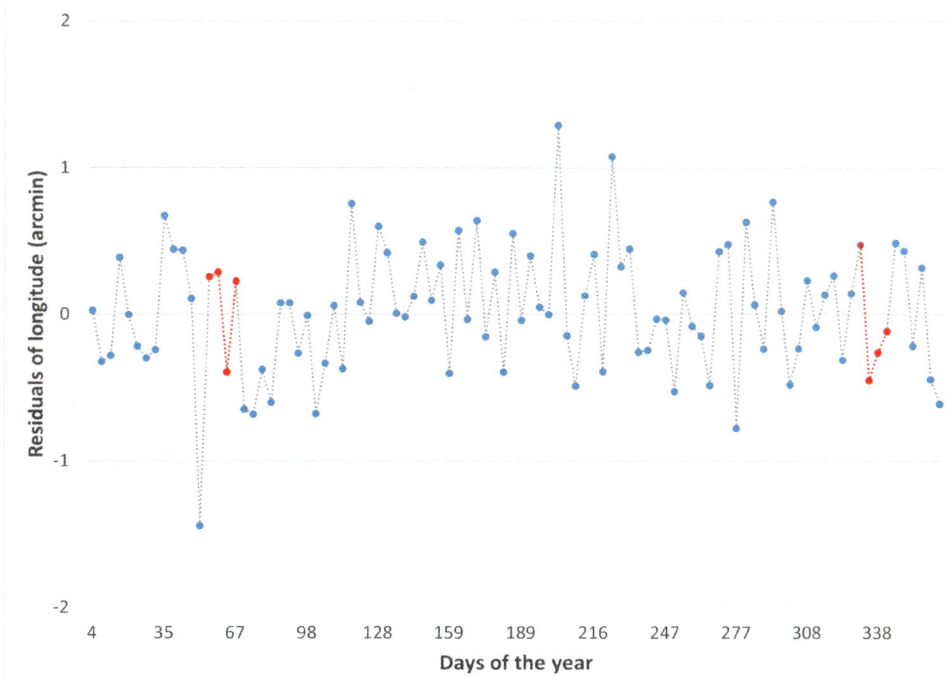

Figure 14.8. Venus in 1338 at four-day intervals, John of Saxony–PAT (forty-eight time minutes east), from Erfurt CA 2° 386, folio 99r. Red points mark places where the mean argument is between eighty and ninety or 270 and 280 degrees.

of mean anomaly to 0;36,59,18°/d (Figure 14.10). This assumes that John computed the first entry for the anomalistic period (where the mean argument is zero) with exact PAT values for the mean motions and then computed the subsequent 583 days of the anomalistic period with a reduced value for the mean argument.[24] Reducing the daily mean argument to three sexagesimal places would match the precision found above in his values for the Sun's daily mean motion in longitude.

I cannot explain why John might have introduced this slight change in a parameter. We might guess that, rather than computing the Venus positions with the PAT, he used the Tables of 1321 authored by John of Murs. These tables employ what has been called a *contratabula* format, presenting the PAT equations in double-entry tables with arguments of the mean center and a period of time equal to a fraction of the anomalistic period (about nine days for Venus). These tables are structured around the times of mean conjunction of a planet with the Sun (when, for the interior planets, the mean argument is zero) and might also have been constructed with the daily increment of mean anomaly slightly

24 In Figure 14.10, I can lower the flat trend line to zero by computing the Venus longitudes for a meridian of forty-five time minutes east. I cannot imagine, however, why John of Saxony would compute for this meridian. Further analysis of John's Venus data is required to explore this feature in the 1338–40 residuals.

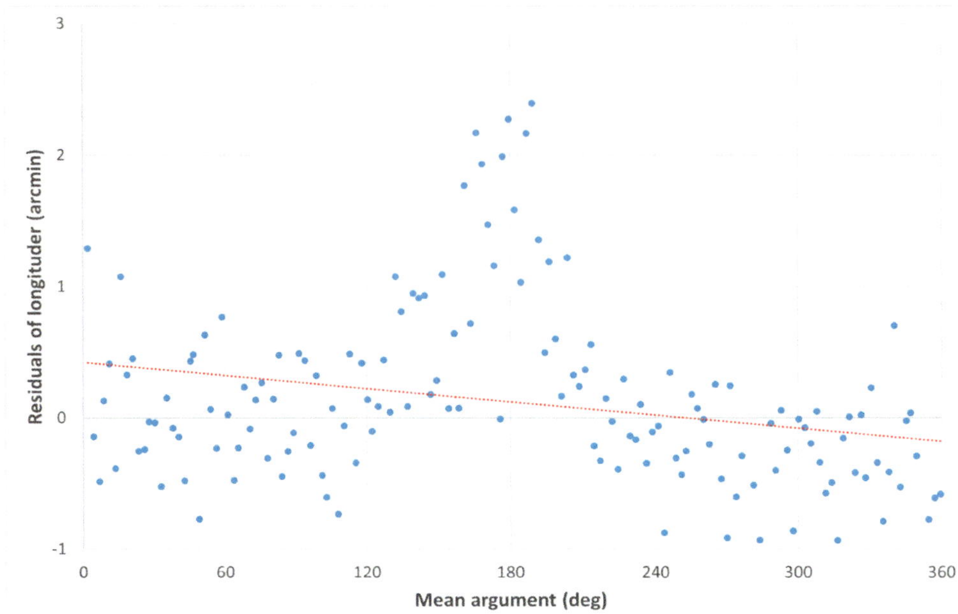

Figure 14.9. Venus, 20 July 1338–20 February 1340 at four-day intervals (one anomalistic period), John of Saxony–PAT (forty-eight time minutes east), from Erfurt CA 2° 386, folio 99^{r}. Two ± 10 arcminute scribal errors corrected.

tweaked from the PAT value.[25] I cannot fully analyse John of Murs's tables for the equations here. However, a comparison of true longitudes for the 1338–40 Venus anomalistic period generated from his *contratabula* against the PAT true longitudes for those times shows considerably more scatter than I find in the residuals of John of Saxony's Venus longitudes (see Figs 14.9 and 14.11).[26] Hence, it does not seem likely that John of Saxony used John of Murs's *contratabula* to generate Venus ephemerides for 1338–40; we do note, however, that John of Murs's Venus equations, for the lower part of the epicycle, are computed at roughly four-day intervals, as are the entries in John of Saxony's Venus ephemerides. Moreover, the residuals in Figure 14.11 show a slight negative peak at the lower part of the epicycle, which suggests that he might have slightly increased the values of the mean anomaly as he computed the Venus *contratabula*.

This preliminary analysis of John of Saxony's ephemerides has revealed several components of his computational practices:

- The meridian seems generally set to forty-eight time minutes east of Toledo (i.e. Paris), but in practice it may, perhaps inadvertently, be shifted by some minutes.

---

25 For a description of the morphology of these tables, see Chabás and Goldstein, 'John of Murs's Tables of 1321'. They do not, however, recompute the tables or compare a full anomalistic period of positions against PAT values. John of Murs constructed the radices of his tables for the meridian of Toledo, not Paris, and I have made my recomputation for Toledo.

26 Note, however, the slight peak in Figure 14.11, centered on a mean anomaly of 180 degrees. This suggests that John of Murs also adjusted the PAT mean anomaly parameter, increasing rather than decreasing its value.

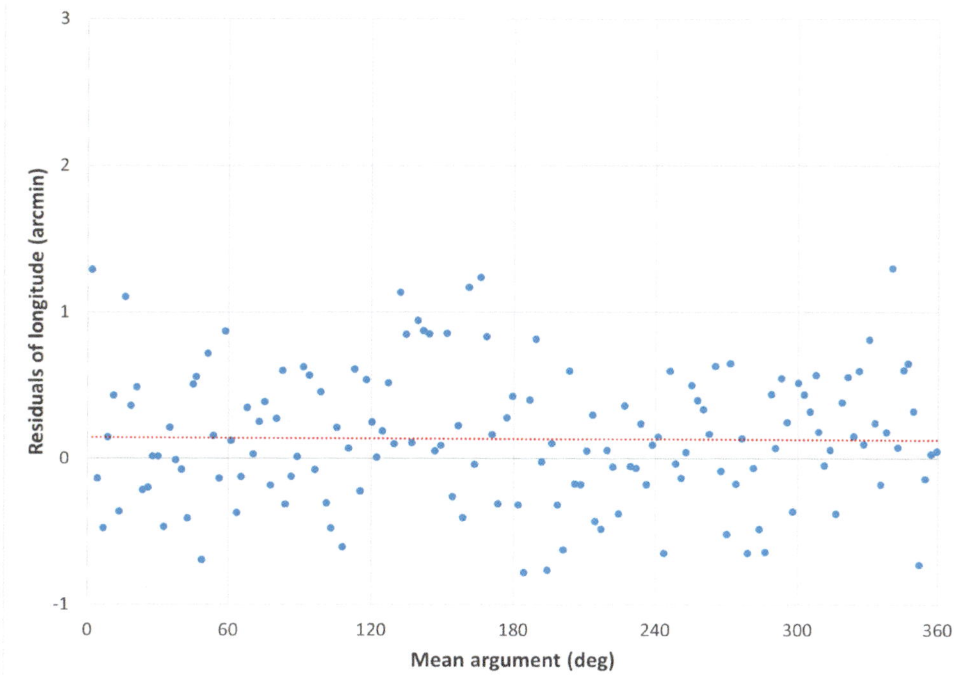

Figure 14.10. Venus, 20 July 1338–20 February 1340 at four-day intervals (one anomalistic period), John of Saxony–PAT (forty-eight time minutes east), from Erfurt CA 2° 386, folio 99r. Two ± 10 arcminute scribal errors corrected, PAT mean argument reduced to 0;36,59,18°/day after the initial computed longitude.

- The true longitudes, computed and copied into the Erfurt manuscript, generally agree with my PAT recomputation to ± 1 arcminute for the Sun, Moon, and Venus, with no more than four or five outliers per year. John of Saxony's sexagesimal arithmetic and my decimal arithmetic apparently do not generate noise that exceeds the precision of the ephemerides (arcminutes).
- Given the accuracy with which John of Saxony worked, we can identify several examples of him employing deviant equation tables, that is, consistently transmitted scribal errors in manuscripts that do not appear in the 1483 *editio princeps*.
- Daily increments of the mean anomaly (or four-day increments for Venus) may have been rounded or truncated to fewer sexagesimal digits than are present in the PAT.
- The lunar and solar positions appear to have been fully computed for each day with no linear interpolation between multi-day computation.
- It seems quite unlikely that John of Murs's *contratabula* were used for Venus.

Hence, the earliest surviving large ephemeris computed with the PAT shows that at least one member of the founding generation of Parisian Alfonsine astronomers was a very competent

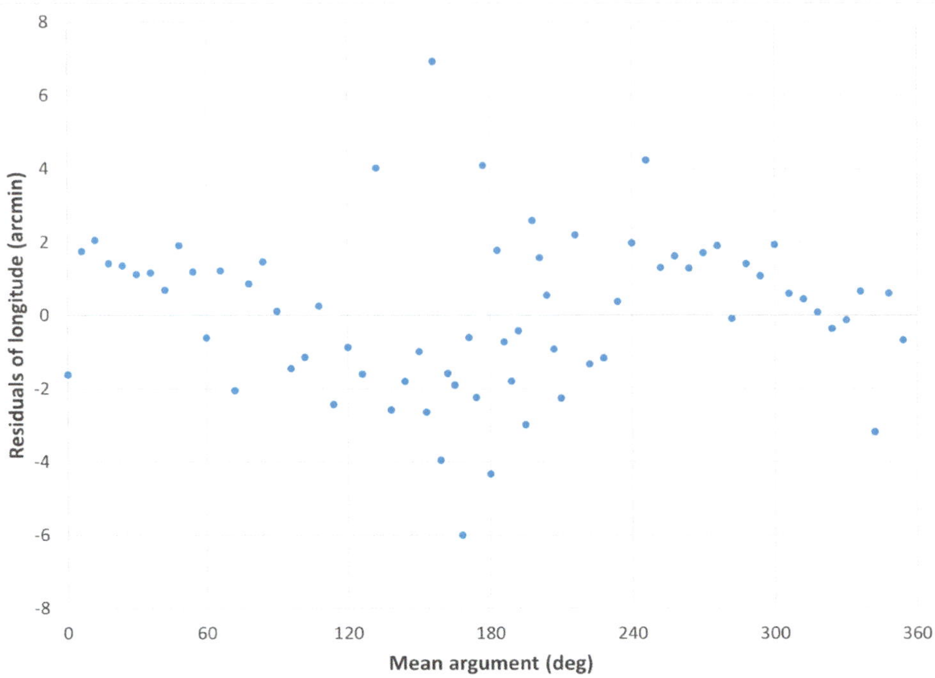

Figure 14.11. Venus, 20 July 1338–20 February 1340 at 9 ¾-day intervals on the upper half of the epicycle, half that length on the lower half (one anomalistic period), John of Murs–PAT (Toledo), Lisbon, Ajuda 52-XII-35, folios 43^{r}–47^{v}. Four ± 10 arcminute scribal errors corrected.

and consistent calculator.[27] Following advice from earlier canons on making ephemerides, John of Saxony reduced the workload by computing daily positions only for the luminaries and multi-day positions for the planets. For the thousands of positions he did compute, John of Saxony consistently followed the steps of the PAT algorithm, as described in his own canon of 1327 (*Tempus est mensura*). The PAT were designed to generate positions to a precision of arcminutes,[28] and their first major user, John of Saxony, produced the first Alfonsine ephemeris to just that precision (except for when Venus was in the lower half of its epicycle). Further analysis, however, may reveal other deviant patterns in the residuals for the other planets.

2. Regiomontanus's ephemerides for 1448–1506

John of Saxony prepared the first major ephemeris in the Alfonsine era; not until Regiomontanus (*c.* 1436–76) would another Alfonsine astronomer compute ephemerides

27 For a discussion of John of Saxony and John of Lignères on precision, see Chabás and Goldstein, 'The Master and the Disciple', p. 95.

28 Mathematical astronomy, in the tradition of Ptolemy, had, of course, long worked to a precision of minutes of time or angles. Not until the mid-sixteenth century would European Latin astronomy begin to employ a precision of seconds.

on a comparable scale. John's data extend for more than forty years; Regiomontanus's cover more than fifty years, thirty-four of which are preserved only in manuscript and thirty-two (from 1475–1506) of which are printed in numerous editions. These latter imprints would circulate widely and be extensively used by astronomers and calendar-makers from the 1480s through the 1510s.[29]

Unlike John's set, preserved as a single, coherent project, the ephemerides of Regiomontanus reflect diverse computational practices and his peripatetic life. Born in Königsberg, a small village northwest of Nuremberg, Regiomontanus may have enrolled at the university in Leipzig in 1447.[30] In 1450, he matriculated at the university in Vienna where he came into contact with the humanist master, astronomer, and court astrologer Georg Peurbach (1432–61) and in 1458 eventually became a Master of Arts. In 1460, the Italian humanist Cardinal Bessarion (*c.* 1395–1472) invited Peurbach and Regiomontanus to return with him to Italy and to write a new abridgment of Ptolemy's *Almagest*. Peurbach died in 1461, but Regiomontanus went to Italy where he perfected his Greek, lectured at the university in Padua, and completed an *Epitome* of the *Amagest*. By 1467 he had moved to the court of Janos Vitéz in Esztergom (Hungary), who commissioned a set of tables for the spherical astronomy of horoscopes. In 1471, Regiomontanus moved to Nuremberg and established a printing shop that issued nine editions between about 1472 and 1475, including a broadside prospectus of some ninety works in mathematics and astronomy that he hoped to print. Corresponding with leading astronomers of Europe, computing new sets of tables for spherical astronomy, writing a comprehensive text on plane triangles, studying Greek mathematical texts, making observations to test the predictions of the PAT, and calling for the 'restoration' of received astronomy, Regiomontanus is usually considered to have been one of the most creative mathematicians and the most reform-minded astronomer of the Latin fifteenth century.[31]

Already as a university student, Regiomontanus had begun computing annual ephemerides. A manuscript now in Vienna, ÖNB Cod. 4988, eventually acquired by the Nuremberg astronomer Johannes Schöner (1477–1547), reveals the development of his ephemeris-making over fifteen years as he expanded the content and experimented with format.[32] Most of the first section of this codex, containing ephemerides for years from 1448

29 For example, nearly 500 editions of broadside almanacs were printed during the incunabula period, from 1462 to 1500. Starting in 1477, a large majority of these editions present astronomical data copied directly from Regiomontanus's printed ephemerides. See Richard L. Kremer, 'Kinetic Almanacs: Astrological Broadsides and Local Markets During the Early Years of Printing', *Cahiers de recherches médiévales et humanistes*, 47 (2024), 497-526.

30 Georg Erler, *Die Matrikel der Universität Leipzig*, 3 vols (Leipzig: Giesecke & Devrient, 1895–1902), 1:161. The name 'Johannes Molitoris' (Johannes Müller) appears in the matriculation record but may or may not refer to Regiomontanus. For the affirmative argument, see Ernst Zinner, *Leben und Wirken des Joh. Müller von Königsberg genannt Regiomontanus*, 2d rev. ed. (Osnabrück: Zeller Verlag, 1968), pp. 13–18; Hans Wussing, 'Regiomontanus and Leipzig', in Ernst Zinner, *Regiomontanus: His Life and Work*, transl. by Ezra Brown (Amsterdam: Elsevier, 1990), pp. 307-11.

31 For an introduction to the massive historiography, see Zinner, *Leben und Wirken*; N. M. Swerdlow, 'Regiomontanus on the Critical Problems of Astronomy', in *Nature, Experiment, and the Sciences*, ed. by T. H. Levere and W. R. Shea (Dordrecht: Kluwer Academic Publishers, 1990), pp. 165–95; Michael H. Shank, 'Regiomontanus Versus George of Trebizond on Planetary Order, Distances, and Orbs (*Almagest* 9.1)', in *Ptolemy's Science of the Stars in the Middlle Ages*, ed. by David Juste, *et al.* (Turnhout: Brepols, 2020), pp. 305–86.

32 The manuscript consists of three sections that circulated independently before being bound, probably by Schöner, sometime after 1525 (the front pastedown bears his printed exlibris). On folio 1^{r}, Schöner wrote that in 1525 the Nuremberg humanist scholar and businessman Willibald Pirckheimer had purchased the

to 1462, is an autograph. The initial year of 1448 (quire 1) includes only the daily planetary longitudes (including the luminaries and lunar nodes) and twenty-two horoscopes for the Sun's entry into the twelve signs and the times of true syzygy (see Figure 14.12). Quires 2 and 3 present years 1451–53. For 1448 and 1451, only longitudes are included in eight columns. For 1452, the leaves contain no numerical data but are ruled in fifteen (Jan-Mar) or fourteen (Apr-Dec) columns. For 1453, a fourteen-column ephemeris provides both daily longitudes and latitudes (except, of course, for the Sun and the nodes). Quire 3 spans the fourteen-column leaves and is in a different hand, probably Peurbach's. The twenty-two horoscopes for 1453 are in Regiomontanus's hand.[33]

For the years, 1454–56, three sixteen-leaf quires each record one year's ephemeris and thirty-eight horoscopes. Arranged in eight columns, the ephemerides now list daily longitudes with only scattered latitudes squeezed between the lines.[34] For 1455 and 1456, syzygy times are jotted into the margins, as are letters indicating occasional lunar aspects. For 1457, two quires are devoted to the ephemeris and forty-four horoscopes (several leaves are blank), and many more lunar aspects are marked with symbols rather than letters; at the head of each column, the direction of the daily motion is denoted ('d[irectus]' or 'r[etrogradus]'). Regiomontanus laid out two quires to present 1458 and 1459, giving the ephemerides as one month per page followed by a year of horoscopes. However, the ephemeris abruptly halts with the words 'Corrige Martem' placed at the bottom of the February 1459 page and the remaining leaves of that quire are filled with horoscopes from January 1458 through February 1459. Apparently, Regiomontanus noticed an error here while writing the manuscript and abandoned his copying of the 1459 ephemeris.

Another two-year set for 1459–60 fills the next two quires, restarting with January 1459 and continuing through March 1460; apparently he had solved the problem with Mars. Now the content has expanded to fill two pages per month, with the longitudes on the verso and the lunar aspects and syzygies on the recto sides of an opening. This would become his standard format for the remainder of the autograph ephemerides in Cod. 4988 and for the 1474 printed edition (Figure 14.13). The final two quires present an ephemeris for April 1461 through March 1462, which suggests that two quires for April 1460–March 1461 have gone missing. Perhaps the shift to starting the year in April rather than January is related to Regiomontanus's move to Italy in 1460.

---

ephemerides for the years 1448 to 1464 [actually 1463] 'from the library of Regiomontanus' and that this material is in Regiomontanus's hand. Schöner recorded on folio 189^r that he had purchased the ephemerides for the years 1464–84 in 1509 in Bamberg. For the fate of Regiomontanus's library, see Zinner, *Leben und Wirken*, pp. 297–371; Monika Maruska, 'Johannes Schöner — "Homo est nescio qualis": Leben und Werk eines fränkischen Wissenschaftlers an der Wende vom 15. zum 16. Jahrhundert', unpublished PhD thesis, Vienna, Universität Wien, 2008; Richard L. Kremer, 'Text to Trophy: Shifting Representations of Regiomontanus's Library', in *Lost Libraries: The Destruction of Great Book Collections since Antiquity*, ed. by James Raven (London: Palgrave Macmillan, 2004), pp. 75–90.

33 See Helmuth Grössing, *Humanistische Naturwissenschaft: Zur Geschichte der Wiener mathematischen Schulen des 15. und 16. Jahrhunderts* (Baden-Baden: Koerner, 1983), pp. 133–35. For Peurbach's hand, see Vienna, ÖNB, Cod. 5203, ff. 67^r–69^v, 88^r–92^r.

34 Latitudes for Mercury at two- or three-day intervals (regularly), for Venus at five- or ten-day intervals, and for the superior planets at ten-day intervals (occasionally) appear for the years 1454–63. I do not analyse latitudes in this chapter.

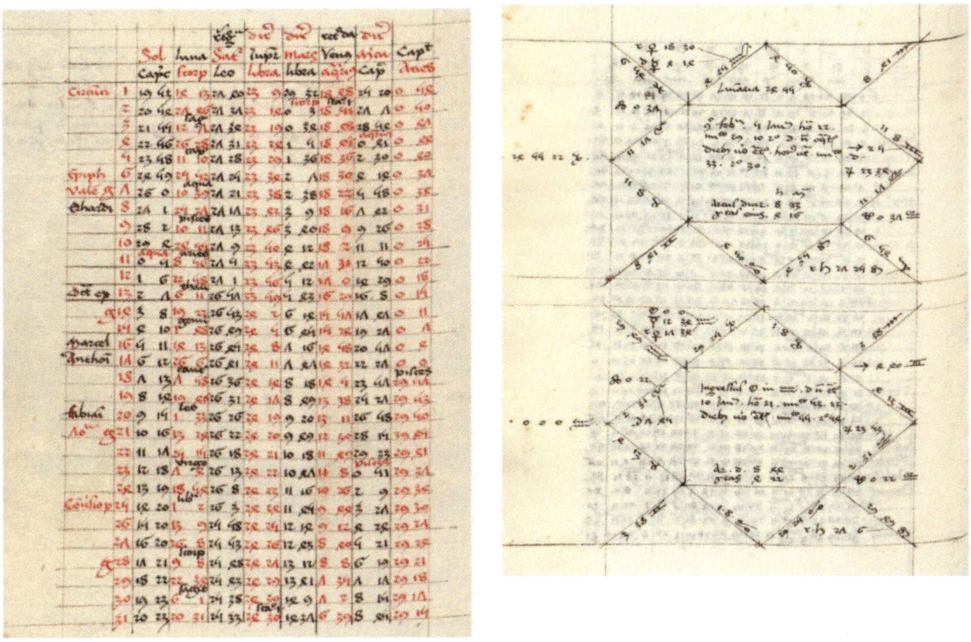

Figure 14.12. Vienna Cod. 4988, folios 1ᵛ, 7ᵛ. Regiomontanus's autograph ephemerides and horoscopes for January 1448. The horoscopes list both mean and apparent times.

The second section of Cod. 4988 contains an autograph ephemeris for 1463, from January to December. It is explicitly labelled as for Ferrara and Bologna, but otherwise follows the two-page-per-month format of the 1459–62 ephemerides. The leaves are smaller (10.8 × 14.5 cm) than in the quires of the first section (14.5 × 20.5 cm) and bear a watermark attested in Naples and dated 1463. Undoubtedly, Regiomontanus copied the 1463 ephemeris in Italy.[35] We note that each of the fourteen quires supporting the ephemerides from 1448 through 1462 generally has only one watermark, and that mark is unique in the codex. These fourteen paper types for the years 1448–62 are attested mostly in Vienna or Nuremberg and are dated from 1449 to 1459. Hence, the paper and the evolving formats in Cod. 4988 suggest that Regiomontanus copied these ephemerides over the course of the 1450s and early 1460s; he did not compute and write them as a unified, coherent project (as had John of Saxony). I know of no other ephemerides, prepared during the Alfonsine era, that show such development over time.

A 1522 inventory of books designated as Regiomontanus's personal library lists an 'Almanach scripta ab anno 1461 usque 1506'.[36] This codex has not been found, but the third section of Cod. 4988, folios 189ʳ–314ᵛ, may contain longitudes extracted from the lost

35 Zinner, *Leben und Wirken*, pp. 297–99 speculated that this quire may have been intended as a gift for Giovanni Bianchini (d. after 1469) in Ferrara, with whom Regiomontanus began corresponding in the summer of 1463. See Armin Gerl, *Trigonometrisch-astronomisches Rechnen kurz vor Copernicus: Der Briefwechsel Regiomontanus-Bianchini* (Stuttgart: Steiner, 1989).

36 Zinner, *Leben und Wirken*, p. 298.

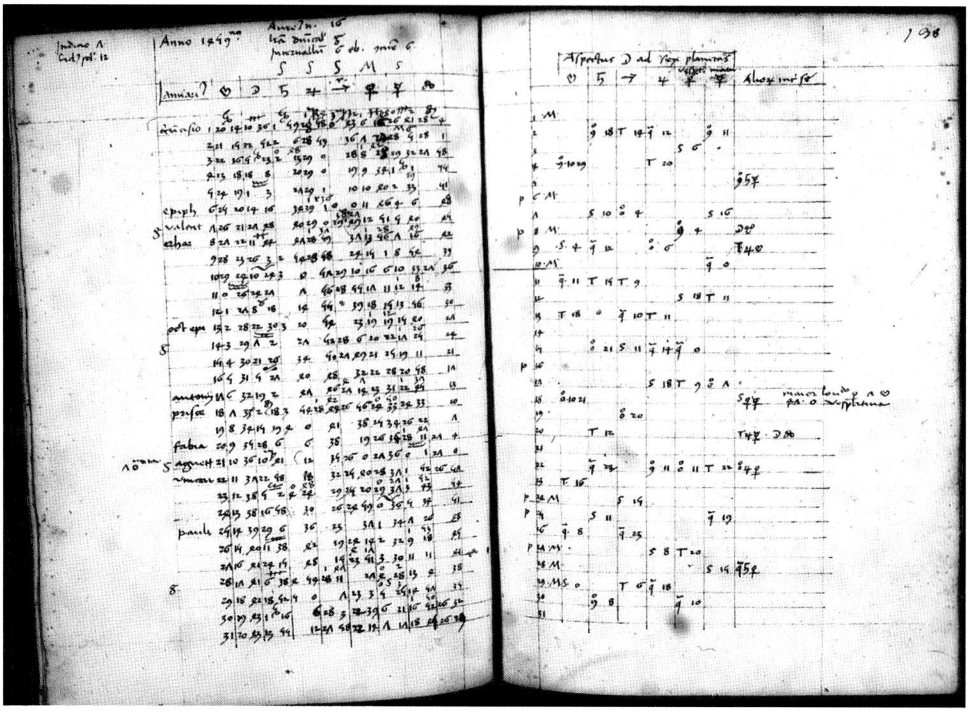

Figure 14.13. Vienna Cod. 4988, folios 137[v]–138[r], for January 1459. Regiomontanus's final layout for a daily ephemeris. The 1463 Ferrara manuscript and 1474 printed edition display an identical format.

codex for the years 1464–84. An identical copy for the years 1470–71 is preserved in Vat. Pal. lat. 1392, folios 1[r]–12[r], as is a copy for the year 1472 (often considered a Regiomontanus autograph) in Munich BSB Rar. 299a.[37] Only the latter manuscript, however, follows what I call the standard format; the first two manuscripts list the longitudes with scattered aspects squeezed into the margins rather than systematically filling the recto pages.

The remaining ephemerides attributed to Regiomontanus appear in the printed edition and comprise the most extensive Alfonsine ephemerides of the fifteenth century. This set of daily positions (only longitudes; latitudes are described by month merely as 'northern' or 'southern') and eclipses covers thirty-two years, from 1475–1506.[38] Unlike the development shown in Cod. 4988, the layout of the printed edition remains fixed over the thirty-two years. By 1471, Regiomontanus had already announced his intention to publish a large ephemeris in a letter to Christian Roder (*c.* 1410–1478), a mathematician at the university in Erfurt:

---

37 Zinner, *Leben und Wirken*, p. 298. As we shall note below, the 1472 copy contains syzygy times for a different meridian from that found in Vienna Cod. 4988.

38 Richard L. Kremer, 'Printing Sacrobosco in Leipzig, 1488 — ca. 1521: Local Markets and University Publishing', in *Publishing Sacrobosco's De Sphaera in Early Modern Europe: Modes of Material and Scientific Exchange*, ed. by Matteo Valleriani and Andrea Ottone (Cham: Springer, 2022), pp. 409–57. Regiomontanus's ephemerides were the first large set of astronomical tables to be printed.

Figure 14.14. Vienna Cod. 4988, book block, showing an early user's (Schöner?) index of the ephemerides, year-by-year, not distinguishing between the Regiomontanus autograph (1448–63) and later (1464–84) sections.

> ... what I want now from you I will also request from other observers of the heavens and from all centers of learning, to the extent that it is possible, so that somewhat better numerical tables may be produced by means of diverse, certain observations that have been collected. When this is done, with God's help I will compose and print, for all my friends, ephemerides of the planets, called an almanac, for thirty years or more. For even if we doubt whether in our time a general science can be created, it must nevertheless with all force be attempted, lest we be accused of spending our life in indolent silence.[39]

By 1472, Regiomontanus began printing what would become a very large book of thirty-two quires, filled with more than half a million sexagesimal digits. The colophon reads '*Explicitum est hoc opus anno Christi domini MCCCCLXXIIII ductu Ioannis de Monteregio*'.[40] Thus, the visible signatures of Regiomontanus's ephemerides fall into two eras: a period of experimentation from 1448 to 1459, followed by a period of stability as he produced the first large, printed edition of numbers.

We now turn to Regiomontanus's less visible computational practices. A two-page canon opens the printed edition, but, as is generally the case with this genre, the user instructions reveal nothing about how the ephemerides were computed. I cannot here explore the full range of computational practices illustrated in Regiomontanus's ephemerides. Rather, I shall consider the following three questions:

- Did he employ 'somewhat better numerical tables', as reported in the 1471 letter to Roder, to calculate the positions?
- For which local meridian(s) did he make the ephemerides?
- How did he seek to reduce the labor in his computations?

First, I have found no evidence that Regiomontanus used tables whose parameters had been 'improved' by recourse to modern observations of planetary positions. It is well known that Peurbach and Regiomontanus had, since the late 1450s, complained about inaccuracies in the predictions of the PAT (e.g., the 1471 letter to Roder quoted above); later authors reported that Peurbach had drafted 'partly corrected and emended' tables of the planetary equations and had 'calculated an almanac for many years'.[41] These tables, however, have not been found, and, as far as I can tell by recomputation with Alfonsine-based

39 Quod autem ex te nunc desidero, ab aliis quoque celi spectatoribus ac generalibus studiis omnibus, quantum fieri potest, exposcam, ut congestis variis ac certis observationibus numeri tabulares meliusculi reddantur. Quo facto, si deus aspirabit, planetarum ephemerides, vocant almanach, ad triginta vel plures annos charactere pressili pro amicis universis componam. Nam, etsi universe arti instaurande etatem nostram dubitemus suffecturam, conandum tamem est pro viribus, ut aliquid saltem veritati propinquius investigetur, ne vitam inerti silentio transegisse criminentur. Maximilian Curtze, 'Der Briefwechsel Regiomontans mit Giovanni Bianchini, Jacob von Speier und Christian Roder', *Abhandlungen zur Geschichte der mathematischen Wissenschaften*, 12 (1902), 187–336 (p. 327); Zinner, *Leben und Wirken*, pp. 163–65.

40 Richard L. Kremer, 'Controlling Errors in the First Printed Book of Astronomical Tables: Regiomontanus's *Ephemerides* (Nuremberg 1474)', in *Printing and Misprinting: A Companion to Mistakes and In-House Corrections in Renaissance Europe (1450–1650)*, ed. by Geri Della Rocca de Candal, Anthony Grafton and Paolo Sachet (Oxford: Oxford University Press, 2023), pp. 295-324.

41 Jerzy Dobrzycki and Richard L. Kremer, 'Peurbach and Maragha Astronomy? The Ephemerides of Johannes Angelus and Their Implications', *Journal for the History of Astronomy*, 27 (1996), 187–237 (pp. 222, 235); Georg Tanstetter Collimitius (ed.), *Tabulae eclypsium Magistri Georgij Peurbachij. Tabula primi mobilis Joannis de Monte Regio* (Vienna: Johann Winterburger, 1514), sig. aa4^{r}. In his autograph of the *Almagest* (Nuremberg,

tables, none of the extant ephemerides prepared by Regiomontanus show any systematic deviations from PAT-generated positions.[42]

On the other hand, it is known that by about 1457 Regiomontanus was computing with the astronomical tables composed circa 1442 by Bianchini.[43] As shown by Chabás and Goldstein, Bianchini's tables, although arranged in a different format, retain all the parameters of the PAT. Presumably they would not have been the 'somewhat better' tables described in the letter to Roder. However, below we shall look for evidence of Bianchini's tables in Regiomontanus's various ephemerides.

Second, recomputation suggests that Regiomontanus's ephemerides were not all computed for a consistent local meridian. Rather (and not unexpectedly) the meridians track Regiomontanus's movements across Europe. Nowhere in the manuscript copies of the ephemerides are meridians explicitly recorded. The 1474 printed ephemerides and calendars include a *Tabula regionum* listing for sixty-two European towns their distances in time from a central meridian set for Braunschweig, Nuremberg, Ulm, and Milan and located eighty-four time minutes east of Toledo. Ferrara is not included in this table, but some astute readers noticed that Bianchini's astronomical tables also set the meridian of Ferrara eighty-four time minutes east of Toledo and assumed that Regiomontanus must also have used that meridian for his printed ephemerides.[44] Regiomontanus presumably had found this *Tabula regionum* somewhere south of the Alps, for copies appear in earlier Italian manuscripts and its longitudes generally do not match those glossed in manuscript copies of the PAT (of provenance generally north of the Alps).[45] But, as we shall see, the times extracted from Regiomontanus's printed ephemerides do not consistently match a

---

Stadtbibliothek Cent. III, 25, f. 80^{r}, dated Vienna *c.* 1460), Regiomontanus wrote the following beneath the tables of equations for Jupiter and Mars in Bk 11: 'Correctiones invenies in libro magistri Georgij'. Presumably, however, Peurbach had corrected scribal errors and had not recomputed the equations with different parameters.

42 Already in the mid 1970s, Owen Gingerich had programmed an electronic computer to generate daily ephemerides based on the parameters of the PAT for the entire fifteenth and sixteenth centuries. Although he did not publish this work, Gingerich showed that the longitudes in Regiomontanus's printed edition closely tracked the computer-generated positions. See Poulle and Gingerich, 'Application du calcul électronique'. For examples of systematic deviation from PAT positions in later ephemerides for 1510 and 1512, see Dobrzycki and Kremer, 'Peurbach and Maragha Astronomy?'.

43 Peurbach to Johann Nihil, *c.* 1457, in Albin Czerny, 'Aus dem Briefwechsel des grossen Astronomen Georg von Peurbach', *Archiv für österreichische Geschichte*, 72 (1888), 281–304 (p. 302); Zinner, *Leben und Wirken*, p. 33; José Chabás and B. R. Goldstein, *The Astronomical Tables of Giovanni Bianchini* (Leiden: Brill, 2009).

44 For example, in 1499 Amerigo Vespucci used Regiomontanus's ephemerides and the lunar distance method to determine his longitude on the northern coast of what is now Brazil. Writing to Lorenzo de Medici, Vespucci indicated that Regiomontanus had made his almanac for the meridian of Ferrara. See Frederick J. Pohl, *Amerigo Vespucci Pilot Major* (New York: Columbia University Press, 1944), pp. 80–81, 222; Chabás and Goldstein, *Bianchini*, p. 37.

45 Paris, BnF lat. 7316A, f. 187^{v} (first half fourteenth century, Italy or Spain; 62 places, same order as Regiomontanus, prime meridian Denmark and Naples), Milan, Ambrosiana C 278 bis Inf., f. 12^{v} (second half of the fifteenth century, Bologna; 63 places adding Bologna, times converted to degrees of longitude with Denmark and Naples at 40 degrees east and Toledo at 10 degrees east, ordered by degrees of longitude), London, British Library Add 22317, f. 36v (identical to Ambrosiana C 278 bis inf. but lacks Danzig; in a version of the Tabule resolute [title in italics] for Bologna, made by Nicolaus Wodka of Kwidzyn in 1477). See Lynn Thorndike, 'Notes on Some Astronomical, Astrological and Mathematical Manuscripts of the Bibliothèque nationale, Paris', *Journal of the Warburg and Courtauld Institutes*, 20 (1957), 112–72 (pp. 137–38); Kremer and Dobrzycki, 'Alfonsine meridians', p. 196; David Juste, *'MS Paris, Bibliothèque nationale de France, lat. 7316A'* (updated: 11.06.2022), *Ptolemaeus Arabus et Latinus. Manuscripts*, URL = http://ptolemaeus.badw.de/ms/126.

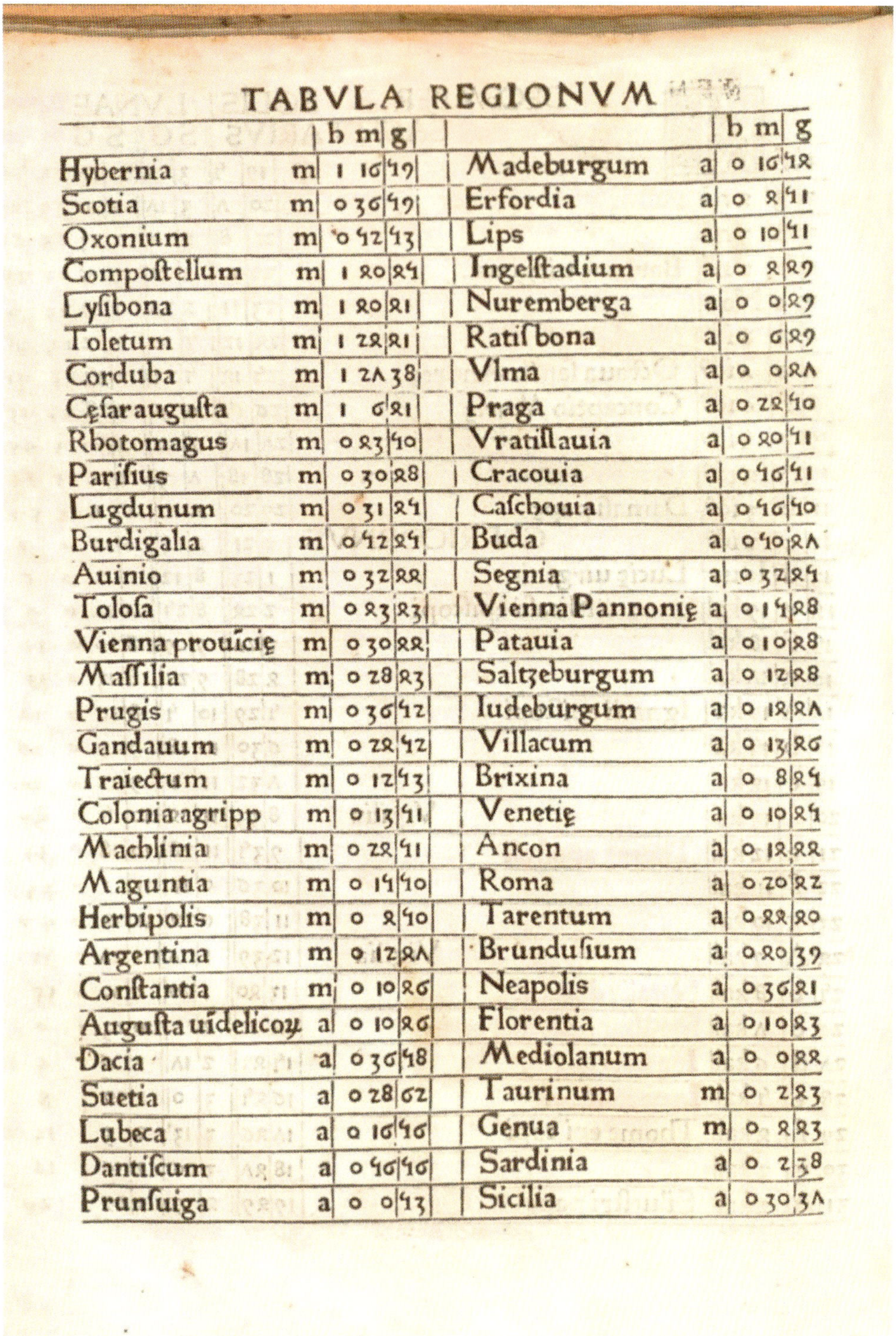

TABVLA REGIONVM

		b	m	g			b	m	g
Hybernia	m	1	16	57	Madeburgum	a	0	16	54
Scotia	m	0	36	59	Erfordia	a	0	4	51
Oxonium	m	0	52	53	Lips	a	0	10	51
Compostellum	m	1	40	45	Ingelstadium	a	0	4	49
Lysibona	m	1	40	41	Nuremberga	a	0	0	49
Toletum	m	1	24	41	Ratisbona	a	0	6	49
Corduba	m	1	27	38	Vlma	a	0	0	47
Cęsaraugusta	m	1	6	41	Praga	a	0	24	50
Rhotomagus	m	0	43	50	Vratislauia	a	0	40	51
Parisius	m	0	30	48	Cracouia	a	0	56	51
Lugdunum	m	0	31	45	Caschouia	a	0	56	50
Burdigalia	m	0	52	45	Buda	a	0	50	47
Auinio	m	0	32	44	Segnia	a	0	32	45
Tolosa	m	0	43	43	Vienna Pannonię	a	0	15	48
Vienna prouīcię	m	0	30	44	Patauia	a	0	10	48
Massilia	m	0	28	43	Saltzeburgum	a	0	12	48
Prugis	m	0	36	52	Iudeburgum	a	0	14	47
Gandauum	m	0	24	52	Villacum	a	0	13	46
Traiectum	m	0	12	53	Brixina	a	0	8	45
Colonia agripp	m	0	13	51	Venetię	a	0	10	45
Machlinia	m	0	24	51	Ancon	a	0	14	44
Maguntia	m	0	15	50	Roma	a	0	20	42
Herbipolis	m	0	4	50	Tarentum	a	0	44	40
Argentina	m	0	12	47	Brundusium	a	0	40	39
Constantia	m	0	10	46	Neapolis	a	0	36	41
Augusta uīdelicorum	a	0	10	46	Florentia	a	0	10	43
Dacia	a	0	36	58	Mediolanum	a	0	0	44
Suetia	a	0	28	62	Taurinum	m	0	2	43
Lubeca	a	0	16	56	Genua	m	0	4	43
Dantiscum	a	0	56	56	Sardinia	a	0	2	38
Prunsuiga	a	0	0	53	Sicilia	a	0	30	37

Figure 14.15. Table of sixty-two places from Regiomontanus, *Kalendarium*, 1474, sig. [b5v]. Munich, Bayerische Staatsbibliothek, 4 Inc.s.a. 1552.

meridian of eighty-four time minutes east of Toledo. Regiomontanus may have inserted the Italian *Tabula regionum* into his printed ephemerides not to indicate the meridian for which he had computed the positions but merely to enable readers to shift the computed syzygy and eclipse times to their local meridians.

Meridians in Alfonsine ephemerides can be extracted by recomputation, either directly by comparing the times of syzygy against those computed for the PAT meridian of Toledo, or indirectly by comparing the daily longitudes against those computed from the PAT for a meridian that minimizes the least squares of the residuals. The latter procedure is more robust for the faster moving bodies. For the slower moving bodies, sums of the squared residuals do not change appreciably when I shift my meridian by five or even ten minutes of time, and the effects of random scribal (or printing) errors, that I cannot detect, can shift least-squares results.

The horoscopes for true syzygies found in Cod. 4988 were generally cast for a meridian of eighty time minutes east of Toledo. This is the common meridian for Vienna as recorded in glossed manuscript copies of the PAT.[46] The meridians of the syzygies for 1463, however, are set to eighty-four time minutes east, that is, the meridian of Ferrara, as explicitly stated in Bianchini's PAT-based tables. This strengthens our suggestion that Regiomontanus prepared the first section of Cod. 4988 in Vienna and the second section, the 1463 ephemerides, while he was in Italy. In most of the syzygy horoscopes in Cod. 4988, Regiomontanus wrote both the mean and the apparent (i.e. corrected by Peurbach's equation of time) times in the central squares. When he listed syzygies in the margins or aspects pages, he gave only apparent times.

The true syzygy times for 1464–84 in Cod. 4988 range from eighty-four to eighty-seven time minutes east, always corrected with Peurbach's equation of time. They also include ascendents for each syzygy time, computed for a geographical latitude of forty-five degrees, that is, the latitude of Venice specified in Regiomontanus's *Tabula regionum* (Figure 14.15). These data imply that Regiomontanus computed the 1464–84 ephemerides during his sojourn in Italy. For the manuscript copy of 1472 and the printed ephemerides, true syzygy times are adjusted with Peurbach's equation of time for a meridian of sixty-five or sixty-six time minutes east of Toledo, the common Alfonsine meridian for Nuremberg found in manuscript copies of the PAT.[47] Hence, the syzygy meridians generally correspond to the locations where they were computed.

Meridians deduced from Regiomontanus's lunar ephemerides, however, do not always match the syzygy meridians. In Cod. 4988, the meridian extracted from the lunar positions for 1448 is eighty time minutes east, matching the syzygy meridians. However, for 1451–63, the lunar meridians vary from eighty-four to eighty-eight minutes east. The lunar meridian in the 1472 manuscript ephemerides is eighty-seven minutes east, and the lunar meridians in the printed edition vary between eighty-four and eighty-six time minutes east (except for 1485, where the lunar meridian is eighty-two time minutes east).

---

46 Kremer and Dobrzycki, 'Alfonsine Meridians', p. 196. Peurbach set the meridian of his eclipse tables to Vienna at eighty time minutes east of Toledo. See Tanstetter Collimitius (ed.), *Tabulae eclypsium*, sig. bb1^r, a1^r.

47 See Kremer and Dobrzycki, 'Alfonsine meridians', p. 194. We might guess that the 1472 ephemerides was copied after Regiomontanus had arrived in Nuremberg, prompting him to shift its syzygy times from the values he had computed in Italy.

Scattered scribal or typographical errors might account for these slight variations by year.[48] Of the forty years of Regiomontanus's ephemerides that I have examined, twenty-three show lunar meridians of eighty-four to eighty-seven time minutes east, i.e., the meridian of Bianchini's astronomical tables (Ferrara) and of Nuremberg in Regiomontanus's *Tabula regionum*. On the basis of these least-squares meridians, we might conclude that from 1451 onward Regiomontanus used Bianchini's tables to compute the daily lunar longitudes.[49] By presenting his ephemerides with a *Tabula regionum* showing Nuremberg and Milan (read Ferrara) on the same meridian, he could use data originally computed for the Italian location for ephemerides that his readers (to this day) assumed were made for Nuremberg. This would mean that the syzygy and lunar longitude data of the printed edition were generated in separate computational campaigns for different meridians.

Meridians extracted from the Mercury positions show considerably more scatter, ranging from eighty time minutes east in 1448 to sixty-eight time minutes in 1454, ninety time minutes in 1463, 101 time minutes in 1464, eighty-four time minutes in 1475 (edition), ninety time minutes in 1476 (edition), seventy-one time minutes in 1495, and eighty-nine time minutes in 1506. Random computational or typographical errors, however, could shift the slower moving Mercury's least-squares meridians by minutes, so these less robust results do not necessarily imply that Regiomontanus was intentionally shifting the Mercury meridians year by year. The standard deviations of the Mercury residuals also fluctuate by year and do not correlate with the extracted meridians. This might suggest that Regiomontanus was experimenting with different labor-saving computational procedures for the Mercury longitudes or that different persons computed some of the years. In any case, the daily Mercury positions do not enable reliable meridians to be extracted by least-squares methods.

Finally, we explore the labor-saving procedures Regiomontanus may have used in generating all the numbers for his ephemerides, beginning again with the times of true syzygy. For the years 1448, 1454, and 1456, Regiomontanus apparently used an iterative method to compute the times of true syzygy, for his times for those years vary by no more than ± 1 minute from those I compute with the iterative method (the standard deviations of the residuals are about half a minute for a year of twenty-four or twenty-five syzygies).[50] However, for the other years in Cod. 4988, standard deviations of the syzygy residuals rise to two or more minutes, and several residuals each year vary by ± 5 or 6 minutes. These results suggest that for these years, Regiomontanus used a labor-saving but less precise double-entry table for true syzygy times.

48 Given the Moon's mean motion of about twelve degrees/day or one arcminute/two time minutes and the precision of Regiomontanus's computations (standard deviations of a year's residuals of about two-to-three arcminutes), a least-squares extraction of lunar meridians is fairly robust, even with occasional scribal or computational errors approaching ten arcminutes.

49 For evidence that Regiomontanus used Bianchini's tables to compute the lunar longitudes of the 1474 edition, see Kremer, 'Controlling Errors', pp. 316-17.

50 For Ptolemy's iterative method of finding true syzygy times, see José Chabás and B. R. Goldstein, 'Nicholaus de Heybech and His Table for Finding True Syzygy', *Historia mathematica*, 19 (1992), 265–89; Richard L. Kremer, 'Thoughts on John of Saxony's Method for Finding True Syzygy', *Historia mathematica*, 30 (2003), 264–77. In this method, true positions of the Sun and Moon are computed for times shifted until the respective longitudes agree to the nearest arcminute.

By the 1450s, two such tables would have been readily available to Regiomontanus: one in the astronomical tables of Bianchini, the other in the eclipse tables recently completed by Peurbach, expanding the *Tabulae permanentes* created in the 1340s by John of Murs and Firmin of Beauval.[51] Both tables eliminate the need to compute true positions of the luminaries. Bianchini provided calendrical tables to find the time of the first mean syzygy (separately for conjunction and opposition) of a calendar year and the values of the mean solar and lunar anomalies at that time. Separate single-entry tables then provide, for the next twelve consecutive conjunctions or oppositions, the true longitudes of the Sun and Moon and their respective velocities. Users must then divide the difference in longitudes by the difference in the luminaries' velocities at each syzygy to find the time of true syzygy. Peurbach offered tables giving the times and anomalies for each mean syzygy. Users then interpolate in a double-entry table to find the time correction to be added to the mean syzygy time. To my computing fingers, Peurbach's method requires considerably less labor than Bianchini's since the intervals for most of the Peurbach double interpolations are small enough to be solved by mental arithmetic (estimation) without pencil-paper sexagesimal arithmetic using proportional parts.[52] In Table 1, I compare the performance of Bianchini's and Peurbach's syzygy tables, both of which use lunar velocities to approximate the time correction, against the more laborious but exact results iterated from the PAT. As can be seen in Table 1, Peurbach's tables estimate the time correction somewhat more accurately than Bianchini's do; however, when the time between mean and true syzygy exceeds twelve hours, Peurbach's tables yield a characteristic error of three-to-six time minutes.[53]

Table 1 also suggests that Regiomontanus used Peurbach's tables to determine the times of true syzygy in the 1455 ephemerides. Most of these time corrections match my computation with Peurbach to ± 2 minutes.[54] For the four cases where the time corrections exceed twelve hours and the errors in Peurbach's estimates reach three to six minutes (marked red in Table 1), Regiomontanus's times match those of Peurbach. Scattered checks of Regiomontanus's true syzygy times for the large time corrections, both in Cod. 4988 and in the printed edition, confirm that he consistently used Peurbach's tables for those

51 For details, see Chabás and Goldstein, *Bianchini*, pp. 85–91; Richard L. Kremer, 'Cracking the *Tabulae Permanentes* of John of Murs and Firmin of Beauval with Exploratory Data Analysis', in *Editing and Analysing Astronomical Tables: Towards a Digital Information System for the History of Astral Sciences*, ed. by Matthieu Husson, Clemency Montelle, and Benno van Dalen (Turnhout: Brepols, 2022), pp. 363–422. In Vienna *c.* 1452, Regiomontanus had copied the *Tabulae permanentes* (now Nuremberg, Stadtbibliothek Cent. VI, 23, ff. 99^{v}–111^{r}); his autograph of Peurbach's eclipse tables including the true syzygy table, prepared in Vienna *c.* 1460, is now in Nuremberg, Stadtbibliothek Cent. V, 57, ff. 108^{r}–153^{v}. See Ingeborg Neske, *Die Handschriften der Stadtbibliothek Nürnberg, Bd. V: Die lateinischen mittelalterlichen Handschriften, Varia, 13.-15. und 16–18. Jahrhundert* (Wiesbaden: Harrossowitz Verlag, 1997), pp. 91, 181.

52 See M. Husson, in this volume, 'Proportional Parts and Interpolation Practices in Alfonsine Astronomy: Astronomers' Work with a Mathematical Procedure'.

53 This imprecision arises from both computational noise in Bianchini's tables and the fact that users must round the anomalies to the nearest degree before entering the tables. If one were to interpolate between the entries provided at intervals of one degree of anomaly, the errors in Bianchini's time corrections would be somewhat reduced but at a cost of more than doubling the computational labor. Bianchini's canons instruct users to round, not interpolate.

54 For 16 April 1455, the manuscript reads 12;48 hours, a value I have emended to 12;38 hours.

Date	Delta T			True syzygy
	Iterated (hrs)	Peur-Itera. (mins)	Bian-Itera. (mins)	Peur-ÖNB 4988 (mins)
3 Jan	10;26	1	1	-79
17 Jan	-6;52	0	-7	-80
2 Feb	13;10	-6	-4	-80
16 Feb	-5;34	-2	-2	-80
3 Mar	12;49	-5	-8	-83
17 Mar	-3;20	0	3	-80
2 Apr	9;42	0	-1	-79
16 Apr	-0;37	-1	0	-78
1 May	4;46	1	-1	-79
16 May	1;59	0	2	-79
30 May	-1;03	0	-3	-79
14 Jun	4;05	0	6	-79
29 Jun	-6;44	-1	-7	-81
14 Jul	5;25	0	5	-79
28 Jul	-11;18	1	-12	-82
12 Aug	5;48	0	3	-80
26 Aug	-13;46	4	-12	-81
11 Sep	5;18	-1	-3	-80
25 Sep	-13;22	3	-8	-80
10 Oct	3;57	-2	-2	-82
25 Oct	-9;51	0	-3	-80
9 Nov	1;59	1	-5	-77
23 Nov	-3;56	0	25	-79
8 Dec	-0;15	1	-8	-79
23 Dec	3;03	1	2	-78

Table 1. Computed times from mean to true syzygy for 1455, using PAT for iterated (i.e. exact) times and the syzygy tables of Peurbach and Bianchini. Since the 1495 edition of Bianchini's tables is filled with typographical errors, I have used a copy found in Paris, BnF lat. 7269. Times larger than twelve hours are printed in red.

computations.[55] I estimate that this tabular procedure would have more than halved the time required to compute true syzygies by iteration with the PAT.

What labor-saving procedures might Regiomontanus have employed to find the daily planetary positions? Did he use the PAT to compute daily positions, or might he have computed from the PAT at intervals of five or ten days (as John of Saxony had done) and then found the daily positions by linear interpolation? As noted above, computing a planetary longitude with the PAT generally requires seven distinct steps with five separate

55 Regiomontanus's autograph copy of Bianchini's tables, in Nuremberg, Cent. V, 57, ff. 21r–107r, does not include the true syzygy table; we might guess that, even before copying, Regiomontanus had decided against using Bianchini's method for approximating true syzygy times. See Chabás and Goldstein, *Bianchini*, p. 26.

single-argument tabular interpolations. Reducing this work by eighty percent (five-day intervals) or by ninety percent (ten-day intervals) might well have been very appealing and, indeed, many of the medieval canons for making ephemerides recommend such a labor-saving move. For example, one canon accompanying the Toledan Tables recommends that positions for the fastest moving body (the Moon at roughly twelve degrees/day) be computed at daily intervals. Bodies with the slowest daily accelerations (Sun, lunar nodes) are to be computed at fifteen-day intervals. With their slow mean motions and small epicycles, Saturn and Jupiter can be computed at eight-day intervals; Mars with a larger epicycle at five-day intervals; and the inner planets, with fast motions around large epicycles (i.e. large daily accelerations), are to be computed at three-day intervals.[56] If such interpolations over intervals were used, we might expect to find larger residuals where the body's daily acceleration is greatest since linear interpolation assumes zero acceleration across the intervals. Plotting residuals against daily accelerations would thus offer a means of exploring whether linear interpolation has occurred across multi-day intervals in the ephemerides.[57]

We begin this investigation with the lunar ephemerides, where we might expect Regiomontanus to have computed daily positions. For their first year of 1448, however, Regiomontanus computed lunar longitudes at two-day intervals and linearly interpolated for the intermediate day. As can be seen in the residuals for the first half of that year (Figure 14.16), the computed longitudes match my PAT values to about ± 1 arcminute with very little noise. The interpolated longitudes, however, show residuals of up to ± 10 arcminutes, with their signs correlated to the changing daily lunar accelerations. Already in his earliest ephemerides, the young Regiomontanus sought to reduce labor by computing only half of the lunar longitudes from the tables.

For 1453, the ephemerides preserved (possibly) in Peurbach's hand, the lunar residuals for the first and second halves of the year reveal different patterns and different meridians. For January-June (Figure 14.17), the residuals again suggest exact computation at two-day intervals, although the noise has increased to about ± 13 arcminutes.[58] For July-December (Figure 14.18), the two-day pattern is gone and the magnitude of the residuals is reduced to about ± 5 arcminutes, lightly correlated with the daily accelerations. These patterns could be explained if we assume that the computer used Bianchini's tables for the second half of 1453, as these give precisely computed daily positions for the time of mean syzygy for each month and each subsequent day of that month at that same time.[59] This hypothesis is nicely confirmed if we consider the anomalistic month of 16 September–13 October 1453

56 Pedersen, *Toledan Tables*, p. 542. A mid-fifteenth-century canon by Laurentius de Laurentiis recommends ten-day intervals for the superior planets, five-day intervals for the Sun and inferior planets, and three-day intervals for the Moon. I thank Nicholas Jacobson, who is editing these tables, for this information.

57 In the graphs below, I determine the body's daily velocity by taking second differences of my computed daily longitudes and then smoothing by making a running average of three consecutive values: $a_i = (a_{i-1} + a_i + a_{i+1})/3$. I seek exploratory data analysis here, not robust statistical tests. See Kremer, 'Cracking the *Tabulae Permanentes*'.

58 This hypothesis is strengthened by three large residuals (-32, -57, -37 arcminutes, not visible in Fig. 15) for days 88 to 91, generated by a one-degree computational error for day 89.

59 In 1452 Bianchini had presented his astronomical tables to Emperor Frederick III in Ferrara. See Chabás and Goldstein, *Bianchini*, p. 15. Perhaps Peurbach and Regiomontanus acquired a copy in 1453; see n. 43 above.

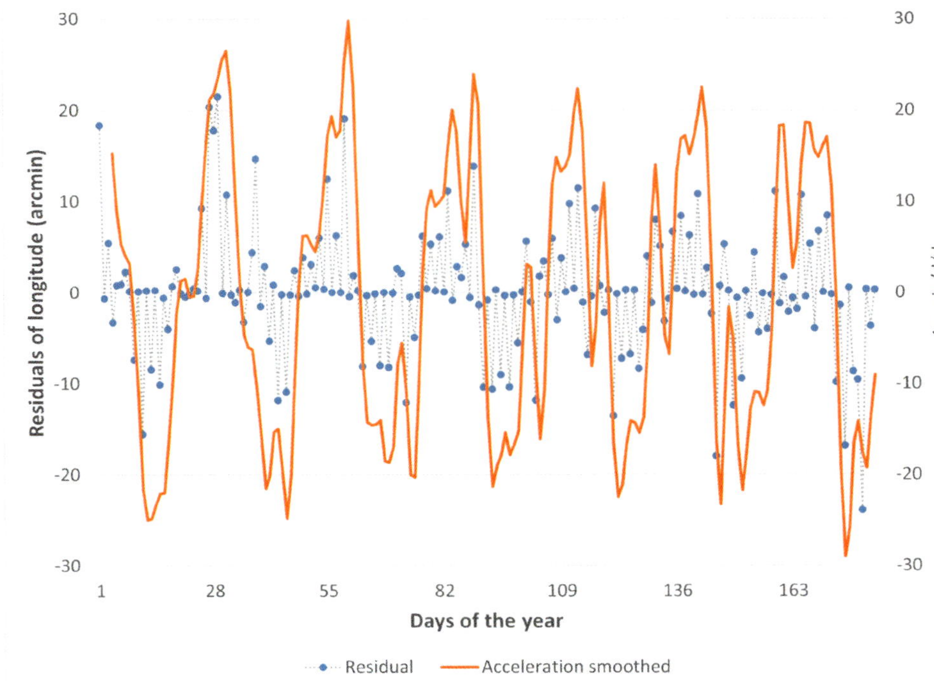

Figure 14.16. Computed residuals (eighty time minutes east) and daily accelerations for the Moon, January-June 1448, from Vienna Cod. 4988, folios 1^v–7^r. One ten-minute scribal error has been corrected.

(Figure 14.18, red points). The residual on day four of this month falls to -20 arcminutes, undoubtedly generated by the twenty-arcminute error in Bianchini's lunar tables for this day (a double elongation of 340 degrees).[60] The ephemeris then gives four lunar longitudes with +10 arcminute residuals, followed by fifteen longitudes with -10 arcminute residuals. In this month, the ephemeris longitudes for days zero, one, two, three, five, ten, and twenty-six match Bianchini's to ± 1 arcminutes; the remaining days show large residuals. I cannot explain this pattern, but surely the computer of the second half of the 1453 ephemeris (Peurbach himself?) committed systematic errors in deploying Bianchini's lunar tables.[61]

The remainder of Regiomontanus's lunar ephemerides that I have examined, in Vienna Cod. 4988 and in the printed edition, show residuals mostly similar to those found in Figure 14.19. It seems quite likely that, after 1453, Regiomontanus consistently used Bianchini's tables to generate the thousands of daily lunar positions required for the ephemerides. Rather than laboriously performing the sequence of steps required by the PAT, Regiomontanus would simply look up the mean motions for each successive mean syzygy (i.e. anomalistic month) in Bianchini's tables and then copy out, for the

60 Both the 1495 edition and Regiomontanus's autograph copy, Nuremberg Cent. V, 57, f. 29^v, read 35;32, whereas the correct value is 35;52 (a later hand has corrected the entry in Nuremberg Cent. V, 57).

61 Since this anomalistic month began (mean lunar anomaly of zero) at 23;26 hours past noon, that is, very close to noon or the time of the entries in the 1453 ephemeris, the September-October residuals cannot have been created by linear interpolation across the time shift from Bianchini's tables to ephemeris noon.

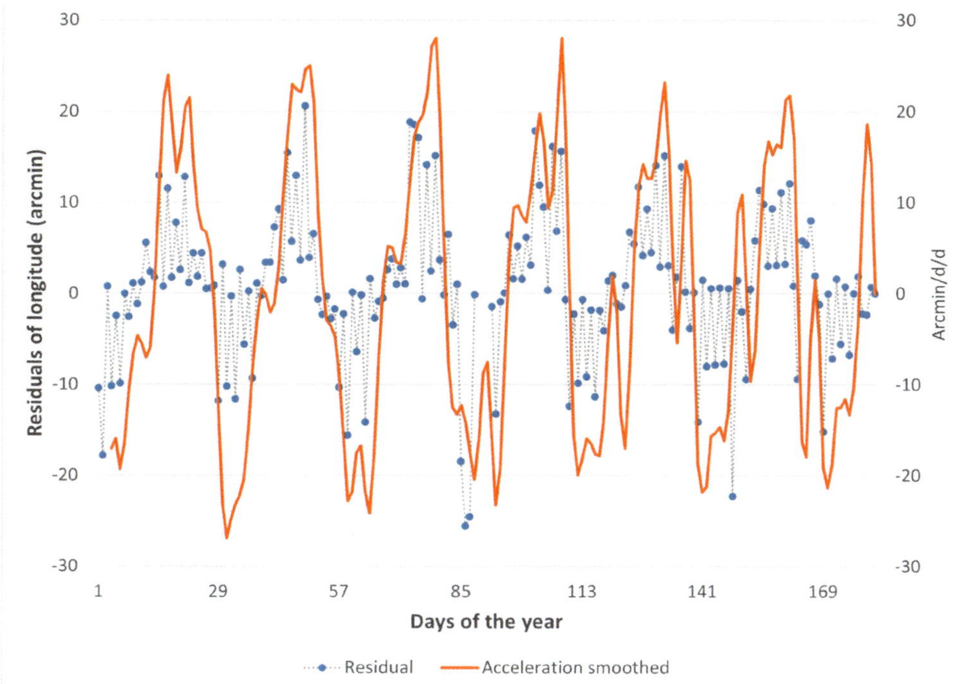

Figure 14.17. Computed residuals (eighty-eight time minutes east) and daily accelerations for the Moon, January-June 1453, from Vienna Cod. 4988, folios 35r–40v. Three outliers (-32, -57, and -37 arcminutes) dropped.

next twenty-eight days, the true longitudes and the proportional parts needed to shift the times to local noon and the double mean elongation to its exact value. Essentially, only several simple additions and multiplications would be required to generate any daily lunar longitude.[62]

We find a similar development in Regiomontanus's computational procedures for the planets. For Mercury, he began by computing exact positions at five-day intervals. Again, his computed values differ from mine by only ± 1–2 arcminutes, but the intermediate interpolated residuals for this rapidly moving planet can reach up to forty arcminutes (Figure 14.20). For the 1451 Mercury positions, Regiomontanus appears to have shifted to Bianchini's tables, which provide exact longitudes for that planet at two- or three-day intervals (Figure 14.21). The three anomalistic periods extending across 1451 show residuals of varying magnitudes, depending on the number of hours between local noon and the

62 In the printed ephemerides, we find several examples of scribes, in Regiomontanus's print shop, adding manuscript instructions to the printed leaves for correcting lunar longitudes across an entire anomalistic month, each of which suggests that Regiomontanus had incorrectly copied the initial values for the month from Bianchini's tables. See Kremer, 'Controlling Errors'. For another example of Regiomontanus using Bianchini's tables, see E. Andriani and S. Zieme, in this volume, 'Regiomontanus on the Equation of Time: Computational Practices of an Alfonsine Astronomer'.

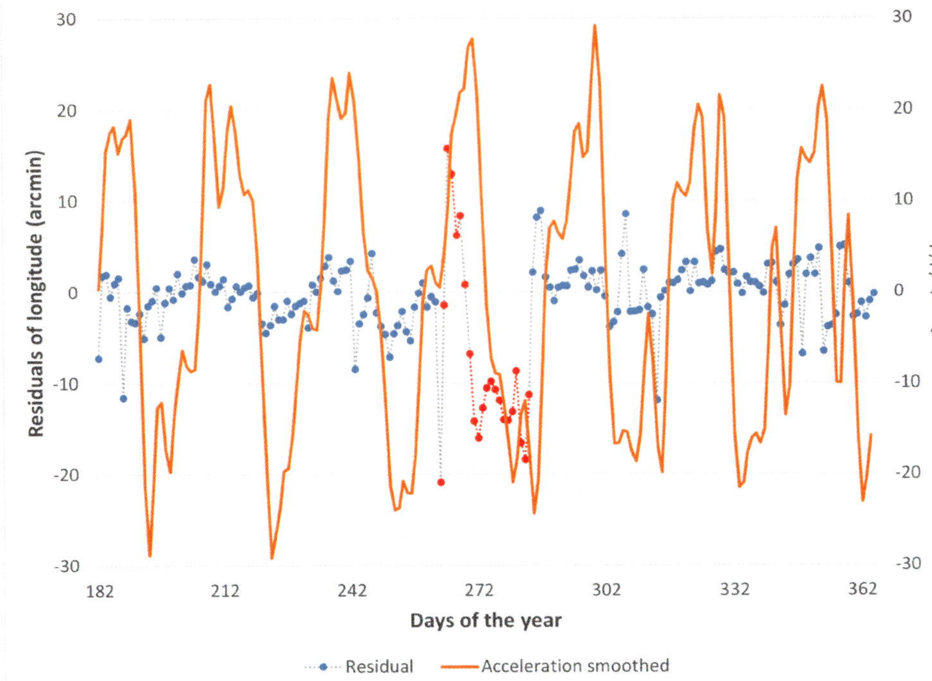

Figure 14.18. Computed residuals (eighty-eight time minutes east) and daily accelerations for the Moon, July-December 1453, from Vienna Cod. 4988, folios 35^r–40^v. Red points mark residuals for 20 September–11 October.

time of the initial mean conjunction, that is, the beginning of the planet's anomalistic period.[63] The maximal magnitude of the residuals reaches about ± 10 arcminutes.

Regiomontanus continued using the Bianchini tables to generate Mercury positions, but he did not always control the procedure. In 1458–59, the Mercury residuals twice exceed the usual magnitudes of ± 10 arcminutes, reaching twenty arcminutes in one anomalistic period and forty arcminutes in another (Figure 14.22). In the former case, the larger residuals result from the fact that for this cycle, the time correction to reach local noon is more than nine hours. Residuals occur because Bianchini's method uses linear interpolation to adjust the daily times to local noon, that is, he assumes the daily acceleration of the planet is zero. When, however, the daily acceleration reaches eight arcminutes/day/day, over nine hours, Bianchini's linear interpolation will err by (9/24) x eight arcminutes/day/day, which is about three arcminutes per day or six arcminutes every two days (the interval of his Mercury tables), as we see in Figure 14.22. In the latter case, I have discovered that Regiomontanus failed to switch the sign of one entry from Bianchini's tables, that is, the 'Equatio' or the correction needed when the mean center falls between the ten-degree intervals of the tables. For the mean center of 5s 50°, the 'Equatio' is initially negative, becomes positive for days thirty-six to forty-six, is negative for days

63 For the four anomalistic periods ('Bianchini days' in Fig. 14.21) of 1451, the time shifts were, respectively, 4;11, 1;26, 1;39, and 4;33 hours.

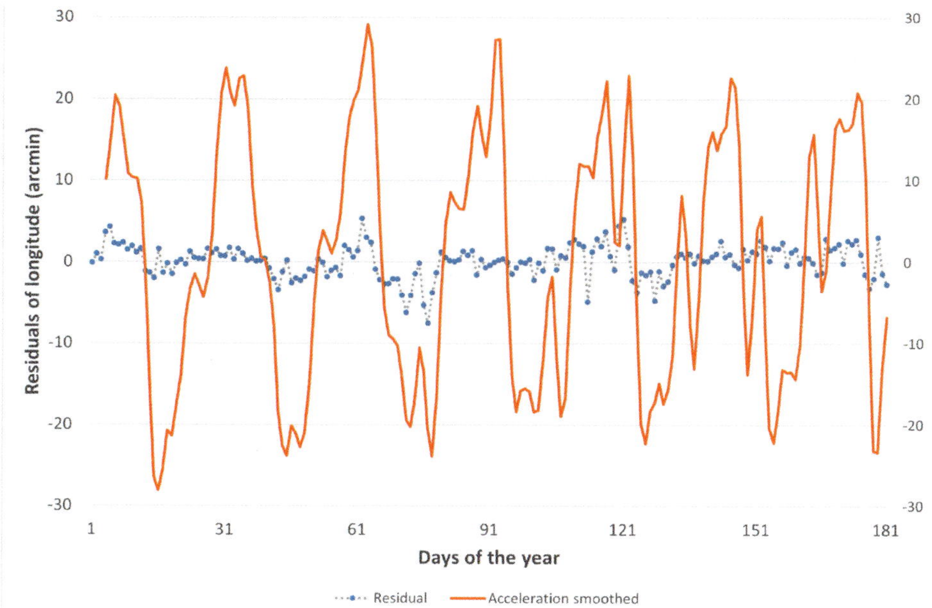

Figure 14.19. Computed residuals (eighty-five time minutes east) and daily accelerations for the Moon, January-June 1506, from *Ephemerides*, 1474.

forty-seven to 106, and is positive for the final portion of the 115-day cycle. For whatever reason, Regiomontanus used a positive 'Equatio' to compute days forty-seven to 106, thereby generating a systematic error of forty arcminutes in the third cycle of 1458–59.[64]

As can be seen in Figure 14.22, the green residuals of the third cycle that I have prepared by laboriously computing (incorrectly, as did Regiomontanus) with Bianchini's tables are nearly flat. In that third cycle, I find the average absolute error (green residuals) to be 1.4 arcminutes and the standard deviation of the 115 values to be 1.8 arcminutes. Despite the mistake in the algebraic sign, Regiomontanus otherwise wielded Bianchini's tables with considerable accuracy.

For the several internal years I have examined in Regiomontanus's printed ephemerides (1476, 1485, 1495), the Mercury residuals generally present the Bianchini pattern, reaching maximums of about ± 5 arcminutes, following the peaks and valleys of the daily acceleration. For the initial and final years (1475, 1506), however, the Mercury residuals go nearly flat (Figure 14.23). I suspect that for these years, Regiomontanus computed the Mercury positions directly from the PAT for each day of the year. The average of the absolute values of the residuals is 1.1 arcminutes, the standard deviation being 1.5 arcminutes. Might we

64 In Regiomontanus's autograph copy of Bianchini's tables (Nuremberg Cent. V, 57), column headings for the mean center $5^s 50°$ table correctly label the additions and subtractions for 'Equatio'. In the 1495 printed edition, however, two of the column headings on the final page of the $5^s 50°$ table are reversed, which would generate nearly the same errors seen in Figure 14.22. I have not examined all the manuscript copies of Bianchini's tables and cannot assess how these errors in the 'Equatio' headings circulated. I have found, however, that the 1495 edition overflows with typographical errors; for the various tables I have recomputed, Nuremberg Cent. V, 57 offers a much cleaner copy than the *editio princeps*.

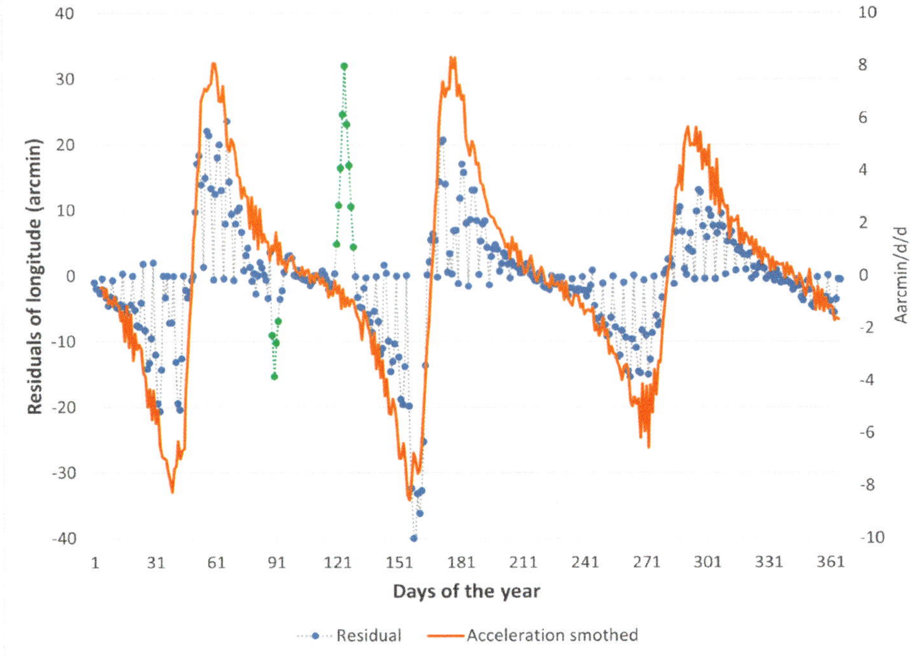

Figure 14.20. Computed residuals (eighty time minutes east) and daily accelerations for Mercury, 1448, from Vienna Cod. 4988, folios 1ᵛ–7ʳ. Notice the large computational errors at days 90 and 125, marked in green.

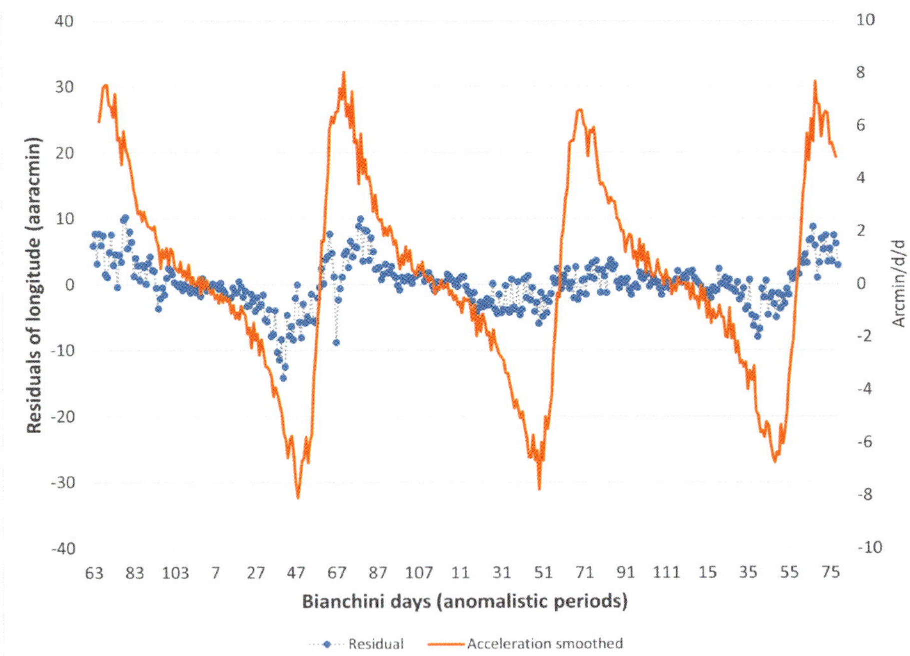

Figure 14.21. Computed residuals (ninety-three time minutes east) and daily accelerations for Mercury, 1451, from Vienna Cod. 4988, folios 13ᵛ–19ʳ.

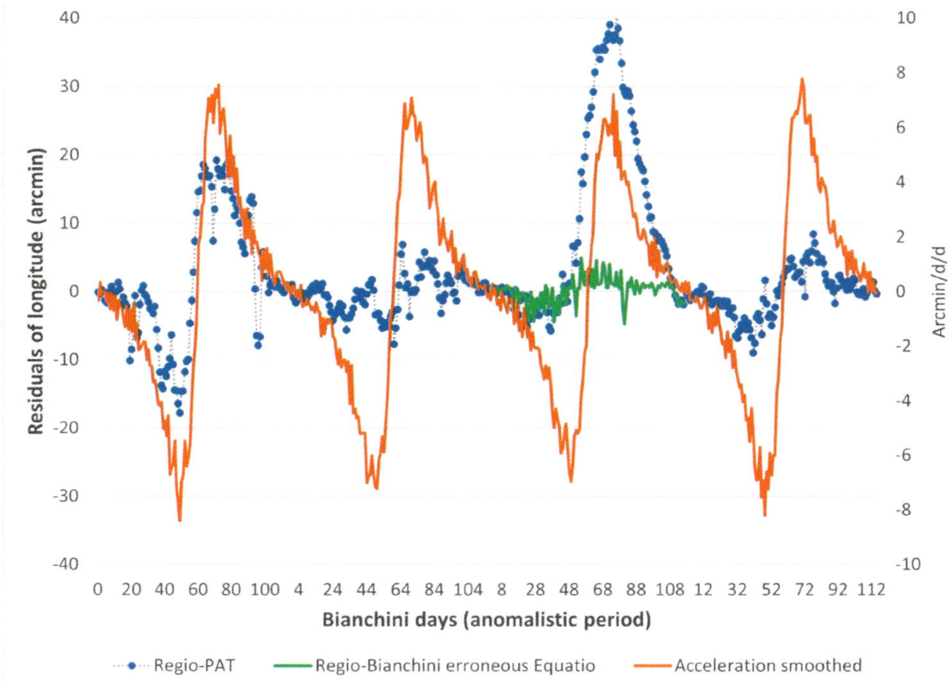

Figure 14.22. Computed residuals (eighty-four time minutes east) and daily accelerations for Mercury for four anomalistic periods, from 15 February 1458–24 May 1459, from Vienna Cod. 4988, folios 114r–119r, 137v–141v. In the third cycle, the green residuals are produced by computing longitudes with Bianchini's tables and erroneously not switching the sign of the 'Equatio' correction over days 48–106.

understand the Mercury residuals of these two years as the mark of a proud craftsman, signing his extensive work with a flourish that, until now, no one has noticed? If true, this would suggest that Regiomontanus knew that Bianchini's tables, which he used to reduce computational labor, invariably introduce errors of several arcminutes or more into longitudes when the bodies are rapidly accelerating.

What Regiomontanus did not know, however was that a predecessor, the first calculator of a large Alfonsine ephemeris John of Saxony, had surpassed the Nuremberg printer and astronomer in both consistency and accuracy. John's Mercury positions, computed at four-day intervals for 1336, yield residuals the absolute value of which average 0.45 arcminutes, with a standard deviation of 0.70 arcminutes. However, did John of Saxony know that expanding four-day positions with linear interpolation to daily intervals would introduce errors of up to twenty arcminutes in the Mercury longitudes (Figure 14.24)? Judging simply from his ephemeris, we cannot know the level of precision that he expected from his computational practices with the PAT.

This preliminary analysis of more than fifty years of ephemerides prepared by Regiomontanus has uncovered several features of computational practices in what was truly a massive project of Alfonsine data production:

- Despite his rhetoric about 'reforming' astronomy, Regiomontanus never adjusted parameters of the PAT as he computed ephemerides.

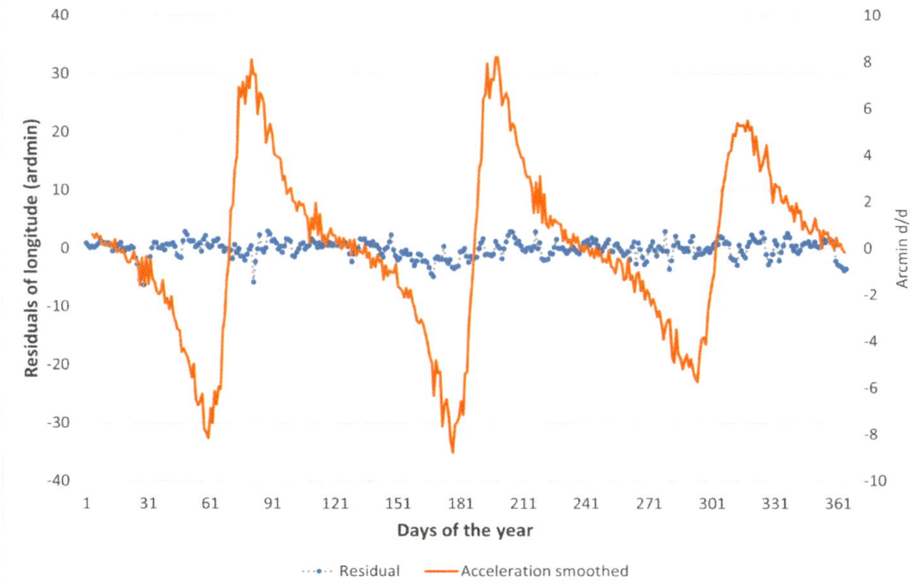

Figure 14.23. Computed residuals (eighty-nine time minutes east) and daily accelerations for Mercury, 1506, from *Ephemerides*, 1474.

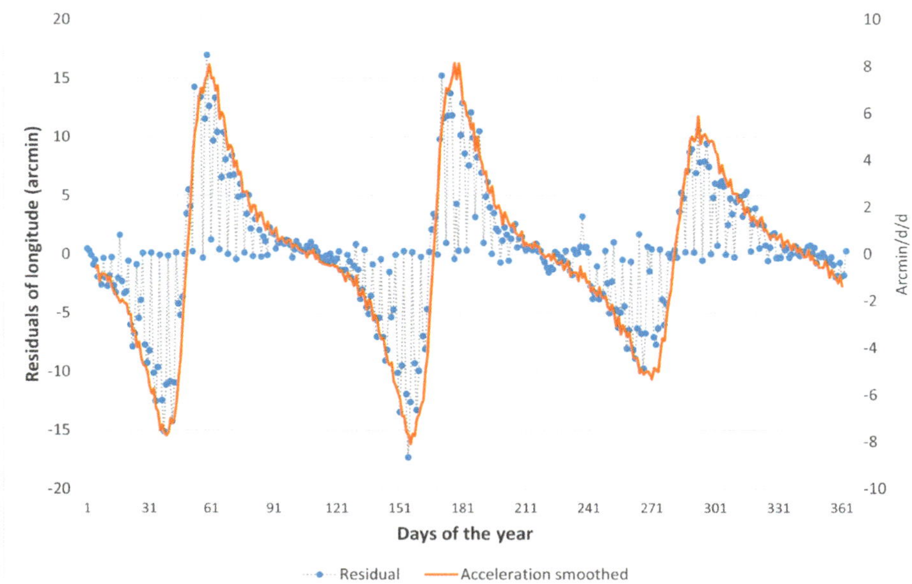

Figure 14.24. John of Saxony's four-day positions expanded by linear interpolation to one-day positions for Mercury, 1336, from Erfurt CA 2°386, folio 103ᵛ, several ± 10 or ± 20 arcminute scribal errors corrected, two outliers (twenty-one and twenty arcminutes) dropped. The residuals (forty-eight time minutes east) near zero represent John's computed positions.

- Inconsistent meridians can be found in his ephemerides, reflecting the radices of his tables (eighty time minutes east for Peurbach's eclipse tables, eighty-four time minutes east for Bianchini's tables) and local traditions (sixty-five time minutes east for Nuremberg).
- In the 1450s, Regiomontanus experimented with different tools for computing true syzygy times, but eventually settled on Peurbach's double-entry tables.
- Regiomontanus also experimented with labor-saving techniques by computing positions for intervals greater than one day; by 1453, he would use Bianchini's tables (which provide exact longitudes at different intervals for the various bodies). However, Regiomontanus realized that linear interpolation in Bianchini's procedures can introduce significant deviations from exactly computed positions.
- Regiomontanus occasionally erred in using Bianchini's tables, writing down incorrect initial values for an anomalistic period or not managing the algebraic signs, which produced systematic patterns in the residuals of his computed positions.
- As the first printer of astronomical tables, Regiomontanus standardized a format for ephemerides that had long existed in manuscript traditions (going back to Greek and Arabic sources); the layout of Regiomontanus's 1474 edition would be consistently repeated in the dozens of ephemerides editions printed over the next two centuries and beyond.[65]

Conclusions

This preliminary study has explored what modern historians can learn by recomputing medieval ephemerides with the tools medieval computers used and comparing the results (the residuals). Earlier scholars have productively compared the performance of astronomical models and parameters, realized by modern computation, against historical ephemerides. However, the goal of this chapter has been to understand how the medieval computational tools (i.e., astronomical tables) generated a set of computational practices that we moderns can uncover. During the Alfonsine period spanned by the protagonists in my story, neither the astronomical models nor the parameters changed; but the computational habits of late medieval astronomers did not remain fixed.

The most significant finding of this study is that the residuals generally remain small, at least for well-known astronomers such as John of Saxony and Regiomontanus. As I perform the sequence of steps, using decimal arithmetic, to compute a true position with the PAT, my results generally match the medieval results, achieved with sexagesimal arithmetic, to within several arcminutes. Computational noise, which might result if the medieval computers inconsistently rounded or if their procedural steps differed from mine, is quite low, usually within one-to-two arcminutes. This means that systematic patterns in the residuals can be identified and sometimes explained. Such patterns allow us to know whether the medieval tables had, in certain places, values different from those in the 1483 *editio princeps*, arising from scribal errors in the manuscript transmission before 1483. We

---

65 For example, see the *Connoissance* [sic] *les temps* for 1680, produced by the Paris Observatory. Guy Boistel, *'Pour la gloire de M. de la Lande': Une histoire matérielle, scientifique, institutionnelle et humaine de la Connaissance des temps, de 1679 à 1920* (Paris: IMCCE, 2022).

can also discover where medieval computers miscalculated a single value that affected computational results for a chain of true positions (e.g., an anomalistic period).

The residuals also allow us to identify the meridians of the medieval ephemerides, the particular sets of PAT-based tables they used to compute the positions, various labor-saving shortcuts the medieval computers developed such as rounding mean motions, and how they batched the computational labor (computing sequences of positions for calendar years or anomalistic periods). And by watching how they vacillated between labor-saving and full PAT-based procedures, we can occasionally gain a sense of the medieval computers' criteria for computational 'precision' in their ephemerides.

Finally, this preliminary study has demonstrated the importance of considering all the data available in historical ephemerides as we seek to extract details about their makers' computational practices. Sampling their positions at arbitrarily selected intervals (e.g. five days) may allow the modern computer to recognize if the medieval computer employed non-Ptolemaic models; yet, such sampling may well miss many subtle features of their computational practices.[66] For comprehensive analyses, the modern historian must recompute all of the historical data available. And this chapter has shown that to recognize examples of linear interpolation in historical ephemerides it can be very helpful to plot smoothed second differences, that is, daily accelerations, of the daily recomputed positions. Knowing when a planet is accelerating rather than moving uniformly can help us understand otherwise puzzling systematic patterns in the residuals.

John of Saxony and Regiomontanus were the two most prolific Alfonsine ephemeris-makers, computing daily positions for periods extending over decades. Another fifty or so Alfonsine ephemerides have been preserved, often covering only a single year or a few years. However, these smaller products, like the large ephemerides, display both levels of agency by the maker, i.e., decisions about external format and content and decisions concerning which tables to use and how to compute the daily positions. Analysing the residuals in the smaller ephemerides will provide rich information about what historian John D. North has called the 'calculating classes' of medieval society and will let us compare the computational practices of the leading Alfonsine astronomers, such as John of Saxony and Regiomontanus, against those of the more quotidian Alfonsine practitioners who wrote and collected the 900+ extant manuscripts and 100+ printed editions that document what we call Alfonsine astronomy.[67] To return to the question posed in the title of this chapter, recomputing Alfonsine ephemerides has the potential to reveal considerable information about massive computation in late-medieval Europe.[68]

66 For an earlier investigation of historical ephemerides at only five-day intervals, see Dobrzycki and Kremer, 'Peurbach and Maragha astronomy?'.

67 J. D. North, 'The Alfonsine Tables in England', in *Prismata, Naturwissenschaftsgeschichtliche Studien: Festschrift für Willy Hartner*, ed. by Y. Maeyama and W. G. Saltzer (Wiesbaden: Steiner, 1977), pp. 269–301 (290–91).

68 In the history of earlier mathematics or astronomy, very few computational projects can be identified that involved literally millions of individual arithmetic procedures. For example, Ulugh Beg's sine table (early fifteenth century) presents sines to five sexagesimal places (six between eighty-seven and ninety degrees) for every arcminute, independently computed at multiples of five arcminutes, but his computational procedures have not been fully reconstructed. In what Charette has called the 'largest trigonometric tables of the Middle Ages', a 400,000 entry, triple-argument table solving the spherical triangle for three given arcs from one to ninety degrees was constructed *c.* 1300 in Cairo by the Muslim timekeeper and astronomer Najm al-Dīn al-Miṣrī.

Manuscript Sources

Brussels, Bibliothèque royale, 10117-26
Erfurt, Universitätsbibliothek, CA 2° 386
Lisbon, Biblioteca da Ajuda, 52-XII-35
London, British Library, Add 22317
Milan, Biblioteca Ambrosiana, C 278 bis Inf.
Munich, Bayerische Staatsbibliothek, Cgm 737
Munich, Bayerische Staatsbibliothek, Rar. 299a
Nuremberg, Stadtbibliothek, Cent. III, 25
Nuremberg, Stadtbibliothek, Cent. V, 57
Nuremberg, Stadtbibliothek, Cent. VI, 23
Paris, Bibliothèque nationale de France, lat. 7269
Paris, Bibliothèque nationale de France, lat. 7316A
Vatican, Biblioteca Apostolica Vaticana, Pal. lat. 1392
Vatican, Biblioteca Apostolica Vaticana, Pal. lat. 1403
Vienna, Österreichische Nationalbibliothek, Cod. 4988
Vienna, Österreichische Nationalbibliothek, Cod. 5203

Bibliography

Boffito, J., and C. Melzi d'Eril, eds, *Almanach dantis aligherii, sive Profhacii Judaei Montispessulani Almanach perpetuum ad annum 1300 inchoatum* (Florence: Olschki, 1908).

Boistel, Guy, *'Pour la gloire de M. de a Lande': Une histoire matérielle, scientifique, institutionnelle et humaine de La Connaissance des temps, de 1679 à 1920* (Paris: IMCCE, 2022).

Boudet, Jean-Patrice, and Laure Miolo, 'Alfonsine Astronomy and Astrology in Fourteenth-Century Oxford: The Case of MS Bodleian Library Digby 176', in *Alfonsine Astronomy: The Written Record*, ed. by Richard L. Kremer, Matthieu Husson, and José Chabás (Turnhout: Brepols, 2022), pp. 57–105.

Chabás, José, and B. R. Goldstein, 'Nicholaus de Heybech and His Table for Finding True Syzygy', *Historia mathematica*, 19 (1992), 265–89.

———, and ———, *The Alfonsine Tables of Toledo* (Dordrecht: Kluwer, 2003).

———, and ———, *The Astronomical Tables of Giovanni Bianchini* (Leiden: Brill, 2009).

———, and ———, 'John of Murs's Tables of 1321', *Journal for the History of Astronomy*, 40 (2009), 297–320.

———, and ———, 'The Master and the Disciple: The Almanac of John of Lignères and the Ephemerides of John of Saxony', *Journal for the History of Astronomy*, 50 (2019), 82–96.

———, and ———, 'The Almanac of Jacob Ben Makhir', in *Editing and Analyzing Numerical Tables: Towards a Digital Information System for the History of the Astral Sciences*, ed. by

See Benno van Dalen, 'Ulugh Beg', in *Biographical Encyclopedia of Astronomers*, ed. by Thomas Hockey (New York: Springer, 2007), 2:1159–61; François Charette, 'A Monumental Medieval Table for Solving the Problems of Spherical Astronomy for all Latitudes', *Archives internationales d'histoire des sciences* 48 (1998), 11–64 (p. 11).

Matthieu Husson, Clemency Montelle, and Benno van Dalen (Turnhout: Brepols, 2022), pp. 53–78.

Charette, François, 'A Monumental Medieval Table for Solving the Problems of Spherical Astronomy for all Latitudes', *Archives internationales d'histoire des sciences*, 48 (1998), 11–64.

Curtze, Maximilian, 'Der Briefwechsel Regiomontans mit Giovanni Bianchini, Jacob von Speier und Christian Roder', *Abhandlungen zur Geschichte der mathematischen Wissenschaften*, 12 (1902), 187–336.

Czerny, Albin, 'Aus dem Briefwechsel des grossen Astronomen Georg von Peurbach'. *Archiv für österreichische Geschichte*, 72 (1888), 281–304.

Dalen, Benno van, 'Ulugh Beg', in *Biographical Encyclopedia of Astronomers*, ed. by Thomas Hockey (New York: Springer, 2007), pp. 1159–61.

Dobrzycki, Jerzy, and Richard L. Kremer, 'Peurbach and Maragha Astronomy? The Ephemerides of Johannes Angelus and Their Implications', *Journal for the History of Astronomy*, 27 (1996), 187–237.

Erler, Georg, *Die Matrikel der Universität Leipzig*. 3 vols, Leipzig: Giesecke & Devrient, 1895–1902.

Gerl, Armin. *Trigonometrisch-astronomisches Rechnen kurz vor Copernicus: Der Briefwechsel Regiomontanus-Bianchini* (Stuttgart: Steiner, 1989).

Gingerich, Owen, 'Applications of High-Speed Computers to the History of Astronomy', *Vistas in Astronomy*, 9 (1967), 229–36.

———, 'The Role of Erasmus Reinhold and the Prutenic Tables in the Dissemination of Copernican Theory', *Studia Copernicana*, 6 (1973), 43–62, 123–25.

———, 'Crisis' Versus Aesthetic in the Copernican Revolution', *Vistas in Astronomy*, 17 (1975), 89–95.

Gingerich, Owen, and Barbara L. Welther, 'The Accuracy of the Toledan Tables', in *Prismata: Festschrift für Willy Hartner*, ed. by Y. Maeyama and W. G. Saltzer (Wiesbaden: Steiner, 1977), pp. 151–63.

———, and ———, *Planetary, Lunar, and Solar Positions, New and Full Moons, A.D. 1650–1805. Memoirs of the American Philosophical Society*, 59S (Philadelphia: American Philosophical Society, 1983).

Grössing, Helmuth, *Humanistische Naturwissenschaft: Zur Geschichte der Wiener mathematischen Schulen des 15. und 16. Jahrhunderts* (Baden-Baden: Koerner, 1983).

Juste, David, *'MS Paris, Bibliothèque nationale de France, Latin 7316A'* (updated: 11.06.2022), *Ptolemaeus Arabus et Latinus. Manuscripts*, URL = http://ptolemaeus.badw.de/ms/126.

Kennedy, E. S., 'The Digital Computer and the History of the Exact Sciences', *Centaurus*, 12 (1968), 107–13.

Kokott, Wolfgang, 'Regiomontans Ephemeriden für die Jahre 1475 bis 1506', *Sterne und Weltraum*, 36 (1997), 446–51.

Kremer, Richard L., 'Thoughts on John of Saxony's Method for Finding True Syzygy', *Historia mathematica*, 30 (2003), 263–77.

———, 'Text to Trophy: Shifting Representations of Regiomontanus's Library', in *Lost Libraries: The Destruction of Great Book Collections since Antiquity*, ed. by James Raven (London: Palgrave Macmillan, 2004), pp. 75–90.

———, 'Cracking the *Tabulae Permanentes* of John of Murs and Firmin of Beauval with Exploratory Data Analysis', in *Editing and Analysing Astronomical Tables: Towards a Digital Information*

System for the History of Astral Sciences, ed. by Matthieu Husson, Clemency Montelle, and Benno van Dalen (Turnhout: Brepols, 2022), pp. 363–422.

———, 'Printing Sacrobosco in Leipzig, 1488–ca. 1521: Local Markets and University Publishing', in *Publishing Sacrobosco's De Sphaera in Early Modern Europe: Modes of Material and Scientific Exchange*, ed. by Matteo Valleriani and Andrea Ottone (Cham: Springer, 2022), pp. 409–57.

———, 'Controlling Errors in the First Printed Book of Astronomical Tables: Regiomontanus's Ephemerides (Nuremberg 1474)', in *Printing and Misprinting: A Companion to Mistakes and In-House Corrections in Renaissance Europe (1450–1650)*, ed. by Geri Della Rocca de Candal, Anthony Grafton and Paolo Sachet (Oxford: Oxford University Press, 2023), pp. 295-324.

———, 'Kinetic Almanacs: Astrological Broadsides and Local Markets During the Early Years of Printing', *Cahiers de recherches médiévales et humanistes*, 47 (2024), 497-526.

———, and Jerzy Dobrzycki. 'Alfonsine Meridians: Tradition Versus Experience in Astronomical Practice c. 1500', *Journal for the History of Astronomy*, 29 (1998), 187–99.

Maruska, Monika, 'Johannes Schöner — "Homo est nescio quails": Leben und Werk eines fränkischen Wissenschaftlers an der Wende vom 15. zum 16. Jahrhundert', unpublished PhD thesis, Universität Wien, 2008.

Neske, Ingeborg, *Die Handschriften der Stadtbibliothek Nürnberg, Bd. V: Die lateinischen mittelalterlichen Handschriften, Varia, 13.-15. und 16–18. Jahrhundert* (Wiesbaden: Harrossowitz Verlag, 1997).

North, J. D., 'The Alfonsine Tables in England', in *Prismata, Naturwissenschaftsgeschichtliche Studien: Festschrift für Willy Hartner*, ed. by Y. Maeyama and W. G. Saltzer (Wiesbaden: Steiner, 1977), pp. 269–301.

———, *God's Clockmaker: Richard of Wallingford and the Invention of Time* (London: Hambledon and London, 2005).

Pedersen, Fritz S., *The Toledan Tables: A Review of the Manuscripts and the Textual Versions with an Edition* (Copenhagen: C. A. Reitzels Forlag, 2002).

Pohl, Frederick J., *Amerigo Vespucci Pilot Major* (New York: Columbia University Press, 1944).

Poulle, Emmanuel. *Astronomie théorique et astronomie practique au Moyen Âge* (Paris: Palais de la découverte, 1967).

———, 'Jean de Murs et les tables alphonsines', *Archives d'histoire doctrinale et littéraire du moyen âge*, 55 (1980), 241–71.

———, ed., *Les Tables alphonsines avec les canons de Jean de Saxe: Édition, traduction et commentaire* (Paris: Éditions du Centre national de la recherche scientifique, 1984).

———, 'Les Astronomes parisiens au XIVe siècle et l'astronomie alphonsine', *Histoire littéraire de la France*, 43 (2005), 1–54.

———, and Owen Gingerich, 'Les Positions des planètes au Moyen Âge: Application du calcul électronique aux tables alphonsines', *Académie des inscriptions & belles-lettres, Comptes rendus* (1968), 531–48.

Schum, Wilhelm, *Beschreibendes Verzeichniss der Amplonianischen Handschriftensammlung zu Erfurt* (Berlin: Weidmannsche Buchhandlung, 1887).

Shank, Michael H, 'Regiomontanus Versus George of Trebizond on Planetary Order, Distances, and Orbs (*Almagest* 9.1)', in *Ptolemy's Science of the Stars in the Middlle Ages*, ed. by David Juste *et al.* (Turnhout: Brepols, 2020), pp. 305–86.

Stahlman, W. D., and Owen Gingerich, *Solar and Planetary Longitudes for Years -2500 to +2000* (Madison: University of Wisconsin Press, 1963).

Swerdlow, N. M, 'Regiomontanus on the Critical Problems of Astronomy', in *Nature, Experiment, and the Sciences*, ed. by T. H. Levere and W. R. Shea (Dordrecht: Kluwer Academic Publishers, 1990), pp. 169–95.

Tanstetter Collimitius, Georg, ed., *Tabulae eclypsium Magistri Georgij Peurbachij. Tabula primi mobilis Joannis de Monte Regio* (Vienna: Johann Winterburger, 1514).

Thomann, Johannes, 'Reverse-Engineering Applied to Ephemerides: Analysis and Edition of the Arabic Ephemeris of 1327/7 CE (Ms Cairo, Dār Al-Kutub, Mīqāt 817)', in *Editing and Analyzing Numerical Tables: Towards a Digital Information System for the History of Astral Sciences*, ed. by Matthieu Husson, Clemency Montelle, and Benno van Dalen (Turnhout: Brepols, 2022), pp. 469–510.

Tuckerman, Bryant, *Planetary, Lunar, and Solar Positions, 601 B.C. to A.D. 1649*. 2 vols (Philadelphia: American Philosophical Society, 1962–64).

Wilson, Curtis, 'Predictive Astronomy in the Century after Kepler', in *The General History of Astronomy*, ed. by René Taton and Curtis Wilson (Cambridge: Cambridge University Press, 1989), pp. 161–206.

Wussing, Hans, 'Regiomontanus and Leipzig', in Ernst Zinner, *Regiomontanus: His Life and Work*, transl. by Ezra Brown (Amsterdam: Elsevier, 1990), pp. 307-11.

Zinner, Ernst, *Leben und Wirken des Joh. Müller von Königsberg genannt Regiomontanus*. 2d rev. ed. (Osnabrück: Zeller Verlag, 1968).

ALEXANDRE TUR

From Astronomical Computation to Astrological Interpretation

Figurae Caeli *and 'Astrological Squares' in Fifteenth-Century Annual Prognostications**

Introduction

While it has long been a controversial topic among historians of science, it is nowadays widely accepted that medieval astronomy and astrology were so closely linked that they are often indistinguishable in the literature. Astronomy was required in order to practice astrology, and, vice versa, astrology was the natural application of astronomical knowledge. They were considered the same discipline, and while practitioners might specialize either in mathematical computation or in astrological counsel, they were trained and usually fluent in both. Furthermore, most (medieval) scientific advancements in the field of astronomy, including the Copernican revolution, seem to result from the need to obtain more accurate astrological predictions — failures often being considered the result of a defect in the astronomical source data used.[1]

In practice, the point of contact between these two elements making up the 'science of the stars' is the horoscope, which was usually referred to as *figura caeli* (figure of the sky) in the late Latin Middle Ages.

* This work was supported by the ERC project ALFA: Shaping a European Scientific Scene (COG 723085, 2017–2022, PI Matthieu Husson). The author thanks the project's members for their valuable feedback and comments over the course of this research project, and particularly Eleonora Andriani, Jean-Patrice Boudet, José Chabás, Alena Hadravova, Matthieu Husson, and Richard L. Kremer, who reviewed this paper in its entirety at different stages of its composition.

1 On Copernicus, see the remarkable study by Robert S. Westman, *The Copernican Question: Prognostication, Skepticism, and Celestial Order* (Berkeley: University of California Press, 2011). In that symbolic case, the defended hypothesis is disputed. More broadly, on the role of astrology in the medieval history of sciences, Lynn Thorndike's work remains a reference: Lynn Thorndike, *A History of Magic and Experimental Science*, 8 vols (New York: Columbia University Press, 1928–58), vols I–IV for the medieval period.

Alexandre Tur • Bibliothèque nationale de France

Alfonsine Astronomy: Computational Practices, ed. by Matthieu Husson, José Chabás, and Richard L. Kremer, Turnhout, 2026 (*Alfonsine Astronomy: Studies and Sources*, 3), p. 487-550
 10.1484/M.ALFA-EB.5.137920

Horoscope, or figura caeli

The word horoscope is a tricky one as it has evolved from the Greek ὡροσκόπος, one who observes (σκοπός) the sky at the hour (ὥρα) of birth, with different meanings. In medieval Latin, *horoscopus* is synonymous with *ascendens*, designating the part of the sky (or, more specifically, the point of the ecliptic) rising over the horizon at a precise time. The contemporary English meaning of both the celestial configuration at a given moment and the chart representing it is covered by the medieval expression *figura caeli*. The astrological prediction based on the interpretation of that configuration is usually called *judicium* (judgement), although other, less widely used designations coexist.[2]

The medieval horoscope — the *figura caeli* — summarizes all the astronomical data needed to form an astrological prediction. It can be established through observation, using instruments such as astrolabes, but it is much more frequently computed ahead (or in retrospect) of a specific event using tables (or similar tools) and methods such as the one found in the Alfonsine sets. Based on a precise place and time, the *figura caeli* gathers the following:

- the true ecliptic longitude of each of the seven planets (Moon, Mercury, Venus, Sun, Mars, Jupiter, Saturn), the lunar nodes (known as *Caput* and *Cauda Draconis*), and in some (rarer) cases a selection of 'fixed stars'; and
- a mathematical division of the sky into twelve houses identified by the longitude of their intersection with the ecliptic, known as 'cusps' (*cuspides*; the cusp of the first house is the Ascendant itself).

The conjunction of these two sets of data is the basis for astrological analysis and interpretation, which can include the production of intermediary data such as planetary aspects (noteworthy zodiacal angles between planets) or lots (*partes* — geometrical points resulting from the projection of an angle from another point of the ecliptic).

A judgment is made by interpreting this data according to recognized rules and authorities, some being more general (Ptolemy's *Quadripartitum*; Alcabitius' *Liber introductorius*, etc.) and others more specialized, for different types of predictions (Aomar's *De nativitatibus*; Messahallah's *De revolutionibus annorum mundi*, etc.).[3] Different categories of astrological predictions differ by their goals, the authorities used, and mostly by their *figurae caeli* depicting different celestial events. Nativities are based on the celestial configuration at the birth (and anniversaries) of

---

2 A medieval Latin glossary of astronomical and astrological technical terms has long been a *desideratum*. For mathematical astronomy, José Chabás makes a significant contribution with a glossary to be published in a subsequent book of the present collection. No such synthesis has yet been published (to my knowledge) for medieval astrology, but it may be of use to browse André Le Boeuffle, *Astronomie, astrologie: lexique latin* (Paris: Picard, 1987), keeping in mind that it describes classical (ancient) practices.

3 Al-Qabīṣī (Alcabitius), *The Introduction to Astrology*, ed. by Charles Burnett and Keiji Yamamoto (London: Warburg Institute, 2004). Most astrological treaties translated from the Arabic are described in Francis James Carmody, *Arabic Astronomical and Astrological Sciences in Latin Translation: a Critical Bibliography* (Berkeley: University of California Press, 1956); Charles Burnett and David Juste are preparing an updated version of this reference bibliography, including translations from the Greek and Hebrew: see David Juste and Charles Burnett, 'A New Catalogue of Medieval Translations into Latin of Texts on Astronomy and Astrology', in *Medieval Textual Cultures: Agent of Transmission, Translation and Transformation*, ed. by Faith Wallis and Robert Wisnovsky (Berlin: W. de Gruyter, 2016), pp. 63–76.

their subject; annual prognostications (also called *revolutiones anni*, 'revolutions of the year') are based on the sky at the Spring equinox; interrogations look for the answer to a specific question in the figure at the place and time the question was asked; elections go the other way around and try to identify an auspicious celestial configuration in the future to begin a specific initiative.[4] Other remarkable celestial events, such as eclipses, planetary conjunctions, apparitions of comets, and so on, also merit their own horoscopes and judgments.

Displaying the astronomical source of the prediction

Such horoscopes are at the core of the resulting judgments, but they are not always revealed by the astrologer. Like a modern computer program, a 'compiled' prediction can be delivered without producing its 'code' (which can, however, be reverse engineered — to a certain extent — by another skilled astrologer). Horoscopes can even be considered political secrets and/or valuable data, the possession of which may provide a competitive advantage over other astrologers. That is especially true in the case of the birth horoscopes of political rulers, at a time when a precise date of birth was a rare and precious piece of information. These could be found in private workbooks or astrological 'files' that an astrologer could come back to without having to memorize or recalculate anything. Several such manuscripts have been preserved to the present day: they do not include astrological predictions, only a collection of base horoscopes for significant events and births.[5]

On the other hand, demonstrating the method used is also proof of competence and a way to exonerate oneself if things don't go according to plan — which can be attributed either to the incompetence of the astrologer or more imponderable factors such as free will or divine omnipotence. Much like modern scientists, astrologers tended to share their raw (astronomical) data and their (astrological) method of deduction to justify their result. This is especially true among university exercises and demonstration samples such as annual prognostications. Very widespread in the fifteenth century, this type of astrological prediction for the year to come mostly described general phenomena such as the weather, agriculture, epidemics, and warfare on the basis of well-known recurring astronomical events, mainly the spring equinox.[6] Some universities required their lecturer in astrology to publish, and publicly defend, an almanac and an annual prediction.[7] Many astrologers

4 On the different types of astrological predictions and their medieval designations, see Charles Burnett, 'Astrology', in *Medieval Latin: an Introduction and Bibliographical Guide*, ed. by Franck Anthony Carl Mantello and Arthur George Rigg (Washington: The Catholic University of America Press, 1996), pp. 369–82. An example of the fifteenth-century practicing of these different predictions is given by Helena Avelar de Carvalho in *An Astrologer at Work in Late Medieval France: The Notebooks of S. Belle* (Leiden: Brill, 2021).

5 Our case B below is copied in one such 'astrological file' in the manuscript Paris, BnF, lat. 7443. Other famous examples include Lichtenberger's recently discovered collection, Wolfenbüttel, Herzog August Bibliothek, Cod. Guelf. 115 Noviss. 4° or S. Belle's Paris, BnF, NAL 398 (Avelar de Carvalho, *An Astrologer at Work*...). While the diagram described below is a common way to 'save' a horoscope in a synthetic and visually meaningful form, it coexists with textual or tabular descriptions.

6 On annual prognostications, see Alexandre Tur, 'Hora introitus solis in Arietem: Les prédictions astrologiques annuelles latines dans l'Europe du XVe siècle (1405–84)', unpublished doctoral thesis, Université d'Orléans, 2018, http://www.theses.fr/2018ORLE1163) and its extensive bibliography.

7 This practice is well attested in Bologna, Louvain, and Cracow, but it was probably common in other universities. See Tur, 'Hora introitus', pp. 282–86 and, more specifically on the Bologna case, Albano Sorbelli, 'Il 'Tacuinus' dell'Università di Bologna e le sue prime edizioni', *Gutenberg Jahrbuch*, 33 (1938), 109–14; Tommaso Duranti,

also sent their own prediction to political rulers — and potential patrons — free of charge to demonstrate their expertise. However, if they hoped to impress students and/or future patrons (often advised by their current astrologers), the authors needed to exhibit their skills both in procuring adequate astronomical data and in analysing it.

An easy way to do so was to include justifications in the text. Those would appear just after each conclusion and give astronomical data, the interpretation mechanism used, and which authority provided them (sometimes the exact quotation). This is known today as 'cooked data'. For instance, Pietro de Monte Alcino wrote the following in his prediction for 1421 (our case A below):

> This next year, there will be an epidemic with high mortality, especially in lands depending on the triplicity of Aries, Leo, and Pisces, such as Florence, Perugia, Venice and Genoa. This conclusion comes from the [planet] ruling the Ascendent, in the House of death at the beginning of the revolution, which also takes part in an aspect with Mars. For this reason, diseases will be raging with contagiousness and rapidly deadly. This conclusion agrees with Haly [Habenragel] in the 8th part of his *Judiciarium*, chapter 31.[8]

An even more 'open' way (complementarily to in-text justifications) is to add a complete description of the horoscope used at the beginning of the prediction. It would detail the astronomical event chosen, its calculated time and place, longitudes of the planets, cusps of the houses; and, in some cases, the result of a basic mathematical analysis, including planetary aspects and lots and key astrological interpretations, such as determining the dominant planet ('lord of the year' in annual predictions).

That is the case in Pietro de Monte Alcino's prediction, although the whole chapter entitled 'on the layout of the sky' (*de dispositione celi*) has been cut from one of the two preserved copies (see below). This technical information was clearly not valued by every reader, and some less interested copyists did not hesitate to omit it to spare paper. As we rarely preserve several copies of these ephemeral practical texts, it can be difficult to assess whether an astrologer decided not to include this astronomical information in their prediction, or whether we have a copy abbreviated by the scribe.[9]

This base set of data, from which all other conclusions are inferred, is the *figura caeli*, an abstract representation of an astronomical event to use for one (or several) astrological predictions. It can be written down (or not) in the predictions or compiled in files or horoscope collections if it is likely to be reused. The form can differ; Pietro de Monte

---

Mai sotto Saturno: Girolamo Manfredi, medico e astrologo (Bologna: CLUEB, 2008), p. 159.

8 Oxford, Bodleian Library, Ashmole 357, f. 186v (Chapter 3, 1st conclusion): 'In anno futuro erit pestis et mortalitas magna, et fortius in terris de trigono Arietis, Leonis et Piscium, prout est Florentia, Perusium, Venecie et Janua. Hec conclusio colligitur ex domino Ascendentis in principio revolutionis in domo mortis, habente etiam participationem in aspectu cum Marte. Ratione cujus egritudines erunt furiose cum contagio cito interficientes. Hanc conclusionem probat Hali in parte 8a sui *Judiciarii* capitulo 31o.' The last sentence refers to a usual authority, Haly Habenragel, *De judiciis astrorum* (translated from the Arabic by Aegidius de Thebaldis after 1254): book VIII is dedicated to annual prognostications and Chapter 31 to agriculture and diseases. See Haly Abenragel, *De judiciis astrorum* (Venice: Erhard Ratdolt, 1485, ISTC ih00004000), VIII, 31, f. 147v and Tur, 'Hora introitus', pp. 11–14.

9 Of the 111 manuscript Latin prognostications between 1405 and 1484 known today, 101 are preserved in only one copy, eight are in two copies, one is in three, and one is in four. See Tur, 'Hora introitus', pp. 256–62.

Alcino chose to write a textual description, but several astrologers recapitulated this data in a table. The Oxford copyist of Monte Alcino opted for a more graphical way of displaying this data in the margin, with a classical diagram called an astrological square often being provided instead of a textual description.

The diagrammatic representation of the astrological square

A standard way of arranging this astronomical data in a geometrical figure is indeed taught to medieval astrologers, and it is often used in predictions and horoscope collections to synthesize the *figura caeli*. In fact, a specific sense of the locution *figura caeli* (as well as the modern *horoscope*) would come to coexist with the broader abstract representation described so far. This double meaning makes it sometimes difficult to interpret a phrase such as 'as you can see in the figure...' (*ut videtur in figura*), which is often found in justifications; does it refer specifically to the diagram — and could it indicate, for instance, that the astrologers intended to include a diagram even though there isn't one in our preserved copy — or to the abstract representation of the celestial configuration? Despite the probable loss, of the 111 known annual predictions preserved in manuscript form, 25 include an astrological square in at least one of their copies.[10]

The medieval layout of the astrological squares differs from the circular diagram found in ancient sources and from present-day 'sky charts'. Still, they are rooted in a tradition that Johannes Thomann traced back to China, which reached Europe through Arabic and Hebraic astrology, maybe following the Silk Road.[11] In his reference work on the subject, John D. North provided a visual typology of known layouts, revealing their limited variability.[12] In fourteenth- and fifteenth-century Europe, North's 'Type e' is largely hegemonic, although variations sometimes appear.

Unlike contemporary astrologers, who usually depict a horoscope through a circle representing the ecliptic, equally divided into signs with the cusps of the houses distributed on this axis of longitude, medieval astrologers preferred to base their diagram on a representation of equal houses, depicted as successive triangles, often around a central square (serving as a cartel). The planets (and other significant astronomical objects) are distributed among the twelve triangles depending on their longitudes. The houses' cusps are copied, as well as the planets' longitudes, but there is no scaled depiction of the ecliptic itself. If — as it is safe to assume — this normative layout was the first mental representation of a horoscope for most medieval astrologers, it carried a hierarchy in the

10 Tur, 'Hora introitus', catalogue pp. 357–811. The astrological square is rarer in early printed predictions, probably for technical reasons. On that matter, see our case E below.

11 Johannes Thomann, 'Square Horoscope Diagrams in Middle Eastern Astrology and Chinese Cosmological Diagrams: Were These Designs Transmitted through the Silk Road?', in *The Journey of Maps and Images on the Silk Road*, ed. by Philippe Forêt and Andreas Kaplony (Leiden: Brill, 2008), pp. 97–117. While the layout of Chinese hemerology diagrams may have inspired Arabic and Hebraic astrological squares (transferred to Europe through the translation of scientific treaties), they did not have the same content and meaning, with notions such as 'astrological house' not being used in Chinese astrology.

12 John D. North, *Horoscopes and History* (London: The Warburg Institute, 1986). In the first part of this book, North examines the different mathematical methods of computing the astrological houses — crucial in any horoscope. The second part details 28 historical case studies from ancient Greek horoscopes to the seventeenth century, comparing their calculations as well as their diagram layouts.

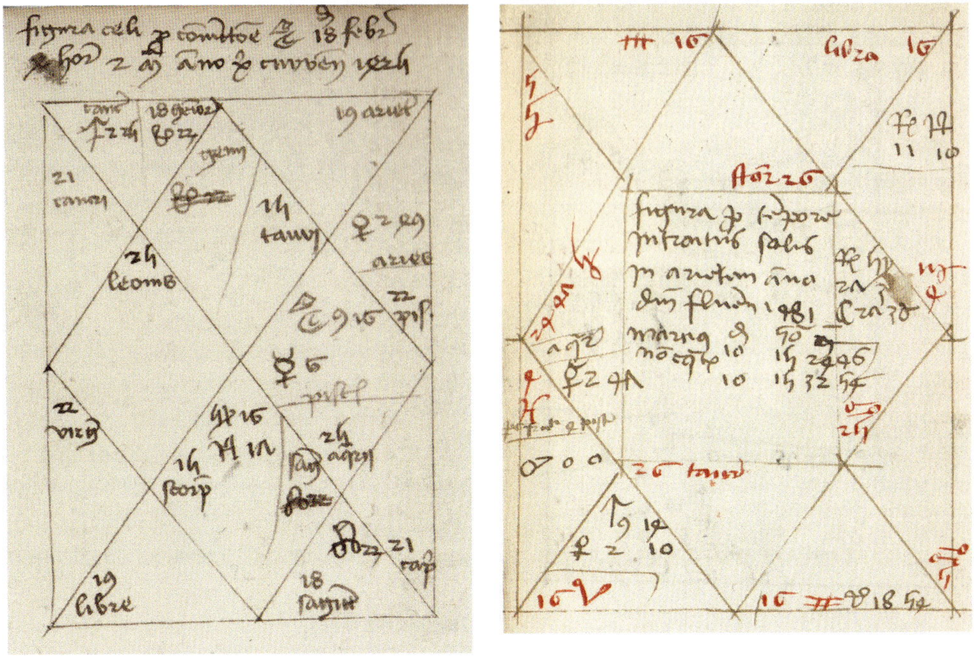

Figure 15.1. Most common 15th-century astrological square layouts, type d (left) and e (right) in John D. North's classification. Paris, BnF, lat. 7443, f. 125^{r} and lat. 7427, f. 269^{r}.

possible layers of interpretation, favouring, for instance, the relation between planets and houses at the expense of less evident geometrical aspects between planets.

Five case studies in fifteenth-century prognostications

To further investigate the astronomical practices of Alfonsine astrologers while building their *figura caeli*, here we detail five case studies from the corpus of fifteenth-century annual prognostications and examine what the preserved sources tell us about the context of production, the role of the figure in the prediction — both intellectually and as a diagram — and the accuracy (and precision) of the astronomical calculations.

This approach is inspired by North's case studies, and more recently by Stephan Heilen's meticulous analysis of John of Lübeck's 1474 prediction on the coming of the Antichrist, based on the 1504 Saturn-Jupiter conjunction.[13] In that work, Heilen carefully examines the different known witnesses of the prediction and their astronomical data and proof-computes it to highlight a probable manipulation by the astrologer in order to ensure his desired interpretative outcome (the coming of the Antichrist). While such a

13 Stephan Heilen, *Konjunktionsprognostik in der Frühen Neuzeit: Die Antichrist-Prognose des Johannes von Lübeck (1474) zur Saturn-Jupiter-Konjunktion von 1504 und ihre frühneuzeitliche Rezeption* (Baden-Baden: Verlag Valentin Koerner, 2020). Two more volumes dedicated to other fifteenth-century astrological predictions on Saturn-Jupiter conjunctions are in preparation.

daring manipulation is probably an exception among medieval prognostications, it reaffirms the historical interest of simultaneously studying the astronomical data, the astrological interpretation, and the context of production.

1. Methodology

The five cases below were selected among the 111 known annual predictions for the years between 1405 and 1484 preserved in manuscript copies, and, more precisely, from the 25 of them that include astrological squares. The choice of annual predictions was made for the sake of convenience, as it made it possible to lean on previous research; and, as explained in the introduction, this type of prediction was also the most likely to include detailed astronomical information.[14] The decision to only consider predictions bearing astrological squares (although several others include textual or tabular *figurae celi*) allowed a more detailed graphical comparison and technical considerations related to the production of this diagram.

In selecting these five cases, we strive to achieve a certain diversity in the apparent link between text and diagram. In case A, the diagram is not provided by the astrologer (who, however, includes a textual *figura caeli*) but reconstituted in the margins by a scribe. Case B displays a double square, while case C draws three different squares side by side (including two for the same horoscope). Case D is a rare representative of a formal prediction, designed to be addressed to the king, and the astrological square is both an astronomical diagram and a decoration. Case E includes the first (known) printed astrological square, a technical leap for fifteenth-century printers, demonstrating the importance of this diagram for astrologers instead of a textual or tabular *figura*.

The following analysis was only performed after the initial selection of the cases, and in some instances it reveals that the selection might not be representative of all the investigated criteria. For example, all five cases probably use the standard domification method, yet North's broader study would have led us to expect some more diversity in fifteenth-century predictions.

For each of these cases, we posed the following questions:

- What manuscript or printed witnesses have been preserved and what do we know of them? We examined the composition of these exemplars, as well as their history, as these might explain the current state of the prediction (maybe selected by a collector or abbreviated by a copyist, etc.).
- What was the context of production? What do we know of the astrologer and the context in which he was writing? This information might come from the text of the prediction (which countries are concerned?) or from external sources (what was the political spectrum at the time?).
- What role does the horoscope (in a broad sense) play in the prediction? How many *figurae caeli* are referred to in the prediction and which ones are they? Do all or only some

14 This consideration also applies to other 'public' predictions, based on astronomical events such as planetary conjunctions, eclipses, or comet sightings.

of them have a diagram? Does the diagram duplicate textual or tabular information? Are there explicit references to the diagram in the text?
- How is the diagram displayed? Does it use astronomical symbols to abbreviate the name of the planets and/or zodiacal signs? How precise are the longitudes, and do these match the data found in the text?

In each case, we tried to reproduce and transcribe the main *figurae caeli*, that is, both the text describing the horoscope (when it exists) and the associated diagram.[15]

To assess the accuracy of these data and spot a potential attempt to manipulate it, we compared the planetary longitudes to the theoretical Alfonsine result using Lars Gislen's *Astromodels* spreadsheets.[16] These recomputed planetary positions are based on the theoretical time of the intended event in Toledo (found by successive approximation through *Astromodels*). The lots were not proof computed, except for the *pars fortune*, which is the most meaningful for astrological interpretation and defined as the projection of the angular distance between the Sun and Moon (by day; or between the Moon and Sun by night) from the ascendent.[17]

Similarly, to check the houses' cusps and identify which of the mathematical domification methods was used, we employed a migrated version of John D. North's *Horosc* software.[18] It is important to remember that, unlike *Astromodels*, *Horosc* uses modern mathematical functions to compute the expected value based on the geometrical model. Medieval astrologers did not have access to such solutions, and they computed their house boundaries using tables adapted (more or less precisely) for their latitude. Depending on

15 In the pursuit of brevity, only the annual *figurae caeli* — the Spring equinox and/or its *prevention*, the syzygy immediately preceding it — have been analysed, even when other horoscopes are mentioned and displayed in the prediction (as in cases D and E). These figures are thought to be the most important for an annual prediction, although others may be required or used in addition in some circumstances. On that technical choice, see further details in Tur, '*Hora introitus*', pp. 31–48.

16 In 2008, Lars Gislen provided a series of MS Excel spreadsheets simulating the computation of planetary longitudes at a given time with different sets of tables. The spreadsheet used here is 'AlfonsoTab', based on the 1483 edition of the Alfonsine tables. This tool is freely downloadable on Gislen's website: home.thep.lu.se/~larsg/ <accessed 18.09.2021>.

17 For instance, by Alcabitius, *Liber introductorius*, V, 3: 'An example of this is the calculation of the Lot of Fortune. You take what is between the Sun and the Moon, and you cast it out from the ascendant; that is, you add the degrees of the ascendant to it, and you cast that out from the beginning of the sign of the ascendant, and where the number is exhausted is the position of the Lot of Fortune.' (ed. and transl. from the Arabic by C. Burnett and K. Yamamoto, p. 141; compare with the Latin version, pp. 350–51: 'Cuius exemplar est opus Partis Fortune, quia accipis quicquid est inter Solem et Lunam ex gradibus, et addis super eos gradus signi ascendentis, et quod collectum fuerit proicis eum .xxx. .xxx. gradus. Et incipis proicere ab initio signi ascendentis, et quo finitur numerus, ibi est Pars Fortune; et in nocte econtra a Luna in Solem, et addes desuper gradus signi ascendentis, et proicis hoc .xxx. .xxx.'). The Pars Fortune and its computation are mentioned by Ptolemy in the third book (on the nativities) of his *Quadripartitum* (III, 10, f. 66^{r} in the 1493 *princeps* edition).

18 North, *Horoscopes and History*, Appendix 4 pp. 197–218, provides the MS-DOS Pascal code for a program with two main functions: computing all theoretical cusps according to the seven 'classical' methods identified by North given an Ascendent and place's latitude; or, given all the cusps of a horoscope, deducing the theoretical latitude (with a range of uncertainty) and quality coefficients assessing the probability that each of the methods were used. The Pascal language can prove difficult to compile nowadays; I am very grateful to François J. Tur for porting the code in JavaScript and providing an easy-to-use Google spreadsheet. Any inexactitude resulting from that process of course remains my sole responsibility.

the demanding nature of the practitioner and their mathematical skills, they would use ready-made house tables or compute the cusps independently using oblique ascension and trigonometric tables. In any case, the several steps involved leave space for rounding and interpolation approximations, and the final result can't be expected to match the geometrical one exactly. It was also possible to read the cusps directly on an adapted astrolabe, which is a less precise but much swifter way to obtain the complete domification, subject to reading approximations that cannot be reproduced.

In assessing the accuracy of the provided data, the consequences of rounding need to be considered. While mathematical rounding follows strict rules nowadays, medieval mathematicians could round up and down alternatively. Therefore, a longitude precise to the degree (which is the most common case) should be considered exact when it is one degree above or below the theoretical value. In the comparison tables, we coloured distances superior to one degree in yellow, and discrepancies of more than five degrees in red. The same goes for the domification tables, but we only coloured the column(s) of the most probable domification method(s).

In some cases, the source includes neither the exact time nor the place used to compute the horoscope. When the time was missing, we approximated it using the actual time of the astronomical event (after *Astromodels*). Restituting an approximate place based on the text of the prediction was often relatively easy, while determining the exact coordinates used in the calculation was less so.

When provided, we can establish that the time of the figure is given for the local meridian, and thus we can compute a theoretical longitude by comparing our computed Toledan time with the time given in the horoscope. However, such a conversion does not account for the astrologer's approximation in computing the time of the equinox or preceding syzygy, which is known to be a difficult operation. In several cases, the inaccuracy can be spotted through the variability of the longitudes across several horoscopes of the same prediction, but there isn't enough data to assess it in detail (see, for instance, case B). There was little consensus with regard to geographical longitudes before the seventeenth century; thus, comparing with known medieval (or modern) values is of limited significance.[19]

Geographical latitudes were less variable and were crucial for the domification computations. Provided with the set of resulting cusps and an expected precision, the *Horosc* software is able to compute a range of possible underlying latitudes. However, assuming a rounding to the degree, this range is often too wide to be truly informative; it is also easily disrupted by computation or copy mistakes resulting in erroneous cusps. When possible, we compared this estimation with the longitude and latitude tables of the Parisian Alfonsine Tables 1483 edition or the geographical coordinates collected by North from several fifteenth-century lists.[20] In the absence of any match, we resorted to any other historical indications available (see, for instance, case E).

19 See Richard L. Kremer and Jerzy Dobrzycki, 'Alfonsine Meridians: Tradition versus Experience in Astronomical Practice *c.* 1500', *Journal for the History of Astronomy*, 29 (1998), 187–99.

20 Alphonsus, Rex Castellae, *Tabulae astronomicae* (Venice: Erhard Ratdolt, 1483, ISTC: ia00534000), sig. m5^r-m6^r; North, *Horoscopes and History*, Appendix 2, pp. 186–95.

2. Case A. Pietro de Monte Alcino, prediction for 1421

2.1. Witnesses

The 1421 prediction by Pietro de Monte Alcino is known by two manuscript witnesses:

Oxford, Bodleian Library, Ashmole 357, folios 185^{r}–191^{r}.[21]

This composite manuscript from the Ashmolean collection — its earlier provenance is not well established — is made of two codicological units. The first one (ff. 1–178) dates from the fourteenth century and contains several mathematical, astronomical, and astrological treatises, such as Anianus' *Computus manualis* (ff. 1^{r}–5^{v}), Alcabitius's *Liber introductorius* translated by John of Seville (ff. 27^{r}–52^{v}), the *Centiloquium Hermetis* (ff. 53^{r}–57^{v}), or Euclid's *Elementa* translated by Adelard of Bath (ff. 71^{r}–75^{v}). A second codicological unit made of four fifteenth-century astrological predictions (ff. 179–216) was later added to the manuscript; it contains three annual predictions by Pietro de Monte Alcino (for 1419, 1421, and 1430) and an anonymous prediction for 1448.[22]

Vatican, Biblioteca Apostolica Vaticana, Pal. lat. 1438, folios 91^{r}–94^{r}.[23]

This composite volume, compiled by the humanist and Palatine chancellor Matthias of Kemnat (*c.* 1430–*c.* 1476) contains several manuscript and printed astrological predictions concerning the 1472 comet in particular, as well as isolated horoscopes (in the form of astrological squares), almanacs, and extracts of prophetical treatises or predictions. The prediction for 1421 by Pietro de Monte Alcino is copied in an abbreviated version that omits most of the astronomical data and astrological justifications.

Its incipit is *Si animadverto viri hoc nostrum judicium inspecturi, quia miro ordine atque ineffabili sapientia Deus noster omnipotens mundi pro gubernatione celi connexum suis intelligentiis…* [24]

21 'Ms Ashmole 357', *A Catalogue of Western Manuscripts at the Bodleian Libraries and Selected Oxford Colleges*, https://medieval.bodleian.ox.ac.uk/catalog/manuscript_332 <accessed 18.07.2021>; William R. Black, *A Descriptive and Critical Catalogue of the Manuscripts Bequeathed unto the University of Oxford by Elias Ashmole* (Oxford: University Press, 1845), 266–68; David Juste, 'MS Oxford, Bodleian Library, Ashmole 357', *Ptolemaeus Arabus et Latinus: Manuscripts*, http://ptolemaeus.badw.de/ms/423 <accessed 18.07.2021, last update 23.08.2018>.

22 The prediction for 1448 (f. 206^{r}–216^{r}) used to be attributed to Pietro de Monte Alcino as well, although there is no mention of the astrologer in the text. The comparison to the three other predictions, and to Pietro de Monte Alcino's very different prediction for 1448 preserved in Munich (Bayerische Staatsbibliothek, Clm 14504, f. 236^{r}–237^{v}), makes the attribution very unlikely. See Tur, *Hora introitus*, pp. 448 and 525–28.

23 Ludwig Schuba, *Die Quadriviums-Handschriften der Codices Palatini Latini in der Vatikanischen Bibliothek* (Wiesbaden: Dr L. Reichert, 1992), pp. 229–36, https://digi.ub.uni-heidelberg.de/diglit/schuba1992/0263 <accessed 20.09.2021>. The manuscript is digitized and available online from the Universitätsbibliothek Heidelberg (https://digi.ub.uni-heidelberg.de/diglit/bav_pal_lat_1438) and from the Biblioteca Apostolica Vaticana (https://digi.vatlib.it/mss/detail/Pal.lat.1438).

24 Lynn Thorndike and Pearl Kibre, *A Catalogue of Incipits of Mediaeval Scientific Writings in Latin* [1937] (2nd edition, Cambridge: The Mediaeval Academy of America, 1963), col. 1441. The incipit is actually only found in the Oxford manuscript, as the prologue doesn't appear in the Vatican manuscript, where the prediction begins: "Incipit judicium magistri Petri quodam Bernardi de Monte Alcino. Anno Domini MºCCCCºXXºI die XI Martii post meridiem…'. See also Lynn Thorndike, *A History of Magic and Experimental Science* (New York: Columbia University Press, 1934), IV, 91.

2.2. Context of production

This prediction for 1421 is the third (that we know of) made by the prolific Italian astrologer Pietro de Monte Alcino (fl. 1401–48) during his time as Professor of Astrology at the University of Pavia (1418–27). His predictions for 1418 and 1419 are preserved, as well as later prognostications for 1430 and 1448, which were probably issued while he was serving as one of the physicians of the Duke of Milan Filippo Maria Visconti (1412–47). It is not known whether the University of Pavia required its professor of astrology to publish an annual prediction, but the practice was common in other Italian universities as early as 1405.[25]

All preserved predictions by Pietro de Monte Alcino share the same thematic plan, beginning with the presentation of the *figura caeli* (*in dispositione celi*), then meteorology (*in dispositione temporali*), diseases (*in esse sanitatis et infirmitatis*), agriculture (*in esse sterilitatis et fertilitatis*), war (*in esse pacis et guerris universaliter*), and finishing with particular predictions to princes and Italian cities (*in significatione particulari*: *in esse Ecclesie*; *imperatoris*; *regis Francorum*; *regine Apulie*; *Romanorum*; *Florentinorum*; *Venetorum* and *Perusinorum*).[26]

2.3. Role of the horoscope in the prediction

This prediction, as well as all the others known of Pietro de Monte Alcino, is based on the horoscope of the spring equinox (the 'revolution of the year'). In 1421, some conclusions were also drawn from the lunar eclipse of August 13. The first chapter of the prediction is supposed to contain the description of those astronomical grounds.

It doesn't seem, however, that the author provided astrological squares to summarize this data. The copyist of the sole Oxford manuscript added an approximate diagram collecting all the astronomical information described in the text in the margins of the three predictions by Pietro de Monte Alcino. The 1421 astrological square is drawn in the bottom margin of folio 186^{r}, a little after the end of the textual description (See Fig. 15.2).[27]

Both planetary longitudes and houses' cuspids match the textual data. Positions are given to the nearest degree and to the nearest minute for Saturn, Jupiter, and the Dragon. Retrograde planets are identified both in the text and in the diagram, and several lots (*pars ordei*, *pars tritici*, *pars ovarum*, *pars milii*) are also included in both.

Both the text and the diagram use Arabic numerals and give the textual denomination of zodiacal signs, but the names of the planets are abbreviated with their symbols in the diagram. The symbols for the Head and Tail of the Dragon are inverted from their contemporary use (☊ for Caput Draconis; ☋ for Cauda Draconis).

---

25 See note 7 above. The duty of performing an annual prediction was, for instance, well established in the 1405 statutes of the Faculty of Arts at the University of Bologna: Carlo Malagola, *Statuti delle Università e dei collegi dello studio Bolognese* (Bologna: Nicola Zanichelli, 1888), Chapter LX, p. 264.

26 Tur, 'Hora introitus', 448 and 476–77.

27 While the text of the prediction also describes and uses the figure of the lunar eclipse, only the revolution is summarized as such in an astrological square. The same goes for the other predictions in Ashmole 357; other figures are used — in 1419, the solar eclipse of March 26; in 1430 the prevention — but a marginal square is only added for the revolutions. The last prediction in this manuscript, for 1448, although seemingly copied in the same hand, does not include a marginal square. However, there is a blank space on f. 207^{v} after the textual description of the figures to be used that might have been reserved for any future drawing of an astrological square (which would have been, in this case, inserted in the text and not in the margins).

2.4. Figurae Caeli

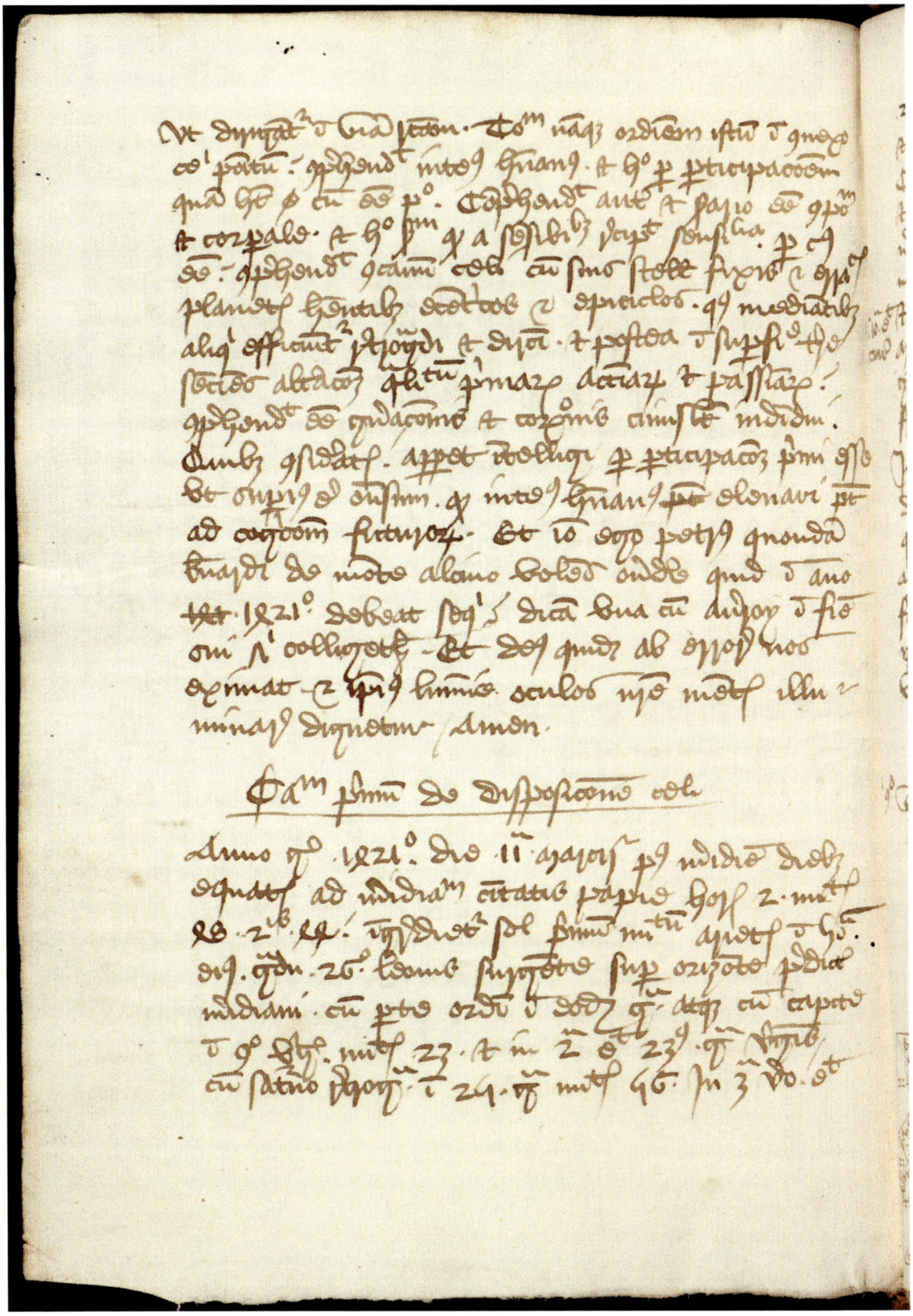

Figure 15.2. Pietro de Monte Alcino, prediction for 1421. Oxford, Bodleian Library, Ashmole 357, ff. 185v-186r.

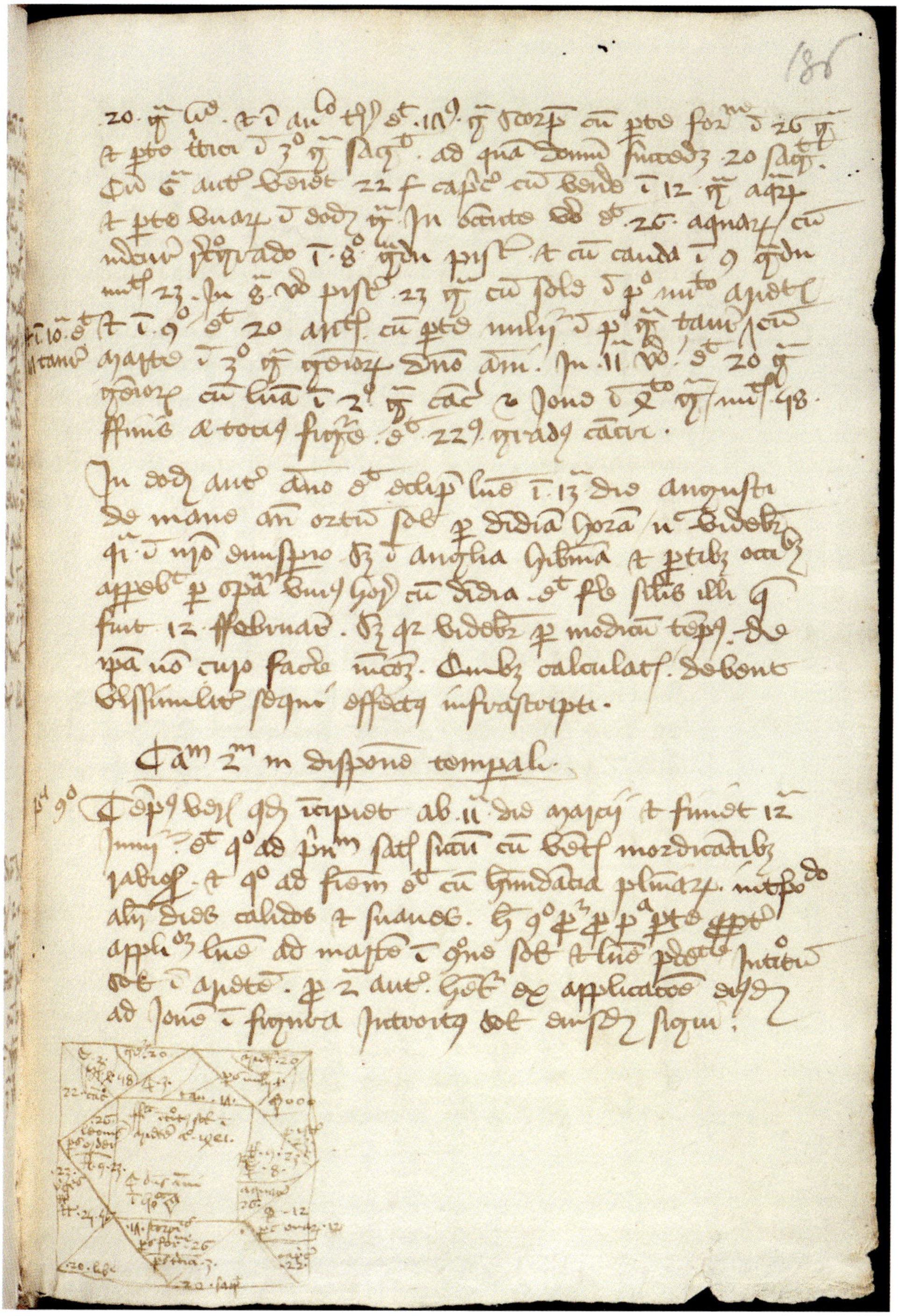

Judicium ejusdem [clarissimi astrorum que doctissimi viri magistri Petri de Monte Alcino] anno Domini 14210 incompleto
[The text begins with a cosmological prologue, f. 185$^{r\text{-}v}$]
Capitulum Primum de Dispositione Celi

Annus igitur 1421 die 11^{a} martii post meridiem diebus equatis ad meridianum civitatis Papie horis 2 minutis 48 secundis 44 ingredietur Sol primum minutum Arietis in hora ejus, gradu 26° Leonis surgente super orizonte predicti meridiani, cum parte ordei in eodem gradu atque cum Capite in 9° Virginis minutis 23. Et in 2^{a} erit 23us gradu[s] Virginis cum Saturno retrogrado in 25 gradu minutis 56. In 3^{a} vero erit *[186^{r}]* 20 gradu[s] Libre. Et in angulo terre erit 17us gradu[s] Scorpionis cum parte fortune in 26 gradu et parte tritici in 3° gradu Sagittarii. Ad quam domum succedet 20 Sagittarii. Cum 6^{a} autem veniet 22 [gradus] Capricorni cum Venere in 12 gradu Aquarii et parte uvarum in eodem gradu. In occidente vero erit 26 Aquarii, cum Mercurio retrogrado in 8° gradu Piscium, et cum Cauda in 9 gradu minutis 23. In 8^{a} vero Piscium 23 gradu[s] cum Sole in primo minuto Arietis et in 9° erit 20 Arietis cum parte milii in primo gradu Tauri < in 10^{a} erit [?] Tauri *add. in margine* > cum Marte in 3° gradu Geminorum, domino anni. In 11^{a} vero erit 20 gradu[s] Geminorum cum Luna in 2° gradu Cancri et Jove in 4to gradu minutis 58. Finis et totius figure erit 22us gradus Cancri.

(…)

[Follow the description of the figure for the lunar eclipse of August 13, then 'capitulum 2^{m} in dispositione temporali'. The astrological square is in the bottom margin of f. 186^{r}.]

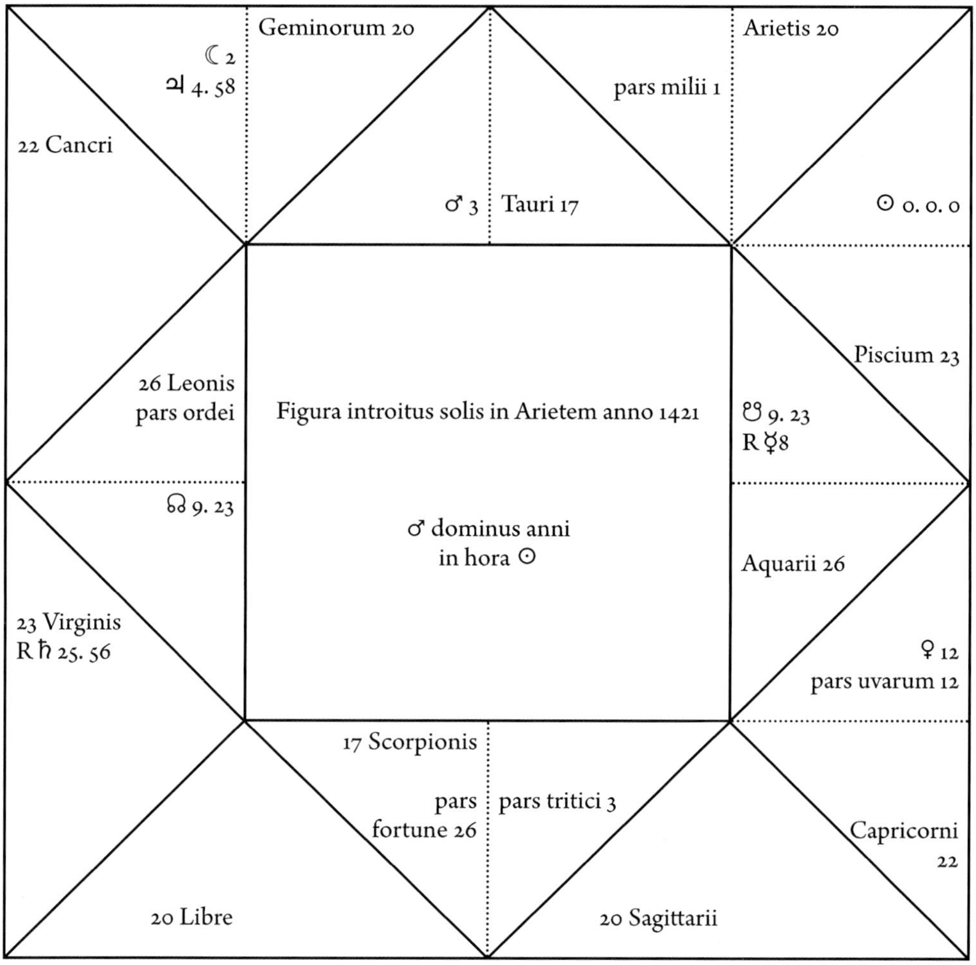

2.5. Recomputation

The text of Chapter 1 clearly states the date and place for which the horoscope has been calculated: March 11, 1421, *diebus equatis*, at 2h 48m 44s after noon, at the meridian of Pavia.[28] The astrologer doesn't explain, however, which geographical longitude and latitude he assumes for Pavia.

The fifteenth-century geographical lists compiled by North do not include coordinates for Pavia. Neither does the table of latitudes and longitudes printed in the 1483 edition of the Alfonsine Tables, but it does have a set of coordinates for Milan, the political centre

28 Precision of the time to the second is unusual and seems exaggerated given the precision chosen for the longitudes. It probably results from the addition of the equation of time (in seconds) to the mean time (in minutes). I am grateful to Stefan Zieme for pointing out this practice, which is also attested in Conrad Heingarter's prediction (case C below).

associated with the University of Pavia: 30°20' E (of the further west inhabited region, that is, 19°20' east of Toledo) and 44°50' N.[29]

Given the values of the cusps of the houses stated in the prediction, and supposing a rounding to the degree, North's *Horosc* software suggests they should have been computed for a latitude between 43°8' N and 48°51' N.

In the following recomputation of the domification, we assumed the widespread value of 45°, for which many tables were available.[30]

According to *Astromodels*, the theoretical Alphonsine equinox in Toledo was to happen on March 11, 1421, at 1h 21m pm. Supposing that Pietro de Monte Alcino indeed computed that value, the switch from Toledo to Pavia is of 1h 27m 44s to the east, placing Pavia at a longitude of 21°56'' E.

Compared with true positions provided by the Alfonsine tables (as recomputed by the *Astromodels* spreadsheets), Pietro de Monte Alcino is accurate to the rounding degree for all planets, except Venus, which is given at 12° Aquarius instead of 18°. This doesn't affect the astrological signification of this planet and should probably be considered a computation mistake or perhaps a misreading at some point. Given with a better precision (to the minute), the longitudes of Saturn and the lunar nodes are indeed rather accurate, with, respectively, ± 3' and ± 1'; Jupiter's position is slightly less accurate (± 30').

Planet	**Figura Caeli (diagram and text)**	**Alfonsine Tables**
	11/03/1421 2h48m44s pm Pavia	11/03/1421 1h21m pm Toledo
Moon	2° Cancer (XI)	1°46' Cancer
Sun	0°0'0'' Aries (VIII)	0°0'1'' Aries
Mercury	8° Pisces (VII)	8°39' Pisces
Venus	12° Aquarius (VI)	17°55' Aquarius
Mars	3° Gemini (X)	2°38' Gemini
Jupiter	4°58' Cancer (XI)	5°32'16'' Cancer
Saturn	25°56' Virgo (II)	25°53'13'' Virgo
Caput	9°23' Virgo (I)	9°21'57'' Virgo
Cauda	9°23' Pisces (VII)	9°21'57'' Pisces

The recomputation of the theoretical cusps according to the different domification methods, through North's *Horosc* software for a latitude of 45°N and an Ascendant at 26° Leo, leads us to suppose that methods 1 (Standard method) or 2 (Dual Longitude method) were used with an accuracy to the rounding degree (except for one set of cusps: Houses III and IX if method 1 was used; or Houses VI and XII if method 2 was used).

29 Dobrzycki and Kremer, 'Alfonsine Meridians', does not include either of those meridians. Contemporary coordinates for Pavia are 45°11'N 09°09'E (East of Greenwich, that is, 13°10' east of Toledo, which lies 4°1' west of Greenwich); for Milan, 45°28'N 9°11'E (13°12' east of Toledo).

30 Should 44°50' N have been used instead, the difference between the (theoretical) resulting cusps would be less than 6'5'', no matter the method.

The quality coefficients calculated by the *Horosc* software given the entire set of cusps (and independently of the latitude chosen) give a slight preference to the Standard method.

	Figura	North 0	North 1	North 2	North 3	North 4	North 5	North 6
I	26° Le	26°0' Le	26°0' Le	26°0' Le	26°0' Le	26°0' Le	26°0' Le	26°0' Le
II	23° Vi	17°21' Vi	23°18' Vi	23°10' Vi	25°19' Vi	18°9' Vi	28°7' Vi	26°0' Vi
III	20° Li	14°16' Li	21°3' Li	20°20' Li	21°16' Li	13°26' Li	0°24' Sc	26°0' Li
IV	17° Sc	17°30' Sc	17°30' Sc	17°30' Sc	17°30' Sc	17°30' Sc	0°27' Sa	26°0' Sc
V	20° Sa	24°6' Sa	20°18' Sa	20°20' Sa	17°32' Sa	27°48' Sa	28°25' Sa	26°0' Sa
VI	22° Cp	27°44' Cp	22°4' Cp	23°10' Cp	22°0' Cp	1°9' Aq	26°15' Cp	26°0' Cp
VII	26° Aq	26°0' Aq	26°0' Aq	26°0' Aq	26°0' Aq	26°0' Aq	26°0' Aq	26°0' Aq
VIII	23° Pi	17°21' Pi	23°18' Pi	23°10' Pi	25°19' Pi	18°9' Pi	28°7' Pi	26°0' Pi
IX	20° Ar	14°16' Ar	21°3' Ar	20°20' Ar	21°16' Ar	13°26' Ar	0°24' Ta	26°0' Ar
X	17° Ta	17°30' Ta	17°30' Ta	17°30' Ta	17°30' Ta	17°30' Ta	0°27' Gm	26°0' Ta
XI	20° Gm	24°6' Gm	20°18' Gm	20°20' Gm	17°32' Gm	27°48' Gm	28°25' Gm	26°0' Gm
XII	22° Cn	27°44' Cn	22°4' Cn	23°10' Cn	22°0' Cn	1°9' Le	26°15' Cn	26°0' Cn
QC		**5.498**	**0.229**	**0.250**	**1.741**	**7.330**	**7.044**	**4.750**

3. *Case B. anonymous, prediction for 1443*

3.1. Witnesses

An anonymous prediction for 1443 is copied in the following manuscript, its only (known) witness:
Paris, Bibliothèque nationale de France, Latin 7443, folios 117^{v}–119^{v}.[31]
This manuscript is a compendium of quite diverse astrological material, including extracts of theoretical treatises, some of the earliest known annual predictions, and several 'astrological files', as Jean-Patrice Boudet calls them, which are collections of texts and astrological squares related to political events in the Armagnac-Burgundian Civil War. The volume as we know it was probably bound late in the fifteenth century, either by the astrologer Simon de Phares or before being acquired by him, from files originating from three main fifteenth-century astrologers, that might be identified with Simon of Boesmare (1380–1438), Jean Halbout of Troyes, and, perhaps, Roland L'Escripvain. The prediction for 1443, however, has been copied by another hand on leaves left blank by the latter in the middle of a 1427 'file' on the imminent confrontation between Charles VII and Henri VI.[32]

31 'Latin 7443', *BnF Archives et Manuscrits*, https://archivesetmanuscrits.bnf.fr/ark:/12148/cc666669 <accessed 05.08.2021>. A detailed description is given in Jean-Patrice Boudet, *Lire dans le ciel: la bibliothèque de Simon de Phares, astrologue du xve siècle* (Brussels: Centre d'étude des manuscrits, 1994), pp. 113–51. See also David Juste, *Les manuscrits astrologiques latins conservés à la Bibliothèque nationale de France*, Catalogus codicum astrologorum latinorum, II (Paris: CNRS éditions, 2015), pp. 162–67. This manuscript has been digitized under the ALFA project and is available online on Gallica: https://gallica.bnf.fr/ark:/12148/btv1b52515212f <accessed 03.01.2022>.

32 Here, we followed Boudet, *Lire dans le ciel*, pp. 113–15 and 139–42. The astrological file on 1427, presumably by Roland L'Escripvain, has been critically edited by Jean-Patrice Boudet and Thérèse Charmasson, 'Une consultation astrologique princière en 1427', in *Comprendre et maîtriser la nature au Moyen âge: mélanges d'histoire des sciences offerts à Guy Beaujouan* (Paris: Champion, 1994), pp. 255–78, republished in Jean-Patrice Boudet, *Astrologie et politique entre Moyen Âge et Renaissance* (Florence: SISMEL — Edizioni del Galuzzo, 2020), pp. 247–71.

The text's incipit (not considering the astrological square at the beginning of the prediction, see below) is *Anno domini 1443 fluente, sol primum minutum Arietis ingreditur 11 die martii hora 10ª noctis cum 38 minutis ad latitudinem Parisiensem ...*[33]

3.2. *Context of production*

This very general prediction doesn't leave many clues as to its author. Its conclusions cover both Europe and North Africa. It follows a classical thematic plan, not unlike that of Pietro de Monte Alcino in 1421. A short prologue synthesizes the horoscopes used and their main characteristics; then, the first chapters *de aeris dispositione in generali* and *in speciali* give meteorological previsions for each season. In the second chapter on agriculture (*de esse victualium*), the astrologer uses not only direct astrological interpretation, but also — and this was quite rare in fifteenth-century annual predictions — his previous meteorological conclusions; for example, because of the wet summer, wheat might be scarce. The following chapters, on war (*de esse guerra et pace*) and on the epidemics (*de esse infirmitatum*) remain very general, and there are no 'particular' chapters or formal conclusions, which may be down to a lack of space in which to copy them.

3.3. *Role of the horoscope in the prediction*

Unlike Pietro de Monte Alcino's predictions, there is no detailed textual description of the horoscopes involved, only a brief presentation in the prologue, giving the exact times and ascendant of the two main figures, with the equinox and the syzygy immediately preceding them.[34] The main astrologically significant characteristics are reiterated (aspects, *dominus anni* and *significator rerum*), but it seems adequate to say that the planets are 'disposed as in the figure' (*planetis dispositis ut in figura*), that is, on the astrological square.

The entire verso of folio 117 (the recto being part of the 1427 astrological file, with unrelated textual *figurae caeli*) is dedicated to the double astrological square, and it doesn't seem absurd to say that it might have circulated independently from the prediction, which begins on folio 118ʳ. The design of the double square, with the interior compartments dedicated to the prevention and the exterior to the revolution (in this case, but more frequently the other way around) is not unprecedented, but it remains unusual in fifteenth-century annual predictions.[35] It is its only appearance in this manuscript: both Roland L'Escripvain

33 Thorndike and Kibre, *A Catalogue...*, col. 103; Thorndike, *HMES*, IV, 99.

34 The spring equinox, astronomically the entry of the Sun in the sign of Aries, is sometimes called the 'revolution' (of the year), while the immediately preceding syzygy or conjunction or opposition of the Sun and the Moon is called the 'prevention'. While most astrological authorities, and most notably Ptolemy, insist on drawing predictions only from the latter (entirely respected by Conrad Heingarter in 1476), it is very common to have both as *figurae caeli* in an annual prediction (in the present case as well as in Richard Trewythian's prediction for that same year), or even only the revolution (as in Pietro de Monte Alcino's predictions). In 1479, Johannes Laet relied mostly on the revolution, but he calculated the prevention and cites it occasionally.

35 In almanacs and annual predictions, the revolution (entry of the Sun in Aries or in other signs) is often at the center of the diagram, and the prevention is described on the rim. In other contexts, the same layout was used to relate other astrologically connected horoscopes; Bernard Goldstein and David Pingree, for instance, found it in the Cairo Genizah for anniversary horoscopes (connecting the nativity and its revolution). On that matter, see Bernard R. Goldstein and David Pingree, 'Horoscopes from the Cairo Geniza', *Journal of New*

and his fellow astrologers prefer simple, more common astrological squares (or textual descriptions). One of the downsides of this layout is obviously the difficult task of writing all information in the small inner diagram, which might explain why it is rather uncommon.

The cartel is used for the title of the prevention, while the revolution's title is written at the top of the leaf. Both include the detailed date, as in the textual prologue, but not the place. Astrological symbols are used for the planets, in the diagram and in the text of the prediction (although more randomly), but not for the name of the signs. Longitudes are given to the nearest degree. The outer diagram includes several lots: the *pars fortune*, *pars pani*, *pars guerre*, *pars vini*, *pars tritici*, and *pars aquae vel pluvie*. It is not clear whether they are absent from the prevention figure because they would be less meaningful there, or just for want of space.

It should be noted that while the revolution and its prevention are the main figures of the judgement, the only one explicitly mentioned and the only one described with an astrological square, some meteorological conclusions implicitly rely on the other horoscopes, probably of both equinoxes and solstices.

Eastern Studies, 36 (1977), 113–44; North, *Horoscopes and History*, II.6, pp. 91–96. In the fifteenth century, Richard Trewythian is known to use that design in both ways, and even to connect three or four figures. John D. North devoted his twenty-second historical case to what he calls 'compound horoscopes' (North, *Horoscopes and History*, II.22, pp. 149–53).

3.4. Figurae Caeli

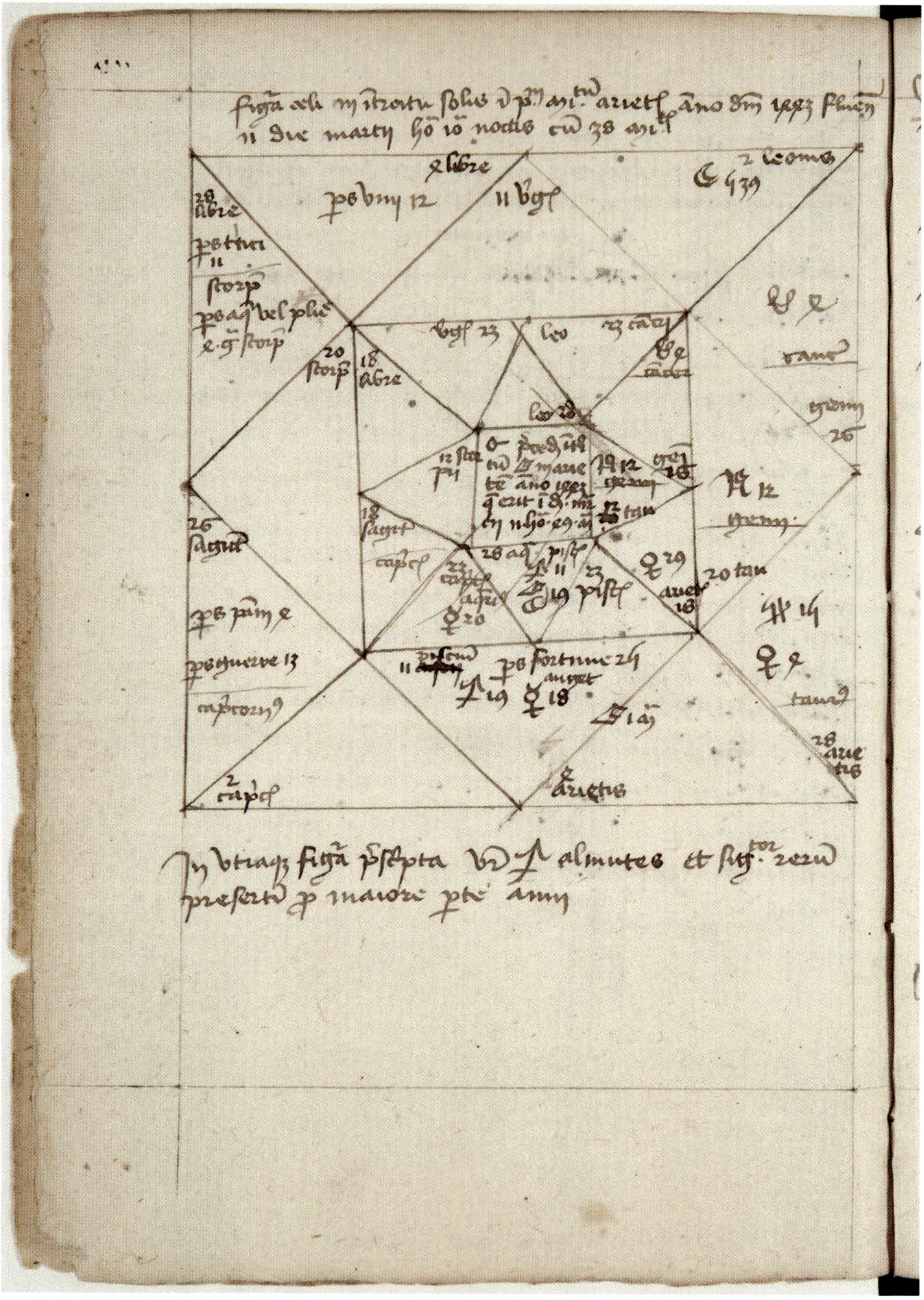

Figure 15.3. Anonymous prediction for 1443. Paris, BnF, Latin 7443, fols 117^{v}-118^{r}.[36]

36 This title would probably have been the cartel of the exterior square. For practical reasons, we edited the interior square (prevention of the year) and the exterior square (revolution) separately.

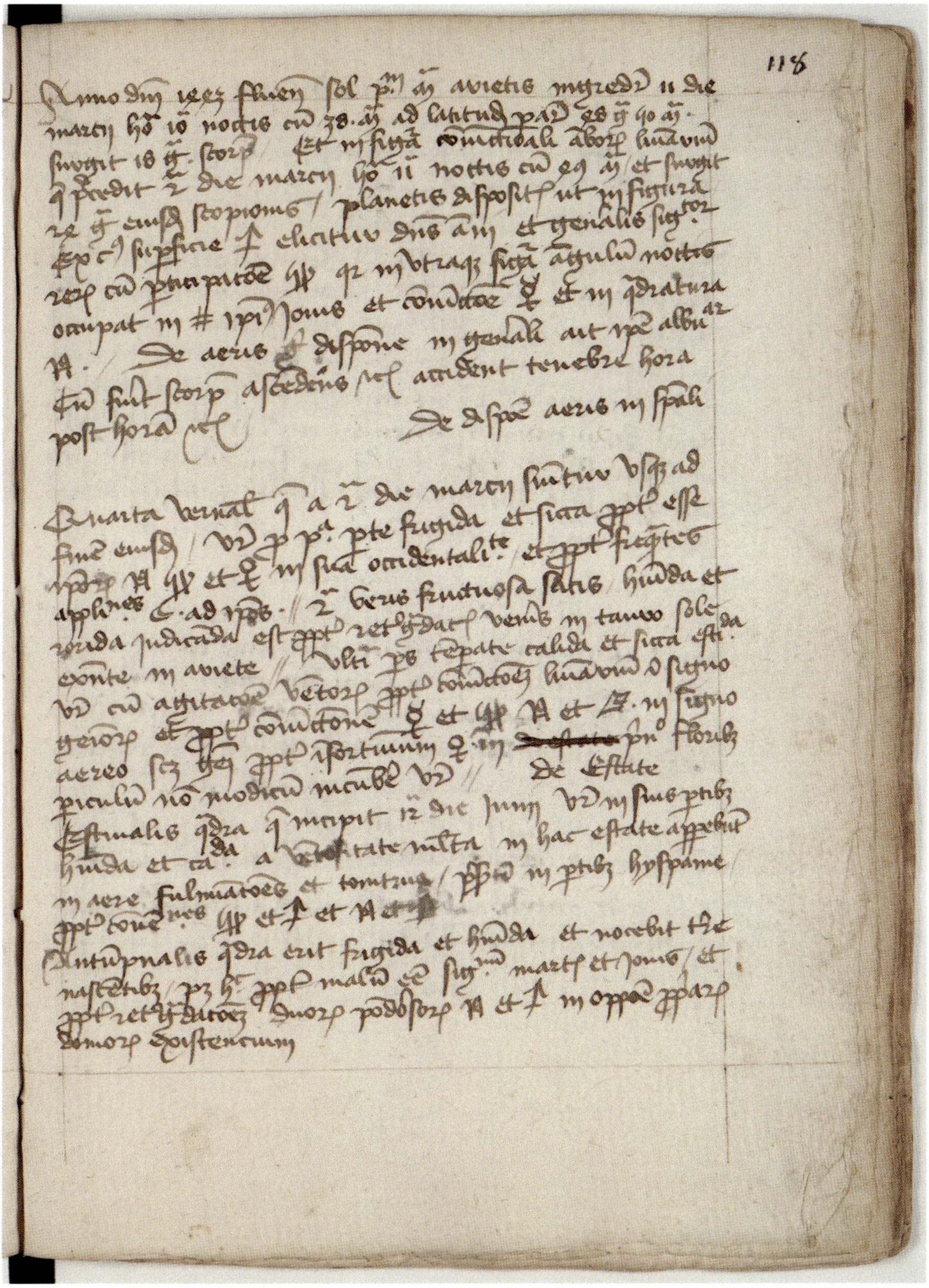

[177v] **“Figura celi in introitu Solis in primum minutum Arietis anno Domini 1443 fluente 11 die martii hora 10a noctis cum 38 minutis.”**

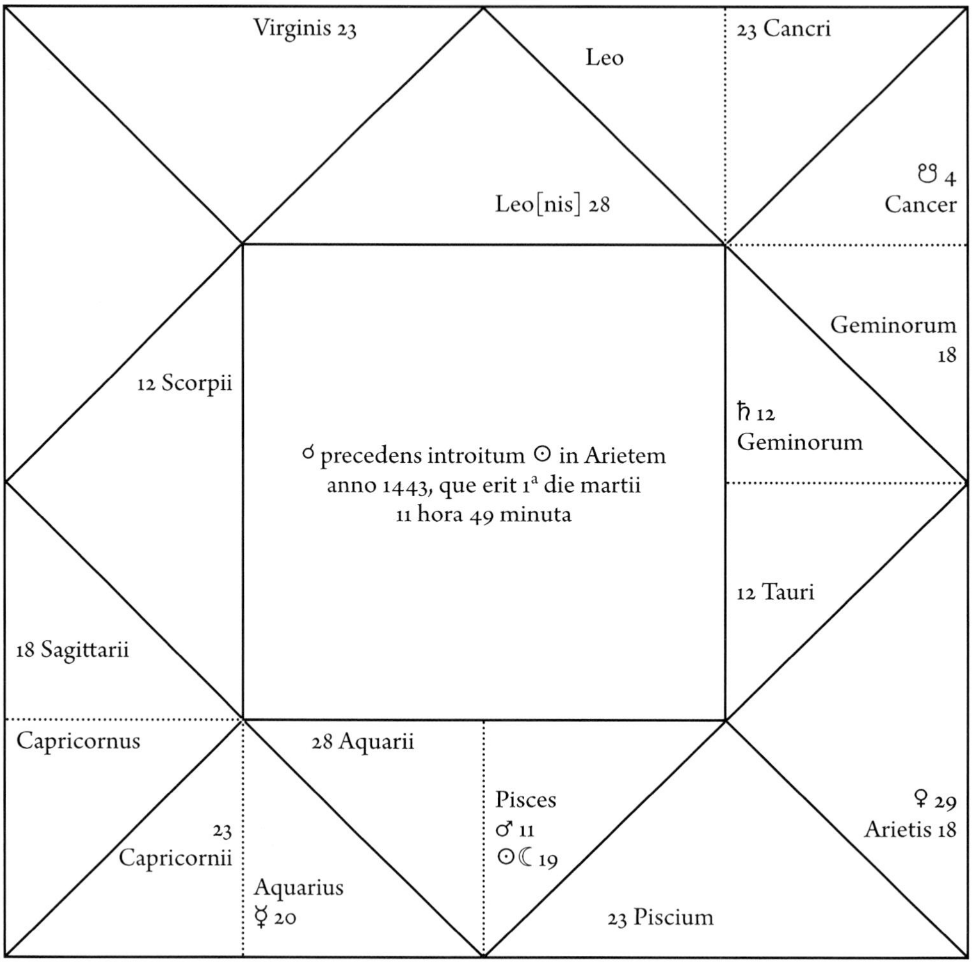

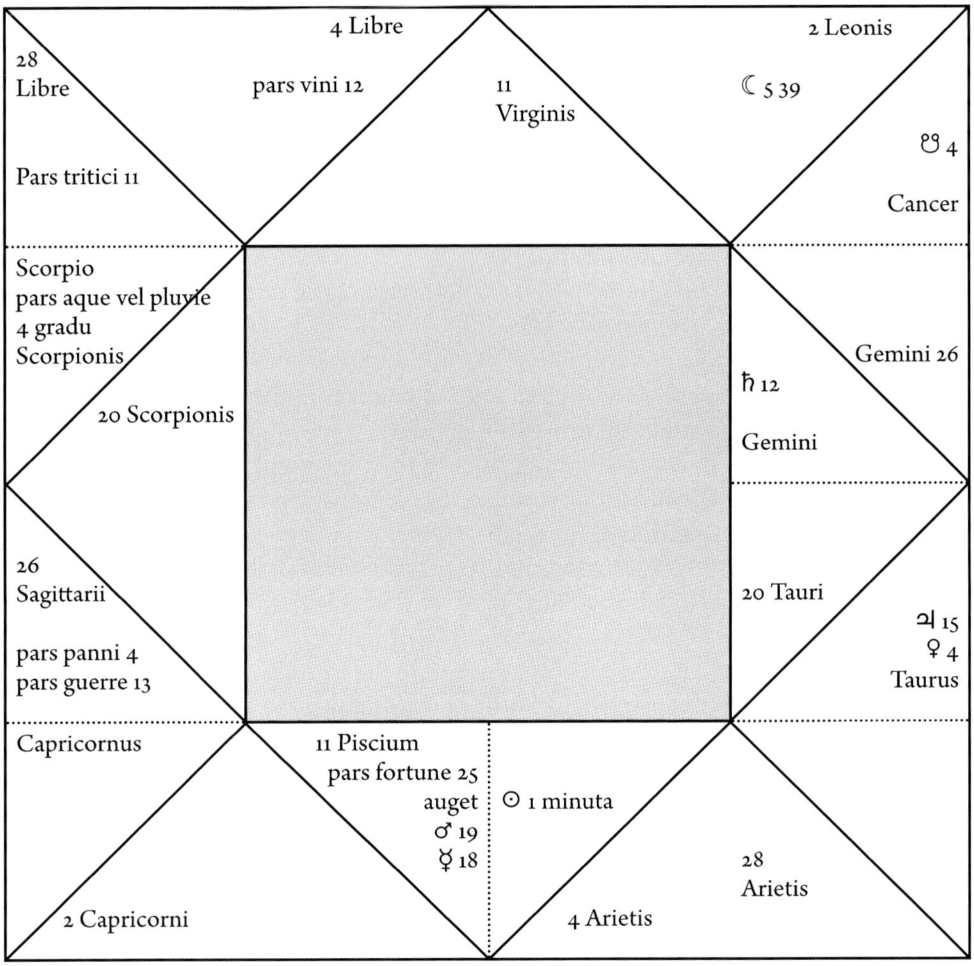

In utraque figura prescripta videtur Mars *almutes* et significator rerum presertim pro majore parte anni.

[118ʳ]

Anno Domini 1443 fluente Sol primum minutum Arietis ingreditur 11 die martii hora 10ª noctis cum 38 minutis; ad latitudinem Parisiensem, 48 gradu 50 minutis, surgit 18 gradu Scorpionis. Et in figura conjunctionali amborum luminarium que precedit, 2ª [*sic*[37]] die martii hora 11ª noctis cum 49 minutis, et surgit 24 gradu ejusdem Scorpionis, planetis dispositis ut in figura. Ex cujus superficie, ♂ elicitur dominus anni et generalis significator rerum cum participatione ♃, quia in utraque figura angulum noctis occupat in # ipsius Jovis et conjunctor ☿ et in quadratura ♄.

37 Probably a copy mistake for *1ª die martii*, as stated in the astrological square's diagram, which is a much more astronomically accurate date.

(...)

[Follow meteorological chapters.]

3.5. Recomputation

Both the text of the prologue and the titles of the astrological square give an adequate date and hour: March 11, 1443, at 10h38m p. m.[38]

Only the prologue mentions the place, but in a precise and detailed manner: *ad latitudinem Parisiensem, 48 gradu 50 minuta.*

The table of longitudes and latitudes of the 1483 Alfonsine tables places Paris at 48° N and 22°20' E, but 48°50' is also one of the admitted values compiled by North (along with 48°, 48°35', 48°40', 48°50', and very diverse longitudes).[39]

There is no reason why any latitude other than the one announced would have been used, which is why the approximations calculated by the *Horosc* software based on the provided cusps — between 51°10' and 54°8' for the prevention and between 49°53' and 53°39' for the revolution — probably reveal computation mistakes or even the use of faulty tables.[40]

Astromodels gives the Toledan equinox on March 11 1443 at 9h25m p. m., 1h13m before the adopted time, theoretically placing Paris 18°15' east of Toledo.[41] The prevention, however, was to happen in Toledo at 10h55m, only 54m before the figure's time, placing Paris 13°30' east of Toledo. This discrepancy obviously doesn't mean that the astrologer adopted two different longitudes for Paris, but rather that his determination of either the syzygy time and/or the revolution time was not very accurate.

All in all, the astronomical data are among the least precise of the five predictions (along with Conrad Heingarter's prediction for 1476). Compared to the Alfonsine Tables (as emulated by *Astromodels*), the longitudes of three planets, including Saturn and Mars, diverge by more than one degree in the revolution and five in the prevention. Given in minutes in the revolution, the lunar longitude has an inaccuracy of 33'. Except for prevention Mercury, however, which is given at 20° Aquarius instead of 1° Pisces (a divergence by 11°, which would make it change sign and house), the inaccuracy doesn't exceed two degrees.

The divergence between the theoretical and computed value of the *pars fortune* is more puzzling. By night, the astrologer should have computed the distance from the Moon (at 5°39' Leo) to the Sun (at 0°1' Aries), that is 234°39', and cast it from the Ascendant, given at 20° Scorpio in the diagram, resulting at 14°22' Cancer (or 13°47'54" Cancer with *Astromodels* theoretical values).[42] Here, it seems he kept to the day formula, instead using the distance from the Sun to the Moon (125°38') resulting in a final longitude of 25°38' Pisces (or 26°12'06" with *Astromodels* results). This might have been an honest mistake,

---

38 That is to be understood as 10h38m *after noon*, meaning night-time in March in Paris (which 10h38 after midnight would not be).

39 North, *Horoscopes and History*, p. 194. Longitudes range from 20° E to 67° E (!).

40 This divergence affects the quality coefficients: see below.

41 Dobrzycki and Kremer, 'Alfonsine Meridians', p. 194, collated diverse longitudes for Paris in fifteenth-century manuscripts, ranging from 11°05' E (of Toledo) to 18°06' E. Both the 1483 Alfonsine tables and North's sources (mentioned above) adopt larger values. Modern coordinates for Paris are 48°51' N 2°21' E (6°22' east of Toledo).

42 See note 17.

forgetting to consider the time of day, but we should not ignore the fact that the resulting *pars fortune*, in House IV (associated with family and legacy), in near-conjunction with the Sun, happens to be in a much more favourable position than it would have been had the formula for a night horoscope been followed, in Cancer, in House VIII (associated with death and assassination), and in conjunction with *Cauda Draconis*, which might have been purposefully avoided.

However, neither the longitude of Mercury during the prevention nor the position of the *pars fortune* in the revolution is actually interpreted for its astrological meaning in the course of the prediction; the consequence of these manipulations (if they are indeed manipulations) is thus very discrete — unless of course only part of the text is preserved.

Prevention

Planet	**Figura Caeli (inner diagram)**	**Alfonsine Tables**
	1/03/1443 11h49m pm Paris	1/03/1443 10h55m pm Toledo
Moon	19° Pisces (IV)	20°10' Pisces
Sun	19° Pisces (IV)	20°10' Pisces
Mercury	20° Aquarius (III)	0°48' Pisces
Venus	29° Aries (VI)	0°10' Taurus
Mars	11° Pisces (IV)	12°41' Pisces
Jupiter	*missing*	13°35' Taurus
Saturn	12° Gemini (VII)	12°38' Gemini
Caput	4° Cancer (VIII)	4°23' Cancer

Revolution

Planet	**Figura Caeli (outer diagram)**	**Alfonsine Tables**
	11/03/1443 10h38m pm Paris	11/03/1443 9h25m pm Toledo
Moon	5°39' Leo (IX)	6°12'07" Leo
Sun	0°1' Aries (IV)	0°0'1" Aries
Mercury	18° Pisces (IV)	19°3' Pisces
Venus	4° Taurus (VI)	3°42' Taurus
Mars	19° Pisces (IV)	20°24' Pisces
Jupiter	15° Taurus (VI)	15°26' Taurus
Saturn	12° Gemini (VII)	13°12' Gemini
Caput	4° Cancer (VIII)	3°52' Cancer
pars fortune	25° Pisces (IV)	13°48' Cancer (26°12' Pisces by day)

While the prologue offers a precise value of the geographical latitude (used in the following recomputation), it also gives ascending degrees (18° Scorpio for the revolution and 24°

Scorpio for the prevention) that are different from the longitude of the first cusps in the diagram (respectively 20° and 12° Scorpio).[43] The following recomputation is based on the values found in the diagram.

It should be noted that the prevention cusps of Houses III and IX, respectively 2° Capricorn and 2° Leo, are not opposed on the ecliptic as they should be. This is probably due to a copy mistake, either for 2° Cancer or (more likely) 2° Aquarius.[44] Despite this, methods 1 (Standard method) and 2 (Dual Longitude method) seem to have been favoured for casting theses horoscopes, although with a low level of accuracy (1–3° divergence in the revolution and 1–4° in the prevention, overlooking the discrepancy between Houses III and IX).

Because of the difference between the declared latitude and the one proposed by *Horosc* given the known cusps (as stated above), the quality coefficients relate to domifications different to those computed below.[45]

Prevention

	Figura	North 0	North 1	North 2	North 3	North 4	North 5	North 6
I	12° Sc	12°0' Sc	12°0' Sc	12°0' Sc	12°0' Sc	12°0' Sc	12°0' Sc	12°0' Sc
II	18° Sa	11°4' Sa	16°51' Sa	16°35' Sa	18°47' Sa	6°51' Sa	11°7' Sa	12°0' Sa
III	23° Cp	17°11' Cp	20°12' Cp	21°10' Cp	26°14' Cp	12°55' Cp	8°45' Cp	12°0' Cp
IV	28° Aq	25°46' Aq	25°46' Aq	25°46' Aq	25°46' Aq	25°46' Aq	7°7' Aq	12°0' Aq
V	23° Pi	28°10' Pi	21°9' Pi	21°10' Pi	19°41' Pi	28°17' Pi	7°51' Pi	12°0' Pi
VI	18° Ar	22°52' Ar	17°3' Ar	16°35' Ar	13°18' Ar	21°19' Ar	10°23' Ar	12°0' Ar
VII	12° Ta	12°0' Ta	12°0' Ta	12°0' Ta	12°0' Ta	12°0' Ta	12°0' Ta	12°0' Ta
VIII	18° Gm	11°4' Gm	16°51' Gm	16°35' Gm	18°47' Gm	6°51' Gm	11°7' Gm	12°0' Gm
IX	23° Cn	17°11' Cn	20°12' Cn	21°10' Cn	26°14' Cn	12°55' Cn	8°45' Cn	12°0' Cn
X	28° Le	25°46' Le	25°46' Le	25°46' Le	25°46' Le	25°46' Le	7°7' Le	12°0' Le
XI	23° Vi	28°10' Vi	21°9' Vi	21°10' Vi	19°41' Vi	28°17' Vi	7°51' Vi	12°0' Vi
XII	18° Li	22°52' Li	17°3' Li	16°35' Li	13°18' Li	21°19' Li	10°23' Li	12°0' Li
QC		**6.248**	**0.600**	**0.500**	**3.960**	**8.126**	**10.975**	**8.500**

43 North describes an attested (although rare) method of domification where the cusps are given a longitude of 5° before the usually calculated value, no matter the method. This does not seem to be the case here. According to *Horosc*, increasing the cusp's longitude by 5° results in worse quality coefficients (although not much: 0.750 instead of 0.500 for method 2 in the revolution; 0.690 instead of 0.248 for method 1 in the prevention). As seen above, an Ascendant of 20° Scorpio was assumed to compute the *pars fortune* during the revolution.

44 Quality coefficients have been computed in *Horosc* with the House IX hypothesis, that is, supposing the cusp of House III is at 2° Aquarius.

45 North's software can either compute the theoretical domification after the geographical latitude and ascendant (which results in the tables given here) or, given a set of cusps, find the best domification possible and affect quality coefficients to the different methods. See North, *Horoscopes and History*, Appendix 4 pp. 197–218.

Revolution

	Figura	North 0	North 1	North 2	North 3	North 4	North 5	North 6
I	20° Sc	20°0' Sc	20°0' Sc	20°0' Sc	20°0' Sc	20°0' Sc	20°0' Sc	20°0' Sc
II	26° Sa	20°23' Sa	25°10' Sa	25°48' Sa	28°30' Sa	15°45' Sa	18°32' Sa	20°0' Sa
III	2° Cp	28°30' Cp	29°43' Cp	1°36' Aq	8°7' Aq	24°16' Cp	16°9' Cp	20°0' Cp
IV	11° Pi	7°24' Pi	7°24' Pi	7°24' Pi	7°24' Pi	7°24' Pi	15°2' Aq	20°0' Aq
V	4° Ar	8°31' Ar	2°5' Ar	1°36' Ar	29°55' Pi	7°57' Ar	16°26' Pi	20°0' Pi
VI	28° Ar	1°47' Ta	26°40' Ar	25°48' Ar	22°3' Ar	29°40' Ar	19°1' Ar	20°0' Ar
VII	20° Ta	20°0' Ta	20°0' Ta	20°0' Ta	20°0' Ta	20°0' Ta	20°0' Ta	20°0' Ta
VIII	26° Gm	20°23' Gm	25°10' Gm	25°48' Gm	28°30' Gm	15°45' Gm	18°32' Gm	20°0' Gm
IX	2° Le	28°30' Cn	29°43' Cn	1°36' Le	8°7' Le	24°16' Cn	16°9' Cn	20°0' Cn
X	11° Vi	7°24' Vi	7°24' Vi	7°24' Vi	7°24' Vi	7°24' Vi	15°2' Le	20°0' Le
XI	4° Li	8°31' Li	2°5' Li	1°36' Li	29°55' Vi	7°57' Li	16°26' Vi	20°0' Vi
XI	28° Li	1°47' Sc	26°40' Li	25°48' Li	22°3' Li	29°40' Li	19°1' Li	20°0' Li
QC		**4.881**	**0.248**	**1.000**	**6.186**	**6.576**	**12.466**	**10.000**

4. Case C. Richard Trewythian, prediction for 1443

4.1 Witnesses

Richard Trewythian's prediction for 1443 is only known by one manuscript witness:

London, British Library, Sloane MS 428, folio 7ᵛ.[46]

This manuscript is known as Richard Trewythian's mostly autographical notebook, which was studied by Sophie Page in a decisive paper.[47] Probably unbound in the fifteenth century (and maybe in a different order from the one offered by the contemporary binding), it contains almanacs and horoscopes used by Trewythian in his own practice and as a source for his astrological (probably mostly private) services.[48] The first part of the manuscript, folios 1–114, contains monthly ephemerides for 1442–58, including occasional astrological squares in the margins for meaningful astronomical events (such as the spring equinox, an eclipse, or a planetary conjunction), accompanied in several cases by brief interpretation notes. Most of these ephemerides seem to have been issued year by year, with a paper cover

46 'Mag. Richard Trewythian, Medical Practitioner, Astrologer and Money Lender in London: Entry book: 1442–1458. Henry VI of England: Notes of public events by Magister R. Trewythian: 1442–1458. Astrology: Horoscopes and astrological predictions: 1442–1458', *Explore Archives and Manuscripts*, http://searcharchives.bl.uk/IAMS_VU2:IAMS040-002112777 <accessed 19.08.2021>; Samuel Ayscough, *A Catalogue of the manuscripts preserved in the British Museum hitherto undescribed, consisting of five thousand volumes, including the collections of Sir Hans Sloane, Bart. the Rev. Thomas Birch, D. D. and about five hundred volumes bequeathed, presented or purchased at various times* (London: J. Rivington, 1782), vol. 1 p. 409 and vol. 2 p. 865.

47 Sophie Page, 'Richard Trewythian and the Uses of Astrology in Late Medieval England', *Journal of the Warburg and Courtauld Institutes*, 64 (2001), 193–228. One of John D. North's horoscope case studies is taken from Richard Trewythian's workbook: North, *Horoscopes and History*, II.22, pp. 149–54.

48 An earlier set of horoscopes written in Trewythian's hand is preserved in Oxford, Bodleian Library, MS Ashmole 189, ff. 201–02 (see Page, 'Richard Trewythian …', p. 197), but no 'published' prediction of his conception (in manuscript or print) is known.

often used to write down the annual prediction. Folio 7 is the first of such separating leaves, between ephemerides for 1442 and 1443, and contains a prediction for 1442 on the recto and this prediction for 1443 on the verso. Like several separating leaves in the notebook, folio 7 suffered damage, probably while it was serving as a cover for the ephemerides, and parts of the text and diagrams are now missing.

The second part of the book, folios 115–38, contains a series of horoscopes, including annual revolutions, interrogations, nativities, and elections, often centred around an astrological square.[49] In total, the notebook contains 13 annual predictions for 1429–58, making it one the largest repositories of such material.[50]

4.2. Context of production

Richard Trewythian's principal activity in London was that of a doctor, who was consulted by citizens, artisans, and merchants. Much like money-lending, astrological consultation was something of a side activity, perhaps one he himself had a specific interest in (as suggested by the number of horoscopes concerning him or members of his family).[51] While it is probable that he used his notebook during astrological consultations, and maybe took advantage of his double and triple astrological squares to make an impression on clients, most of the ephemerides and figures of recurring astronomical events were mainly there as sources to draw from in later predictions.[52] The prediction for 1443 is complementary to the prediction for 1442 written on the recto of the same leaf, and to the immediately following almanac for 1443. Unlike most of the preserved astrological prognostications, it is not a literary product, but it could resemble a first draft even recognized astrologers would have made before writing down the definitive judgement, and as such it is very valuable to our understanding of the process of conceiving an astrological prediction. In particular, the frequent use of astrological squares, while it might be a personal preference by Trewythian, seems to have a certain advantage for synthesizing astronomical information (rather than writing it down as a text or a table, for instance) and perhaps, for astrologers who grew used to it, even as a graphical way to spot astrologically meaningful connections.

4.3. Role of the horoscope in the prediction

The prediction for 1443 is based on three simple horoscopes, completed by relatively brief interpretative notes (see reproduction). The first two squares at the top of the leaf are dedicated, respectively, to the conjunction of the luminaries preceding the entry of the Sun

49 Page, 'Richard Trewythian', Appendix I, pp. 217–20.

50 Tur, 'Hora introitus', p. 451. Page edited Trewythian's prediction for 1452 (f. 60rv), one of the most complete, as well as one of his nativities: Page, 'Richard Trewythian', Appendix II, pp. 220–28.

51 Page, 'Richard Trewythian', p. 197 and *passim*.

52 Page, 'Richard Trewythian', p. 218. Trewythian often used double or even triple 'compound horoscopes' in the last part of his notebooks (see above such a diagram in the Parisian anonymous prediction for 1443); one such case is documented by North, *Horoscopes and History*, II.22, pp. 149–53. The use of a visually impressive medium by physicians to convince patients, notably in late medieval England, has been suggested in other cases, such as bat-book medical almanacs. See, for instance, Chelsea Silva, 'Opening the Medieval Folding Almanac', *Exemplaria*, 30 (2018), 49–65.

in Aries (the prevention, not dated), and to the entry of the Sun in Aries itself, here dated of the year 1442 completed, that is, 'after the end of the year 1442', and thus March, 1443.[53] These figures are not textually described, but the next three lines give a rapid analysis to determine which planet is the 'lord of the year' (*dominus anni*).

These lines appear to have been (slightly) crossed out, and a new astrological square below has been filled for the entry of the Sun in Aries 'in the year 1443 incomplete' (*anno Christi 1443 imperfecto*) — which is the same date. The hour is given as 10h instead of 10h43m, but this seems to be an omission of the minutes rather than a real change. This third figure is obviously very close to the previous one, but there is no denying that the calculation was made anew. Switches in the degrees of the planetary positions might indicate a different rounding method. The cusps of the Houses also vary by no more than two degrees, but that is enough to significantly increase the distance from the theoretical cusps, as estimated by North's quality coefficient (see below). This inaccuracy is particularly suspect coming from Trewythian, whom North accepts as usually very accurate in his domification.[54]

The first notes next to the figure attribute it a different lord of the year: Jupiter (still at 16° Taurus, but now in the important House VII instead of VI) instead of the Moon (House IX in both figures).[55] It is noteworthy that the Sun is now also in House V instead of IV. Might Trewythian have decided to change the figure by falsifying his cusps' longitudes by commodity (to have a more evident lord of the year) or because he had a certain interpretation in mind?

The following conclusions, although more detailed than other predictions in Trewythian's notebook, aren't enough to validate such a hypothesis, but they do not contradict it; they are mainly optimistic, drawing evidently from Jupiter's good nature in a good position, while the initial figure, with Saturn in House VII, might have been less positive.

The predictions are not a progressive interpretation of the horoscope (as, for instance, in Trewythian's prediction for 1442 on the recto of that same leaf); they are organized by class of population (monks, women, merchants, etc.), a plan that became more influential in the second half of the fifteenth century and might demonstrate an intent to turn these notes into a more complete prediction.

53 *Introitus Solis in Arietem anno Christi 1442 completo, duobus mensibus 11 diebus 10 horis 43* minutis (cartel). As it appears from the given planetary positions, this is to be understood as March 11, (sic) 1443, 10h43m p. m. This curious mix of 'complete' and 'incomplete' might be a reason why Trewythian decided to make a new figure. It was not uncommon for fifteenth-century astrologers to explain the equivalence of a complete and incomplete year in a prologue, but the 'complete' style is very rarely used in computed data.

54 North, *Horoscopes and History*, p. 149: 'Whether or not Richard Trewythian had university training, he certainly appears to have had a much better knowledge of astronomy than most of those with that advantage. One cannot be certain that he calculated the ephemerides his book contains, but he clearly calculated the horoscopes, accurately, and on the Standard system'.

55 In the last astrological square, Jupiter appears in House VII, although its longitude (16° Taurus) should place it in House VI (before 20° Taurus). Likewise, in the second astrological square (1442 completo), the cusp of House VII is 18° Taurus, two degrees after Jupiter, which, however, appears correctly in House VI at 16° Taurus. The astrological conclusions refer to Jupiter in House VII several times, which means that the third square is used, including the (perhaps voluntary) mispositioning of Jupiter in House VII.

4.4. *Figurae caeli*[56]

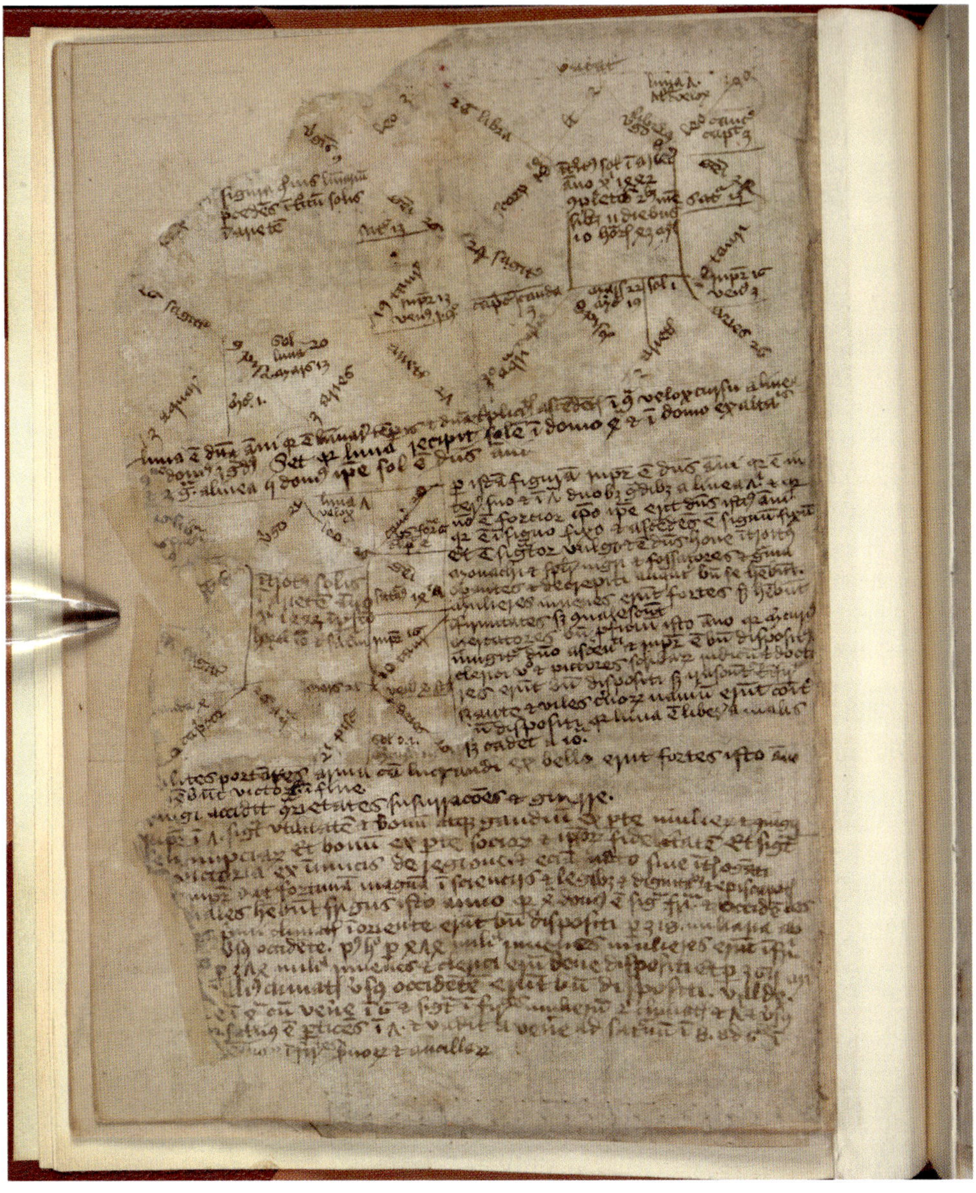

Figure 15.4. Richard Trewythian, prediction for 1443. London, British Library, Sloane MS 428, f. 7v.

56 Richard Trewythian seems to express the cusp longitudes using the nominative and genitive cases of the signs alternatively. This (dubious) grammatical choice has not been corrected in the edition below; only confusing abbreviations have been arbitrarily extended as genitive.

[7v][57]

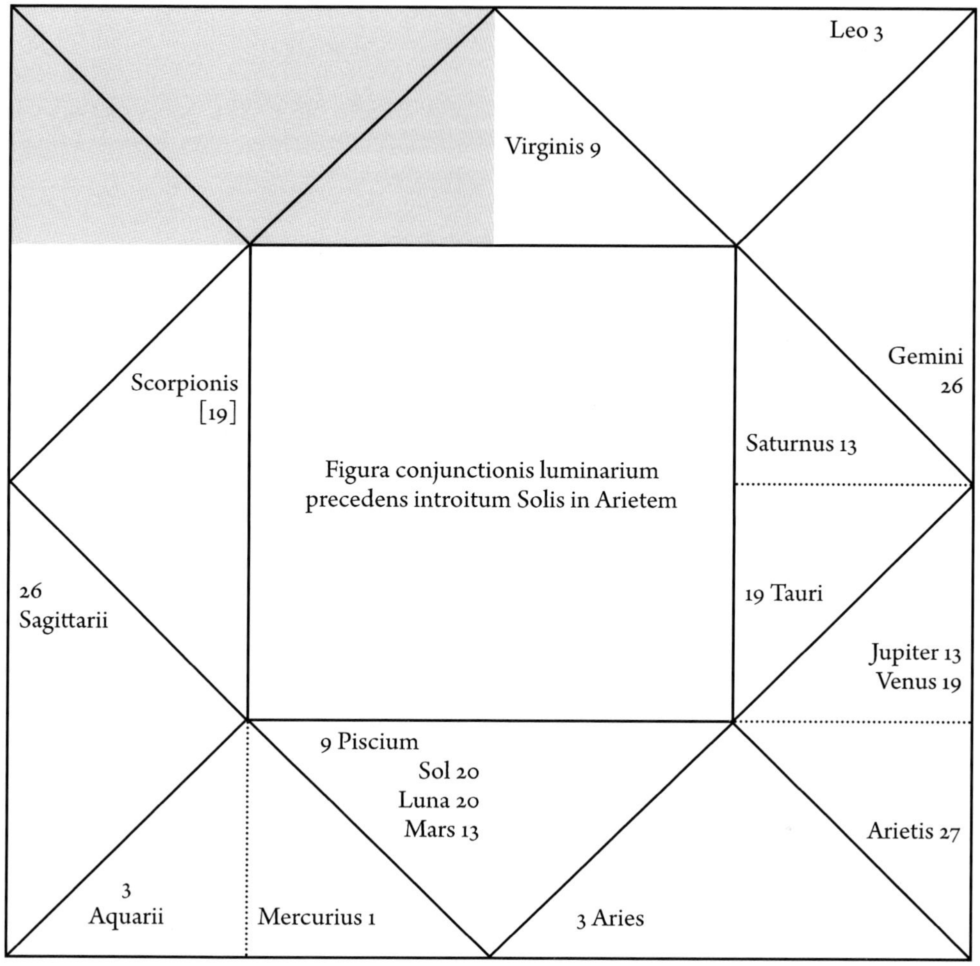

57 The first two figures are visible on the same line (see reproduction). Note that this leaf is damaged; unreadable parts of the astrological squares are rendered in grey.

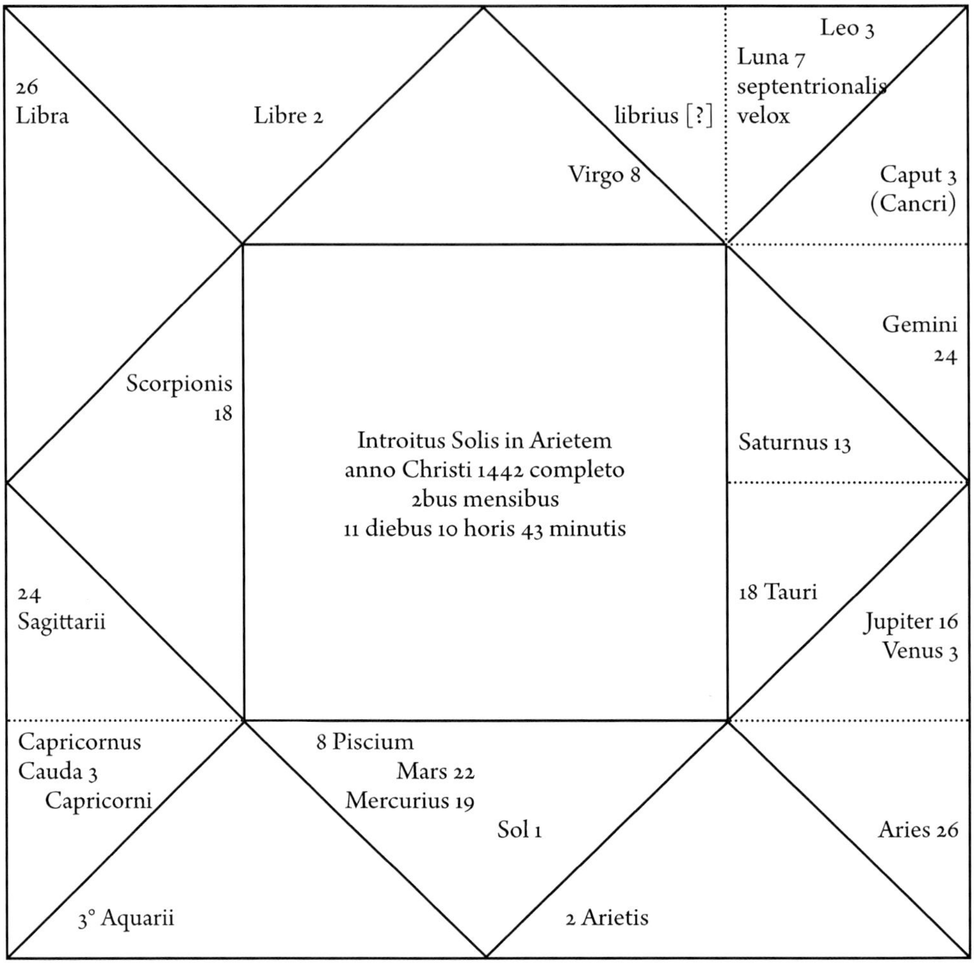

[Cancelled:] Luna est domina anni quia est luminarium temporis et domina triplicitatis ascendentis in 9^{a} velox cursu a linea 9^{ne} domus 3 gradibus. Set quia Luna recipit Solem in domo 4^{a} et in domo exaltationis et 3 gradu a linea 5 domus, ipse Sol est dominus anni.

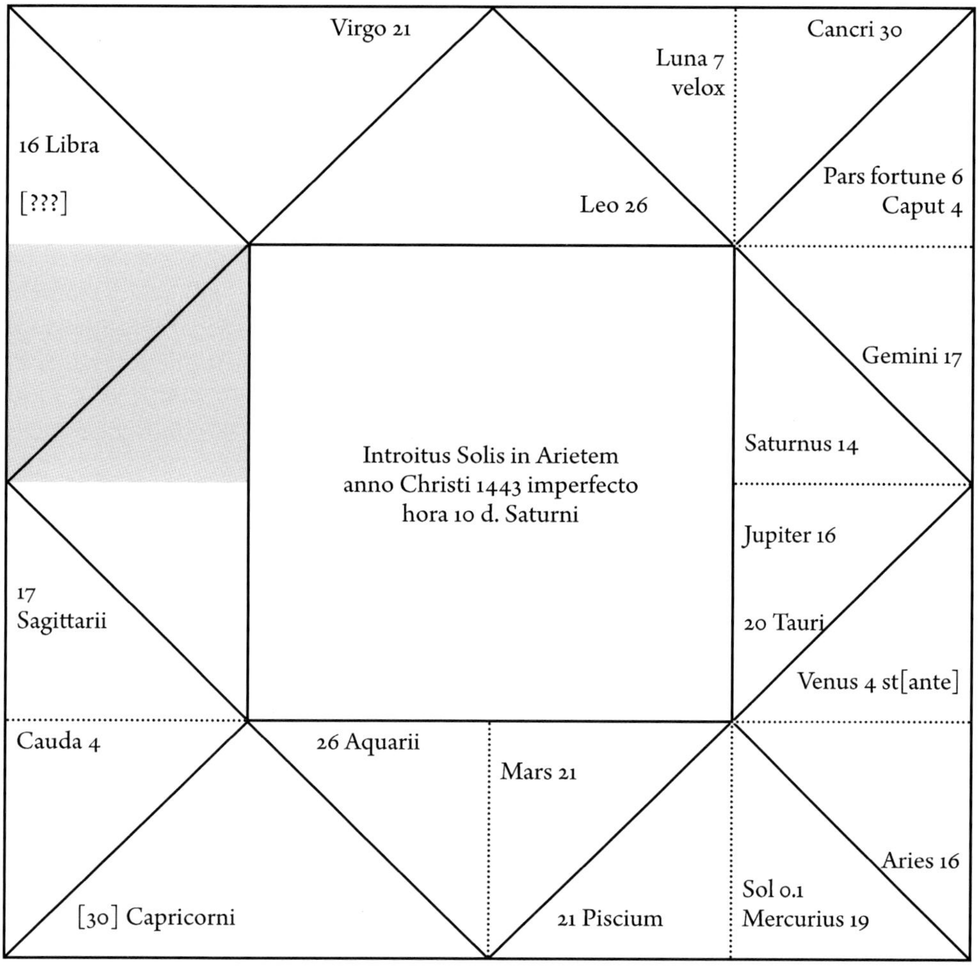

Per istam figuram, Jupiter est dominus anni quia est in termino suo et in 7[a], duobus gradibus a linea 7[i]; et quia [nullus] est fortior ipso, ipse erit dominus istius anni, quia est in signo fixo et ascendens est signum fixum.

Et est significator vulgi et est dominus hore introitus.

Monachi et specialiter nigri et fossatores… [*Astrological interpretations follow.*]

4.5. *Recomputation*

Neither the text nor the diagram clearly indicates the time and place used for computation, and that is part of the ambiguity that occurs when differentiating between the second and third figures.

The first figure (prevention) does not include a calculated date for the event, but the planetary longitudes closely match the actual conjunction emulated by *Astromodels* on March 1, 1443, at 10h55m p. m. (Toledan time). All planetary longitudes except that of Venus are accurate to the degree, which is a very good performance for data rounded to the degree. The position of Venus, however, with a 2–19° inaccuracy (depending on the reading), probably denotes a misreading or computational error.

The same performance is evident for the two figures of the revolution; longitudes are all accurate to the degree with *Astromodels*' restitution of the Alfonsine tables (with the exception of Mars in the second figure, given at 22° Pisces instead of 20°26', but at 21° in the third figure). Differences between the two figures do not exceed one degree in longitude and can in all cases be considered rounding decisions. All things considered, the planetary positions in Trewythian's three figures are all more accurate than those computed by the author of the Bibliothèque nationale de France, Latin 7443 1443 prediction (studied above). Only the third figure includes the *pars fortune*; unlike the astrologer of the previous prediction, Richard Trewythian rightfully used the night formula, but oddly he ends with the value of 6° Cancer (instead of 13°1' Cancer with the third figure's data, or 12° Cancer following the second figure). This change doesn't seem of astrological significance and is probably a mistake, despite a relatively easy computation.

Prevention

Planet	**Figura Caeli #1**	**Alfonsine Tables (Toledan conjunction)**
	?	1/03/1443
	?	10h55m pm
	London	Toledo
Moon	20° Pisces (IV)	20; 10° Pisces
Sun	20° Pisces (IV)	20; 10° Pisces
Mercury	1° Pisces (III)	0; 48° Pisces
Venus	(1)9° Tauri (VI)[58]	0; 10° Taurus
Mars	13° Pisces (IV)	12; 41° Pisces
Jupiter	13° Tauri (VI)	13; 35° Taurus
Saturn	13° Gemini (VII)	12; 38° Gemini
Caput		4; 23° Cancer

58 Reading is uncertain for the latitude of Venus in House VI. It may be 9° Tauri or 19° Tauri (see the reproduction). It would be tempting to read 1.9° (for 1°9'), an astronomically more accurate value, but that would be the only precision to the minute (except for the Sun in the third figure). It is also worth noting that the anonymous prediction of BnF, lat. 7443 for 1443 (our previous case study above) places Venus at a longitude of 29° *Aries* in the same prevention.

Revolution

Planet	Figura celi #2	Figura Caeli #3	Alfonsine Tables (Toledan equinox)
	11/03/1443 10h43m pm London	[11/03/]1443 10h [?] pm London	11/03/1443 9h25m pm Toledo
Moon	7° Leo (IX)	7° Leo (IX)	6°12' Leo
Sun	1° Aries (IV)	0°1' Aries (V)	0°0'1" Aries
Mercury	19° Pisces (IV)	19° Pisces (V)	19°3' Pisces
Venus	3° Taurus (VI)	4° Taurus (VI)	3°42' Taurus
Mars	22° Pisces (IV)	21° Pisces (IV)	20°24' Pisces
Jupiter	16° Taurus (VI)	16° Taurus (VII)	15°26' Taurus
Saturn	13° Gemini (VII)	14° Gemini (VII)	13°12' Gemini
Caput	3° Cancer (VIII)	4° Cancer (VIII)	3°52' Cancer
Cauda	3° Cancer (II)	4° Cancer (II)	3°52' Capricorn
pars fortune		6° Cancer (VIII)	13°48' Cancer[59]

Although we expect Trewythian made his computations in London, he did not clarify which coordinates he used. In studying other horoscopes of the same manuscript, North found them (mathematically) to yield a latitude of approximately 51°25' N.[60] This value is coherent with the estimated latitude of London; geographical lists compiled by North include latitudes such as 51°34', 51°39', and 51°40'.[61] With the cusps computed by Trewythian, the *Horosc* software deduces a compatible base latitude between 50°39' and 53°44' for the first figure (prevention) and between 51°11' and 54°17' for the second one. The third figure, however, gives much stranger results with a deduced latitude between 27°22' and 32°44', which clearly indicates a significant computation approximation, if not data falsification.

Only the second figure provides a detailed time, 10h43m p. m., which is 1h18m after the Alfonsine equinox in Toledo, placing London 19°30' east of Toledo. Although the value is clearly overestimated — it lies east of case B's Paris — it is not shocking for the time and not enough to confirm either a copy mistake or a significant approximation in computing of the time of the equinox.[62]

Even considering only the two first figures, Trewythian's accuracy in his domification is not as good as North experienced it in other horoscopes of the notebook or compared to our other cases. If Trewythian used the Standard method, as North postulated for other horoscopes, four cusps out of twelve in the prevention are inaccurate by more than one

59 This result is valid for an ascendant at 20° Scorpio declared in the third figure. Should the ascendant be 18° (as in the second figure), the casted distance of 233°48' between the Moon and the Sun would place the *pars fortune* at 11°48' Cancer.

60 North, *Horoscopes and History*, section II.22, pp. 149–53. We used this value in the following computations through *Horosc*.

61 North, *Horoscopes and History*, p. 194.

62 North's geographical list mentioned above includes longitudes ranging from 16°00' E to 19°26' E (with a marginal mention of 59°30' E). The 1483 edition of the Alfonsine tables does not include London in its longitudes and latitudes table, but it gives Oxford at 17°20' E and 53° N. Modern coordinates for London are 51°30' N; 0°08' W (3°57' east of Toledo), and for Oxford they are 51°45' N; 1°15' W (2°46' east of Toledo).

degree, plus six out of twelve in the second figure (ten out of twelve in the third, with six inaccurate by more than 10°).[63] In this case as well, the Dual Longitude method cannot be excluded and even gives slightly better quality coefficients.

Prevention

	Figura	North 0	North 1	North 2	North 3	North 4	North 5	North 6
I	[19°] Sc	19°0' Sc	19°0' Sc	19°0' Sc	19°0' Sc	19°0' Sc	19°0' Sc	19°0' Sc
II	26° Sa	19°16' Sa	24°47' Sa	25°24' Sa	29°7' Sa	14°19' Sa	17°36' Sa	19°0' Sa
III	3° Aq	28°23' Cp	29°54' Cp	1°49' Aq	9°44' Aq	23°44' Cp	15°13' Cp	19°0' Cp
IV	9° Pi	8°13' Pi	8°13' Pi	8°13' Pi	8°13' Pi	8°13' Pi	14°2' Aq	19°0' Aq
V	3° Ar	9°7' Ar	2°17' Ar	1°49' Ar	29°43' Pi	8°28' Ar	15°21' Pi	19°0' Pi
VI	27° Ar	1°40' Ta	26°14' Ar	25°24' Ar	21°3' Ar	29°25' Ar	17°56' Ar	19°0' Ar
VII	19° Ta	19°0' Ta	19°0' Ta	19°0' Ta	19°0' Ta	19°0' Ta	19°0' Ta	19°0' Ta
VIII	26° Gm	19°16' Gm	24°47' Gm	25°24' Gm	29°7' Gm	14°19' Gm	17°36' Gm	19°0' Gm
IX	3° Le	28°23' Cn	29°54' Cn	1°49' Le	9°44' Le	23°44' Cn	15°13' Cn	19°0' Cn
X	9° Vi	8°13' Vi	8°13' Vi	8°13' Vi	8°13' Vi	8°13' Vi	14°2' Le	19°0' Le
XI	[3° Li]	9°7' Li	2°17' Li	1°49' Li	29°43' Vi	8°28' Li	15°21' Vi	19°0' Vi
XII	[27° Li]	1°40' Sc	26°14' Li	25°24' Li	21°3' Li	29°25' Li	17°56' Li	19°0' Li
QC		**5.652**	**1.082**	**0.750**	**5.103**	**7.358**	**13.228**	**10.750**

Revolution

Figura Caeli #2

	Figura	North 0	North 1	North 2	North 3	North 4	North 5	North 6
I	18° Sc	18°0' Sc	18°0' Sc	18°0' Sc	18°0' Sc	18°0' Sc	18°0' Sc	18°0' Sc
II	24° Sa	18°4' Sa	23°44' Sa	24°14' Sa	27°50' Sa	13°11' Sa	16°40' Sa	18°0' Sa
III	3° Aq	26°52' Cp	28°40' Cp	0°28' Aq	8°8' Aq	22°12' Cp	14°16' Cp	18°0' Cp
IV	8° Pi	6°42' Pi	6°42' Pi	6°42' Pi	6°42' Pi	6°42' Pi	13°1' Aq	18°0' Aq
V	2° Ar	7°48' Ar	0°53' Ar	0°28' Ar	28°24' Pi	7°15' Ar	14°15' Pi	18°0' Pi
VI	26° Ar	0°33' Ta	25°1' Ar	24°14' Ar	19°57' Ar	28°22' Ar	16°51' Ar	18°0' Ar
VII	18° Ta	18°0' Ta	18°0' Ta	18°0' Ta	18°0' Ta	18°0' Ta	18°0' Ta	18°0' Ta
VIII	24° Gm	18°4' Gm	23°44' Gm	24°14' Gm	27°50' Gm	13°11' Gm	16°40' Gm	18°0' Gm
IX	3° Le	26°52' Cn	28°40' Cn	0°28' Le	8°8' Le	22°12' Cn	14°16' Cn	18°0' Cn
X	8° Vi	6°42' Vi	6°42' Vi	6°42' Vi	6°42' Vi	6°42' Vi	13°1' Le	18°0' Le
XI	2° Li	7°48' Li	0°53' Li	0°28' Li	28°24' Vi	7°15' Li	14°15' Vi	18°0' Vi
XII	26° Li	0°33' Sc	25°1' Li	24°14' Li	19°57' Li	28°22' Li	16°51' Li	18°0' Li
QC		**5.817**	**1.110**	**1.083**	**5.200**	**7.573**	**13.237**	**10.750**

63 North, *Horoscopes and History*, p. 152.

Figura Caeli #3

	Figura	North 0	North 1	North 2	North 3	North 4	North 5	North 6
I	[20° Sc]	20°0' Sc	20°0' Sc	20°0' Sc	20°0' Sc	20°0' Sc	20°0' Sc	20°0' Sc
II	17° Sa	20°28' Sa	25°50' Sa	26°35' Sa	0°25' Cp	15°28' Sa	18°32' Sa	20°0' Sa
III	[30°] Cp	29°55' Cp	1°8' Aq	3°9' Aq	11°20' Aq	25°16' Cp	16°9' Cp	20°0' Cp
IV	26° Aq	9°44' Pi	9°44' Pi	9°44' Pi	9°44' Pi	9°44' Pi	15°2' Aq	20°0' Aq
V	20° Pi	10°25' Ar	3°41' Ar	3°9' Ar	1°1' Ar	9°41' Ar	16°26' Pi	20°0' Pi
VI	16° Ar	2°47' Ta	27°26' Ar	26°35' Ar	22°8' Ar	0°27' Ta	19°1' Ar	20°0' Ar
VII	20° Ta	20°0' Ta	20°0' Ta	20°0' Ta	20°0' Ta	20°0' Ta	20°0' Ta	20°0' Ta
VIII	17° Gm	20°28' Gm	25°50' Gm	26°35' Gm	0°25' Cn	15°28' Gm	18°32' Gm	20°0' Gm
IX	30° Cn	29°55' Cn	1°8' Le	3°9' Le	11°20' Le	25°16' Cn	16°9' Cn	20°0' Cn
X	26° Le	9°44' Vi	9°44' Vi	9°44' Vi	9°44' Vi	9°44' Vi	15°2' Le	20°0' Le
XI	20° Vi	10°25' Li	3°41' Li	3°9' Li	1°1' Li	9°41' Li	16°26' Vi	20°0' Vi
XII	16° Li	2°47' Sc	27°26' Li	26°35' Li	22°8' Li	0°27' Sc	19°1' Li	20°0' Li
QC		**10.783**	**9.686**	**9.750**	**9.702**	**9.377**	**6.066**	**6.750**

5. *Case D. Conrad Heingarter, prediction for 1476*

5.1 Witnesses

The 1476 prediction by Conrad Heingarter is known by one manuscript witness:

Paris, Bibliothèque nationale de France, Latin 7450, folios 3^{r}–21^{v}.[64]

This manuscript is a presentation copy of this prediction alone. Emmanuel Poulle identified the same hand that copied Heingarter manuscripts Bibliothèque nationale de France, Latin 7232, Latin 7446 (ff. 1–14), and Latin 11232, suggesting it might be Heingarter's own formal handwriting (as opposed to his cursive annotation style in manuscripts intended for himself such as BnF Latin 7197).[65] It was probably intended to be sent by Conrad Heingarter to its dedicatee, King Louis XI of France. This apparently never happened, as it was handed over with other books from Heingarter's library, such as Bibliothèque nationale de France, Latin 7197 and Latin 7295A, to Claude Hardy, Etienne Baluze, and Jean-Baptiste Colbert, before entering the Royal Library in 1732.

The prediction is entitled *Judicium anni millesimi quadringentesimi septuagesimi sexti currentis*, and its incipit is *Quom curialibus laboribus et negociis familiaribus aut omnino aut*

64 'Latin 7450', *BnF Archives et Manuscrits*, https://archivesetmanuscrits.bnf.fr/ark:/12148/cc66676h <accessed 05.08.2021>. See Juste, *Les manuscrits astrologiques latins conservés à la Bibliothèque nationale de France*, pp. 96–97; Charles Samaran and Robert Marichal, *Catalogue des manuscrits en écriture latine portant des indications de date, de lieu ou de copiste*, vol. II, *Bibliothèque Nationale, fonds latin: Nos 1 à 8000* (Paris: CNRS, 1962), p. 517. This manuscript is digitized and available online on Gallica: https://gallica.bnf.fr/ark:/12148/btv1b100202499q.

65 Emmanuel Poulle, 'Catalogue des manuscrits en écriture latine portant des indications de date, de lieu ou de copiste, par Charles Samaran et Robert Marichal', recension in *Revue d'histoire des sciences et de leurs applications*, 16 (1963), 267–69.

magna ex parte aliquando liberatus, retuli me, rex christianissime, ad ea studia astrologie... (f. 3^{r}).[66]

5.2. Context of production

Conrad Heingarter, originally from Zurich, worked as personal physician and astrologer of the Duke of Bourbon, Jean II (1426–56–88). He seems to have been a trusted political advisor and probably served as occasional ambassador to the court of France under Louis XI (1461–83) and Charles VIII (1483–92). While he might have been serving the latter after the death of Jean II, there is no doubt he was still in the service of the duke in 1476 when he wrote this prediction. It is thus surprising, and admittedly meaningful, that he would dedicate it to Louis XI instead.

The French political chessboard remained quite uncertain at the end of 1475, and one may consider Louis XI to have been in a position to appreciate astrological advice.[67] From 1474–75, Louis XI managed to proclaim several strategic victories, but whether or not they would last was probably unclear at the time. After managing to buy the withdrawal of Edward IV of England — and the end of the Hundred Years' War — in August 1475 in Picquigny, Louis XI was still in open conflict with Charles the Bold of Burgundy. Likewise, the peace with François II of Britain, reinstated in September, could have been considered precarious, and the 1474 seizure of Anjou, though not (yet) militarily challenged by the official Duke René, made the latter a potential enemy of the Crown. In that context, it is not surprising that Jean II, former member of the protest League of the Public Weal (1465), might have wished to reassure the king of his allegiance by providing astrological counsel as a gift.

Whether or not he — or his political advisor, Heingarter himself — influenced and censored the content of the prediction is difficult to know, especially as this is the only preserved annual prediction by Heingarter.[68] In fact, this prediction — that we must remember was probably never sent, for an unknown reason — mentions only very vaguely such political intrigues. On the contrary, it is often very technical, with lengthy explanations of astrological methods, and

66 Thorndike and Kibre, *A Catalogue...*, col. 288; Ernst Zinner, *Verzeichnis der astronomischen Handschriften des deutschen Kulturgebietes* (Munich: C. H. Beck, 1925), n°4082; Thorndike, *HMES*, VI, 360–61 and 365. This prediction is well known and has been thoroughly studied in the last decades; it was critically edited by Maxime Préaud in his 1969 École des chartes dissertation, *Les méthodes de travail d'un astrologue du xve siècle, Conrad Heingarter* (unpublished dissertation, École nationale des chartes, 1969), pp. 197–292, and was part of Philippe Contamine's prognostications corpus in 'Les prédictions annuelles astrologiques à la fin du Moyen Age: genre littéraire et témoin de leur temps', in *Histoire sociale, sensibilités collectives et mentalités: Mélanges Robert Mandrou* (Paris: Presses Universitaires de France, 1985), pp. 191–204 (prediction H). See also Tur, *Hora introitus*, pp. 669–71 and 'À l'attention du roi Louis XI: une prédiction astrologique sérieuse pour l'année 1476', *Manuscripta: manuscrits médiévaux conservés à la BnF*, 20 April 2020, https://manuscripta.hypotheses.org/3398 <accessed 31.08.2021>.

67 On Louis XI's own astrological habits, see Jean-Patrice Boudet, 'Les astrologues et le pouvoir sous le règne de Louis XI', in *Observer, lire, écrire le ciel au Moyen Age*, ed. by Bernard Ribémont (Paris: Klincksieck, 1991), pp. 7–61 (republ. in Boudet, *Astrologie et politique*, pp. 293–349).

68 As a personal astrologer, Heingarter probably supplied the duke with secret, personalized advice, rarely copied and thus unknown to us. Annual predictions were an academic exercise that could be used by astrologers in search of new patrons to illustrate their skills, or in some cases by political leaders to propagate well-chosen predictions in the population. See Tur, 'Hora introitus', part III and *passim*. For an example of (a suspicion of) propaganda through annual predictions, see Monica Azzolini, *The Duke and the Stars: Astrology and Politics in Renaissance Milan* (Cambridge: Harvard University Press, 2013).

seems to be addressed more to the Kings' astrologers than Louis XI himself.[69] In that prediction, which might have been the only one he had actually written in some time, Heingarter champions an astrological innovation of humanistic influence, the predilection for Ptolemy's methods over those suspected to be attributable to later astrologers such as Albumasar and Hermes.

The prediction is organized in four chapters (*tractatus*), one per season, each one with the same six parts: astronomical causes of the events to come (*de causis*); places concerned (*de terrarum locis*); timespan of these events (*de inceptione atque duratione eventuum*); people and objects particularly affected (*de rebus magis huic influentiis aptis*); and, finally, the events themselves (*qualis res in generali accidat*). A final part is dedicated to meteorological predictions (*in speciali aeris dispositione*).

5.3. Role of the horoscope in the prediction

His wish to 'return to Ptolemy' brought Heingarter to abandon the figures of revolution (equinoxes and solstices) in favour of their prevention, that is, the syzygy immediately preceding. In spring (March 10) and autumn (September 3), those happen to be lunar eclipses. Each of the four tracts is organized around one of these figures, beginning with spring. They are presented and analysed in the first part, but only the astrological squares hold the entirety of the astronomical data. Heingarter actually uses the word *figura* to describe them on several occasions: *secundum quod figura subjecta ostendit* (f. 5^{v}), ... *et reliqua omnia que in figura subjecta patebunt* (f. 12^{r}), and so on.

These astrological squares are thus a crucial and carefully executed part of the argumentation. They also act as illustrations in a formal, but not illuminated, manuscript.[70] They are meticulously drawn as a vignette of around half the text's width in the first part of each tract. Heingarter uses three different inks: brown (the same as the text) for the signs, which are textually abbreviated; blue for the symbols of the planets; and red for the degrees in Arabic numerals. The layout is traced with the same ink used for the ruling, but in-house sign shifts are traced at the quill with the red ink of the degrees (and text rubrication).

Heingarter employed the central cartel to remind us not only of the nature of the event (*oppositio Solis et Lune*) and its date and time (to the minute), but also the corresponding equation of time and the right ascension of the ascendent (in minutes of degree). Cusps and planetary longitudes are given in degrees, except for the cardinal houses of the spring prevention (only), precise to the minute.

Although Heingarter strived to respect 'Ptolemy's method' and mainly used those figures, he still resorted to the other monthly syzygies in the meteorological part of each tract; however, he only describes them superficially and does not draw astrological squares.

---

69 In 1475, two astrologers were officially at the service of the court of France: Pierre Choinet from 1470, and Jacques Loste from 1469 (Boudet, *Les astrologues et le pouvoir*, p. 304). It is not surprising that Louis XI, upon receiving an astrological prediction, would entrust its analysis to the professionals at his service, and, as a regular ambassador, Heingarter probably knew the court and its astrologers well. Whether he wrote as such on purpose or got carried away in the technical description of his 'return to Ptolemy' cannot be said with certainty.

70 The astrological squares of this manuscript are, however, much less aestheticized than the sky chart of another of Heingarter's predictions, his nativity for Jean de la Goutte as found in manuscript Paris, BnF, lat. 7446, f. 2r. The latter is rendered as a painted illumination, representing zodiacal signs and houses, but from which all precise astronomical data is missing. It is noteworthy that this is one of the rare fifteenth-century circular depictions of a horoscope. There are probably clues to be found concerning the differing visualizations of celestial configurations between astrologers and non-astrologers.

5.4. Figurae caeli

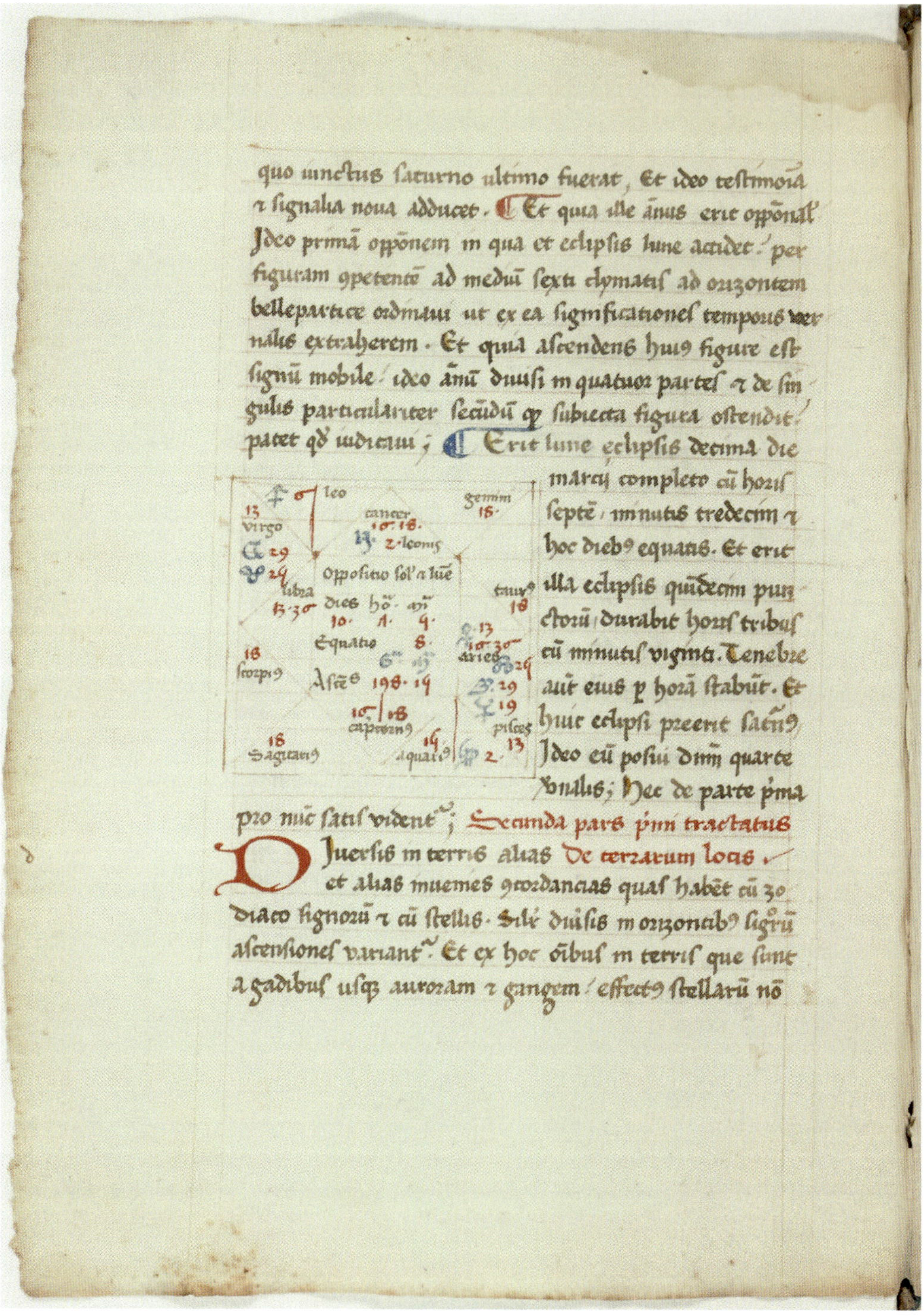

Figure 15.5. Conrad Heingarter, prediction for 1476. Paris, BnF, Latin 7450, fol. 5v.

[3v]

Tractatus Primus de Quarta Vernali in Partes Sex Secatur. Quarum Prima Est de Causis.

[Begins with discussions of the astrological methods and particularly the use of syzygies instead of revolutions, according to Ptolemy's teachings]

[5r]

(…)

Consideratione dignum videtur quod Jupiter hoc anno ad locum celi perveniet in [5v] quo junctus Saturno ultimo fuerat. Et ideo testimonia et signalia nova adducet. § Et quia ille annus erit oppositionalis, ideo primam oppositionem in qua et eclipsis lune accidet per figuram competentem ad medium sexti clymatis ad orizontem Bellepartice ordinavi, ut ex ea significationes temporis vernalis extraherem. Et quia Ascendens hujus figure est signum mobile, ideo annum divisi in quatuor partes, et de singulis particulariter secundum quod subjecta figura ostendit, patet quod judicavi.

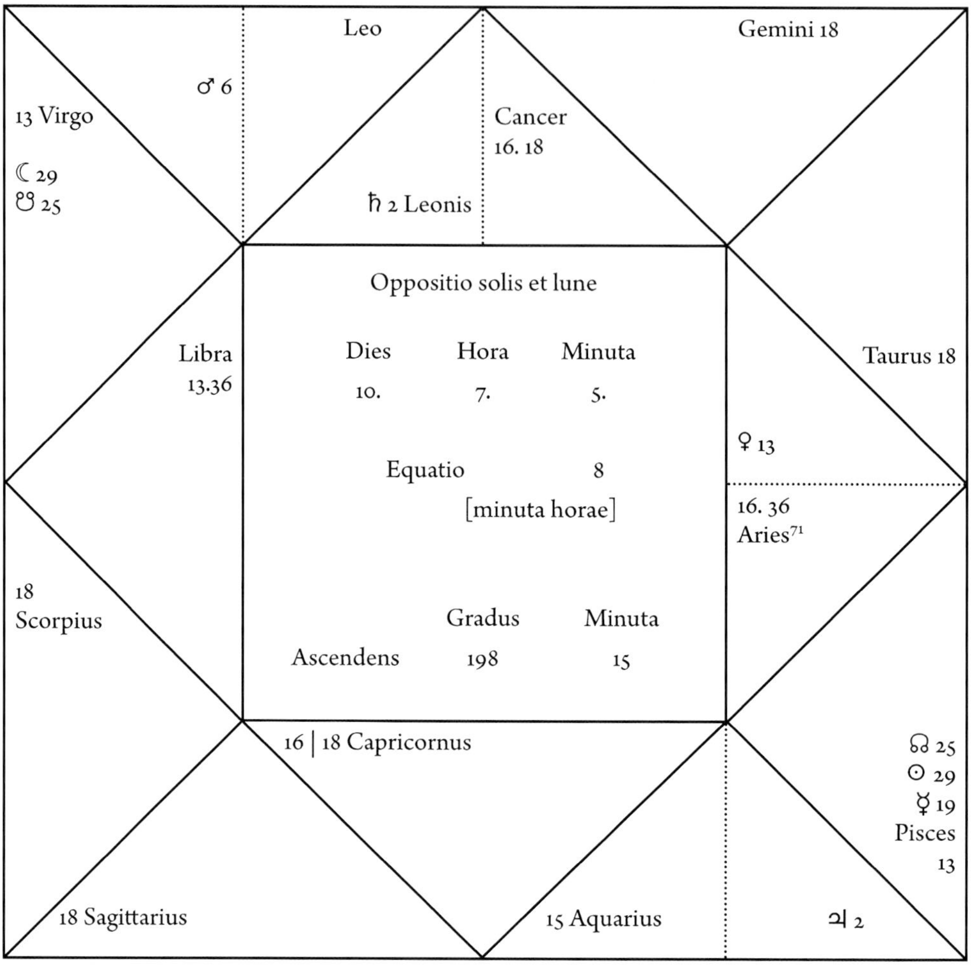

Erit lune eclipsis decima die martii completo cum horis septem, minutis tredecim et hoc diebus equatis. Et erit illa eclipsis quindecim punctorum, durabit horis tribus cum minutis viginti. Tenebre autem ejus per horam stabunt. Et huic eclipsi preerit Saturnus.

Ideo eum posui dominum quarte vernalis. Hec de parte prima pro nunc satis videntur.

Secunda Pars Primi Tractatus: de Terrarum Locis

(...)

5.5. Recomputation

Unlike most prognosticators, Heingarter's base data for computing the horoscope is overly complex. In the prologue, he gives the coordinates of Belleperche, for which he draws the following figure: *Hoc mihi ad nostra tempora atque ad nostrum clyma sextum, ad longitudinem*

71 This should probably be read as Cusp of House VII: 13°36' Aries (opposite the ascendant); Mercury at 16° Aries.

23 gradus cum quarto unius, ad latitudinem vero 46 graduum et semis ad orizontem Bellepartice verificandum putavi. (f. 3r).[72]

Similarly, the date and time of each of the figures are given to the minute; for instance, in the case of the spring prevention, for March 10, 7h 13m p. m.: *Erit lune eclipsis decima die martii completo, cum horis septem, minutis tredecim, et hoc diebus equatis. Et erit illa eclipsis quindecim punctorum, durabit horis tribus cum minutis viginti, tenebre autem ejus per horam stabunt.* (f. 5v).[73]

Maxime Préaud studied this figure in his *École des chartes* thesis, recomputing it by following the Alfonsine canons of John of Saxony.[74] He thereby reconstituted Heingarter's time for the eclipse — 6h 13m 39s p. m. at Toledo — that is, 7h 2m 39s in Belleperche (instead of 7h 5m), and its duration of 3h 20m (14s).

Préaud used the 1483 edition of the Alfonsine tables to check the ascendant given by Heingarter in the diagram; while the right ascension in the cartel matches his own calculations, he finds the corresponding (oblique) ascendant at that latitude should be 13°49' Libra (instead of 13°36'). Such a difference probably means that Heingarter was adjusting his results to the horizon of Belleperche, while Préaud used the table of oblique ascensions for the seventh climate (computed for the Paris latitude).[75]

In spite of Heingarter's reputation, and his highly precise range of base values, the accuracy of his results is not outstanding compared to our other case studies. This might, however, be a sign that he did more recomputation instead of using pre-existing tables. Equating his data for Belleperche instead of Paris, for instance, probably imposed several more steps, roundings, and opportunities for error.

Three of the seven planets' longitudes have an inaccuracy of more than one degree. This does not change the astrological signification of any of these planets, and it cannot be considered an intentional manipulation.

72 In his recomputation, Préaud considers that the longitude is given east of the Fortunate Isles, which are situated 11° west of Toledo, and thus he accepts the longitude as 12°15' east of Toledo (Préaud, 'Les méthodes de travail', vol. 1, p. XLVI and vol. 3, note 1 p. 200). This matches the time difference of 49 minutes between the true eclipse found by *Astromodels* at the meridian of Toledo and the time given for Belleperche, which is a surprisingly good performance. The *Horosc* estimation of the latitude is not very helpful in that case; for a rounding to the degree, the base latitude is given between 35°40' N and 51°07' N. Belleperche castle was too small a locality to appear in major fifteenth-century geographical lists; it is reputed to have been located near the modern village of Bagneux, 46°40' N; 3°12' E (7°13' east of Toledo).

73 Préaud, 'Les méthodes de travail', vol. 3, p. 206.

74 Préaud, 'Les méthodes de travail', vol. 1, pp. XLV–LX. Préaud notably recomputed the mean and true time of the eclipse according to the Alfonsine canons of John of Saxony. See also his critical edition of the prediction in the third volume.

75 Préaud used the latitude of Paris for his recomputation because he assumed Heingarter was using the table of oblique ascensions for the seventh climate found in early editions of the Alphonsine tables.

Planet	Figura Caeli	Alfonsine Tables (Toledan conjunction)
	10/03/1476 7h5m pm Belleperche	1/03/1443 6h14m pm Toledo
Moon	29° Virgo (XII)	29°52' Virgo
Sun	29° Pisces (VI)	29°52' Pisces
Mercury	19° Pisces (VI)	20°13' Pisces
Venus	(13° or) 16° Aries (VII)	14°3' Aries
Mars	6° Virgo (XI)	6°14' Virgo
Jupiter	2° Pisces (VI)	3°0' Pisces
Saturn	2° Leo (X)	2°28' Leo
Caput	25° Virgo (XII)	25°37' Virgo
Cauda	25° Pisces (V)	25°37' Pisces

Comparisons of the domification and *Horosc* quality coefficient are not conclusive as to whether the Standard method (Nr 1 by North) or the Prime Vertical method (Nr 3 by North, often referred to as Campanus' method) was used. This is the only one of our cases where this method might have been used. It is not chronologically improbable, as the method gained publicity in 1468–69, when Regiomontanus criticized it, and Heingarter's propension to try new methods might support this hypothesis.[76] It is surprising, however, that he would have used such a new method without announcing and defending it in the text of the prediction.

	Figura	North 0	North 1	North 2	North 3	North 4	North 5	North 6
I	13°36' Li	13°36' Li	13°36' Li	13°36' Li	13°36' Li	13°36' Li	13°36' Li	13°36' Li
II	18° Sc	9°50' Sc	16°54' Sc	14°40' Sc	16°32' Sc	7°23' Sc	15°0' Sc	13°36' Sc
III	18° Sa	11°16' Sa	17°25' Sa	15°43' Sa	18°4' Sa	7°53' Sa	13°53' Sa	13°36' Sa
IV	16°18' Cp	16°47' Cp	16°47' Cp	16°47' Cp	16°47' Cp	16°47' Cp	11°29' Cp	13°36' Cp
V	15° Aq	21°5' Aq	13°49' Aq	15°43' Aq	14°6' Aq	23°18' Aq	10°2' Aq	13°36' Aq
VI	13° Pi	20°10' Pi	13°4' Pi	14°40' Pi	12°28' Pi	20°48' Pi	11°1' Pi	13°36' Pi
VII	13°36' Ar	13°36' Ar	13°36' Ar	13°36' Ar	13°36' Ar	13°36' Ar	13°36' Ar	13°36' Ar
VIII	18° Ta	9°50' Ta	16°54' Ta	14°40' Ta	16°32' Ta	7°23' Ta	15°0' Ta	13°36' Ta
IX	18° Gm	11°16' Gm	17°25' Gm	15°43' Gm	18°4' Gm	7°53' Gm	13°53' Gm	13°36' Gm
X	16°18' Cn	16°47' Cn	16°47' Cn	16°47' Cn	16°47' Cn	16°47' Cn	11°29' Cn	13°36' Cn
XI	[15°] Le	21°5' Le	13°49' Le	15°43' Le	14°6' Le	23°18' Le	10°2' Le	13°36' Le
XII	13° Vi	20°10' Vi	13°4' Vi	14°40' Vi	12°28' Vi	20°48' Vi	11°1' Vi	13°36' Vi
QC		**6.473**	**0.958**	**2.000**	**0.988**	**8.515**	**3.517**	**2.700**

76 On 'Campanus' method' and its (re)appearance in 1468, see North, *Horoscopes and History*, I.9, pp. 27–30.

6. Case E. Johannes Laet, prediction for 1479

6.1. Witnesses

The 1479 prediction by Johannes Laet is known (in its Latin version) by three witnesses.

(i) *Incunabulum London, British Library, IB 4240, the only known exemplar of an edition by Johann Guldenshaff in Cologne (ISTC il00022140).*[77]

This edition is the fifth preserved edition of an annual prediction by Johannes Laet, after those printed by Johannes de Westfalia in Louvain for 1476 (ISTC il00022100, 2 exemplars preserved) and 1477 (ISTC ip01005700, 1 exemplar preserved), by Richard Blandin and by Guillaume Février in Paris for 1477 (same prediction, ISTC il00022120, 1 exemplar preserved), and by Colard Mansion in Bruges for 1477 (prediction in French, ISTC il00022125, lost fragment documented).[78]
It is (together with the edition by Nicolaus Götz below) the first known printed prognostication to include an astrological square and the first (known) astrological print of Johann de Guldenshaff.[79]
This exemplar was acquired by the British Library from the Library of the Earls of Leicester at Holkham Hall in 1951, which probably provided the blue morocco bibliophile binding (gilded *Johannis Laet de Borchloen prenosticata sine loco aut anno* on the spine); it does not seem to have been compared to the existing fragment IB 3813 at the time.

(ii) *Incunabulum London, British Library, IB 3813, a one-leaf fragment and the only known exemplar of an edition by Nicolaus Götz in Cologne (ISTC ip01005800).*[80]

Unidentified so far, this fragment is part of an otherwise unknown edition of the same prediction by Johannes Laet for 1479. Nowadays bound the wrong way around (verso first), it only contains part of the prologue (inc. *invidentibus scientibus que in hoc se*

77 Laet, Johannes, *Prognosticationes de anno 1479* (Cologne: Johann Guldenschaff, [1478], ISTC il00022140 <https://data.cerl.org/istc/il00022140, accessed 31.08.2021>, GW M1665510 <https://www.gesamtkatalogderwiegendrucke.de/docs/M1665510.htm, accessed 31.08.2021>). 'Begin. [f. 1 recto:] Scire et intelligere gloriosũ est. q ōnis sciēcia et sapiēcia a deo e, etc. End. [f. 6 recto:] Expliciunt prenosticata Iohannis laet de Horchloen. [Astrological predictions for the year 1479.] G.L.', *Explore Archives and Manuscripts*, http://explore.bl.uk/BLVU1:LSCOP-ALL:BLL01020005123 <accessed 31.08.2021>. This incunable is digitized and available online: http://access.bl.uk/item/viewer/ark:/81055/vdc_100108907985.0x000001.

78 Beside this edition, two others are known for 1479: the edition by Nicolaus Götz presented below (same prediction) and an edition by Lukas Brandis in Lübeck (Low German translation, ISTC ip00949500, fragments preserved).

79 His only other similar editions are broadside almanacs for 1485 and 1489. This sort of material being very ephemeral, it wouldn't be surprising that he printed others now lost to us.

80 *Prognosticatio pro anno 1478 vel 1479* ([Cologne: Nicolaus Götz, about 1478], ISTC ip01005800 <https://data.cerl.org/istc/ip01005800, accessed 31.08.2021>, GW M35594 <https://www.gesamtkatalogderwiegendrucke.de/docs/M35594.htm, accessed 31/8/2021>). The identification of the printer was first given by Robert Proctor, *An Index to the Early Printed Books in the British Museum: from the Invention of Printing to the Year 1500* (London: Kegan Paul, Trench, Trübner, 1898), vol. 1, n° 1111 (p. 89); Ernst Voulliéme, *Der Buchdruck Kölns bis zum Ende des fünfzehnten Jahrhunderts* (Bonn, 1903), n°978 (p. 430); *Catalogue of Books Printed in the XVth Century, now in the British Museum*, part I, *Xylographica and Books Printed with Types at Mainz, Strassburg, Bamberg and Cologne*

male facturos...), the first astrological square (for the entry of the Sun into Capricorn in December 1478) and the beginning of the meteorological predictions (expl. ... *et venient tonit[rua et fulgura ex] frigiditate et hoc mirabile super [terram])*.

The astrological square includes the same data, but the amateurish layout strongly suggests it was drawn by hand (while Guldenshaff's seems to have been a woodcut print). The information inside, however, is printed, including Arabic numerals and astrological symbols that do not appear in Guldenshaff's diagram.

(iii) Manuscript Munich, Bayerische Staatsbibliothek, Clm 648, folios 1ʳ–18ᵛ.[81]

This manuscript is the second volume (after Clm 647) of a collection of astrological predictions curated (and partially copied) by the Nuremberg physician and humanist Hartmann Schedel (1440–1514).[82] While working in Amberg between 1477 and 1482, the future author of the *Nuremberg Chronicle* specifically collected and copied annual predictions as they were published (or more probably, as he was able to access them through his bibliographical network). The two volumes contain 17 predictions in total, including the retrospective copying of a 'prediction' (in fact, rather a didactic example of medical astrology) by Augustine of Trento for 1340 at the beginning of Clm 647.[83] Most of them, including Laet's prediction for 1479, are in Schedel's own handwriting with his monogram.[84]

There are some similarities evident in all those predictions. None of them are anonymous, for instance, and Johannes Laet is a followed astrologer (with his predictions for 1477, 1478, and 1479), much like Jan of Glogau (1476 and 1480) or Girolamo Manfredi (1478 and 1481); they are often lengthy predictions based on several *figurae caeli* and give detailed meteorological and economical predictions using numerous astrological authorities, but it is difficult to see whether those result from strategical collection decisions on Schedel's part or by coincidence.[85] Only two include astrological squares: the initial prediction by Augustine of Trento and this prediction by Johannes Laet for 1479. Two inserted slips, one at the end of Clm 647 and one in Paul of Middelburg's prediction for

(London: British Museum, 1908), pp. 238–39; 'Prognosticatio pro anno 1478 vel 1479', *Explore Archives and Manuscripts*, http://explore.bl.uk/BLVU1:LSCOP-ALL:BLL01019976514 <accessed 31.08.2021>. This fragment is digitized and available online: http://access.bl.uk/item/viewer/ark:/81055/vdc_100107247272.0x000001.

81 'Judicia de anno 1479 et quidem Johannis de Laet, Marci Scribanarii [u.a.] — BSB Clm 648', *BSB-Katalog OPACplus*, https://opacplus.bsb-muenchen.de/search?id=796186563 <accessed 31.08.2021>; Carl Halm, Georg Laubmann et al., *Catalogus codicum latinorum bibliothecæ regiæ monacensis* (Munich: sumpt. Bibliothecæ Regiæ, 1868–81), I.1, pp. 168–69; David Juste, *Les manuscrits astrologiques latins conservés à la Bayerische Staatsbibliothek de Munich*, Catalogus codicum astrologorum latinorum, I (Paris: CNRS éditions, 2011), pp. 79–80. A microfilm reproduction of this manuscript is digitized and available online: http://mdz-nbn-resolving.de/urn:nbn:de:bvb:12-bsb00073052-1.

82 Alexandre Tur, 'Hartmann Schedel, collectionneur et copiste de prédictions astrologiques annuelles', *Bulletin du bibliophile*, 2 (2015), 278–96.

83 Schedel seems to have considered this text he copied from another manuscript in his possession, now Munich, Bayerische Staatsbibliothek, Clm 276 (ff. 87ʳ–91ʳ; astrological squares f. 96ʳ) as an early archetype for annual prognostications. See the critical edition: Agostino da Trento, *Astrologia e medicina*, ed. by Domenico Gobbi (Trento: Civis, 2009).

84 The last four predictions of manuscript Clm 648 are in another, less 'humanistic' hand, probably that of a scribe hired by Schedel to copy them under his supervision. They are still dated Amberg and 'signed' with Schedel's monogram. See Tur, 'Hartmann Schedel...', fig. 3, p. 291.

85 Tur, 'Hartmann Schedel...', pp. 281–86 and Tur, 'Hora introitus', pp. 266–72.

1480 (Clm 648, between f. 34 and f. 35) contain reconstitutions by Schedel (or at least in his handwriting) of the astrological squares for the figures of the spring equinox, respectively in 1478 and 1480, with a rather low level of precision (sign).[86]

Bought by Hans Jakob Fugger after Schedel's death, his library joined the collection of Albert V, Duke of Bavaria in 1571 and is now (for the most part) in the Bayerische Staatsbibliothek.

The incipit of the prediction is *'Scire et intelligere gloriosum est, quia omnis scientia et sapientia a Deo est': ergo laudes copiose cum infinito honore sint ipsi Deo glorioso…* [87]

6.2. Context of production

This prediction for 1479 is chronologically the fourth preserved by Johannes Laet, either in a manuscript or in print, all dedicated to Louis of Bourbon, bishop of Liège (1438–82). However, at least one earlier prediction, which is referred to in his prognostication for 1476, remains lost. Residing in Louvain but, as far as we know, not associated with the flourishing university of the town, Johannes Laet of Borchloen (*c.* 1410–*c.* 1490) is the first in a dynasty of astrologers that would continue with his son, Jaspar (or Gaspar) Laet the Elder and grandson Jaspar Laet the Younger, both physicians and prolific authors of prognostications.[88] He seems to have been a resolute proponent of the printing of prognostications, and his supposed role in making printers incorporate astrological squares corroborates it. He is also one of the first to publish vernacular translations of his predictions, in several languages, such as the Low German version of this prediction for 1479.[89]

In 1479, Laet followed the same thematical organization he adopted in previous prognostications; he begins with a dedication to Louis of Bourbon and a prologue revisiting his predictions for the last year, and notably a verified war between France and the northern

86 The last prediction before the slip at the end of Clm 647 is by Battista Piasio for 1478 and known only through this copy; however, there is no evidence that the square was designed for this prediction and not for another one in the collection, or even for all 1478 predictions, for none of which do the other preserved manuscript copies or incunable editions include diagrams. Likewise, the only known printed edition of Paul of Middelburg's prediction for 1481 is without diagrams.

87 Thorndike and Kibre, *A Catalogue…*, col. 1406; see Zinner, *Verzeichnis*, n°6297; Thorndike, *HMES*, vol. 4, p. 451; Tur, 'Hora introitus', pp. 704–06. The initial quotation is the beginning of Aegidius de Thebaldis' preface to his translation of Ptolemy's *Quadripartitum*: *Scire et intelligere gloriosum est quia omnis sapientia est a Deo, eius igitur est anima gloriosa que intelligit et discernit…* (ed. by David Juste, 'Ptolemy, *Quadripartitum* (tr. Aegidius de Tebaldis)', *Ptolemaeus Arabus et Latinus: Works*, http://ptolemaeus.badw.de/work/47 <accessed 11.09.2021; last update 20.07.2021>).

88 Eustace Bosanquet, *English Printed Almanacks and Prognostications: A Bibliographical History to the Year 1600* (London: Chiswick Press, 1917), pp. 18–19. See Tur, 'Hora introitus', pp. 417–18 (on Johannes Laet) and p. 413 (on Jaspar Laet the Elder).

89 This edition by Luckas Brandis in Lübeck (ISTC ip00949500) is only known in fragments; see Carl Meltz, 'Die Drucke der Michaelisbrüder zu Rostock, 1476 bis 1530', *Wissenschaftliche Zeitschrift der Universität Rostock*, 5 (1955–56), 229–62, 21 and figs 6–7. The first translation of Laet's predictions is one of the first known altogether: a French version of his 1477 prediction published by Colard Mansion in Bruges. There is unfortunately no surviving copy, but it was studied and partially edited (based on material that has since been lost) in Lotte Hellinga and Wytze Hellinga, 'A Prognostication Printed by Colard Mansion (CA 1444)', *Gutenberg-Jahrbuch*, 1972, 79–85. Based on the *Incunabula Short Title Catalogue*, no vernacular prognostication is preserved before 1477. Another is preserved for that year, an Italian translation of Pietro Bono Avogario (Ferrara: Severinus Ferrariensis, ISTC ia00057080). This process became common in the 1490s, and by the sixteenth century the Latin prediction was no longer systematically published. The Laet dynasty played a significant role in that transformation.

lands (presumably the predictable conflict between Louis XI and Maximilian of Habsburg). He then provides lengthy meteorological predictions based on the figures (and astrological squares) of the revolutions of winter and spring, as well as the solar eclipse of July 19, 1478. The following chapters are dedicated to agriculture, diseases, particular predictions by planet and by land, and general predictions on war.[90] His predictions encompass all of Europe, with a clear focus on (modern-day) Belgium in certain cases (*patria Geltensi, terra Leodiensi, Brabancia, Lovanio, civitas vel oppidum Bruxellense*, etc.).

6.3. Role of the horoscope in the prediction

Throughout the prediction, Laet frequently relies on the annual (spring) prevention and revolution, but he does not describe them textually. There is in fact a gap between the figures highlighted through an astrological square in the meteorological section — the conclusions of which rely on all seasonal revolutions and preventions, and even on monthly syzygies — and their actual astrological use. Likewise, it seems that the winter revolution is drawn to make a point, maybe to stand out from the competition by beginning the year in winter and not in spring, as it is rarely quoted in the following justifications.[91] However, he seems to have been committed to providing astrological squares to 'make clear to everyone how the sky stands with all the planets at said time'.[92] So much so, in fact, that he probably convinced first-time printers to make the necessary technical investment to be able to print such diagrams.[93] Earlier predictions by Johannes Laet do not include astrological squares, but there is one in his next known prediction, for 1481, which was published by another printer, Johannes de Westfalia, in his hometown of Louvain.[94] This practice deserves to be highlighted here, as it was truly unusual in the 1480s.[95]

90 Tur, 'Hora introitus', pp. 704–06.

91 In the fifteenth century, most astrologers adopted the astronomical year, beginning with the spring equinox (from March to March), rather than local civil or ecclesiastical styles (such as Christmas or Easter being the start of a year).

92 *Hoc anno volo tenere et ponere figuram cum Sol intrabit in Capricornum. Et hoc facio ad hunc finem ut unusquisque clare sciat qualiter et quomodo totum celum illa hora stat cum planetis.* (f. 1v in British Library IB 4240).

93 None of Guldenschaff's or Götz's earlier preserved prints suggest they might already have had experience in printing astrological squares (although many such ephemeral early prints are probably lost). I was not able to find any earlier printed prognostication including an astrological square (consulting most — but not all — of the editions referred to in the *Incunabula Short Title Catalogue*). For instance, another influential prognosticator, Girolamo Manfredi, left a 1479 prediction known in four different editions, none of which include astrological squares.

94 Johannes Laet, *Prognosticationes de anno 1481*, ([Louvain: Johannes de Westfalia, 1480], ISTC il00022180) is known by only one fragment (including a partial square for the entry of the Sun in Libra), now preserved in Frankfurt (Universitätsbibliothek Johann Christian Senckenberg, Inc. Fragm. Löw. I; digitization available here: urn:nbn:de:hebis:30:2–221859). The same astrological square also appears in the Dutch version of this prediction, printed in Audenarde by Arend de Keysere (ISTC il00022150), which is now lost, but was edited in the nineteenth century by Johan Hendrik Van Dale, 'Ontdekking van een oud Vlaamsch drukwerk: Prenosticaten van Johannes Laet van Borchloen, voor het jaer 1481', *Bijdragen tot de Oudheidkunde en Geschiedenis, inzonderheid van Zeeuwsch-Vlaanderen*, 1 (1856), 263–74 and 282–86 (digitized by the New-York Public Library on Internet Archive: ark:/13960/t. onsoxk88). Johannes de Westfalia had already printed Laet's predictions for 1476 and 1477 without astrological squares; it would make sense that he also printed the predictions for 1478, 1479, and 1480, but these editions are lost. Johannes Laet's later preserved predictions for 1484 and 1485 (by other printers) are all fragmentary; they may or may not have included astrological squares.

95 In addition to the previously mentioned prognostications, some noteworthy exceptions are the following: Konrad Rengart, prediction for 1483, printed in Augsburg by Christmann Heyny (ISTC ir00145600; Augsburg,

Printing the astrological square comes with technical constraints, especially with regard to typographical composition. This is especially visible in Guldenshaff's edition, probably the one copied by Schedel. The astronomical data in each 'House triangle' is composed as a separate paragraph, with Roman numerals, and not line by line. In the spring equinox square, for instance, the whole House VIII is pivoted by 90°, and caesura is dictated by layout optimization and not by astronomical meaning (one line per planet). House VIII thus reads *Mer. XV Mar. XXIIII Ari / es VI Sol i[n]t[ra]t Ari / etem in d. m / or.* (Cusp: Aries 6°. Mercury at 15° Aries; Mars 24°. Sun enters Aries in the House of death.).[96] In Clm 648, Hartmann Schedel switched to Arabic numerals and split the data by line, but he didn't reorganize it (to have the cusp in the right place, for instance), nor did he investigate obvious irregularities, such as the entry of the Sun in Aries in House VIII, while this house begins at 6° Aries.[97] Both Guldenshaff and Schedel use textual abbreviations for planets and signs.

Nicolaus Götz, on the other hand, seems to have been more thorough in printing the astrological square, but there is suspicion that he drew the borders manually, which was a more costly but less technically challenging choice. As the spring equinox is missing, we can only compare the first square for the winter solstice (see the reproduction). There, Götz uses astrological symbols for the signs, the Sun, and the Moon, thus duplicating information that he also carefully prints in its entirety. He also tries to break lines between each planet. While it is visible that he has not yet mastered the exercise, this gives a much more finished look to his diagram compared to that of Guldenshaff.

While Schedel used paper unsparingly, dedicating one full page to each square, Guldenshaff and Götz were both concerned about economical sustainability and printed smaller squares. Guldenshaff even justifies the text around the last two astrological squares, a typography exercise in which he was likely an expert.[98] The result is not unlike the layouts of some rather elaborate manuscripts, such as Conrad Heingarter in his 1476 prediction as copied in Bibliothèque nationale de France, Latin 7450 (see above).

Staats- und Stadtbibliothek, 2° Ink 587, digitized online: urn:nbn:de:bvb:37-dtl-0000000804); Marcus Schinnagel, prediction for 1487, printed both as a broadside (Augsburg: Erhard Ratdolt, ISTC is00334600; see, for instance, Munich, Bayerische Staatsbibliothek, Einbl. Kal. 1487 f: urn:nbn:de:bvb:12-bsb00101301-5) and in the classical in-4° (Strasbourg: Johann Grüninger, ISTC is00334650, Basel, Universitätsbibliothek, Aleph A IV 17:1, https://www.e-rara.ch/id/5193828); and, probably, Jaspar Laet, prediction for 1488 (Anvers: Gerard Leeu, ISTC il00019900), according to the description in *Bod-Inc Online*, L-016, http://incunables.bodleian.ox.ac.uk/record/L-016 (I did not consult this edition).

96 The entry of the Sun in Aries in House VIII (*domo mortis*) is obviously conflicting with a cusp at 6° Aries. The Sun would enter Aries ahead of it, in House VII (associated with adversaries and women). House II (of which the cusp should be opposite House VIII, at 6° Libra) is only given as 'Libra', with no precision of the degree.

97 Schedel seems to have spotted the anomaly, though, as he wrote 'Aries VI' in Roman numerals instead of switching to Arabic.

98 He only does so for the last two diagrams (spring equinox and solar eclipse), and only the first of Götz' diagrams is preserved. Thus, we cannot rule out that Götz did the same in the next part of his edition, perhaps copying the manuscript presentation (although it seems unlikely).

6.4. Figurae caeli

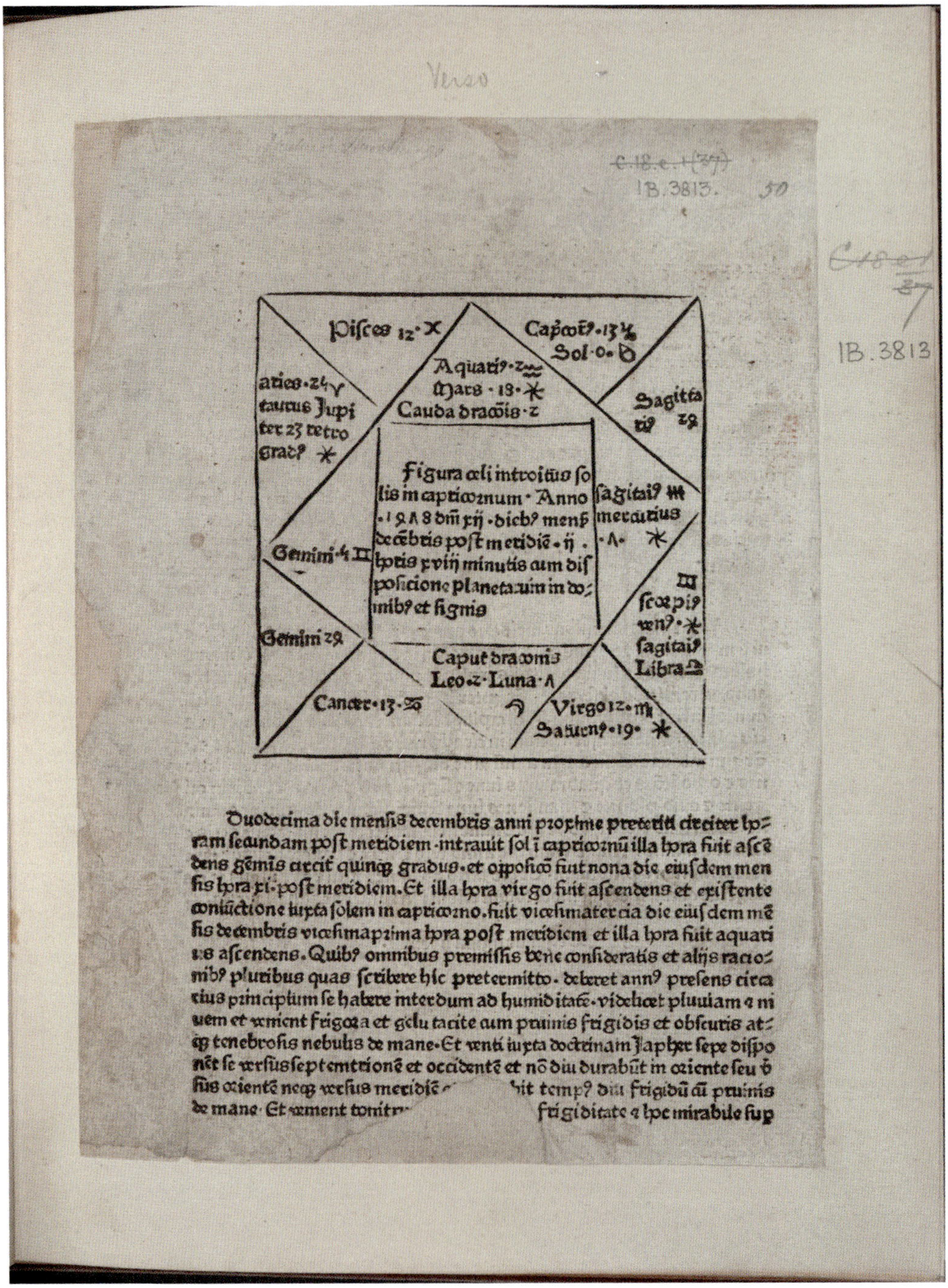

Figure 15.6a. Johannes Laet, prediction for 1479, Cologne: Nicolaus Götz, astrological square for the winter solstice. London, British Library, IB 3813.

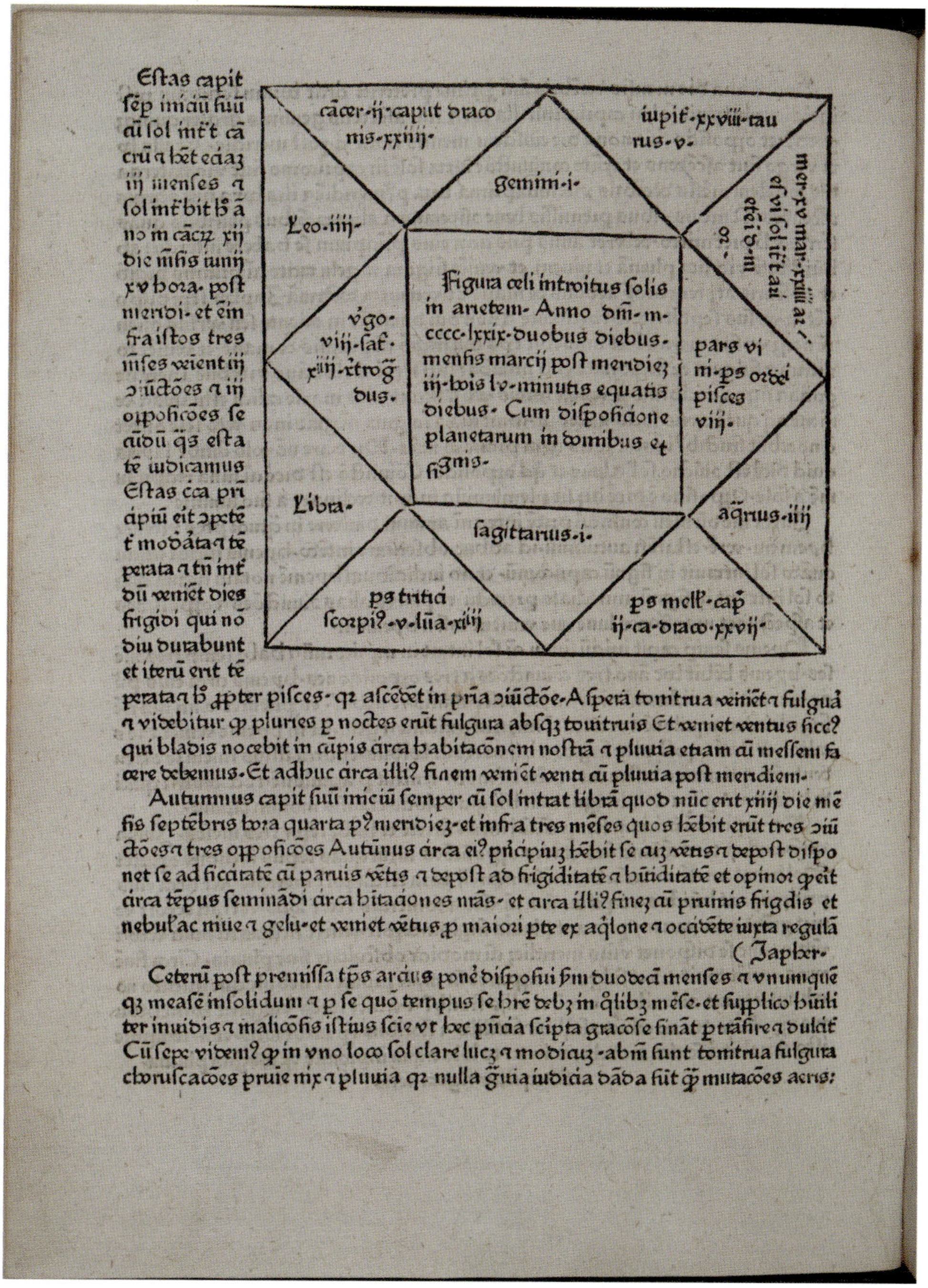

Estas capit sēp iniciū suū cū sol intrat cā crū et hēt eciaȝ iij menses et sol intrabit hō ā no in cācrȝ xij die mēsis iunij xv hora post meridi et ēim fra istos tres mēses vēient iij cōiūctōes et iij oppositiōes se cūdū q̄s esta tē iudicamus Estas circa pri cipiū erit cōpetē ter modāta et tē perata et tn̄ int dū veniēt dies frigidi qui nō diu durabunt et iterū erit tē

perata et hō ꝓpter pisces· q2 ascēdet in pria cōiūctōe· Aspera tonitrua veniēt et fulgurā et videbitur q̄ pluries p noctes erūt fulgura absq3 tonitruis Et veniet ventus sicc9 qui bladis nocebit in cāpis circa habitacōnem nostrā et pluuia etiam cū messem fa cere debemus· Et adhuc circa illi9 finem veniēt venti cū pluuia post meridiem·

Autumnus capit suū iniciū semper cū sol intrat librā quod nūc erit xiiij die mē sis septēbris hora quarta p9 meridiez· et infra tres mēses quos hēbit erūt tres cōiū ctōes et tres oppositiōes Autūnus circa ei9 pncipiū hēbit se cū vētis et depost dispo net se ad siccitatē cū paruis vētis et depost ad frigiditatē et hūiditatē et opinor q̄ erit circa tēpus seminādi circa hitaciones nrās· et circa illi9 finez cū pruinis frigdis et nebul ac niue et gelu· et veniet vētus p maiori pte ex aq̄lone et occidēte iuxta regulā (Japher·

Ceterū post premissa tp̄s circuus ponē disposiciū ꝟm duodecī menses et vnumquē q3 mēsē in solidum et p se quō tempus se hrē deb3 in q̄lib3 mēse· et supplico hūili ter inuidis et malicōsis istius sciē vt hec pn̄cia scipta graciōse sināt ptrāsire et dulcit Cū sepe videm9 q̄ in vno loco sol clare lucz et modicuȝ ·abīn sunt tonitrua fulgura chorusciacōes pruie nix et pluuia q2 nulla ḡnia iudicia dāda sūt q̄ mutacōes aeris:

Figure 15.6b. Johannes Laet, prediction for 1479, Cologne: Johann Guldenshaff. London, British Library, IB 4240, fol. 2ᵛ.

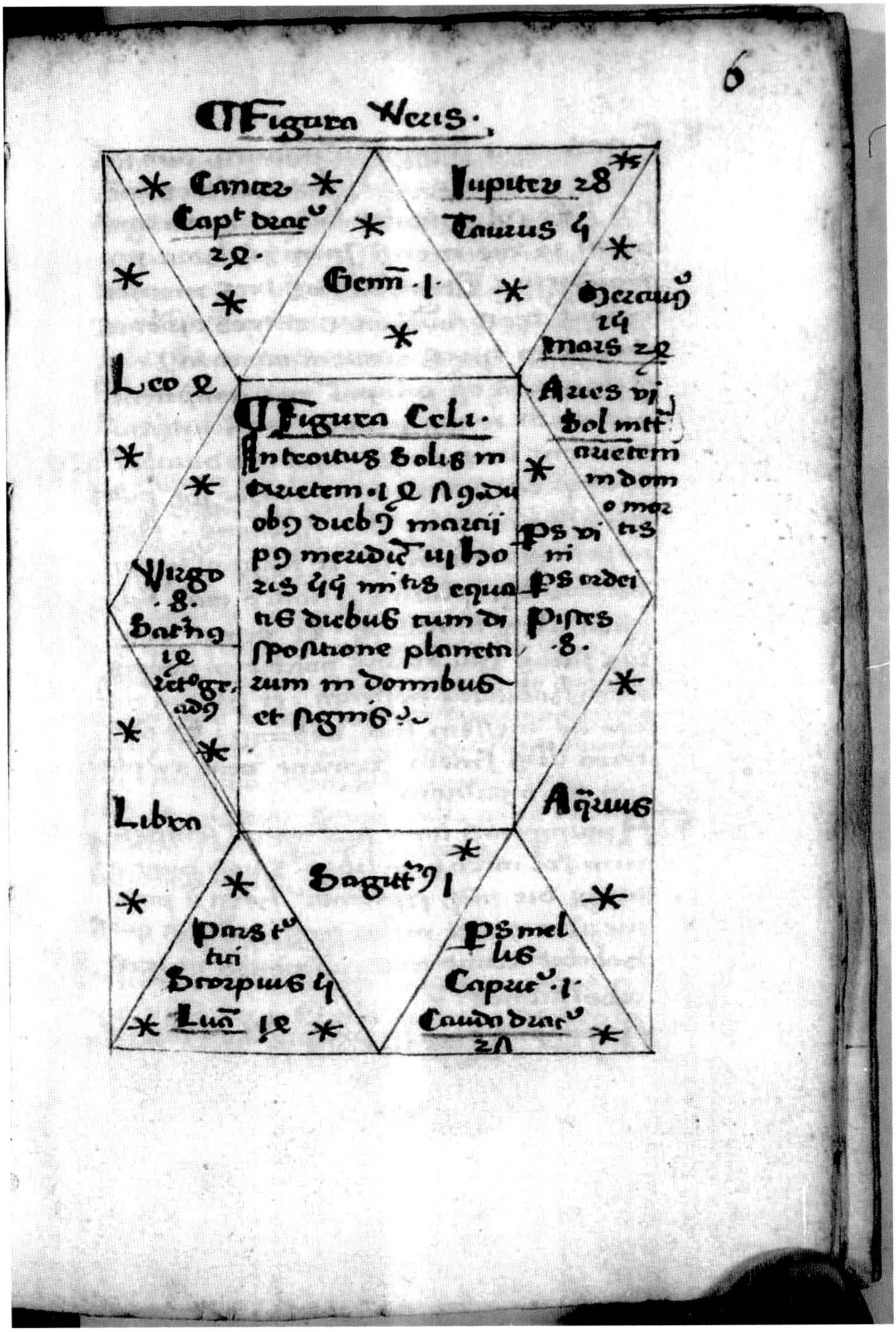

Figure 15.6c. Johannes Laet, prediction for 1479, manuscript copy by Hartmann Schedel. Munich, Bayerische Staatsbibliothek, Clm 648, fol. 6r.

[IB 4240, f. 2^r; Clm 648, f. 5^r]

Ver capit suum initium semper cum sol intrat Arietem, et habet semper tres menses. Infra verem veniunt tres conjunctiones et tres oppositiones juxta quas tempus judicare oportet.

Sol intrat Arietem undecima die mensis marcii hora tercia post meridiem. Ver circa ejus principium erit humidus et tempestuosus cum vento et tonitruis in certis locis, et statim de post veniet grando minuta et nix parva aut pluvia frigida. Depost veniet [Clm 648, 5^v] siccitas … *[followed by general meteorological predictions]*
Sequitur figura veris:

[IB 4240, f. 2^v; Clm 648, f. 6^r]

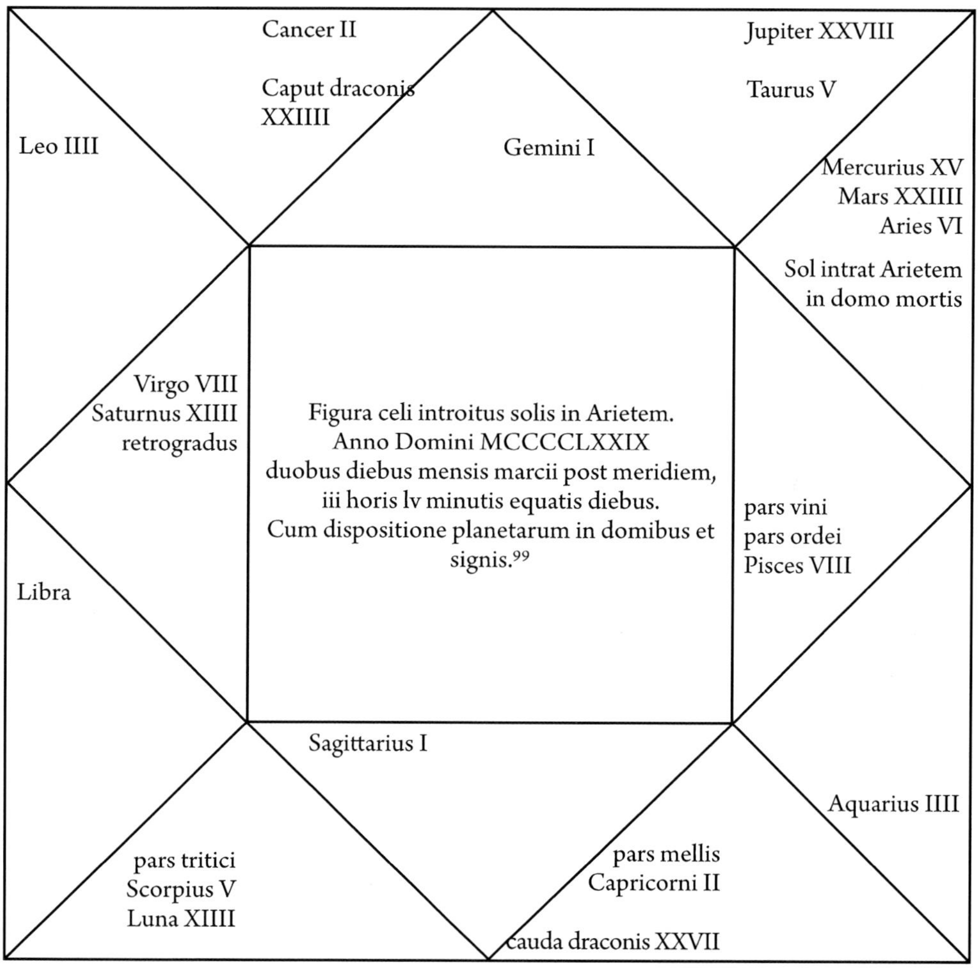

99 Schedel's copy in manuscript Clm 648 displays the following differences: cusp of House V: Capricorni 1; cusp of House VI: *Aquarius* (no degree); cusp of House XI: *Cancer* (no degree).

6.5. *Recomputation*

As in most of our cases, this prediction highlights the date much more than the location.

The cartel gives the time as follows: *Anno Domini MCCCCLXXIX duobus diebus mensis marcii post meridiem, iii horis lv minutis equatis diebus*. Hartmann Schedel copied it faithfully, only switching the Roman into Arabic numerals. Given the planetary longitudes, it seems obvious, however, that a misreading during the printing process is responsible for changing *11 diebus* to *duobus diebus*. Such a small mistake is typical of the difficulties astrologers may have had in working with early printers employing different 'digital' traditions.

No textual mention of the place is to be found either in the text or in the cartel, not to mention the exact coordinates used. Laet is usually associated with Louvain, and a note in the manuscript copy of his 1476 prediction gives his address in the town (*in die Ridder strate*).[100] In this prediction's prologue, however, he describes himself as 'exiled' (but he probably would not change his base location for calculations pertaining to the same region). It is also possible he would have used Liège for predictions dedicated to bishop Louis of Bourbon, although that would probably mean computing new tables instead of benefiting from the scientific activity of the Louvain university.

None of these towns appear either in the list of places compiled by North or in the 1483 Alfonsine Tables. A few decades later, however, an adaptation of the Oxford Tables printed in Louvain in 1528 by Henri Baers uses the geographical latitude 51° N for its table of astrological houses.[101] It is relatively safe to assume that this value (or a close one) was already accepted in Johannes Laet's time.[102] Given the cusp calculated by Laet, the *Horosc* software deduces the compatible range of 47°8'–54°53' N.

The 56-minute difference between the Toledan equinox established by *Astromodels* at 2h59m p. m. and the time declared on the figure (3h55m p. m.) implies a reasonable longitude of 14° E.

Except for Venus, which is missing from the astrological square, all planetary longitudes are in accordance with the *Astromodels* recomputation, within the one-degree accuracy rounding. Only the Dragon's Head, given at 24° Cancer, in contradiction with the opposite lunar node at 27° Capricorn, is thereby inaccurate by three degrees, but this is probably a copy or printing mistake.[103]

---

100 Brussels, KBR, MS 1086–115, f. 136v.

101 Henri Baers or Vekenstyl, *Les tables astronomiques de Louvain de 1528*, ed. by Emmanuel Poulle and Antoine De Smet (Brussels: Ed. Culture et civilisation, 1976); José Chabás and Bernard R. Goldstein, 'Adaptations of the Oxford Tables to Paris, Mantua, and Louvain', *Journal for the History of Astronomy*, 49 (2018), 99–115.

102 Actual coordinates of Liège and Louvain are, respectively, 50°38' N; 5°34' E (9°35' east of Toledo) and 50°53' N; 4°42' E (8°43' east of Toledo).

103 In spite of this obvious contradiction, Hartmann Schedel copied the two values without trying to correct them.

Planet	Figura Caeli (incunabulum)	Alfonsine Tables (Toledan equinox)
	11/03/1479 3h55m pm Louvain (?)	11/03/1479 2h59m pm Toledo
Moon	14° Scorpion (III)	14°15' Scorpio
Sun	0° Aries (VII or VIII)	0°0' Aries
Mercury	15° Aries (VIII)	15°23' Aries
Venus	*missing*	25°41' Pisces
Mars	24° Aries (VIII)	24°23' Pisces
Jupiter	28° Taurus (IX)	27°39' Taurus
Saturn	14° Virgo (I)	14°21' Virgo
Caput	24° Cancer (XI)	27°35' Cancer
Cauda	27° Capricorn (V)	27°35' Capricorn

Both comparison with the theoretical cusp values for a 51° N latitude and *Horosc*'s quality coefficient give a clear preference for the Standard method, with all proposed longitudes being within the one degree rounding accuracy. This result is obtained by adjusting the cusp of House II (only given in Libra) with the opposite degree of House VIII (at 6° Aries). This confirms that 6° Aries is the correct value for the cusp of House VIII, and that the Sun should be found in the preceding House while entering Aries.

	Figura	North 0	North 1	North 2	North 3	North 4	North 5	North 6
I	8° Vi	8°0' Vi	8°0' Vi	8°0' Vi	8°0' Vi	8°0' Vi	8°0' Vi	8°0' Vi
II	[6°] Li	29°7' Vi	6°38' Li	5°41' Li	8°56' Li	29°11' Vi	10°32' Li	8°0' Li
III	5° Sc	26°31' Li	4°47' Sc	3°22' Sc	5°29' Sc	24°44' Li	12°9' Sc	8°0' Sc
IV	1° Sa	1°4' Sa	1°4' Sa	1°4' Sa	1°4' Sa	1°4' Sa	11°15' Sa	8°0' Sa
V	2° Cp	8°48' Cp	2°17' Cp	3°22' Cp	29°30' Sa	12°55' Cp	8°53' Cp	8°0' Cp
VI	4° Aq	11°35' Aq	3°45' Aq	5°41' Aq	2°57' Aq	14°50' Aq	7°15' Aq	8°0' Aq
VII	8° Pi	8°0' Pi	8°0' Pi	8°0' Pi	8°0' Pi	8°0' Pi	8°0' Pi	8°0' Pi
VIII	6° Ar	29°7' Pi	6°38' Ar	5°41' Ar	8°56' Ar	29°11' Pi	10°32' Ar	8°0' Ar
IX	5° Ta	26°31' Ar	4°47' Ta	3°22' Ta	5°29' Ta	24°44' Ar	12°9' Ta	8°0' Ta
X	1° Gm	1°4' Gm	1°4' Gm	1°4' Gm	1°4' Gm	1°4' Gm	11°15' Gm	8°0' Gm
XI	2° Cn	8°48' Cn	2°17' Cn	3°22' Cn	29°30' Gm	12°55' Cn	8°53' Cn	8°0' Cn
XII	4° Le	11°35' Le	3°45' Le	5°41' Le	2°57' Le	14°50' Le	7°15' Le	8°0' Le
QC		**7.476**	**0.346**	**1.250**	**1.775**	**9.754**	**5.455**	**3.750**

7. Synthesis table

Prediction	**Witness**	**Graphical type of diagrams**	**Horoscopes used in the prediction**	**Place**[104]	**Precision and accuracy (planets)**	**Precision and accuracy (domification)**	**Domification method(s)**
A. Pietro de Monte Alcino for 1421	Ashmole 357	Cursive marginal addition	Spring equinox	Pavia [43°08' - 48°51' N] [21°56' E]	Minutes (Sun, Jupiter, Saturn, Caput); degrees < 30" (7/8) Venus ± 6°	Degrees < 1° (10/12) < 2° (2/12)	North 1 or 2
	Pal. lat. 1438	No astrological square	No data				
B. Anonymous prediction for 1443	Latin 7443	Full-page, double astrological square, rule-drawn	Spring equinox	Paris 48°50' N [49°53' – 53°39' N] [18°15' E]	Minutes (Sun and Moon); degrees < 1° (5/8) or 2° (3/8)	Degrees < 1° (6/12) < 2° (4/12) < 3° (2/12)	North 1 or 2
			Preceding conjunction	Paris 48°50' N [51°10' – 54°08' N] [13°30' E]	Degrees < 1° (2/7) or 2° (4/7) Mercury ± 10° Jupiter missing	Degrees < 1° (4/12) < 2° (4/12) < 3° (4/12)	North 1 or 2

104 Name of the place for which the ***figura caeli*** is computed (in brackets when it is not plainly stated in the text or diagram); geographical coordinates mentioned in the prediction; in brackets: recomputed latitudes' range (after ***Horosc*** given the houses' cusps), recomputed longitude (after the local time difference with the astronomical event in Toledo computed with ***Astromodels***). All longitudes are based on the Toledan meridian.

Prediction	Witness	Graphical type of diagrams	Horoscopes used in the prediction	Place[104]	Precision and accuracy (planets)	Precision and accuracy (domification)	Domification method(s)
C. Richard Trewythian for 1443	Sloane 428	Vignette, cursive (draft)	Prevention (#1)	[London?] [50°39' – 53°44' N]	Degrees < 1° (6/7) Venus ± 9° or 19°	Degrees < 1° (8/12) < 2° (2/12) < 4° (2/12)	North 1 or 2
		Vignette, cursive (draft)	Revolution (#2)	[London?] [51°11' – 54°17' N] [19°30' E]	Degrees < 1° (8/9) or 2° (1/9)	Degrees < 1° (6/12) < 2° (4/12) < 5° (2/12)	North 1 or 2
		Vignette, cursive (draft)	Revolution (#3)	[London?] [27°22' – 32°44' N]	Degrees < 1° (9/9)	Degrees < 2° (4/12) < 8° (2/12) > 10° (6/12)	Inconclusive
D. Conrad Heingarter for 1476	Latin 7450	Vignette, meticulous (3 colours, decorative)	Spring prevention	Belleperche 46°30' N; 12°15' E [35°40' – 51°07' N] [12°15' E]	Degrees < 1° (6/9) or 2° (3/9)	Minutes (I, IV, VII, X) or degrees < 1° (8/12, incl. IV ± 29') < 2° (4/12)	North 1 or 3
E. Johannes Laet for 1479	IB 3813 (incunabulum)	Full length, manuscript layout around printed data	Fragmentary: the spring horoscope is missing				
	IB 4240 (incunabulum)	Vignette, printed astrological square	Spring equinox	[Louvain?] [47°8' – 54°53' N] [14° E]	Degrees <1° (except Caput, inconsistent) Venus missing	Degrees < 1° (12/12)	North 1
	Clm 648	Full page, decorative					

Conclusion

These five case studies make up just a small selection of fifteenth-century annual predictions and cannot be considered representative of the entire Alfonsine astrological production. However, their analytical study allows a few assertions.

First of all, they illustrate the importance of the astrological square diagram as a practical way to synthesize (A) and visualize (B) the data, but also as something more: an aestheticized symbol of the astrological scientific status (D) to be displayed on printed prognostications, despite the cost (E). Unlike some later attempts at astrological illustration, such as planetary personifications, astrological squares do not embody the astrological science at the expense of their actual scientific meaning. They are indeed a support for carefully computed and gathered ground data — although with astrological interpretation in mind (C) — and even a possible substitute for the whole prediction.

While the analysis of precision, accuracy and transformation reveals something about computational practices, there is only so much that can be assumed from these five cases.

While precision is generally low (mostly to the degree), accuracy of planetary longitudes with regard to the Alfonsine Tables is relatively good: 50 out of 65 longitudes are less than 1° from the theoretical position, and only four diverge by more than 2°. Major incoherencies can usually be attributed to reading or copying mistakes. The correction of mean times through the equation of time is attested (D) or likely (A) in several cases, and this explains the strange discrepancy between time and longitude precisions. This shift in precision — a time calculated to the minute 'gains' precision to the second — is very peculiar and merits further analysis in the context of Alfonsine practices.[105]

Domification is somewhat poorer, or at least less in phase with the theoretical values generated by *Horosc*: fourteen out of ninety-six *cuspides* diverge from *Horosc's* best proposition by more than 2°. We can speculate with several possible explanations. Firstly, unlike *Astromodels*, *Horosc* is an analytic tool and is not based on historical tables that may have been used (with their own flaws); secondly, such tables were much more diverse and less united than the Alfonsine Tables, as each one had to be adapted to a specific latitude. Table mistakes would have been more difficult to identify and correct. In several of our cases, we had to guess the exact latitude assumed (although that does not account for such a difference in the resulting data). There is also a possibility that some astrologers resorted to a graphical method to compute the house boundaries, using a moving instrument such as an astrolabe. Surviving astrolabes and their textual instructions attest that this use was well known. For astrologers living alone at a tableless latitude — let's say their own domain in Belleperche, 23°15'E 46°30'N — it would make sense to have a rete engraved rather than computing a whole set of tables. Rounding to the degree would also be coherent with reading on a rete scale. It is surprising, however, that this use of the astrolabe would not be more publicized in astrological predictions, all the more so as astrolabes were, at the time, symbolic attributes of astronomers and astrologers, even more than astrological squares.

105 On that question, see, in the present volume, Eleonora Andriani and Stefan Zieme, 'Regiomontanus on the Equation of Time'. Adding the equation of time allows us to shift from mechanical clock time to sundial time, two technologies widely used in fifteenth-century Europe.

In all of our cases, the analysis of the domification points towards the use of the same domification method, the so-called Standard method (although the use of the Dual Longitude method in cases A, B, and C, and the Campanus method in case D, cannot be entirely excluded). While we might have expected a little more diversity in the fifteenth century, this is not entirely surprising. The Standard method is the one prescribed by John of Lignères in his widespread Canons to the Alfonsine Tables.[106] While North was very prudent when it came to the possible use of several different domification methods at any period, and a large mathematical survey of horoscopes drawn in the fourteenth and fifteenth centuries — an enormous research effort that would, however, undoubtedly prove worthwhile if such a database could be built — might show otherwise; there is not much evidence that any other method was widely used before the Regiomontanus controversy of 1467. Regiomontanus himself, in his 1456 annual prediction, uses the Standard method.[107]

Despite all this emphasis on the astronomical computation as the source for the astrological prediction, it is not above these astrologers to manipulate astronomical data with astrological interpretation in mind (B, C, and E), and we cannot say that there has been much effort (at least in these documents) to hide it (although they probably did not expect them to be checked with computer tools such as *Horosc*).[108] In each case, the agenda does not seem to be political (in order to justify predetermined conclusions) in the way Stephan Heilen demonstrates for John of Lübeck's Antichrist prognostication of 1474. The motivation seems to be to obtain a set of data that would allow a coherent astrological reading (and avoid obvious contradictions).[109] There again, estimating how frequent such a manipulation happened could only be done by carefully proof-computing a large corpus of known horoscopes, but it would be very instructive of the actual mindset of the astrological community at the end of the Middle Ages. Was such a practice frowned upon in the name of astronomical precision, that same ideal that might have prompted the Copernican revolution? Or was there another group of the opinion that it was perfectly acceptable to adjust one's computed astronomical data if such an adaptation would allow a much more coherent astrological interpretation? Would it not actually be proof that the

106 John of Lignères, *Canones primi mobilis*, Chapter 37: 'Duodecim domos aequare', critically edited by Marie-Madeleine Saby-Rousset, 'Les canons de Jean de Lignères sur les tables astronomiques de 1321: Édition critique, traduction et étude', unpublished dissertation, Paris, Ecole Nationale des Chartes, 1987, chapter 37, pp. 152-55.

107 The analysis of Regiomontanus' prognostication is not included in this paper as it deserves a more detailed study in its own right. While the endeavour of detailed and precise computation is visible, analysis through *Horosc* definitely points towards the Standard method (quality coefficient of 0.037 for the revolution and 0.019 for the prevention). This is not very surprising historically, and it would tend to confirm that Regiomontanus only supported the Equatorial method towards the end of his career and mostly employed it as a reaction to the so-called Campanus Prime Vertical method. See North, *Horoscopes and History*, pp. 27–30.

108 Richard L. Kremer has highlighted the historical importance of Regiomontanus' ephemerides, first printed in 1474 and so widespread they would have given easy access to planetary positions even to non-mathematicians. In theory, this should have led to a significant change in this practice of altering astronomical data, providing both a reassurance for an astrologer unsure of their computations and a source for easy external criticism. Then again, this would need to be statistically confirmed, as trends in late-fifteenth- and early-sixteenth-century astrological practices divert from interpretative rigour in favour of popularity. Richard L. Kremer, *The Ephemerides: Massive Hand Computations in Alfonsine Astronomy* (forthcoming).

109 Stephan Heilen, *Die Antichrist-Prognose*, chapter IV–D, pp. 163–89. In his 1474 written prediction on the Saturn-Jupiter conjunction of 1504, John of Lübeck substantially alters the longitudes of Mars and Venus, without regard for astronomical consistency, in support of his main theory: the birth of the Antichrist.

original data was flawed — by tables or computation mistakes, excessive rounding, or even a slightly defective model — and that a little 'tweak' would render it more trustworthy?

It would be beneficial to challenge all of these hypotheses with a much more extensive survey of preserved horoscopes, including both astrological squares and the sometimes much scarcer (and difficult to analyse) textual justifications.

Manuscript sources

References preceded by an asterisk are the primary source for one of the case studies.
Bruxelles, KBR, ms 1086–115
*London, British Library, Sloane MS 428
Munich, Bayerische Staatsbibliothek, Clm 276
Munich, Bayerische Staatsbibliothek, Clm 647
*Munich, Bayerische Staatsbibliothek, Clm 648
Munich, Bayerische Staatsbibliothek, Clm 14504
*Oxford, Bodleian Library, Ashmole 357
*Paris, Bibliothèque nationale de France, lat. 7443
Paris, Bibliothèque nationale de France, lat. 7446
*Paris, Bibliothèque nationale de France, lat. 7450
Paris, Bibliothèque nationale de France, NAL 398
*Vatican, Biblioteca Apostolica Vaticana, Pal. lat. 1438
Wolfenbüttel, Herzog August Bibliothek, Cod. Guelf. 115 Noviss. 4°

Early printed sources

References for incunabula editions mention ISTC and GW numbers. ISTC numbers refer to the *Incunabula Short Title Catalogue* (https://data.cerl.org/istc/), an international database administered by the British Library; GW numbers to the *Gesamtkatalog der Wiegendrucke* (https://www.gesamtkatalogderwiegendrucke.de/), administered by the Staatsbibliothek zu Berlin — Preussischer Kulturbesitz. The titles are generally those attributed retrospectively by one of these bibliographic databases, with the exceptions of annual prognostication, for which a homogenous retrospective title has been forged. References preceded by an asterisk are the primary source for one of the case studies.

Alphonsus, Rex Castellae, *Tabulae astronomicae* (Venice: Erhard Ratdolt, 1483, ISTC ia00534000, GW 1257)

Avogario, Pietro Bono, *[Annual prognostication for 1477]*, Italian translation by Antonio Cornazzano ([Ferrara: Severinus Ferrariensis, 1477], ISTC ia00057080, GW 232)

Claudius Ptolemaeus, *Quadripartitum* (Venice: Bonetus Locatellus, 1493, ISTC ip01089000, GW M36394)

Hali filius Abenragel, *De judiciis astrorum* (Venice: Erhard Ratdolt, 1485, ISTC ih00004000, GW 12117)

Laet, Johannes, *[Annual prognostication for 1476]*, Latin (Louvain: Johannes de Westfalia, [1475], ISTC il00022100, GW M16653)

Laet, Johannes, *[Annual prognostication for 1477]*, Latin (Louvain: Conradus de Westfalia, [1477], ISTC ip01005700, GW M16637)

Laet, Johannes, *[Annual prognostication for 1477]*, Latin (Paris: Richard Blandin and Guillaume Février, [1477], ISTC il00022120, GW M16654)

Laet, Johannes, *[Annual prognostication for 1477]*, French ([Bruges: Colard Mansion, 1476–77], ISTC il00022125, GW M1665410): no known extant copy

*Laet, Johannes, *[Annual prognostication for 1479]*, Latin (Cologne: Johann Guldenschaff, [1478], ISTC il00022140, GW M1665510): London, British Library, IB 4240

*Laet, Johannes, *[Annual prognostication for 1479]*, Latin ([Cologne: Nicolaus Götz, about 1478], ISTC ip01005800, GW M35594): London, British Library, IB 3813

Laet, Johannes, *[Annual prognostication for 1479]*, Low German ([Lübeck: Lucas Brandis, ca. 1478–79], ISTC ip00949500, GW M35161)

Laet, Johannes, *[Annual prognostication for 1481]*, Latin ([Louvain: Johannes de Westfalia, 1480], ISTC il00022180, GW M1665710)

Laet, Johannes, *[Annual prognostication for 1481]*, Dutch (Audenarde: [Arend de Keysere, 1480], ISTC il00022150, GW M16656)

Rengart, Konrad, *[Annual prognostication for 1483]*, Latin ([Augsburg: Christmann Heyny], 1483, ISTC ir00145600, GW M3514510)

Schynnagel, Marcus, *[Annual prognostication for 1487]*, Latin (Augsburg: Erhard Ratdolt, [about 1486–87], ISTC is00334600, GW M40840)

Schynnagel, Marcus, *[Annual prognostication for 1487]*, Latin ([Strasbourg: Johann Grüninger, about 1487], ISTC is00334650, GW is00334650)

Bibliography

Agostino da Trento, *Astrologia e medicina*, ed. by Domenico Gobbi (Trento: Civis, 2009).

Al-Qabiṣī (Alcabitius), *The Introduction to Astrology*, ed. by Charles Burnett and Keiji Yamamoto (London: Warburg Institute, 2004).

Avelar de Carvalho, Helena, *An Astrologer at Work in Late Medieval France: The Notebooks of S. Belle* (Leiden: Brill, 2021).

Azzolini, Monica, *The Duke and the Stars: Astrology and Politics in Renaissance Milan* (Cambridge: Harvard University Press, 2013).

Bosanquet, Eustace, *English Printed Almanacks and Prognostications: A Bibliographical History to the Year 1600* (London: Chiswick Press, 1917).

Boudet, Jean-Patrice, 'Les astrologues et le pouvoir sous le règne de Louis XI', in *Observer, lire, écrire le ciel au Moyen Age*, ed. by Bernard Ribémont (Paris: Klincksieck, 1991), pp. 7–61, republ. in Jean-Patrice Boudet, *Astrologie et politique entre Moyen Âge et Renaissance* (Florence: SISMEL — Edizioni del Galuzzo, 2020), pp. 293–349.

———, *Lire dans le ciel: la bibliothèque de Simon de Phares, astrologue du xve siècle* (Bruxelles: Centre d'étude des manuscrits, 1994).

———, and Thérèse Charmasson, 'Une consultation astrologique princière en 1427', in *Comprendre et maîtriser la nature au Moyen âge: mélanges d'histoire des sciences offerts à Guy Beaujouan* (Paris: Champion, 1994), pp. 255–78, republ. in Jean-Patrice Boudet, *Astrologie et politique entre Moyen Âge et Renaissance* (Florence: SISMEL — Edizioni del Galuzzo, 2020), pp. 247–71.

Burnett, Charles, 'Astrology', in *Medieval Latin: an Introduction and Bibliographical Guide*, ed. by Franck Anthony Carl Mantello and Arthur George Rigg (Washington: The Catholic University of America Press, 1996), pp. 369–82.

Carmody, Francis James, *Arabic Astronomical and Astrological Sciences in Latin Translation: a Critical Bibliography* (Berkeley: University of California Press, 1956).

Chabás, José, and Bernard R. Goldstein, 'Adaptations of the Oxford Tables to Paris, Mantua, and Louvain', *Journal for the History of Astronomy*, 49 (2018), 99–115.

Contamine, Philippe, 'Les prédictions annuelles astrologiques à la fin du Moyen Age: genre littéraire et témoin de leur temps', in *Histoire sociale, sensibilités collectives et mentalités: Mélanges Robert Mandrou* (Paris: Presses Universitaires de France, 1985), pp. 191–204.

Dale, Johan Hendrik van, 'Ontdekking van een oud Vlaamsch drukwerk: Prenosticaten van Johannes Laet van Borchloen, voor het jaer 1481', *Bijdragen tot de Oudheidkunde en Geschiedenis, inzonderheid van Zeeuwsch-Vlaanderen*, 1 (1856), 263–74 and 282–86.

Duranti, Tommaso, *Mai sotto Saturno: Girolamo Manfredi, medico e astrologo* (Bologna: CLUEB, 2008).

Goldstein, Bernard, and David Pingree, 'Horoscopes from the Cairo Geniza', *Journal of New Eastern Studies*, 36 (1977), 113–44.

Heilen, Stephan, *Konjunktionsprognostik in der Frühen Neuzeit: Die Antichrist-Prognose des Johannes von Lübeck (1474) zur Saturn-Jupiter-Konjunktion von 1504 und ihre frühneuzeitliche Rezeption* (Baden-Baden: Verlag Valentin Koerner, 2020).

Hellinga, Lotte, and Wytze Hellinga, 'A prognostication printed by Colard Mansion (CA 1444)', *Gutenberg-Jahrbuch*, 1972, 79–85.

Henri Baers or Vekenstyl, *Les tables astronomiques de Louvain de 1528*, ed. by Emmanuel Poulle and Antoine De Smet (Brussels: Ed. Culture et civilisation, 1976).

Juste, David, *Les manuscrits astrologiques latins conservés à la Bayerische Staatsbibliothek de Munich* (Paris: CNRS éditions, 2011).

———, *Les manuscrits astrologiques latins conservés à la Bibliothèque nationale de France* (Paris: CNRS éditions, 2015).

———, and Charles Burnett, 'A New Catalogue of Medieval Translations into Latin of Texts on Astronomy and Astrology', in *Medieval Textual Cultures: Agent of Transmission, Translation and Transformation*, ed. by Faith Wallis and Robert Wisnovsky (Berlin: de Gruyter, 2016), pp. 63–76.

———, 'Ptolemy, *Quadripartitum* (tr. Aegidius de Tebaldis)', *Ptolemaeus Arabus et Latinus: Works*, http://ptolemaeus.badw.de/work/47 <accessed 11.09.2021; last update 20.07.2021>

Kremer, Richard L., and Jerzy Dobrzycki, 'Alfonsine Meridians: Tradition versus Experience in Astronomical Practice c. 1500', *Journal for the History of Astronomy*, 29 (1998), 187–99.

———, *The Ephemerides: Massive Hand Computations in Alfonsine Astronomy* (forthcoming).

Le Boeuffle, André, *Astronomie, astrologie: lexique latin* (Paris: Picard, 1987).

Malagola, Carlo, *Statuti delle Università e dei collegi dello studio Bolognese* (Bologna: Nicola Zanichelli, 1888).

Meltz, Carl, 'Die Drucke der Michaelisbrüder zu Rostock, 1476 bis 1530', *Wissenschaftliche Zeitschrift der Universität Rostock*, 5 (1955–56), 229–62.

North, John D., *Horoscopes and History* (London: The Warburg Institute, 1986).

Page, Sophie, 'Richard Trewythian and the Uses of Astrology in Late Medieval England', *Journal of the Warburg and Courtauld Institutes*, 64 (2001), 193–228.

Poulle, Emmanuel, 'Catalogue des manuscrits en écriture latine portant des indications de date, de lieu ou de copiste, par Charles Samaran et Robert Marichal', recension in *Revue d'histoire des sciences et de leurs applications*, 16 (1963), 267–69.

Préaud, Maxime, *'Les méthodes de travail d'un astrologus du XVe siècle, Conrad Heingarter'*, unpublished thesis, Paris, École Nationale des Chartes, 1969.

Saby, Marie-Madeleine, 'Les canons de Jean de Lignères sur les tables astronomiques de 1321: édition critique, traduction et étude', unpublished thesis, Paris, École Nationale des Chartes, 1897.

Schuba, Ludwig, *Die Quadriviums-Handschriften der Codices Palatini Latini in der Vatikanischen Bibliothek* (Wiesbaden: Dr L. Reichert, 1992).

Silva, Chelsea, 'Opening the Medieval Folding Almanac', *Exemplaria*, 30 (2018), 49–65.

Sorbelli, Albano, 'Il "Tacuinus" dell'Università di Bologna e le sue prime edizioni', *Gutenberg Jahrbuch*, 33 (1938), 109–14.

Thomann, Johannes, 'Square Horoscope Diagrams in Middle Eastern Astrology and Chinese Cosmological Diagrams: Were These Designs Transmitted through the Silk Road?', in *The Journey of Maps and Images on the Silk Road*, ed. by Philippe Forêt and Andreas Kaplony (Leiden: Brill, 2008), pp. 97–117.

Thorndike, Lynn, *A History of Magic and Experimental Science*, 8 vols (New York: Columbia University Press, 1928–1958).

———, and Pearl Kibre, *A Catalogue of Incipits of Mediaeval Scientific Writings in Latin* [1937], (2nd edition, Cambridge: The Mediaeval Academy of America, 1963).

Tur, Alexandre, 'Hartmann Schedel, collectionneur et copiste de prédictions astrologiques annuelles', *Bulletin du bibliophile*, 2 (2015), 278–96.

———, 'Hora introitus solis in Arietem: Les prédictions astrologiques annuelles latines dans l'Europe du xve siècle (1405–1484)', unpublished doctoral thesis, Université d'Orléans, 2018, http://www.theses.fr/2018ORLE1163.

———, 'À l'attention du roi Louis XI: une prédiction astrologique sérieuse pour l'année 1476', *Manuscripta: manuscrits médiévaux conservés à la BnF*, 20 April 2020, https://manuscripta.hypotheses.org/3398.

Westman, Robert S., *The Copernican Question: Prognostication, Skepticism, and Celestial Order* (Berkeley: University of California press, 2011).

Zinner, Ernst, *Verzeichnis der astronomischen Handschriften des deutschen Kulturgebietes* (Munich: C. H. Beck, 1925).